Acoustical Imaging
Volume 23

Acoustical Imaging

Recent Volumes in This Series:

A Continuation Order Plan is available for this series. A continuation order will bring delivery of
each new volume immediately upon publication. Volumes are billed only upon actual shipment.
For further information please contact the publisher.

Acoustical Imaging
Volume 23

Edited by

Sidney Lees

Forsyth Dental Center
Boston, Massachusetts

and

Leonard A. Ferrari

Virginia Polytechnic Institute and State University
Blacksburg, Virginia

SPRINGER SCIENCE+BUSINESS MEDIA, LLC

The Library of Congress cataloged the first volume of this series as follows:

International Symposium on Acoustical Holography.
Acoustical holography; proceedings. v. 1–
New York, Plenum Press, 1967–

v. illus. (part col.), ports. 24 cm.
    Editors: 1967– . A. F. Metherell and L. Larmore (1967 with H. M. A. el-Sum)
    Symposium for 1967– held at the Douglas Advanced Research Laboratories, Huntington
Beach, Calif.

    1. Acoustic holography—Congresses—Collected works. I. Metherell. Alexander A.,
ed. II. Larmore, Lewis, ed. III. el-Sum, Hussein Mohammed Amin, ed. IV. Douglas
Advanced Research Laboratories. v. Title.

QC244.5.I.5   69-12533

Proceedings of the 23rd International Symposium on Acoustical Imaging,
held April 13–16, 1997, in Boston, Massachusetts

ISSN 0270 5117

ISBN 978-1-4613-4640-1   ISBN 978-1-4419-8588-0 (eBook)
DOI 10.1007/978-1-4419-8588-0

© 1997 Springer Science+Business Media New York
Originally published by Plenum Press, New York in 1997
Softcover reprint of the hardcover 1st edition 1997

http://www.plenum.com

10 9 8 7 6 5 4 3 2 1

23rd International Symposium on Acoustical Imaging

Chairman: Sidney Lees

Executive Advisory Council
Joie P. Jones   James F. Greenleaf Hua Lee Lawrence W. Kessler
John P. Powers Glen Wade Piero Tortoli

International Scientific Advisory Board
Leonard A. Ferrari, Chairman

Pierre M. Alais	(France)	Hua Lee	(USA)
Michael Andre	(USA)	Sidney Leeman	(UK)
Walter Arnold	(Germany)	Sidney Lees	(USA)
Jeffrey C. Bamber	(UK)	Frederick L. Lizzi	(USA)
A.A. Beex	(USA)	Leonard Masotti	(Italy)
Andrew Biggs	(UK)	Andrzej Nowicki	(Poland)
Valentin A. Burov	(Russia)	Veernon L. Newhouse	(USA)
Richard C. Cobbald	(Canada)	William D. O`Brien Jr.	(USA)
John Duke	(USA)	Jonathan Ophir	(USA)
Kenneth Erikson	(USA	Song Bai Park	(Korea)
Helmut Emmert	(Germany)	Peder C. Pedersen	(USA)
Mathias Fink	(France)	E. J. Pisa	(USA)
Woon Siang Gan	(Singapore)	John P Powers	(USA)
James F. Greenleaf	(USA)	Piero Tortoli	(Italy)
Michael Insana	(USA)	Peter N.T. Wells	(UK)
Joie P. Jones	(USA)	Robert Waag	(USA)
Lawrence W. Kessler	(USA)		

Local Arrangements Committee
Paul Barbone   Peder C Pedersen   Thomas L. Szabo

ACKNOWLEDGMENTS

The organizers of the 23rd International Symposium on Acoustical Imaging gratefully acknowledge the financial support by

Acoustical Society of America

Advanced Technology Laboratories

Hewlett-Packard Company

Materials Systems Inc

RITEC, Inc

Bradley Department of Electrical Engineering
Virginia Polytechnic Insitute

Department of Aerospace and Mechanical Engineering
Boston University

PREFACE

The contents of this volume are the proceedings of the 23rd International Symposium on Acoustical Imaging which took place 13-16 April, 1997, in Boston, Massachusetts. The first Symposium met 25 years ago. Originally the Symposium met in California, then elsewhere within the United States but beginning in 1988 the Symposia began to meet outside of the United States as well. It is now being held about every eighteen months, alternately in the United States and then outside. The present pattern is to hold one meeting in East Asia, then in the USA, then in Europe and again in the USA. However, for scheduling reasons the next Symposium will be in Santa Barbara, California, followed by England and then East Asia.

It is to be noted that the Symposium is a free standing institution, not associated with any other organization. Each meeting is the total responsibility of its chairman with the advice of past chairmen. Papers are submitted in response to the call for them and reviewed by an International Scientific Advisory Board.. The quality depends entirely on the response to the call. It is gratifying to note that the Symposium has attained the status that attracts high quality contributions despite (or perhaps because of) the loose structure. Two factors that have appeal are that there is only one session and that there is time during the meeting for extensive discussion. It is the opportunity to meet one's scientific peers, to exchange views, to be informed of the work at the forefront, to be recognized and to present one's work for evaluation and recognition, which can best be found at a small meeting.

The meeting was attended by about 130 persons who came from 20 countries, attesting to the international character of the Symposium. Because the Symposium is limited to three days, and the large number of submitted papers, about half the papers were presented orally and the rest by posters. Both types are regarded of equal merit and no distinction appears in this volume of proceedings.

A prize for the best poster was awarded to Erikson et al. The prize for the next best poster was awarded to Bridal et al. Each prize was a copy of the renowned Paul Revere bowl, who was a silver smith as well as a famous American patriot.

We wish to thank the members (past chairmen) of the Executive Advisory Council who provided guidance based on their past experience for many questions that arose during the planning and organizing phases. The International Scientific Advisory Board, chaired by Professor Ferrari, reviewed all the abstracts attesting to their relevance and quality. We thank the members of the Board for their patience and care.

The three members of the Local Arrangements Committee did yeoman's work before the Symposium and during it. Many people in the Boston area attended the meeting because they were recruited by these three. Professors Barbone and Pedersen brought their students to operate the projectors and when more attendees arrived than expected they pitched in to provide more books of abstracts. The Symposium proceeded smoothly without a hitch because of the efforts of this supporting crew. We are grateful to them.

Finally we wish to thank those who contributed financially and thereby helped make the Symposium solvent. With this support the International Symposium on Acoustical Imaging continues to be a free standing institution and a platform for reporting high quality research in this rapidly advancing field of acoustics

Sidney Lees
Leonard A. Ferrari

CONTENTS

Acoustic Microscopy

Biomedical

Bone and Cartilage

Cardiology

Components and Systems

Doppler

Mathematics and Physics

Nondestructive Evaluation

Oceanography

SYSTEM RESOLUTION ANALYSIS OF THE
SCANNING TOMOGRAPHIC ACOUSTIC MICROSCOPE

S. Davis Kent and Hua Lee

Department of Electrical and Computer Engineering
University of California, Santa Barbara
Santa Barbara, California 93106–9560

INTRODUCTION

The scanning tomographic acoustic microscope (STAM) is an imaging device capable of producing tomograms of the acoustic properties of the internal structures of thick, optically opaque materials. Its usefulness extends to many different areas, such as in the study of biological specimens where the determination of acoustic velocity and attenuation indicate the health of tissues, and in nondestructive evaluation of electronic components and composite materials where voids and other flaws are easily visualized using acoustic energy.[1, 2]

The STAM is a complicated imaging device that employs many different components to permit high-resolution reconstructions. In addition to making imaging possible, they also present limits on the ability of the overall device to form accurate images. To understand their influence on system performance, a resolution-limit analysis has been performed on an automated STAM system. This study considers each component individually to determine its effects and then combines these results to determine the performance of the overall system. This resolution analysis is useful in several ways. First, the analysis aids in the interpretation of reconstructions. Second, it can be used to determine weaknesses in the system, permitting effort to be directed to improving problem areas. Third, this resolution analysis illuminates the capabilities and limitations of the STAM, to determine better its applicability to different imaging situations.

This paper presents the resolution analysis of the STAM using the technique of spectral coverage. First, the STAM image formation process is described. Then, the spectral coverage technique is described, and a theoretical analysis of the STAM is provided. To verify the results of this study, experimental results are presented that show close agreement between the predicted and measured resolution-limit.

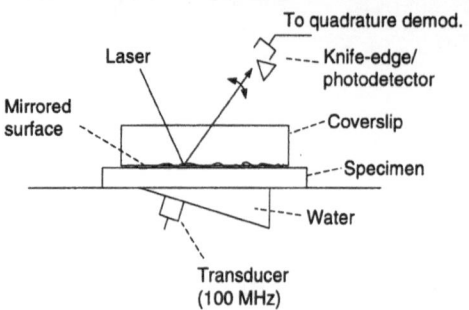

To quadrature demod.

Laser

Knife-edge/
photodetector

Mirrored
surface

Coverslip

Specimen

Water

Transducer
(100 MHz)

Figure 1: Geometry of Scanning Tomographic Acoustic Microscope Stage.

IMAGE FORMATION

The STAM is a transmission-mode device. As illustrated in Fig. 1, a transducer produces ultrasonic plane waves that propagate upward through a water coupling medium. They enter the specimen, propagate through it, and become modulated by its structure. The waves exit the top of the specimen and impinge upon the mirrored bottom of a plastic coverslip, creating a dynamic ripple. A raster-scanned laser beam reflects off of the reflective surface at an angle related to the dynamic ripple. A knife-edge/photodiode detector system converts this angular motion into an electrical signal that is quadrature demodulated, producing both the in-phase (I) and quadrature-phase (Q) components of the signal present on the coverslip. These signals are digitized and stored for computer processing.

In STAM imaging, plane-to-plane propagation is used to form subsurface reconstructions. From the I and Q components, the magnitude and phase of the coverslip wave field can be computed. Knowledge of the phase is crucial since it allows backpropagation, or focusing, of the received wave field to the desired depth, Δz, by application of the backpropagation filter of Eq. (1).

$$H(f_x, f_y; \Delta z) = \begin{cases} \exp\left\{-j2\pi\Delta z\sqrt{1/\lambda^2 - f_x^2 - f_y^2}\right\}, & f_x^2 + f_y^2 < 1/\lambda^2 \\ 0, & otherwise \end{cases} \tag{1}$$

Here, f_x and f_y are the spatial frequency indices in the x and y directions respectively, and λ is the wavelength of insonification.[3, 4, 5] In holographic reconstructions, demodulation of the resulting wave field yields the final image.

To improve upon the reconstruction quality of holographic imaging, the multiple-frequency and multiple-angle tomography are used. In tomographic imaging, many received wave fields are captured, each corresponding to different frequencies or angles of insonification. These techniques increase the spectral coverage of the specimen which improves system resolution, and increases the signal-to-noise ratio. In multiple-angle tomography, physical constraints require that the specimen be rotated inside of the microscope. During the reconstruction process, this rotation requires a subsequent numerical rotation of the received wave fields to provide a common orientation and proper superposition of projections. Using the received wave fields, a minimum mean-squared error estimate of the transmission profile at depth $z = z_s$ of a single slice through the specimen is computed to be

$$\tilde{t}(x, y) = \frac{\sum_{m=1}^{M} u^{(m)^*}(x, y; z_s) v^{(m)}(x, y; z_s)}{\sum_{m=1}^{M} u^{(m)^*}(x, y; z_s) u^{(m)}(x, y; z_s)} \tag{2}$$

where $*$ represents the complex conjugate operator and $v^{(m)}(x, y; z_s)$ and $u^{(m)}(x, y; z_s)$ are estimates of the mth wave fields above and below the plane at $z = z_s$. The wave field $v^{(m)}(x, y; z_s)$ is obtained by backpropagating and aligning the mth received wave field. $u^{(m)}(x, y; z_s)$ is either estimated iteratively or assumed known from other measurements.

SPECTRAL COVERAGE TECHNIQUE

The spectral coverage technique specifies the resolving capability of an imaging system in terms of its spatial frequency coverage. In general, the larger the spectral coverage, the better the resolving capability. This relationship yields resolution measurements consistent with the Rayleigh resolution criterion.[5, 6] The spectral coverage technique for resolution analysis exploits the reciprocal-spreading relationship between the space and spatial frequency domains. For example, a point-scatterer imaged by an ideal diffraction-limited system yields received wave fields bandlimited to the circular region $f_x^2 + f_y^2 + f_z^2 \leq 1/\lambda^2$ where λ is the wave length of insonification. The corresponding point-spread function is the Bessel function

$$h(\rho) = J_1(2\pi\rho/\lambda)/\rho \tag{3}$$

where ρ is a radius parameter. As $1/\lambda$ increases, corresponding to greater spectral coverage, the width of the main lobe of the point-spread function decreases, indicating improved resolution.

The relationship between the resolution–limit and the spectral coverage is described by

$$\Delta S \cdot W = c \tag{4}$$

where ΔS is the resolution cell size, W is the spectral coverage or bandwidth, and c is a system and resolution-definition dependent constant.

In order to apply Eq. (4) to a system, a baseline or reference must be established by determining c for a known system. Using the ideal, diffraction limited system described earlier, the bandwidth is taken to be $W = 2/\lambda$. The corresponding 3–dB width of the point-spread function is measured to be 0.5145λ. The constant c is thus calculated to be

$$\begin{aligned} c &= (0.5145\lambda)(2/\lambda) \\ &= 1.029. \end{aligned} \tag{5}$$

THEORETICAL APPLICATION

The STAM has many factors that limit the system's spectral coverage, including (1) the system electronics, (2) coverslip effects, (3) propagation effects, (4) the finite aperture, and (5) the detector system.[7] Although each of these components influences the resolution-limit of the microscope, under normal operation only the last three play a significant role in system performance, with the detector system dominating. In this section, the resolution-limit of the STAM is predicted using the spectral coverage technique.

The scanning laser's cross-section and the knife-edge detector combine to yield the detector system's response characteristics. Assuming a Gaussian intensity profile of the laser beam with a spot-size of radius r_0 meters, the detector's transfer function is[8]

$$M(f_x, f_y) = \mathrm{erf}\left(\frac{f_x r_0 \pi}{\sqrt{2}}\right) \exp\left(\frac{-\pi^2 r_0^2 (f_x^2 + f_y^2)}{2}\right). \tag{6}$$

The STAM is designed to operate at the peak of $M(f_x, f_y)$. The exponential roll off of all frequency components beyond a small distance from the peak, however, greatly restricts the detector's passband. Assuming a 3–dB cutoff, the passband in the $f_x - f_y$ plane is limited to the small shaded region seen in Fig. 2. Also included in Fig. 2 are the $1/\lambda$ propagation cutoff frequencies for water and acrylic specimens assuming an insonification frequency of 103.5 MHz. Since the spectral components of the received wave fields must satisfy the dispersion relationship

$$f_x^2 + f_y^2 + f_z^2 = 1/\lambda^2 \tag{7}$$

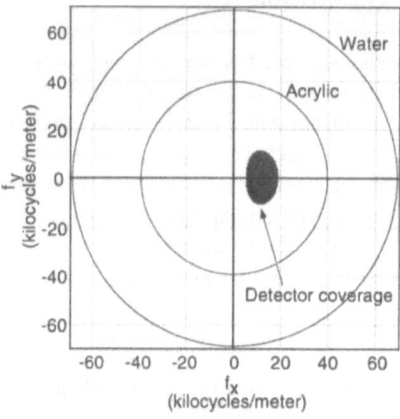

Figure 2: Spectral coverage in the $f_x - f_y$ plane as allowed by the STAM's detector system. Also shown are the propagation cutoff frequencies for water and acrylic (PMMA) specimens.

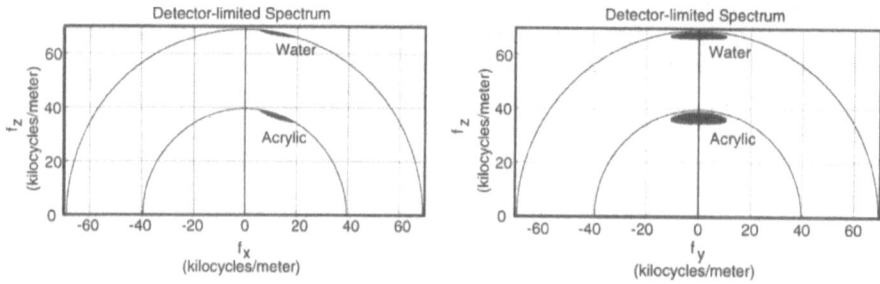

Figure 3: Spectral coverage in the $f_x - f_z$ and $f_y - f_z$ planes as allowed by the STAM's detector system. Also shown are the propagation cutoff frequencies for water and acrylic specimens.

the spectrum of the received wave field resides on the positive-half of the hemisphere described by Eq. (7). The regions of spectral coverage in the $f_r - f_z$ and $f_y - f_z$ planes are illustrated in Fig. 3. The small amount of coverage in the f_z direction is caused by the small rate of curvature of the hemisphere in the vicinity of the detector's passband. It is interesting to note that the coverage is greater for the acrylic specimen than for the water specimen, as a result of the longer wavelength in acrylic. This result is counter intuitive, and indicates that superior range resolution is obtained for longer wavelengths.

Spectral coverage for the case of tomographic imaging is determined by superimposing the regions of spectral coverage of many holographic projections. Considering the case of multiple-angle tomography, each projection whose spectrum is represented in Figs. 2 and 3 is backpropagated, demodulated, and rotated to a common orientation. Backpropagation does not affect the spectral coverage. However, demodulation shifts the spectra to the origin, and rotation rotates the spectra of the received wave fields about the $z-$axis. Superposition of these spectra produces regions of spectral coverage illustrated in Fig. 4.

The bandwidths of the spectra of Fig. 4 are measured to be $W_{cr} = 21350\ Hz$ and $W_r = 4178\ Hz$ in cross-range and range, respectively. Applying these values to Eq. (4), the resolution-limit in cross-range and range are predicted to be

$$\Delta r_{cr} = 24.1\ \mu m \qquad (8)$$

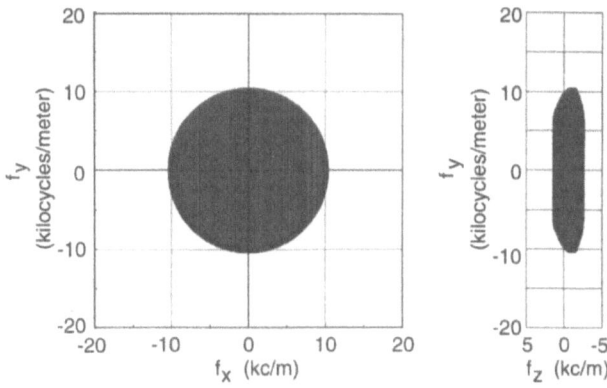

Figure 4: Spectral coverage in the $f_x - f_y$ and $f_y - f_z$ planes as allowed by the STAM's detector system.

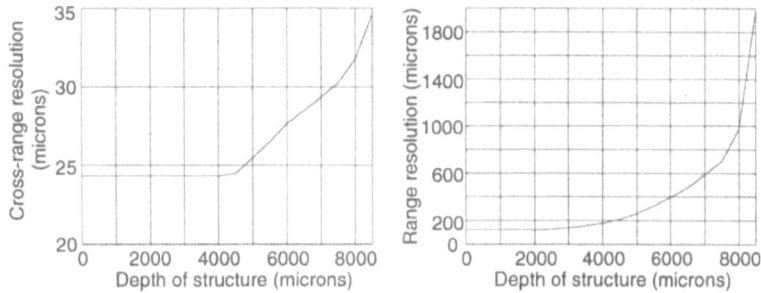

Figure 5: Cross-range and range resolution-limit as influenced by the finite aperture and detector system.

and

$$\Delta r_r = 123.2 \, \mu m. \tag{9}$$

where resolution is defined as one-half the resolution cell size.

For most imaging situations, the effects of the finite aperture can be ignored. However, for thick specimens the finite aperture limits spectral coverage, lowering the resolving ability. As the depth of structures increases, the angle subtended by the aperture decreases, introducing a lower spatial frequency cutoff. For deep structures, this cutoff frequency encroaches upon the passband allowed by the detector system. The effect of the finite aperture is seen in Fig. 5, with worsening resolution as the finite aperture increasingly excludes the detector's passband.

A similar effect is observed with increasing wavelength, with the exception that, initially, the range resolution improves with increasing wavelength. However, when the wavelength exceeds approximately 55μm the $1/\lambda$ propagation cutoff frequency restricts the detector system's passband, and the resolving ability decreases.

EXPERIMENTAL RESULTS

To evaluate the accuracy of the theoretical predictions, a single-layered specimen consisting of a 25μm-thick copper grid encased in acrylic at a depth of 1090μm was insonified from 36 different directions, spaced evenly 10 degrees apart. The frequency of insonification was 103.5 MHz. Multiple-angle reconstructions in both cross-range and range were formed. The reconstructions were then used to estimate the point-spread functions of the system. The resolution-limits were measured from these estimates to be

$$\Delta r_{cr,exp} = 29.9 \, \mu m$$
$$\sigma_{cr,exp} = 2.1 \, \mu m$$

(10)

and

$$\Delta r_{r,exp} = 158 \, \mu m$$
$$\sigma_{r,exp} = 31.9 \, \mu m$$

(11)

in cross-range and range, respectively. Differences between theoretical and experimental results can be attributed to the low signal-to-noise ratio for this specimen which restricts the usable bandwidth of the detector. If a 2–dB signal bandwidth is adopted for the passband of the knife-edge detector, for example, the theoretical predictions for the resolution-limit increase to

$$\Delta r_{cr,2dB} = 29.5 \, \mu m$$

(12)

and

$$\Delta r_{r,2dB} = 157 \, \mu m,$$

(13)

which are in closer agreement with the experimental results. Also significant is the consistency of the ratios of the range to cross-range resolutions for both experimental and theoretical evaluations. Both ratios indicate that the range resolution is 5.3 times worse than the cross-range resolution.

CONCLUSION

In this paper, the spectral coverage technique for determining the resolution-limit of an imaging system was described. This technique was applied to an automated STAM system, and the theoretical results were compared with experimentally determined values. For a high noise environment, theoretical and experimental results indicate cross-range and range resolution-limits on the order of 29.5 and 157μm, respectively. The effects of the system's finite aperture and the wave length of insonification were presented, also.

ACKNOWLEDGMENT

This research is supported by the National Science Foundation under Grant No. MSS-9020556.

REFERENCES

1. J. F. Havlice and J. C. Taenzer. Medical ultrasonic imaging: An overview of principles and instrumentation. *Proceedings of the IEEE*, 67(4):620–641, Apr. 1979.

2. L. W. Kessler and D. E. Yuhas. Acoustic microscopy — 1979. *Proceedings of the IEEE*, 67(4):526–536, Apr. 1979.

3. Z. C. Lin, H. Lee, and G. Wade. Back-and-forth propagation for diffraction tomography. *IEEE Transactions on Sonics and Ultrasonics*, SU-31:626–634, Mar. 1984.

4. R. Y. Chiao. *Signal Processing and Image Reconstruction for Scanning Tomographic Acoustic Microscopy*. PhD thesis, University of Illinois, Urbana-Champaign, July 1990.

5. J. W. Goodman. *Introduction to Fourier Optics*. McGraw-Hill, New York, 1968.

6. H. Lee. Formulation for quantitative performance evaluation of holographic imaging. *Journal of the Acoustical Society of America*, 84(6):2103–2108, Dec. 1988.

7. S. D. Kent. *Image Formation and System Analysis of the Scanning Tomographic Acoustic Microscope*. PhD thesis, University of California, Santa Barbara, June 1997.

8. Z.-C. Lin. *A planar ultrasonic tomographic imaging system*. PhD thesis, University of California, Santa Barbara, Aug. 1984.

VISUALIZATION OF LIVING CELLS BY ACOUSTIC MICROSCOPY

Yoshifumi Saijo[1], Hidehiko Sasaki[1], Hiroaki Okawai[1], Shin-ichi Nitta[1], and
Motonao Tanaka[2]

[1]Department of Medical Engineering and Cardiology,
 Institute of Development, Aging and Cancer, Tohoku University.
 4-1 Seiryo-machi, Aoba-ku, Sendai 980-77, JAPAN.
[2]Tohoku Kosei-nenkin Hospital
 1-12-1 Fukumuro, Miyagino-ku, Sendai 983, JAPAN.

INTRODUCTION

With the development of molecular biology, there is an increasing need to observe the morphology of the cultured cells, which are known to change their properties or to transform themselves in the process of cellular culture. The standard staining techniques of pathology, which exterminate the intracellular organella, can not be applied to the visualization of living cells. We believe that the acoustic microscopy is one of the best methods to visualize the morphology of the cells, since the method does not require the fixation or staining.

Acoustic microscopy not only visualize the cells, but also provides physical information of the living cells. To understand the physical properties of the cells is important in the pathology. For example, the metastasis of the cancer cells may have some relationships to the physical properties of the cells.

The objective of the present study is to visualize the cultured cells by acoustic microscopy. This paper also includes the assessment of the acoustic properties of the internal organella of the cell.

METHODS

Figure 1 is a block diagram of the SAM system[1-4]. The acoustic focusing element comprises a ZnO piezoelectric transducer with a sapphire lens. The ultrasonic frequency is variable over the range of 100 to 210 MHz and the beam width at the focal volume ranges from 5 μm (at 210 MHz) to 10 μm (at 100 MHz). The focusing element is mechanically scanned at 60 Hz, in the lateral direction (X) above the specimen, which remains stationary on the specimen holder, while the holder is scanned in other lateral direction (Y) in 8 sec, thus providing two-dimensional scanning. The mechanical scanner is so arranged that the ultrasonic beam is transmitted for every 1 μm interval over a 0.5 mm width. The number of sampling points is 480 in one scanning line and 480×480 points make one frame. Both amplitude and phase images are obtained in a field of view 0.5mm×0.5mm.

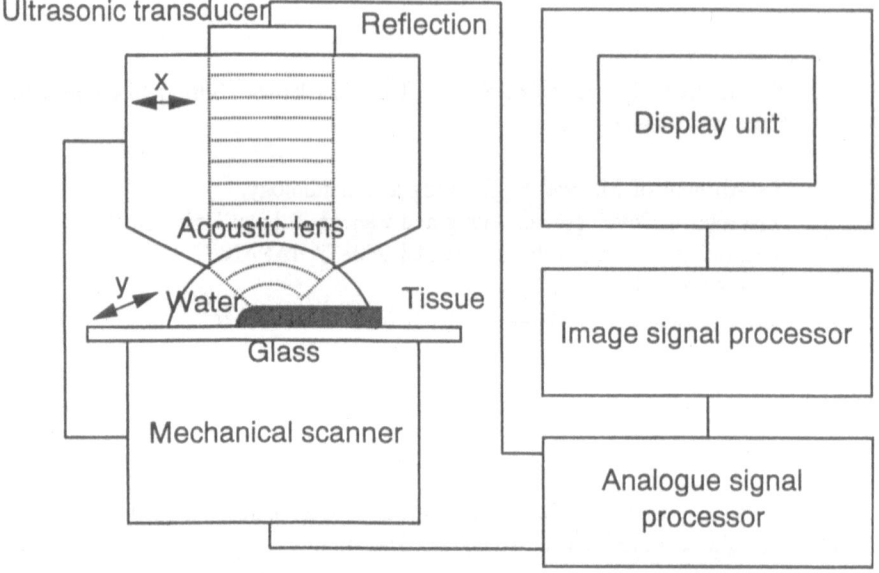

Figure 1. Block diagram of the scanning acoustic microscope (SAM) system.

A–549 cells, a cell line of adenocarcinoma of the lung, were cultured in the EMEM medium supplemented with 10% heat–incubated fetal bovine serum at 37°C. Figure 2 is the optical image of A–549 cells obtained with an inverted microscopy. Both spindle–shaped and round–shaped cells. The change of the shape indicated the cell cycle. The ratio of the nucleus (N) / cytoplasm (C) also indicate the cell cycle; N / C ratio is high when the nucleus is divided. The characteristic behavior of A–549 cells is to adhere the glass surface and to grow in monolayer form. A single cell is approximately 15 μm in size, and 2 μm in thickness[5].

Phosphate buffer saline, which was maintained at 37°C, was used for the coupling medium between the transducer and the specimen.

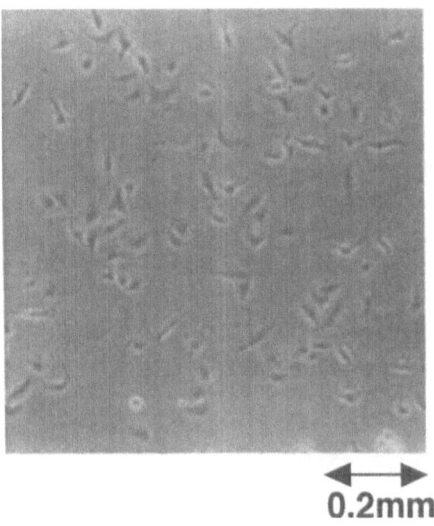

0.2mm

Figure 2. An inverted microscopy image of A–549 cells. ($\times 100$)

RESULTS

Figure 3 shows the acoustic microscopy images of A–549 cells. The left picture is the amplitude image obtained at 210 MHz, and the right one is the phase image. The outer silhouette of the cells are represented by both amplitude and phase images.

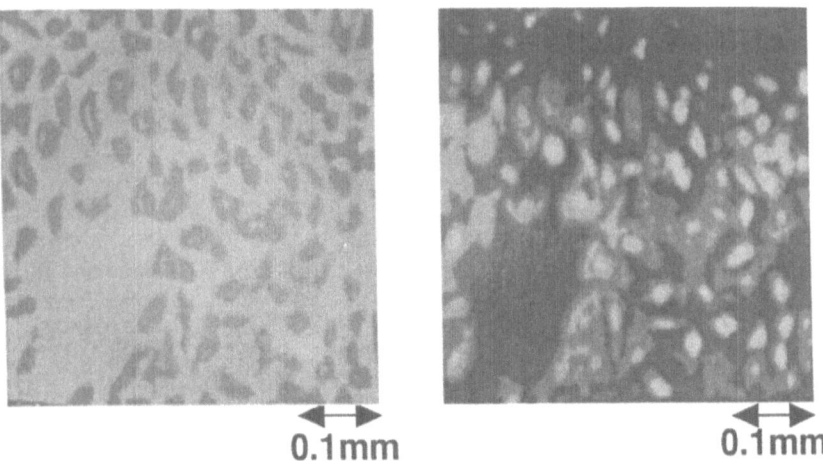

0.1mm   0.1mm

Figure 3. Acoustic microscopy images of A–549 cells. (left: amplitude image, right: phase image)

Although both images exhibit the structure of the cell, the phase image exhibited the intracellular structure more clearly than the amplitude image. The nucleus of the A–549

9

cells are shown in the center part of the cell.

The intensity is higher in the nucleus than in the cytoplasm, in both amplitude and phase images.

A–549 cells are still living in the phosphate buffered saline during the acoustic microscopy study. The SAM system equipped in the present study requires 8sec to make one frame, since the Y–direction scan needs 8sec.

In order to test the stability of the cells during SAM measurement, two consecutive scans were done. Figure 4 shows the results. The second scan (upward scan) was done soon after the first scan (downward scan) was done. So the time interval of the two scans was 8sec.

There is no significant difference between two consecutive scans, although the image was completely changed after 30min of the first scan.

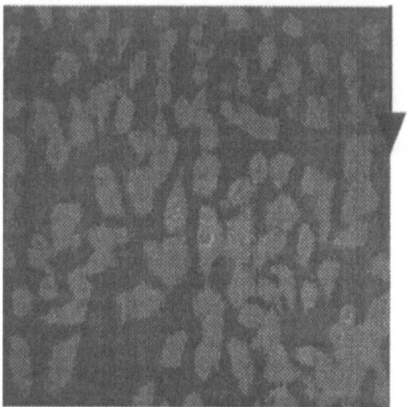 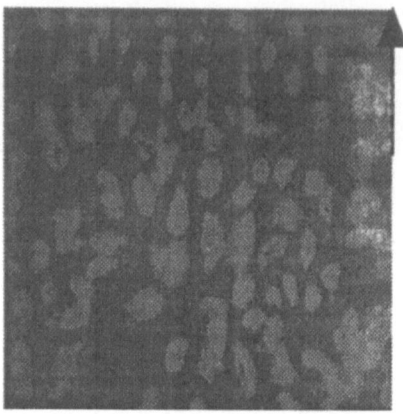

Figure 4. The amplitude images of A–549 cells. Two consecutive scans of A–549 cells were represented. The time interval of two scans was 8sec. (left: downward scan, right: upward scan)

DISCUSSION

The acoustic lens focuses the ultrasonic beam to 5μm at the focal point and the mechanical scanner enabled the every 1μm interval's transmission. These specification may be enough to observe the A–549 cells, since the size of the cell is approximately 15μm.

The ultrasonic frequency used in the present study is 100–210 MHz. If the sound speed in the cell is 1550 m/s, the wavelength of the ultrasonic beam is calculated as 7.5μm. Although the precise measurement of the thickness was not performed in the present study, the thickness of the A–549 cell is considered to be approximately 2μm. As the ultrasonic beam passes through the specimen twice in the transmission–reflection method used in the present study, the whole distance is 5μm at most. Then only the vague intracellular structure of the A–549 cells might be exhibited by the amplitude method.

The phase image clearly exhibited the nucleus of the cell. In the previous study, the attenuation and sound speed of the nucleus were considered to be higher than those of the cytoplasm, because the average values were high where the N / C ratio is high[2]. In the present study, the nucleus showed higher phase shift than the cytoplasm. The result indicates that the sound speed of the nucleus is higher than that of the cytoplasm. The nucleus is consisted of nucleic acid and the cytoplasm is mainly consisted of protein. The molecular weight of the nucleus is known to be larger than the cytoplasm. So the result of the present study is considered to be ideal from the view point of relationship between the molecular weight and sound speed.

Figure 5 is the phase image of A–549 cells and the digital zoom up image of the region of interest (ROI). The phase shift of the nucleus is larger than that of the cytoplasm, however, the phase shift of the cytoplasm close to the nucleus is smaller than that of the ordinary cytoplasm. This finding may proof that the center part of the cell is the nucleus, of which specific acoustic impedance is different from that of cytoplasm.

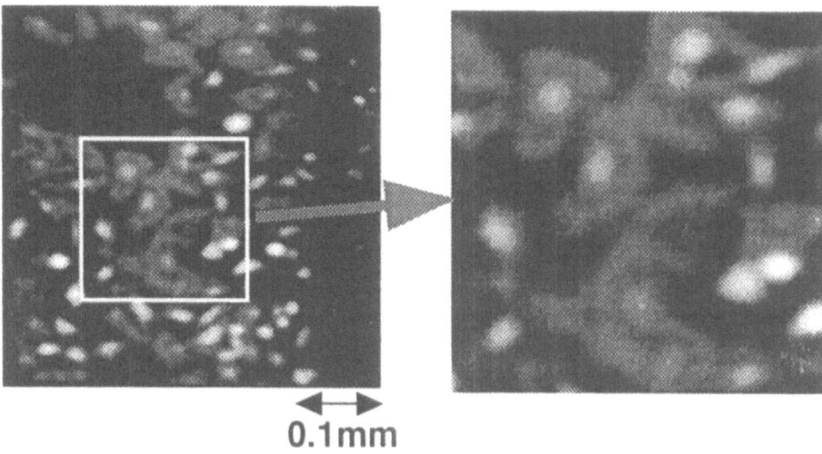

0.1mm

Figure 5. Phase image of A–549 cells (left) and the digital zoom up image of the ROI (right).

Figure 6 is the schematic illustration of the propagation of the ultrasonic beam in the cell. If the high phase shift is determined only from the thickness of the cell, the phase shift should be gradually increased toward the center portion. The small phase shift area near the center portion indicates that some kinds of reflections or scattering are occurred at the interface between two different impedance materials. Thus the nucleus exhibited larger phase shift than the cytoplasm, which indicated the sound speed would be higher in the nucleus than in the cytoplasm.

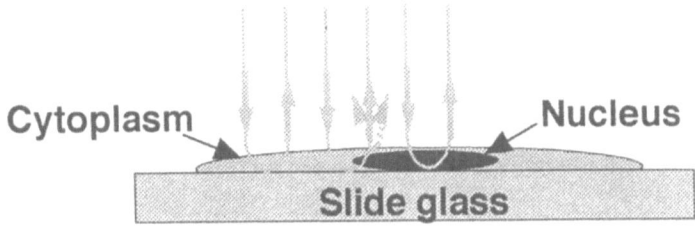

Figure 6. Schematic illustration of the ultrasonic beam propagation in the cytoplasm and the border between the cytoplasm and the nucleus.

CONCLUSION

The morphology of the living cells were visualized by acoustic microscopy, without any pathological treatments. The method should be applied for cellular biology, since the physical properties of the living cells can be obtained.

REFERENCES

1. Okawai H, Tanaka M, Dunn F, Chubachi N, Honda K, Qualitative display of acoustic properties of the biological tissue elements, *Acoustical Imaging* 17:193 (1988).
2. Saijo Y, Tanaka M, Okawai H, Dunn F, The ultrasonic properties of gastric cancer tissues obtained with a scanning acoustic microscope system, *Ultrasound in Med and Biol* 17:709 (1991).
3. Saijo Y, Sasaki H, Okawai H, Tanaka M, Dunn F, Intravascular ultrasound and acoustic microscopy evaluation of aortic wall, *Acoustical Imaging* 21:423 (1994).
4. Saijo Y, Sasaki H, Okawai H, Tanaka M, Development of ultrasonic spectroscopy for biomedical use, *Acoustical Imaging* 22:335 (1995).
5. Kobayashi S, Fujimura S, Characterization and the clinical application of cultured human pulmonary carcinoma cells, *Tohoku J Exp Med* 168:375 (1992).

SURFACE ACOUSTIC WAVE DISPERSION STUDIES ON THE LAYERED CU/SI STRUCTURE USING SCANNING ACOUSTIC MICROSCOPY

Zuliang Yu [1], Siegfried Boseck [2]

[1] Laboratory of Information Technology, University of Hannover
  D-30167 Hannover, Germany
[2] Department of Physics, University of Bremen
  D-28359 Bremen, Germany

ABSTRACT

Surface acoustic wave (SAW) dispersion studies on the layered Cu/Si structures for the elastic characterization have been performed by scanning acoustic microscopy (SAM). The bonding quality at the interface has been also evaluated by examining the SAW dispersion changes. We have proposed an interface layer modeling approach to interpret the experimental results and to examine the elastic properties at the interface. The studies have shown that the behaviors of Sezawa wave is very sensitive to the interface properties and the bonding conditions in the used samples.

INTRODUCTION

Scanning acoustic microscopy (SAM) has proven to be a useful technique for the characterization of a thin film deposited on an elastic substrate [1]. In this work we have used SAM technique to carry out the surface acoustic wave (SAW) dispersion studies on thin layered Cu/Si structures. In the silicon integrated-circuits technology, copper is a material of high potential for ULSI multilevel metallization [2]. Usually, Rutherford backscanning spectroscopy (RBS), transmission electron microscopy (TEM), Auger electron spectroscopy (AES) and scanning electron microscopy (SEM) are the major tools for the characterization of the Cu/Si structures to analyze microstructures and interface structures of thin films as well as to determine the compositions. However, there has been to date little study on the elastic characterization, although it should be an important subject. The aim of our studies is twofold. First of all to compare the SAW dispersion measurements performed on the samples with a predicted good adhesion at the interface, with the theoretical predictions. Its goal is examining the elastic properties of the formed interface phase structure. Secondly to analyze the SAW dispersion changes performed on the samples with a predicted inperfect bonding at the interface. The goal is the use of the metrology mode of SAM to evaluate quantitatively the quality of the bond.

SPECIMENS AND SAW DISPERSION MEASUREMENTS

The Cu layers deposited on Si(100) wafer were prepared by sputtering process. The four specimen groups I, II, III and IV with the Cu layerthicknesses of 109 nm, 156 nm, 219 nm and 343 nm, respectively, were designed for obtaining a wide range of frequency-layerthickness of the studied samples. In order to examine the bonding properties at the interface, the used specimens were made with different processes that would predict good as well as bad adhesions at the interface. So, each of the specimens has two kinds of the bonding properties. The specimen groups I, II and III with bad adhesion were prepared simply with substandard sputtering process. The structure mode of the sample IV with bad bonding was specifically designed, by introducing a thin TiN film, approximately 20 nm, at the interface between the top Cu-layer and the Si-substrate. Usually, Cu layers have poor adhesion on oxides and nitrides by sputtering technique[2]. Thus, it can be predicted that the prepared sample Cu/TiN/Si(100) has inperfect bonding at the interface between the Cu-layer and the TiN-layer.

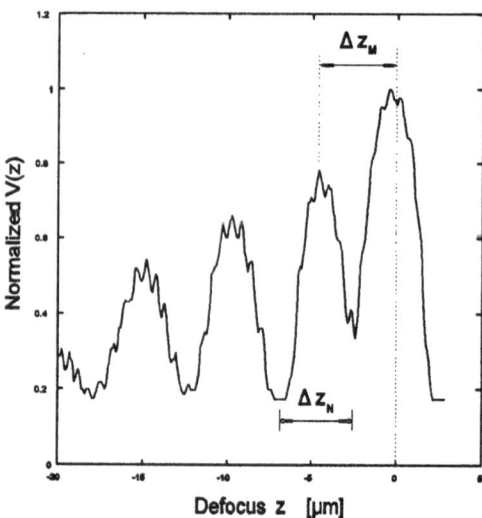

Figure 1. Measured V(z) curve with water ripple, Δz_N and Δz_M

The SAW dispersions were measured by Leica ELSAM from V(z) curve. Both Δz_N and Δz_M measurements [3,4] have been carried out in this work to extend the upper range of the measured SAW velocity. Δz_N measures the dip spacing of the oscillations in V(z) curve, and Δz_M measures the separation along the lens axis between focal plane and first V(z) peaks, see Fig.1. The SAW velocities are deduced by the analytical equation (1) and the empirical equation (2), respectively,

$$v_{LSAW} = v_W \; / \; [\,1 - (1 - v_W / (\,2 f \Delta z_N\,)\,)^2\,]^{\,1/2} \;, \tag{1}$$

$$v_{LSAW} = v_W \; / \; \sqrt{[\,1 - (1 - \frac{3 v_W}{4 f \Delta z_M})^2\,] / \alpha} \;, \tag{2}$$

where v_W is the sound velocity in water and f is the operating frequency. α is an empirical parameter, which depends largely on the acoustic lens material.

EXPERIMENTAL RESULTS AND DISCUSSION

The measured dispersion data points are shown in Fig.2, denoted by "$\star$" and "o" for good and bad adhesions, respectively. The theoretically predicted Rayleigh and Sezawa dispersion curves of a single-layered Cu/Si system are also superimposed in Fig.2 for comparison.

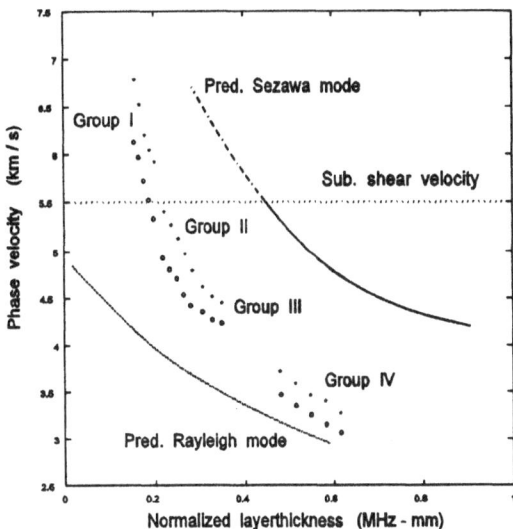

Figure 2. Measured SAW dispersion results, superimposed by theoretically predicted curves. "$\star$" and "o" indicate the measured data points for the predicted good and bad bonding structures, respectively; the lines are the predicted SAW dispersion curves.

In appearance, Fig.2 reveals that the experimental data for the specimen groups I, II and III tend towards the predicted Sezawa dispersion, though they deviate still obviously from Sezawa dispersion. The measured data for the specimen IV are already closer to the Rayleigh dispersion. But, they lie also above the theoretical curve. Thus, all of the measured results cannot agree with the theoretical predictions and provide anomalous SAW dispersion behaviors for the used specimens. Furthermore, all of the measured data for the bad adhesions are always less than that of the corresponding good adhesions.

Discussion: Usually, Rayleigh wave is the fundamental SAW mode excited by SAM in a thin metallic layered structure, whereas Sezawa mode may be excited only under a rather restrictive condition [1]. It is not clear why the experimental results deviate from the Rayleigh dispersion, but rather tend towards the Sezawa dispersion. It is necessary to make effort to interpret this anomaly. We consider that the reasons should result from the specific excitation mechanism of relevant SAW modes in the used specimens, caused by a possible formed intermediate layer at the interface between the Cu layer and Si substrate, somehow producted in the sample preparation. Usually, in sputtering process, Cu can diffuse into active regions of devices where it is known to form a silicide compound Cu_xSi_{1-x} at the Cu/Si interface [2]. After growth of an intermediate phase structure the measured SAW dispersions should be considered as the comprehensive elastic properties of the Cu/Si structure containing the resulting intermediate, which thus cannot agree with the theoretical predictions of the single-layered Cu/Si structure. In the bad specimens the formed intermediates should be coarser due to the substandard process, which largely changes the SAW dispersion properties compared with that in the good bonding.

INTERFACE LAYER MODELING

We have proposed a modeling approach to characterize the elastic properties of the formed intermediate structures, by considering the entire interface between the Cu layer and the Si substrate, as well as its vicinity, as a modeling interface layer. Introducing a modeled interface layer can narrow the agreement of the experimental results with the theoretical predictions, and allows to evaluate quantitatively the interface properties. The effective parameters of the modeling interfacial layer, i.e., the thickness, the density, the longitudinal wave and shear wave velocities, are determined through SAW dispersion matching by simulation test. Usually, the shear wave velocity has the greatest influence on the SAW dispersion, especially on Sezawa mode. The density and the thickness also have some influences, while the longitudinal wave velocity has very little influence and may be taken as same as the top Cu layer. We have firstly carried out the modeling approach for the specimens with good adhesion. For the specimen groups I, II and III, the approach has been made to match the SAW dispersions on Sezawa mode with the measured results, since the measured data seem to tend towards the Sezawa dispersion. The parameters of the modeling layer obtained by the numerical simulation are shown in Table 1, from which the calculated Sezawa dispersion was well agreed with the experimental results, see Fig.3. Similar approach also applied to the specimen IV. We have started with trying to match the SAW dispersion on Rayleigh mode, since the measured results appear to be closer to the Rayleigh dispersion. However, no matter how the parameters have been chosen, the calculated Rayleigh dispersion cannot agree with the experimental data. We have then returned to make the dispersion matching on Sezawa mode. With reduced shear velocity and less density, as well as with thicker thickness, compared to that of the specimen groups I, II and III, the calculated Sezawa dispersion was also well agreed with the experimental results, see Table 1 and Fig.3. This implies that the measured results for the specimen IV should be still Sezawa dispersion. The modeling approach suggests that Sezawa mode was dominantly excited by the acoustic microscope beams in all specimens. The thinner intermediates were indeed formed at the interface between the Cu layer and the Si substrate.

Table 1. Effective parameters of the modeled layers for good and bad specimens, the listed values indicate the corresponding ratios to that of the top Cu-layer.

Sample	Predicted bonding	Layer thickness	S-wave velocity	Density
I,II,III Cu/Si	good	0.10	0.25	0.40
	bad	0.20	0.22	0.38
IV Cu/Si and Cu/TiN/Si	good	0.35	0.18	0.35
	bad	0.38	0.14	0.70

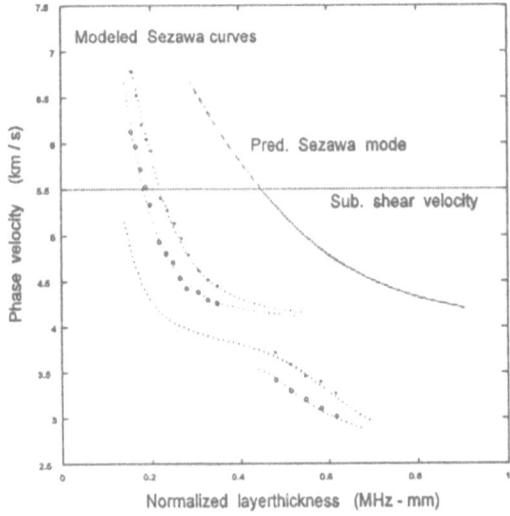

Figure 3. The SAW dispersion curves with the modeled interface layers, superimposed by the experimental results. The dotted lines indicate the matched SAW dispersion curves.

Next, we get into the bonding quality evaluation using the modeling approach. For all of the specimens with bad bonding, the excited SAWs by the acoustic microscope beams appear to be still the Sezawa mode. But, Their velocities are significantly perturbed and always less than that in the good adhesion. This implies that the inperfect bonding conditions change the interface structures and affect the SAW propagation properties. For the specimen groups I, II and III, the introduced modeling interface layers lay still at the interface between the Cu layer and the Si substrate. While, for the Cu/TiN/Si sample, the modeled layer was introduced at the interface between the top Cu layer and the TiN layer from the preparation knowledge. After the numerical experiments two sets of the modeling parameters were determined for the specimen groups I, II and III, as well as for the Cu/TiN/Si, respectively, see Table 1. Whose calculated Sezawa dispersions were well agreed with the experimental results, see Fig.3. The modeling approach shows that the inperfect bonding conditions change the elastic properties of the formed intermediates, with more reduced shear velocities and larger layerthicknesses compared to that in the good adhesion. The result also shows that in the Cu/TiN/Si sample an intermediate layer was indeed formed at the interface between the Cu layer and the TiN layer.

SUMMARY

The SAW dispersion study has enabled estimates of the Cu diffusion-induced changes of the elastic properties of the used samples. In the Cu/Si(100) specimens, prepared by sputtering process, Cu can diffuse into active regions of devices and forms an intermediate phase structure with the Si substrate. Furthermore, the inperfect bonding conditions change the SAW dispersion properties compared with that in the good bonding, from which the bonding quality has been evaluated by examining the dispersion changes. The studies have shown that the behaviors of Sezawa wave are very sensitive to the intermediate structures and the bonding conditions. Large reduction of the propagation velocity of Sezawa mode indicates that either the produced intermediate at the interface is thick or the reductions of the shear velocity and the density are large, compared with that of the top deposited layer. Thus, Sezawa wave appears to be more suitable than Rayleigh wave to assess the interface properties in layered structure by acoustic microscopy techniques.

REFERENCES

1. Z.Yu and S.Boseck, Scanning acoustic microscopy and its applications to material characterization, *Reviews of Modern Physics.* **67**:863 (1995).
2. J.Li, Y.Shacham-Diamand and J.W.Mayer, Copper deposition and thermal stability issues in copper-based metallization for ULSI technology, *Materials Science Reports.* **9**:1 (1992).
3. R.D.Weglein, Acoustic micro-metrology, *IEEE Trans. Sonics Ultrason.* **SU-32**:225 (1985).
4. R.D.Weglein, Acoustic material signature extension, *Electronics Letters.* **32**:30 (1996).

SCANNING NEAR FIELD ACOUSTIC MICROSCOPES FOR THE EVALUATION OF POLYCRYSTALLINE MATERIALS

B. Y. Zhang,[1] X. X. Liu,[2] M. Maywald,[2,3] Q. R. Yin,[1] L. J. Balk[2]

[1] Shanghai Institute of Ceramics, Chinese Academy of Sciences,
    Lab. of Functional Inorganic Materials, Shanghai 200050, P. R. of China
[2] Bergische Universität Gesamthochschule Wuppertal,
    Lehrstuhl für Elektronik, 42119 Wuppertal, Germany
[3] Present address: Siemens AG, HL QA 3, Postfach 801709,
    81617 München, Germany

INTRODUCTION

Analysis of functional inorganic ceramics necessitates techniques enabling a detailed knowledge of thermal, mechanical, and - if applied for electrical engineering - electronic features. Due to the typical size of material-relevant structures the spatial resolution has to be clearly beyond a micrometer, in difficult cases even in the lower nanometer region. Whereas conventional acoustic microscopes lack sufficient resolution, so-called scanning near-field acoustic microscopes can overcome the situation. In general, any microscope system can be understood in this manner that uses the impact or generation of sound in or at the direct vicinity of the sample surface with an interaction volume being much smaller in size than the wavelength of the acoustic signal used for the detection of the material properties. Techniques in this respect are photo, electron[1], and ion acoustic microscopes as well as just recently probe acoustic microscopes[2].

In this paper, scanning electron acoustic microscopy (SEAM) and scanning probe acoustic microscopy (SPAM) are compared as those have the highest spatial resolution capability. SEAM relies either on an intermediate generation of thermal waves[3] due to the impact of a modulated electron beam or on a direct coupling mechanism[4,5] such as piezoelectricity. SPAM, as discussed in this paper, uses a modulated force between a tip and the sample as a direct sound source. The detection scheme for both can be made identical, such as by adapting a transducer to the sample's rear surface. The comparison is demonstrated for two different polycrystalline materials. One is microwave CVD diamond with its extreme thermal conductivity and mechanical stiffness, the other barium titanate with its pronounced ferroelectric behavior.

EXPERIMENT

SEAM can be realized by modification of a scanning electron microscope. Its imaging mechanism is as follows: When an electron beam whose intensity is modulated at a certain frequency injects into a sample, it will be partially absorbed by the sample and generates acoustic waves by various effects (such as electronic, thermal, elastic). The electron acoustic signal can then be detected with a PZT transducer that is in intimate contact with the rear surface of the sample. A block diagram for SEAM is shown in Fig. 1[6].

Similarly SPAM is derived from a commercial scanning force microscope. In this paper a modulation of the force between a mechanical tip and the surface of the sample is used for sound generation. Detection is usually identical to the SEAM set up(Fig. 2(a)). However, for the detection of domains, a sheer wave detector is used, too (Fig. 2(b)). If the sample is piezoelectric on its own, the voltage between the bottom surface and the tip can be used to determine the acoustically induced voltage (Fig. 2(c)).

ANALYSIS OF POLYCRYSTALLINE DIAMOND FILMS

In both techniques a pronounced dependence of the signal on the operation frequency has to be mentioned. Only by adjusting the preferential frequencies, valuable information can be obtained. Different quality of adhesion of various grains becomes visible in the SEAM result of Fig. 4(c), whereas in Fig. 4(b) a replica of the secondary electron(SE) topography image of Fig. 4(a) is to be seen only. In Fig 5(b) the SEAM contrast within the front face of a single crystal indicates the presence of two twin boundaries not being visible in the topography micrography of Fig. 5(a).

The experiment results by SPAM to the polycrystalline diamond films are shown in Fig. 6(a) and 6(b). Here both features are visible at once: the adherence difference within the different grains as above, and inhomogeneities within the single grains due to higher defect concentrations at the circumference of the grains. Complementary TEM investigation showed that there are more twins and dislocations in darker areas[7].

ANALYSIS OF BARIUM TITANATE CERAMICS

From the SE topography of Fig. 7(a), only the grain structures of the sample can be observed, whereas the SEAM picture of Fig. 7(b) clearly depicts the complex ferroelectric domain structures without being perturbed by the topography.

An attempt to image these structures by means of SPAM and by using the longitudinal wave detection mode failed. Moreover, considering that in SPAM the sheer wave signal should be affected by the electric surface field of the domains, the according experiment showed comparable results of SPAM and SEAM, though up to now with a worse signal to noise by SPAM(Fig. 8).

A feature not visible in SEAM is the local variation of piezoelectric property of the ceramics. In SPAM piezoelectric inhomogeneities become evident. Force modulation causes an induced voltage between tip and back surface of the sample, whereas a voltage modulation between these two electrodes should yield an acoustic signal. Fig. 9(b) and Fig.9(d) are the corresponding results as taken with the set ups of Fig. 3 and Fig. 2(c), both delivering identical information. The standard SPAM image of Fig. 9(c) is here dominated by the topography(Fig. 9(a))

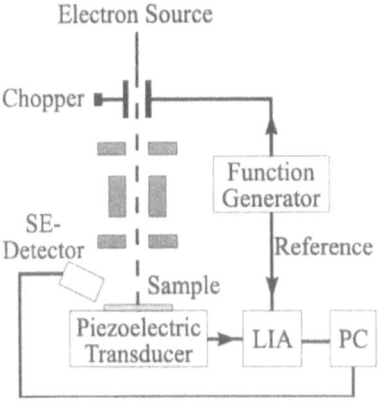

Fig. 1: Experiment setup of SEAM

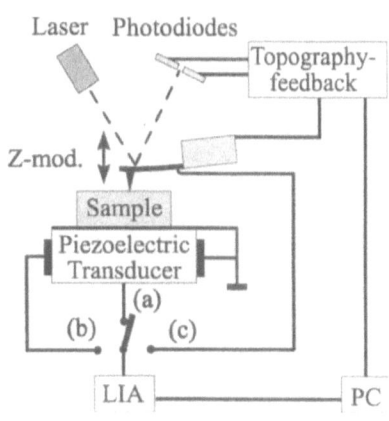

Fig. 2: SPAM setup: (a) Longitudinal wave, (b) Sheer wave, (c) Voltage image

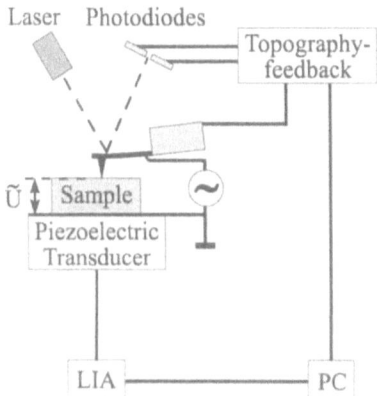

Fig. 3: SPAM setup for voltage stimulation

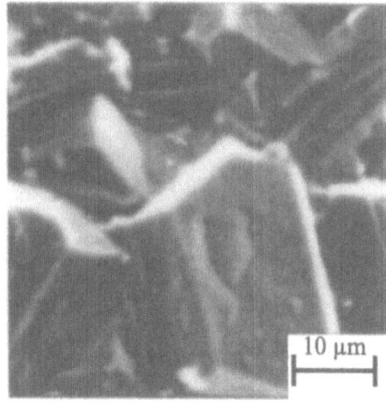

Fig. 4(a): SE image of a diamond film

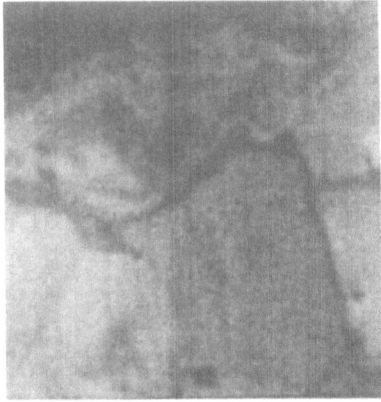

Fig. 4(b): SEAM image of the diamond film at f=143.3kHz

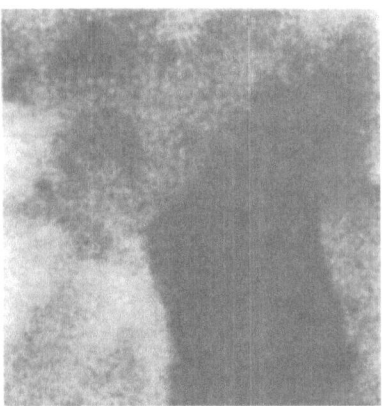

Fig. 4(c): SEAM image of the diamond film at f=148.8kHz

Fig. 5(a): SE image of a diamond grain

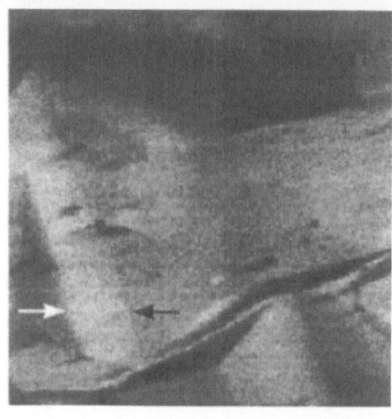

Fig. 5(b): SEAM image of the diamond grain at f=98.4kHz

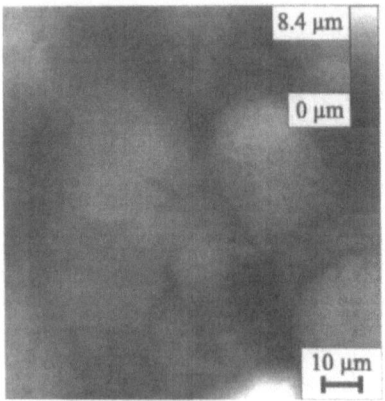

Fig. 6(a): Topography of a diamond film by SFM

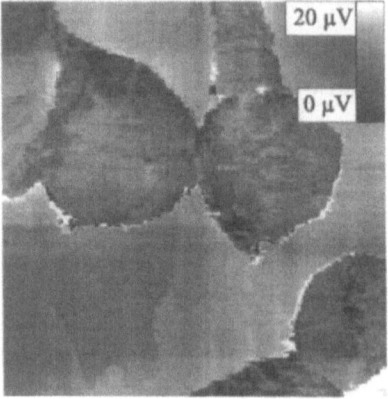

Fig. 6(b): SPAM image of the diamond film at f=102.4kHz

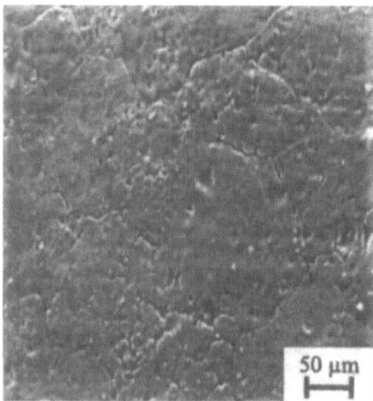

Fig. 7(a): SE image of a barium titanate ceramics

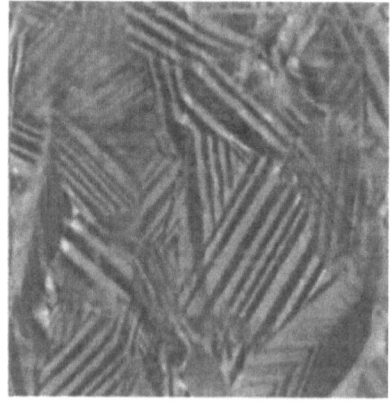

Fig. 7(b): SEAM image of the barium titanate ceramics at f=138.1kHz

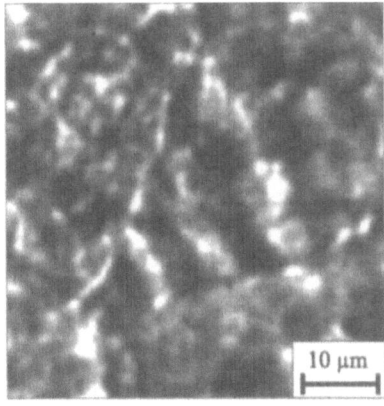

Fig. 8(a): SPAM image of domains in BaTiO$_3$ at f=102.4kHz (sheer wave)

Fig. 8(b): SEAM image of domains in BaTiO$_3$ at f=169kHz

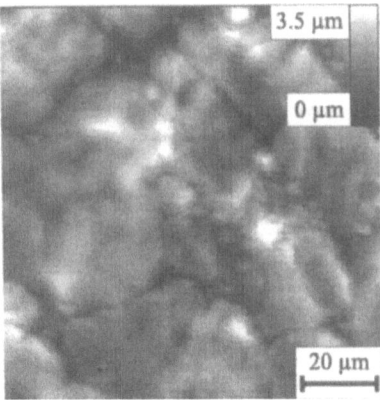

Fig. 9(a): Topography of BaTiO$_3$ ceramics by SFM

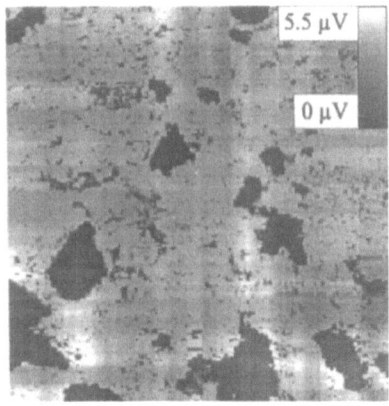

Fig. 9(b): Acoustic image with ac voltage modulation at f=64.9kHz

Fig. 9(c): SPAM image at f=64.9kHz (longitudinal wave)

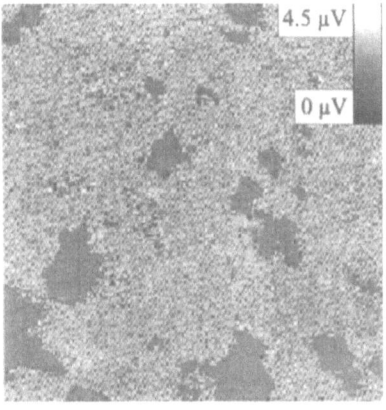

Fig. 9(d): Voltage image due to force modulation at f=64.9kHz

DISCUSSION

The diamond analysis showed that on principle a similar information can be gained by both techniques, stiffness and adhesion properties can be visualized with about equal quality. The barium titanate results are less comparable due to the strongly different generation mechanisms in both cases. Whereas domains show up clearly in SEAM, they are only visualized in SPAM by the new mode of sheer wave detection. On the other hand piezoelectric properties become visible in SPAM.

These similarities and differences can become a great help in understanding the complex signal generation mechanisms of SEAM which can be thermal, mechanical as well as electrical or electronical. In this manner the domain contrast could be clarified not to be of thermal or mechanical origin.

Both methods have the advantage over many other microscopical techniques[8] that they need nearly no specimen preparation. This allows observation of structures even for dynamic experiments, as could be done for the analysis of domain behavior in barium titanate. Further they have the advantage to be realized without large costs out of a standard SEM or SFM. Finally they are complementary tools: the SEAM is good for getting an overview of a sample surface, but is limited to a spatial resolution of, say, 0.5 to 1.0 µm, SPAM can up to now only view at a maximum area of 150 µm × 150 µm, on the other hand yields resolution values of down to some tens of nanometers.

A more detailed comparison of both techniques should involve new detection schemes, especially an improvement of the sensitivity of the sheer wave detection. The improvement should further allow experiments at high operation frequencies and should consider the use of other modes then solely imaging the signal magnitude.

ACKNOWLEDGEMENTS

The authors would like to thank Dr. P. K. Bachmann, Philips Research Labs. Aachen, for supply of the diamond samples, Dr. Xue Jiming, Shanghai Institute of Ceramics, for supplying the Barium Titanate samples, Dr. P. Pongratz, TU Wien, for complementary TEM investigation, Dr.-Ing. R. Heiderhoff for helpful discussions and Dipl.-Ing. R. E. Stephan for the experimental help and discussions.

REFERENCES

1. L. J. Balk and N. Kultscher, Techniques for Scanning Electron Acoustic Microscopy, *Inst. Phys. Conf. Ser.* **67**(1983), pp387-392.
2. M. Maywald and L. J. Balk, Evaluation of Local Thermoelastic Features by Means of Scanning Force Microscopy Based Techniques, *Proc. of the 9th Inter. Conf. on Photoacoustic and Photothermal Phenomena.*. ppS-107, S. Y. Zhang, ed., Science in China Press, Nanjing, Jun. 27-30(1996).
3. J. Opsal and A. Rosencwaig, Thermal Wave Depth Profiling: Theory, *J. Appl. Phys.,* **53**(1982), pp4240-46
4. L. J. Balk, Scanning Electron Acoustic Microscopy, in: *Advances in Electronic and Electron Physics,* **71**:pp1-73, Peter W. Hawkes, ed., Harcourt Brace Jovanovich Publishers, Toulouse, France(1988).
5. B. Y. Zhang and Q. R. Yin, Piezoelectric Coupling Effect of Ferroelectrics in Scanning Electron Acoustic Imaging, *Proc. of the 9th Inter. Conf. on Photoacoustic and Photothermal Phenomena,* ppS-123, S. Y. Zhang, ed., Science in China Press, Nanjing, Jun. 27-30(1996).
6. B. Y. Zhang, F. M. Jiang, Y. Yang and Q. R. Yin, Piezoelectric Electron Acoustic Study of Domain Structures in Ferroelectric Ceramics $BaTiO_3$, *Ferroelectrics Letters,* **22**(1996), pp21-25.
7. P. Wurzinger, M. Joksch and P. Pongratz, Defects and Microstructure of CVD Diamond as Observed with the TEM, *Inst. Phys. Conf. Ser.,* **146**(1995), pp523-528.
8. V. J. Tennery and F. R. Anderson, Examination of the Surface and Domain Structure in Ceramic Barium Titanate, *J. Appl. Phys.,* **29**(1958), pp755-758.

NON-INVASIVE TEMPERATURE IMAGING USING

ULTRASOUND ECHO STRAIN: PRELIMINARY SIMULATIONS

J C Bamber, P Meaney, M M Doyley, R L Clarke, G R ter Haar

Joint Department of Physics,
Institute of Cancer Research and Royal Marsden NHS Trust,
Downs Road, Sutton, Surrey, SM2 5PT, U.K.

INTRODUCTION

Focused ultrasound surgery (FUS) is currently being studied and developed for application to the local control of cancer and the treatment of various benign pathologies. The role of non-invasive imaging methods in facilitating the these applications of FUS can be described as:

1. *planning,* which involves localization of the pathology to be treated,
2. *targeting* which, in addition to (1), requires localizing the FUS lesion prior to its formation,
3. *treatment monitoring*, which requires the ability to image with high contrast the location and extent of regions of ablated tissue,
4. *monitoring of response*, which among other things involves non-invasive assessment of changes in tissue architecture and blood flow.

It has been suggested (Seip et al, 1996) that the temperature dependence of the tissue's apparent echo displacement, which is due both to the temperature dependence of sound speed and to thermal expansion of the tissue, can be used for non-invasive real-time spatial temperature estimation and control of hyperthermia. Such a technique may have application in requirement 2 of the above list since, by running the FUS system at a non-ablative intensity, reversible heat-induced apparent echo displacements might be used to predict where a lesion will form at higher intensities. The same phenomenon, observed by means of the time decorrelation of radio frequency (r.f.) echo signals (equivalent to colour Doppler imaging) that occurs as the tissue cools following lesioning, has been shown to be capable of depicting the location and extent of FUS lesions (Malcolm, 1996). In this study it appeared that the observed effect was dominated by sound speed variations and that thermal expansion represented the smaller contribution. The relative contribution of these components, however, remains unkown as far as reversible heating of living tissue is concerned.

This paper describes preliminary work on a computer simulation that we are developing to study the behaviour of the apparent echo displacement and strain under conditions where FUS might be employed. At present only changes due to the contribution from the temperature dependence of sound speed have been modelled. In due course these results will be compared with experimental measurements and, if necessary, the model will be extended to account for thermal expansion.

Seip et al (1966) provided an experimental demonstation that their method could provide real-time noninvasive temperature estimates, both with a rubber phantom and a sample of beef liver, but using temperature changes induced to vary from room temperature to about $30^{0}C$. The sound speed variations over this temperature range, particularly for the rubber, may have produced an optimistic result by comparison with that which would be expected from known temperature variations of sound speed of tissues over the more realistic range of $37\ ^{0}C$ to $42\ ^{0}C$ (Bamber and Hill, 1979; Bamber, 1996). The possible exception to this is fat which, like the rubber, has a strongly negative temperature coefficient of sound speed, but there are relatively few regions of the body where there are sufficiently homogeneous regions of fat for this to be useful. More frequently, the sound propagation path will include complicated patterns of mixed fat and non-fatty tissue, which would lead one to anticipate that non-invasive temperature monitoring of such tissues will be difficult. We are therefore particularly interested in understanding whether this type of approach is likely to enable us to visualize the focal lesioning volume for a realistic FUS system heating a medium with a realistic temperature dependence of sound speed over an achievable temperature range.

THEORY AND METHODS

Temperature contrast discrimination

The smallest temperature variation that can be made visible is determined by the contrast discrimination of the measurement, which is dependent on the contrast-to-noise ratio,

$$CNR_T = \frac{\Delta S}{\sigma_s} \tag{1}$$

where ΔS is the contrast or range in the "signal" (i.e., the quantity measured and assumed to be directly related to the variation in sound speed, c, and hence to temperature variation) and σ_s is the standard deviation of the noise present on this measured quantity. A more complete discussion of contrast discrimination is provided by Hill et al (1993). In the present case the signal S is obtained from one of a number of possible r.f. echo processing techniques, each with their own advantages and disadvantages, as discussed below. The contrast, ΔS, is determined firstly by the starting temperature in combination with the greatest reversible temperature rise that can be induced by the focused ultrasound therapy system, then by the speed of sound change corresponding to this temperature range and finally by the alteration in the measured signal, S, that the sound speed change will cause. High magnitudes for both dS/dc and dc/dT are therefore desirable, so as to maximize ΔS. The noise may have contributions from various sources, both biological and technical. For example, variations in dc/dT across the tissue being studied will contribute to the intrinsic biological variance whilst extrinsic variations, such as the mean of dc/dT for a given tissue, will determine the degree to which absolute temperature rise can be quantified. The intrinsic and extrinsic technical variances will have contributions, for example, from errors inherent in the signal processing methods used to extract the signal S from the echo waveforms and in the statistical nature of the echoes from ensembles of scatterers.

Diffraction field of therapy transducer

A program that employs the Fresnel-Kirchoff diffraction integral for a spherical bowl continuous wave source is used to calculate the three dimensional intensity distribution in front of the therapy transducer.

Temperature and sound speed distributions

The intensity distribution is the starting point for a finite element calculation (based on a direct implementation of the bioheat equation) of the heat diffusion that takes place after absorption of the ultrasonic energy, resulting in a spatially and temporally varying temperature. The temperature distribution (for example, at steady state) is then mapped to a sound speed distribution via a polynomial fit to data on the temperature dependence of sound speed in specific tissues (Bamber and Hill, 1979).

Echo Imaging

A central two dimensional section through the sound speed distribution is then used to calculate a distribution of acoustic pathlengths for pulse echo imaging. An "apparent" displaced distribution of acoustic scatterers is then calculated, described in terms of the inhomogeneous acoustic pathlengths in the heated tissue, in a model of the scattering structure of the tissue, as would be seen by an ultrasound pulse-echo system imaging through this "heat haze". This is accomplished using a modified form of the two dimensional simulation described by Bamber and Dickinson (1980), implemented in MATLAB^{TM}, in which the tissue's scattering structure is modelled as a random inhomogeneous continuum of bulk adiabatic compressibility values, the second spatial differential of which produces a tissue scattering impulse response that must be convolved with the simulated backscatter imaging systems r.f. point response function to obtain an r.f. pulse-echo image of the tissue.

For the formation of r.f. backscatter images of the tissue before heating, an homogeneous sound speed at a given baseline temperature is assumed. At present, the simulation neglects possible wave distortion due to the inhomogeneous sound speed in heated tissue and calculates the echo images of tissues after heating by redistributing the tissue compressibility values (i.e. in effect, the positions of the acoustic scatterers) according the acoustic pathlengths calculated from the array of speed of sound values. In practice, aberration and refraction will further degrade the temperature imaging performance.

Displacement distributions

Consider a two dimensional discrete matrix of compressibility values, $\beta\,(x,y)$, that will be imaged using a linear scan format where y is the direction of sound propagation divided into depth increments (pixels) of Δy, and x is the direction of scanning. It is assumed that the imaging transducer that both launches sound pulses and detects the echoes is positioned at $y = 0$. For a uniform sound speed, c, and any fixed value of x, the pulse-echo transit time to the kth element in depth will be

$$t_k = \frac{2k\Delta y}{c} \tag{2}$$

and for a spatially variant sound speed $c(x,y)$ at fixed position x it will be

$$t'_k = 2\Delta y \sum_{i=1}^{k} \frac{1}{c_i} \tag{3}$$

where c_i is the sound speed value for the ith element in the acoustic path between the transducer and the region of tissue under consideration. Therefore one can define an apparent displacement of the kth tissue element in time from the position in the unheated tissue to the position in the heated tissue as

$$\delta t_k = t'_k - t_k = 2\Delta y \left(\sum_{i=1}^{k} \frac{1}{c_i} - \frac{k}{c} \right) \tag{4}$$

To an imaging system that displays echo information assuming that distance is linearly related to propagation time via an assumed uniform sound speed, c_o, this temporal displacement will translate to a spatial displacement of the tissue element in the y direction, that is given by

$$\delta y_k = \frac{c_o \delta t_k}{2} = c_o \Delta y \left(\sum_{i=1}^{k} \frac{1}{c_i} - \frac{k}{c} \right) \tag{5}$$

For fat δy is positive, indicating that tissue scatterers will appear to move away from the transducer during heating and for non-fatty tissue δy is negative, an apparent foreshortening of the transducer-scatterer distance.

Implementation of the above within the simulation involves calculating a matrix of apparent tissue axial displacement values by applying Eq. 5 to every element in the sound speed matrix. Bilinear interpolation is then used to obtain an axial displacement value, and hence new spatial coordinates, for each element of the tissue compressibility matrix. Linear interpolation of these displaced tissue compressibility values along the y direction is then used to calculate samples of the distorted tissue compressibility function at the spatial coordinates of the original compressibility matrix elements. A final r.f. backscatter image of the heated tissue is obtained by scanning this distorted compressibility function by convolution with the imaging system point response, as described above for the case of uniform sound speed.

Tissue displacement and temperature imaging

The final stage involves the application of either time-delay tracking or autocorrelation processing methods to detect and image this apparent internal displacement distribution. The methods employed at this stage are essentially the same as those employed to measure tissue internal displacements in elasticity imaging, except that in the case of elasticity imaging the displacements are real and in the present model the displacements are due only to changes in acoustic pathlength induced by temperature mediated variations in sound speed.

28

From further inspection of Eq. 5 it may be seen that images of relative sound speed distribution (and hence temperature distribution) in the heated tissue may be recovered by differentiating the images of measured echo displacement, $\delta y'$, with respect to y. The result may then be compared with the original temperature distribution.

RESULTS

For the purposes of this preliminary report specific examples of results will be given, based on a spherical bowl therapy transducer of diameter 8.4 cm and 15 cm radius of curvature heating a sample of human liver and another of bovine peritoneal fat. A water path coupling was assumed with the focal spot being positioned 1 cm below the tissue surface. A steady state in temperature was acheived after about 4.5 minutes of FUS exposure with a maximum intensity of 62 Wcm^{-2} at the spatial peak, starting at an ambient temperature of 21°C. This case was chosen to produce a peak temperature rise of 35°C; representing for the moment, the optimistic case used by Seip et al (1996). Sections through the resulting sound speed distributions are shown in Fig. 1.

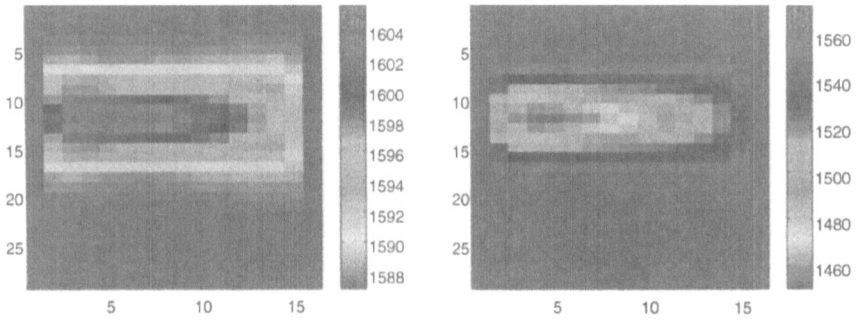

Figure 1. Predicted sound speed in heated human liver (left) and bovine peritoneal fat (right). The gray scale bars show sound speed (ms^{-1}). Each region is 30 mm by 30 mm.

Predicted apparent tissue displacement distributions due to heating were then calculated, using Eq. 5, from the sound speed distributions for the fat and each of the non-fatty tissues. To study the dependence on the diagnostic-therapy beam angle, displacement distributions were calculated both for a beam angle of 90 degrees (Fig. 2) and for parallel beams (Fig. 3).

Figure 4 shows example r.f. echo scans of the tissue compressibility matrix before and whilst imaging through the "heat haze distortion" created by the therapy beam.

Examples of echo displacement images obtained from the synthesized backscatter images of heated tissue are shown in Figs. 5 and 6. As may be seen, these resemble the predicted apparent displacement images but with varying amounts of noise present. This noise arises only from the errors inherent in the signal processing methods used to estimate echo displacement, which will be smaller in this simulation than in practice, where wave aberration due sound speed inhomogeneities and additive electronic noise with frequency dependent attenuation are present.

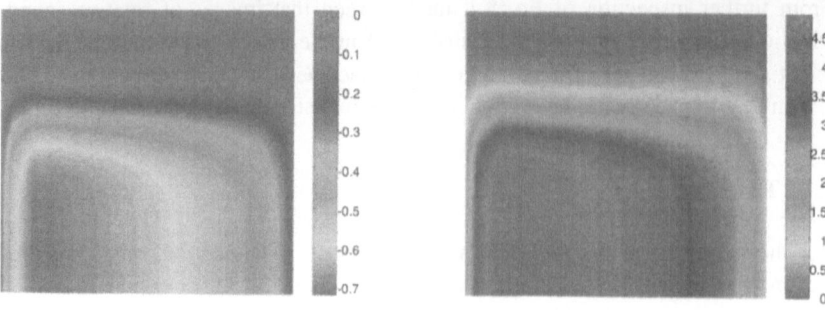

Figure 2 Predicted apparent tissue displacement due to heating of human liver (left) and bovine peritoneal fat (right), when the therapy beam is 90^{0} to the diagnostic beam.

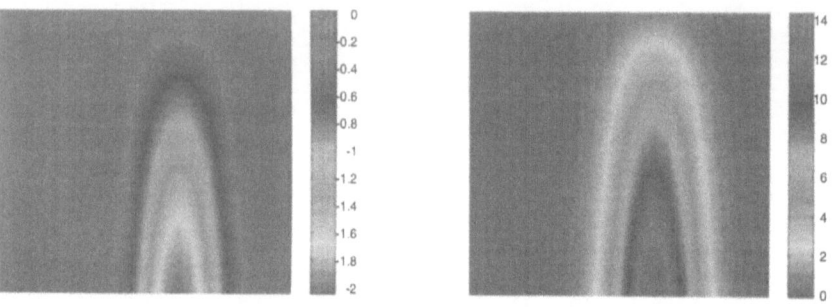

Figure 3. Predicted apparent tissue displacement due to heating of human liver (left) and bovine peritoneal fat (right), when the therapy beam is parallel to the diagnostic sound beam

Figure 4 r.f. echo image of before (left) and after heating (right) bovine peritoneal fat when the therapy beam is parallel to the diagnostic sound beam, as in Fig. 3.

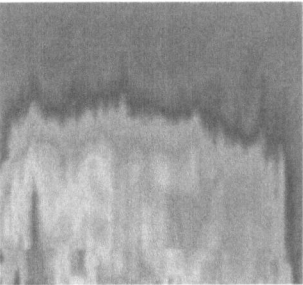 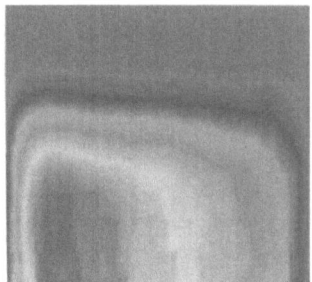

Figure 5 Echo displacement maps of human liver imaged perpendicular to the therapy beam, as in Fig. 2, derived by time-domain tracking of envelope detected (left) and r.f. (right) echo images.

Figure 6 Echo displacement maps of bovine peritoneal fat imaged perpendicular to the therapy beam, as in Fig. 4, derived by tracking envelope detected (left) and r.f. (right) echo images.

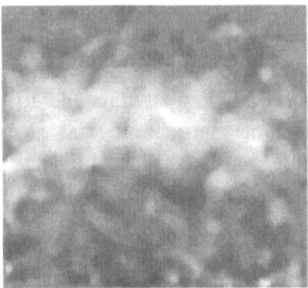 

Figure 7 Echo strain (relative temperature) maps of human liver imaged perpendicular to the therapy beam and derived from envelope (left) and r.f. (right) echo tracking.

 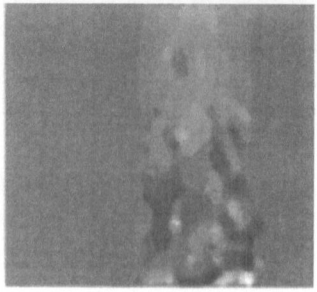

Figure 8 Echo strain (relative temperature) maps of bovine peritoneal fat imaged parallel to the therapy beam and derived from envelope (left) and r.f. (right) echo tracking.

Note the poorer performance of the echo displacement estimator when the therapy and diagnostic beams are parallel, especially in regions where there is a large gradient of sound speed in the x-direction, which causes echo decorrelation and is a known effect from elasticity imaging. The gradient of echo displacement in the y-direction, or "echo strain", shown in Figs. 7 and 8 depicts measured relative sound speed and may be compared with the error-free counterparts in Fig. 1.

DISCUSSION

This study has confirmed the feasibility of using r.f. tracking of apparent echo displacement to visualize temperature distributions in specific tissues, *viz.*, normal liver and fat, heated from room temperature. Envelope detected B-scan images can be used for the same purpose and it is desirable to maintain an angle of 90° between the heating and diagnostic sound beams For r.f. tracking $CNR_{peak} \sim 10.8$ and for envelope tracking $CNR_{peak} \sim 4.9$ but the simulation does not model all sources of noise. Also, as may be seen from Fig. 8, envelope tracking results in greater spatial distortion of the imaged temperature distribution.

From published temperature dependencies of sound speed we anticipate greater difficulty in visualizing temperature distributions when heating from body temperature and for tissues such as breast or fatty liver, which have either a low value or a large biological variance for dc/dT in the temperature range of interest. Thermal expansion is currently neglected in the simulation but is expected to cause additional measurement noise. *In vivo* application of the method will require compensation for bulk tissue motion, for example due to respiration. This will generate a further source of noise.

Further work in the immediate future will include (a) the study of systems and temperature rises more relevant to *in vivo* FUS treatment situations, and (b) further quantitative analysis of the resulting relative sound speed images, both in terms of the detectability of the temperature rises and by comparing the measured distributions with the true temperature and sound speed distributions.

REFERENCES

J.C. Bamber and C.R. Hill, Ultrasonic attenuation and propagation speed in mammalian tissues as a function of temperature, *Ultrasound Med Biol.* **5**:149-157 (1979).

J.C. Bamber, R.J. Dickinson, Ultrasonic B-scanning: a computer simulation, *Physics Med Biol.* **25**:463-479 (1980).

J.C. Bamber, Acoustical characteristics of biological media, pp. 1703-1726 in: *Encyclopedia of Acoustics*, M.J. Crocker, ed., John Wiley and Sons, New York, (1997).

C.R. Hill, J.C. Bamber and D.O. Cosgrove, Performance criteria for quantitative ultrasonology and image parameterization, *Clin Phys Physiol Meas.* **11**(Supp.A):57-73 (1990).

A. Malcolm, *An investigation into ultrasonic methods of imaging the tissue ablation induced during focused ultrasound surgery*, PhD Thesis, University of London (1996).

R. Seip, P. VanBaren, C. Cain and E. S. Ebbini, Noninvasive real-time multipoint temperature control for ultrasound phased array treatments, *IEEE Trans UFFC.* **43**:1063-1073 (1996).

ACKNOWLEDGEMENTS

This research was supported in part by the Cancer Research Campaign, The Institute of Cancer Research, The Royal Marsden NHS Trust and NATO.

MEDICAL ULTRASOUND IMAGING USING PULSE COMPRESSION AND DEPTH-DEPENDENT MISMATCHED-FILTERING

A. R. Brenner, K. Eck, W. Wilhelm, T. G. Noll

Chair of Electrical Engineering and Computer Systems
RWTH Aachen, University of Technology
52062 Aachen
Germany

INTRODUCTION

In conventional medical ultrasonic imaging, the signal-to-noise ratio (SNR) as well as the penetration depth are insufficient for many diagnostic cases. In pulse-echo ultrasonic imaging tissue heating is negligibly small, whereas there is a potential for acoustic cavitation at peak intensities currently used in ultrasonic imaging[1]. Therefore improving the SNR by increasing the peak intensity of the acoustic signal is not possible. This restriction can be circumvented by using elongated, coded signals with high average energy at moderate peak intensity[2,3]. Frequency-modulated chirps are suitable signals for this purpose.

The frequency-dependent attenuation of tissue substantially limits the advantages of coded excitation: tissue attenuation causes energy loss and a frequency-dependent distortion of the echo spectrum. This leads to an echo with undesired sidelobe characteristics after matched-filter pulse compression. In this paper, we present a mismatched-filter pulse compression scheme which substantially suppresses sidelobes compared to a matched-filter pulse compression. An estimation of the requirements for a VLSI implementation of the depth-dependent mismatched filter is given.

CHOICE OF SUITABLE CHIRP SIGNALS

The aim of the design of chirp signals is to get a time-limited signal with almost constant envelope and a predefined spectrum. Since it is not possible to reach both of these goals simultaneously, chirp signal design mainly consists of a trade-off between signal energy and a suitable shape of the magnitude spectrum at limited signal length. Proposed solutions are the apodized linear chirp[1], the pseudo-chirp[3], and the nonlinear chirp[4].

The linear frequency-modulated chirp combines maximum energy and maximum effective bandwidth. The signal after matched-filter compression shows a narrow mainlobe and high sidelobes at -13 dB. To suppress the undesired sidelobes, either a time-domain apodization or a nonlinear frequency modulation can be used. In the case of the apodized linear chirp, a weighting function is multiplied in the time domain to obtain a chirp signal with the desired spectrum. The disadvantage of this method is the reduced energy of the compression result compared to the case of a linear chirp with constant envelope function. In the case of the nonlinear frequency-modulated chirp, the desired spectrum is approximated by varying the instantaneous frequency of the rectangular envelope chirp signal.

MISMATCHED-FILTERING

Matched-filter compression is the standard pulse compression method. It is well known that the matched-filter pulse compression yields the highest SNR in case of white additive noise. However, matched-filter compression does not guarantee a suitable sidelobe level of the compressed signal and therefore limits the dynamic range of the image. Pulse compression with a filter that differs in the magnitude of its transfer function (*mismatched filter*) can shape the compression result into a desired mainlobe-sidelobe characteristic, but lowers the SNR[5]. The resulting dynamic range is given by the difference between mainlobe height and sidelobe level. The SNR for mismatched-filter pulse compression depends on the similarity between the spectrum of the echo and the transfer function of the mismatched filter. The difference of the SNR of matched- and mismatched-filter pulse compression is typically smaller than 1 dB for mismatched-filtering of proper designed nonlinear chirps in the non-attenuated case.

The Design of Mismatched Filters

Our approach of designing a mismatched-filter for enhanced pulse compression is basically as follows: Suppose a time-discrete channel impulse response x_k of finite-length L, i.e. k=0, ..., L-1 and a mismatched filter with finite impulse response of c_k; k=0, ..., N-1. Then the compressed echo signal becomes

$$y_k = \sum_{j=0}^{N-1} c_j \cdot x_{k-j}; \quad \text{with } k = 0, ..., N+L-2. \tag{1}$$

We define a time domain template according to the desired mainlobe width and the desired sidelobe level. Similar to the frequency domain approximation process in filter design, we can now approximate a Chebyshev characteristic in the time domain. Discretization in time leads to the desired compressed echo signal $\tilde{y}_k$. The mismatched filter can then be fitted according to the minimum mean-square error criterion

$$E[e_k^2] = \frac{1}{N+L-2} \sum_{k=0}^{N+L-2} (\tilde{y}_k - y_k)^2. \tag{2}$$

Using (1) this is equivalent to minimize a distortion D

$$D = \sum_{k=0}^{N+L-2} \left(\tilde{y}_k - \sum_{j=0}^{N-1} c_j \cdot x_{k-j} \right)^2 \rightarrow min, \tag{3}$$

requiring

$$\nabla_c D = 0 \tag{4}$$

or

$$\frac{\partial}{\partial c_j}D = 2\left\{-\sum_{k=0}^{N+L-2} \tilde{y}_k \cdot x_{k-j} + \sum_{i=0}^{N-1} c_i\left(\sum_{k=0}^{N+L-2} x_{k-i} \cdot x_{k-j}\right)\right\} = 0. \tag{5}$$

Introducing the abbreviations

$$a_{ij} = \sum_{k=0}^{N+L-2} x_{k-i} \cdot x_{k-j} \quad \text{and} \quad r_j = \sum_{k=0}^{N+L-2} \tilde{y}_k \cdot x_{k-j}, \tag{6}$$

(5) becomes

$$0 = -r_i + \sum_{i=0}^{N-1} c_i \cdot a_{ij}. \tag{7}$$

Defining the $N \times N$ matrix $A = (a_{ij})$, the vectors $c = (c_0, c_1, ..., c_{N-1})^T$ and $r = (r_0, r_1, ..., r_{N-1})^T$, the coefficients of the mismatched filter c_j can be calculated by solving

$$A \cdot c = r. \tag{8}$$

Figure 1 shows results of matched-filter and mismatched-filter pulse compression of linear and nonlinear chirps with rectangular envelopes. According to the Chebyshev characteristic, the mismatched-filter compression result shows nearly equiripple sidelobes.

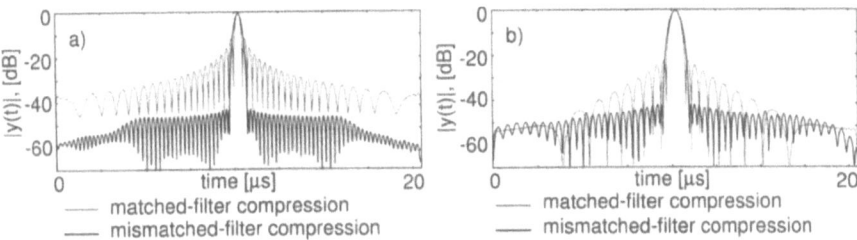

Figure 1. Matched-filter and mismatched-filter compression results of a) linear and b) nonlinear rectangular envelope chirps.

DEPTH-DEPENDENT MISMATCHED-FILTERING

The application of mismatched-filtering and the optimization of the chirp signal to ultrasonic imaging must furthermore take into account the transfer function of the tissue. The transfer function of a round-trip propagation for an ideal reflector at depth z in an attenuating medium is given by

$$H_m(f, z) = \exp\left\{-2z\left[\alpha(f) + j\frac{2\pi f}{v}\right]\right\} \tag{9}$$

with the tissue dependent attenuation $\alpha(f)$ and the sound velocity v. The effect of $H_m(f, z)$ on the imaging process is a severe energy loss and a distortion of the spectrum of x_k which results in high sidelobes of y_k after pulse compression[1]. Conventional matched-filter pulse compression techniques do not take into account this frequency-dependent attenuation. Prefiltering of the chirp signal to compensate for the effect of attenuation is possible, but only for one imaging depth at a time[4], and therefore it under- or overcompensates attenuation in all other depths.

Depth-dependent mismatched-filtering, however, consists of applying a set of mismatched filters each adjusted to a certain depth. For each of these filters the altered effective center frequency and bandwidth of an echo at this depth has to be taken into account.

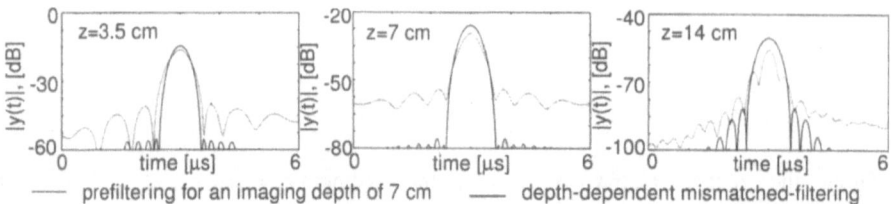

Figure 2. Comparison of compression results of depth-dependent mismatched-filtering and prefiltering approach for three different imaging depths (attenuation 0.6 dB / MHz cm).

Figure 2 shows compression results of the combination of prefiltering for a depth of 7 cm with matched-filtering and compression results of the mismatched-filtering approach. It can be seen that even for a depth of 7 cm for which the prefiltering was optimized the depth-dependent mismatched filter features significant lower sidelobes.

At depths of 3.5 cm and 14 cm the under- and overcompensation by the prefilter leads to poor compression results, whereas the depth-dependent mismatched-filter compression again shows significantly better results.

Trading Mainlobe Width for Sidelobe Suppression

Depth-dependent mismatched-filtering offers an additional degree of freedom for the three signal characteristics mainlobe height, mainlobe width and sidelobe level, which can be traded for each other up to a certain account. Figure 3 shows compression results optimized with regard to mainlobe width and sidelobe suppression.

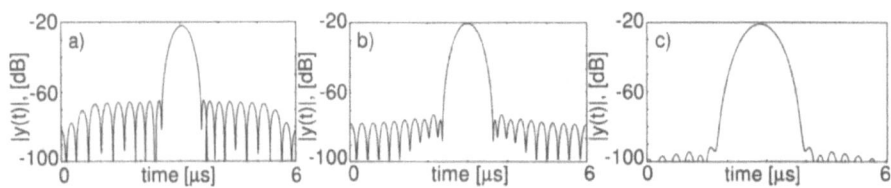

Figure 3. Depth-dependent mismatched-filter compression results for $z = 7$ cm. Optimization criterion a) tight mainlobe, b) dynamic range > 50 dB, c) maximum sidelobe suppression

FIRST EXPERIMENTAL RESULTS

First experimental results confirm that ultrasonic imaging using nonlinear chirp signals and depth-dependent mismatched-filtering results in improved image quality and penetration depth compared to pulse-echo excitation with the same signal peak intensity. Figure 4 shows A- and B-Scans of a tissue mimicking phantom (ATS 539) with an attenuation of 0.5 dB / MHz cm. The experiment was performed using a curved array (Pie Medical CA 80) with a center frequency of 3.75 MHz. Time gain correction was applied. Further experimental investigations will compare the depth-dependent mismatched-filter approach to standard compression and prefiltering techniques.

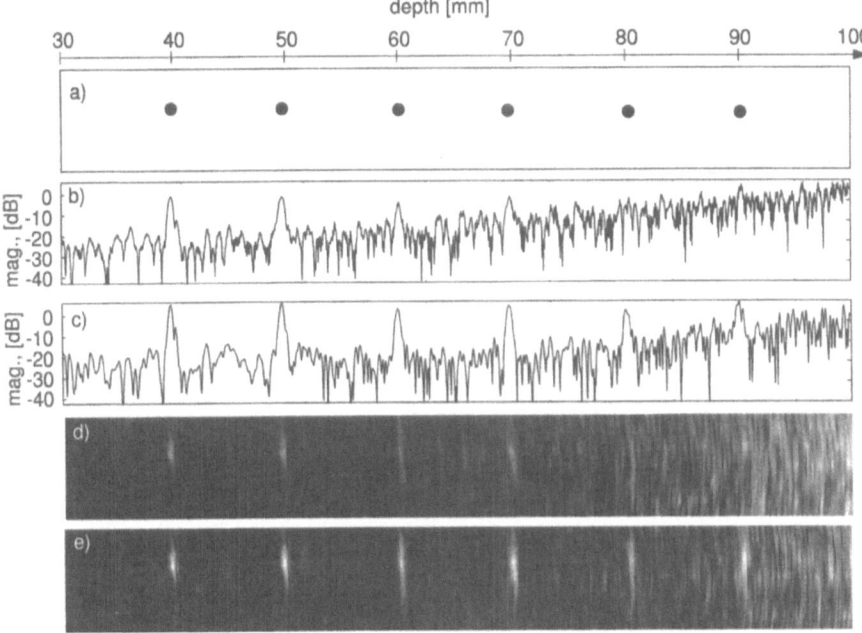

Figure 4. Experimental results of conventional pulse-echo excitation and mismatched-filter pulse compression technique: a) phantom sketch, b) A-Scan of the phantom (pulse-echo), c) A-Scan of the phantom (20 μs nonlinear chirp with mismatched filter compression), d) corresponding B-Scan to b), e) corresponding B-Scan to c)

VLSI IMPLEMENTATION

The progress in CMOS technology leads to a permanent miniaturization. Additionally, the maximum clock-frequency of VLSI circuits increased dramatically in the past and will increase further. If the speed requirements of a system to be implemented are lower than the speed potential of the used technology, trading speed for silicon area becomes possible by time-sharing of hardware resources. By that, either the silicon area and therefore costs for constant functionality can be decreased, or the complexity on constant area can be increased. For example in the case of the application discussed here a sampling rate of typically 14 MSample/s is sufficient, while a 0.5 μm CMOS technology enables clock frequencies of 200 MHz for optimized designs, allowing a time-sharing factor of 14.

Further, a mismatched filter with approximately 500 taps (filter order) and a wordlength of 10 bit for the input signal and the coefficients allows to reach the SNR required for ultrasonic imaging. Designing such a filter according to todays semi-custom design strategies[7] based on synthesis using predesigned standard cells leads to an unacceptable silicon area of about 300 mm^2 in 0.5 μm CMOS technology. An optimized full-custom design style enables the implementation of this filter on an area of about 50 mm^2. Applying a constructive mapping technique called bit-level folding[8] with a time-sharing factor of 14 in a full-custom design results in a silicon area of less than 9 mm^2.

Aspiring a fully integrated digital frontend consisting of a digital beamformer, a pulse compression filter and further advanced signal processing units, e.g. for phase aberration correction[9], a mismatched filter of 9 mm^2 occupies only a small part of the total silicon area.

SUMMARY

By the means of depth-dependent mismatched-filtering, the dynamic range and resolution of coded excitation imaging in an attenuating medium can substantially be improved. Depth-dependent mismatched-filtering gives full control over the properties (mainlobe height, mainlobe width, sidelobe characteristic) of the compressed echo. A VLSI implementation of a depth-dependent mismatched filter was shown to be feasible. Further work will be done concerning experimental evaluation and verification as well as the simultaneous estimation of attenuation.

ACKNOWLEDGMENT

The authors would like to thank J. Geijsen and his colleagues at Pie Medical for valuable discussions and hardware support.

REFERENCES

1. N.A.H.K. Rao, Investigation of a pulse compression technique for medical ultrasound: a simulation study, *Medical & Biological Engineering & Computing*, 32:181 (1994).

2. B. Haider, P.A. Lewin and K.E. Thomenius, Pulse elongation and deconvolution filtering for medical ultrasonic imaging, *IEEE Proc. UFFC*, 1303 (1995).

3. M. O'Donnell, Coded excitation system for improving the penetration of real-time phased-array imaging systems, *IEEE Trans. UFFC*, 39(3):341 (1992).

4. M. Pollakowski and H. Ermert, Chirp signal matching and signal power optimization in pulse-echo mode ultrasonic nondestructive testing, *IEEE Trans. UFFC*, 41(5):655 (1994).

5. H.D. Lüke, *Korrelationssignale*, Springer-Verlag Berlin (1992).

6. C. Passmann and H. Ermert, A 100-MHz ultrasound imaging system for dermatologic and ophthalmologic diagnostics, *IEEE Trans. UFFC*, 43(4):545 (1996).

7. S. Mita, R. Shimokawa, T. Matsuura, H. Sawaguchi, S. Miyazawa and K. Hikasa, A 150Mb/s PRML chip for magnetic disk drivers, *IEEE International Solid-State Circuits Conference*, 39:62 (1996).

8. W. Wilhelm and T.G. Noll, A new mapping technique for automated design of highly efficient multiplexed FIR digital filters, To be published in: *IEEE International Symposium on Circuits and Systems* (1997).

9. A.R. Brenner, K. Eck, G. Engelhardt and T.G. Noll, Phase aberration correction using dynamic time warping, *IEEE Proc. UFFC*, 1361 (1995).

ULTRASONIC, SPECTRUM-ANALYSIS, TISSUE-TYPING IMAGES FOR PROSTATE-BIOPSY GUIDANCE AND STAGING

E.J. Feleppa,[1] W.R. Fair,[2] T. Liu,[1] W. Larchian,[2] A. Kalisz, V. Reuter,[2] and A. Rosado [1]

[1] Riverside Research Institute, New York, NY 10036
[2] Memorial Sloan-Kettering Cancer Center, New York, NY 10021

INTRODUCTION

Prostate cancer is diagnosed using needle biopsies guided by conventional ultrasonic images. However, conventional ultrasound has inadequate sensitivity and specificity to direct biopsies reliably into cancerous regions. Urologists use conventional images to guide the needle systematically, but "blindly," into six or more selected regions of the gland. Unless a distinctly palpable or ultrasonically visible nodule is present (e.g., as a hypoechogenic region) to raise suspicion, the urologist simply samples tissue from the left and right base, mid, and apical regions of the gland, and possibly from the transition zone or seminal vesicles. Because of the blind nature of this procedure, initial biopsies miss about half the cancers, which results in a high incidence of false-negative biopsies; in addition, more than two thirds of the initial biopsies sample benign tissue, which results in an unfortunately high incidence of true negative biopsies.

Tissue-typing images based on normalized (system-corrected) spectrum analysis of radio-frequency (rf) ultrasonic echo signals[1-3] are showing encouraging sensitivity and specificity for prostatic cancer.[4-7] We generate images of prostate glands based on spectrum analysis by computing spectra within a sliding window along each scan line (or vector) of rf data from an ultrasound scan, then comparing computed spectral-parameter values at each point in the scan plane with a range of values known to indicate cancer. If unknown tissue presents values within the cancerous range at a specific window site, then the pixels at that site are encoded or "stained," e.g., depicted in a color, such as red, or in a saturated grey-scale. The pattern of stained pixels throughout a scan provides a guide to regions warranting biopsy. (Images also can be generated to depict parameter values, discriminant-function values, or scatterer properties such as effective scatterer size.) Comparisons of spectrum-analysis methods (using leave-one-out analysis) to levels of suspicion (LOSs) assigned by the urologist on a scan-by-scan basis showed an ROC curve area of 79% for spectrum analysis versus 60% for conventional-image determinations. We are expanding our data base and developing new, statistically more-powerful means of generating tissue-typing images for guiding biopsies, including interactive 3-D volume renderings to better visualize suspicious tissue in the prostate.

METHODS

Rf echo-signal data were acquired during standard, trans-rectal ultrasound examinations involving ultrasound-guided prostate biopsies. Acquired rf data were used to generate grey-scale images equivalent to conventional B-mode images; these data were subjected to normalized, spectrum analysis 1) to generate a data base of spectral parameters associated with specific prostate tissue types; 2) to generate two-dimensional (2-D), tissue-type images of potential value in biopsy guidance; and 3) to generate three-dimensional (3-D), tissue-type images for cancer staging. True tissue type was determined histologically. Classification by spectrum analysis used "nearest-neighbor" algorithms, and its efficacy was assessed using standard "leave-one-out" and ROC methods. Tissue typing of 2- and 3-D images was based on pre-defined parameter-value ranges associated with cancerous tissue. The methods used in these studies continue to be applied in our on-going investigations, and are summarized below.

Data Acquisition

We acquire rf data by capturing echo signals from 318 scan lines in the 112-degree sector of a B&K Medical Systems Model 3535 scanner and 8551 probe. Rf signals are digitized after front-end gain, but before any non-linear processing such as compression or envelope detection are applied. (Rf echo signals, along with pulse-trigger signals are accessible from a back-panel port.) Data acquisition timing is controlled by a custom interface module and a desktop computer with custom digitization software. The sampling frequency is 50 MHz, and 2500, 8-bit samples are acquired along each scan line; accordingly, each line spanned a radial segment of approximately 36 mm after an 8-mm delay. Rf data for an entire scan plane are digitized and buffered in a digital sampling oscilloscope (DSO), then transferred to computer memory. These data are retained in computer memory until a scan series is complete, then are transferred onto the computer disk. Rf data for each scan are placed in a uniquely identified file that contains various scan variables in its header. For each patient, the scan series consists of 1) a set of 10 to 20, closely spaced, parallel, transverse scans spanning the prostate from the base to the apex, and 2) the biopsy-guiding scan associated with every biopsy; rf data in biopsy-guidance scans are acquired the instant before the biopsy needle is inserted and is identified by prostate region, e.g., left mid, right base, transition zone, etc., for subsequent correlation with histology results.

We maintain an examination log that contains patient and examination information, such as patient age, PSA level, etc. A key element of that information is the examining urologists assessment of the likelihood of cancer in the biopsied region of the gland; this assessment, termed the level of suspicion (LOS) is made on a scale of 1 to 5 with 1 indicating near certainty of no cancer, 3 indicating an indeterminate assessment, and 5 indicating near certainty of cancer. Our efficacy assessments are based on a comparison of the LOS assignments with classifications made using spectrum analysis.

Each biopsy specimen is identified by prostate region, e.g., left mid, right base, transition zone, etc., for subsequent correlation of histoligical tissue-type determinations with spectrum-analysis results.

Data Analysis

Rf data are analyzed according to the methods described in the references.[1-5] In summary, an operator defines region of interest (ROI) that matches the image region corresponding to the biopsied tissue to select rf data for normalized spectrum analysis. The analysis provides power spectra that are corrected for the system transfer function over the effective, noise-limited bandwidth of the system; in the present case, this bandwidth is 4.5 MHz and extends from 3.5 MHz to 8.0 MHz Since the ROI contains several scan-lines, the spectrum represents the ensemble average for the set of included echo-signal segments. Spectral parameters are derived from the computed spectrum by linear regression, and the values of the regression slope, intercept, and amplitude at the band center (mid-band value) are compared to biopsy tissue type to establish the basis for tissue classification. As described in the references, cancer of the prostate appears to present significant lower values of intercept and mid-band parameters than non-cancerous tissues present. The differences in intercept and mid-band values for cancerous and non-cancerous prostate tissue serves as the basis of tissue typing in 2-D and 3-D images.

Biopsy specimens are histologically examined by the collaborating pathologist who defines tissue type for each biopsy core. This identification is made specifically for each core, which enables us to explicitly relate spectral parameter values to histologically determined tissue type on a core-by-core basis. Our data base now contains spectral-parameter and tissue-type data for over 600 biopsies.

We investigated the application of linear discriminant analysis and nearest-neighbor methods for classifying cancerous and non-cancerous prostate tissue. To date, the nearest-neighbor methods incorporated in commercially available ModelWare™ software has proven to be the most effective classification tool. We evaluated the classification efficacy of this approach using the "leave-one-out" or "jackknife" tools within ModelWare™, and presented the results using standard receiver-operator characteristics (ROC) methods. We generated ROC curves from the leave-one-out computations using the LABROC software of Metz.[8]

To serve as a reference for the classification efficacy of spectrum-analysis; we generated ROC curves depicting the classification efficacy of current clinical methods. We generated an ROC curve for the LOS assignments made by the urologist for each of the biopsied regions of tissue using the ROCFIT software of Metz.[8]

The "gold standard" for comparing these two classification methods was the tissue type determined histologically by the urological pathologist

Image Generation

We generate B-mode grey-scale images from rf data through digital rectification and envelope detection. Such images are used to guide ROI placement, as described in the references.[4-6] We generate parameter type images by performing normalized power-spectrum analysis along the full length of each scan line using a sliding, 64-point, Hamming window. Typically, we generate grey-scale images depicting parameter values with pixel values ranging from 0 to 255. As discussed in the references, slope and mid-band values are affected by attenuation.[1-3] Accordingly, their values are corrected for an assumed attenuation coefficient of 0.5 dB / MHz-cm.[4,5] For tissue-type encoding or "staining" images, we compare local values of parameters to reference values for each tissue type, e.g., for prostate cancer, we compare the intercept and mid-band values at each window location with the values associated with cancer. If the values at a given location fall within the range associated with cancer, the pixel at that location is given a distinct value, e.g., a value corresponding to red if color is used or a saturated white in grey scale; otherwise, the pixel value retains its normal grey-scale value.

RESULTS

Our initial classification results indicated that spectral-parameter values were markedly superior to conventional image-feature criteria for distinguishing cancerous from non-cancerous prostate tissue. Therefore, we utilized ranges of parameter values (computed using sliding windows for the rf data along each scan line of the entire scan plane) to determine which pixels in 2-D and 3-D images should be stained to indicate a locally high likelihood of cancer.

ROC Results

For 31 biopsied patients, we obtained 147 tissue cores with histologically determined tissue types, and 147 corresponding independent ROIs for spectrum analysis. Of these 147 tissue samples, 23 were histologically determined to contain cancerous tissue; 124 were histologically determined to consist of benign prostatic tissue. We excluded tissue types that were not present in a statistically significant number of samples, e.g., chronic or acute inflammation, and intraepithelial neoplasia. The ROC-curve area computed by LABROC for spectrum analysis, as determined using leave-one-out methods applied to a nearest-neighbor analysis, was 0.79 ± 0.07. The corresponding ROC-curve area computed by ROCFIT for LOS assignments made by the examining urologist at the time of biopsy was 0.60 ± 0.22

2-D Images

Using the match of local intercept and attenuation-corrected mid-band values with the ranges of values associated with cancer as the criterion for tissue-type staining, we generated 2-D images for a series of biopsy-proven cases. Because of their quantitative, reduced-speckle qualities, we applied tissue-type encoding to mid-band parameter images rather than to computer-generated B-mode grey-scale images. Examples of a biopsy-proven benign scan plane and a biopsy-proven cancerous scan plane are shown in Figs. 1 and 2, respectively. Figure 1 shows a computer generated B-mode image next to a stained mid-band image for the benign case; Fig. 2 shows the same types of images for the cancerous case. Both figures show the longitudinal scan plane used for biopsy; the apex of the sector shows a small portion of the rectal wall; the seminal vesicles are to the left. In both, the stain encoding for cancer is a saturated (white) grey-scale value. The benign case of Fig. 1 shows no staining for cancer, even in hypoechoic portions of the peripheral zone adjacent to the rectal wall; however, the cancerous case shows significant staining in the peripheral zone where cancer was detected.

3-D Images

We generated 3-D images from stained 2-D images by morphing pairs of images in the set of 2-D images corresponding to the transverse scan planes digitized at the time of ultrasonic examination and biopsy guidance. Morphing, which is performed by standard, commercially available software, fills in the volume between acquired scan planes. Prior to morphing, we manually demarcate the capsule of the gland, then mask off the region of each image that lies outside the gland. After morphing, we render the entire set of scan planes into a 3-D volume using commercially available VoxelView™ software in an SGI Onyx® imaging workstation.

Figure 3 shows an example of a 3-D rendering using two volumetric presentations of a biopsy-proven cancerous case. This rendering consists of 91 component planes morphed from 10 original scan planes. Although we generated several presentations of this rendering, only the two shown here are suitable for grey-scale reproduction. Figure 3A depicts the volume rendering of the region suspected of being cancerous based on spectrum analysis; it spatially corresponds to the region that was histologically positive for cancer. Figure 3B depicts the lesion within the capsule of the gland and shows its proximity to the gland margins and its distribution along the peripheral glandular zone of the prostate in which cancer originates. The gland is viewed from its apex in both depictions; the rectal wall corresponds to the indentation at the lower right of 3B.

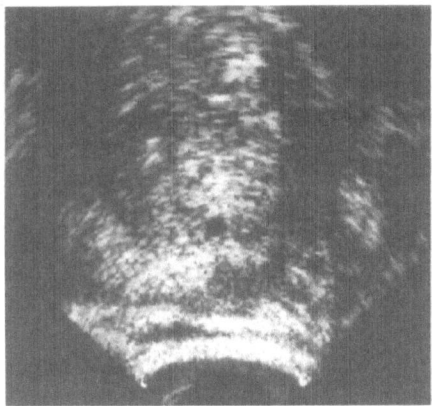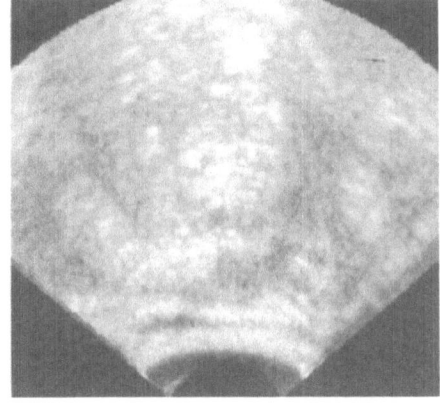

Figure 1 A biopsy-proven cancer-free scan plane. A. Computer-generated B-mode image. B. Stained image using saturated white to indicate the suspected presence of cancer.

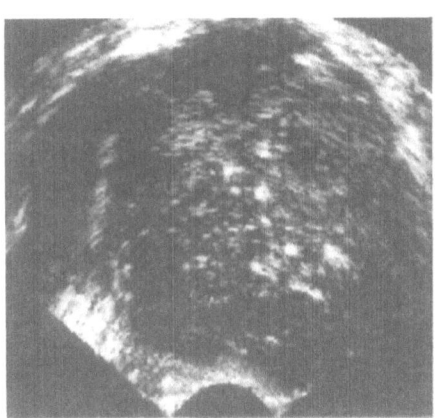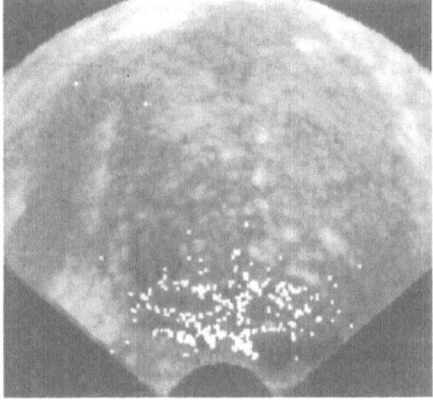

Figure 2 A biopsy-proven cancer-containing scan plane. A. Computer-generated B-mode image. B. Stained image using saturated white to indicate the suspected presence of cancer.

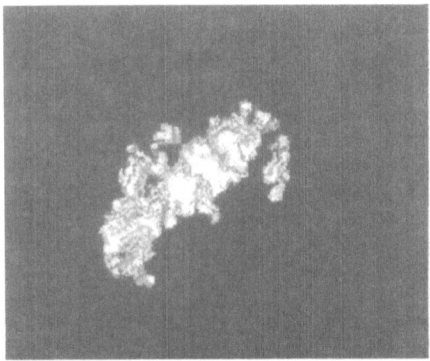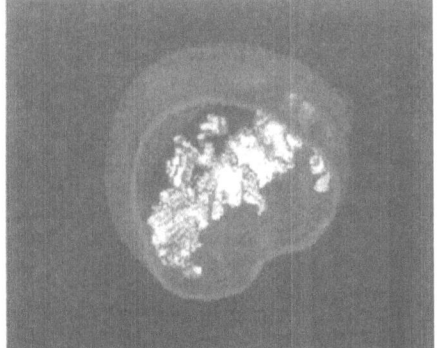

Figure 3. 3-D rendering of a biopsy-proven cancerous lesion viewed from the apex of the gland. A. The lesion alone; B. The lesion within the prostate capsule.

DISCUSSION

The results summarized above suggest that spectrum analysis may provide a means of imaging cancer of the prostate that is significantly superior to conventional imaging methods. This superiority has important implications for biopsy guidance, i.e., for detection and diagnosis of cancer, and for staging, i.e., for assessment of tumor volume and spatial distribution required for planning and monitoring treatment.

The improved specificity and sensitivity offered by spectrum-analysis-based images may permit more-effective sampling of prostate tissue resulting in fewer missed cancers and fewer unnecessary negative biopsies. Inferences drawn from published studies on incidence and efficacy combined with the implications of the ROC analyses of our data suggest that 100,000 more prostate cancers might be detected annually in the U.S. if these methods prove to be as effective as current results suggest.[9,10]

A key prognostic variable for prostate cancer is tumor size. Yet no reliable imaging means of volume estimation exists. The choice between surgical and non-surgical treatment depends on clinical determination of tumor extent; surgery is appropriate only if cancer is organ-confined. Yet typically 30% of excised glands show positive margins indicating that surgery was inappropriate. Neo-adjuvant treatment prior to surgery may benefit markedly from an accurate means of gland and tumor volume assessment; the efficacy of conformal radiation, implanted-seed radiation, cryosurgery, and thermal methods under development can be enhanced by reliable assessment of cancer distribution in 3-D; chemotherapeutic therapies have toxic side effects that must be balanced with tumor regression and can benefit from a means of determining tumor response. Yet no reliable imaging means of pre-treatment assessment and treatment monitoring exists. The described 3-D imaging methods offer a possible, effective, imaging technique 1) for evaluating tumor size, shape and location, to aid in planning treatment, and 2) for quantitatively evaluating changes in response to treatment.

ACKNOWLEDGEMENT

This research is supported in part by NIH/NCI Grant CA53561.

REFERENCES

1. F.L. Lizzi, M. Greenebaum, E.J. Feleppa, M. Elbaum and D.J. Coleman, Theoretical framework for spectrum analysis in ultrasonic tissue characterization, *J. Acoust. Soc. Am.* 73:1336-1373 (1983).
2. E.J. Feleppa, F.L. Lizzi and D.J. Coleman, Diagnostic spectrum analysis in ophthalmology: a physical perspective, *Ultrasound Med. & Biol.* 12:623-631 (1986).
3. F.L. Lizzi, D.L. King, M.C. Rorke et al., Comparison of theoretical scattering results and ultrasonic data from clinical liver examinations, *Ultrasound Med. & Biol.* 14:377-385 (1988).
4. E.J. Feleppa, W.R. Fair, A. Kalisz, et al., Typing of prostate tissue by ultrasonic spectrum analysis, *IEEE Trans. Ultrason. Ferro. Freq. Cont.* 43:609-619 (1996).
5. E.J. Feleppa, T. Liu, A. Kalisz, M.C. Shao, W.R. Fair, N. Fleshner and V. Reuter, Ultrasonic spectral-parameter images of the prostate, *Int. J. Imaging Sys. Tech.* 8:11-25 (1997).
6. N. Fleshner, E.J. Feleppa, T. Liu, A. Kalisz, M. O'Sullivan, V. Reuter and W.R. Fair, Spectrum analysis: novel technique for ultrasonic prostate imaging and biopsy guidance, *Molec. Urol.* 1:21-28 (1998)
7. E.J. Feleppa, W.R. Fair, A. Kalisz, W. Larchian, T. Liu and V. Reuter, Spectrum analysis and 3-dimensional imaging for prostate evaluation, *Molec. Urol.* (in press).
8. C.E. Metz, Some practical issue of experimental designed data analysis in radiological ROC studies, *Invest. Radiol.* 24:234-245 (1989).
9. S.L. Parker, T. Tong, B.A. Bolden and P.A. Wingo, Cancer statistics, *CA Cancer J. Clin.* 46:5-27 (1996).
10. W.J. Ellis and M.K. Brawer, Repeat prostate needle biopsy: who needs it?, *J. Urol.* 153:1496-1498 (1995).

PERFUSION OF PARENCHYMA - IN VITRO SIMULATION USING CONTRAST AGENTS AND 2nd HARMONIC IMAGING

Bernhard Gassmann,[1] Wolfram Wermke [2]

[1] Klinikum Berlin-Buch
Institute for Medical Physics
13125 Berlin Germany
[2] Humboldt-Universität Berlin
Universitätsklinikum Charité
IV. Medical Clinc Dept. Gastroenterology
10117 Berlin Germany

INTRODUCTION

The aim of contrast agents for ultrasound diagnostic is the increase of blood flow signal returning from vessels inside the body. Better signal to noise ratio is fundamental to distinguish between blood flow and noise. Many limitations for detecting very small flow and very low flow velocities are well known, but with the help of contrast agents together with harmonic imaging it should be possible to overcome some of these limitations. Artifacts are visible by using contrast agent. With harmonic imaging some of these artifacts will decrease. With the help of a new type of capillary phantom the simulation of effects using contrast agent in parenchymal organs can be demonstrated. The discussion of the resulting images is in progress, the same problems with interpretation like the in vivo-images.

EXPERIMENTAL

The diameter of the capillary phantom is about 3 cm to 4 cm. The diameter of only one capillar tube is less than 200 µm. The number of capillaries bundled together depends on the required outer diameter of the phantom. It is possible to conduct more than one phantom in a line to produce opposite flow direction. The full phantom is placed in a watertank and different types of transducers placed in a fixed, but adjustable position for measurements. The ATL HDI 3000 is used for measurements with fundamental and 2nd harmonic imaging, connected with Covex (C4-2, C7-4) and linear (L10-5) transducers. Imaging the phantom was done in B-mode, color- and power-Doppler as well as spectral-Doppler (PW). Levovist® from Schering AG, Berlin-Germany, is used to demonstrate the effects by using contrast agents. The solution with Levovist® was isotonic and ph-stabilized with phosphate (standard solution). A roller pump is used for adjustable flow. By the construction of the phantom the equalized flow in all capillaries is accomplished. The flow volume was measured by collecting the outflow volume.

RESULTS

The images from the phantom with and without contrast agent, in fundamental and 2nd harmonic mode show the potential of the phantom to simulate perfusion. The b-mode images look like images from parenchymal organs, for instance liver. Because frozen images not show the effect of contrast filling it is difficult to show this in frozen images - it´s a dynamic effect and only to see in b-mode. Therefore the main demonstration of the usefulness of the phantom was done in color- and power Doppler-mode, respectivly spectral Doppler. To demonstrate the ability of the phantom, some images below show the phantom and the effect of adding contrast agent in fundamental and 2nd harmonic imaging (b-mode).

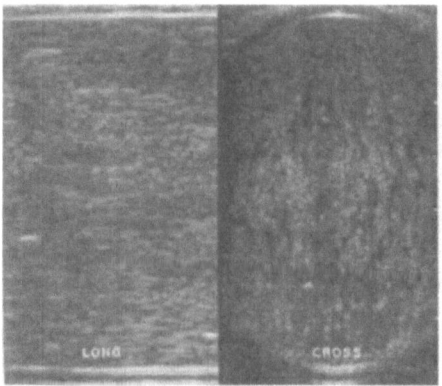 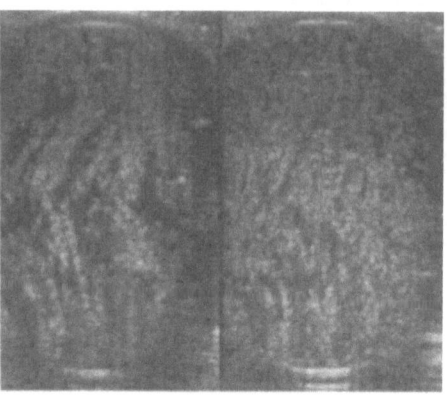

a) fundamental b-mode
wihtout contrast agent,
longitudinal plane and

b) fundamental b-mode
without contrast agent,
angle dependent backscatter

Figure 1. B-mode images from the phantom in different planes.
a) The phantom imaged in fundamental b-mode and without contrast agent (nativ mode). On left is the longitudinal plane and on the right is the transvers plane. Espacially in the cross section the interference of the scattered waves is clear to see.
b) The phantom is imaged in fundamental b-mode and withoet contrast agent. The two images show the phantom in transverse plane but with different angles. The angle dependent backscatter is a problem by using this type of capillary material.

The phantom is biult from a couple of capillaries. These capillary are not pressed together in a tube; each capillary is surrounded by water in the tank. Therefore angle dependent backscatter my be a problem at low backscatter intensities. With the use of contrast agent in the lumen of the capillaries the amplitude of the scattered ultrasound is much more higher than the intensity scattered from the capillaries. Under use of 2nd harmonic imaging the angle dependent backscatter decreases and the images are much more homogenius.

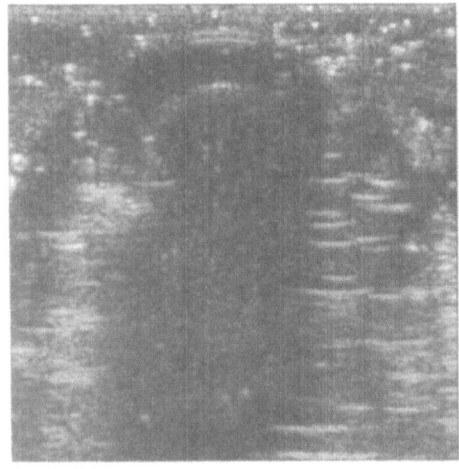 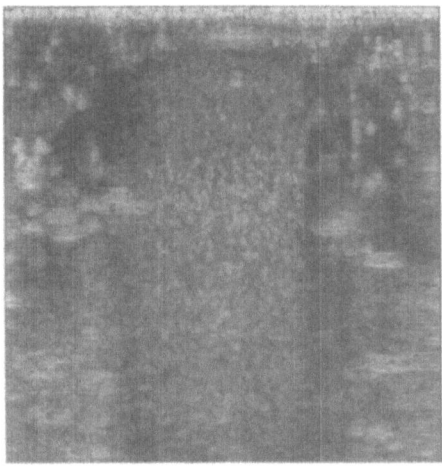

a) fundamental b-mode wiht contrast agent

b) 2nd harmonic imaging in b-mode, with contrast agent

Figure 2. B-mode images from the phantom in different modes.
a) The phantom imaged in fundamental b-mode and with contrast agent (nativ mode). The lumen inside the phantom is the dark area, only small backscatter intensity.
b) 2nd harmonic b-mode with contrast agent, the lumen of the phantom is filled with signals coming from the bubbles of the contrast agent.

The effect of 2nd harmonic imaging with contrast agent is very clear to see with the phantom for artifacts like "blooming", "bubble noise" (acoustic emission), shadowing and overestimation of maximum Doppler-shift.
It s easy to handle and to look for effects depending on the transmit power and on the position of the focal zone. Signal detection in 2nd harmonic mode is more dependent on the position of the focal zone than in fundamental mode. Decreasing bubble noise is dependent on the transmit power. Sometimes it is very helpful to allow reperfusion, to work with triggered images.

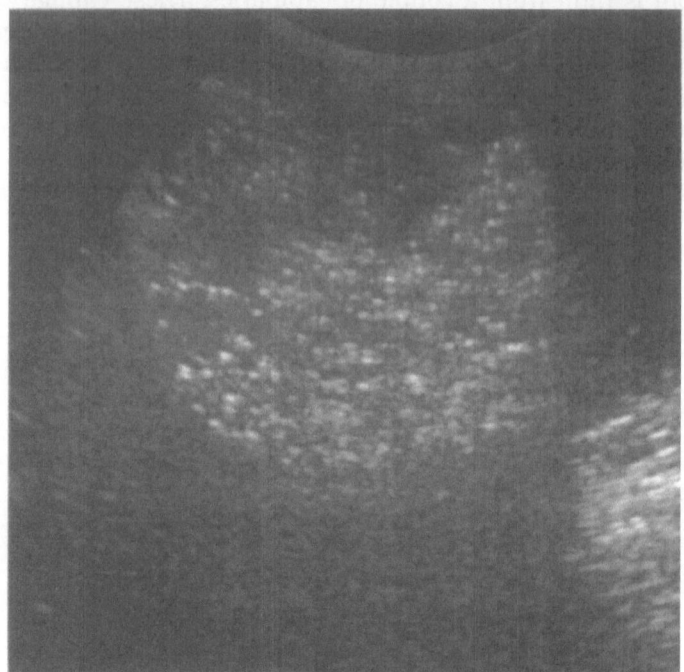

Figure 3. Fundamental b-mode with contrast agent. The noise on the edge of the phantom is clearly identified as noise from multiple backscattered signals form the contrast agent bubbles. The edge detection is difficult to do, the image looks unsharp and overloaded with noise. Under use of 2nd harmonic imaging the noise is decreasing.

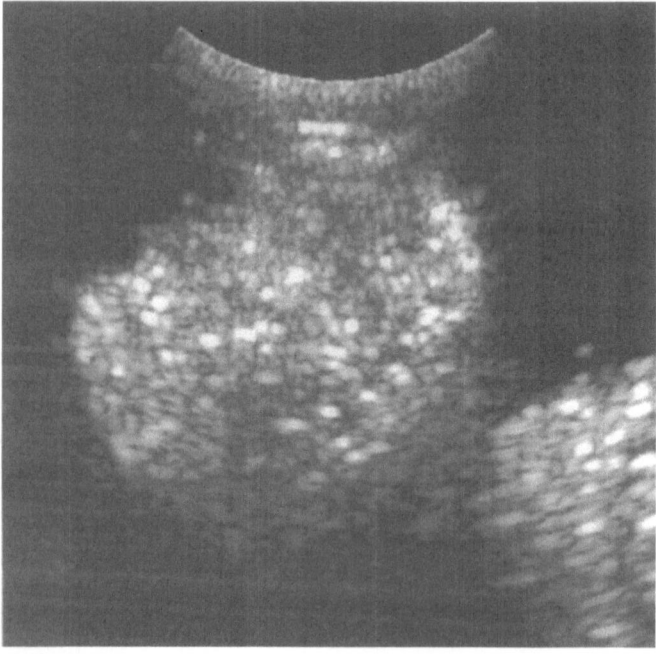

Figure 4. 2nd harmonic b-mode with contrast agent. The edge detection of the phantom is easy to handle, low noise but reduction in spatial resolution.

CONCLUSION

Perfusion of parenchyma is still not to see but harmonic imaging has the potential to visualize the concentration of contrast agents in parenchyma at very low velocities. To simulate parenchyma the capillary phantom is useful and looking like parenchyma in b-mode imaging. Perfusion of the phantom filled with isotonic solution and contrast agent was measured at different velocities and concentrations of the agent. With the help of this phantom it is possible to demonstrate advantages as well as disadvantages of 2nd harmonic imaging. It´s easy to simulate different types of perfusion defects inside the phantom. Visualisation of these perfusion defects will increase the use of contrast agents. The diagnostic liability is to detect hypo- or hyperperfused areas in parenchymic organs. Interference between scattering effects from the structure of phantom and clutter noise of moving contrast agent makes it difficult to interpret the images. The phantom is also useful to distinguish between technical effects from the equipment and artifacts, caused by the contrast agent. The problems of prototype phantom are clear to see. The differences between mimicking phantoms and inhomogeneous and anisotropic tissue are well known. This phantom has the ability to get a feeling, how perfusion defects in parenchyma were imaged.

NON-STATIONARY PARAMETRIC SPECTRAL ESTIMATION FOR ULTRASOUND ATTENUATION

Jean-Marc Girault[1], Frederic Ossant[1], Abdeldjalil Ouahabi[1], Christelle Guittet[2], Denis Kouamé[1] and Frederic Patat[1].

[1]LUSSI / GIP Ultrasons-EIT, 7 avenue Marcel Dassault, BP 407, 37204 TOURS Cedex 3 FRANCE.
[2]INSERM U316, Fac. Médecine, 2 bis bd Tonnelé, 37032 TOURS Cedex FRANCE.

INTRODUCTION

Great progress[1] has been made recently in high frequency ultrasound imaging. In particular, several works have shown the interest of echographic exploration for frequencies ranging from 20 to 100 MHz, in dermatology and ophthalmology. These results are extending the field of application of tissue characterization. The measurement of acoustic attenuation tissues has received much interest in the field of ultrasound tissue characterization. Indeed, several clinical applications have shown a correlation between attenuation values and pathological states.

Attenuation caused by scattering and absorption affects the instantaneous spectrum of echoes in downward direction. Consequently, in order to estimate the ultrasound attenuation, we propose to analyze the interest of new spectral analysis tools allowing an efficient estimation of the ultrasound attenuation in this high frequencies domain.

Several techniques[2, 3, 4] can be used to determine the spectrum of backscattered signal. The Fourier analysis and the parametric spectral estimation commonly employed in the tissue characterization field, give similar results in weakly attenuating media. These two methods are based upon global and local stationary signal assumptions and use a sliding window technique. In the reflection mode and only in highly attenuating media, it has been shown[4] that the parametric spectral analysis provides a better estimation of attenuation in terms of relative error than the Fourier approach. In order to obtain better results, than conventional parametric and Fourier approaches, both in weakly and highly attenuating media, we sought to test another parametric spectral analysis: the nonstationary autoregressive (AR) spectral analysis.

In this communication, we propose two new approaches of parametric spectral analysis for the attenuation estimation: an adaptive recursive method and a time-varying recursive method.

Thanks to these two techniques, it is possible to obtain instantaneous frequency determination and evolution of the spectrum of the maximum energy frequency. These two new methods applied to the backscattered simulated signals are compared for different attenuation coefficient values (1 to 5 dB/cmMHz) with a 45 MHz ultrasonic transducer central frequency.

ATTENUATION MODELING AND DATA SIMULATION

Attenuation modeling: A reasonably accurate model for backscattered signal by a biological tissue consists of a function of components representing the contributions:

On the one hand, the measuring system impulse response (basic ultrasonic wavelet).

On the other hand, the scattering and absorption function of the explored medium.

Consequently, the reflected signal $x(t)$ from a given area of a medium can be represented as a function of the ultrasound wave $e(t)$ on a time-varying impulse response $h(t, \tau)$ from the explored tissue:

$$x(t) = F\{e(t), h(t, \tau)\} \tag{1}$$

In our model, we do not take into account the diffraction effects in order to state the problem in simpler terms.

It is well-known that for most biological tissues, the attenuation α (dB/cm) appears to be linearly dependent[5] on frequency.

Attenuation information appears in all depth-varying frequencies of the spectrum and particularly for the maximum energy frequency. We assume that the Power Spectral Density (PSD) of emitted signal $e(t)$ has a Gaussian form[6]; this implies that its spectral variance is a constant versus depth.

The attenuation coefficient can be calculated with the following relation:

$$\beta_{(dB/cmMHz)} = \frac{-4.34}{c\sigma^2} \frac{df_{max}(t)}{dt} \tag{2}$$

where σ^2 the variance of PSD which is related with transducer bandwidth, c: speed of ultrasound. Note that 1 dB corresponds to 8.68 Nepers.

Accordingly, by using a linear regression, we can derive the frequency slope and then the slope of the attenuating coefficient.

Data simulation: Simulated uncorrelated A-lines are computed by the following method. Several assumptions are considered on the simulated medium,

1) scatterers with random spatial distribution are punctual,
2) attenuation is homogeneous,
3) single scattering is considered.

Concerning the transducer, we suppose that its transfer function has a Gaussian form and diffraction effects are neglected.

Accordingly, a spectrum of the backscattered signal by L scatterers is given by:

$$X_{(f)} = E_{(f)} \sum_{i=1}^{L} \xi_i e^{-\frac{\alpha(f)ct_i}{8.68}} e^{-2\pi f t_i} \tag{3}$$

where α is the ultrasonic attenuation, $E_{(f)}$ is the transfer function of the transducer and ξ_i is a random variable ($\xi_i \in [0,1]$) which weights the backscattering for each scatterer.

An homogeneous medium is simulated with attenuation ranging from *1* to *5* dB/cmMHz. The sampling frequency f_s is *400* MHz, the transducer center frequency f_0 is *45* MHz and the standard deviation σ of the DSP is *7.46* MHz. Each simulation contains *256* A-lines of *1024* samples; with a speed of ultrasound of *1530* m/s, the pulse duration is about *2.56* µs corresponding of *2* mm explored tissue.

NONSTATIONARY AUTOREGRESSIVE MODELING

The echographic signal $x(n)$, which is strongly nonstationary in highly attenuating media, is modeled as the output of a linear filter driven by a white Gaussian noise $u(n)$ with zero mean and variance σ_u^2. It is given by Kay and Marple[7]:

$$x(n) = -\sum_{i=1}^{p} a_i(n)x(n-i) + u(n) \qquad (4)$$

where $a_i(n)$ are AR parameters at each time n which is related to distance by: $d = nT_sc/2$, c is the speed of ultrasound, T_s is the sampling interval and p is the order of the AR model.

Nonstationary RF signals modeling will not require a high AR order because the aim is to identify one time-varying frequency and not all the spectrum which will require a high AR order. In this study we will adopt a second order AR model[4, 8].

If we suppose that the time-varying parameters are linear combinations of a set of basis time-varying functions $F_g(n)$, then we transform a linear nonstationary[9] problem into a linear stationary one by replacing a scalar process with a vector process. In this case, time-varying AR parameters are expressed by:

$$a_i(n) = \sum_{g=0}^{m} a_{i,g} F_g(n) \qquad (5)$$

where $a_{i,g}$ are AR constant coefficients and m is the dimension of time functions basis $F_g(n)$.

The AR time-varying model[9, 10] for a nonstationary sample signal $x(n)$ is:

$$x(n) = -\sum_{i=1}^{p}\sum_{g=0}^{m} a_{i,g} F_g(n-i)x(n-i) + u(n) \qquad (6)$$

The number of unknowns is multiplied by $(m+1)$ but this seems a small price to pay, compared to the benefit of keeping the problem linear.

Several bases functions[10] have been used in different fields for example in a speech processing. Among all these bases, we decided to introduce one of the most commonly used: The Power of time functions.

The Power of time functions are usually used in a quasi-linear time-varying parameters evolution. They are given by:

$$F_g(n) = \frac{1}{g!}\left(\frac{n}{N}\right)^g \qquad (7)$$

where $F_0(n) = 1$.

In practice, the choice of basis functions is obtained by keeping the basis in which order criteria or other criteria are the smallest. Concerning the choice of both the dimension basis functions m and the AR order p, we will take the same criteria as in the basis functions choice. Recently, decomposing the representation[11] of the basis coefficients using the discrete Karhunen-Loeve transformation makes it possible to find an appropriate basis in a more optimal way.

AR PARAMETER ESTIMATION

In this communication the AR parameters estimation is realized by using the Least-Squares algorithms[7, 12, 13].

If long data sequences are available, it is possible to use time-variant identification method. The algorithm which allows us to obtain a new set of parameters each time a new sample is available. This is accomplished by updating the previously evaluated sample on the basis of the prediction error, and weighting by means of a *forgetting factor* λ. In this case, the cost function becomes

$$J_k = \sum_{n=1}^{k} \lambda^{k-n} \left(x(n) - \hat{x}(n) \right)^2 \tag{8}$$

where in practice $0.95 < \lambda < 1$ and k is the index of the last sample considered. The general expression of the recursive weighted algorithm is given by:

$$\begin{bmatrix} \text{parameter vector} \\ \text{at time n} \end{bmatrix} = \begin{bmatrix} \text{parameter vector} \\ \text{at time (n-1)} \end{bmatrix} + [\text{gain}] \begin{bmatrix} \text{observation} \\ \text{vector} \end{bmatrix} \begin{bmatrix} \text{prediction} \\ \text{error} \end{bmatrix} \tag{9}$$

$$\hat{\theta}_n \quad = \quad \hat{\theta}_{n-1} \quad + \quad P_n \quad \phi_n \quad \varepsilon_n \tag{10}$$

$$P_n = \frac{P_{n-1}}{\lambda} (1 - \frac{P_{n-1} \phi_n \phi_n^T P_{n-1}}{\lambda + \phi_n^T P_{n-1} \phi_n}) \tag{11}$$

$$\varepsilon_n = x(n) - \phi_n^T \hat{\theta}_{n-1} \tag{12}$$

where:

$$\phi_n^T = [-x(n-1), \quad \cdots \quad , -x(n-p)] \tag{13}$$

$$\theta = [a_1, \quad \cdots \quad , a_p]^T \tag{14}$$

Extending the recursive weighted algorithm to a time-varying algorithm can be performed by: multiplying the observation vector by the basis functions, increasing the AR order p by $p(m+1)$ and setting the forgetting factor to one. By using (6), the algorithm becomes:

$$x(n) = \varphi_{n-1}^T \hat{\theta} + u(n) \tag{15}$$

$$\varphi_{n-1}^T = \phi_{n-1}^T V \tag{16}$$

$$V = [F_0(n-1), \quad \cdots, \quad F_m(n-p)]^T \tag{17}$$

$$\theta = [a_{1,0} \quad \cdots \quad a_{1,p} \quad \cdots \quad a_{m,0} \quad \cdots \quad a_{m,p}]^T \tag{18}$$

At the beginning of parameters estimation, both the parameters vector and the gain are initialized. It is possible in this way to calculate a power spectrum at each successive sample signal and, hence to study the spectral characteristics even under nonstationary conditions. Fig. 2 illustrates the frequency evolution versus depth by using an adaptive algorithm for simulated medium. We note that frequency estimation is biased and that is becomes meaningful after transient area. Fig. 3 shows the maximum energy frequency evolution versus depth in a time-varying case for simulated medium. We can establish that frequency evolution is weakly distorted and feebly biased.

SPECTRAL ESTIMATION

If AR parameters are known, the power spectral density (PSD) $S_{xx}(f,n)$ of a nonstationary AR modeling is given by (19). Extending the latter expression to the time-varying AR modeling, the PSD becomes (20).

$$S_{xx}(f,n) = \frac{\sigma_u^2(n)}{\left| 1 + \sum_{i=1}^{p} a_i(n) e^{-j2\pi fi} \right|^2} \tag{19}$$

$$S_{xx}(f,n) = \frac{\sigma_u^2(n)}{\left| 1 + \sum_{i=1}^{p} \sum_{g=0}^{m} a_{i,g} F_g(n) e^{-j2\pi fi} \right|^2} \tag{20}$$

where f is the normalized frequency $-0.5 \leq f \leq 0.5$, n is the time and $\sigma_u^2(n)$ is the noise power.

Attenuation information which appears in the time-varying spectrum can be derived by many particular frequencies as the centroid frequency, the maximum energy frequency or the resonating frequency ...

In this paper, we will focus on the maximum energy frequency[2, 14]. It is obtained by differentiating (19) or (20) with respect to f and setting it to zero.

$$f_{max}(n) = \frac{f_s}{2\pi} \cos^{-1}\left(\frac{-a_1(n)}{4}\left(1 + \frac{1}{a_2(n)}\right)\right)$$

(21)

where f_s is the sampling frequency, $a_1(n)$ and $a_2(n)$ are the two considered AR parameters.

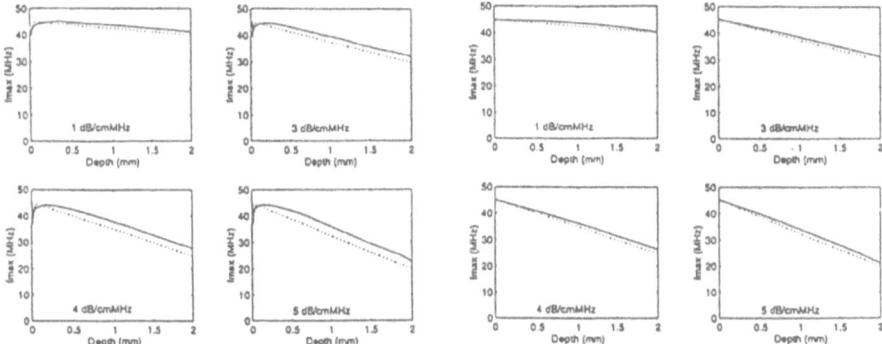

Fig. 2. Evolution of the Maximum Energy frequency versus depth in Adaptive approach (solid line) for 2, 3, 4 and 5 dB/cmMHz, (dashed line) theoretical value.

Fig. 3. Evolution of the Maximum Energy frequency versus depth in Time-varying approach (solid line) for 2, 3, 4 and 5 dB/cmMHz, (dashed line) theoretical value.

RESULTS AND DISCUSSION

Previous operations before estimation: The necessary operations to use the algorithms in optimal conditions are listed below:

1) We have verified that a second AR order is suitable and gives an efficient attenuation estimation. What is more, we can mark that a two dimensions function is sufficient to characterizing attenuation phenomenon.

2) We have searched the forgetting factor λ that offers the best trade off between a good accuracy and a feebly statistical fluctuation. The best value that we have found is $\lambda = 0.98$.

3) The initialization of parameters vector and gain is realized in substituting information at the i line estimation by collected information at the $i+1$ line estimation. This original operation permits an improvement of the algorithm convergence speed and then an improvement of the attenuation coefficient estimation.

Statistical analysis of Estimates: In order to characterize estimates, we focus our attention on the bias and the variance of estimation. Moreover, to take into account the stochastic variation of the signal, these two statistical characteristics are examined at a fixed depth.

Measurements regarding the frequency estimation emphasize four interesting points:

1) Excepting transient area, the bias and the variance of the frequencies estimators are independent of the explored medium depth. This phenomenon can be simply explained thanks to the homogeneity of the medium.

2) The systematic error on the frequency estimation is dependent on the attenuation while the variance appears very feebly dependent (see Fig. 4). The attenuation-dependence of

the frequency bias does not disturb attenuation estimation because its slope constitutes relevant information (since the frequency bias is constant versus depth).

3) Fig. 4 dwells on the fact that the time-varying approach is more efficient than adaptive technique (low variance and small bias).

4) Adaptive and especially time-varying methods are powerful techniques to evalu attenuation upon the basis of only one A mode echo line compared to the short time method. This accuracy of the time-varying estimation leads us to conclude that few A m echo lines are sufficient to obtain a good estimation. Therefore, it implies that element investigated volume tissue will be smaller than one realized with classical methc Accordingly, it becomes possible to realize quantitative imaging with a very h resolution.

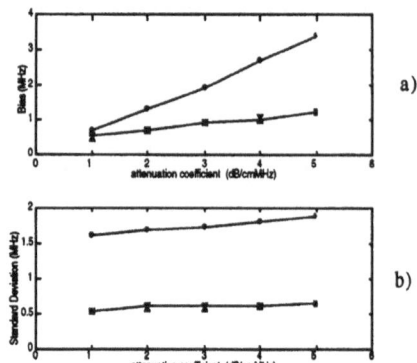

Fig. 4. Bias (a) and standard deviation (b) of frequency estimation versus attenuation coefficient. Adaptive (o) and Time-varying (*) approaches.

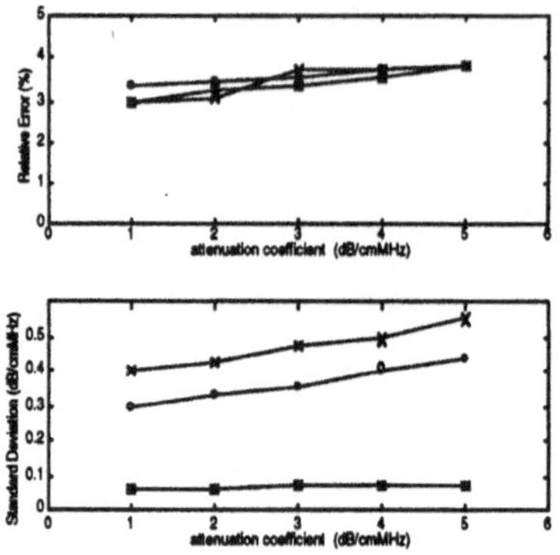

Fig. 5. Relative error (a) and standard deviation attenuation coefficient versus attenuation coeffici Batch (x), Adaptive (o) and Time-varying (*) approaches.

The attenuation estimation highlights two important points (see Fig. 5).

1) Relative error is below *4 %* and standard deviation which is weakly dependent attenuation is below 0.1 dB/cmMHz. Consequently, the *a posteriori* knowledge of rela error means that it becomes possible to easily correct this error on the attenuat estimation.

2) Time-varying method is better than adaptive and batch methods in terms of stand deviation while the relative error is nearly the same in all these methods (see Fig. 5). using a time-varying method, we can improve on the standard deviation by a factor Finally, as expected, the time-varying method is more efficient than the adaptive method.

CONCLUSION

Ultrasound attenuation is an interesting tool in the biological tissue characterizat One mean to evaluated the attenuation coefficient is the spectral shift method. In the reflec mode and for weakly attenuating media, the Short Time Fourier Analysis is sufficient, for highly attenuating media, it is not the case. To overcome this drawback the Short Ti Parametric Analysis has been introduced.

In this communication we have proposed two new approaches for the ultrasor attenuation estimation approaches which give better results than the classical methods b in weakly and highly attenuating media. The second order AR adaptive method, as wel the second order AR time-varying approach whose basis is a two dimensions power ti function provide bias and variance on spectral estimation that are independent from explored tissue depth. However, to implement these algorithms special cares must be tak

The first issue relates to initialization of both gain and parameters of algorithms and the second concern implies researching the best basis function and its best dimension.

Consequently, the joint use of high frequency and time-varying analysis is a promising way which has to be confirmed by both in vitro and in vivo measurements. It would permit to contemplate the following applications: high-resolution quantitative imaging, quantitative identification in biological tissues of superposed layers.

REFERENCES

[1] M. Lethiecq, G. Feuillard, M. Berson, and F. Patat,"Principles and Applications of High Frequency Medical Imaging" *Advances in Acoustic Microscopy*, Ed. A. Briggs, W. Arnold, Plenum Press New-York, vol. 2, pp. 39-102, 1996.
[2] R. Kuc and H. Li," Reduced-Order Autoregressive Modeling for Center Frequency Estimation," *Ultrasonic Imaging*, pp. 244-251, 1985.
[3] G. Berger, P. Laugier, M. Fink, and J. Perrin," Optimal Precision in Ultrasound Attenuation Estimation and Application to the Detection of Duchenne Muscular Dystrophy Carriers," *Ultrasonic Imaging*, vol. 9, pp. 1-17, 1987.
[4] T. Baldeweck, P. Laugier, A. Herment, and G. Berger," Application of Autoregressive Spectral Analysis for Ultrasound Attenuation Estimation: Interest in Highly Attenuation Medium," *IEEE Transactions on U. F. F. C.*, vol. 42, no. 1, pp. 99-110, 1995.
[5] M. Fink, F. Hottier, and J. F. Cardoso," Ultrasonic Signal Processing for in Vivo Attenuation Measurement: Short Time Fourier Analysis," *Ultrasonic Imaging*, vol. 5, pp. 117-135, 1983.
[6] L. Ferrari, J. P. Jones, V. Gonzalez, and M. Behrence,"In Vivo Measurement of Attenuation Based on the Gaussian Pulse Propagation," *Ultrasonic Imaging*, vol. 5, pp. 95-116, 1983.
[7] S. M. Kay and S. L. Marple," Spectrum Analysis-A Modern Perspective," *proceeding of IEEE*, vol. 69, no. 11, pp. 1380-1419, 1981.
[8] K. A. Wear, R. F. Wagner, and B. S. Garra," High Resolution Ultrasonic Backscatter Coefficient Estimation Based on Autoregressive Spectral Estimation Using Burg's Algorithm," *IEEE Transactions on Medical Imaging*, vol. 13, no. 3, pp. 500-507, 1994.
[9] T. S. Rao," The Fitting of Non-Stationary Time-Series Models with Time-Dependent Parameters," *Journal of the Royal Statist. Soc.*, Series B, vol. 32, no. 2, pp. 312-322, 1970.
[10] Y. Grenier," Time-Dependent ARMA Modeling of Non-Stationary Signal," *IEEE Transactions on A. S. S. P.*, vol. 31, no. 4, pp. 899-911, 1983.
[11] J. J. Rajan and P. J. W. Rayner," Generalized Feature Extraction for Time-Varying Autoregressive Models," *IEEE Transactions on Signal Processing*, vol. 44, no 10, pp. 2498-2507, 1996.
[12] L. Ljung," Recursive Identification: Theory for the Users Prentice Hall," New Jersey, 1987.
[13] L. Ljung," Adaptation and Tracking in System Identification. A Survey", *Automatica*, vol. 26, no. 1, pp. 7-21, 1990.
[14] T. Wang, J. Saniie, and X. Jin," Analysis of Low-Order Autoregressive Models for Ultrasonic Grain Signal Characterization," *IEEE transactions on U. F. F. C.*, vol. 38, no. 2, pp. 116-124, 1991.

FDTD SIMULATION OF TRANSCRANIAL FOCUSING USING ULTRA-SONIC PHASE-CONJUGATE ARRAYS

Ibrahim M. Hallaj[1], Robin O. Cleveland[1], Steven G. Kargl[1], and Ronald A. Roy[2]

[1] Applied Physics Laboratory
University of Washington, Seattle, WA 98105

[2] Department of Aerospace and Mechanical Engineering
Boston University, Boston, MA 02215

INTRODUCTION

Focusing ultrasonic fields onto selected targets within the body with good accuracy and sufficient intensity to allow controlled necrosis of a region of tissue is a topic of great interest in medical acoustics. Several techniques have been demonstrated through simulation and laboratory experiment to achieve promising results [1, 2, 3]. The purpose of the present paper is to demonstrate the use of acoustic phase conjugation, or more narrowly in this context, time reversal as a valid focusing technique in biological tissue, which does not require *a priori* knowledge of the inhomogeneous propagation medium's properties. Another goal of this paper is to demonstrate the effectiveness of the finite-difference time-domain (FDTD) method for the simulation of linear acoustic propagation in an inhomogeneous propagation medium.

DESCRIPTION OF THE PROBLEM

It is desired to use a sparse linear acoustic array to back-propagate a signal scattered from a point-like source deep within a modeled human head. A finite-aperture array is modeled as 64 point source elements outside the head in a water bath used as a coupling medium.

An arbitrary acoustic signal, here a pulse from a spark source (lithotripter), is used as the model waveform for illuminating the target. Any pulse shape can be used, but the illustration of the concept without using ideal narrowband signals was to demonstrate the utility of the time reversal technique for arbitrary waveforms using the FDTD code.

Brief Description of the Time Reversal Method

The theory and operation of time reversal mirrors (TRM) is described by Fink [4]. A brief description of the time reversal technique follows: Consider a scatterer or specular reflector within an inhomogeneous medium. The scatterer is our target in this case, and is assumed to have an acoustic impedance mismatch causing it to scatter an incoming signal that could be originated at some element of the array for example. The scattered signal is distorted as it passes through the inhomogeneous propagation medium and undergoes boundary reflections. We record the arriving field at each of the array elements digitally, and store these into buffers during the *receive mode* of the operation. Once the main scattered signal is stored, the array is switched into its *transmit mode*. In transmit mode the recorded data from each channel buffer is reemitted *time reversed*, and amplified if needed. The unchanged medium will back-propagate the signals from all the array elements to focus onto the original scattering target.

It is the time-invariance of the lossless wave equation and the medium's Green's function that allow the use of linear acoustic time reversal. The reciprocity of the system and the medium are necessary for acoustic phase conjugation to be an exact solution. Note that this receive/transmit cycle can be performed more than once (iteratively) to achieve better focusing, and can allow temporal windowing and target selection as well. In our example we assume a single scatterer exists in the medium.

Certain caveats apply to the phase conjugation method. Foremost perhaps is that a well-defined scatterer can be found within the illuminated region. The other main requirement for perfect phase conjugation is that the propagation medium be reciprocal. Strong attenuation, dynamic medium, and dispersion can have detrimental effects on the effectiveness of a real time reversal mirror. Other effects such as phase distortion, nonlinear transfer functions, or jitter in the phase of the signals can also degrade the focusing of the TRM.

DESCRIPTION OF THE SIMULATIONS

The simulations are carried out in two dimensions (x, y). The example presented uses a 1024 x 1024 spatial grid, with a uniform square grid spacing of 0.13-mm, and the time stepping is done using 8-nsec steps. Absorbing boundary conditions were used to minimize reflection from the outer boundaries of the computational domain. The 2-dimensional head model, inspired by the Shepp and Logan phantom for tomography [5], was composed of 21 ellipses that form the major cross-sectional features. Data for the ultrasonic properties were obtained from empirical values given in the literature, as in Goss [6].

The Wave Equation

The inhomogeneous linear acoustic wave equation for a lossy medium is

$$\nabla^2 p - \frac{1}{c^2}\frac{\partial^2 p}{\partial t^2} - \frac{1}{\rho}\nabla p \cdot \nabla \rho + \frac{\delta}{\rho c^4}\frac{\partial^3 p}{\partial t^3} = 0 \tag{1}$$

which is obtained from the equations of fluid mechanics, and solved for the acoustic pressure, p. With the exception of the the finite-amplitude terms, equation (1) retains all terms from the Westervelt equation, described by Tjøtta and Tjøtta [7], and Cleveland [8]. The attenuation comes from the sound diffusivity, δ, and accounts for viscous and thermal effects. The attenuation term implicitly contains a second power dependence on the central pulse frequency, f_c, that can be seen if the equation is rewritten using nondimensional time and space variables, $\hat{t}$ and $\hat{x}$, in one dimension for clarity

$$\frac{\partial^2 p}{\partial \hat{x}^2} - \frac{\partial^2 p}{\partial \hat{t}^2} - \frac{1}{\rho}\left(\frac{\partial p}{\partial \hat{x}}\frac{\partial \rho}{\partial \hat{x}}\right) + \frac{\delta}{\rho c^3}f_c^2\frac{\partial^3 p}{\partial \hat{t}^3} = 0 \tag{2}$$

where the non-dimensional variables are defined to be $\hat{x} = x/\lambda_c$, and $\hat{t} = tf_c$. The wave equation is solved to second-order accuracy in space and time using the finite-difference time-domain (FDTD) method described by Yee [9]. Note that five time frames of the pressure field are required to be stored at all times due to the third-order time derivative in (1).

The Loss Term

A detailed study of the relaxation mechanisms of tissue is beyond the scope of the present study, however Khokhlova, *et al.* [10] and Averkiou, *et al.* [11] have shown that the propagated field with a frequency to the η power dependence gave similar results for η between 1.1 and 2. The studies cited were done for finite-amplitude waveforms, so the results would be conservative when used for our linear case due to the absence of excessive energy in the higher harmonics. Thus, effective attenuation coefficients can be used for the linear simulations. The sound diffusion in (1) is directly related to the absorption coefficient by

$$\alpha = 2\frac{\delta f^2}{\rho c^3} \tag{3}$$

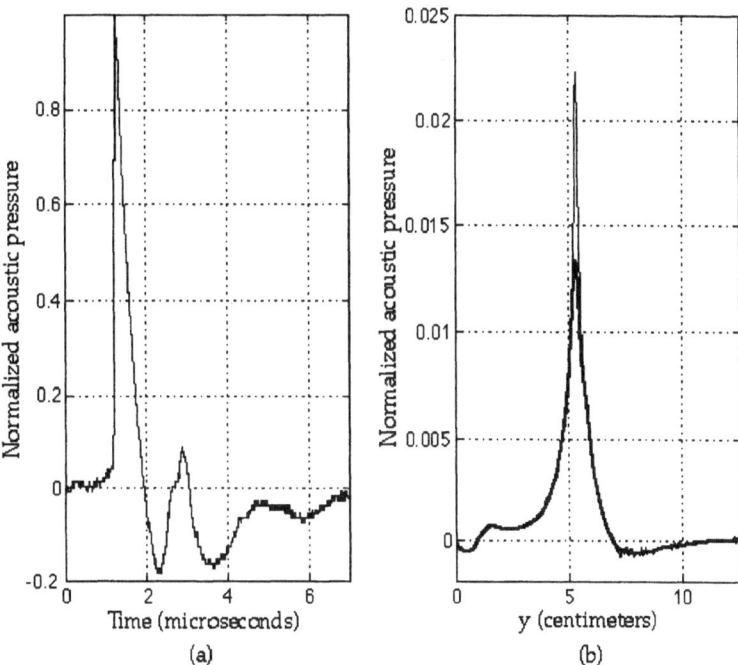

Figure 1: (a) The illuminating pressure waveform, normalized to its maximum value (b) Acoustic pressure slices through the focal spot normalized to illuminating signal maximum. The FWHM extent for the case with attenuation (bold line) is approximately twice the FWHM for the non-attenuating case (fine line).

SIMULATION RESULTS

Output files are generated throughout the simulation to create images such as those shown in figure 2. These frames show snapshots of the acoustic pressure field normalized to the peak pressure of the illuminating signal. An outline of the head model is overlayed on the fields for reference. A small circle is drawn around the location of the scatterer in the receive mode frames. The vertical line to the left of the head shows the position and aperture of the linear array. The large feature within the head in the Receive mode frames is the wavefront reflected by the inner surface of the skull, while the fainter circular traces outside are the transmitted wavefronts. Figure 2 compares the case $\delta = 0$ (no attenuation) with the case having modeled attenuation for the human head during the transmit mode. It is clear that the TRM is capable of focusing acoustic energy back onto the scatterer as predicted by the theory for the reciprocal case. What is not so obvious is what effect the reciprocity-violating factors would have upon the focusing ability of the array. Any terms containing odd-order time derivatives, such as the attenuation term, will reduce the effectiveness of the time reversal system.

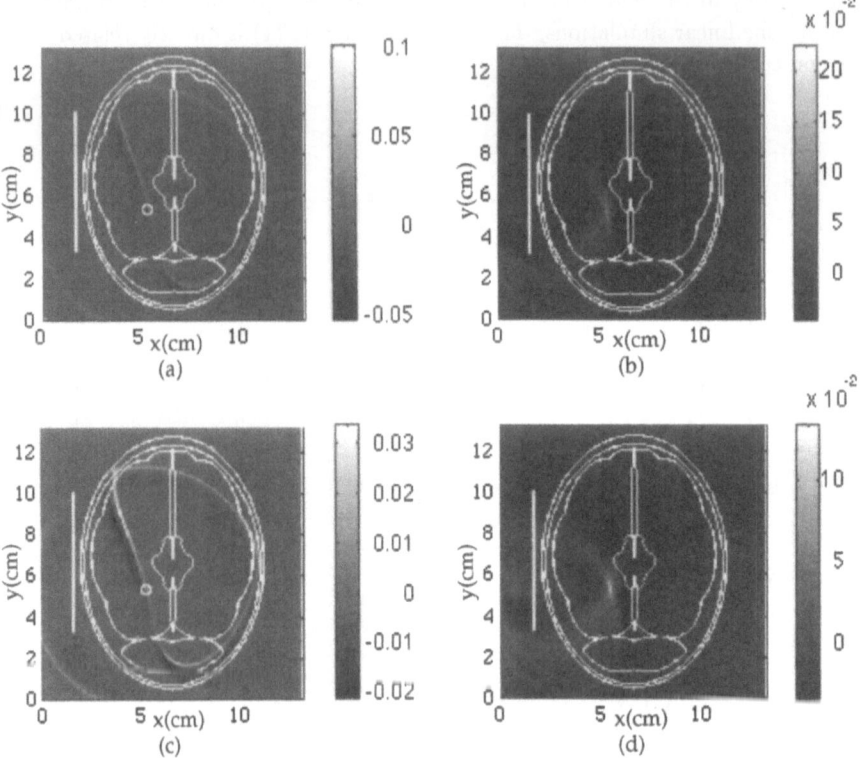

Figure 2: Acoustic pressure field normalized to the peak pressure of the illuminating waveform. The outline of the major modeled structures is also shown. The vertical line to the left of the head is the aperture of the array. (a) Receive mode pressure at some time, t_o, without attenuation (b) Transmit mode pressure at the time of maximum focusing, t_{max}, without attenuation (c) Receive mode pressure at time, t'_o, with attenuation (d) Transmit mode pressure at the time of maximum focusing, t'_{max}, with attenuation. Note that each frame is scaled individually to make full use of the grayscale color map.

LIMITATIONS OF THE TIME-REVERSAL METHOD

Quantitative knowledge of the limitations of the time reversal technique is required to determine whether the technique can be effectively used for medical purposes, especially in the brain. The present study does not attempt to serve as proof of the feasibility of using a TRM for brain tumor ablation. The use of high-intensity phase-conjugate systems is probably more appropriate at this stage for kidney stone comminution and other, less critical, procedures. Thomas and Fink [12] have shown experimentally that the attenuating skull has a detrimental effect on a TRM focus, and suggested a method for amplitude distortion compensation.

Simulations for the human head provide a challenge for imaging and focusing techniques, and as such, serve as a rigorous test of the techniques. This is due to the presence of the highly attenuating and reflecting skull bone, and other internal inhomogeneities. Undesired energy deposition in the tissue other than in the region of interest can result from propagation through the bone. The removal of a section of the skull before applying the ultrasound has been suggested, and was performed on dogs by Fry, *et al.* [13], and is not a difficult or unusual procedure.

In another study [14] the authors investigated the effect of temporal jitter in the initial phase of the signals. As one would expect from a method based on phase, the initial phase of the signals was found to play an important role in the ability of the array to obtain a tight focus. Temporal jitter greater than 5 to 10 percent of one period of a narrowband system will noticeably degrade the focus of the array. Such phase jitter will occur in all systems due to sampling resolution errors as well as transient mechanical errors, which can result from the motion of the array or the target between the receive and the transmit stages of the time reversal process. These phase errors can be important for the megahertz frequencies used in medical ultrasound.

CONCLUSIONS

An explicit FDTD code to solve for the inhomogeneous acoustic wave equation with attenuation was used to simulate the use of a time reversal mirror for focusing ultrasound in a modeled human head cross-section. The head poses special challenges for imaging and focusing techniques because of the attenuation and reflection from the skull bone. The present study modeled a finite-aperture linear array to focus retrodirectively onto a well-defined scatterer deep within the modeled brain tissue. For the case of no attenuation, the time reversal method is exact, and very good focusing was obtained as expected. For the waveform used in this example, the acoustic pressure FWHM points were about 4 mm wide for the attenuation-free case. When attenuation was included, the focusing ability of the TRM was degraded, and the FWHM grew to 9mm. The FDTD is well-suited for simulations such as this, and was found in benchmark tests to be in excellent agreement with calculated focal zone widths down to one wavelength. The propagation through an inhomogeneous medium with sharp interfaces was demonstrated. Further tests can be made to test the code for use as a qualitative prediction tool for hyperthermia applications.

ACKNOWLEDGEMENT

The authors wish to acknowledge the Office of Naval Research (Code 321TS) for its support of this research.

References

[1] R. J. Lalonde, A. Worthington, J. W. Hunt, "Field conjugate acoustic lenses for ultrasound hyperthermia," *IEEE Trans. Ultrason., Ferroelect., Freq. Contr.*, 40, pp 592-602 (1993).

[2] C. J. Diederich, K. Hynynen, "The feasibility of using electrically focused ultrasound arrays to induce deep hyperthermia via body cavities," *IEEE Trans. Ultrason., Ferroelect., Freq. Contr.*, 38, pp 207-219 (1991).

[3] H. Wan, P. VanBaren, E. S. Ebbini, C. A. Cain, "Ultrasound surgery: Comparison of strategies using phased array systems," *IEEE Trans. Ultrason., Ferroelect., Freq. Contr.*, 43, pp 1085-1098 (1996).

[4] M. A. Fink, "Time reversal of ultrasonic fields- Part I: Basic principles," *IEEE Trans. Ultrason., Ferroelect., Freq. Contr.*, 39, pp 555-566 (1992).

[5] A. C. Kak, M. Slaney, *"Principles of Computerized Tomographic Imaging,"* IEEE Press (1988) pp 52-56.

[6] S. A. Goss, R. L. Johnston, F. Dunn, "Comprehensive compilation of empirical ultrasonic properties of mammalian tissues," *J. Acoust. Soc. Am.*, 64, pp 423-457 (1978).

[7] J. N. Tjøtta, S. Tjøtta, "Nonlinear equations of acoustics, with application to parametric acoustic arrays," *J. Acoust. Soc. Am.*, 69, pp 1644-1652 (1981).

[8] R. O. Cleveland, *Propagation of Sonic Booms Through a Real, Stratified Atmosphere*, PhD Dissertation, University of Texas at Austin, (1995).

[9] K. S. Yee, "Numerical solution of initial boundary value problems involving Maxwell's equations in isotropic media," *IEEE Trans. Antennas and Propagation*, AP-14, pp 302-307 (1966).

[10] V. A. Khokhlova, O. V. Sapozhnikov, M. A. Averkiou, L. A. Crum, "Modified spectral solution of a Burgers-type equation for the description of shock wave propagation in biological media," in Proceedings:*World Congress on Ultrasonics*, (1995) pp 1099-1102.

[11] M. A. Averkiou, L. A. Crum, V. A. Khokhlova, O. V. Rudenko, "Nonlinear waveform distortion and energy attenuation of intense acoustic waves in biological tissue," in 14th *Symposium on Nonlinear Acoustics: Nonlinear Acoustics in Perspective*, R. Wei, ed., Nanjing University Press (1996) pp 463-468.

[12] J.-L. Thomas, M. A. Fink, "Ultrasonic beam focusing through tissue inhomogeneities with a time reversal mirror: Application to transskull therapy," *IEEE Trans. Ultrason., Ferroelect., Freq. Contr.*, 43, pp 1122-1129 (1996).

[13] F. J. Fry, *et al.*, "A focused ultrasound system for tissue volume ablation in deep seated brain sites," in Proceedings:*IEEE Ultrasonics Symposium*, (1986) pp 1001-1004.

[14] I. M. Hallaj, S. G. Kargl, R. A. Roy, "Penalties of retrodirective arrays containing signal uncertainties: Implications for mine countermeasures," to appear in Proceedings:*High Frequency Acoustics in Shallow Water*, (1997).

A METHOD FOR IMAGING THE SPATIAL COHERENCE OF PULSED ULTRASOUND FIELDS

Andrew J. Healey, Sidney Leeman and Mark Betts

Department of Medical Engineering and Physics,
King's College School of Medicine and Dentistry,
Kings College Hospital (Dulwich)
London, SE22 8PT, U.K.

INTRODUCTION

The pulsed ultrasound fields employed in conventional medical ultrasound systems are considered to be 'highly' coherent. The development of concepts and definitions of coherence stem from the field of optics, and are well known. However, there are differences between optical fields and the acoustic fields employed in medical ultrasound applications, and the devices available for their measurement, which make direct application of the optical concepts to the acoustic case problematic. Abbott and Thurstone[1] comment that the concept of coherence has been applied to acoustic energy without explanation and the definition of coherence is best taken *"simply as the ability to interfere"* Gehlbach[2] notes that *"the concept of incoherence in acoustic imaging is difficult to define due to the short pulse lengths involved"*.

The second order coherence theory of scalar wave fields is first briefly stated because of its direct relevance to the ultrasound case. The problems of applying these optics concepts directly to ultrasound is then discussed. Application to pulsed wave ultrasound fields is then proposed, along with methods of measurement. Emphasis is placed on a novel method for directly measuring spatial coherence. Results are presented for typical pulsed fields employed in medical imaging systems.

OPTICAL COHERENCE AND THE MUTUAL COHERENCE FUNCTION

The properties of strict coherence and incoherence (or noncoherence) are a mathematical idealisation which are not physically realisable, and may be treated as limiting cases of partial coherence. Hence it is more accurate to say that waves possess a 'degree' of coherence which can be considered quantitatively under a single framework via the mutual coherence function (MCF). Let $V^{(r)}(\mathbf{r},t)$ denote a real (optical) field variable which is a function of a position vector, $\mathbf{r}$, and time, t. It is convenient to consider not the real variable, but rather the associated complex analytic signal $V(\mathbf{r},t)$. Due to the fact that all optical fields found in nature have associated with them certain statistical features, they represent a statistical dynamic system. Thus a statistical description of the field, $V(\mathbf{r},t)$, is used, and the statistical properties of the field manifest themselves in the coherence properties. Hence it is usual to define a mutual coherence function[3], $\Gamma(\mathbf{r}_1,\mathbf{r}_2,t_1,t_2)$, in terms of an ensemble average, in the optics case, as

$$\Gamma(\mathbf{r}_1,\mathbf{r}_2,t_1,t_2) = \langle V_1(\mathbf{r}_1,t_1)V_2^*(\mathbf{r}_2,t_2)\rangle_E \equiv \lim_{N\to\infty} \frac{1}{N}\sum_{i=1}^{N} V_{1i}(\mathbf{r}_1,t_1)V_{2i}^*(\mathbf{r}_2,t_2), \tag{1}$$

where the angle brackets symbolise the ensemble average. The function $\Gamma(\mathbf{r}_1,\mathbf{r}_2,t_1,t_2)$ is the cross-correlation function (in space and time) of the random processes $V_1(\mathbf{r},t)$ and $V_2(\mathbf{r},t)$ and represents the correlation that exists between the fields at the points $\mathbf{r}_1$ and $\mathbf{r}_2$ at times t_1 and t_2. The notion of coherence can also be applied to a single field. Here the two interfering waves, V_1 and V_2 arise from two spatial points of the field V. The cross-correlation function is then replaced with the autocorrelation function, $\Gamma(\mathbf{r}_1,\mathbf{r}_2,t_1,t_2) = \langle V(\mathbf{r}_1,t_1)V^*(\mathbf{r}_2,t_2)\rangle_E$. The following is phrased in terms of interference effects between the two fields V_1 and V_2, but is directly applicable to the concept of coherence of a single field. If it is appropriate to equate time and ensemble averages (ergodic theorem) then the ensemble average in Equation (1) may be determined from a temporal average,

$$\Gamma(\mathbf{r}_1,\mathbf{r}_2,\tau) = \langle V_1(\mathbf{r}_1,t+\tau)V_2^*(\mathbf{r}_2,t)\rangle_t \tag{2}$$

An appropriately normalised version of $\Gamma(\mathbf{r}_1,\mathbf{r}_2,\tau)$, is termed the complex degree of coherence, and is defined as

$$\gamma(\mathbf{r}_1,\mathbf{r}_2,\tau) \equiv \frac{\Gamma(\mathbf{r}_1,\mathbf{r}_2,\tau)}{\sqrt{\Gamma(\mathbf{r}_1,\mathbf{r}_1,0)\Gamma(\mathbf{r}_2,\mathbf{r}_2,0)}} \tag{3}$$

Using the form

$$\gamma(\mathbf{r}_1,\mathbf{r}_2,\tau) = |\gamma(\mathbf{r}_1,\mathbf{r}_2,\tau)|\exp[i\varphi(\mathbf{r}_1,\mathbf{r}_2,\tau)] \tag{4}$$

the normalisation is such that $|\gamma(\mathbf{r}_1,\mathbf{r}_2,\tau)|$ lies between 0 and 1. When $|\gamma(\mathbf{r}_1,\mathbf{r}_2,\tau)|$ assumes the limiting value 0 it is associated with incoherence and the value 1 with coherence. Space and time coherence are concepts which are often considered in the literature. They arise from an attempt (in optics) to separate the effects of spectral width of the radiation from those associated with the finite spatial dimensions of the primary source[3]. These types of coherence may both be expressed in terms of $\gamma(\mathbf{r}_1,\mathbf{r}_2,\tau)$, with spatial coherence being associated with $\gamma(\mathbf{r}_1,\mathbf{r}_2,0)$, and temporal coherence with $\gamma(\mathbf{r},\mathbf{r},\tau)$.

ACOUSTIC COHERENCE

There are a number of important differences between the optical and ultrasound cases :
(1) Optical detectors are usually sensitive only to the intensity of the optical field. Because of the relatively high frequency of the optical field, optical periods are in the order of 10^{-15}s.

68

The resolving time of typical optical detectors is in the order of 10^{-9} s, hence it is not generally possible to study the rapid fluctuations of the optical field, although it is possible to measure the correlations of the field at two or more space-time points. In contrast, the usual ultrasound detectors provide direct measurement of the ultrasound pressure wave, $p(\mathbf{r},t)$, providing 'instantaneous' amplitude and phase information. An empirical knowledge of $p(\mathbf{r},t)$, is thus, in principle, directly available, and recourse does not have to be made to measuring correlations of the field by necessity.

(2) There is usually no inherent random nature associated with the ultrasound source. Hence there is no need to refer to an *eight* dimensional function, $\gamma(\mathbf{r}_1,\mathbf{r}_2,t_1,t_2)$ (or seven dimensions for $\gamma(\mathbf{r}_1,\mathbf{r}_2,\tau)$ if the field is assumed stationary in time), for a description of the field when it is already completely specified by $p(\mathbf{r},t)$.

(3) The pulses employed by pulse-echo imaging systems are effectively bounded in both space and time

Let us first consider the concept of temporal coherence. The temporal signal recorded at the spatial location, $\mathbf{r}$, is $x(t)$. This may be regarded as a stochastic signal where every realisation is identical to every other. The autocorrelation, $\Gamma(t_1,t_2)$, of the real pressure field, $x(t)$, is defined by the expectation of the product $x(t_1)x(t_2)$,

$$\Gamma(t_1, t_2) \equiv \langle x(t_1)x(t_2)\rangle_E \equiv \int x_1 x_2 p_2[x_2(t_2), x_1(t_1)]\, dx_1 dx_2 \tag{5}$$

where $p_2[x_2(t_2),x_1(t_1)]$ is the two-fold joint probability density of the variates x_1 and x_2 at the two time parameters t_1 and t_2. For a deterministic signal the expectation value, $\langle x(t_1)x(t_2)\rangle_E$, simply becomes the product $x(t_1)x(t_2)$. Normalisation of this autocorrelation function via

$$\gamma(t_1, t_2) \equiv \frac{\Gamma(t_1, t_2)}{\sqrt{\Gamma(t_1, t_1)\Gamma(t_2, t_2)}} \tag{6}$$

provides the result that $|\gamma(t_1,t_2)| = 1$, and hence the concept of incoherence, or a degree of partial coherence, is not meaningful in this context. Hence it is appropriate to use the usual definition of the normalised autocorrelation function for deterministic signals,

$$\Gamma(\tau) = \frac{\int x(t+\tau)x(t)dt}{\int x(t)x(t)dt} \tag{7}$$

Thus an appropriate definition for the complex degree of temporal coherence for a deterministic wave, $p(\mathbf{r},t)$, may be given via,

$$\gamma_{pt}(\mathbf{r}, \tau) \equiv \frac{\int p(\mathbf{r}, t+\tau)p^*(\mathbf{r}, t)dt}{\int p(\mathbf{r}, t)p^*(\mathbf{r}, t)dt} \tag{8}$$

where the subscript 'pt' indicates that this is the pulse temporal coherence function. The function $\gamma_{pt}(\mathbf{r},\tau)$ may be calculated directly from a signal recorded from an ideal point hydrophone receiver placed at a location, $\mathbf{r}$, in the field. Note that $\gamma_{pt}(\mathbf{r},\tau)$ may vary rapidly with a change in spatial location, $\mathbf{r}$, due to diffraction effects, especially in the near field.

Similarly it is appropriate to define the concept of spatial coherence for a deterministic pulse via

$$\gamma_{sp}(\mathbf{R}, t) \equiv \frac{\int d^3\mathbf{r}\, p(\mathbf{r}+\mathbf{R}, t)p^*(\mathbf{r}, t)}{\int d^3\mathbf{r}\, p(\mathbf{r}, t)p^*(\mathbf{r}, t)} \tag{9}$$

The subscript 'sp' indicates that it is the pulse spatial coherence function. Note that for the optical case spatial coherence, $\gamma(\mathbf{r}_1, \mathbf{r}_2, t)$, (a seven dimensional function) provides the 3D spatial extent of correlations in the field about every point in 3D space for every time instant, whereas $\gamma_{sp}(\mathbf{R}, t)$ provides a single 3D averaged value of the spatial correlation of the field, for every time instant. This function has a number of useful properties, given in the following section, and will be termed the complex degree of 'pulse' spatial coherence. Note that it has the advantage of being a function of only four dimensions. Consider a (four dimensional) ultrasound (pressure) wave, $p(\mathbf{r}, t)$, propagating linearly in a loss-less medium according to the canonical wave equation. The pressure field may be expressed as,

$$p(\mathbf{r}, t) = \frac{1}{(2\pi)^3} \int\int\int_{-\infty}^{\infty} d^3\mathbf{k}\, F(\mathbf{k})\exp(i\{\mathbf{k} \bullet \mathbf{r} - \omega t\}) \tag{10}$$

where $F(\mathbf{k})$ is the spatial Fourier transform of the field at some time, say $t=0$.

$$F(\mathbf{k}) = \int\int\int_{-\infty}^{\infty} d^3\mathbf{r}\, p(\mathbf{r}, 0)\exp(-i\{\mathbf{k} \bullet \mathbf{r}\}) \tag{11}$$

$F(\mathbf{k})$ is comprised of a set of travelling continuous plane waves and is termed the 'directivity spectrum' and $|F(\mathbf{k})|$ the 'directivity signature'[5]. Note that $F(\mathbf{k})$ does not depend on either $\mathbf{r}$ or t. From the Wiener-Khintchine theorem it follows that the pulse spatial coherence of $p(\mathbf{r}, t)$ (the spatial autocorrelation of the deterministic field $p(\mathbf{r}, t)$), is independent of t, as it is related (via an inverse Fourier transform) to $|F(\mathbf{k})|^2$ (the energy density spectrum of $p(\mathbf{r}, 0)$). Hence the interesting property holds that propagation of the pulse spatial coherence *is independent of t* for propagation in a uniform homogenous loss-less medium. This indicates that a measurement of the pulse spatial coherence properties prior, and subsequent, to propagation in a medium may directly afford a quantification of the change in pulse spatial coherence properties of the pulse due to propagation in the medium.

MEASUREMENT OF COHERENCE PROPERTIES

The measurement of temporal coherence is naturally and directly performed with a point hydrophone. The nature of such devices and typical signals for medical ultrasound fields are well known. A method of conveniently and directly measuring spatial coherence properties with a point hydrophone is more problematic. However spatial coherence may be directly measured with a 'Large Aperture Hydrophone' (LAH). The LAH consists of a thin, uniformly poled and electroded PVDF membrane, stretched flat on a supporting frame[6]. As long as the device is large enough to encompass the entire cross-section of the ultrasound field, the output corresponds to a projection of the field, the device effectively integrating over a plane. Consider the projection onto the (z, t) plane,

$$P(z, t) \equiv \int\int_{-\infty}^{\infty} dx\,dy\, p(x, y, z, t) = \int\int dx\,dy \int\int\int d^3\mathbf{k}\, F(k_x, k_y, k_z)\exp(i[k_x x + k_y y + k_z z - \omega t]) \tag{12}$$

Prudent use of the delta function provides,

$$P(z,t) = \frac{1}{2\pi} \int dk_z \, F(0,0,k_z) \exp\left(i[k_z z - k_z \tau]\right) \equiv P(z - \tau) \tag{13}$$

where $\tau = ct$, and c is the velocity of the wave. If $P(z,t)$ is measured at some initial location, $z=0$, then the value of the directivity spectrum, along a line in $\mathbf{k}$ space, intersecting the origin and oriented orthogonal to the plane of projection, can be directly obtained via a simple Fourier Transform,

$$F(0,0,k_z) = \frac{1}{2\pi} \int d\tau \, P(0,\tau) \exp\left(ik_z \tau\right). \tag{14}$$

This results holds for all orientations of the projection plane, and thus a complete measurement knowledge of $F(\mathbf{k})$, and hence the spatial coherence function, may be obtained by angulation of the LAH.

RESULTS

Figure 1(a) shows a 2D slice through the 3D directivity signature (in $\mathbf{k}$-space) which intersects the origin and is oriented in the direction of the acoustic axis. The pulsed field is generated by a K&B Aerotech 5MHz centre frequency, 1-4 cm focal zone, circular aperture transducer, and the resulting field posses cylindrical symmetry. Figure 1(b) shows a 2D slice through the 3D spatial coherence function which intersects the origin and the entire acoustic axis. A measure of spatial coherence in the axial direction was obtained from the second moment of the one dimensional values of the spatial coherence function, $|\gamma_{sp}(R_z,t)|$, via

$$\frac{1}{2}\left[\frac{\int dR_z \, |\gamma_{sp}(R_z,t)|^2 R_x^2}{\int dR_z \, |\gamma_{sp}(R_z,t)|^2} \right]^{\frac{1}{2}} \tag{15}$$

A similar expression gives a measure for the lateral directions R_y and R_x. These definitions, based on $\gamma_{sp}(\mathbf{R},t)$ provide single time dependent measures, e.g. a single value for the spatial coherence lengths at each time instant. Moreover for propagation in an ideal uniform homogenous loss-less medium, $\gamma_{sp}(\mathbf{R},t)$ is time invariant and thus provides global measures for the pulse. The axial measure provides 0.0893 cm (3 s.f., ~ 2.9 λ) and 0.211 cm (3 s.f., ~ 6.7 λ) for each the lateral directions (where λ is the wavelength associated with the centre frequency of the pulse). The cylindrical symmetry of the field, due to the circular symmetry of the transducer, implies that the coherence area may be given by π times the square of half the lateral spatial coherence length, 0.0350 cm^2 (3 s.f.), and the coherence volume by the axial spatial coherence length times the coherence area, 3.13 mm^3 (3 s.f.). These numbers represent an attempt to quantitate the degree of spatial coherence of this ultrasound pulse.

At first glance it appears that the value of 3.13 mm^3 for the coherence volume does not suggest an extremely 'high' degree of spatial coherence, bearing in mind that the wavelength at the centre frequency is approximately 0.3 mm. However, as the pulse is bounded in both space and time it is instructive to compare the (spatial) extent of the pulse to the (spatial) coherence lengths. If the extent of the pulse is of the same order of magnitude as the coherence length, then this implies that the field is very highly correlated over its extent. Figure 2 presents an indication of the spatial extent of the focused pulse, a volume measurement was obtained by determining the -6 dB contour, which enclosed 95 % of the energy of the pulse. For the locations presented in Figure 2(a), (b) and (c) these volumes

71

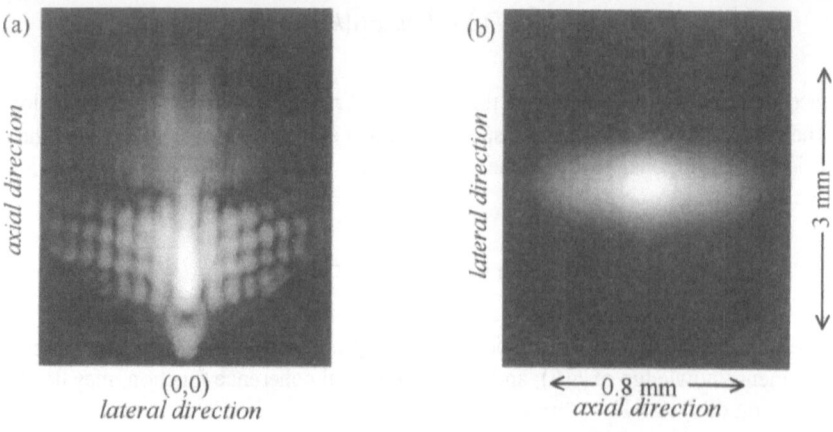

Figure 1 : (**a**) A 2D slice through $|F(\mathbf{k})|$ in **k**-space. (**b**) A 2D slice, $|\gamma_{sp}(R_x, R_y, t_0)|$, through the spatial coherence function for a 5 MHz centre frequency, focused, pulsed field.

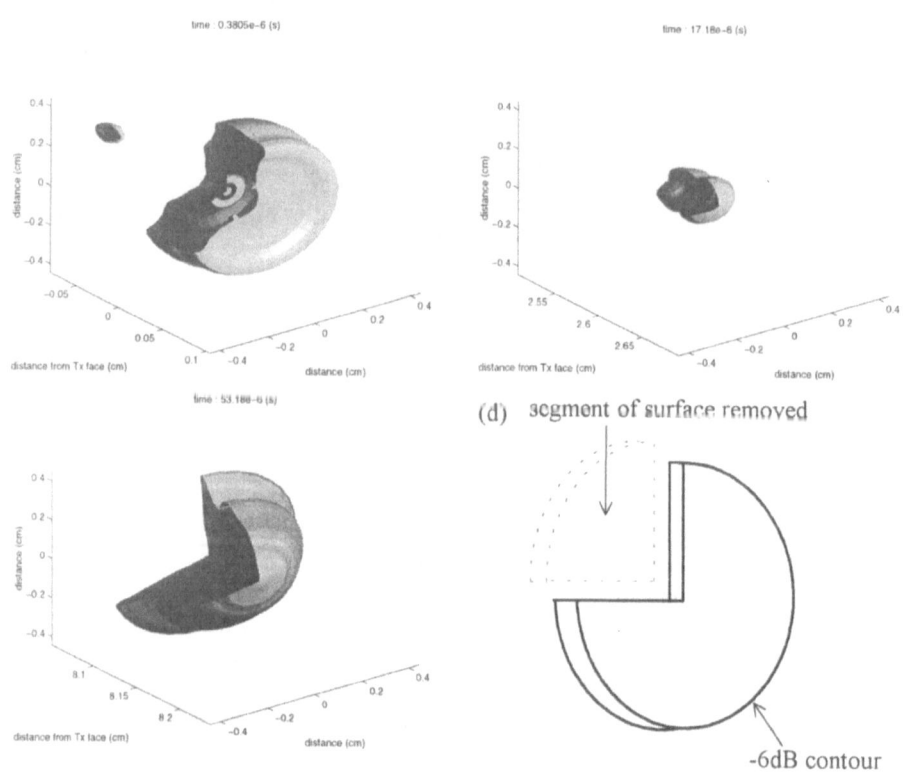

Figure 2 : Graphics showing the -6 dB spatial contour surfaces (in 3D) of the 5 MHz focused pulse at different time instants, in order to convey an indication of the spatial extent of the field. (**a**) Close to the transducer face, (**b**) in the focal zone and (**c**) into the far field. A segment of the surface has been removed in these graphics, see (**d**), in order to aid visualisation.

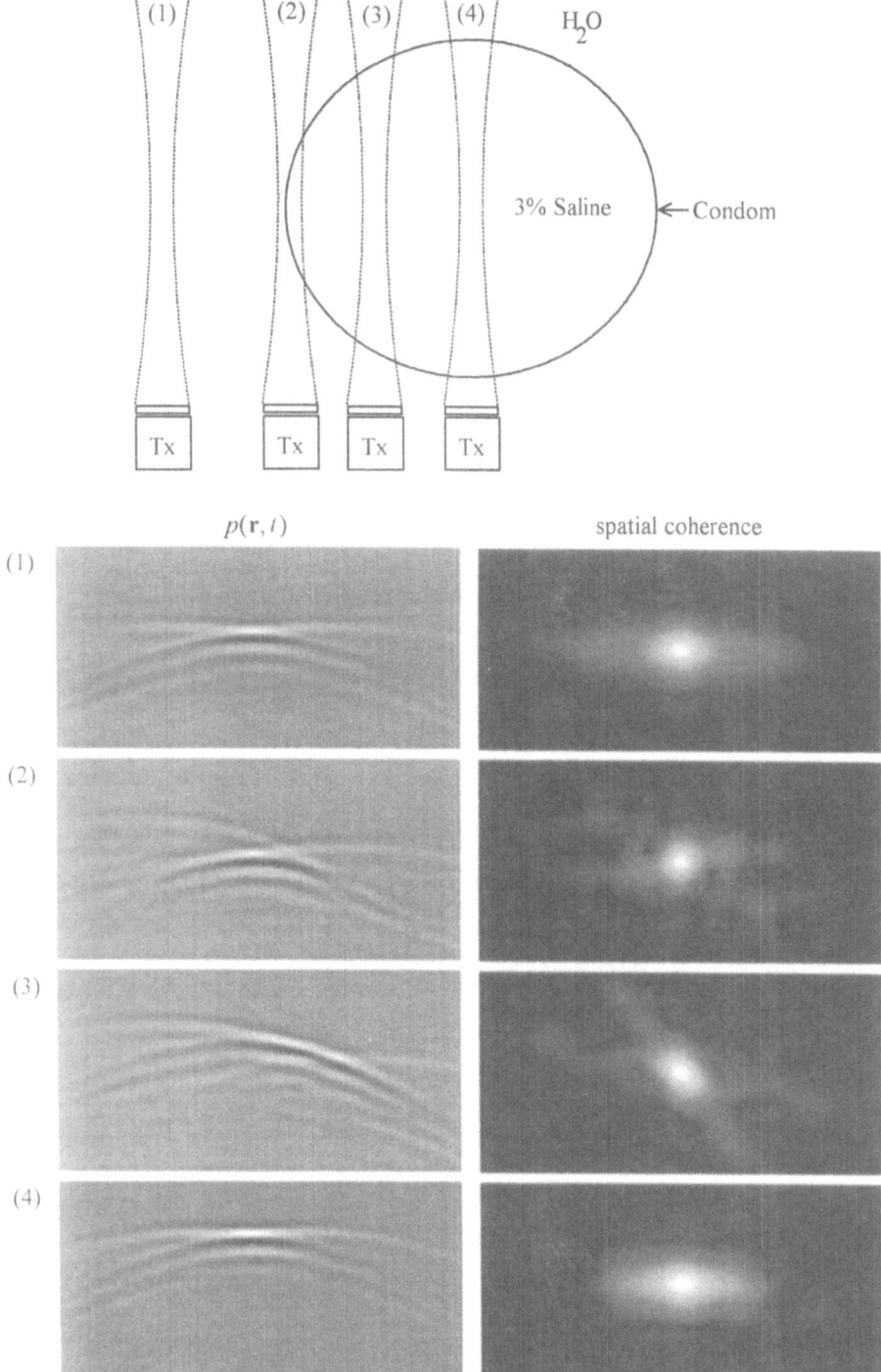

Figure 3 : Projections of the pressure field $p(\mathbf{r},t)$. (1) water path measurement and (2), (3), (4), recorded after propagation through a thin walled cylinder (condom) of 47 mm diameter containing 3% saline solution.

were calculated to be 20.8, 3.5 and 15.7 mm³ respectively. Note that in the focal zone (Figure 2(**b**)) the volume of the pulse is close to that of the coherence volume.

Figure 3 shows 2D projections of a 3.5 MHz centre frequency (9 cm focus) pulse, and associated projections of the spatial coherence function, after propagation through a 47 mm diameter cylinder containing 3% saline solution, immersed in a water bath. The spatial coherence is greatest when the pulse is oriented in position (1), which corresponds to a straight water path. In positions (2), (3) and (4) the pulse encounters various portions of the cylinder. For these paths there is an obvious associated loss of spatial coherence. For propagation through parallel sided 1 cm thick *in-vitro* soft tissue samples comprising animal liver, and animal kidney, a change in the coherence volume of approximately 7% is observed.

CONCLUSIONS

A measure of the spatial coherence for a deterministic bounded ultrasound pulse, $\gamma_{sp}(\mathbf{R},t)$, has been shown to be time independent for linear propagation in an homogenous loss-less fluid. The measurement of $\gamma_{sp}(\mathbf{R},t)$ may be directly achieved with a Large Aperture Hydrophone. Typical focused pulsed fields exhibit a 'high' degree of spatial coherence in the focal zone. Preliminary *in-vitro* measurements of spatial coherence indicate that there is a general loss of spatial coherence of typical medical ultrasound fields when propagating in biological soft tissues.

REFERENCES

1. Abbot J. G. and Thurstone F. L., 1979, Acoustic speckle : theory and experimental analysis, *Ultrasonic Imaging*, 1, pp 303-324.
2. Gehlbach S. M., 1983, *Pulse reflection imaging and acoustic speckle*, PhD dissertation, Stanford university.
3. Beran M. J., and Parrent G. B., 1964, *Theory of partial coherence*, Prentice Hall, Inc., Englewood Cliffs N.J.
4. Wolf E., 1958, *Proc. Phys. Soc* (London) **71**, 257.
5. Leeman S., Seggie D. A., Ferrari L. A., Sankar P. V., and Doherty M., 1985, Diffraction-free attenuation estimation. In : *Ultrasonics International 85*, Ed.: Novak Z. [Butterworth, Guilford], pp 128-132.
6. Costa E. T. and Leeman S., 1990, "A measurement cell for ultrasound attenuation estimation in liquids", *Revista Brasileira de Engenharia, Caderno de Engenharia Biomedica*, 7/1, 459-464.

COMPUTER PHANTOMS FOR SIMULATING ULTRASOUND B-MODE AND CFM IMAGES

Jørgen Arendt Jensen and Peter Munk

Department of Information Technology, Build. 344,
Technical University of Denmark,
DK-2800 Lyngby, Denmark

ABSTRACT

Programs capable of simulating ultrasound images have recently been developed. This opens the possibility of evaluating transducers and focusing schemes not only from their point spread function, but also from an imaging point of view. The calculation of the ultrasound field is based on linear acoustics using the Tupholme-Stepanishen method for calculating the spatial impulse response. Any transducer can be simulated by splitting the aperture into rectangular or triangular sub-apertures, and the calculation can include any transducer excitation and apodization. The acoustic settings can be controlled in the entire image through dynamic apodization and focusing. The transmit and receive apertures can be defined independently of each other. Frequency dependent attenuation can also be included in the simulation.

The B-mode images are generated by specifying a number of independent scatterers in a file that defines their position and amplitude. Adjusting the number of scatterers and their relative amplitude yields the proper image.

Five different computer phantoms are described. The first one consists of a number of point targets. It is used for studying the point spread function as a function of spatial position, and can give an indication of sidelobe levels and focusing abilities. The second phantom contains a number of cysts and point tagets along with a homogeneous speckle pattern. This is used for investigating image contrast, and the system's ability to detect low-contrast objects. The third phantom is for realistic clinical imaging. It contains the image of a 12 week old fetus, where the placenta and the upper body of the fetus is visible. This phantom gives an indication of the whole system's capability for real imaging. The current fetus phantom is only two-dimensional, as it is constant in reflection amplitude in the elevation direction. The program, however, can handle the full three-dimensional simulation, and the whole body could in principle be simulated. An example for a simulated kidney is also shown.

The last phantom is used for color flow mapping and is a combination of static and moving scatterers. A model with stepwise movement of the scatterers in an artery with a parabolic flow profile surrounded by tissue is used. The signal from the scatterers is recorded between each movement and then they are propagated for the next image acquisition. The phantom can be used to study both spectral estimation techniques and color flow mapping estimators.

The scatterer description is three-dimensional and so is the acoustic field calculation. This approach allows the evaluation of an imaging system design to be performed very fast and early in the development process. Typically a single phantom simulation takes less than 12 hours of simulation time.

Transducer designs can be optimized and the practical implementation reduced to only one trial. Also different signal processing approaches can be evaluated realistically.

1 INTRODUCTION

One of the first steps in designing an ultrasound system is selecting the appropriate number of elements for the array transducers and the number of channels for the beamformer. The focusing strategy in terms of number of focal zones and apodization must also be determined. These choices are often not easy, since it is difficult to determine the effect in the resulting images of increasing the number of channels and selecting more or less advanced focusing schemes. It would therefore be beneficial to have easy access to programs that can quantify the image quality. The first approach has been to make simulation programs that can calculate the ultrasound fields and the point spread function for the imaging system. Our first version [1] used the Tupholme-Stepanishen theory for calculating fields for arbitrary transducer geometries, excitations, focusing and apodization schemes. The program is menu driven, and can handle a fixed focus and fixed apodization, and thereby the point spread function at a single position in the image can be determined. This makes it possible to evaluate and design appropriate point spread functions, but it is difficult to evaluate the influence on the image, especially with dynamic focusing and apodization schemes. The program was rewritten to interface more closely with Matlab to handle time varying focusing and apodization as described in [2] and [3]. This has paved the way for doing realistic imaging with multiple focal zones for transmission and reception and for using dynamic apodization. It is hereby possible to simulate ultrasound imaging for all image types including flow images, and the purpose of this paper is to present some standard simulation phantoms that can be used in designing and evaluating ultrasound transducers, beamformers and systems. The phantoms described in this paper can be divided into ordinary string/cyst phantoms, artificial human phantoms and flow imaging phantoms. The ordinary computer phantoms include both a string phantoms for evaluating the point spread function as a function of spatial positions as well as a cyst/string phantom. Two artificial human phantoms are presented; one for a left kidney in a longitudinal scan, and a fetus in the third month of development. The last phantom generates signals from a carotid artery with a parabolic flow profile, and can be used for generating color flow mapping (CFM) images with colors superimposed on the normal anatomic B-mode image. The phantom is, thus, suited for testing new blood velocity estimation algorithms. All the phantoms can be used with any arbitrary transducer configuration like single element, linear, convex, or phased array transducers, with any apodization and focusing scheme.

2   SIMULATION MODEL

A powerful approach to calculating ultrasound fields has been jointly devised by Tupholme [4] and Stepanishen [5]. The pressure generated by the transducer is described by the spatial impulse response as found from the Rayleigh integral:

$$h(\vec{r},t) = \int_S \frac{\delta(t - \frac{|\vec{r}|}{c})}{2\pi|\vec{r}|} dS \tag{1}$$

in which $\vec{r}$ indicates the position of the field point in space, c is the speed of sound, and S is the surface of the transducer. The emitted pressure field is then

$$p(\vec{r},t) = \rho \frac{\partial v_n(t)}{\partial t} * h(\vec{r},t) \tag{2}$$

where $v_n(t)$ is the surface velocity of the transducer and ρ is the density of the medium. The spatial impulse response describes how the transducer shape emits sound in space, and can be seen as the impulse response for the linear system at a particular point in space. Since linear acoustics is used the effect of apodization of the transducer surface can readily be included, and responses from different transducer elements can be directly added for array transducers.

The scattered field and the received response can also be found from spatial impulse response. This has been done in [6] and the received signal from the transducer is:

$$p_r(\vec{r},t) = v_{pe}(t) \underset{t}{\star} f_m(\vec{r}) \underset{r}{\star} h_{pe}(\vec{r},t) \tag{3}$$

where $\underset{r}{\star}$ denotes spatial convolution. v_{pe} is the pulse-echo impulse, which includes the transducer excitation and the electro-mechanical impulse response during emission and reception of the pulse. f_m accounts for the inhomogeneities in the tissue due to density and propagation velocity perturbations which give rise to the scattered signal. h_{pe} is the pulse-echo spatial impulse response that relates the transducer geometry to the spatial extent of the scattered field. Explicitly written out these terms are:

$$v_{pe}(t) = \frac{\rho}{2c^2} E_m(t) \underset{t}{\star} \frac{\partial^3 v(t)}{\partial t^3}, \qquad f_m(\vec{r}_1) = \frac{\Delta\rho(\vec{r})}{\rho} - \frac{2\Delta c(\vec{r})}{c}, \qquad h_{pe}(\vec{r},t) = h_t(\vec{r},t) * h_r(\vec{r},t) \tag{4}$$

So the received response can be calculated by finding the spatial impulse response for the transmitting and receiving transducer and then convolving with the impulse response of the transducer. A single RF line in an image can be calculated by summing the response from a collection of scatterers, in which the scattering strength is determined by the density and speed of sound perturbations in the tissue. Homogeneous tissue can thus be made from a collection of randomly placed scatterers with a scattering strength with a Gaussian distribution, where the variance of the distribution is determined by the backscattering cross-section of the particular tissue. This is the approach taken in this paper.

The phantoms typically consist of 100,000 or more scatterers, and simulating 50 to 128 RF lines can take several days depending on the computer used. It is therefore beneficial to split the simulation into concurrently run sessions. This can easily be done by first generating the scatterer's position and amplitude and then storing them in a file. This file can then the be used by a number of workstations to find the RF signal for different imaging directions, which are then stored in separate files; one for each RF line. These files are then used to assemble an image. This is the approach used for the simulations shown in this paper in which 3 Pentium Pro 200 MHz PCs can generate one phantom image over night using Matlab 4 and the Field II program.

3   SYNTHETIC PHANTOMS

The first synthetic phantom consists of a number of point targets placed with a distance of 5 mm starting at 15 mm from the transducer surface. A linear sweep image is then made of the points and the resulting image is compressed to show a 40 dB dynamic range. This phantom is suited for showing the spatial variation of the point spread function for a particular transducer, focusing and apodization scheme.

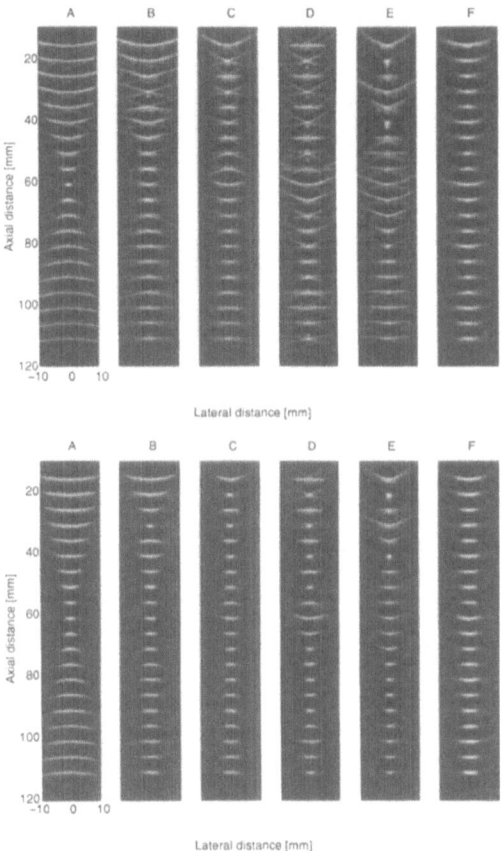

Figure 1: Point target phantom imaged for different set-up of transmit and receive focusing and apodization. See text for an explanation of the set-up.

Twelve examples of using this phantom is shown in Fig. 1. The top graphs show imaging without apodization and the bottom graphs show when a Hanning window is used for apodization in both transmit and receive. A 128 elements transducer with a nominal frequency of 3 MHz was used. The element height was 5 mm, the width was a wavelength and the kerf 0.1 mm. The excitation of the transducer consisted of 2 periods of a 3 MHz sinusoid with a Hanning weighting, and the impulse response of both the emit and receive aperture also was a two cycle, Hanning weighted pulse. In the graphs A – C, 64 of the transducer elements was used for imaging, and the scanning was done by translating the 64 active elements over the aperture and focusing in the proper points. In graph D and E 128 elements were used and the imaging was done solely by moving the focal points.

Graph A uses only a single focal point at 60 mm for both emission and reception. B also uses reception focusing at every 20 mm starting from 30 mm. Graph C further adds emission focusing at 10, 20, 40, and 80 mm. D applies the same focal zones as C, but uses 128 elements in the active aperture.

The focusing scheme used for E and F applies a new receive profile for each 2 mm. For analog beamformers this is a small zone size. For digital beamformers it is a large zone size. Digital beamformer can be programmed for each sample and thus a "continuous" beamtracking can be obtained. In imaging systems focusing is used to obtain high detail resolution and high contrast resolution preferably constant for all depths. This is not possible, so compromises must be made. As an example figure F shows the result for multiple transmit zones and receive zones, like E, but now a restriction is put on the active aperture. The size of the aperture is controlled to have a constant F-number (depth of focus in tissue divided by width of aperture), 4 for transmit and 2 for receive, by dynamic apodization. This gives a more homogeneous point spread function throughout the full depth. Especially for the apodized version. Still it can be seen that the composite transmit can be improved in order to avoid the increased width of the point spread function at e.g. 40 and 60 mm.

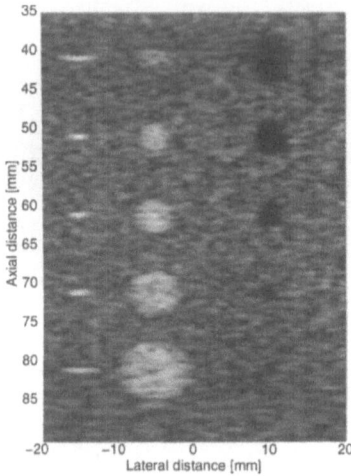

Figure 2: Computer phantom with point targets, cyst regions, and strongly reflecting regions.

The next phantom consists of a collection of point targets, five cyst regions, and five highly scattering regions. This can be used for characterizing the contrast-lesion detection capabilities of an imaging system. The scatterers in the phantom are generated by finding their random position within a $60 \times 40 \times 15$ mm cube, and then ascribe a Gaussian distributed amplitude to each scatterer. If the scatterer resides within a cyst region, the amplitude is set to zero. Within the highly scattering region the amplitude is multiplied by 10. The point targets has a fixed amplitude of 100, compared to the standard deviation of the Gaussian distributions of 1. A linear scan of the phantom was done with a 192 element transducer, using 64 active elements with a Hanning apodization in transmit and receive. The element height was 5 mm, the width was a wavelength and the kerf 0.05 mm. The pulses where the same as used for the point phantom mentioned above. A single transmit focus was placed at 60 mm, and receive focusing was done at 20 mm intervals from 30 mm from the transducer surface. The resulting image for 100,000 scatterers is shown in Fig. 2. A homogeneous speckle pattern is seen along with all the features of the phantom.

4   ANATOMIC PHANTOMS

The anatomic phantoms are attempts to generate images as they will be seen from real human subjects. This is done by drawing a bitmap image of scattering strength of the region of interest. This map then determines the factor multiplied onto the scattering amplitude generated from the Gaussian distribution, and models the difference in the density and speed of sound perturbations in the tissue. Simulated boundaries were introduced by making lines in the scatterer map along which the strong scatterers were placed. This is marked by completely white lines shown in the scatterer maps. The model is currently two-dimensional, but can readily be expanded to three dimensions. Currently, the elevation direction is merely made by making a 15 mm thickness for the scatter positions, which are randomly distributed in the interval.

Two different phantoms have been made; a left kidney in a longitudinal scan, and a fetus in the third month of development. For both was used 200,000 scatterers randomly distributed within the phantom, and with a

Gaussian distributed scatter amplitude with a standard deviation determined by the scatter map. The phantoms were scanned with a 5 MHz 64 element phased array transducer with $\lambda/2$ spacing and Hanning apodization. A single transmit focus 70 mm from the transducer was used, and focusing during reception is at 40 to 140 mm in 10 mm increments. The images consists of 128 lines with 0.7 degrees between lines.

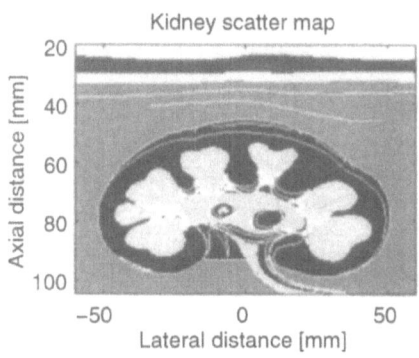

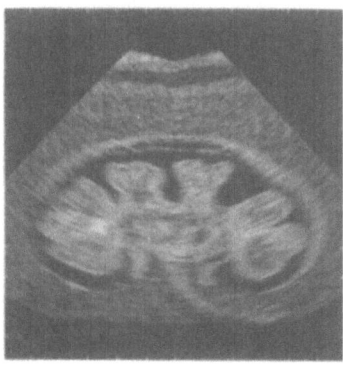

Figure 3: Simulation of artificial kidney.

Fig. 3 shows the artificial kidney scatterer map on the left and the resulting image on the right. Note especially the bright regions where the boundary of the kidney is orthogonal to the ultrasound, and thus a large signal is received. Note also the fuzziness of the boundary, where they are parallel with the ultrasound beam, which is also seen on actual ultrasound scans.

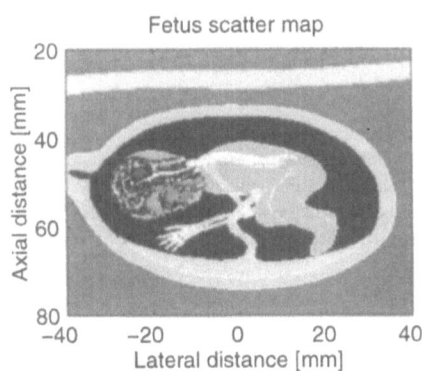

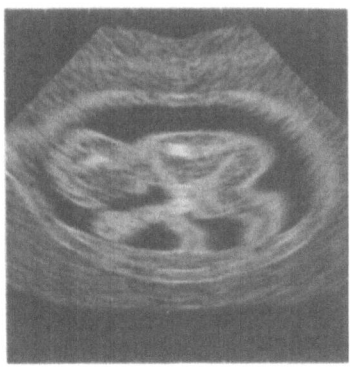

Figure 4: Simulation of artificial fetus.

Fig. 4 shows the fetus. Note how the anatomy can be clearly seen at the level of detail of the scatterer map. The same boundary features as for the kidney image is also seen.

The images have many of the features from real scan images, but still lack details. This can be ascribed to the low level of details in the bitmap images, and that only a 2D model is used. But the images do show great potential for making powerful fully synthetic phantoms, that can be used for image quality evaluation.

5   FLOW PHANTOMS

The last phantom is used for evaluating color flow imaging. It generates data for flow in vessels with properties like the carotid artery, The velocity profile is close to parabolic, which is a fairly good approximation during most of the cardiac cycle [7] for a carotid artery. The phantom generates 10 files with positions of the scatterers at the corresponding time step. From file to file the scatterers are then propagated to the next position as a function of their velocity and the time between pulses. The ten files are then used for generating the RF lines for the different imaging directions and for ten different times. A linear scan of the phantom was made with a 192 element transducer using 64 active elements with a Hanning apodization in transmit and receive. The element height was 5 mm, the width was a wavelength and the kerf 0.05 mm. The pulses where the same as used for the point phantom mentioned above. A single transmit focus was placed at 70 mm, and receive focusing was done at 20 mm intervals from 30 mm from the transducer surface. The resulting signals have then been used in a standard autocorrelation estimator [7] for finding the velocity image.

The resulting color flow image is shown in Fig. 5. Note how the vessel is larger at the bottom than the top.

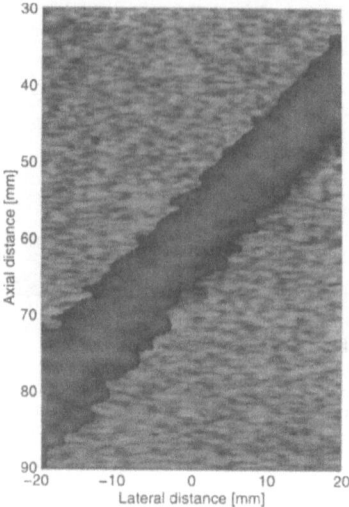

Figure 5: Color flow image of vessel with a parabolic flow profile.

6   CONCLUSION

This paper has shown that it is possible to generate realistic simulated ultrasound images. The rather rough scatterer map images generate ultrasound images with a reasonable resemblance to actual ultrasound images. Increasing the maps detail and the number of scatterers should make it possible to get even more detailed images.

Version 1.41 of the Field II program used for this paper can be found at the web-site: http://www.it.dtu.dk/bme under Research. The program can be downloaded for PCs and the most common workstations. The code for the examples and scatterer maps can also be found at the web site.

References

[1] J. A. Jensen and N. B. Svendsen. Calculation of pressure fields from arbitrarily shaped, apodized, and excited ultrasound transducers. *IEEE Trans. Ultrason., Ferroelec., Freq. Contr.*, 39:262–267, 1992.

[2] J. A. Jensen. Field: A program for simulating ultrasound systems. *Med. Biol. Eng. Comp.*, 10th Nordic-Baltic Conference on Biomedical Imaging, Vol. 4, Supplement 1, Part 1:351–353, 1996b.

[3] J. A. Jensen. Ultrasound fields from triangular apertures. *J. Acoust. Soc. Am.*, 100(4):2049–2056, 1996a.

[4] G. E. Tupholme. Generation of acoustic pulses by baffled plane pistons. *Mathematika*, 16:209–224, 1969.

[5] P. R. Stepanishen. Transient radiation from pistons in an infinte planar baffle. *J. Acoust. Soc. Am.*, 49:1629–1638, 1971.

[6] J. A. Jensen. A model for the propagation and scattering of ultrasound in tissue. *J. Acoust. Soc. Am.*, 89:182–191, 1991.

[7] J. A. Jensen. *Estimation of Blood Velocities Using Ultrasound: A Signal Processing Approach*. Cambridge University Press, New York, 1996.

CONTRAST RESPONSE FOR MEDICAL ULTRASOUND ANALYSIS

Richard K. Johnson

Palinurus Associates
20035 SE 27th Place
Issaquah, WA 98029

ABSTRACT

Several investigators have described the contrast performance of systems using the concept of averaged response to anechoic or weakly echoic regions. Their work has demonstrated the importance of average sidelobe level in contrast resolution. This paper discusses the use of a similar measure, called contrast response, which quantifies the ability of a system to accurately image the contrast of an anechoic region surrounded by a uniform scattering medium. The contrast response is defined as the ratio of the system output when the anechoic target is not present to the minimum output when it is present, evaluated over a range of target sizes. Contrast response analysis is useful for evaluating system operating parameters. Examples are presented which demonstrate the effects of pitch, number of active elements, aperture shading and elevation focus.

INTRODUCTION

In medical ultrasound imaging, much of the clinically important information for tissue characterization and differentiation is contained in the lower strength echoes. High performance imaging requires that these weak echoes be clearly presented despite the presence of strong scatterers in the vicinity. To achieve this goal, a system must have good detail resolution (a narrow mainlobe width), and it also needs good contrast resolution.

Contrast resolution has proven difficult to explain and to quantify. It has been studied using low and high contrast cysts in phantoms. Smith et al.[1], Bly et al.[2], Turnbull et al.[3], Liu and Waag[4] and Vilkomerson et al.[5] compare the average echo from a cyst to the average echo from its neighborhood as a measure of contrast performance. Their work demonstrates the value of this approach for analysis of experimental images and for system simulation. The results in these papers indicate that both mainlobe width and sidelobe level are important to contrast resolution.

The method proposed here, which is called contrast response[6], is an extension of this previous work. It includes the contrast resolution effects of both mainlobe width and sidelobe level on image presentation in terms of the system response to anechoic cysts having a range of diameters.

DEFINITION

The response of a linear system to a target is the convolution of the system spatial impulse response (three dimensional beam pattern) and the target scattering function[7]. Expressing the problem in polar coordinates outward from the face of the transducer, the output O from a system with spatial impulse response R in a medium with scattering function T can be expressed by the convolution integral

$$O(r,\theta,\phi) = \int_{r_1}^{r_2} \int_{-\pi/2}^{\pi/2} \int_{0}^{2\pi} R(r',\theta',\phi')T(r-r',\theta-\theta',\phi-\phi')r'dr'd\theta'd\phi' \tag{1}$$

where (r_1,r_2) spans the active volume. Note that this equation assumes coherent summation from all of the scatterers in the medium. Consider a spherical anechoic region V surrounded by a region with unity echo strength such that

$$T(r,\theta,\phi) = 0 \quad (r,\theta,\phi) \in V$$
$$T(r,\theta,\phi) = 1 \quad (r,\theta,\phi) \notin V \tag{2}$$

The contrast response is defined as the ratio of the system output when the anechoic region has zero volume to the minimum system output when the anechoic region is present. Thus a higher contrast response means better system performance. The contrast response can be calculated as

$$\frac{O(r,\theta,\phi)\, for V = 0}{\min(O(r,\theta,\phi))\, for V > 0} \tag{3}$$

Contrast response will be expressed in decibels for easy comparison with image dynamic range.

EXAMPLES

The examples in this paper are based on a linear system using linear or phased array transducers. The operating frequency is 3.5 MHz, and both the transmit and receive apertures are based on an f number of 2. The transducers have an elevation length of 1.5 cm and an elevation focus at 7 cm. The calculations are simplified by assuming coherent

82

summation of the echoes from the scattering medium and attenuation is not included. Implementation issues which can also affect perceived image contrast such as noise, beamformer errors, edge enhancement filtering and grayscale mapping are not included.

Elevation Focus

Most transducers in use today have an elevation focus at a single depth which is determined by a mechanical lens. The effect of elevation focus depth on cyst presentation may be seen in Figure 1. The anechoic cyst fills in with echoes from the surrounding region due to the finite mainlobe width and the sidelobes. The amount of cyst filling increases with distance of the cyst from the elevation focus depth. This cyst filling distorts the shape of the cyst, and echo filling near the top of the cyst is common (Figure 1a).

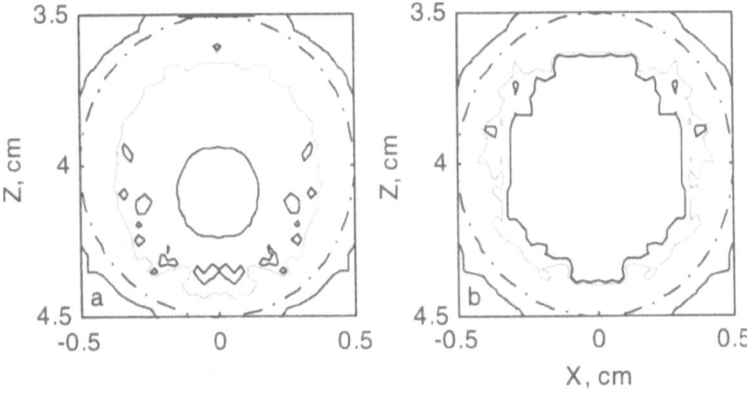

Figure 1 Contour plot of system response using a 128 element phased array with elevation focus at a) 7cm or b) 4cm to an anechoic cyst at a depth of 4cm. Contours are at 0, -10, -20 and -30 dB. The cyst position is shown with a dash dot linestyle.

In this example the image of the cyst presented by the system is narrower than the actual cyst diameter and has a minimum intensity level of at best -35 dB. Because of the finite mainlobe width, the system output from the cyst does not have the sharp square sides of the actual cyst. The effect of this intensity tapering on human visual performance does not appear to have been studied. The scattering floor or finite minimum output intensity near the center of the cyst is determined by the integrated sidelobe level of the system beam pattern.

The detail in Figure 1, while useful for image artifact analysis is not necessary for system characterization. The contrast response metric uses a single number (the minimum output) from each case.

Depth

The response of this phased array system to anechoic regions which range in depth from 4 to 18 cm is shown in Figure 2. As would be expected, the contrast response decreases as cyst diameter decreases or depth increases. This simulation is consistent with the subjective imaging experience that contrast resolution decreases with depth more quickly than detail resolution. Over this depth range, the system output is affected by both elevation and lateral beampattern. The sidelobes of the beam pattern are limiting system performance, and these limitations become more significant with depth.

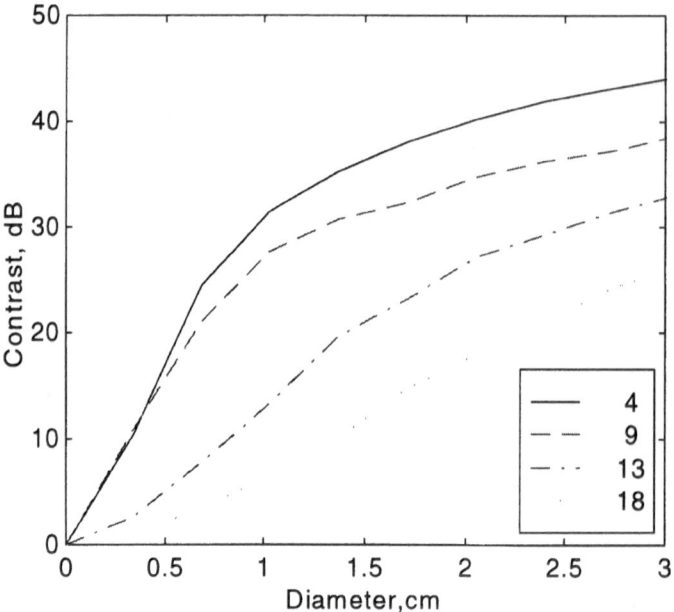

Figure 2 Contrast response of a 128 element phased array system using Hamming shading for a range of cyst diameters vs. depth in cm.

This contrast - diameter curve (Fig. 3) represents the system output to defined inputs. It is different from the similarly named contrast - detail curves [1] which are ROC (receiver operating characteristic) curves characterizing the performance of a human or computer observer for lesion detection in speckled images.

The presentation in Figure 2 becomes too cluttered when more than a few curves are plotted. To display a larger set of parameters, the figures in the subsequent examples are contour plots of contrast response against depth and diameter.

Pitch

Early 128 channel systems used linear array transducers with 1.5 or 2 wavelength pitch. More recent systems use 1 wavelength pitch when the resulting field of view is large enough. Figure 3 shows that linear arrays with these three pitches have similar performance for small cysts that are just resolved. For bigger cysts however, the contrast response decreases dramatically as the pitch increases.

Number of Active Elements

Since the pitch of a phased array is limited to no more than ½ wavelength, it might be expected that the number of active elements would have a big impact on performance. Figure 4 indicates that the performance increase from 64 to 128 active elements should be visible, but the step to 256 would be marginal. This simulation is consistent with recent experience that 256 element phased arrays do not have significantly better contrast resolution than 128 element phased arrays.

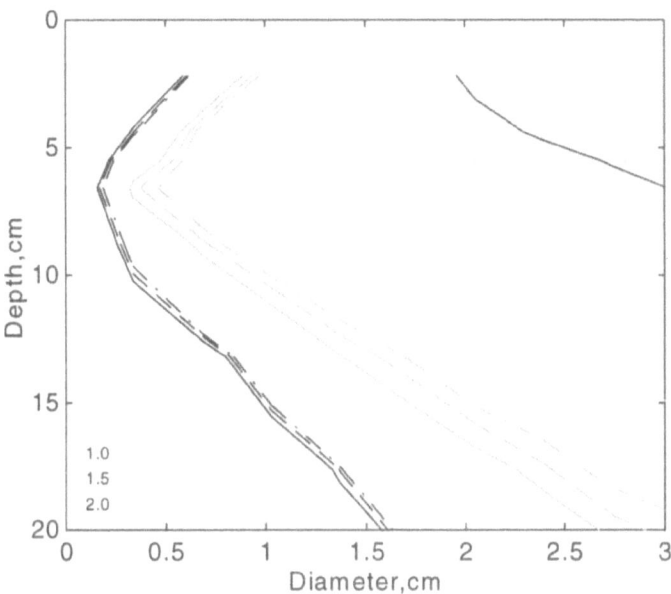

Figure 3 Contour plot of contrast response for transducers with array pitch of 1 (solid line), 1 ½ (dashed line) or 2 (dash dot line) wavelength array pitch. Contours are at 10, 20, 30 and 40 dB.

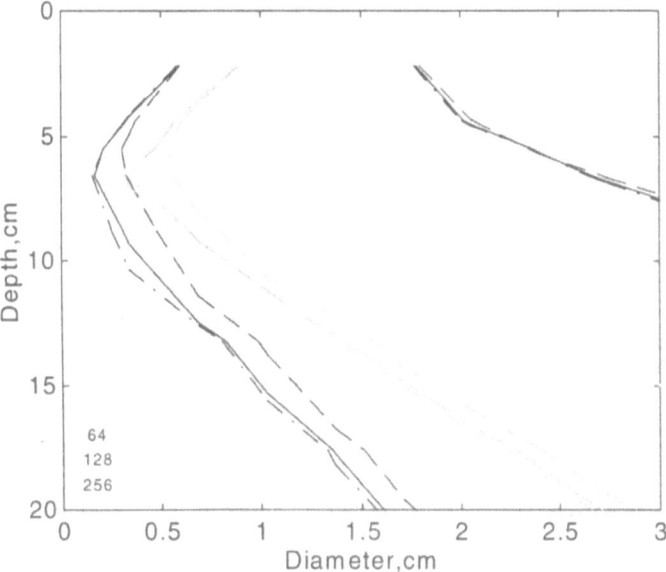

Figure 4 Contrast response for phased arrays with 64 (dashed line), 128 (solid line) or 256(dash dot line) active elements operating with Hamming shading and f number of 2 for transmit and receive. Contours are at 10, 20, 30 and 40 dB.

Beamformer Shading

Shading or aperture apodization can have a profound effect on the system beam pattern -- particularly the sidelobe levels. Experience has shown that nonrectangular shading functions are required for superior contrast performance. Reduced sensitivity is a necessary cost of these shading functions. The effect of aperture shading on the system output from anechoic cysts can be seen in Figure 4. The use of rectangular shading for transmit *or* receive leads to a modest reduction in contrast response for the bigger cysts. The use of rectangular shading for transmit *and* receive leads to a significant reduction in contrast response which would not be acceptable for a high performance system. For nonrectangular shading, the cyst images are slightly narrowed, but much more contrast is produced for most cyst sizes. Shading functions with extremely low sidelobe levels, such as Blackman, are popular in digital signal processing but little used in medical ultrasound because they produce essentially the same contrast response as Hamming with less sensitivity.

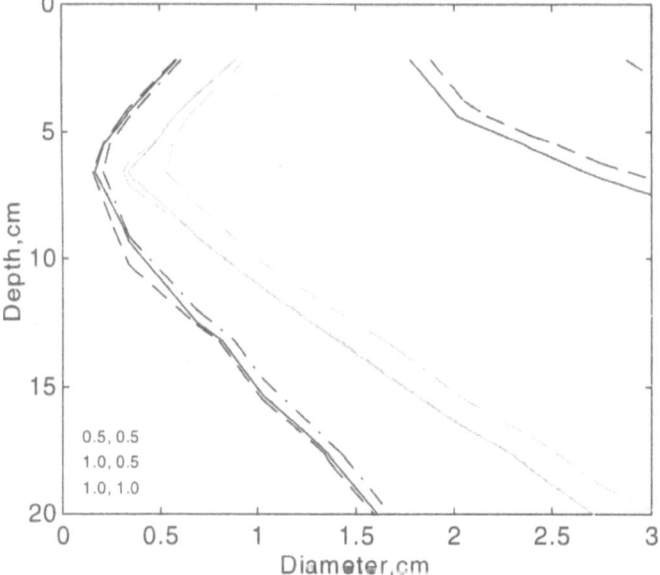

Figure 5 Contrast response of a 128 element phased array with Hamming shading for transmit and receive (solid line), for receive only (dashed line) or for neither (dash dot line). The unshaded aperture is uniformly excited. Contours are at 10, 20, 30 and 40 dB.

Elevation Shading

The previous examples suggest that the sidelobe levels of the elevation beampattern are a major limitation on contrast resolution. Shading in the elevation direction is not used for average transducers because it somewhat difficult to achieve. The contrast response simulation in Figure 6 indicates that some amount of elevation shading is very desirable for contrast resolution.

CONCLUSION

The contrast response metric quantifies the ability of a system to accurately portray weak scattering regions in the presence of strong scatterers. It can be used to determine the effect of system operating parameters on contrast resolution for a range of the sizes and locations of the weak scattering regions. The examples suggest that the judicious use of elevation shading can significantly improve contrast resolution.

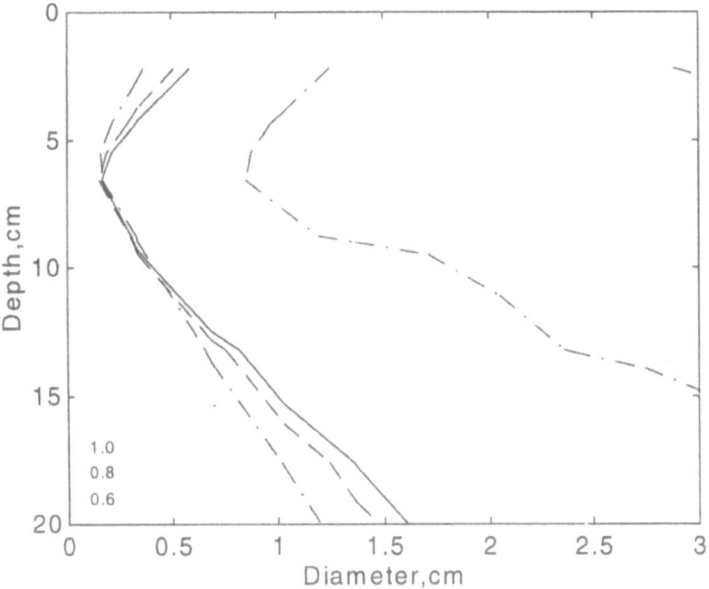

Figure 6 Contrast response of a Hamming shaded 128 element phased array no elevation shading (solid line), intermediate elevation shading (dashed line) or Hamming elevation shading (dash dot line). Contours are at 10, 30 and 50 dB.

REFERENCES

1. S.W. Smith, R.F. Wagner, J.M. Sandrik and H. Lopez, "Low contrast detectability and contrast/detail analysis in medical ultrasound," *IEEE Trans. Sonics and Ultras*, 30, 164-173 (1983).
2. S. Bly, D. Lee-Chahal, D. Foster, M. Patterson, F. Foster and J. Hunt, "Quantitative contrast measurements in B-mode images," *Ultra in Med. & Biol.*, 12, 197 (1986).
3. D. Turnbull, P. Lum, A. Kerr, F. Foster, "Simulation of B-scan images from two-dimensional transducer arrays," *Ultrasonic Imaging*, 14, 323-343 (1992).
4. D. Liu and R. Waag, "Time-shift compensation of ultrasonic pulse focus degradation using least-mean-square error estimates," *J. Acoust. Soc. Am.*, 95, 542-555 (1994).
5. D. Vilkomerson, J. Greenleaf and V. Dutt, "Towards a resolution metric for medical ultrasonic imaging," in *IEEE Ultrasonics Symposium*, 1405-1410, (1995).
6. R.K. Johnson, "Contrast response analysis for medical ultrasound imaging," *IEEE Trans. Sonics and Ultras*, Scheduled for July 1997.
7. 7. B.D. Steinberg, *Principles of Aperture and Array System Design*. New York: John Wiley, 1976, pp. 198-199.

NON-CONTACT ULTRASONIC IMAGING FOR THE EVALUATION OF BURN-DEPTH AND OTHER BIOMEDICAL APPLICATIONS

J. P. Jones,[1] D. Lee,[1] M. Bhardwaj,[2] V. Vanderkam,[3] and B. Achauer[3]

[1]Department of Radiological Sciences and [3]The UCI Burn Center
University of California Irvine
Irvine, CA 92697-5000   USA
[2]Ultran Laboratories
Boalsburg, PA 16827 USA

INTRODUCTION

Conventional wisdom suggests that ultrasonic imaging of the body cannot be accomplished without direct contact (or at least contact via water coupling). Thus, non-contact imaging (i.e., through air) is traditionally viewed as impossible. The present study shows that non-contact imaging **is** possible, certainly for superficial body regions, provided judicious choices of piezoelectric materials and matching layers are made. In preliminary experiments reported here, reflections from the dermal/fat interface in human skin were clearly seen using non-contact transducers operating in the low MHz frequency range. Such measurements are sufficient to determine burn-depth which, in turn, is sufficient to provide, for the first time, a quantitative and non-invasive method for burn evaluation and treatment specification.

In the present paper we (1) describe the development of a transducer and an ultrasound system specifically for non-contact imaging of human skin, (2) discuss the initial laboratory experiments using this system with appropriate test phantoms and human volunteers, and (3) detail preliminary clinical studies at the UCI Burn Center. In all cases, non-contact ultrasonic imaging provided accurate estimates of burn-depth which were consistent with biopsy measurements.

PROBLEM DEFINITION

The evaluation of burns and the determination of burn severity is an unmet medical need. At the present time, burns are assessed by simple visual observation in both combat and civilian situations. Unfortunately, such clinical evaluation of burns, even by an experienced physician, has only the same probability as chance in predicting outcome. A more aggressive management and treatment of burn patients has been shown to significantly reduce infection and associated problems and, therefore, to shorten hospital stay as well as the duration of the illness. Improved function and appearance have also been noted. Such management requires the early measurement of burn depth to determine the clinical course or treatment plan to be followed. For example, in a full thickness burn, no residual dermis survives (requiring the removal of the dead skin followed by a skin transplant), while, in a partial thickness burn, there is a variable thickness of viable dermis beneath a layer of coagulated necrosis (requiring simply the application of appropriate creams and bandages for survival). Please see Figure 1.

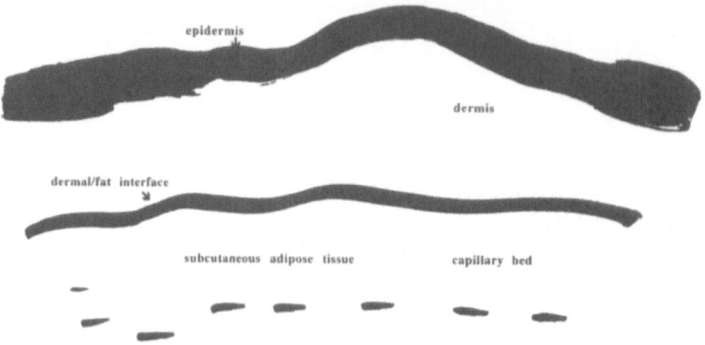

Figure 1. Simplified schematic of the human skin. In a partial thickness burn, the dermis and the capillary bed below remain viable, allowing a new epidermis layer to form. In a full thickness burn, the dermis is destroyed thereby removing contact with the capillary bed and requiring surgical intervention to reestablish the skin.

Although a plethora of technologies have been developed for the evaluation of burns (including biopsy, vital dyes, fluorescein fluorometry, laser Doppler flowmetry, thermography, ultrasound, magnetic resonance imaging, and light reflectance), they are, with the exception of ultrasound, either inaccurate or unsuitable for use in either an emergency situation or a remote location. Although ultrasound, with its inherent ability to differentiate soft tissue structures, appears to be the appropriate technology for the evaluation of burns, conventional contact ultrasound is difficult and time-consuming to use. Non-contact ultrasonic imaging would offer a fast and reliable method for burn evaluation and would permit, for the first time, effective triage of burn victims.

METHODS

The challenges (as well as the utility) of non-contact imaging are easily appreciated from the results of a simple calculation of the insertion loss experienced by an ultrasonic pulse when reflected from skin. For example, the signal reflected from the dermal/fat interface using a water or direct coupled PZT transducer (i.e., a transducer with a quarter-wavelength matching layer which matches the PZT to water) is about 20 dB down from the original pulse. If this same transducer were operating in air, the signal loss would be about 180 dB, which explains why non-contact imaging is considered impossible. However, if the PZT transducer is now matched to air, the loss is about 80 dB -- still significant but certainly not impossible to overcome, especially with superficial tissue structures. In our experiments, described below and using far from optimized materials, we observed losses between 100 and 120 dB, clearly demonstrating the feasibility of using this technology for burn evaluation.

Clearly the key to any non-contact system is the air-matched transducer. For some years Ultran Laboratories has offered commercially air-matched transducers operating in the kHz frequency range. Such transducers normally used PZT as the active element with a special ceramic coating which matched to air. Unfortunately, such coatings proved to be no longer feasible when the frequency was increased to the MHz range. Working together, UCI and Ultran have developed a proprietary polymer which serves as a matching layer from PZT to air. Although this polymer does not provide an exact match to air, it has proven adequate for the preliminary studies described here. Ultran provided a series of such air matched transducers for the experiments described here. The transducers had center frequencies of 1.0, 2.25, and 5.0 MHz, respectively, and each was an unfocused, broad-banded, plane-piston source with an active diameter of about 1 inch.

Initial experiments at UCI used a simple A-mode scanner consisting of these non-contact transducers with a Panametrics pulser/receiver and a Tektronix digital oscilloscope. This A-mode system was applied to a number of specially constructed test phantoms which verified the capabilities of such a system to make accurate spatial measurements with only air between the transducer and the test object. Applying this non-contact scanner to a series of human volunteers, reflections from the dermal/fat interface were clearly seen using both the 2.25 and the 5.0 MHz transducers.

In order to test the feasibility of non-contact imaging for burn evaluation, we constructed a special self-contained data acquisition system for use in the clinic. This system, shown in Figure 2 consists of a simple pulser/receiver, a 50 MHz ADC, and 32 MB of storage which can retain over 1000 A-line records. The unit is powered by a self contained battery pack and was designed to operate with the 5.0 MHz air-coupled transducer. The operation of the device is quite simple: the on/off switch is engaged, the transducer is held about one inch above the region of interest, and the data collect button is pushed. Engaging the data collect button causes a single ultrasonic pulse to be generated and the backscattered signal to be recorded for a spatial extent of about 2 cm. A front panel dial allows a specific time delay to be set before the ADC begins operation, thereby insuring that only relevant data are recorded. The RF A-line data recorded by the clinical data acquisition system are off-loaded to a laboratory computer (Power Macintosh 9500) for evaluation and analysis.

Figure 2(a) The self contained clinical data acquisition system.

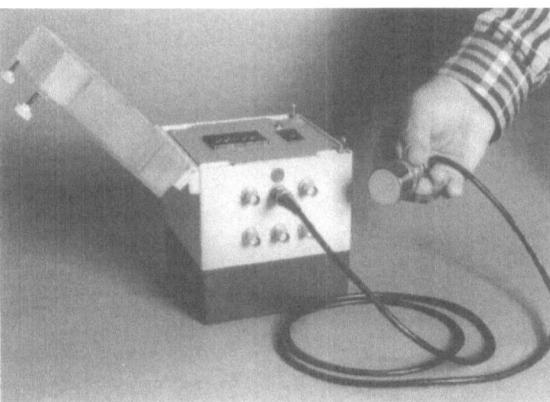

Figure 2(b) The system as utilized in the clinic. The dark bottom of the device is the integrated battery pack. Controls on the top panel consist of an on/off switch, a dial to select the time delay before data acquisition, and a button which engages the generation of a single ultrasonic pulse and the recording of the backscattered signal.

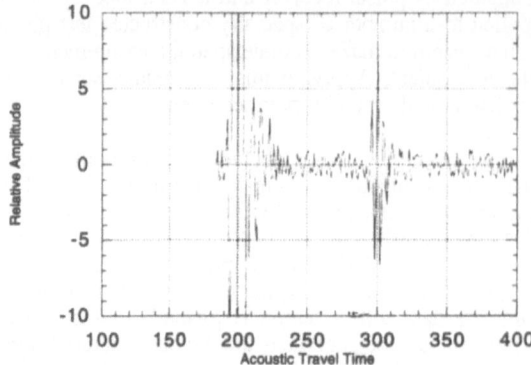

Figure 3a. A-line of normal skin (upper right arm) taken with the clinical data acquisition system using a non-contact (air coupled) transducer operating at a center frequency of 5 MHz. From left to right the major echo structures are the skin surface (and epidermis) and the dermal/fat interface.

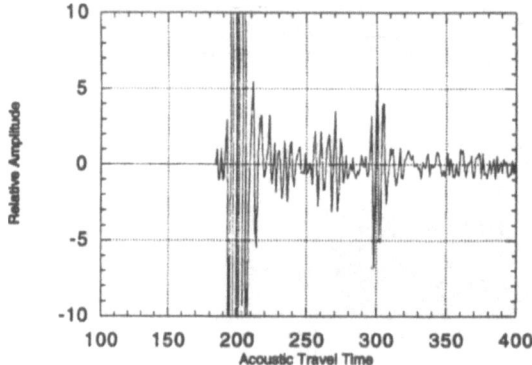

Figure 3b. A-line of burned skin (upper left arm) showing a partial thickness burn. Note that the dermal/fat interface is still intact.

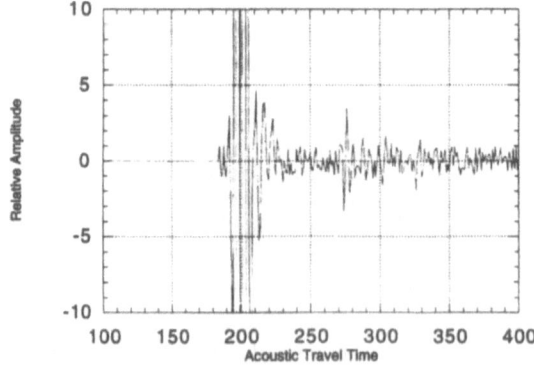

Figure 3c. A-line of burned skin (upper left arm) showing a full thickness burn. Note that the dermal/fat interface has been destroyed severing contact with the underlying capillary bed.

RESULTS

The clinical data acquisition system was first used on the same set of test phantoms and human volunteers to which the A-mode lab system had been applied. Obtaining similar and consistent results, we moved on to a series of clinical studies at the UCI Burn Center. Although these studies are still in progress, some preliminary results can be presented here. Figure 3 gives a series of A-line scans taken on a patient with significant burns to his left arm. All of these scans were recorded with the clinical data acquisition and off-loaded to the laboratory computer. Some minor time adjustments were made to the three A-lines presented in Figure 3 for ease of comparison.

Figure 3a is an A-line taken on the normal right arm of the subject with burns on his left arm. The large echo structure at time units 200 is the skin surface and epidermis. The generally echo free region between time units 200 and 300 is the dermis while the echo structure at time units 300 is the dermal/fat interface. This A-line is typical of normal skin structure.

Figure 3b is an A-line from a particular section of the left (burned) arm of the same patient. Note that the echo structure is similar to Figure 3a with the additional reflectors in the dermis. These additional echoes, confirmed by biopsy, are regions of coagulated necrosis associated with a partial thickness burn. Note that the dermal/fat interface is still intact.

Figure 3c is an A-line from a different section of the same left arm. Note that here the dermal/fat interface is no longer present which indicates a full thickness burn. Reflections from what is probably coagulated necrosis is seen in the dermis but the important link with the capillary bed is severed.

CONCLUSIONS AND FUTURE DEVELOPMENTS

We believe the simple A-lines shown in Figure 3 clearly demonstrate the proof of concept that non-contact ultrasound operating through air can quantitate burn depth and severity. Work now in progress seeks to collect a larger data base from burn patients using our clinical data acquisition system so as to determine clinical efficacy and to better define a practical burn evaluation system. Ongoing work also seeks to develop more optimal materials for matching the piezoelectric to air.

One of the major advantages of this technology is its potential applicability to large scale triage as might be required in a significant armed conflict or a terrorist act as well as in a forward combat zone. Other potential application areas for non-contact ultrasonic imaging include the evaluation of wounds and the wound healing process as well as the detection and assessment of pressure sore development.

DETERMINATION OF GALLBLADDER STONE COMPOSITION WITH ULTRASOUND - IN VIVO AND IN VITRO STUDIES

C. Kocherscheidt[1], U. Schmidt[1], W. Albert[1], J. Racky[2], M. Pfeiffer[2], M. Pandit[2]

[1]Westpfalz Klinikum Kaiserslautern , 67653 Kaiserslautern
[2]Universität Kaiserslautern, 67653 Kaiserslautern, Germany

INTRODUCTION

The determination of the composition of gallbladder stones on the basis of (non-invasive!) ultrasound scans has been a subject of intense research. One sound reason for the interest is that it is advantageous to select the therapy on the basis of the composition. Especially, detection of calcification is important in order not to employ lithotripsy, as the pulverized calcium compounds cannot be dissolved and washed away with solvent medicines. Opinion is divided as to whether a reliable method of composition determination is possible at all with ultra-sound especially on the basis of in vivo B-scans. Thus /1/ and /2/ claim that it is possible to detect calcification whereas, /3/ and /4/ say that it is not so.

Setting out from this point we undertook to check some of the methods published in the literature and to devise and try out new methods for detection of calcification in gallbladder stones. To make the investigation as comprehensive as possible, both the high frequency signal and the B- scans were examined, in the latter case in vivo as well as in vitro. Finally, calcification was established using x-ray imaging.

At first there seemed to be differences in the signal parameters which could be used as criteria for classification. However, these differences proved to be indicative of the surface geometry rather than calcification.

EXPERIMENTAL METHODS AND APPARATUS

Tests were carried out at the Westpfalz Klinik and the University of Kaiserslautern. First, patients were examined with the ultrasound scanner before undergoing cholestystectomy (in vivo features). After surgery the removed gallbladder stones were scanned a second time (in vitro features) to acquire the rf signal and the B mode image. Finally the gallbladddder stones were photographed with x-rays to determine a conclusive classification. Details of the tested material are given in Table 1

Table 1. Details of tested material

100 GBS of 100 patients undergoing cholecystectomy at the
Westpfalz Klinikum Kaiserslautern July 1995 to September 1996

	in vivo	in vitro
Ultrasound (B-Mode)	29 GBS	100 GBS
Ultrasound (RF-Signal)		20 GBS
X - Ray	34 GBS	100 GBS

a. The stones were suspended in a water tank and the 3 mhz rf echo signal obtained with a Toshiba Sonolayer-L SAL-32 B instrument was fed to a sampling and data acquisition and processing unit. The rf signal was sampled at a rate of 21 mhz. The samples are stored and processed in a PC. Fig.1 shows the experimental set-up and typical signal runs.

LT parameters and strain*	infinite	infinite	8 bits	6 bits	4 bits
m_1 (1.0)	0.9993 ± 0.0016	0.9715 ± 0.0116	0.9968 ± 0.01	0.9945 ± 0.0014	0.9902 ± 0.0131
$\Delta_{11}(0)$ mean	$-0.7 \pm 1.6 \; x10^{-3}$	$-29 \pm 11.6 \; x10^{-3}$	$-3.2 \pm 10 \; x10^{-3}$	$-5.5 \pm 1.4 \; x10^{-3}$	$-9.8 \pm 13.1 \; x10^{-3}$
s.d.	$2.2 \pm 1.0 \; x10^{-3}$	$15.4 \pm 6.7 \; x10^{-3}$	$13.5 \pm 5.9 \; x10^{-3}$	$1.9 \pm .8 \; x10^{-3}$	$17.4 \pm 7.6 \; x10^{-3}$
m_2 (0)	0.0004 ± 0.0017	0.0031 ± 0.0064	0.0022 ± 0.0054	0.0018 ± 0.0128	0.0020 ± 0.009
$\Delta_{12}(0)$	$0.4 \pm 1.7 \; x10^{-3}$	$3.1 \pm 6.4 \; x10^{-3}$	$2.2 \pm 5.4 \; x10^{-3}$	$1.8 \pm 12.8 \; x10^{-3}$	$2 \pm 9.0 \; x10^{-3}$
s.d.	$2.3 \pm 1.0 \; x10^{-3}$	$8.5 \pm 3.7 \; x10^{-3}$	$7.2 \pm 3.2 \; x10^{-3}$	$17 \pm 7.5 \; x10^{-3}$	$11.9 \pm 5.2 \; x10^{-3}$
m_3 (.0349)	0.0348 ± 0.0002	0.0351 ± 0.0009	0.0351 ± 0.0014	0.0362 ± 0.0004	0.0367 ± 0.0009
Δ_{21} (34.9 $x10^{-3}$)	$34.8 \pm .2 \; x10^{-3}$	$35.1 \pm .9 \; x10^{-3}$	$35.1 \pm 1.4 \; x10^{-3}$	$36.2 \pm .4 \; x10^{-3}$	$36.7 \pm .9 \; x10^{-3}$
s.d.	$.2 \pm .09 \; x10^{-3}$	$1.2 \pm .5 \; x10^{-3}$	$1.9 \pm .8 \; x10^{-3}$	$.5 \pm .2 \; x10^{-3}$	$1.3 \pm .57 \; x10^{-3}$
m_4 (1.0)	0.9999 ± 0.00008	1.0000 ± 0.0002	1.0001 ± 0.00008	1.0011 ± 0.0008	1.0006 ± 0.0003
$\Delta_{22}(0)$	$-0.1 \pm .08 \; x10^{-3}$	$0 \pm .2 \; x10^{-3}$	$0.1 \pm .08 \; x10^{-3}$	$1.1 \pm .8 \; x10^{-3}$	$0.6 \pm 0.3 \; x10^{-3}$
s.d.	$.1 \pm .04 \; x10^{-3}$	$.3 \pm .1 \; x10^{-3}$	$0.1 \pm .04 \; x10^{-3}$	$1.0 \pm .4 \; x10^{-3}$	$0.4 \pm 0.18 \; x10^{-3}$

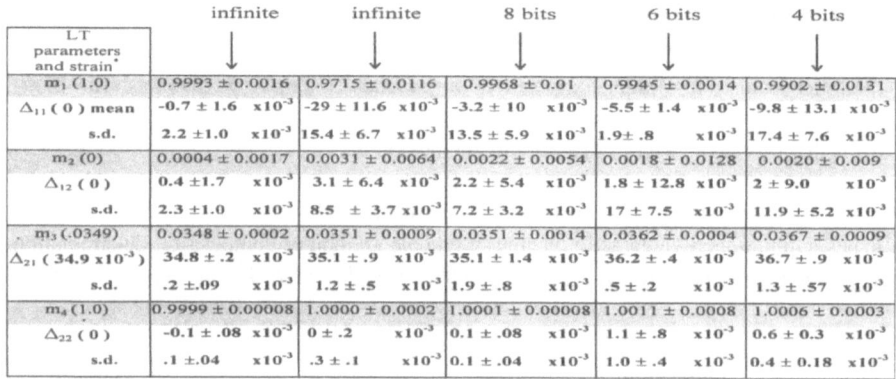

Fig. 1 Experimental set-up and signal run

b. Using an Acuson 128 XP/10 scanner with a 3.5 mhz curved array transducer the in vivo and in vitro B mode images were obatained (Fig.2 a/b).

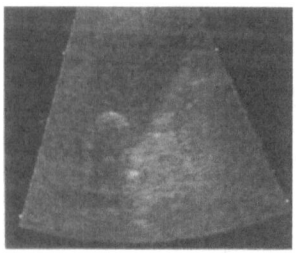
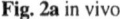
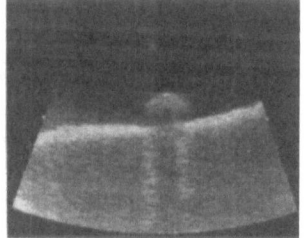

Fig. 2a in vivo   Fig. 2b in vitro

c. The stones were checked for calification using x-rays. Fig. 3 shows images of various calcified and uncalcified gallbladder stones.

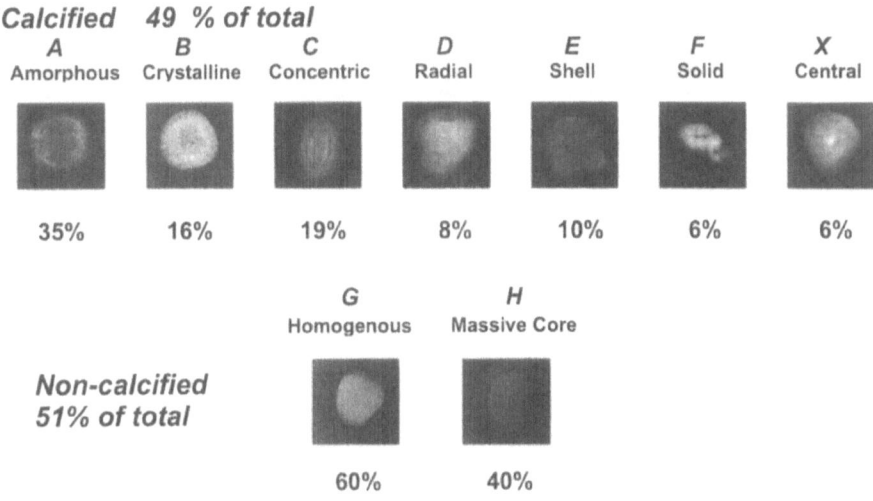

Fig. 3   x-ray photographs of calcified and uncalcified gallbladder stones with various structures

METHODS OF ANALYSIS OF B MODE AND RF SIGNALS AND RESULTS

Rf signal analysis

One method which is cited /5/ indicates that the velocity of propogation of ultra-sound can be used for assesing the composition. An indirect method of measuring the velocity consists of determining the group delay given by the expression:

$$\tau[\omega] = \frac{T_a}{|S[\omega]|^2}(X[\omega] \cdot U[\omega] + Y[\omega] \cdot V[\omega])$$

where X and Y represent the real and imaginary part of the diskrete Fourier-transform of the sampled rf echo signal S over the normalized frequency . U and V represent the real and imaginary part of $n \cdot S$, where n denotes the diskrete time. Finally T_a denotes the sampling time.

For the ultra-sound signal this was calculated using the short term FFT with a hanning windows. A disadvantage is the short pulse duration which gives rise to large variations of the calculated value. Table 2 shows typical values obatined for representative stones.

Table 2 Group velocity calculated for a set of stones

Calcified stones			Uncalcified stones		
1	*2*	*3*	*4*	*5*	*6*
27±1	29±2	30±3	29±1	35±5	40±6

B mode image analysis

First the image of the gallstone has to be extracted from the acquired ROI. To do this, a segmentation algorithm is employed using a threshold value set at

$$t = m + s,$$

where m is the mean gray level and s is the standard deviation. Using morphological closing by a 7x7-circle small structures are suppressed and gaps in the shape are closed. Based on the fact that most stones have a convex surface, we assume the surface of the stone to be convex. The shape which is obtained by the above algorithm is not convex, thus it has to be modified. This is done by finding the convex hull of the shape.

Next the presence or lack of the scythe - like shape is detected by constructing a region which would constitute the scythe and checking whether this region is distinctly brighter than the rest of the image. Based on experiments, the leading 20% of the gallstone image is taken as the region under investigation. The various steps of image processing are shown in Fig. 4.

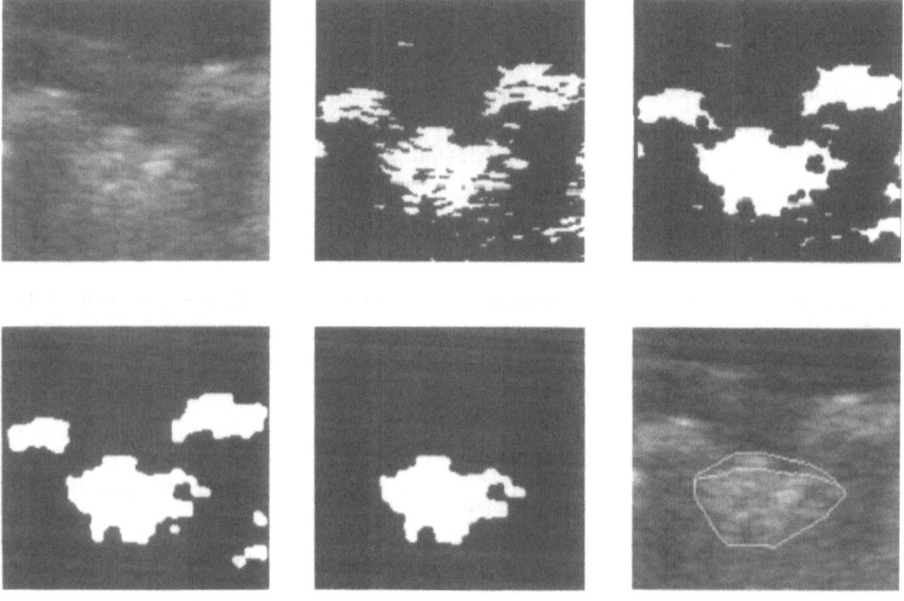

Fig. 4 Steps of image processing

As a first approach the mean gray level and the standard deviation is calculated. Results are presented below.

Table 3 Results obtained by B mode analysis

#	m front	m inner	s front	s inner	calcium
1	193.9	198.9	47.3	44.8	C
2	203.8	209.1	45.2	46.8	N
3	193.7	215.9	41.8	35.2	N
4	194.2	220.9	19.3	25.7	C
5	204.1	215.1	35.4	28.3	C
6	185.6	208.0	47.5	39.5	N
7	199.4	208.7	39.6	40.5	N
8	195.2	206.6	45.1	30.3	N
9	207.1	211.5	39.6	33.1	N
10	214.5	214.6	28.3	25.8	N
11	183.1	210.8	59.0	36.5	N
12	189.8	208.9	50.9	37.3	N
13	216.8	223.0	38.7	24.7	N
14	166.9	203.9	67.8	38.9	N
15	170.9	202.8	71.1	41.1	N

DISCUSSIONS AND CONCLUSIONS

In the case of rf signal analysis, the group delay does not turn out to be an indicator of calcification as can be seen from Table 2. The intraclass variations are of the same magnitude as the interclasss variations. However, this is not conclusive because of the limited pulse duration and the ensuing unsatisfactory singnal to noise ratio.

The B mode image analysis, summarized in Table 3, reveals, that the results are not consistent and therefore can not be used to distinguish between calcified and uncalcified galbladder stones.

In conclusion it can be said that the tested methods are not reliable for assessing the composition of gallbladder stones. Visual inspection of the gallbladder stones and review of the signal parameters indicate that the signal parameters are dependent on the geometry of the stone surface. Crescentness could be traced back to a jagged surface rather than high calcium content.

REFERENCES

1. B.A. Carroll, Gallstones: In vitro comparison of physical, radiographic and ultrasonic characteristics, *Am. J. Roentgen* 153, 223 - 226, 1978

2. R. Gladisch, H.K. Deininger, Die Darstellung von Gallensteinen im Sonogamm und Radiogramm, *Fortschr. Roentgenstr.* 139, 3, 249 - 255, 1983

3. W. Swobodnik, H.Wortmann, J.G.Wechsler et al, Sonographie von Gallenblasensteinen : Möglichkeiten und Grenzen der Auswahl konservativ lysierbarer Steinträger, *Ultraschall* 3, 117 - 122, 1986

4. C. Jakobeit, L. Greiner, S. Rebensburg et al, Sonographie und biliäre extrakorporale Stoßwellenlithotripsie (ESWL), *Ultraschall in Med.* 13, 255 - 262, 1992

5. A. Goedegebure, A. Van der Steen, J.M. Thijssen, In vitro classification of gallstones by quantitative echography, *Ultrasound in Med. & Biol.* Vol. 18, Nos. 6/7, 1992

ULTRASONIC BREAST IMAGING ASSISTED BY ACOUSTIC VELOCITY RECONSTRUCTION

M. Krueger, A. Pesavento, H. Ermert

Dept. of Electrical Engineering, Ruhr University
44780 Bochum, Germany

INTRODUCTION

Approximately 9% of all women in the United States are affected by breast cancer[1]. The chance of survival depends strongly on the stage of the cancer at the date of the diagnosis. Consequently, an early-stage detection of breast cancer is very important. Since the mortality of breast cancer has not changed in the past 60 years[2] the success of the current screening (X-ray mammography and palpation) has to be seen in a very critical way. Therefore, additional diagnostics are investigated. Several approaches have been made in the last two decades: (a) imaging of further (ultrasonic, optical, functional) parameters, (b) enhanced imaging of a presently 'manually' investigated parameter, i. e. elasticity (elastography), and (c) processing of diagnostic images or raw data.

In the scope of diagnostic ultrasound quantitative imaging of three parameters (reflectivity, attenuation, and sound velocity) promises a very good separation of different tissue types[3]. While reflectivity and attenuation can be determined from ultrasonic echo data the determination of the acoustic velocity needs a tomography approach.

In the following sections a review on past imaging systems is given. Then we propose a modified time-of-flight tomography concept for ultrasonic breast imaging which requires a different reconstruction algorithm. This algorithm is adjusted with a simulation using test data. We will then present the reconstruction results of a simple phantom. The results are discussed and an outlook on future objectives is given.

TIME-OF-FLIGHT TOMOGRAPHY USING A CIRCULAR APERTURE

The time-of-flight tomography concept is derived from X-ray computer tomography[1]. In this concept a water bath is necessary and the breast is not fixed. The geometry is presented in Figure 1. Either two linear arrays are moved on a circular line or a special ring array is used. Both concepts provide time-of-flight measurements sampling the complete

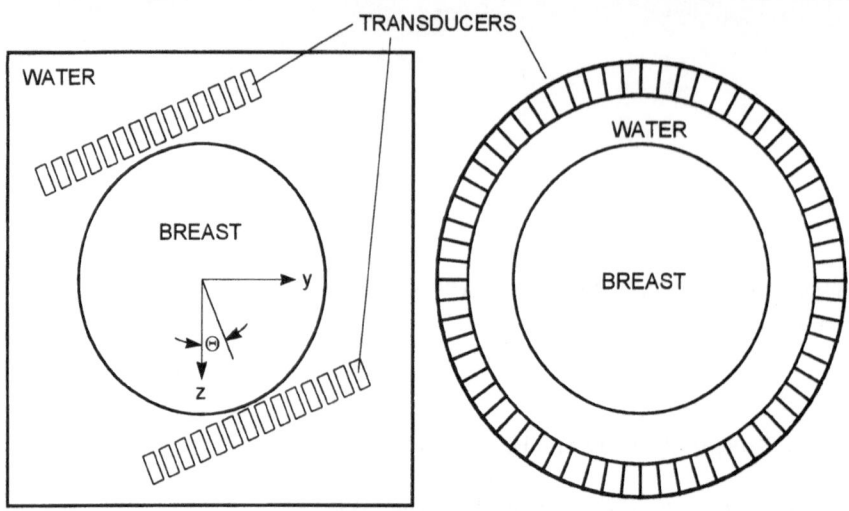

Figure 1. Geometries of time-of-flight tomography concepts using circular apertures. Two linear arrays are moved on a circular line (left) or a special ring array is used (right). The transducers are symbolized by small rectangles ('□') and their aperture is always towards the breast tissue.

angular range of $0 \leq \Theta \leq 2\pi$. Hence, an isotropic resolution can be obtained. Also, in contrast to mammography the breast is not compressed.

However, in this way the compatibility to the clinically well-established X-ray mammography is lost. First, the reconstructed cross-sectional plane is perpendicular with respect to the projection plane obtained by mammography. This could be overcome by reconstructing planes in various elevations followed by image processing. Second, the position of the breast is different and not as well-defined as in mammography. To support breast cancer screening a concept compatible to mammography which preserves the fixation of the breast should be applied.

TIME-OF-FLIGHT TOMOGRAPHY USING A LINEAR APERTURE

Richter[4] proposed to acquire a sequence of parallel displaced B-mode images of the female breast compressed between a Plexiglas plate and a plane metal reflector (see Figure 2, left). The strong echo of the reflector would appear as a bright straight line in the bottom of the B-scan of the breast if the sound velocity were homogeneous. In case of a lesion with a different sound velocity the line will be no longer straight. Hence, the displacement of the line contains time-of-flight information. The displacement as a function of the lateral coordinate y is called a CARI-line (CARI: clinical amplitude/velocity reconstructive imaging). All CARI-lines obtained in different elevations x (i. e. B-scan planes) comprise a so-called CARI-image. Besides the displacement the variation of the brightness as an indirect measure of the attenuation is considered. In Richter's clinical study 88% of all carcinomas gave rise to a displacement in the CARI-line. The specificity of the combination of B-Scans, velocity and amplitude CARI-images was 78%.

Obviously, in the CARI approach the ultrasonic wave propagation has been considered in a very simple way. Therefore, a modification of the concept should promise a better resolution as well as improved quantitative information on the sound velocity. We decided to use multiple channel echoes for a time-of-flight tomography concept with a linear aperture (see Figure 2, right). The times-of-flight are obtained from multistatic synthetic aperture

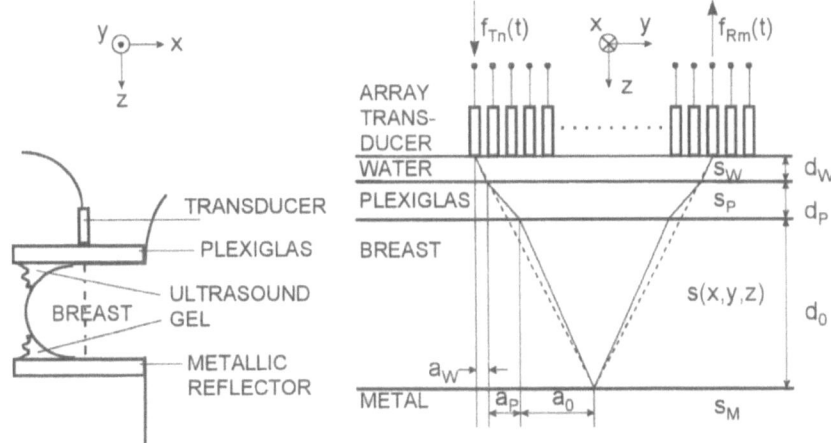

Figure 2. Left: Sonographic concept proposed by Richter[4]. The breast is compressed like in X-ray mammography. The B-scan is taken perpendicular to the reflector (dash-dotted line) and the transducer is moved in elevation direction (x). Right: The same combination of a conventional transducer and the compression plates can be used for a modified time-of-flight tomography concept. Note the changed orientation of the coordinate systems. Symbols: $f_{Tn}(t)$: transmit signal, $f_{Rm}(t)$: signal received by the m-th channel ($n < m = 1,...,M$), d_W, d_P, d_0: thicknesses of water layer, Plexiglas layer, and compressed breast, respectively, a_W, a_P, a_0: intercepts of the propagation path, s: slowness of each layer (the slowness is reciprocal to the acoustic velocity). The dashed line describes the acoustic path with shear waves. The considered acoustic path without shear waves is given by the solid line.

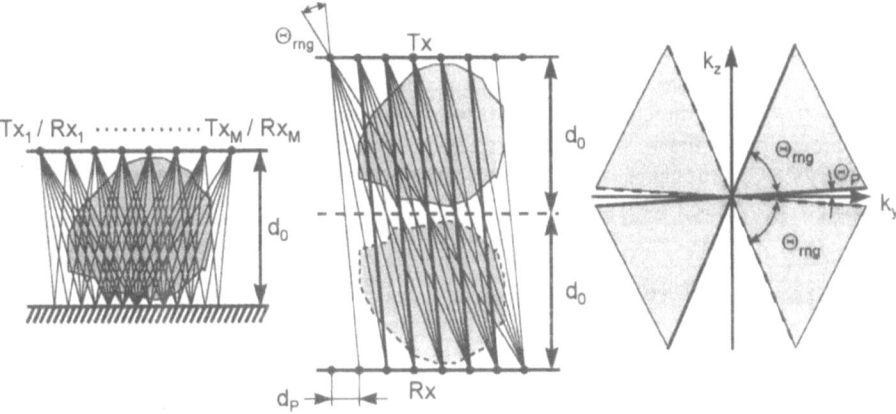

Figure 3. Limited angle tomography. Left: geometry of the modified time-of-flight tomography concept, center: equivalent transmission tomography system, right: frequency response of the resulting spatial filter.

data. Each element is transmitting a short ultrasonic pulse while all elements which are located right to this transmitting element (according to Figure 2) are receivers. The arrival times of the first echo of the metallic reflector provide the raw data for the time-of-flight tomography.

In case of a homogeneous breast layer the arrival time can be derived by Fermat's principle. Concerning the Plexiglas layer two propagation modes have to be considered. A longitudinal wave with a phase velocity of 2720 ms^{-1} (solid line in Figure 2) and a transversal wave with a phase velocity of 1300 ms^{-1} (dashed line). Since the velocity of the longitudinal mode is significantly larger than the shear wave velocity the arrival time of the first echo is determined by the time-of-flight of the ray containing the longitudinal wave path. Therefore, the shear wave mode may remain unconsidered. However, in B-mode and Doppler Imaging this mode can lead to imaging problems due to multiple foci[5].

To simplify the following equations the 'slowness' s as the reciprocal quantity of the acoustic velocity c

$$s = c^{-1} \qquad (1)$$

is introduced. According to Fermat's principle the shortest time-of-flight along a path from the n-th element to the m-th element (Figure 2) is given by

$$t_0^{mn} = 2 \left[s_W (a_W{}^2 + d_W{}^2)^{1/2} + s_P (a_P{}^2 + d_P{}^2)^{1/2} + s_0 (a_0{}^2 + d_0{}^2)^{1/2} \right]. \qquad (2)$$

The intercepts a_W, a_P, and a_0 are determined by minimizing t_0^{mn}. The minimization is constrained by the distance of the transmitting and the receiving element.

Since the variations of the sound velocity inside the tissue are sufficiently small, refraction can be neglected here. Consequently, in case of an inhomogeneous medium the difference of the time-of-flight is given by:

$$t^{mn} - t_0^{mn} = \sqrt{a_0{}^2 + d_0{}^2} \left[\int_0^{d_0} \left\{ s' \left(0, \frac{a_0 z}{d_0}, z \right) - s_0 \right\} dz + \int_0^{d_0} \left\{ s' \left(0, -\frac{a_0 z}{d_0}, z \right) - s_0 \right\} dz \right]. \qquad (3)$$

This equation describes the relationship of the raw data (time-of-flight) and the reconstruction data (slowness) for our modified time-of-flight tomography concept. In contrast to the concept using total circular apertures, the reconstructed object is spatially filtered. Figure 3 demonstrates that, according to the Fourier-Slice-theorem, only a limited angular range is recorded by this tomography concept. Especially, there is no information with respect to the z-direction. Therefore, two fundamental reconstruction limits appear: (a) the projection of the reconstruction result in y-direction is independent from the real slowness data; (b) the axial resolution (z-direction) is much coarser than the lateral resolution (y-direction).

RECONSTRUCTION ALGORITHM

Time-of-flight tomography using a circular aperture allows the application of the reconstruction algorithms which were developed for X-ray computer tomography: filtered back-projection or algebraic algorithms. The limited angle as well as the varying spatial sampling of the projections which occur in the new concept suggest the use of a parametric approach. By sampling the object and collecting the representatives of the slowness into a slowness vector **s** and by collecting the times-of-flight into vectors **t** and **t₀**, respectively, Equation 3 can be expressed by a matrix equation:

$$\mathbf{t} - \mathbf{t_0} = \mathbf{L} \, (\mathbf{s} - \mathbf{s_0}) + \mathbf{z}. \qquad (4)$$

L represents the linear relationship described by Equation 3 while **z** represent errors of the time-of-flight measurements. As the number of measurements is larger than the number of representatives Equation 4 is over-determined. On the other hand the rank of the matrix is less than its maximum value because of components which cannot be reconstructed. For this kind of singular equation systems the singular value decomposition[6] is a powerful tool to achieve an optimal solution in a least square sense. The optimization of this algorithm using a simulation tool is described in Krueger et al.[7] In the simulations the axial resolution was

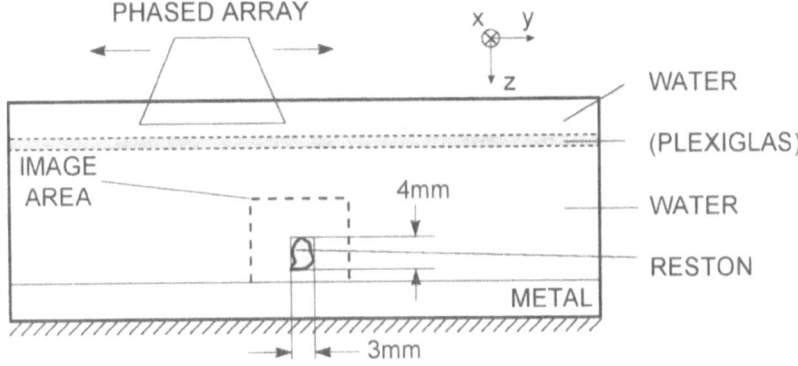

Figure 4. Experimental setup. The reconstructed object was a cylindrical piece of RESTON (Trademark of 3M) surrounded by water. The velocity contrast was approximately 20 ms^{-1}.

3 mm and the lateral resolution was 0.5 mm. The velocity resolution was estimated as 5 ms^{-1}. The applied transducer was a phased array with a center frequency of 3.5 MHz. The time-of-flight estimation was performed using the maximum of the RF-signal. Correlation of neighbored propagation paths were used to improve the signal-to-noise ratio to the time-of-flight measurements. An oblique orientation of the reflector was compensated numerically.

RECONSTRUCTION OF A PHANTOM

Encouraged by the simulation results an experimental setup was designed. Figure 4 shows the phantom that was constructed to test the algorithm in practice. To increase the aperture the phased array was moved mechanically. The phased array is connected to a multistatic synthetic aperture focusing system which is described in details in Krueger et. al.[7]

The time-of-flight measurement was not affected by the Plexiglas layer as demonstrated two sections before. Multiple reflection modes and shear wave modes could be clearly separated from the first order echo. Therefore, for these experiments the mechanical construction could be simplified by omitting the Plexiglas layer. The reconstruction result is presented in Figure 5.

The properties of the time-of-flight tomography concept as derived from the simulations are confirmed by the experiment: a coarse axial resolution (3 mm), a much better lateral resolution (0.5 mm), and side lobes resulting from the missing projections in lateral direction. Considering the axial blurring of the maximum, the quantitative resolution is quite good, as well. The distribution as presented in the right-handed plot in Figure 5 with a maximum of 2.4 µsm^{-1} corresponds to a rectangular distribution with a maximum of 7.3 µsm^{-1} and a width of 4 mm. Hence, the reconstructed velocity contrast is 16 ms^{-1}. Therefore, a real quantitative resolution of 5 ms^{-1} can be estimated.

CONCLUSION

A modified time-of-flight algorithm to reconstruct the acoustic velocity of breast tissue has been introduced. In contrast to ring apertures a conventional X-ray mammography system and a clinical B-mode imaging system allowing single channel data recording can be used. The breast is in a fixed position which is compatible to X-ray mammography. Corresponding B-mode images can be recorded without any further efforts. Disadvantages are a coarse axial resolution and side lobes in lateral direction.

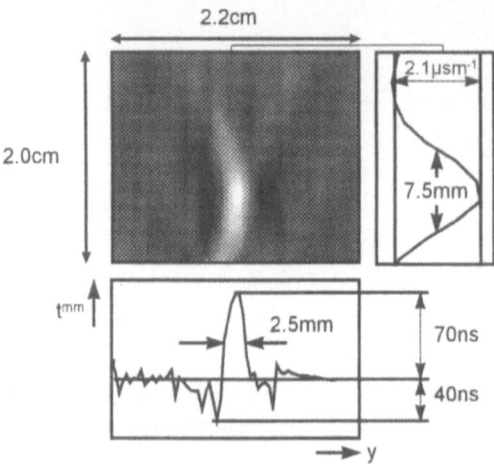

Figure 5. Reconstruction results. Gray-scale image: reconstructed slowness distribution, lower plot: axial projection (t^{mm}: times-of-flight in axial direction), right-handed plot: axial slice containing maximum $((1470 \text{ ms}^{-1})^{-1})$ of slowness.

Due to the fundamental limits the algorithm does not provide a high-resolution quantitative image. However, compared to other tissue characterization systems the segment size is not extraordinarily large. In our future activities we want to apply different array types. The center frequency will be higher (7MHz) and the aperture will be larger (linear array). Furthermore, a commercial clinical B-scan system will be modified to decrease the data acquisition time considerably and to allow clinical studies. In conjunction with B-mode imaging and attenuation estimation the time-of-flight reconstruction has the potential to improve breast cancer diagnostics significantly.

ACKNOWLEDGEMENT

The simulations and experiments were partially supported by Siemens Medical Systems, Ultrasound Group, Issaquah, WA, USA.

REFERENCES

1. J. F. Greenleaf, J. Ylitalo, J. J. Gisvold. Ultrasonic computed tomography for breast examination. *IEEE Eng. Med & Biol. Mag.*, pp 27-32, Dec. 1987
2. R. McLelland. Screening for breast cancer: Opportunities, status and challenges. In S. Brünner, B. Langfeld (eds.), *Advances in Breast Cancer Detection*, pp 29-38. Springer, Berlin, 1990.
3. F. S. Foster. M. Strban, G. Austin. The ultrasound macroscope: Initial studies of breast tissue. *Ultrasonic Imaging*, 6: 243-261,1994.
4. K. Richter. Clinical Amplitude/Velocity Reconstructive Imaging (CARI) - A New Sonographic Method for Detecting Breast Lesions. *Brit. J. Radiol.*, 68: 375-384, 1995.
5. P. Berti, A. Gubbini, P. Tortoli. Refraction artifacts in Doppler spectra due to dual mode ultrasound propagation. *Acoustical Imaging 23*, S. Lees (ed.), 1997, to be published
6. W.H. Press, S.A. Teukolsky, W. T. Vetterling, B. P. Flannery, Numerical Recipes in C, Cambridge University Press, Cambride (1992)
7. M. Krueger. A. Pesavento. H. Ermert. A modified time-of-flight tomography concept for ultrasonic breast imaging. In *1996 IEEE Ultrason. Symp. Proc.*, pp. 1381-1385, IEEE press, 1996.

VERY-HIGH FREQUENCY ULTRASONIC IMAGING AND SPECTRAL ASSAYS OF THE EYE

Frederic L. Lizzi,[1] Andrew Kalisz,[1] Michael Astor[1], D. Jackson Coleman[2],
Ronald H. Silverman[2], and Dan Z. Reinstein[2]

[1]Riverside Research Institute
330 West 42nd Street
New York, NY 10036
[2]Cornell University Medical College
1300 York Avenue
New York, NY 10021

INTRODUCTION

During the past several years, focused ultrasonic transducers operating at center frequencies near 50 MHz have become available for pulse-echo imaging.[1] The large bandwidths (e.g., 30 MHz) and narrow beamwidths (e.g., 75 μm) afforded by these transducers have greatly increased the resolution attainable for examining superficial segments of the body.

Very-high frequency ultrasound (VHFU) has proven especially useful in imaging the anterior segment of the eye, as initially reported by Pavlin et al.[2] However, as at lower frequencies, B-mode images do not fully convey the information contained in received echo signals. To more fully extract this information, we have developed a VHFU system[3] that digitizes radio-frequency (rf) echoes from parallel-plane scans and includes specialized processing and imaging procedures for specific tissue examinations.

In applying our processing concepts, we have found it useful to divide tissue structures into two basic categories. The first category encompasses deterministic tissues whose properties are well-defined over several beamwidths. This category includes corneal and lenticular surfaces, tumor boundaries, and blood vessels. Such structures constitute coherent scatterers; their echo signals are analyzed by first applying rf pre-processing (to sharpen resolution) and then computing 3-D biometric features such as thickness, curvature and volume.

The second category comprises stochastic tissues whose small internal constituents exhibit random positions or acoustic properties. These tissues give rise to non-coherent scattering. Their echo signals are analyzed by computing calibrated power spectra[4] and applying a theoretical model that relates computed spectral parameters to the effective sizes and acoustic concentration of constituent tissue scatterers.[5] Acoustic concentration is defined as CQ^2 where C is the effective concentration of scatterers and Q is their acoustic impedance relative to the surround.

This report provides a brief overview of our analysis techniques for each tissue category. We will illustrate the analysis of deterministic tissues by describing our procedures for characterizing the outer epithelium layer (50-μm thick) and central stroma (500-μm thick) of the cornea. Precise knowledge of the conformations of these layers is becoming increasingly important for treating corneal opacities and refractive aberrations.[6,7]

Our VHFU procedures for analyzing stochastic tissues[8] are being employed to identify small ocular tumors and to monitor microstructural changes induced by ultrasonic hyperthermia and radiotherapy. Stochastic analysis procedures are also being employed to evaluate ciliary body microstructure in normal and glaucomatous eyes[9]. This report will illustrate how these procedures are employed for analyzing intraocular hemorrhages to detect organized hemorrhagic masses, which can complicate treatment procedures.[10]

SYSTEM OPERATION AND PROCESSING SEQUENCE

The clinical VHFU system runs under control of a Pentium computer. It employs a broadband Panametrics pulser/receiver with focused PVDF and PZT transducers whose useable (-15 dB) bandwidths typically extend from 15 to 60 MHz, as shown below. A LeCroy analog-to-digital converter system is operated at a sampling frequency of at least 200 MHz to acquire 8-bit samples of rf data. The transducer is scanned by a pair of computer-controlled orthogonal stepper motors, and acoustic coupling is provided by means of a shallow saline bath.

In typical VHFU operation, the transducer is linearly scanned in a sequence of 10 to 20 parallel scan planes separated by 0.2 to 0.5 mm. In each plane, 2048 rf samples are typically acquired and buffered along 128 scan lines. Scanning requires about 0.03 sec/plane with interspersed one-second intervals for data transfer to the Pentium. Calibration data are obtained by digitizing rf echoes from a planar water/glass interface placed in the focal plane of each transducer. All spectral amplitudes are specified in dB relative to this reference reflector.

Data processing begins with the synthesis of B-mode images from rf data captured in each scan plane. These images are displayed in rapid succession on a video monitor or as 3-D presentations on a Silicon Graphics workstation. Displayed images are inspected to clarify anatomical relations and to define specific clinical topics to be evaluated. Tissue segments are selected for further analysis by using a mouse to bracket their B-mode locations; customized software is then used to assay deterministic or stochastic structures as described below.

ASSAYS OF DETERMINISTIC STRUCTURES

The analysis of deterministic structures will be illustrated by describing our procedures for characterizing the spatial conformations of the corneal epithelium and stroma.

The most important region of these layers is the 3-mm zone centered along the visual axis. RF data are acquired from parallel planes within this zone after the transducer has been aligned so that its focal point lies near the anterior surface of the epithelium. Acquired RF data are pre-processed to improve resolution and waveform stability. First, inverse filtering is applied to compensate for the system transfer function and broaden the signal bandwidth. In this procedure, rf echoes from the cornea are gated and a Fast Fourier Transform (FFT) algorithm is applied to obtain the complex tissue spectrum T(f) where f denotes temporal frequency. As shown for clinical epithelium echoes in Fig. 1, the magnitude of T(f) exhibits periodic maxima at a frequency interval determined by layer thickness as previously described for detached retina spectra.[4] T(f) is then divided by G(f), the complex calibration spectrum of a planar water/glass interface. Figure 1 shows that the tissue and calibration spectra (offset by 40 dB) exhibit similar shaping due to system and transducer characteristics; these factors limit the -6 dB bandwidth to about 30 MHz. Division of T(f) by G(f) removes this shaping and, as seen in Fig. 1, extends the effective signal bandwidth. The processed spectrum is then multiplied by W(f) to suppress

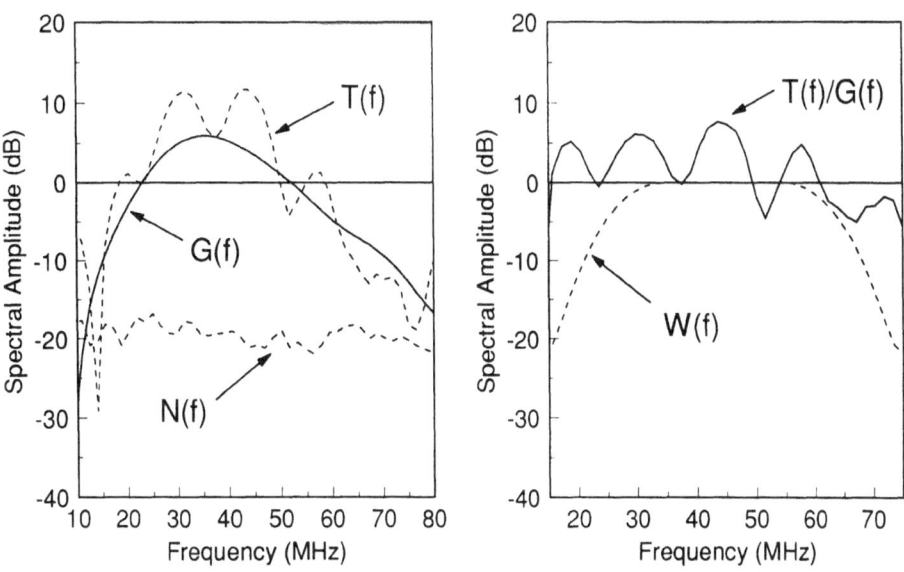

Figure 1. Magnitudes of spectral functions for cornea analysis.

out-of-band noise and obtain a bandwidth exceeding 40 MHz. (The noise spectrum N(f) was computed from signals in the coupling bath anterior to the cornea.)

After these procedures, an inverse FFT of W(f)T(f) / G(f) is computed, yielding the RF echoes shown in Fig. 2a. A Hilbert transformation is then applied to derive the magnitude of the analytic signal, providing a well-defined, stable estimate of the echo envelope as shown in Fig. 2b. Epithelial thickness is computed from the time interval between the peaks of the anterior and posterior echoes; in the illustrated case, a thickness of 56 μm was computed. These procedures have provided a precision better than 1.5 μm

For 3-D corneal assays, these processing techniques are applied to the epithelial and stromal echoes from each scan line in the set of parallel scan planes. For each layer, biometric information is displayed in a set of three images where color depicts local values of: surface height (referenced to a common plane); thickness; and radius-of-curvature.[7] As an example, Fig. 3 presents gray-scale epithelial-thickness images for a normal cornea (left) and for a cornea previously treated with photorefractive keratotomy (right); this laser technique is intended to reshape the cornea and thereby restore normal optical refraction. The treated cornea exhibits an irregular epithelial thickness compared to the normal. In particular, the lower segment comprises a broad thinned (bright) region (37 - 47 μm) containing a thickened (dark) band (47 - 55 μm); this irregularity produced a refractive error in post-treatment vision. (More detailed color biometric images are contained in ref. 7.)

ASSAYS OF STOCHASTIC STRUCTURES

Examination of stochastic structures is based on calibrated power spectra[4]. A sliding Hamming window is used to analyze rf echoes lying within the examined structure. At sequential sites along each scan line, the local power spectrum of gated rf signals is computed and divided by the calibration spectrum to remove system transfer functions (as in Fig. 1). The calibrated power spectrum is then expressed in dB, and linear regression is employed to calculate spectral slope (dB / MHz) and intercept (dB, extrapolation to zero frequency).[5] These procedures are typically employed over the 15 to 60 MHz band, where adequate signal-to-noise ratios are obtained. (The Hamming window length is typically 0.25 mm yielding a -3 dB axial resolution near 70 μm.)

Values of slope and intercept are determined for all scan-line segments in the examined

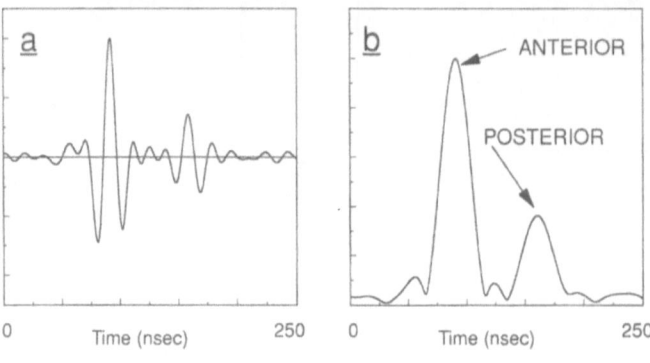

Figure 2. Epithelium echo signals: a. rf; b. analytic signal magnitude.

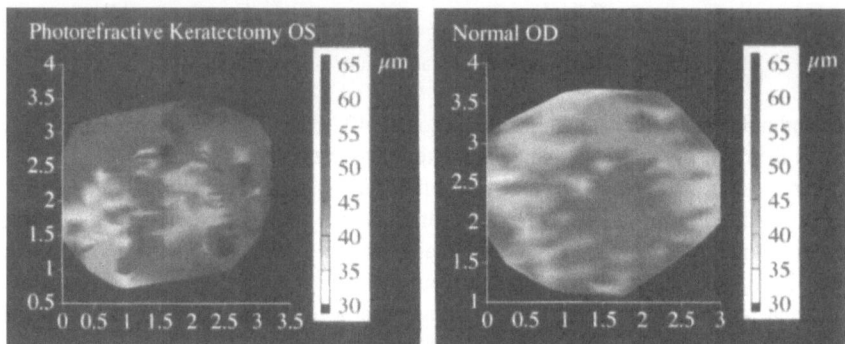

Figure 3. Corneal epithelium thickness (calibration bars) vs. 2-D position (mm).

structure. Then theoretical model results are applied to derive local estimates of effective scatterer diameter D and acoustic concentration CQ^2 throughout the tissue volume.[5] These calculations employ a theoretical model of tissue scattering which has recently been expanded to provide closed-form expressions for arbitrary transducer parameters, center frequencies, and fractional bandwidths.[11] The model treats isotropic scatterer distributions whose acoustic impedance exhibits a Gaussian spatial autocorrelation function (ACF). It assumes that attenuation is negligible or that it has been compensated for. (See ref. 5 and 11 for other ACF's, scatterer morphologies, and discussion of attenuation.)

The theoretical model results in Fig. 4 permit D to be estimated from spectral slope; the smallest sizes that can be estimated depend on the Rayleigh scattering limit where slope is constant (corresponding to f^4 spectra). For the current VHFU frequency band, diameters as small as 10 μm can be estimated. Once D is estimated, the second plot in Fig. 4 is used to estimate CQ^2. This plot shows intercept values for unity CQ^2 as a function of scatterer diameter D; CQ^2 is computed as $I' - I$ where I is the measured intercept and I' is the plotted reference value for the estimated value of D. (CQ^2 is specified in dB relative to $1 / mm^3$.)

Local estimates of D and CQ^2 are displayed in color-coded 2-D or 3-D images. Fig. 5 shows gray-scale reproductions of parameter images for hyphema, an anterior-chamber hemorrhage.[10] In this case, the outer segments of the hemorrhage (H) exhibit low sizes and CQ^2 values, consistent with unorganized blood constituents. The central posterior region of the hyphema exhibits large scatterer sizes and high CQ^2 levels, indicative of blood organization. Such results convey information for planning appropriate treatment, which may be complicated by the presence of large organized masses. (Color and 3-D clinical images for stochastic structures can be found in ref. 8.)

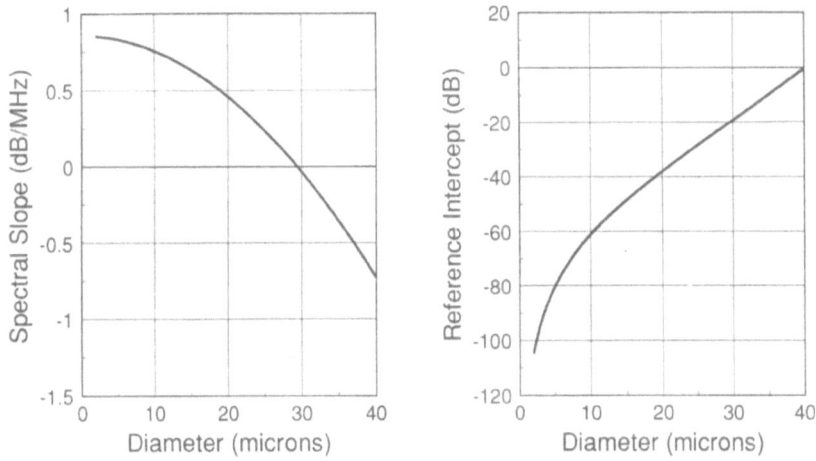

Figure 4. Theoretical scattering model results.

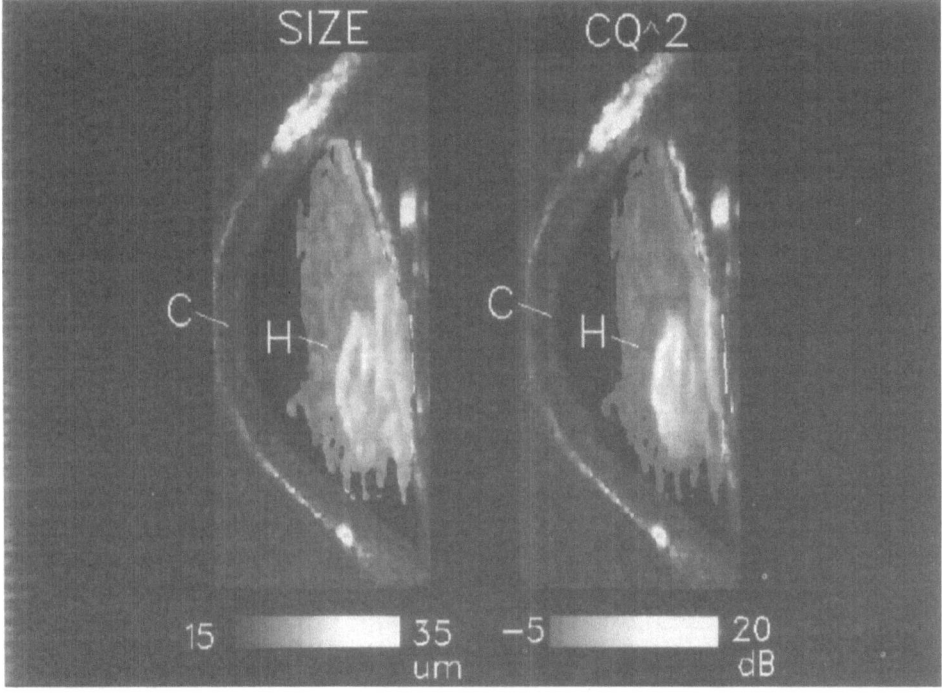

Figure 5. Scatterer property images of hyphema (H) posterior to B-mode image of cornea (C).

SUMMARY

VHFU examinations are providing a new degree of spatial resolution for examining diseases, drug responses, and physiologic changes in ocular microstructure. Digital signal processing can extract key quantitative data that are not conveyed by conventional B-mode images. Digital procedures designed to extract biometric data for deterministic structures are particularly useful for corneal assays before and after laser surgery. Spectrum analysis procedures are providing important information for stochastic tissues including hemorrhages, tumors, and the ciliary body. These methods should continue to improve as VHFU frequency ranges are expanded and as improvements are made in theoretical models that relate measured echo features to underlying tissue microstructure.

Acknowledgement

Portions of this research were supported by NIH Research Grant EY01212, Research to Prevent Blindness, and the St. Giles Foundation.

REFERENCES

1. M.D. Sherar and F.S. Foster, The design and fabrication of high frequency poly(vinylidene fluoride) transducers, *Ultrason. Imaging.* 11:75-94 (1989).
2. C.J. Pavlin, M.D. Shearer, and F.S. Foster, Subsurface ultrasound microscopic imaging of the intact eye, *Ophthalmology.* 97:244:250 (1990).
3. F.L. Lizzi, M.C. Rorke, J.B. Sokil-Melgar, A. Kalisz, and J. Driller, Interfacing very-high-frequency transducers to digital-acquisition scanning systems, *Proc. of the Society for Photo-Optical Instrumentation Engineering (SPIE).* 1844, 313-321 (1992).
4. F.L. Lizzi, M. Greenebaum, E.J. Feleppa, M. Elbaum, and D.J. Coleman, Theoretical framework for spectrum analysis in ultrasonic tissue characterization, *J. Acoust. Soc. of Amer.* 73(4):1366-1373 (1983).
5. F.L. Lizzi, M. Ostromogilsky, E.J. Feleppa, M.C. Rorke, and M.M. Yaremko, Relationship of ultrasonic spectral parameters to features of tissue microstructure, *IEEE Transactions on Ultrasonics, Ferroelectrics, and Frequency Control.* UFFC-34:319-329, (1987).
6. D.Z. Reinstein, I.M. Aslanides, R.H. Silverman, P.A. Asbell, and D.J. Coleman, High-frequency ultrasound corneal pachymetry in the assessment of corneal scars for therapeutic planning. *The CLAO Journal.* 20:198-203 (1994).
7. D.Z. Reinstein, R.H. Silverman, S.L. Trokel, and D.J. Coleman, Corneal pachymetric topography, *Ophthalmology.* 101:432-438 (1994).
8. R.H. Silverman, M.J. Rondeau, F.L. Lizzi, and D.J. Coleman, Three-dimensional high-frequency ultrasonic parameter imaging of anterior segment pathology, *Ophthalmology.* 102: 837-843 (1995).
9. I.M. Aslanides, P.E. Libre, R.H. Silverman, D.Z. Reinstein, G.K. Harmon, D.R. Lazzaro, M.J. Rondeau, and D.J. Coleman, High frequency ultrasound imaging in pupillary block glaucoma, *Brit. J. Ophthalmol.* 79:972-976 (1995).
10. N. Allemann, R.H. Silverman, D.Z. Reinstein, and D.J. Coleman, High-frequency ultrasound imaging and spectral analysis in traumatic hyphema, *Ophthalmology.* 100:1351-7 (1993).
11. F.L. Lizzi, M. Astor, A. Kalisz, T. Liu, D.J. Coleman, R.H. Silverman, R. Ursea, and M. Rondeau, Ultrasonic spectrum analysis of different scatterer morphologies; theory and very high frequency clinical results, *Proc. 1996 Ultrasonics Symp.* pp. 1155-1159. Inst. of Electric and Electronics Engineers, Piscataway, (1997).

TISSUE STRAIN ESTIMATION USING A LAGRANGIAN SPECKLE MODEL

R. L. Maurice, M. Bertrand

Institut de génie biomédical, École Polytechnique, C.P./PO Box 6079, succ. Centre-Ville, Montréal, Québec, Canada H3C 3A7

Institut de Cardiologie de Montréal, 5000 Bélanger est, Montréal, Québec, Canada H3T 1C8

INTRODUCTION

Accurate assessment of tissue motion from spatio-temporal changes in ultrasound speckles is a key factor in computing high signal-to-noise ratio (SNR) elastograms or correlation-based flow profiles. Provided the tissue (or fluid) is subjected to a simple translation movement, reliable sub-wavelength displacement estimates can be obtained through cross-correlation delay computation. On the other hand when the tissue is subjected to more complex movement, displacement estimates are generally found to be less accurate; indeed the changes in speckle patterns that result from such a motion act as the noise source often responsible for most of the displacement estimate variance. In this paper we propose a method to estimate 2-D motion while optimally compensating for speckle decorrelation. The method is based on a Lagrangian speckle model we proposed in [1], and uses a Wiener inverse filtering approach to reduce speckle decorrelation.

THE LAGRANGIAN SPECKLE MODEL (LSM)

We assume the echographic image formation ($I(x,y)$) can be described by a 2-D linear and space-invariant operation on a scattering function $z(x,y)$. Hence the radio-frequency (r.f.) image of the tissue before motion occurs ($I_0(x,y)$) is given by:

$$I_0(x, y) = h(x, y) \otimes z(x, y) \tag{1}$$

where $h(x,y)$ is the imaging system point spread function (PSF). The motion model we assume here is that of a constant strain within the region-of-interest (ROI), that is it can be described by a linear transformation (LT). Assuming the scattering strength at coordinates (x,y) is a material property which is conserved during motion, the image formation after tissue motion can be modeled by:

$$I_1(x,y) = h(x,y) \otimes z_{LT}(x,y) \tag{2}$$

where the subscript LT indicates that the scattering function variables (x,y) are subjected to a linear transformation.

Equation (2) describes the motion in terms of the observer's coordinate system. To study speckle changes, we now define a new r.f. image $I_{Lag}(x,y)$ by performing motion rectification for the tissue displacement over $I_1(x,y)$:

$$I_{Lag}(x,y) = (I_1)_{LT^{-1}}(x,y) \tag{3}$$

where the subscript LT^{-1} indicates the inverse of the linear transformation LT. We call this the Lagrangian speckle image, because the correction of the motion for tissue displacement effectively expresses the speckle dynamics in terms of material point coordinates. In such coordinates, we can consider the speckles only change their "shape", not their position. This motion correction operation leads to an r.f. image that should maximally correlate with the r.f. image before motion occurred (I_0), a property which can be used to determine the tissue motion. It can be easily shown that:

$$I_{Lag}(x,y) = h_{LT^{-1}}(x,y) \otimes z(x,y)|J|^{-1} \tag{4}$$

where $|J|$ is the Jacobean of the LT. In the frequency domain (u,v), equations (1) and (4) lead to:

$$I_0(u,v) = I_{Lag}(u,v)\frac{H(u,v)}{\Im(h_{LT^{-1}}(x,y))}|J| \tag{5}$$

where $\Im(\cdot)$ is a Fourier transform operation. Equation (5) indicates that, in theory, the r.f. image before motion I_0 and the Lagrangian speckle image are filtered version of one another; i.e. the Lagrangian speckle image defined in (3) can be *restored* from the image before motion (I_0) by an appropriate parametric filtering. Notice that the filter parameter is the linear transformation involved in the tissue motion itself. In that case, not only the r.f. image before motion would maximally correlate with the Lagrangian image but it would exactly match its filtered version. The tissue motion estimation method we propose is based on seeking the LT that performs a match of this type.

A LAGRANGIAN SPECKLE TRACKING METHOD

The tissue motion estimator for a linear transformation could be formulated as the following non-linear minimization problem:

$$\underset{LT}{\text{MIN}} \ \left\| I_{Lag}(x,y) - \hat{I}_{Lag}(x,y) \right\|^2 \tag{6}$$

where $\hat{I}_{Lag}(x,y)$ is the estimate of $I_{Lag}(x,y)$ obtained by filtering I_0. In practice, because of the presence of noise components, the filter cannot be implemented using equation (5), and the filtering process would need to be implemented differently. A Wiener filtering that uses an appropriate noise model would be a more suitable restoration technique to provide an estimate of I_{Lag}. Figure 1 illustrates a simple noise model for the case where I_0 is corrupted by an additive noise representing electronic noise or quantization noise for example.

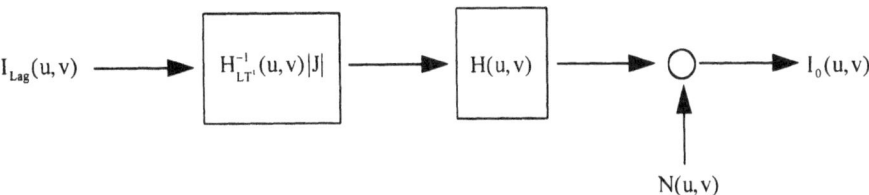

Figure 1: A simple noise model for Lagrangian speckle. Subscript LT^t indicates the linear transformation given by the matrix transpose of LT.

The corresponding Wiener filter used to estimate the Lagrangian speckle image from the image before motion occurred (I_0) is therefore given by:

$$\hat{I}_{Lag}(u,v) = I_0(u,v)\frac{\left(H^{-1}_{LT^t}(u,v)H(u,v)|J|\right)^*}{\left|H^{-1}_{LT^t}(u,v)H(u,v)|J|\right|^2 + \dfrac{S_{NN}(u,v)}{S_{I_{Lag}I_{Lag}}(u,v)}} \qquad (7)$$

where $S_{NN}(u,v)$ and $S_{I_{Lag}I_{Lag}}(u,v)$ are the power spectra of the noise and the Lagrangian image respectively.

RESULTS AND DISCUSSION

We now give some results obtained with the tissue motion estimation method based on the Lagrangian speckle model using simulated r.f. images of a tissue subjected to $2°$ axial shear. For the image formation, we model the scattering function $z(x,y)$ as a 2-D Gaussian white noise subjected to $2°$ shear deformation. The corresponding [LT] matrix and the equivalent displacement field partial derivative matrix $[\Delta_{ij}]$ are given by:

$$[LT] = \begin{bmatrix} m_1 & m_2 \\ m_3 & m_4 \end{bmatrix} = \begin{bmatrix} 1 & 0 \\ \tan\theta & 1 \end{bmatrix}$$

$$\Delta_{ij} = \begin{bmatrix} \Delta_{11} & \Delta_{12} \\ \Delta_{21} & \Delta_{22} \end{bmatrix} = \begin{bmatrix} 0 & 0 \\ \tan\theta & 0 \end{bmatrix} \Rightarrow \varepsilon_{ij} = \begin{bmatrix} \Delta_{11} & \dfrac{\Delta_{12}+\Delta_{21}}{2} \\ \dfrac{\Delta_{12}+\Delta_{21}}{2} & \Delta_{22} \end{bmatrix}$$

where ε_{ij} is the strain tensor and θ, the shear angle; for a $2°$ shear , $\tan(\theta) = .0349$ radian. The imaging system PSF ($h(x,y)$) is a 2-D Gaussian wavelet modeling a 3.5 MHz transducer with a beam width (width at half maximum) of 1.5 mm.

To model the noise term n(x,y), we simulate the quantization noise of an A/D converter. To estimate [LT], we use the Matlab implementation of the Levenberg-Marquardt non-linear minimization algorithm on the objective function defined by equation (6). In this equation, I_{Lag} is provided by implementing equation (3) using bilinear interpolation on the r.f. image after motion, I_1. In order to reduce the bilinear interpolation noise, image I_1 was first oversampled using a band limited interpolator. The results for 2 ROI sizes and quantization noise levels are summarized in Table 1. In each case 10 realizations were used.

Table 1. Mean value and standard deviation (s.d.) of the LT parameters and strain tensor using the proposed LSM, +/- the 95% confidence limits. The reference LT parameters and strain tensor values are shown between parenthesis in the leftmost column of the table.

	ROI sizes				
	1cm x 1cm	5mm x 5mm			
	Resolution				
LT parameters and strain	infinite	infinite	8 bits	6 bits	4 bits
m_1 (1.0)	0.9993 ± 0.0016	0.9715 ± 0.0116	0.9968 ± 0.01	0.9945 ± 0.0014	0.9902 ± 0.0131
$\Delta_{11}(0)$ mean	-0.7 ± 1.6 x10^{-3}	-29 ± 11.6 x10^{-3}	-3.2 ± 10 x10^{-3}	-5.5 ± 1.4 x10^{-3}	-9.8 ± 13.1 x10^{-3}
s.d.	2.2 ± 1.0 x10^{-3}	15.4 ± 6.7 x10^{-3}	13.5 ± 5.9 x10^{-3}	$1.9 \pm .8$ x10^{-3}	17.4 ± 7.6 x10^{-3}
m_2 (0)	0.0004 ± 0.0017	0.0031 ± 0.0064	0.0022 ± 0.0054	0.0018 ± 0.0128	0.0020 ± 0.009
$\Delta_{12}(0)$	0.4 ± 1.7 x10^{-3}	3.1 ± 6.4 x10^{-3}	2.2 ± 5.4 x10^{-3}	1.8 ± 12.8 x10^{-3}	2 ± 9.0 x10^{-3}
s.d.	2.3 ± 1.0 x10^{-3}	8.5 ± 3.7 x10^{-3}	7.2 ± 3.2 x10^{-3}	17 ± 7.5 x10^{-3}	11.9 ± 5.2 x10^{-3}
m_3 (.0349)	0.0348 ± 0.0002	0.0351 ± 0.0009	0.0351 ± 0.0014	0.0362 ± 0.0004	0.0367 ± 0.0009
Δ_{21} (34.9 x10^{-3})	$34.8 \pm .2$ x10^{-3}	$35.1 \pm .9$ x10^{-3}	35.1 ± 1.4 x10^{-3}	$36.2 \pm .4$ x10^{-3}	$36.7 \pm .9$ x10^{-3}
s.d.	$.2 \pm .09$ x10^{-3}	$1.2 \pm .5$ x10^{-3}	$1.9 \pm .8$ x10^{-3}	$.5 \pm .2$ x10^{-3}	$1.3 \pm .57$ x10^{-3}
m_4 (1.0)	0.9999 ± 0.00008	1.0000 ± 0.0002	1.0001 ± 0.00008	1.0011 ± 0.0008	1.0006 ± 0.0003
$\Delta_{22}(0)$	$-0.1 \pm .08$ x10^{-3}	$0 \pm .2$ x10^{-3}	$0.1 \pm .08$ x10^{-3}	$1.1 \pm .8$ x10^{-3}	0.6 ± 0.3 x10^{-3}
s.d.	$.1 \pm .04$ x10^{-3}	$.3 \pm .1$ x10^{-3}	$0.1 \pm .04$ x10^{-3}	$1.0 \pm .4$ x10^{-3}	0.4 ± 0.18 x10^{-3}

The first case is for the "ideal" conditions, i.e. infinite resolution, large (1cm x 1cm) ROI; here we would expect the motion and strain parameters to be exactly recovered. However, we find a standard deviation for the strain that can go up to .0023 (2300 μstrain). This can be accounted for by the interpolation noise introduced when implementing equation 2 and (to a lesser degree) equations 3 and 7. The situation is similar for the smaller 5mm x 5mm ROI except that for Δ_{11} the estimate appears biased (.029 instead of zero) and its standard deviation is eight times larger than with the 1cm x 1cm ROI. This is believed to be due to the algorithm ending in a local minimum, and/or may result from the noise introduced when windowing the ROI; being correlated to the data, such noise can indeed introduce a bias on the estimate. This is being investigated.

For the 8 to 4 bits finite resolution examples, we make the following observations. It appears that the estimate for Δ_{21} becomes biased as the resolution gets below 8 bits; this can be explained by the fact that the quantization noise becomes more correlated to the signal as the quantizer resolution decreases. The bias on Δ_{21} is however quite small, and account for only a 3% error with 4 bits resolution! For Δ_{11}, Δ_{12} and Δ_{22}, the bias is either very small (<.001 strain) or insignificant at the noise level simulated. Finally, the strains which show the largest standard deviation are Δ_{11} and Δ_{12}; these strains are responsible for the lateral motion of the speckles and are thus expected to show larger errors.

This motion estimation model appears promising in that it can take into account and counteract the speckle decorrelation effects which currently probably constitute the most important factor limiting the performance of correlation-based tissue motion estimator using ultrasound signals. We are now working on the implementation of the LSM for experimental data in elastography.

[1] Maurice R.L., Bertrand M.: "Speckle Model in Lagrangian Coordinates for Studying Non-Linear Tissue Deformations in Ultrasonography," Ultrasonic Imaging, Washington, June 1996.

A part of the project upon which this publication is based was performed pursuant to the University of Texas Grant CA64597-01 with the NIH, PHS.

CALIBRATION FOR THE *URTURIP* TECHNIQUE
USING AN ENERGY MINIMIZATION METHOD

Bruno Migeon, Philippe Deforge, and Pierre Marché

Laboratoire Vision et Robotique
63, avenue de Lattre de Tassigny
18020 Bourges Cedex - France

INTRODUCTION

As a part of our project concerned with the development of an ultrasound scanner dedicated to limb study, the URTURIP Technique (Ultrasound Reflection-mode Tomography Using Radial Image Processing) has recently been developed[1]. It consists in using classical B-scan images instead of projections[2-5] and gives qualitative images instead of quantitative images. The final goal of this project is the 2D and 3D reconstruction of anatomical structures at limb level by using echographic image processing. The developed process consists of several successive steps like : multiple reflection removing[6], 2D reconstruction[1], segmentation[7], contour association, contour interpolation[8], 3D reconstruction and visualization[9]. It has been validated by in vitro experiments on anatomical pieces of limbs of new-borns using a simple acquisition system prototype[9].

Now, we wish to validate the reconstruction technique by in vivo experiments with the help of a new acquisition system prototype. In order to use the URTURIP Technique, the exploration plane of the probe must be identical for each angular position, and the subpixel coordinates of the rotation center must be precisely determined before the reconstruction.

The goal of the calibration is to solve these two problems which is the purpose of this paper. After a brief reminder of the acquisition system and the 2D reconstruction principle, the positioning of the probe and the determination of the rotation center using an energy minimization technique are presented.

ACQUISITION AND RECONSTRUCTION

The acquisition system

The acquisition system is composed of two parts : an electronical one and a mechanical one.

The electronical part includes the echograph device with a 5MHz electronical sectorial scanning probe, a graphic card which allows digitalizing the video signal of the echograph device, and a micro-computer (133 MHz Pentium) on which the images are stored and then processed.

The mechanical part is composed of a water tank in which the probe can turn around the object to be studied, and can move with a vertical translation device in order to have access to various slices.

2D reconstruction principle

The Ultrasound Reflection-mode Tomography Using Radial Image Processing (URTURIP Technique) principle is to utilize radial B-scan images instead of projections as most other methods do[2-5]. In comparison, fewer radial directions are needed, it is less time-consuming, but qualitative images instead of quantitative ones are computed.

Let L_i^* i=1,..,N be N radial images obtained from N angulary equidistant directions around the rotation center where $L_i^*(k, l)$ denotes the luminance of the (k, l)-pixel on the L_i^* image.

An adjustment step consists in turning each L_i^* image around the rotation center with its own acquisition angle, to construct N adjusted images L_i in such a way that a pixel (x, y) on each L_i image corresponds to the same real point of the cross-section.

Then, a method based on the URTURIP Technique reconstructs an image L by a combination of the N adjusted images L_i , i.e. $L(x, y) = f(L_i(x,y))$, i=1..N, where f denotes the reconstruction method.

POSITIONING OF THE PROBE

Introduction and principle

During the exploration of a cross-section from different radial directions of investigation, the exploration plane of the probe must correspond to the identical and demanded cross-section for each radial direction of investigation. Due to the mechanical construction of our system, the rotation axis of the probe support is vertical and the rotation of the probe support stays in the horizontal plane. The position of the probe can be characterized by three angles by associating a three axes referential to the probe support and another to the exploration plane of the probe (Figure 1). The positioning stage of the probe consists in setting a zero value to the two first angles in such a way that the exploration plane of the probe always remains horizontal and eventually in setting a zero value to the third angle if we want the exploration area to be maximal.

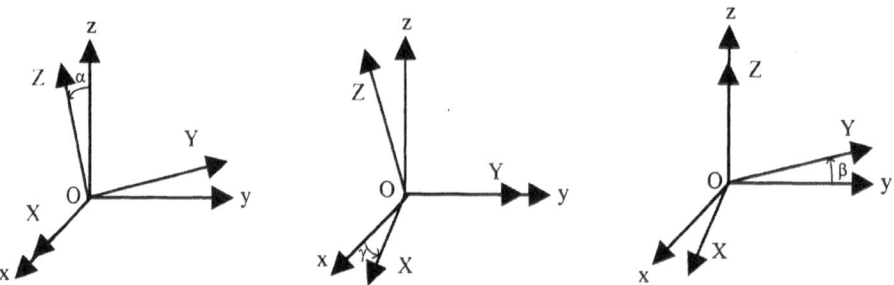

Figure 1. The position of the probe is characterized by the three angles α, γ, and β.

Horizontality of the investigation plane

Setting a zero value to the angle α

In order to set a zero value to the angle α, a horizontal mark (a 0.3 mm diameter taunt nylon thread) perpendicular to the Oy axis of the probe is used. We propose an iterative method which consists in combining the rotation of the probe around its Oy longitudinal axis with the vertical translation of the probe until the mark is correctly displayed on the echograph. The process is described in the following pseudo-algorithm.

While the horizontal mark does not belong to the investigation plane of the probe, do :
- Define the probe sector side where the mark is,
- Turn the probe around its longitudinal axis in the direction which moves the mark away to the outer sector,
- Define the direction of the vertical translation of the probe,
- Move the probe vertically until the mark is close to the median axis sector.

Setting a zero value to the angle γ

In order to set a zero value to the angle γ, a horizontal mark (a 0.3 mm diameter taunt nylon thread) at about 45° with the Ox axis of the probe is used. We propose an iterative method which consists in combining the rotation of the probe around its Ox axis with the vertical translation of the probe until the mark is correctly displayed on the echograph screen. The process is described in the following pseudo-algorithm.

While the horizontal mark does not belong to the investigation plane of the probe, do :
- Turn the probe around its Ox axis in the direction that moves the mark away to the far field until the mark is no longer visible,
- Define the direction of the vertical translation of the probe,
- Move the probe vertically until the mark is in the neighbouring field,

Optimal investigation by Setting a zero value to the angle β

If we want the exploration area to be maximal, the third angle β has to be set to zero too. To adjust this angle, the rotation center must be firstly determined in order to carry out a rotation around the vertical axis until the median of the exploration plane of the probe is aligned with it.

CALIBRATION USING AN ENERGY MINIMIZATION TECHNIQUE

Introduction

Before the 2D reconstruction, the radial images must be readjusted by a rotation around the rotation center[1]. So, the goal of the calibration is to determine the subpixel coordinates of the rotation center, which are identical on each radial image.

The proposed method

Principle

The method presented here to solve this problem is based on an energy minimization technique. Indeed, the real rotation center is such, so that the reconstruction of a cross-section is perfect i.e. all radial information overlaps perfectly with the others. If an energy is able to represent the non-overlapping, the real rotation center minimizes this energy.

Energy minimization

Let $E(x_c, y_c)$ be the energy of a reconstructed image after an adjustment of the radial images by considering the rotation center (x_c, y_c). $E(x_c, y_c)$ is defined by the number of the occupied pixels of the image reconstructed by the *maxima* method[1]. In other words :

$$E(x_c, y_c) = \sum_{(x,y) \in L^{(x_c, y_c)}} \delta^{(x_c, y_c)}(x, y) \quad \text{with} \quad \delta^{(x_c, y_c)} = \begin{cases} 0 & \text{if } L^{(x_c, y_c)}(x, y) = 0 \\ 1 & \text{if } L^{(x_c, y_c)}(x, y) \neq 0 \end{cases}$$

and

$$L^{(x_c, y_c)}(x, y) = \underset{i}{\text{Max}}\, L_i^{(x_c, y_c)}(x, y)$$

where $L^{(x_c, y_c)}$ is the reconstructed image after an adjustment according to the (x_c, y_c) rotation center, i is the index of the radial direction of investigation, $L_i^{(x_c, y_c)}$ is the ith radial image adjusted according to the rotation center (x_c, y_c).

Then, by minimizing this energy with a solid reference object, the subpixel coordinates of the rotation center are determined. To accede to the desired accuracy, each pixel of the images is subdivided into the adequate number of subpixels.

The time consuming aspect

Considering a square area in which the rotation center has to be determined, the expression of the Total Processing Time (TPT) can be expressed by :

$$TPT = \left(\frac{SSL}{RA}\right)^2 \times (RIN) \times (SPT)$$

where SSL is the Length of the Square Side, RA is the Required Accuracy, RIN is the Radial Image Number and SPT is the Step Processing Time which corresponds to the time required to compute the energy at a unique point.

If this method allows obtaining results with an accuracy as much as one could wish, it is in return very time consuming. For example, about 30 minutes are needed to obtain a poor accuracy of 1mm for a 10x10 mm^2 square with 4 radial images.

RESULTS

This method has been applied to a stainless steel hexagonal object. Figures 2, 3, and 4 respectively represent a radial image of the object, a reconstruction by the *maximized average* method with 8 radial images according to an incorrect rotation center, the reconstruction by the *maximized average* method with 8 radial images according to the rotation center determined by the proposed method with a desired accuracy of 0.40 mm. Figure 5 represents the energy function of a part of the image, and the minimum corresponding to the rotation center at position (99,103).

Thanks to this method, we can now use the URTURIP Technique with our system, and Figure 6 represents the first in vivo image we obtained at arm level, by applying the *maximized average* method on 8 radial images.

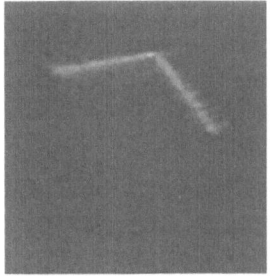 

Figure 2. Radial image.   **Figure 3.** Incorrect rotation center.   **Figure 4.** Good rotation center.

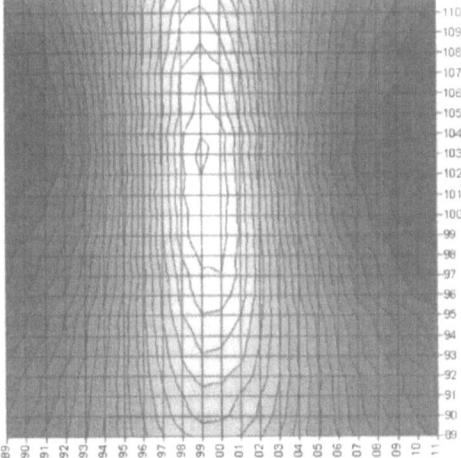

Figure 5. Energy representation for the hexagonal object.

CONCLUSION AND PERSPECTIVES

As a part of our development project of an ultrasound scanner based on the URTURIP Technique, a reference calibration method has been developed. It allows determining the subpixel coordinates of the rotation center which is required, to use the URTURIP Technique, with an accuracy as one could wish.

This method is in return very time-consuming. Thus, the perspective of this work is to develop faster (and may be less accurate) calibration methods, the results of which will be compared to this reference method.

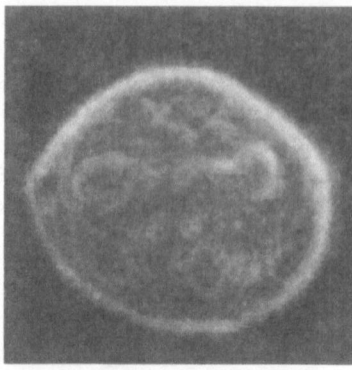

Figure 6. In-vivo arm slice obtained by the *maximized average* method.

REFERENCES

1. B. Migeon and P. Marché, Ultrasound tomography by radial image processing, *Innov. Tech. Biol. Med.,* vol. 13, n°3, pp. 292-304, (1992).
2. C. M. Sehgal, et al., Ultrasound transmission and reflection computerized tomography for imaging bones and adjoining soft tissues, *IEEE Ultrasonic Symp.* Chicago, IL, vol. 2, pp. 849-852, (1988).
3. H. Hiller, H. Hermert, System analysis of ultrasound reflection mode computerized tomography, *IEEE Trans. Sonics Ultrason.*, vol SU 31, pp. 240-250, (1984).
4. M. Friedrich , et al., Computerized ultrasound echo tomography of the breast, *Europ. J. Radiol.*, vol 2, pp. 78-87, (1982).
5. J. Ylitalo, et al., Ultrasonic reflection mode computed tomography through a skullbone, *IEEE Trans. Biomedical Engineering*, Vol. 37, N° 11, Nov. (1990).
6. B. Migeon, P. Vieyres, P. Marché, A simple solution for removing echo bars for URTURIP Technique, Acoustical Imaging, *Acoustical Imaging*, Vol. 22, pp. 543-548, (1995).
7. B. Migeon, V. Serfaty, M. Gorkani, P. Marché, An Adaptive Smoothing Filter for URTURIP Images Applying the Maximum Entropy Principle, *IEEE Engineering in Medicine and Biology*, pp. 762-765, Nov/Dec (1995).
8. B. Migeon, P. Vieyres, P. Marché, Interpolation of star-shaped contours for the creation of lists of voxels : application to 3D visualisation of long bones, *Int. J. of CADCAM and Computer Graphics*, Vol.9, n°4, pp. 579-587, (1994).
9. B. Migeon, P. Marché, Echographic Image Processing for Reconstructing Long Bones, *Proc. of 17th Annual International Conference IEEE-EMBS*, Montreal, (1995).

A METHOD TO IDENTIFY REVERBERATION ECHOES IN MULTILAYERED HOMOGENEOUS MEDIA

Wagner C. A. Pereira,[1] Marcelo A. Duarte,[1] and João C. Machado[1,2]

[1]Biomedical Engineering Program - COPPE/UFRJ
P. O. BOX 68510 - Rio de Janeiro - RJ - Brazil - 21945 - 970
E-mail: wagner@serv.peb.ufrj.br
[2]Currently at Sunnybrook Health Science Centre, Dept. of Medical Biophysics, University of Toronto, 2075 Bayview Av., North York, Ontario M4N 3M5, Canada

INTRODUCTION

In medicine, Ultrasound (US) is generally used for diagnostic purposes through Doppler and Ultrasonography exams. The former type of exam is mainly used, for example, to detect the causes of circulatory complications, such as stenosis. The latter type of exam, based on bidimensional images of biological tissues, provides informations like size and tissue structure of some organs. In this case, the diagnostic accuracy degrades if the image becomes corrupted by some artifact, that can be caused, for instance, by US reverberation - more than one reflection on the same interface (Fish, 1990). The possibility of identifying reverberating echoes may allow image corrections with a chance for improved diagnostic accuracy.

Moreover, in material characterization, it becomes necessary to detect the presence of reverberating echoes to avoid errors in parameter estimation. Examples are in Pereira et alii (1992) and Pereira et alii (1996), for the thickness and wave speed estimate of multilayered media; Coleman et alii (1985) plus Fei and Shung (1985), for US tissue characterization; Derouiche et alii (1995) for heterogeneous media characterization; and Kinra et alii (1994) plus Kinra and Iyer (1995), on the determination of acoustical parameters in multilayered media.

The present work proposes a method to identify reverberating echoes occurring on the US rf-signal reflected from a multilayered homogeneous and lossless media.

THEORY

The method proposed in this work to classify an echo as a reverberating one, or not, assumes the US wave propagating in a lossless medium composed of layers. It relies on the premises that the frequency power spectra of echoes traveled through the same layer are more alike than those that propagate on different layers. Besides this premise, as a reverberating echo follow the same path of a previous one, even though propagating more than once on some of the layers, its Time of flight (TOF) must be a linear combination of the preceding echoes TOF's.

The algorithm to classify an echo as a reverberating one, or not, is based on the following 4 steps:

Step #1: The echoes from the rf-signal are gated out and their frequency power spectra calculated by FFT. The similarity of the frequency power spectrum of two echoes is tested by means of the minimum value for a least square error function, that compares the amplitudes of the same frequency power spectrum components of each echo. This error function, $E(K)$, is defined as:

$$E(K) = \sum_{i=1}^{N} (P_{1[i]} - K \cdot P_{2[i]})^2 \tag{1}$$

with:

- N as the number of frequency power spectrum components;
- i as i^{th} frequency power spectrum component;
- $P_{1[i]}$ as the power of the ith frequency power spectrum component of echo number 1;
- $P_{2[i]}$ as the power of the ith frequency power spectrum component of echo number 2;
- K as a constant that minimizes $E(K)$.

The minimum of $E(K)$ occurs when:

$$K = \frac{\sum_{i=1}^{N} (P_{1[i]} \cdot P_{2[i]})}{\sum_{i=1}^{N} (P_{2[i]})^2} \tag{2}$$

Step #2: The mean (MV) of the minimum values of $E(K)$ for all possible echo pair combinations is determined.

Step #3: For each echo is searched a linear combination (LC), with integer coefficients, expressing its TOF as a function of the preceding echoes' TOF's. A difference, in magnitude, between the result of the linear combination and the TOF of the echo under analysis (so called "reverberation time candidate" - RT) is accepted if less than the transmitted US pulse width. If the linear combination exists, then the echo under analysis becomes a candidate for a reverberating echo (RV). If not, then the present echo becomes classified as a real one (RE).

Step #4: For each RV candidate, the minimum value of $E(K)$, between the present echo and those whose TOF are part of a linear combination, is compared with MV following the time arrival order of the echoes used on the linear expression. If at least one minimum value of $E(K)$ is less than MV, than the present echo becomes classified as RV and considered as reverberation of the echo on the linear combination that provided the minimum of E(K) less than MV. In the case that all values of E(K) are greater than or equal to MV, then the RV candidate becomes classified as RE.

Figure 1 illustrates the algorithm flow chart.

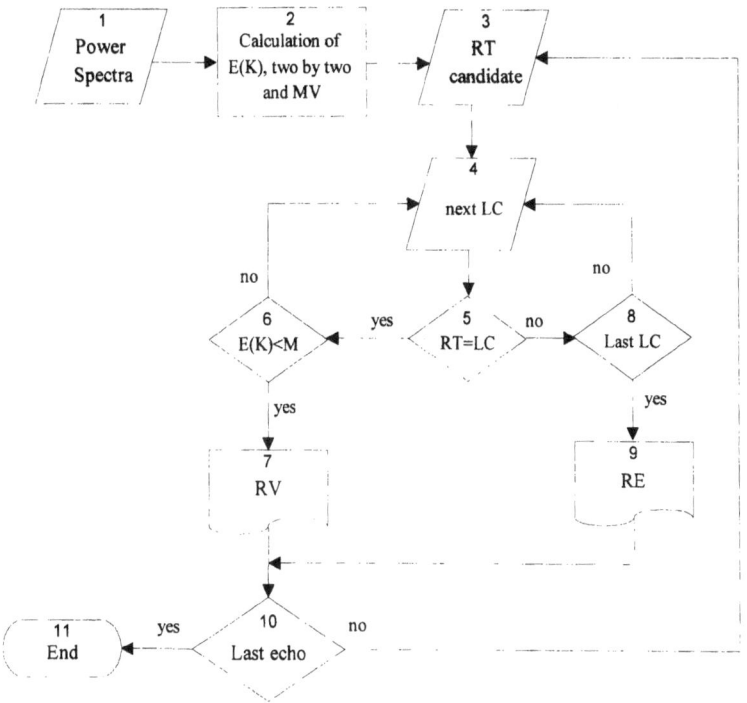

Figure 1. The algorithm flow chart.

EXPERIMENTAL PROCEDURE

In order to test the proposed method, some experiments were carried out in an acoustic tank, according to Figure 2.

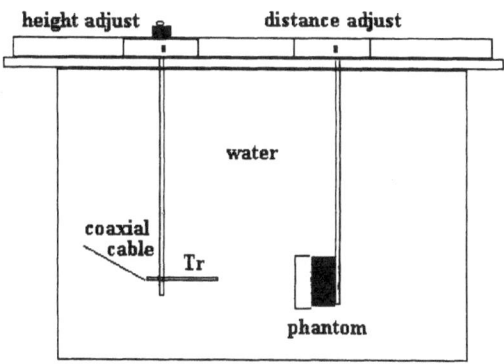

Figure 2. Water tank, transducer (Tr) and phantom.

A US transducer (Tr), 10 mm in diameter and resonating at 7.65 MHz was excited by a sine wave burst generator, TB-1000 - Gated Amplifier Toneburst Plug-in Card by Matec Instruments, Inc., coupled in an IBM PC compatible computer, with amplitude of 100 Vpp, that provided bursts with pulse repetition rate (PRF) of 1 kHz, and width between 0.6 and 1 µs. The transducer, used together with an epoxy spherical lens focused at 27.53 mm from its face and along its central axis, was positioned 35 to 40 mm distant from the phantom, according to the best signal to noise ratio (SNR) of the rf-echo signal returning from the layers that compose the phantom. The received signal passed through an rf amplifier, provided in the same card described above, and sampled on a digital oscilloscope, Tektronix - TDS 420, at a sampling frequency of 500 MHz. Then, the sampled rf echo signal stored on the oscilloscope was sent to the computer for the processing of the echoes.

Four different phantoms, with layers width varying between 4.4 and 60 mm were used:
- 1-layer phantom made with aluminum (Al) (width = 60 mm),
- 1-layer phantom made with glass (G) (width = 9 mm),
- 2-layers phantom made with alcohol and acrylic (AA), with the alcohol layer closer to the Tr (alcohol width = 6.5 mm, acrylic width = 6.3 mm),
- 2-layers phantom made with epoxy and acrylic (EA), with the epoxy layer closer to the Tr (epoxy width = 4.4 mm, acrylic width = 5.0 mm).

RESULTS

Eight experiments were made, performing a total of 80 echoes collected. Among all the experiments, one was conducted with the 1-layer aluminum phantom, five with the 1-layer glass phantom, one with the 2-layers alcohol-acrylic phantom and one with the 2-layers epoxy-acrylic phantom. Detailed results are presented for one of the 1-layer glass phantom experiments, while the remaining results for this and the other seven experiments will only be presented in terms of the number of echoes that were correctly classified.

During all the experiments conducted with the 1-layer glass phantom, an air layer existed behind the glass. The distance between Tr and phantom was 35 mm and Tr was excited with a burst width of 0.7 µs.

From the rf echo signal reflected on the two faces of the glass layer (Figure 3), 11 echoes were collected with sufficient SNR (SNR > 5) to be analyzed. The TOF for each one of these echoes is presented on Table 1.

Table 1. TOF of the echoes from one of the 1-layer glass phantom experiments.

Echo	Time (µs)	Echo	Time (µs)	Echo	Time (µs)	Echo	Time (µs)
1st	46.98	4th	56.90	7th	100.50	10th	143.62
2nd	50.18	5th	93.38	8th	103.62	11th	186.98
3rd	53.54	6th	96.74	9th	140.10	-	-

Figure 4 presents the frequency power spectra of echoes 3. In this experiment, MV was 0.92 power units. The echo classifications are shown in Table 2.

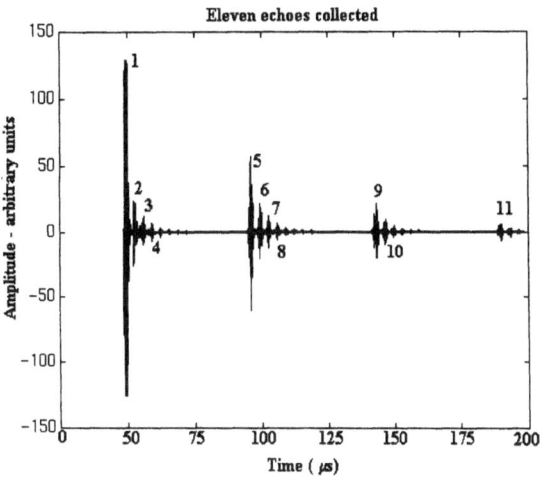

Figure 3. Echoes from the 1-layer glass phantom experiment.

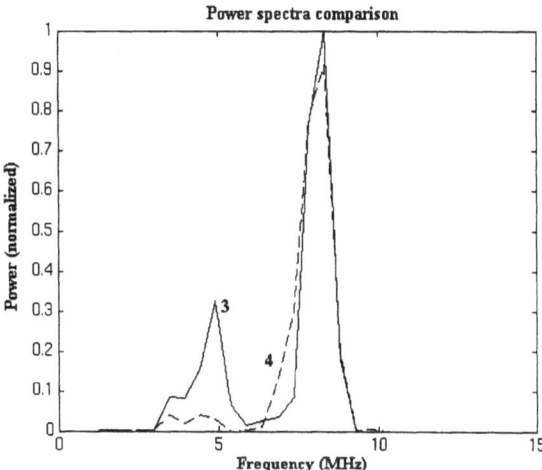

Figure 4. Frequency power spectra of echoes 3 and 4. E(K) = 0.17.

Table 2. Classification of the echoes for one of the 1-layer glass phantom. The last row relates the RV echo with the one from which it is a reverberation. MV = 0.92 power units.

Echo number	1	2	3	4	5	6	7	8	9	10	11
RV candidate ?	NO	NO	YES	YES	YES	YES	YES	YES	YES	YES	YES
E(K) < MV ?	-	-	1.50	0.17	0.68	0.81	0.39	0.51	0.91	0.28	0.72
Classification	RE	RE	RE	RV	RV	RV	RV	RV	RV	RV	RV
RV of echo number	-	-	-	3	1	3	3	3	5	6	9

The overall results obtained in the classification of the 80 echoes from all the 8 experiments are in Table 3.

Table 3. Number of correctly classified echoes for the 8 experiments conducted.

Phantom type	Al	AA	G1	G2	G3	G4	G5	EA	Total
Correctly classified	6	8	11	9	10	10	10	10	74
Incorrectly classified	1	1	0	0	1	1	1	1	6
Total	7	9	11	9	11	11	11	11	80

DISCUSSION

With respect to the experiment results presented in Table 2, for the 1-layer glass phantom, the method correctly identified the 1st and 2nd echoes. They are related to the reflections of the incident wave on the water-glass and glass-air interfaces, respectively. Also, the method correctly classified the echoes from number 4 up to 11 as RV. Only the 3rd echo was incorrectly classified as real. In fact, it is a reverberation of the 2nd one. In this experiment, 10 out of 11 echoes were correctly classified.

For the general classification, presented in Table 3, the method correctly classified 74 (92.5%) echoes out of 80.

The method has still some limitations, such as the inability to deal with a lossy medium, that are the concern for its improvement. Other approaches, like layer transfer function estimates, are also in progress.

ACKNOWLEDGMENTS

The authors would like to thank the Brazilian agencies CAPES, CNPq and FAPERJ for their support.

REFERENCES

COLEMAN, D. J., LIZZI, F. L., SILVERMAN, R. H., HELSON, L., TORPEY, J. H. e RONDEAU, M. J. (1985), "A Model for Acoustic Characterization of Intraocular Tumors", *Investigative Ophthalmology & Visual Science*, Volume 26, Number 4, pages 545-550.

DEROUICHE, Z., DELEBARRE, C., GAZALET, M., ROUVAEN, J. M. e BRIDOUX F. (1995), "Ultrasonic Characterization of Heterogeneous Materials Using a Stochastic Approach", *Journal of Acoustical Society of America*, Volume 97, Number 4, pages 2304-2315.

FEI, D. Y. e SHUNG, K. K.(1985), "Ultrasonic Backscatter from Mammalian Tissues", *Journal of Acoustical Society of America*, Volume 78, Number 3, pages 871-876.

FISH, P. (1990), *Physics and Instrumentation of Diagnostic Medical Ultrasound*, John Wiley & Sons, West Sussex, England.

KINRA, V. K., JAMINET, P. T., ZHU, C. e IYER, V. R. (1994), "Simultaneous Measurement of the Acoustical Properties of a Thin-Layered Medium: The Inverse Problem", *Journal of Acoustical Society of America*, Volume 95, Number 6, pages 3059-3074.

KINRA, V. K. e IYER, V. R. (1995), "Ultrasonic Measurement of the Thickness, Phase Velocity, Density or Attenuation of a Thin-Viscoelastic Plate. Part I: The Forward Problem", *Ultrasonics*, Volume 33, Number 2, pages 95-109.

PEREIRA, W. C. A., SIMPSON, D. A. and MACHADO, J. C. (1992), "An Estimator of Focal Position Based on Geometric Acoustics", *IEEE 1992 Ultrasonics Symposium*, October 20-23, Tucson, U. S. A., Proceedings, Volume 1, pages 323-325.

PEREIRA, W. C. A., GRECO, A. V. D. and MACHADO, J. C. (1996), "Ultrasonic Velocity Mapping of Multilayered Media", Proceedings of the 22nd International Symposium on Acoustical Imaging, Plenum Press, Volume 22, pages 63-68.

SEISMIC SIGNAL PROCESSING

OF ULTRASOUND IMAGING DATA

Thomas L. Szabo and Daniel R. Burns*

Hewlett Packard, Imaging Systems Division
3000 Minuteman Drive
Andover, MA

*Earth Resources Laboratory
MIT
77 Massachusetts Avenue
Cambridge, MA 02139

INTRODUCTION AND OBJECTIVES

Medical ultrasound images often contain spurious signals that interfere with the diagnostically relevant information. The severity of the imaging problems may vary widely and range from overall haze or clutter in the image to unwanted bright reflectors. Even though the physical causes of these artifacts may not be known, the overall problem can be posed as enhancing signal-to-noise. The challenge is that the unwanted signals, some of which may be external to the imaging plane, are spatially variant as well as being time dependent.

In our search to find solutions, we found a method applied in geophysics to similar problems, Velocity-Guided-Median- Filtering (VGMF) (Reiter et al, 1992). The reasons for using VGMF on ultrasonic imaging data are two-fold. First, the filtering process can enhance coherent events relative to incoherent events in the RF data. This enhancement results in an increase in the S/N level in the RF data and the resulting beamformed line.

Second, the filtering process can remove artifacts by means of 'velocity' or angle filtering. By defining the angles (or velocities) over which the alignment search is carried out, we can remove artifacts that are present in the RF data along certain alignments. The effect of this application of the filter is similar to a pie-slice (f-k) filter. By reducing clutter, increasing the S/N, and removing artifacts, the VGMF can improve the final ultrasonic image quality. An improved image may allow finer detail events to be identified and interpreted, and may also improve subsequent border detection applications. After an explanation of VGMF, application of VGMF to cardiac sector scan images will be described.

VELOCITY GUIDED MEDIAN FILTERING (VGMF) METHOD

Overview of the Method

Before considering the method itself, we set the stage for how the data is received for processing. Consider an array of N receivers which are elements in a linear or phased array. Pulse echo signals arriving from various directions are received by each element as a continuous time signal of a certain length determined by the frame rate selected. If we plot the time trace of each element (it resembles an A-Line signal) vertically under the spatial location of its receiving element, the resulting rf-data set presentation is like that of Figure 1. This data presentation is the space-time (x-t) format usually employed in seismic data processing. Within the data set we can see several locations where for several traces the echos are coherent, indicating the presence of a target.

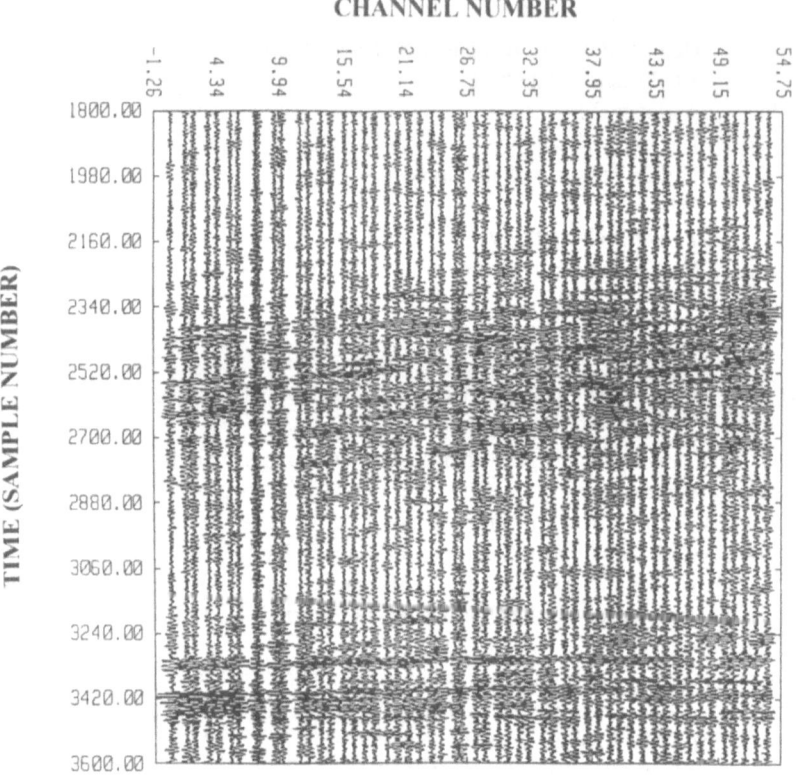

Figure 1. Raw RF data from line 72 (+8.25 degree steer direction) of the AP1 data set . The data are shown for 54 traces of a 64 element transducer (traces 6-59) and for the time sample range of 1800 - 3600 or 90 microseconds. The horizontal events between time samples 3200-3500 correspond to specular tissue reflections, while the events between about 2200-2700 correspond to a lung artifact. Note also the level of incoherent noise in the background. The data have been normalized to the maximum value in the plot (maximum = 65).

The velocity guided median filter (VGMF) acts as a velocity filter as well as a method of improving signal to noise ratio by selectively passing coherent signals. In this case, "velocity" is taken to mean an apparent velocity defined by the slope of a line cutting across several traces, $\Delta x/\Delta t$. Each slanted straight line in the top of Fig. 2. corresponds to a different apparent velocity. The filtering operation is composed of two steps: searching and identifying the dip or apparent velocity of an arrival across an array of receivers, and replacing the actual data point on which the filter is centered by the median value of the traces across the array along the identified slowness alignment.

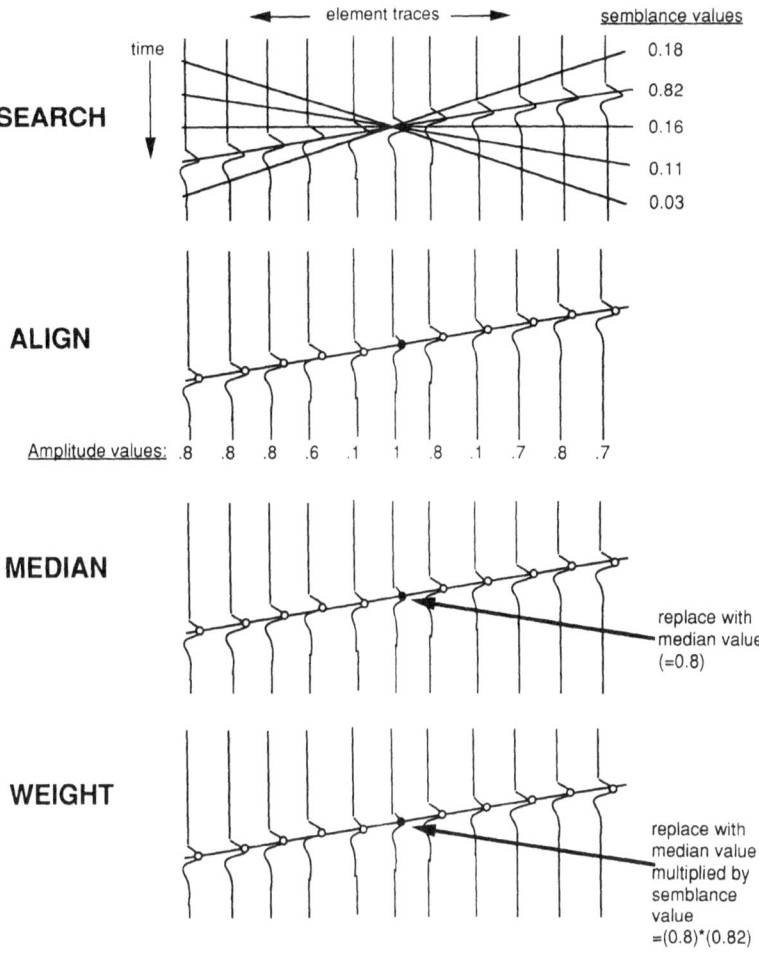

Figure 2. Schematic diagram illustrating the steps involved in velocity guided median filtering (VGMF). At each RF data sample point on which the filter is centered, semblance is calculated along a number of search angles in order to find the optimum alignment direction. A one dimensional median filter is then applied along this optimum alignment direction. This median value can also be weighted by the semblance value itself.

Processing Flow

Event Alignment The first step of the VGMF process is event alignment. This step consists of two parts: (1) the search for an optimum alignment direction which involves the computation of the semblance statistic for a number of linear trajectories (given by slope values or angles in the space-time domain of the RF data), and (2) the selection of the optimum alignment direction which corresponds to the direction with the maximum semblance value. Semblance is the normalized ratio of the energy across the array (in a given time window and along a given alignment direction) to the energy in each trace, and is given by the formula:

$$semblance = \frac{(\sum A_i)^2}{N \sum A_i^2}$$

where: N = number of traces in the array
                A_i = amplitude value for the appropriate time sample of trace i.

The semblance values range between 0 and 1. The slopes over which the alignment search is conducted are selected by the user, as is the number of search directions within the specified slope range. Figure 2 is a schematic diagram showing the search and alignment process. Input to the filter is a local set of 'Nr' traces centered about the trace to be filtered, where 'Nr' is an odd number. Choice of the filter aperture or width 'Nr' depends on the spatial and temporal sampling of the input data and the specific data characteristics to be filtered. Larger filter widths generally provide increased angle or slope resolution but result in greater averaging of the data along the spatial (element) dimension. In the top of Figure 2, the search is carried out over five angles for eleven traces. The maximum computed semblance value defines the optimum alignment direction, shown as the second part of Fig. 2. labeled "align".

Median Filter After the optimum alignment direction (Fig. 2) has been identified, a one-dimensional median filter is applied to the data samples of each trace in the filter aperture. The median value can also be weighted by the semblance value which was computed for the optimum alignment direction. This semblance weighting step can further suppress events which are only weakly correlated.

Time Sequencing Data samples along the optimum alignment trajectory are input into a standard one-dimensional median filter which returns the statistical median of the input. As a result, coherent signals with this alignment are enhanced, while those with different alignments or those which are coherent over less than half of the filter aperture are suppressed. The filter is then moved to the next time sample of the same trace and the semblance values and median filter are computed once again. After filtering a single trace, the local trace gate is centered about the next trace to be filtered. In this way each time sample of each trace of the RF data is subjected to the VGMF process.

Tradeoffs The major problem is the possibility of smearing of arrivals in the data set. Median filtering in general is excellent at maintaining and enhancing 'edges' in images, and in fact this is a main application for median filters in image processing. If, however, the data has breaks in the coherency of an event and those breaks are less than half a filter aperture in width, the VGMF will smear the arrival through that zone. Depending on the type of features which must be imaged, and the nature of the data sets, it is important to optimize the filter aperture and be aware of the implications of the filter aperture parameter.

IMAGING COMPUTATIONS

The major steps in computation were the following:
1. Data acquistion: for each line in a frame a data set consisting of n traces was acquired and put into digital memory. A portion of typical data set for a beamformed line in an image is shown in Fig. 1. For cardiac applications, about 120 lines were taken per frame.
2. VGMF filtering was applied to each line dataset.

3. Beamforming and filtering: This involved fairly standard steps including interpolation, dynamic receive beamforming (delay and sum) and Hilbert transforming.
4. Image Processing: A standard log compression was applied to the resulting line (analytic) envelope data along with scan conversion (Leavitt, 1983).

Cardiac RF data sets were obtained from a 2.5 MHz 64 element sector array. Figure 1 is the raw RF data from an apical 4 chamber image. The arrivals between samples 2350 and 2700 (approximately) are interpreted to be a lung artifact, while the events between samples 3250 and 3500 are specular reflections. Figure 3 shows the same data after application of the VGMF. Much of the incoherent energy has been removed and the signal to noise ratio has been improved. In Figure 3 much of the spatial variability of the coherent events has been maintained, which was not the case when a larger aperture Nr was used in filtering. Longer filter apertures (Nr) produced smearing or averaging effects in the final image.

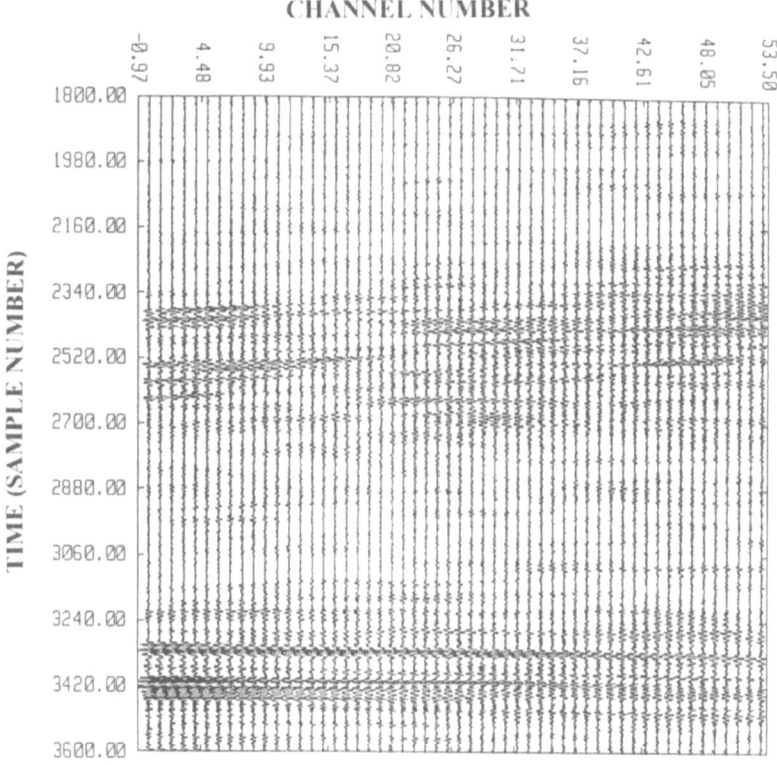

Figure 3. A plot of the same RF data as in Figure 1 after VGMF processing with a filter aperture of 11 traces and semblance weighting. Note the improved signal to noise ratio and the enhancement of events which are coherent and essentially horizontal in alignment. Note also that the spatial variability of the coherent events has been preserved with this short filter aperture. The data have been normalized to the maximum value in the plot (maximum = 30).

IMAGE TESTS

Figure 4 shows an apical four chamber view from a different data set with moving clutter in the ventricles before and after VGMF processing. VGMF processing results in a 'cleaner' image, that is an image with higher S/N and less clutter, especially in the left ventricle. The original, compressed to 50dB, is on the left; the 55dB compressed and filtered image is on the right. The VGMF-processed image, when viewed on a computer screen, reveals a mitral valve hidden in clutter in the original image. Also, the chest wall area of this image is significantly improved (i.e. - more structural definition and less clutter).

In some cases it appears that the VGMF image is similar in appearance to a standard processing result with a smaller compression value, that is the two results have the same S/N appearance. In other words, if the original image was compressed to 45 dB, the S/N is similar to the image on the right but significant tissue structure is lost. The VGMF result retains more tissue structure than the re-compressed standard image. For the apical four chamber view data set of Figs. 1 and 3, VGMF reduces the lung artifact but does not remove it, even with semblance weighting.

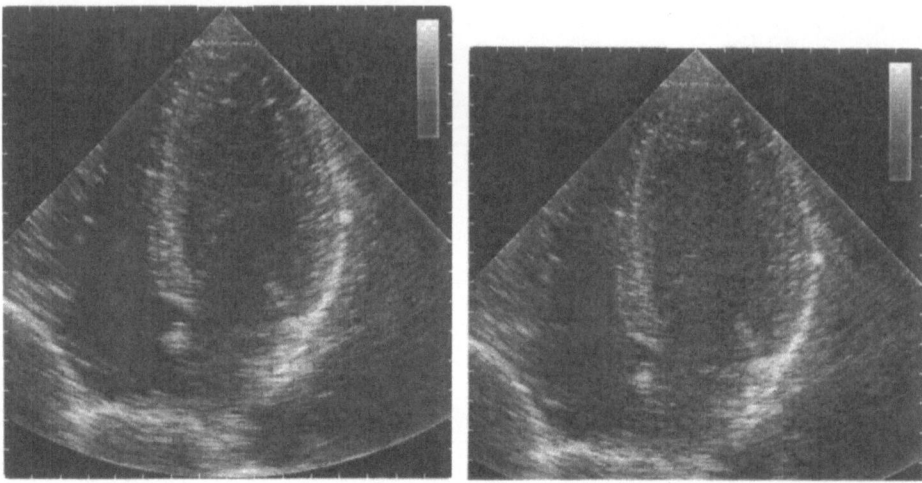

Figure 4. A single frame image which shows a reasonably good apical 4 chamber view with moving clutter in the ventricles (data set AP2). The original, compressed to 50dB, is on the left; the 55dB compressed and filtered image is on the right.

CONCLUSIONS

The general conclusions we have reached from studying processed images is that the VGMF method improves image quality by reducing incoherent clutter, providing some suppression of but not elimination of lung artifacts (and presumably other out of plane events), and clearly improving the shallow depth image quality. These effects can have important consequences: improved shallow area image quality may make identification of thrombi more feasible, suppression of artifacts will aid overall image interpretation, and clutter reduction will improve image interpretation by enhancing fine scale structure and providing improved tissue/blood contrast for border detection algorithms.

REFERENCES

E. C. Reiter, M. N. Toksoz and G. M. Purdy, 1993, A semblance guided median filter, *Geophysical Prospecting* 41:15.
S. C. Leavitt, B. F. Hunt and H. G. Larsen, 1983, A scan conversion algorithm for displaying ultrasound images, *Hewlett-Packard Journal* 34:30.

DETERMINATION OF VELOCITY AND ATTENUATION

USING BROADBAND PULSE TECHNIQUE

Junru Wu

Department of Physics
University of Vermont
Burlington, VT 05405

INTRODUCTION

Velocity and attenuation of ultrasound are the two most important properties of materials used in ultrasonic applications. Many techniques (Sachse and Pao, 1978; Kline, 1984; Bamber, 1986; Zeqiri, 1988) have been developed for measuring velocity and attenuation of longitudinal waves at frequencies ranging from 1 to 10 MHz. Among them, two methods are popular. One is the so-called discrete frequency method; the other is the broadband pulse technique or the ultrasonic spectroscopy method.

The discrete frequency method uses a toneburst, whose duration is chosen long enough so that its center frequency is well-defined, yet short enough to avoid the interference due to multiple reflections within the sample. The detail of this method can be found in our previous publication (Wu, 1996a). Strictly speaking, this method measures the group velocity at the center frequency, as it uses a toneburst. Since most materials used in frequency range between 1 - 10 MHz have very small dispersion, therefore the phase velocity is approximately equal to the group velocity.

In order to determine the phase velocity and attenuation coefficient vs. frequency, several transducers whose center frequencies cover the frequency range of interest are used. As the "wave-shape" changes significantly after it passes a material with high attenuation, it is almost impossible to unambiguously identify the equivalent points in the signals without the sample and with the sample. Therefore, this method is detrimental for materials with high attenuation.

The broadband pulse technique was developed by Sachse and Pao (1978). The phase velocity c_l and attenuation coefficient α_l can be obtained by (Szabo, 1993) comparing the spectra of Fourier Transforms of the signals received by the receiver with and without the sample, i. e.,

$$c_l = \frac{c_w}{1 + c_w(\phi_s - \phi_w \pm 2m\pi)/(2\pi f d)},$$ (1)

$$\alpha_l = \alpha_w + \ln(T_l A_w / A_s)/d,$$

where c_w and α_w are, respectively, the speed of sound and the attenuation coefficient of water, f is frequency, T_l is the transmission coefficient of the sample, i. e., $T_l = 4z_w z_s /(z_w + z_s)^2$, z_w and z_s are the acoustic characteristic impedances of the liquid and the sample respectively, A_w, ϕ_w, A_s, ϕ_s are the amplitude and phase spectra of the Fourier Transform of the case without the sample and with the sample respectively. The term $\pm 2m\pi$ in eq. (1) is to account for the ambiguity of the phase spectrum calculated from the arctangent function, where m is an integer. This technique measures c_l vs. f, and α_l vs. f; it takes one measurement to cover the whole continuous frequency range of interest. Those two transducers have to be broadband transducers; their 6 dB bandwidth should be wider than, or at least equal to, the frequency range of interest. Especially for a sample with large attenuation, for the reason mentioned above, it is superior to the discrete frequency method.

Velocity and attenuation of a shear wave is relatively more difficult to measure compared with its longitudinal wave counterparts. Currently, the discrete frequency method has been used. To generate a shear wave, a shear wave transducer is usually needed. This paper reports a technique developed by the author which measures the phase velocity and attenuation coefficient of shear waves using the broadband technique. This method essentially is an extension of the technique developed by Sachse and Pao. Therefore, it has inherited the advantages of the broadband technique. Furthermore, shear waves in a sample are generated by the mode conversion method and shear wave transducers are not required.

THEORETICAL BACKGROUND

If a sound wave impinges upon a sample from water or other liquids with an angle of incidence θ_i, in addition to the reflected wave in water, in general, a refracted longitudinal wave and shear wave occur in the sample. If $\theta_i > \theta_{cr}$, where θ_{cr} is the critical angle corresponding to the longitudinal wave of the sample material, only a shear wave can propagate through the sample as shown in Fig. 1; θ_t is the angle of refraction for the shear wave. Applying simple trigonometry to the situation of Fig.1, it can be easily shown that the phase velocity c_t and attenuation coefficient α_t can be obtained by

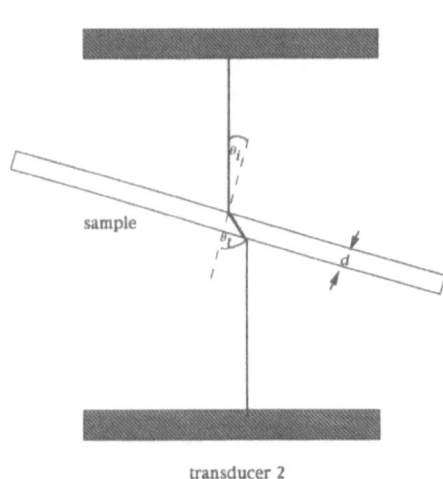

Figure 1. Geometric relation between the transducers and sample.

$$c_t = \frac{c_w}{\sqrt{\sin^2 \theta_i + [\frac{(\phi_s - \phi_w \pm 2m\pi)c_w}{2\pi f d} + \cos \theta_i]^2}},$$ (2)

$$\alpha_t = \alpha_w \cos(\theta_t - \theta_i) + \sqrt{1 - (c_t \sin \theta_i / c_w)^2} \ln(T_t A_w / A_s)/d.$$

Here all the symbols are defined earlier. The amplitude transmission coefficient of a sample for the shear wave is obtained by multiplying those of the two boundaries of the sample given by Brekhovskikh (1980), that is $T_t = |T_1 T_2|$, where

$$T_1 = \frac{-2\rho_w z_{ln}}{\rho_s} \frac{\sin(2\theta_t)}{z_{ln}\cos^2(2\theta_t) + z_{tn}\sin^2(2\theta_t) + z_{wn}},$$

$$T_2 = \frac{\tan\theta_i}{2\sin^2\theta_t}(1 - \frac{z_{ln}\cos^2(2\theta_t) - z_{tn}\sin^2(2\theta_t) + z_{wn}}{z_{ln}\cos^2(2\theta_t) + z_{tn}\sin^2(2\theta_t) + z_{wn}}),\qquad(3)$$

$$z_{wn} = \frac{z_w}{\cos\theta_i}, z_{ln} = \frac{z_l}{\cos\theta_l}, z_{tn} = \frac{z_t}{\cos\theta_t}, z_t = \rho_s c_t.$$

The three angles are related by Snell's Law, that is $\dfrac{\sin\theta_i}{c_w} = \dfrac{\sin\theta_l}{c_l} = \dfrac{\sin\theta_t}{c_t}$.

EXPERIMENTAL SETUP AND RESULTS

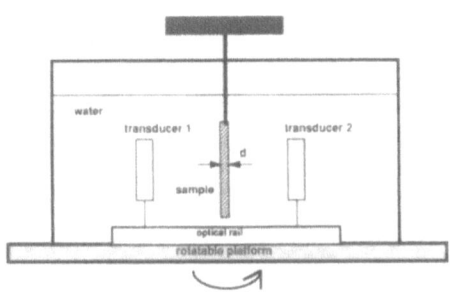

Figure 2. The experimental setup.

A typical experimental setup is very similar to the situation shown in Fig. 1. The above-mentioned broadband transducers are pvdf unfocused transducers (Ultrasonic Sciences Limited, Fleet, Hampshire, England GU13 9RL) whose radii are all 1.22 cm. They were mounted on an optical rail and than immersed in a water tank (26x30x50cm); they can be rotated with the tank by a rotatable platform. A sample, whose lateral dimension/dimensions is/are much larger than -6 dB beamwidth of the ultrasound, was inserted coaxially between the two broadband transducers with its center located at the center of the rotation (Fig. 2). When the platform was rotated, the sample was held stationary and the angle of incidence can be changed with a resolution of 0.1°. Special care was taken in alignment of the transducers and samples; they were initially installed at a normal incidence situation and their surfaces were parallel determined by a vertical laser beam. The sample was placed at 10 cm from transducer 1. The separation of the two transducers was kept to be 24 cm so that the transducer 2 was approximately located at the last axial maximum (the boundary between near-field and far-field) of the sound field in water generated by transducer 1. The pressure amplitude and -6 dB beamwidth at this location were determined to be 22 kPa and 8 mm (that is much smaller than the diameter of the receiver transducer 2). The above-mentioned parameters of the sound field generated by transducer 1 were determined by applying a calibrated pvdf needle hydrophone whose probe diameter is 1 mm.

Figure 3 is a block diagram of the electronic instruments used in the measurement. The synthesizer (HP3325A) and Matec MBS-8000 system generate a narrow electronic pulse (<0.1 μs). This pulse drives transducer 1 (broadband transducer). The -6 dB bandwidth of the sound pulse received by transducer 2 is between 2.2 - 7.6 MHz. The output of transducer 2 is then sent to LeCroy 9310 Digital (12 bits) Oscilloscope for digitizing and stored in a PC. Fourier transformation and other data analysis are done by using Mathematica®, a software package (Wolfram Reseach, Inc., Champaign, Illinois).

Measurements on high density polyethylene and low density polyethylene samples were performed. The samples are all circular discs of diameter 3.8 cm prepared by Hewlett-Packard Co. The samples of various thickness from 2mm to 6 mm were used. The thickness variations across the surfaces of the samples are all smaller than 0.01 mm. The density of samples varies from sample to sample; densities of high density polyethylene and low density polyethylene samples tested are in the range from 0.9155 to 0.9463 and 0.8964 to g/cm^3 respectively.

Figure 4-5 are plots of phase velocities and attenuation coefficients vs. frequency (2.2 - 7.6 MHz) of high density polyethylene (r_s=0.9155 g/cm^3), and low density polyethylene (ρ_s=0.9000 g/cm^3) respectively.

The results of phase velocity and attenuation of longitudinal wave for high density polyethylene are very close to the results obtained by Kline (1984). Since density of the high density polyethylene sample used by Kline was 0.9517 kg/m^3, higher than ours, a perfect agreement was not expected. It was also difficult to perform a careful comparison since there were no individual data provided by Kline (1984) and the scale of the plot provided by him was rather large. However, in general it can be estimated that agreement in phase velocity is better than 3 % and agreement in attenuation coefficient is better than 5 %. Our attenuation coefficient results for this sample in general are slightly higher and phase velocity results are lower than the results of Kline's. It is consistent with our experience with the high density polyethylene samples used since density of our sample is lower.

Comparison for data (attenuation coefficients in particular) of shear waves is not possible since no data in literature are available. The results of both materials exhibit a slight increase in velocity (longitudinal or shear waves) with increasing frequency except that longitudinal velocity has a small decrease at about 2.2 MHz for high density polyethylene. It has been known that many viscoelastic materials such as those samples have this kind characteristic. It should be noted that, in general, results at the extremes of the bandwidth of the transducers were less reliable than results made at the center frequency of the transducers due to the low signal-to-noise ratio for these spectral component at extremes.

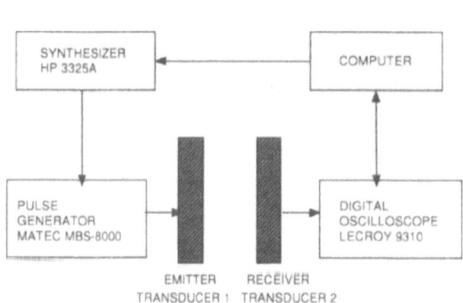

Figure 3. A block diagram of electronics.

Another characteristic of viscoelastic material, i. e., the attenuation coefficient is close to a linear function of frequency, has been demonstrated by those plots as well. They also show that the attenuation coefficient of a shear wave is much greater than its longitudinal counter part for all two samples tested. This is the third characteristic of many viscoelastic materials.

The measurements of longitudinal waves were done using the incident angle $\theta_i = 0$; while those of shear waves were performed letting $\theta_i > \theta_{cr}$.

SUMMARY AND DISCUSSION

It has been shown by this work that the broadband pulse technique can be used in determining phase velocity and attenuation of materials for shear waves in the megahertz

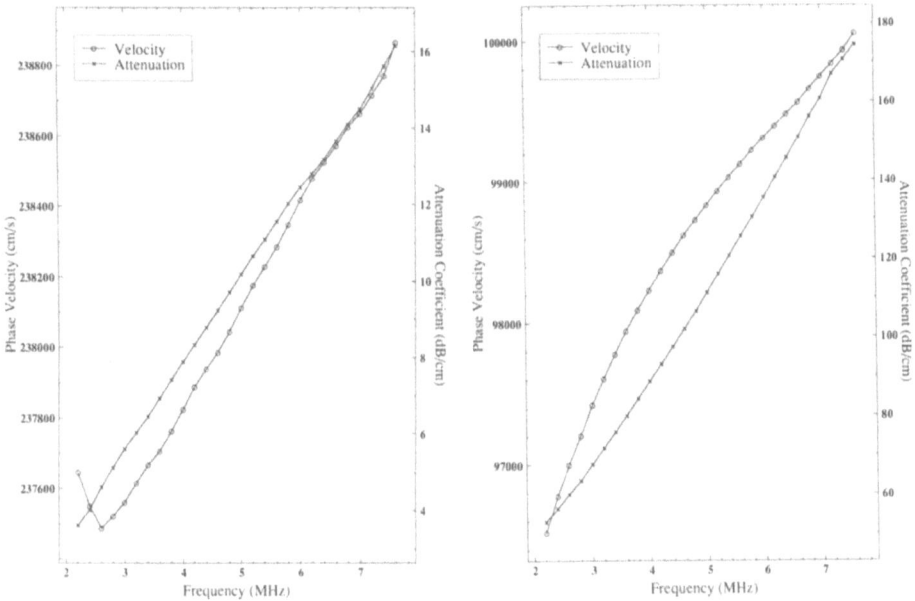

Figure 4. Phase velocity and attenuation coefficient versus frequency for a high-density polyethylene sample.

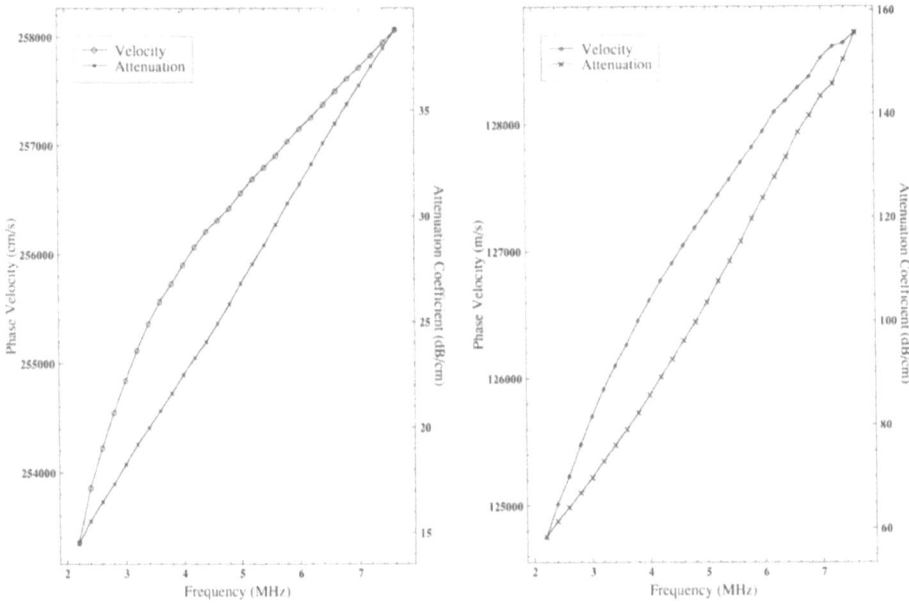

Figure 5. Phase velocity and attenuation coefficient versus frequency for a low-density polyethylene sample.

frequencies. In our view, the broadband technique is especially useful to determine the dispersion (phase velocity and attenuation) for "lossy" (with high attenuation) materials. As pointed earlier, for these material the discrete frequency method often cause unacceptable errors. Diffraction loss is usually a source of error (Szabo, 1993) especially in attenuation measurements. However, errors introduced in our measurements are usually negligibly small. It can be shown as follows. Let us assume that the diffraction losses without and with the sample are DL_w and DL_s, respectively. They are given by (Rogers & Van Buren, 1974; Szabo, 1993)

$$DL_w = 1 - Exp[-(2\pi/s_w)i][J_0(2\pi/s_w) + iJ_1(2\pi/s_w)],$$
$$DL_s = 1 - Exp[-(2\pi/s_s)i][J_0(2\pi/s_s) + iJ_1(2\pi/s_s)], \qquad (4)$$

where $s_w = z\lambda/a^2$, $s_s = s_w\eta$, λ is the wavelength of sound waves in water, $\eta = 1 + \dfrac{d}{z}(\dfrac{c_s}{c_w} - 1)$, z is the distance between the two transducers, a is the radius of the transducers, and c_s is the phase velocity of the relevant waves of the sample. The attenuation coefficient in eq. 2 should be modified as

$$\alpha_t = \alpha_w \cos(\theta_t - \theta_i) + \sqrt{1 - (c_t \sin\theta_i/c_w)^2} \; \ln[(T_t A_w/A_s)|DL_s/DL_w|]/d. \qquad (5)$$

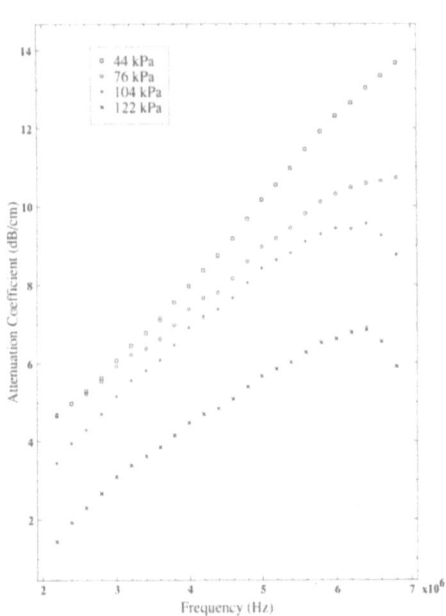

Figure 6. Attenuation coefficient versus frequency for a high-density polyethylene sample. The legends 1,2,3, and 4 correspond to 44, 76, 104, and 122 kPa for peak-peak pressure amplitudes of incident pulses, respectively.

For example, for longitudinal waves of high density polyethylene $c_s = c_f \approx 2,400$m/s, $c_w \approx 1,500$m/s, $d = 2$ mm and $z = 24$ cm, thus $\eta = 1.005$; it is very close to unity. Therefore, $|DL_s/DL_w|$ is also very close to unity; the correction introduced by $|DL_s/DL_w|$ or the difference of the results calculated using eqs. (6) and (7) is less than 0.5%.

Another important source of error is related to nonlinear distortion. This technique is based on the linear wave theory; any nonlinear distortion in wave propagation through water would cause errors. It was found (Wu, 1996b) that attenuation coefficient strongly depends on the pressure amplitude of the incident broadband pulse as shown in Fig. 6. Unaccepted errors may be generated by the nonlinear interaction between the different frequency components of the broadband pulse. Therefore, the acoustic pressure amplitude of the ultrasonic pulse should be kept low in measurements to minimize the errors caused by nonlinear distortion. On the other hand, if the acoustic pressure amplitude is too low, a reasonable signal-to-noise ratio can not be achieved. As a compromise, throughout the experiment, the pressure amplitude was kept 22 kPa. The shock parameter s (Carstensen and Muir, 1989) is used to estimate nonlinear distortion and it is defined as $\sigma = 6.5 \times 10^{-8}$ $p_0 f_c z_s$, when water is the medium, (where p_0 is acoustic pressure amplitude, f_c is the center of the ultrasonic pulse and z_s is the distance from the source to the sample. Generally speaking, if $\sigma < 0.1$, errors caused by nonlinear effects should be very low (Szabo, 1993). At the pressure amplitude used, it is 0.07 if the center freqency 4.8 MHz and $z_s = 10$ cm are used.

REFERENCES

Bamber, J. C., 1986, Attenuation and absorption, in: *Physical Principles of Medical Ultrasonics,* C. R. Hill, ed., John Wiley and Sons, Chichester.

Brekhovskikh, L. M., 1980, *Waves in Layered Media*, 2nd Edition , Academic Press, San Diego.

Carstensen, E. L.; Muir, T. G., 1986, The role of nonlinear acoustics in biomedical ultrasound, in: Tissue Charaterization with Ultrasound, J. F. Greenleaf , ed., CRC, Boca Raton, FL, 1986), Vol. I.

Kline, R. A., 1984, Measurement of attenuation and dispersion using an ultrasonic spectroscopy technique, J. Acoust. Soc. Am. 76: 498.

Rogers, P. H.; Van Buren A. L., 1974, An exact expression for the Lommel diffraction correction integral, J. Acoust. Soc. Am. 55, 724 .

Sachse, W.; Pao, Y. H., 1978, On the determination of phase and group velocities of dispersive waves in solids, J. Appl. Phys. 49: 4320.

Szabo, T. L., 1993, Linear and Nonlinear Acoustic Propagation in Lossy Media, Ph. D thesis, University of Bath, U. K.

Wu, J., 1996a, Determination of velocity and attenuation of shear waves using ultrasonic spectroscopy, J. Acoust. Soc. Am. 99:2871.

Wu, J., 1996b, Effects of nonlinear interaction on measurements of frequency-dependent attenuation coefficients, J. Acoust. Soc. Am. 99:3380.

Zeqiri, B., 1988, An intercomparison of discrete-frequency and broadband techniques for the determination of ultrasonic attenuation, in: *Physics in Medical Ultrasound*, edited by D. H. Evans and K. Martin, ed, IPSM, London.

BONE ELASTOMETRIC IMAGING USING ULTRASOUND CRITICAL-ANGLE

REFLECTOMETRY (UCR)

Peter Antich, Shreefal Mehta, Maithili Daphtary, Billy Smith, Victor Vaguine[1]

Department of Radiology, University of Texas Southwestern Medical Center
5323 Harry Hines Blvd., Dallas, TX 75235

[1]Vitel, Inc., 9786 Skillman St., Dallas, TX 75243

The ability of bone to withstand fracture, or strength, can be affected in normal subjects by exercise, by the environment e.g. in space flight or by the onset of disease, e.g. osteoporosis. The established method for assessing changes in bone properties is gamma- or X-ray densitometry, which measures the quantity of bone mineral.

Bone strength, however, is also dependent on microstructure, orientation of the applied forces, composition, and other physicochemical characteristics of the bone matrix. Interest in the ultrasound measurement of mechanical properties of bone is justified by the fact that not only is there a deterministic relation between the ultrasound velocity V, the density ρ and the elasticity E ($E = \rho \bullet V^2$) but there also is, in general, a strong positive correlation between elasticity and mechanical strength of bone.[1-5]

The measurement of elasticity and anisotropy through ultrasound velocity and density is routinely used in mechanical engineering and material science, and has been tested in bone studies by several researchers.[6-16] Bone material is anisotropic but possesses hexagonal symmetry: to fully characterize it, five independent elasticity coefficients must be measured. Ultrasound has been the technique of choice for determining the five coefficients in isolated bone samples. Ultrasound pressure and shear wave velocities are deterministically related to the elasticity components, which can be precisely measured, nondestructively and noninvasively. To the extent that bone material has not been fundamentally altered, the elasticity so derived is a predictor of the breaking limit or strength measured by destructive mechanical testing.

Of the various methods used to measure ultrasound velocity in bone[17], the UCR (Ultrasound Critical-angle Reflectometry) method has been shown to have several advantages and benefits not available by the other techniques [18,19]. UCR is a method which measures the velocity by detecting the angle for total internal reflection from bone into the medium in which the ultrasound beam is emitted. This technique is described in detail elsewhere[12,18] and here we summarize only points salient to this study.

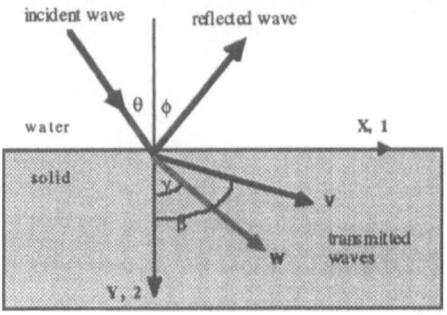

Figure 1: The UCR phenomenon is sketched in a ray diagram of the reflection and refraction at the interface with a solid, of a beam emitted in the calibration medium, typically water.

As is shown in *Figure 1*, a pressure wave (velocity c) propagated through water or tissue, when incident on bone (at angle θ to the surface normal), gives rise to a reflected wave and two waves propagating in bone, a pressure wave with velocity v_p and a shear wave with velocity v_s. As the angle of incidence increases, first the pressure wave and then the shear wave undergo total internal reflection at two critical angles of incidence (θ_p, θ_s). At these two angles, Snell's law is used to obtain the pressure and shear wave velocities from the critical angle and the velocity of sound in water c, $v_i=c/\sin\theta_i$[19]. The velocity of sound in water is accurately known and thus the measurement of critical angle of incidence in water gives by itself the velocity in bone. Following the laws of reflection these critical angles are obtained using clearly recognized features (maxima and discontinuities) in the amplitude and phase spectra in cortical and cancellous bone [12,18,20].

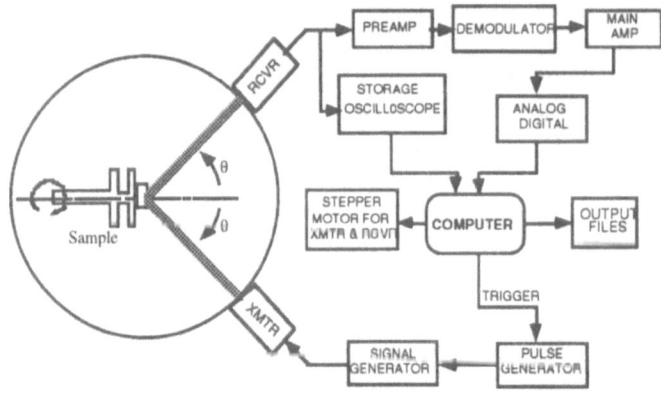

Figure 2: Simplified diagram of the experimental setup for the UCR technique. An ultrasound beam is reflected off the sample at the center and digitized for calculating amplitude and phase, or the demodulated waveform is sampled to give a single value of amplitude, at any given angle of incidence.

The UCR measurement is non-invasive and non-destructive and has been implemented at the University of Texas Southwestern Medical Center at Dallas in collaboration with VITEL, Inc, of Dallas, Texas, at frequencies ranging from 0.5 to 5 MHz for application both in vivo in the ulna and in vitro in bone biopsies [20,21,22].

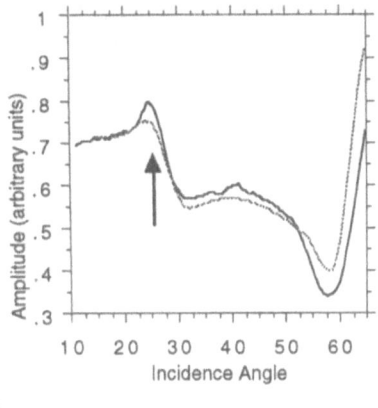
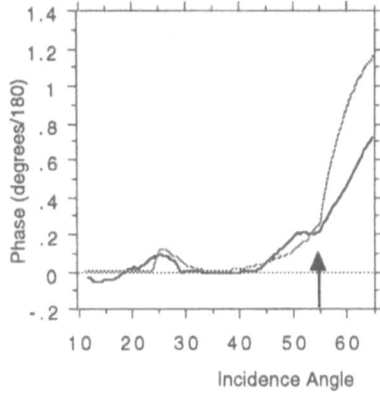

a.   b.

Figure 3 a)Amplitude and b)Phase spectra: experimental (dark line) and simulated (dashed line) for cortical bone Arrows indicate critical angles (**left** graph - pressure and **right** - shear wave critical angles).

The reflected amplitude and phase, for a sample of cortical bone, have a characteristic signature: starting at small angles, the experimental amplitude shows a prominent peak followed by two successive dips and an ultimate rapid increase at large angles. The phase shows a minor signal after the peak in amplitude, but a pronounced hump-and-ascent feature at large angles. Wave reflection theory predicts this behavior rather closely and uniquely identifies the pressure wave critical angle as the first peak, and the shear wave critical angle at the angle at which the phase spectrum has a discontinuity in its first derivative, as shown in *Figures 3a,b*.

Initial studies have demonstrated that UCR can measure the five coefficients of elasticity of a transversally symmetric material from a single surface using only pressure wave transducers[12,18]. To this end, the plane of measurement, identified by the incident beam and the normal (shaded grey in *Figure 4*), is rotated around the normal to the point of measurement and the pressure and shear velocities are obtained at each orientation.

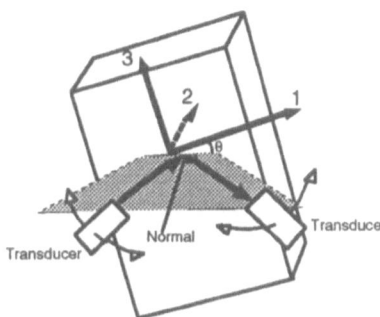

Figure 4. Schematic showing the measurement of anisotropy of pressure and shear velocities in a material. The plane of measurement, shown shaded grey, is defined by the incident beam and the normal to the interface. Rotating this plane, either by rotating the sample or the transducer assembly around the normal, the velocities can be measured at any selected orientation.

This plane can be rotated around the normal to the surface by an angle which defines its orientation. The velocity measured from the first critical angle shows a linear-quadratic dependence upon the square of the cosine of the orientation *(Figure 5)*. That dependence is characteristic of hexagonal symmetry, and allows the unambiguous determination of two elements of the matrix and of a two-term combination from the formula

$$\rho \cdot V_p^2(\theta) = Cos^4(\theta) \cdot c_{33} + Sin^4(\theta) \cdot c_{11} + Cos^2(\theta) \cdot Sin^2(\theta) \cdot (c_{13} + 2 \cdot c_{44})$$

which shows that the first critical angle corresponds to a pure pressure wave velocity along the principal directions of symmetry, or to a quasi-pressure wave at all other orientations.

The change in velocity with orientation is shown in *Figure 5* for a representative sample of bovine cortical bone.

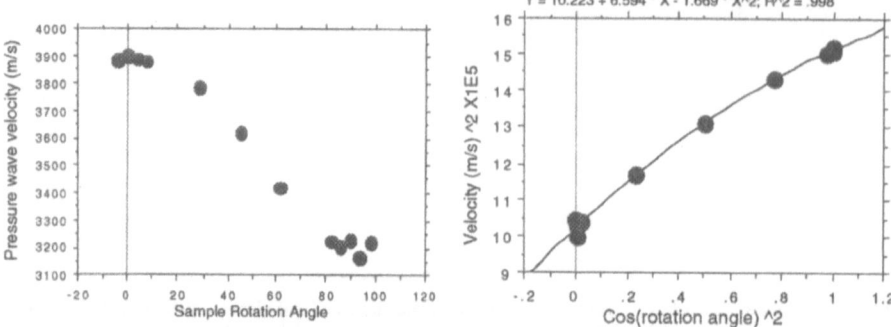

Figure 5 Angular dependence of the pressure wave velocity in a sample of cortical bovine bone, obtained by rotating the plane of analysis over 100°. The graph on the **left** shows the maximum velocity at 0° and the minimum at 90°, 0° being aligned with the long axis of the bone, and the graph on the **right** shows the fitted velocity^2 curve to the Cos^2 function of the rotation angle.

The velocity measured from the second (shear) critical angle exhibits a similar dependence upon orientation. Its behavior is also well fitted by the dependence predicted by hexagonal symmetry, allowing the unambiguous determination of two further elements of the matrix and of a two-term combination[23].

UCR IMAGING

The measurement of anisotropy of velocity at a point can be extended over a surface to visualize the heterogeneity of velocity in a bone sample and also to see the orientations of the maximal stress planes in the sample. An example of this method of functional imaging using UCR is shown in *Figure 6*, where the velocity at different orientations was measured at multiple points over a bovine bone sample.

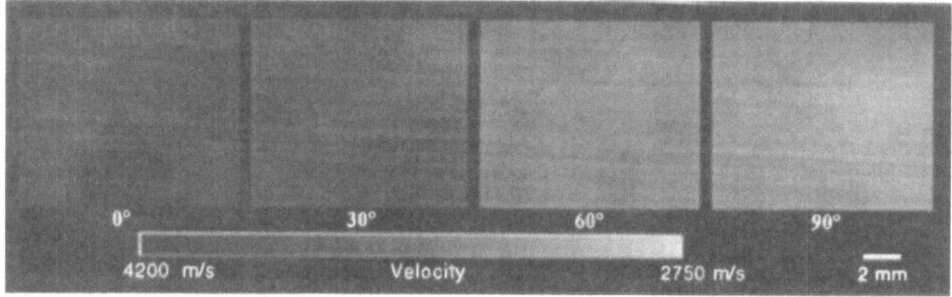

Figure 6. This greyscale image shows the velocity map over the surface of a bovine bone specimen (1.2 cm X 1.2 cm) with plane of analysis at 0°, 30°, 60°, 90° with the longitudinal axis of bone. The average velocity decreases from 4200 to 2750 m/s as the angle of orientation increases from 0° to 90°. The local heterogeneity is ≈ 5% at any given orientation.

For hexagonal symmetry, five independent elements of the matrix of elasticity can completely determine the elastic behavior of bone at any point on the surface of the sample. These measurements of elasticity, taken at a point on a sample, can also be made over a given area by moving the point of measurement as shown in *Figure 6*. The spatial heterogeneity of the elasticity in bone can then be visualized for each principal direction as shown in the two reconstructed images reported below, which display the values of the technical constants and Poisson's ratios across a 2.5 cm x 0.5 cm bone sample.

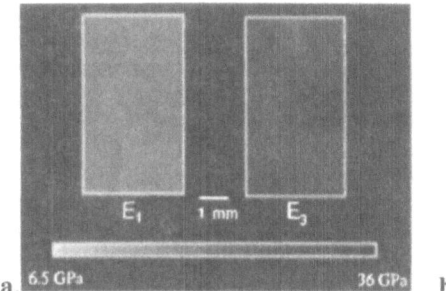 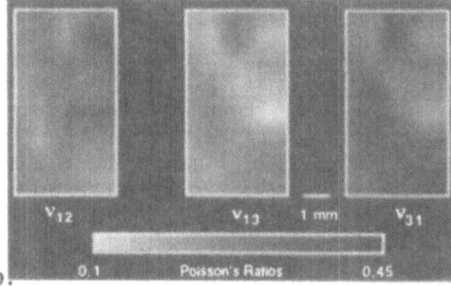

Figure 7 a) Visualization of principal technical constants for a single sample of bone using UCR and the measured density. Here E1<E3: both vary within 10% over the sample's area. **b)**Visualization of Poisson's ratios in the same sample, showing similar variability over this area

The heterogeneity of elastic constants visualized in this manner over a region of interest is putatively indicative of normal structural adaptation to local loads and metabolic requirements. To account for this variability in quantitative applications when such images are not employed, the velocity must be presented as a mean value, averaged over orientation and/or multiple points. This averaging procedure is analogous to processes used in densitometry and other quantitative techniques where numerical results are abstracted from multidimensional data.

Clinical Applications of UCR Imaging

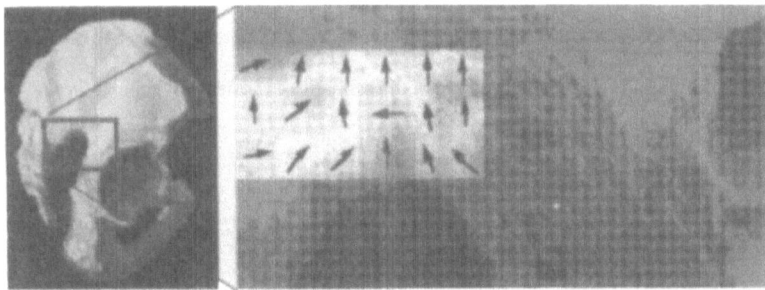

Figure 8 Left hand picture is a side view of a human ilium. The sciatic notch region, shown in expanded view on the right, was scanned using the UCR technique and directions of principal stress were found by measuring the velocity anisotropy over 180° at each location. The overlaid pseudocolor velocity and vector maps show a maximum value at the center of the arch and principal directions of the material properties pointing inwards towards the center of the arch from either side. The confirmation of the UCR measurement of principal directions with visual observations of the material flow in this region, demonstrates the potential of the UCR imaging technique. UCR imaging can be used *in vivo* to determine the directionality of the bone material properties at any location and the behaviour of this material over a period of time.

ACKNOWLEDGEMENTS

This work was supported by NASA grant # NAGW 3582. The authors are grateful to Amar Raheja for his help in image processing.

REFERENCES

1 Pak C, Sakhaee K, Antich P, Zerwekh J, Peterson R, Piziak V, Gonzales J, Hatab M, Poindexter J, 1993, Update on the treatment of osteoporosis with intermittent slow release sodium fluoride and continuous calcium chloride. Ed. Christiansen C, Riis B Proc. *Fourth International Symposium on Osteoporosis on Osteoporosis and Consensus Development Conference.* Aalborg, Denmark. pp. 122-124

2 Vahey J, Lewis J, Vanderby R, 1987, Elastic moduli, yield stress, and ultimate stress of cancellous bone from the proximal tibial epiphysis, *J Biomech* **20**:29-33.

3 Brown T, Ferguson A, 1980, Mechanical property distributions in the cancellous bone of the human proximal femur, *Acta Orthop Scand* **51**:429-437.

4 Turner C, Eich M ,1991, Ultrasonic velocity as a predictor of strength in bovine cancellous bone, *Calcif Tissue Intnl* **49(2)**:116-119.

5 Lafage M, Balena, Battle M, Shea M, Seedor J, Klein H, Hayes W, Rodan G, 1995, Comparison of alendronate and sodium fluoride effects on cancellous and cortical bone in minipigs: one year study. *J Clin Invest* **95**:2127-2133.

6 Ashman R, Cowin S, Van Buskirk W, Rice J, 1984, A continuous wave technique for the measurement of the elastic properties of cortical bone. *J Biomech* **17(5)**:349-361.

7 Bonfield W, Tully A, 1982, Ultrasonic analysis of the Young's modulus of cortical bone, *J Biomed Engin* **4(1)**:23-27.

8 Katz J, 1980, Anisotropy of Young's modulus of bone, *Nature* **283**:106-107.

9 Kim H, Walsh W, 1993, Mechanical and ultrasonic characterization of cortical bone, *Biomimetics* **1**:293-370.

10 Lang S, 1970, Ultrasonic method for measuring elastic coefficients of bone and results of fresh and dried bones, *IEEE Trans, Biomedical Engineering* **17**:101-105.

11 Lees S, Heeley J, Cleary P, 1979, A study of some properties of bovine cortical bone using ultrasound, *Calcif Tissue Intnl* **29**:107-117.

12 Mehta S, 1995, *Analysis of Mechanical Properties of Bone Material Using Nondestructive Ultrasound Reflectometry,* Ph. D. thesis University of Texas Southwestern Medical Center.

13 Rho J, Ashman R, Turner C, 1993, Young's modulus of trabecular and cortical bone material: ultrasonic and microtensile measurements, *J Biomech* **26(2)**:111-119.

14 Van Buskirk W, Cowin S, Ward R, 1981, Ultrasonic measurement of orthotropic elastic constants of bovine femoral bone, *J Biomech Eng.* **103**:67-72.

15 Turner C, Eich M, 1991, Ultrasonic velocity as a predictor of strength in bovine cancellous bone, *Calcif Tissue Intnl* **49(2)**:116-119.

16 Antich P, 1993, Ultrasound Study of bone in vitro, *Calcif Tissue Intnl* **53(Suppl 1)**:S157-S161.

17 Kaufman J, Einhorn T, 1993, Perspectives: Ultrasound assessment of bone, *J Bone Miner Res* **8(5)**:517-525.

18 Antich P, Anderson J, Ashman R, Dowdey J, Gonzales J, Murry R, Zerwekh J, Pak C, 1991, Measurement of mechanical properties of bone material in vitro by ultrasound reflection: methodology and comparison with ultrasound transmission, *J Bone Miner Res* **6(4)**:417-426.

19 Lees S, 1975, Data reduction from critical angle measurements, *Ultrasonics* **13**:213-215.

20 Ashman R, Antich P, Gonzales J, Anderson J, Rho J, 1994, A comparison of reflection and transmission ultrasonic techniques for measurement of cancellous bone elasticity, *J Biomech* **27(9)**:1195-1199.

21 Antich P, Pak C, Gonzales J, Anderson J, Sakhaee K, Rubin C, 1993, Measurement of intrinsic bone quality in vivo by reflection ultrasound: correction of impaired quality with slow-release sodium fluoride and calcium citrate, *J Bone Miner Res* **8(3)**:301-311.

22 Zerwekh J, Antich P, Sakhaee K, Gonzales J, Gottschalk F, Pak C, 1991, Assessment by reflection ultrasound method of the effect of intermittent slow-release sodium fluoride-calcium citrate therapy on material strength of bone, *J Bone Miner Res* **6(3)**:239-243.

23. Antich P, Mehta S, 1997, UCR (ultrasound critical-angle reflectometry): A new modality for functional elastometric imaging, *Physics Med Biol* **In Press**.

NONDESTRUCTIVE EVALUATION OF OSTEOPROGENITOR CELL CULTURES USING HIGH FREQUENCY ACOUSTICAL IMAGING AND IMAGE PROCESSING

D. Callens, F. Lefebvre, Ph. Pernod and B. Nongaillard

Institut d'Electronique et de Microélectronique du Nord
Département d'Opto-Acousto-Electronique (UMR CNRS 9929)
University of Valenciennes BP311
59304 VALENCIENNES - FRANCE

INTRODUCTION

Bone marrow is a source of osteogenic stem cells that is lately used in orthopaedic surgery in order to stimulate the bone formation and to fill some bone deficits[1,2]. Not all individuals demonstrate equal osteogenic potential[3]. In order to understand the effects of the different factors conditionning the effectiveness of the human bone marrow graft, in-vitro studies are currently being developed. Methods of investigation commonly used for this type of characterization have the drawback of degrading the samples and do not permit running investigations. Consequently, an important number of cultures have to be plated in order to study their evolution with incubation time. A new method of characterization is required.

It is well known that ultrasounds constitute a valuable tool to measure mechanical properties of materials without damaging them. Ultrasonic imaging and in particular acoustic microscopy are used for non destructive evaluation in number of industrial applications.

The aim of this study is to apply ultrasonic imaging to bone cell cultures characterization in order to perform a non destructive and quantitative evaluation of cell maturation. Our approach can be divided into several well-defined stages. After acquisition with a classical C-Scan imaging system, image obtained are then subjected to different analysis in order to extract automatically meaningful morphological parameters from each mapping. A preliminary study has first been realized on stopped cultures in order to understand interactions between ultrasounds and cells and to choose the more suitable image processing techniques. The subsequent stage will consist in mapping the inner side of the closed sterile flask in order to perform a dynamic evaluation of cell maturation.

MATERIALS AND METHODS

Culture procedure

Stem cells present in bone marrow are undifferentiated progenitors which can turn into different cellular lineages such as reticular, adipose, fibroblastic or osteoblastic ones. Fibroblasts and osteoblasts derive from a common Colony-Forming Unit-Fibroblastic (CFU-F), differentiation depending on the environmental features. The osteogenic potential of the marrow depends on the presence of CFU-F and on their capacity to differentiate in osteoblastic cells. The purpose of in vitro culture is to evaluate osteogenic potential of bone marrow by counting the number of osteoblastic cell colony derived from each CFU-F present in human bone marrow. The experimental protocol undertaken has two objectives : first, the understanding of the differentiation process and secondly, the evaluation of the osteogenic capacity of bone marow.

In this study, bone marrow was extracted from iliac crest and cells were plated in cultures with two different media in order to favour production of fibroblasts or osteoblasts. Medium was changed every 4 days. Cultures were sacrificed after different incubation times (10, 15, 20, 25 and 30 days) in order to follow their evolution[4].

C-Scan imaging system

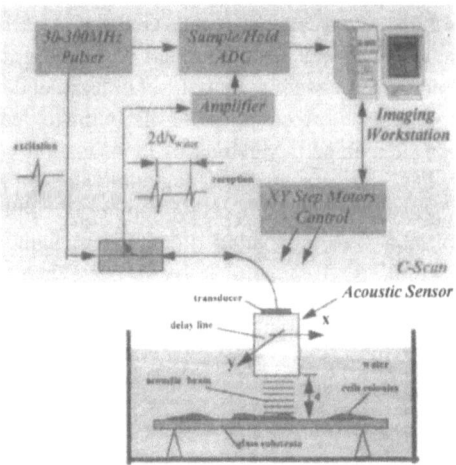

Figure 1.C-Scan imaging system for CSI and SSI investigations.

As it can be seen on figure 1, the C-Scan imaging system we have used to examine the cultures is very classical. It is composed of a wide band pulser, amplifier, sample and hold, analogue to digital converter and a scanning controler. All of these elements are controled by a computer which also serves as an imaging station. For a preliminary study, we have analysed previously stopped cultures, the sterile flask was removed and the culture substrate placed in tank filled with water. Investigations are then done by the cell side with 100 MHz plane waves.

Experimental protocol

We have experimented on three groups of cultures, group 1 and 2 coming from the same individual but prepared with different cell concentration and group 3 coming from an other individual and prepared with the same concentration as group 1. Table 1 shows the cell concentration for each group and the corresponding number of CFU-F evaluated by optical microscopy.

Table 1. Characteristics of each group of cultures.

	Number of CFU-F	Cell Concentration	Evaluated Number of CFU-F
Group 1	$7 / 10^6$ cells	$0.5 . 10^6$ /ml	3 to 4
Group 2	$7 / 10^6$ cells	$2 . 10^6$ /ml	14
Group 3	$20 / 10^6$ cells	$0.5 . 10^6$ /ml	10

Owing to contamination phenomenon that had occured on day 30 of culture for the first group and on day 25 for the second, group 1 and 2 contained less samples.

EXPERIMENTS AND DISCUSSION

Image processing methodology

The first images have been performed on the osteoblastic and fibroblastic lineages of the third group. Raw images obtained by the C-Scan system exhibiting a poor contrast, they have been enhanced in order to expand the dynamic range of gray level. Equalized versions of each lineage are presented on the left part of figure 2. These images really show a good visual quality, other image processings have been applied in order to quantify automatically the cell development. The purpose was to extract on each mapping, morphological parameters of the cultures. Image segmentation was performed in order to partition the image into different regions of interest : the cell colonies and the background[5]. The histogram of the equalized image shows that the differentiation between the features and the background resides in a texture variation instead of intensity variation. For this reason, variance map was computed for each image, in order to select unambigously the binarisation threshold to skeletonize the image. The last stage of image processing has consisted in computing convex hull of each extracted colony[6]. The convex hull is in fact, the minimal convex polygon which emcompasses the boundaries of a colony. It traduces well the entire area occupied by a colony, better than the complex boundaries due to grainy surface. Each stage of the image processing can be followed on figure 2.

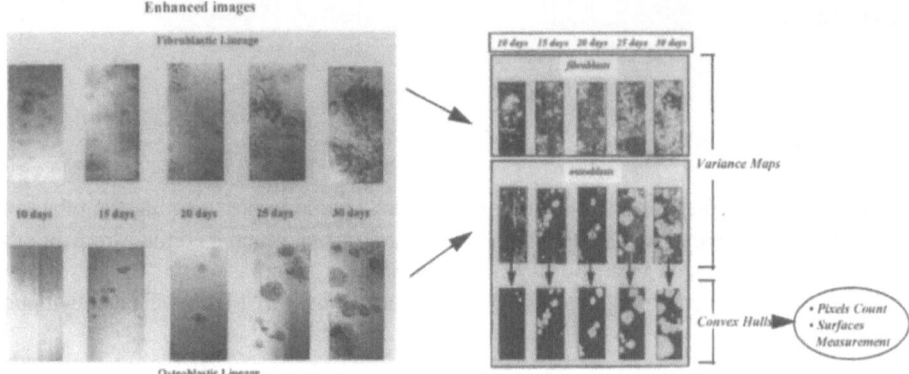

Figure 2. Different stages of the image processing. Application to fibroblastic and osteoblastic lineage of group 3

Preliminary results on stopped cultures

Three morphological parameters which have been found to be characteristic were evaluated from the convex hull maps. These parameters are the number of colonies, their size and the cell recovery rate. Their evolution with incubation time are discussed in the next paragraphs.

Number of colonies

The number of colonies (NOC) is obtained by counting the number of areas in the binary version of the variance map. Figure 3 presents the evolution of this parameter with incubation time for each of the three groups of cultures. NOC increases between the two first incubation stages, i.e. 10 and 15 days and is inclined to decrease for the more advanced stages. Comparison with the values extracted from optical microscopy reveals a good correlation between the acoustical and the optical analysis. At 15 days, the number of colonies acoustically detected fits with the number of CFU-F initially evaluated by optical microscopy. Before this stage, even if all the colonies are present, some of them are too small to be acoustically detected. A limit of detection of colonies can be seen. After this stage, the number evaluated decreases owing to the confluency of colonies.

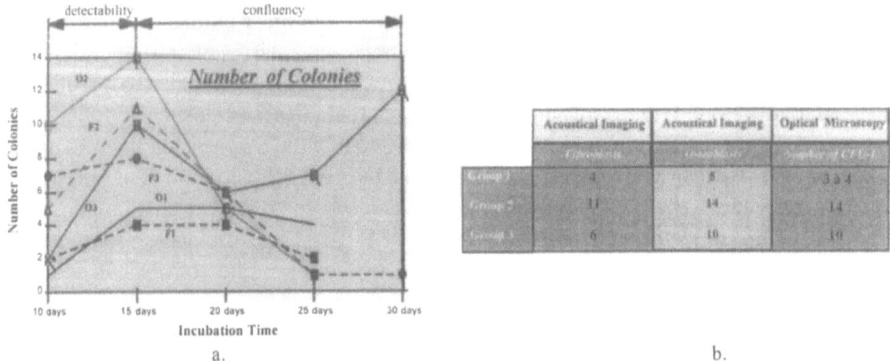

a.

b.

Figure 3. a. Acoustical evaluation of the number of colonies. **b**. Comparison of this parameter acoustically detected with the number of CFU-F.

154

Mean size of colonies

The mean size of colonies has been evaluated for each stage. As it can be seen on figure 4, this parameter increases with time.

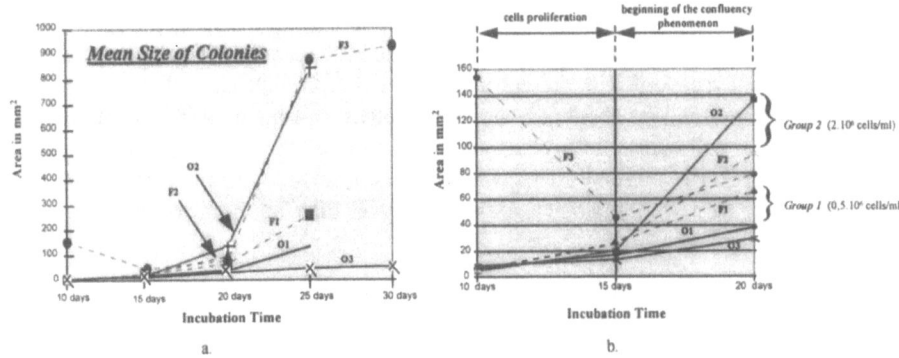

Figure 4. a. Acoustical evaluation of the mean size of colonies. **b.** Focus on the earlier stages.

The evolution is rather smooth for the early stage and increases more rapidly after. This increase is connected to the confluency phenomenon which is related to the cell concentration. The evaluation of the mean size of colonies reveals the rate of cell proliferation when these cells have not reach the stage of confluency and can depict exactly the beginning of this stage.

These two parameters can be fused, producing an other criteria : the total recovery rate, which indicates the cell production obtained from the graft cultured.

Total recovery rate

It represents the ratio of the cell invasion surface to the surface of the substrate. It is a global parameter which traduces at the same time, cell proliferation, confluency and number of CFU-F. Figure 5 shows for each group an increase of the total recovery rate with incubation time. In particular, group 2 which is known to contain an important number of CFU-F presents a rapid evolution of its total recovery rate. The total recovery rate is therefore also a good criteria to evaluate the number of CFU-F produced from a graft.

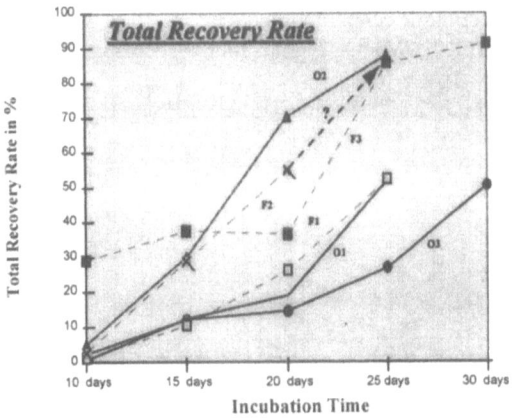

Figure 5. Acoustical evaluation of the cell total recovery rate

155

CONCLUSION

This preliminary study has been implemented to develop a new method of investigation of in vitro bone cell cultures. The results which are presented reveal the great potentialities of acoustic microscopy when associated with appropriate image processing techniques. The three morphological parameters acoustically evaluated : the number of colonies, their mean size and the total recovery rate, present a good correlation with physiological parameters which provide informations about the osteogenic potential of the human bone marrow. The subsequent work will consist in investigating cells through the substrate, without opening the flask, in order to preserve sterility and to perform a dynamic evaluation of cell maturation. The same morphological parameters will then be extracted from acoustical mappings.

Ackowlegdements : This work has been realized in collaboration with Prof. P. Hernigou, Orthopaedic Department, Henri Mondor Hospital, 94010 CRETEIL. It is included in the studies of the regional federative research group entitled : "Fédération des Biomatériaux du Nord Pas de Calais" IRMS Institut Calot 62608 Berck sur Mer. The authors wish to thank F. Beaujan from Blood Transfusion Centre of Val de Marne for preparing bone marrow cell cultures. This research was supported by French Minister of Research and Space found n° 92C0486.

REFERENCES

1. Ph. Hernigou and F. Beaujean, La moelle osseuse, une clé dans la compréhension de certaines pathologies : un potentiel thérapeutique l'autogreffe de moelle, Rev. Chir. Orthop., 79, suppl, pp. 136-137 (1993).
2. J.F. Connolly, Injectable bone marrow preparations to stimulate osteogenic repair, Clin. Orth. Rel. Res., 313, pp.8-18,(1995).
3. S. Palmer, B.S. Strates and J.F. Connolly, Repair of trephine defects in the aged by implants of bone matrix and marrow, Surg Forum, 38, pp.531-533, (1987).
4. D. Callens, Ph. Pernod, E. Radziszewski, B. Nongaillard, JY. Dauphin and Ph. Hernigou, Caractérisation des cultures de progéniteurs osseux par ultrasons haute fréquence. Evaluation de la technique échographique, ITBM, 15, n°4, pp. 391-402, (1994).
5. R.C. Gonzales and R.E. Woods, Digital Image Processing, Addison Wesley Publishing Company, (1993).
6. K. Wu, D. Gauthier, M.D. Levine, Live cell segmentation, IEEE Transactions on Biomedical Engineering, Vol 42, pp.1-11, (1995).

ULTRASONIC ATTENUATION AND DISPERSION OF CANCELLOUS BONE IN THE FREQUENCY RANGE 200 KHZ-600 KHZ

P. Droin, P. Laugier, and G. Berger

Laboratoire d'Imagerie Parametrique URA CNRS 1458
15 rue de l'Ecole de Médecine
75006 Paris FRANCE
e-mail : laugier@idf.ext.jussieu.fr

INTRODUCTION

Measurement of ultrasonic attenuation and velocity of cancellous bone are being applied to aid diagnosis of women with high fracture risk due to osteoporosis (Hans et al., 1996). Ultrasonic measurements in the calcaneus are generally derived using a substitution method, whereby a reference signal is obtained firstly through water alone, and then with the specimen interposed. Velocity measurements are obtained from transit time determinations (usually by detection of the first arrival point, first zero-crossing point or a fixed threshold on the rising front). Phase velocity dispersion in cancellous bone has received little attention up to now. In previous works we have reported a method for characterizing acoustic properties of human cancellous bone using parametric imaging techniques (Laugier et al , 1994). Parametric images of the slope of attenuation coefficient and ultrasonic bone velocity have been compared with bone mass density (BMD) images of the calcaneus (Laugier et al , 1997). The present study was conducted to provide sufficient quantitative information on velocity dispersion of human calcaneus, a preferred site of ultrasonic measurement in clinical practice, so that the relationship between dispersion and BMD could be clarified.

MATERIAL AND METHODS

Trabecular bone specimens were obtained by excision of the calcaneus (heel bone) from cadavers. The calcaneus is a cancellous, subcutaneous, weight-bearing bone. It comprises a thin peripheral cortical layer of compact tissue coating a framework of cancellous (or trabecular) bone. The cortical lateral sides of each calcaneus were sliced off to provide a sample of pure trabecular bone with two parallel faces of approximately 12.5 mm in thickness. The thickness l of the samples was measured using a vernier caliper. The samples were then defatted and finally refilled with water under vacuum.

A matched pair of ultrasonic broadband focused transducers (center frequency of 500 kHz, diameter of 29 mm, focal length of 35 mm) were mounted coaxially, separated by twice the focal length, in a through-transmission normal incidence configuration The supporting electronics comprised a pulse-receiver, amplifier and digitization card (Contrôle US, Orsay, France). Samples were immersed in a water bath at room temperature, suspended by a clamp half way between the transducers. Sample xy displacement was allowed by stepper motors. An ultrasonic scan of each sample was performed with a step of 1 mm in two perpendicular xy directions for a total scan size of 60 mm by 60 mm. Ultrasonic parametric images were obtained from the local estimate of the parameters at each point in the scan. The attenuation and phase velocity were obtained by deconvolving the signal transmitted through the specimen with a reference signal transmitted through water without the sample, following the method described previously (Laugier et al., 1997). The frequency-dependent attenuation coefficient $\alpha(f)$ was derived from the ratio of the magnitude spectra :

$$\alpha(f)l = Ln\frac{|A_r(f)|}{|A(f)|}. \tag{1}$$

where $A_r(f)$ is the complex spectrum of the reference signal, $A(f)$ is that of the signal transmitted through the specimen. As the measured attenuation showed approximately linear behavior over the frequency range (200-600 kHz), the slope of a linear regression fit to $\alpha(f)$ between 0.2 and 0.6 MHz yields the slope of the attenuation coefficient. In clinical practice, it has become known as nBUA (normalized broadband ultrasonic attenuation). The correlation coefficient between attenuation coefficient and frequency is a measure of the degree of closeness of the linear relationship between the two variables. The phase velocity $v(f)$ was derived from the phase of the complex spectra according to:

$$\phi(f) = atan\left[\frac{A(f)}{A_r(f)}\right] = 2\pi hf\left(\frac{1}{v_w} - \frac{1}{v(f)}\right) \tag{2}$$

where v_w is the speed of sound in water. Velocity dispersion was characterized by ultrasound bone velocity (UBV) defined as the value of the phase velocity at the center frequency of initial pulse $UBV=[v(f)]_{f=0.5\,MHz}$ and by a dispersion magnitude index(DMI) defined by $DMI = \Delta V = V(f_{max}) - V(f_{min})$, where $f_{min}=200$ kHz and $f_{max}=600$ kHz are the lower and upper limits of the transducer frequency bandwidth. Provided the phase velocities of the test material and water are reasonably well matched, a requirement comfortably met for measurements on trabecular bone (Laugier et al., 1997), the effect of diffraction on attenuation and velocity estimates is small and can be neglected (Droin, 1997).

BMD was measured with X-ray quantitative computed tomography (QCT). QCT of the specimens was performed using a calibration phantom which contains solutions of K_2HPO_4 encased in Plexiglas. Standard QCT images were obtained using 10 mm thick slices through the center of each specimen. Results were expressed in terms of K_2HPO_4 equivalent in mg/cm^3. With this procedure, approximately the same volume of trabecular bone was investigated with ultrasound and X-ray.

A software was capable of reading both ultrasound and QCT images and of selecting site-matched identical regions of interest on the images. Circular ROIs, 7 mm in diameter were selected. The average value of each parameter was then calculated for each ROI. Several non-overlapping adjacent ROIs were selected for each specimen. A variable number (ranging from 15 to 29) of independent ROIs could be selected according to the size of each specimen.

RESULTS

Typical radiofrequency waveform transmitted through a specimen, frequency dependent attenuation and phase velocity variations are illustrated in figure 1. It shows the nearly linear variation of the attenuation coefficient and negative dispersion. The correlation coefficient between attenuation coefficients and frequency was found constantly to be higher than 0.97, indicating that the attenuation was quasi-linear versus frequency. Negative dispersion was frequently observed. Figure 2 shows the comparison of ultrasound parametric images and QCT Image. The gray scale for DMI image was black (-40 m/s) to white (50 m/s). Substantial variations of the parameter were observed within specimens and also between specimens. DMI values ranged between -42 and 56 m/s. Variations of mean DMI between specimens fell in the range -15 to +2 m/s. On the average, dispersion obtained on all data was negative.

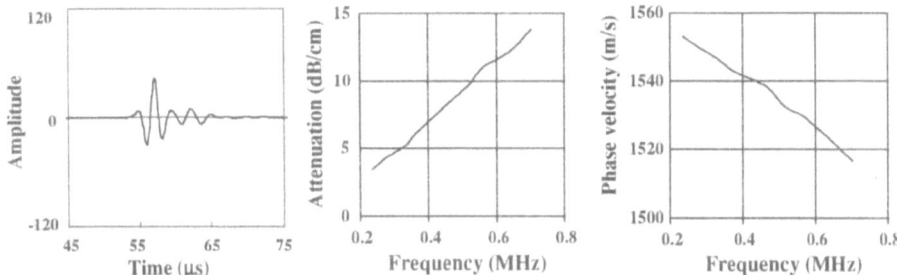

Figure 1 : Typical radiofrequency waveform, frequency dependent attenuation and phase velocity variations measured in trabecular bone.

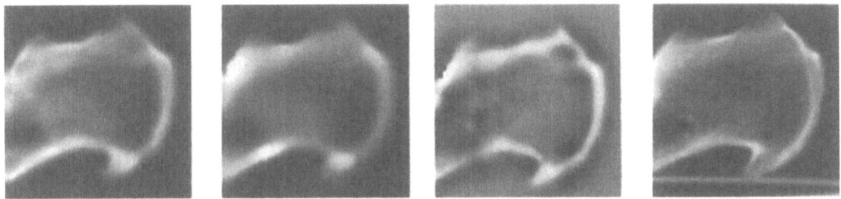

Figure 2: Comparison between nBUA, UBV, DMI, and QCT images (from left to right)

To evaluate the correlation between density measurements, nBUA, UBV, and DMI, the data were analyzed for each specimen as well as grouped together over the entire population. Moderate to non significant correlation were observed and the amount of variance explained were different for different specimens. At the best, BMD could explain only 57% of the variance of DMI. For each specimen, a statistically significant relationship was noted between DMI and nBUA (mean±SD r^2=0.18±0.19, range 0.00-0.56) only for 6 specimens, between DMI and UBV (mean±SD $r^2=0.20\pm0.19$, range 0.00-0.64), only for 7 specimens, between DMI and BMD (mean±SD $r^2=0.22\pm0.20$, range 0.00-0.57), for 8 specimens. When pooling the data, no correlation was found between DMI and the other parameters as illustrated in figure 3.

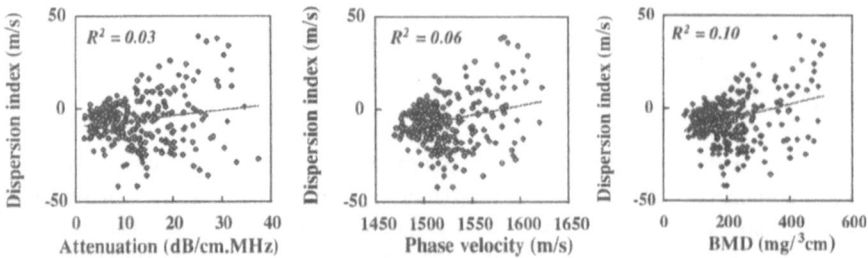

Figure 3: Relationships between dispersion magnitude index and the other parameters

DISCUSSION

Ultrasonic propagation through cancellous bone exhibits marked velocity dispersion as well as high attenuation. Cancellous bone is a highly porous medium composed of a solid matrix (mineralized collagen) of interconnected plates and rods (trabeculae) filled with marrow resulting in dramatic spatial fluctuations of the bulk modulus. Trabecular elements of average size ranging between 50 μm and 150 μm are separated by an average distance of 0.5 mm to 2 mm. Velocity dispersion is likely to be due to bone microarchitecture. Different mechanisms can be invoked including scattering, resonance, viscoelastic absorption, and mode conversion, however no theoretical developments for dispersion has been advanced so far.

Dispersion velocity magnitude in the frequency bandwidth range 200 kHz -700 kHz fell in the range -42 m/s to 56 m/s in the present study, figures which could not be compared to other works, due to missing data. Published data for the ultrasonic velocity dispersion of cancellous bone are very sparse. Fry and Barger (1978) measured an increasing phase velocity with frequency in skull diploe between 0.3 MHz and 2.0 MHz. Emphasis has been recently placed on a comparison of transit time velocity and phase velocity measurements on calcaneus bone specimens. Strelitzki and Evans (1996) found negative dispersion between 600 kHz and 800 kHz in 10 fixed calcaneus bone specimens, while positive and negative dispersion was found in 70 human calcaneae between 200 kHz and 800 kHz by Nicholson et al. (1996) with a general trend for negative dispersion.

Ultrasound parametric imaging provides a method for the assessment of spatial variation of acoustic properties, and in particular that of velocity dispersion. In contrast to the similarity of nBUA, UBV and BMD images (Laugier et al. 1997) the pattern of DMI values was found to differ substantially from that of other parameters. We found moderate to no correlation between DMI and other acoustic parameters or BMD. A second unexpected result was to find such variable dispersion curves, either negative or positive dispersion despite the attenuation coefficient being constantly found to vary quasi linearly with frequency.

The fact that attenuation and dispersion were uncorrelated, as well as the result of negative dispersion, were puzzling and contrasts with other biological or non biological materials where dispersion can be accurately predicted from frequency-dependent attenuation data using causal theories. Generally speaking, there is a relation connecting real and imaginary parts of the wave number, or between attenuation and velocity dispersion. Such a general connection can be expected from the linear, causal and passive properties of the propagation medium. A general relationship between attenuation and dispersion in the area of acoustic was derived by O'Donnell et al. (1981) and is now commonly referred as the Kramers-Krönig relationship. The Kramers-Krönig relationship have been found to hold for a wide class of homogeneous and inhomogeneous media (Weaver and Pao, 1981) which can support plane wave. This include media which are statistically homogeneous, where the usual physical assumption that a spatially local compressibility relating pressure and density exists can be made and where the wave number is that of the average field. For practical use, O'Donnell et al. (1981) further simplified the general Kramers-Krönig relationship to nearly local approximate Kramers-Krönig relationship, under the assumption that the attenuation is much smaller than the wave number or $(\alpha \lambda)^2 << 1$, that the attenuation coefficient is a slowly varying function of frequency, and that the functional form of the attenuation persists beyond the measurement range. The nearly local approximate Kramers-Krönig relationship have been employed in a wide range of experimental circumstances (O'Donnell et al., 1981 ; Lee et al., 1990 ; Pouet and Rasolofosaon, 1993 ; Jeong and Hsu, 1995 ; Szabo, 1995). Excellent agreement has been obtained even for specimens having attenuation coefficient as high as 22 dB/mm (Lee et al., 1990), a value falling in the range of attenuation coefficients observed in cancellous bone (Laugier et al., 1997). Following the Kramers-Krönig relationship, dispersion can be predicted from the frequency-dependent attenuation and for the case of linear variation of the attenuation, the dispersion should be logarithmic with frequency (O'Donnell et al., 1981).

If attenuation and dispersion data are not intrinsic material properties due to competing measurement artifacts, the attenuation dispersion relationship may not apply. To exclude any instrumentation or software error, the system used in this study was carefully calibrated on test materials. Diffraction effect has been analyzed in detail and was found to be negligible (Droin, 1997). It was therefore concluded that attenuation and dispersion data are pure material characteristics of bone and are not due to competing measurements artifacts. The fact that the unexpected results found here cannot be accounted for by system errors would imply that the reason for the observed deviation of dispersion data from theoretical dispersion has to be found in the limit of applicability of the nearly local Kramers-Krönig relationship. This relationship was derived under the condition that the attenuation cannot rise as $f \to \infty$ as fast as a linear function of f. Assuming that the attenuation coefficient is a power function of frequency , $\alpha(f)= \alpha_1 f^n$, n must be less that one (Szabo, 1993). In our case, the correlation coefficient between attenuation and frequency was used as an indicator of goodness of the linear fit, however, the relatively restricted frequency range over which measurements were made limits the confidence which can be placed in the estimates of the parameters for the frequency-dependent attenuation models. As more broadband measurements become available, bone material thought to have linear frequency dependence may be found to be described by an intermediate exponent n>1, or even by a more complex non monotone frequency law. A time domain approach of causality has been proposed by Szabo (1994, 1995), which is more general than the local Kramers-Krönig relationship because it can be used for power law in case where n>1. Time domain causal attenuation-dispersion relationship has been shown to be able to predict negative dispersion as power exponent n approaches an even integer. It remains that, using causal theories, complete dispersion can be predicted from the knowledge of attenuation data. This is in apparent contradiction with the absence of correlation found in this study between nBUA and DMI. However, these theories are based on the assumption that the functional form of the attenuation persists beyond the frequency range (Szabo, 1995), a statement which has still to be documented in cancellous bone.

In summary, velocity dispersion of cancellous bone was curiously found to be either negative or positive and non correlated to the slope of the attenuation coefficient. Since the trabecular network in the calcaneus is neither random, neither perfectly periodic, standard physical models for ultrasound propagation cannot be easily applied. These results and the implication of ultrasound measurements for the assessment of bone strength in osteoporosis is a challenge for further studies. In particular, measurements in extended frequency range should be carried out to document the influence of bone microarchitecture on acoustic wave dispersion and attenuation and to elucidate the apparent contradiction between our findings and the theoretical predictions from causal theories.

REFERENCES

Droin, P. Mesures d'atténuation et de vitesse de propagation ultrasonores dans l'os trabéculaire dans la bande de fréquence 0.2-0.7 MHz, PhD Thesis Report, Université Paris 6, Paris, France, 1997.

Fry FJ, Barger JE, Acoustical properties of the human skull. J Acoust Soc Am 63:1576-1589, 1978.

Hans, D., Dargent-Moline, P., Schott, A. M., Sebert, J. L., Cormier, C., Kotski, P. O., Delmas, P. D., Pouilles, J. M., Breart, G., and Meunier, P. J. Ultrasonographic heel measurements to predict hip fracture in elderly women : the Epidos prospective study. Lancet 348:1996.

Jeong, H., and Hsu, D.K. Experimental analysis of porosity-induced ultrasonic attenuation and velocity change in carbon composites. Ultrasonics 33 :195-203 ;1995.

Laugier, P., Giat, P., and Berger, G. Broadband ultrasonic attenuation imaging : a new imaging technique of the os calcis. Calcif Tissue Int 54:83-86;1994b.

Laugier, P., Droin, P., Laval-Jeantet, A. M., and Berger, G. In vitro assessment of the relationship between acoustic properties and bone mass density of the calcaneus by comparison of ultrasound parametric imaging and QCT. Bone 20:157-1665;1997.

Lee, C.C, Lahham, M., and Martin, B.G. Experimental verification of the Kramers-Krönig relationship for acoustic wave. IEEE Trans Ultrason Ferroelec Contr 37 :286-294 ;1990.

Nicholson, P.H.F., Lowet, G., Langton, C.M., Dequeker, J., and Van der Perre, G.. A comparison of time-domain and frequency-domain approaches to ultrasonic velocity measurements in trabecular bone. Phys Med Biol 41:2421-2435;1996.

O'Donnell, M., Jaynes, E.T., Miller, J.G. Kramers-Krönig relationship between ultrasonic attenuation and phase velocity, J Acoust Soc Am 69 :696-701 ;1981.

Pouet, B.F., and Rasolofosaon, N.J.P. Measurement of broadband intrinsic ultrasonic attenuation and dispersion in solids with laser techniques. J Acoust Soc Am 93 : 1286-1291;1993

Strelitzki, R., Evans, J.A. On the measurement of the velocity of ultrasound in the os calcis using short pulses, Eur J Ultrasound 4 :205-213 ; 1996.

Szabo, T.L. Time domain wave equations for lossy media obeying a frequezncy power law. J Acoust Soc Am 96 : 491-500;1994

Szabo, T.L. Causal theories and data for acoustic attenuation obeying a frequency power law. . J Acoust Soc Am 97 : 14-24;1995

Weaver RL, Pao YH, Dispersion relations for linear wave propagation in homogeneous and inhomogeneous media. J Math Phys 22:1909-1918, 1981

THREE-DIMENSIONAL IMAGING AND ROUGHNESS CHARACTERIZATION OF ARTICULAR CARTILAGE BY 50 MHZ ECHOGRAPHY

A. Saïed[1] , E. Chérin[1], L. Brayard[1], P. Netter[2], G. Berger[1]

[1] Laboratoire d'Imagerie Paramétrique CNRS URA 1458, Faculté de Médecine Broussais Hôtel - Dieu 15 rue de l'École de Médecine75006 Paris
[2] Laboratoire de Pharmacologie CNRS URA 1288 Faculté de Médecine Avenue de la Forêt de Haye BP 184 54505 Vandoeuvre les Nancy, France

INTRODUCTION

Osteoarthritis (OA) is a slowly progressive pathology which affects all tissues of the joint in different ways. The cartilage degeneration is the main characteristic of the disease. In the pathogenic sequence it undergoes morphological, structural, biochemical and functional changes. The earliest alterations of the cartilage are thickness variation and surface roughening (fibrillation and ulceration) followed by fissuring leading to extensive thinning and loss of the tissue at an advanced stage of the disease.[1] At present, the recognition of OA occurs when it is extensive and the characteristic advanced pathological features appear on current diagnostic imaging modalities: radiography, computed tomography, conventional echography (7.5 MHz) and magnetic resonance imaging. These techniques underestimate the cartilage alterations, they are not sensitive to its minor changes and neither permit accurate measure of its thickness.[2]

The development of an imaging technique for early detection and characterization of articular cartilage alterations is of a considerable interest for the clinical diagnosis of OA and the evaluation of pharmacological therapeutics. In a previous work reported on high resolution ultrasound imaging of articular cartilage lesions produced in an experimental model of OA we have shown that 50 MHz echography is sensitive to early degenerative changes of cartilage and permits the evaluation of its thickness variation.[3, 4]

The aim of the current study is to explore the potential of 50 MHz three-dimensional (3-D) echography for imaging and quantifying the initial morphological cartilage changes induced by a model of osteoarthritis. In particular, subtle changes in cartilage thickness and irregularities of its surface were quantitatively evaluated. The estimation of these parameters was made using the 3-D reconstruction of the detected cartilage boundaries. The cartilage surface irregularities or roughness were evaluated using the parameters obtained from a best fit between the studied cartilage surface curvature and the curvature of a model of normal cartilage.

METHODS

Experimental procedure

OA lesions with different degrees of severity were induced in rat knee joints by intra articular injection of 3 mg and 0.1 mg doses of mono-iodo-acetic acid (MIA). In order to assess early lesions and the lesion progression, the animals (n = 41, 5 week old males) were divided into different groups which were sacrificed at 1 week intervals up to 4 weeks after the injection. The injected animal groups are labeled as follows:

Time MIA (mg)	W1	W2	W3	W4
0.1	M01W1 (n=7)*	M01W2 (n=7)	M01W3 (n=6)	M01W4 (n=7)
3	M3W1 (n=7)	M3W2 (n=7)	-	-

* 7 rats received 0.1 mg MIA dose and were sacrificed 1 week after the injection.

In addition to the 41 injected rats, 8 non injected rats which were divided into 4 groups and sacrificed at 1 week intervals up to 4 weeks were also included in the study.

Patellar OA and contralateral control cartilages (intra articular injection of a 0.9% saline solution) were dissected within minutes after the rat sacrifice, mounted on a special holder immersed in a 0.9% saline bath at room temperature and explored *in vitro* using a scanning backscatter acoustic microscope. The system operates with a Panametrics (5052 PRX) pulser receiver which excites a PVDF f/2.5 transducer (Krautkramer IPS 80) to generate a 50 MHz center frequency broadband ultrasound pulse at focus (7.5 mm). The axial resolution was better than 30 microns at focus. To acquire 3-D data from both cartilage and subchondral bone, the transducer was positioned so that it focused on the cartilage surface and the ultrasound beam was perpendicular to the sample at its central part. It was then displaced in the sagittal and transverse planes of the patella (which is approximately 5 mm x 2 mm in size) using high precision (0.1 micron) stepping motors. The inter scan step was 50 microns. At each transducer position the backscattered radio frequency (rf) signals were digitized at 400 MHz, time averaged (N = 50) to improve the signal to noise ratio then transferred to a computer to produce serial contiguous B-scan images of the saddle-shaped rat patella.

Detection of cartilage boundaries

A Sun Spark work station was used to analyze all the B-scan sections and perform an easy detection and 3-D reconstruction of both cartilage surface and cartilage/bone interface.

Cartilage boundaries were detected using a standard contour detection algorithm based on thresholding developed in our laboratory. Figure 1 shows a typical A-scan of an envelope detected backscattered signal from a normal patellar cartilage. The signal exhibits a specular echo at the cartilage/saline water interface (CS) and a high backscatter signal from the cartilage/subchondral bone interface (B). The lower backscatter signal between the 2 interfaces results from the cartilage internal structures. The B-scans acquired in the transverse and sagittal directions were first visually analyzed by the operator to select the optimal values to assign to the different thresholds. The cartilage surface which reflects a specular echo was easy to identify because of the high signal-to-noise ratio at the saline/cartilage interface. The noise maximum value was estimated and the threshold value Ts for cartilage surface detection was selected to be just approximately 6 dB above the maximum level of the noise. The 3-D contour of the cartilage surface was then visualized and the threshold was changed if the surface was judged unsatisfactory. The cartilage subchondral bone interface was less easily detected because most of the cartilage specimens exhibited an echoic internal structure. The depth of this interface was approximately localized by the observer based on *a priori* knowledge that the signal amplitude increases at the cartilage/bone interface. Within this localized region, the interface was detected using a preset threshold Tb such that $Tb > te$, where *te* is the maximum estimated internal structure echo. Due to the low signal-to-speckle noise ratio at the cartilage/bone interface, the value of the threshold Tb was adjusted by the operator after visual inspection of the detected 3-D contour. To eliminate artefacts due to echos from the internal structure of the cartilage and avoid pixels which are not located on the resulting expected edge, a smoothing was applied: the position in depth of the pixel out of the contour was compared to the mean value found in its eight nearest neighbours and was taken into account only if its depth was within the standard deviation of this mean value. If this condition was not satisfied, the pixel was substituted with the mean value.

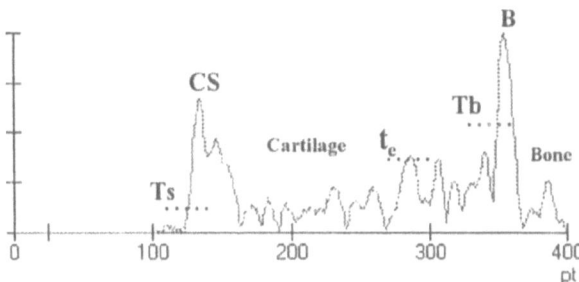

Figure 1. A-scan of an envelope detected backscattered signal from control cartilage. CS: cartilage surface, B: cartilage/bone interface. Ts, te, and Tb are the pre-set thresholds for the detection of cartilage boundaries.

Estimation of cartilage surface irregularities

The non injected patellar cartilage has a saddle shape. In order to quantitatively evaluate the subtle irregularities of cartilage surface and provide an index of its roughness the curvature of the normal cartilage surface in the transverse plane (Figure 2) was described by a parabolic fit (polynomial of degree n):

$$y = \sum_{k=1}^{n} a_k x_i^k \tag{1}$$

For the quantitation of cartilage surface irregularities or roughness the detected water/normal cartilage interface in each B-scan was fit using a least squares polynomial regression. For each B-scan section a merit function χ^2 was defined:

$$\chi^2 = \sum_{i=1}^{m} \left[\frac{y_i - \sum_{k=1}^{n} a_k x_i^k}{\sigma_i} \right]^2 \tag{2}$$

where m is the number of pixels i (x_i, y_i) of the contour on the B-scan, y is the position in depth (in the ultrasound propagation direction), σ_i is the measurement error (standard deviation) of the ith pixel, presumed to be equal to 1.

The parameters of the polynomial model were adjusted such that they minimize χ^2. χ^2 was small if the surface was relatively smooth and large if the detected points were randomly distributed about the fit parabola. Thus, $\chi_s^2 = \chi^2/m$ was used to provide an index of the cartilage section surface irregularities. The index of 3-D cartilage surface roughness R can be expressed in microns by:

$$R = \frac{v}{2Fe}\sqrt{\chi_c^2} \qquad (3)$$

where χ_c^2 is the 3-D cartilage surface merit function obtained by averaging the values of χ_s^2 obtained from all the acquired sections, Fe is the sampling frequency and v is the speed of sound which was equal to 1660 m/s.[5]

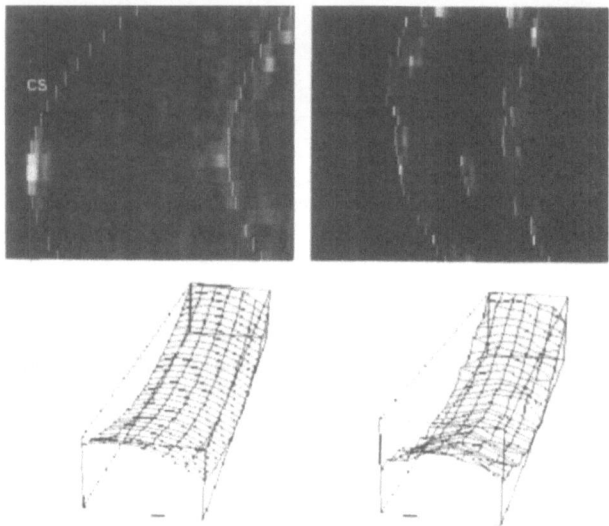

Figure 2. (Top): Transverse B-scans of control and OA cartilages showing detected contours.(dotted lines). The surface curvature (cs) of control cartilage (left) is well described by a parabolic fit. **(Bottom):** 3-D representation of detected (left) and modeled (right) cartilage surface.

RESULTS

In Figure 3 are displayed ultrasound B-scan images of control and osteoarthritic rat patella cartilages at different stages of the pathology. The image of control cartilage shows an echoic and smooth surface and a less regular interface corresponding to the cartilage subchondral bone interface. The cartilage overall thickness demarcated by the two boundaries varies along the patella. The average thickness estimated in the central part of the tissue on several B-scans ranged between 250 and 400 microns. Analysis of B-scan images of OA patellas showed that early lesions induced by a low dose of MIA were mainly characterized by cartilage hypertrophy, mild roughening of the tissue surface and subchondral bone remodeling. Higher dose of MIA induced a cartilage surface ulcerations, a proliferation of fibrous tissue in the deep zone of the cartilage and an extensive bone remodeling. Figure 4 shows 3-D representations of control and OA cartilage of a rat of the M01W4 group. The images show the shape of the rat patellas viewed from two different angles. The grey levels on the surface of the cartilage are indicative of the cartilage thickness variation. This representation permits to localize the lesions and either analyse them visually or quantify them. The grey level variation of the control cartilage surface is due to the irregularities of the osteochondral interface as was shown by echographic B-scan in Figure 3.

Regarding the cartilage roughness characterization which is reported in Figure 5, normal cartilages (non injected) and control cartilages (saline injected) which have a smooth surface exhibited approximately the same values of mean difference between detected surface and the parabolic model. This value was of the order of 2 microns. It demonstrates the sensitivity of the ultrasound system used for this study. For OA cartilages only results obtained on rats sacrificed up to 3 weeks after the 0.1 mg MIA injection are shown since at later stages the cartilage was mostly replaced by a fibrous tissue. The roughness level increased with the injected MIA dose and as a function of the lesion progression.

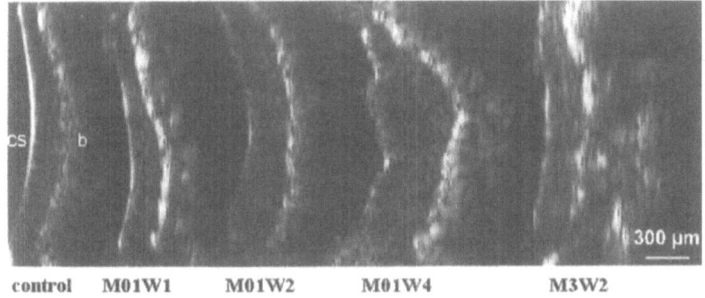

Figure 3. B-scan images made in the sagittal plane of rat patellas showing the evolution in time of cartilage and subchondral bone OA lesions produced by intra articular injection of 0.1 mg MIA dose. cs: cartilage surface, b: bone. Scanned region: 3 mm (vertical axis).

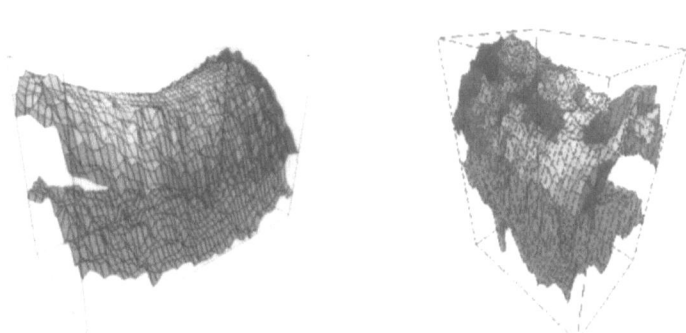

Figure 4. 3 D representation of cartilage surface (upper contour) and osteochondral interface (lower contour) of control (left) and OA (right) patellas of a rat from the M01W4 group. The grey levels on the surface are representative of the cartilage thickness variation. The distance between the pixels corresponds to the scanning step (50 microns). OA cartilage (490 microns in thickness) exhibits surface ulcerations and severe focal erosion of osteochondral interface. Scanned region size: 1.3 mm x 2.3 mm.

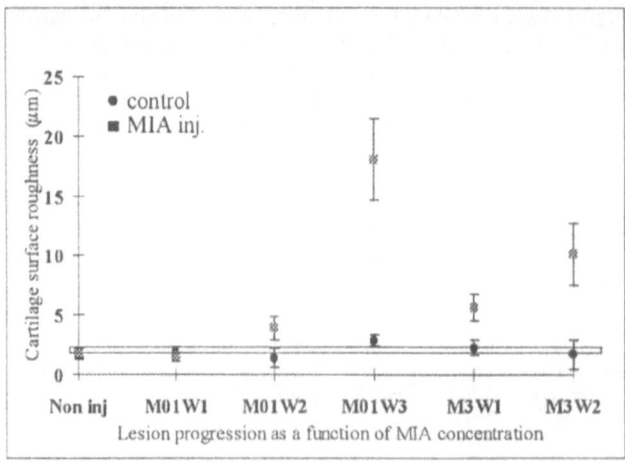

Figure 5. Variation of cartilage surface irregularities or roughness measured in the studied different groups as a function of MIA doses and lesion progression. The bars represent standard deviation. The horizontal line delimits the zone of the surface characteristics corresponding to normal cartilages.

CONCLUSION

High resolution ultrasonography is a promising tool for intra articular direct imaging of 3-D volumetric information of the cartilage and for accurate quantitative evaluation of the subtle and initial alterations of the tissue surface and thickness. 3-D image reconstruction is important for OA diagnosis since it provides information on lesion location, size and extent. Our preliminary results are encouraging. However, routine clinical applications will require real-time 3-D reconstruction which may be achieved by using automatic contour detection. This method is under development in our laboratory. Our approach of cartilage surface irregularities characterization is limited by the orientation of the patella relative to the transducer displacement. In fact, if the patella axis are angled relative to the scanning axis of the transducer the B-scan curvature will be deformed and the resulting χ_S^2 will not represent an exact index of surface roughness. Better appreciation of cartilage morphology and in particular optimization of cartilage surface analysis require control of cartilage orientation relative to the motorized transducer displacement and analysis of 3-D data.

REFERENCES

1. Aubray JH Jr, Sokoloff L, Pathology of osteoarthritis. In: Arthritis and Allied Conditions. Eleventh edition. Edited by DJ McCarty. Philadelphia, Lea & Febiger, 1571 (1989).

2. Adams ME, Wallace CJ, Quantitative imaging of osteoarthritis. Sem. Arth. Rheum. 20: 26 (1989).

3. Saïed, A., Chérin, E., Gaucher, H., Laugier, P., Netter, P., Berger, G., In vitro quantitative assessement of early and progressive lesions in a model of experimental osteoarthritis by high resolution ultrasound, IEEE Ultrasonics Symposium, 95CH35844, 1295 (1995).

4. Saïed, A., Cherin, E., Gaucher, H., Laugier, P., Gillet, P., Netter, P., Berger, G., In Vitro 50 MHz Echography of early and progressive articular cartilage and subchondral bone changes in a chemically induced osteoarthritis (in press).

5. Agemura DH, O'Brien WD Jr, Olerud JE, Chun LE, Eyre DE, Ultrasonic propagation properties of articular cartilage at 100 MHz. J. Acoust. Soc. Am. 87: 1786 (1990).

IMPROVEMENT ON QUALITY OF ECHOCARDIOGRAMS

Yukihiro Abiko[1], Takashi Ito[2], and Masato Nakajima[1]

[1] Department of Electrical Engineering, Faculty of Science
and Technology, Keio University, Japan

[2] Aloka Co.,Ltd.,Japan

INTRODUCTION

Recently, since resolution of echo images have gradually improved more accurate tissue characterization utilizing echo images is anticipated. However, quantitative tissue characterization using echocardiograms is difficult in the existing circumstances, because significant deterioration is prevalent in echocardiograms. The primary cause for deterioration in image quality of echocardiograms is speckle noise components that appear on echo images. Since speckle noise is produced by random interference of ultrasonic pulse wave reflected from the body, it does not necessarily coincide with the minute structure of noise within the organic tissue construction itself. Therefore, it is necessary as a preprocess for tissue characterization to improve the quality of images by reducing speckle components without deteriorating the specular components that reflect the organic tissue structure.

One of the most effective methods in speckle noise reduction is the compounding method[1,2]. This method reduces only the randomly distributed speckle pattern by summing up small echo images that have weak correlation to the speckle pattern in the same cross section. In other words, the compounding method is able to reduce speckle noise components without deteriorating the specular components. However, when applying existing method to the heart, since the heart is constantly moving rapidly and it is enclosed within the lungs, sternum and costae, it is difficult to acquire many appropriate images. On the other hand, since cardiac motion is nearly periodic, it is possible to obtain multiple images in the same cross section. Hence, if it is possible to obtain synchronization echocardiograms, then it becomes possible to apply the Synchronized Summing Method which is used in improving SNR (Signal to Noise Ratio) of the periodic signal in the area of signal processing. It can be said that Synchronized Summing Method is a compounding method. Thereby, it is the purpose of this paper to improve the image quality of echocardiograms by compounding as the pre-process for the tissue characterization using echocardiograms.

METHOD OF IMPROVEMENT ON QUALITY OF ECHOCARDIOGRAMS

Temporally Compounding Method

Synchronized Summing Method is a method utilized in signal processing, since it is effective in improving SNR of the weak periodic signals that are buried noise. In order to briefly explain the concept of the principle of the noise reduction, an illustration of mono-dimensional periodic signal is described below. We discuss a case whereby it is possible to obtain synchronization of such signals. When the synchronized received signals $f_k(t), \{k|0, \ldots, N-1\}$, are represented using periodically transmitted signals $s(t)$ and the noise components $n_k(t)$ as

$$f_k(t) = s(t) + n_k(t), \tag{1}$$

the average for received signals shall be

$$\frac{1}{N} \sum_{k=0}^{N-1} f_k(t) = s(t) + \frac{1}{N} \sum_{k=0}^{N-1} n_k(t). \tag{2}$$

If it is given that these noise are random or gaussian, when $N \to \infty$, then second term of equation (2) closely approaches 0. Consequently, most of originally transmitted signal can be obtained. Assuming that speckle pattern occurs at random, they can be obtained. The synchronization that will be required can be easily achieved by ECG (electrocardiogram) gated imaging using the R-wave as a synchronizing impulse. When processing echocardiograms, multiple periods of images from the same phase are obtained and compounded as shown in Figure 1. This process shall be termed the "Temporally Compounding Method".

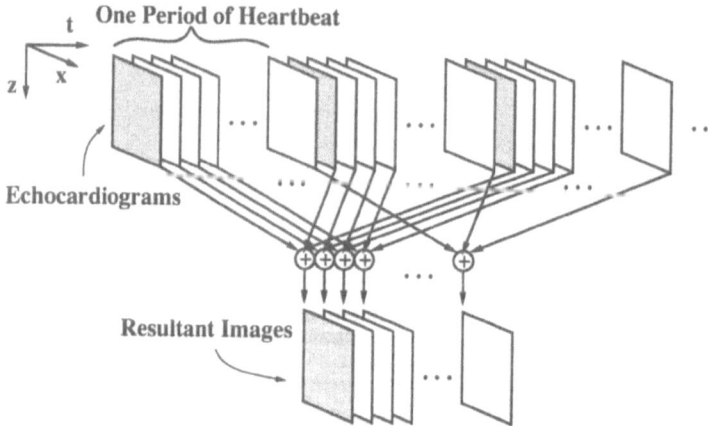

Figure 1. Temporally Compounding Method. Multiple periods of images from the same phase are obtained by synchronizing with the R-wave on ECG and compounded.

Correction in the Time Domain

Although, it is said that heartbeat is periodic, cardiac period is never absolutely constant and temporal fluctuations to some degree. Such fluctuations clearly blur the result images when compounded. Hence, it is becomes necessary to correct fluctuations in the time direction. Furthermore, when looking closely at cardiac motion, you will find local fluctuations which are describe to be corrected. Therefore, such fluctuations are corrected as below.

Fitting. In order to apply the Temporally Compounding Method without causing blurring to the images, images showing identical cardiac motion must be obtained from the same cross section. For this reason, it is necessary to fit the images beforehand. In time correction, one cardiac cycle of the acquired echo images is selected as the standard data against which other sets of the images are corrected and fitted in the follow method.

Since a great number of images are necessary in order to acquire adequate speckle reduction effect according to the above principle of Compounding Method, it is desirable that the correction process is as minimal as possible. Hence, the M-mode images represented two-dimensionally by time and depth axes are used for fitting. Because M-mode images have the characteristic to represent cardiac motion as a locus, it is considered most effective for performing corrections. M-mode images used in the procedure are computerized from B-mode images obtained as shown in Figure 2.

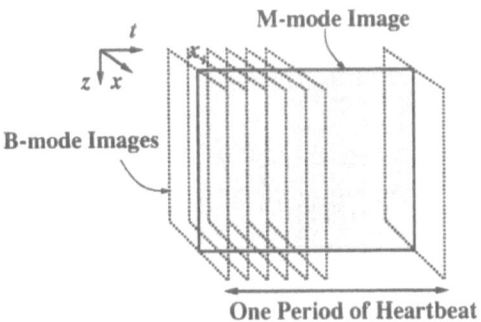

Figure 2. Method of making a M-mode image of B-mode images of a cardiac period. line images in the depth direction on a period of B-mode images in the same cross section are connected sequentially in the time domain.

In these M-mode images, each frame from B-mode images are represented as a line in the depth direction. On the whole, there are no significant differences in cardiac motion of various periods. Therefore, standard data images are not fitted to every phase but rather, set of phases $tm_k, \{k|0,\ldots,K-1\}$, are selected at appropriate intervals. Here, k indicates the number of samples. Set of phases that are fitted to tm_k are represented as td_k. In order to detect td_k, line images (size: $1 \times Z$ pixel) composed from standard data on phase tm_k in M-mode images are utilized as template images T_k for template matching on pre-inputted M-mode images as shown in Figure 3. Specifically speaking, template images are scanned two-dimensionally within the restricted searching area to calculate the normal correlation coefficient $C_k(t,z)$ from template images T_k, and overlapping partial images between the inputted M-mode images and template images. Thus, td_k is determined as the position of T_k in the time domain where $C_k(t,z)$ reaches the maximum. However, when conducting template matching, it is feared that td_k can be wrongly fitted if the searching area is not restricted appropriately. One reason accountable for this is the possible close resemblance between phase images of heart in diastolic state and those images of heart in systolic state. On the other hand, with regard to cardiac motion, the interval in systole is nearly constant as opposed to the interval in diastole which is variant. Hence, it is believed that fitting error will be decreased if searching area is restricted by systolic and diastolic intervals for every input data. For variations in cardiac motion of each cycle, since fluctuations are finite in the time domain, searching area is restricted by the following equation.

$$\text{In systole state:} \quad tm_k \cdot \frac{t'_{es}}{t_{es}} \pm w_s$$

$$\text{In diastole state:} \quad (tm_k - t_{es}) \cdot \frac{t'_{ed} - t'_{es}}{t_{ed} - t_{es}} + t'_{es} \pm w_d$$

In equation (3), t_{es} and t'_{es} represent the standard data and the phase of contraction term for the input data respectively. t_{ed} and t'_{ed} represent the second phase of diastole in the standard data and the input data respectively. w_s and w_d respectively represent the range of searching area in the time domain for the systolic and diastolic state which are then assigned appropriate values determined by the intervals in tm_k. Then, the range of searching area in the depth domain was set at ± 5 pixel.

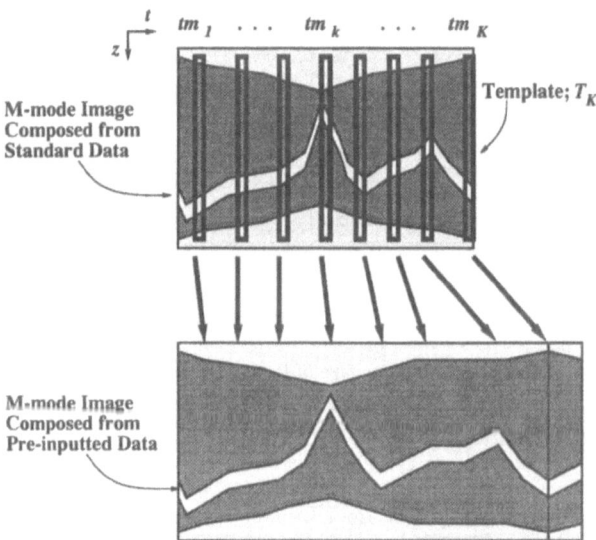

Figure 3. Fitting locally in the time direction. Set of phases tm_k to fit are selected at appropriate intervals. Next, line images T_k (size: $1 \times Z$ pixel) on phase tm_k in M-mode image composed from standard data are utilized as template images for template matching on pre-inputted M-mode images.

Local Expansions/Contractions. Phases tm_k and td_k are utilized to conduct local correction so that input data is fitted to standard data. In this procedure, the range of local area in the time domain that is caught between phases td_k and td_{k-1} in the input data is expanded and/or contracted equally in the time domain so that $td_k - td_{k-1}$ becomes $tm_k - tm_{k-1}$. Upon conducting as many expansions/contractions as there are local areas (K-1), images of expanded/contracted local areas are connected so that they become sequential in the time domain as shown in Figure 4.

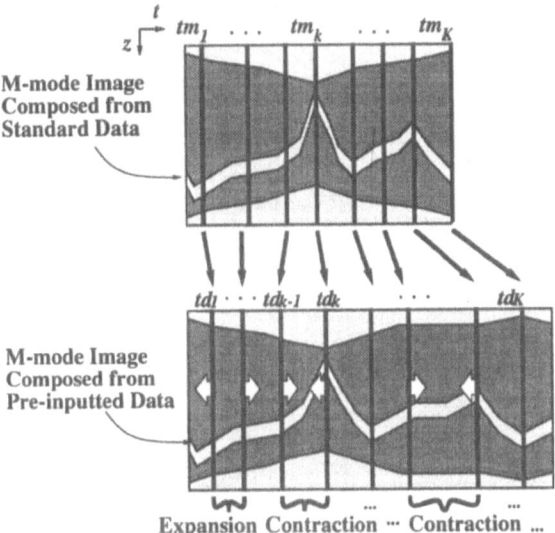

Figure 4. Expansions/contractions on the local areas in the time direction. the range of local area in the time domain that is caught between phases td_k and td_{k-1} in the input data is expanded and/or contracted equally in the time domain so that $td_k - td_{k-1}$ becomes $tm_k - tm_{k-1}$, and are connected so that they become sequential in the time domain. In fact, this procedure is utilized B-mode images.

Correction in the Depth Direction

When compounding, it is necessary to pre-position the subject of each echo image in the same phase so that it coincides with itself. Factors accounting for causing discrepancies in the positioning are, among others, variations in the pressures applied to the probe, the breathing of subjects, and irregular cardiac variations. Hence, images are transferred so that subjects of input data overlap with standard data in each phase when conducting correction. To transfer images, any discrepancies in positions between the standard data and input data must be detected in every phase beforehand. Incidentally, it is common for range resolution to be much more precise than azimuth resolution in echo images. Therefore, it is believed that discrepancies in positions in the depth direction have great influence on the blurring of result images when applying the Temporally Compounding Method.

Detecting Discrepancies in Positions. In order to detect these discrepancies, template matching is performed using M-mode images similarly to time correction. As shown in Figure 5, line images (size: $1 \times Z$ pixel) of all phases of M-mode images composed from standard data images are used as template images T_t. Subsequently, each template is used to perform template matching in the depth direction. In template matching, T_t is scanned in an one-dimensionally (only in the depth direction) restricted searching area of M-mode images composed from input data. Normal correlation coefficient $C_t(t, z)$ is then calculated from T_t and partial images that overlap with T_t on the inputted M-mode image at various positions of T_t . The distance d_t in which the template is transferred when $C_t(t, z)$ reaches a maximum within the searching area shall be detected as the width of discrepancy. The range of searching area in the depth direction shall be ± 5 pixel.

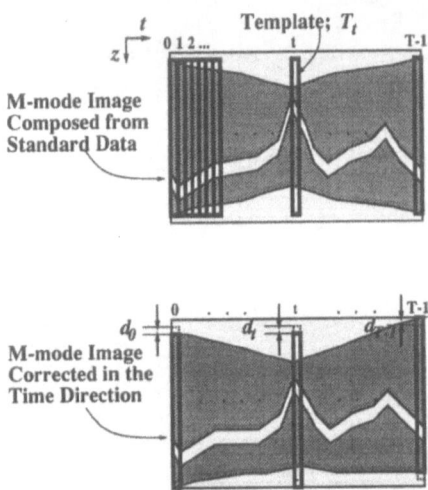

Figure 5. Detecting discrepancies in positions, d_t , in the depth direction. line images (size: $1 \times Z$ pixel) of all phases of M-mode images composed from standard data images are used as template images T_t. Subsequently, each template is used to perform template matching in the depth direction. The distance d_t in which the template is transferred when $C_t(t, z)$ reaches a maximum within the searching area shall be detected as the width of discrepancy.

Transferring images in the Depth Direction.   Next, the echo images of the inputted data is parallelly transferred to d_t in the depth direction to various phases as shown in Figure 6.

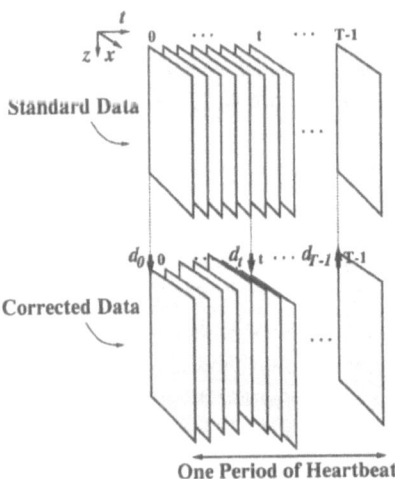

Figure 6.   Transferring images in the depth direction. Inputted B-mode images corrected in the time domain are parallelly transferred to d_t in the depth direction on various phases.

RESULT

To verify the effectiveness of this method, it is applied to 100 periods of echocardio-grams. Figure 7 (a) is an original echo image within the processing area, and (b) and (c) is an improved result image compounded from 20 and 100 echo images from the same phase using this method, respectively. As shown in (b), speckle pattern that appears in echocardio-grams has nearly disappeared and the very intense speckle reducing effect of the Temporal Compounding Method is recognized.

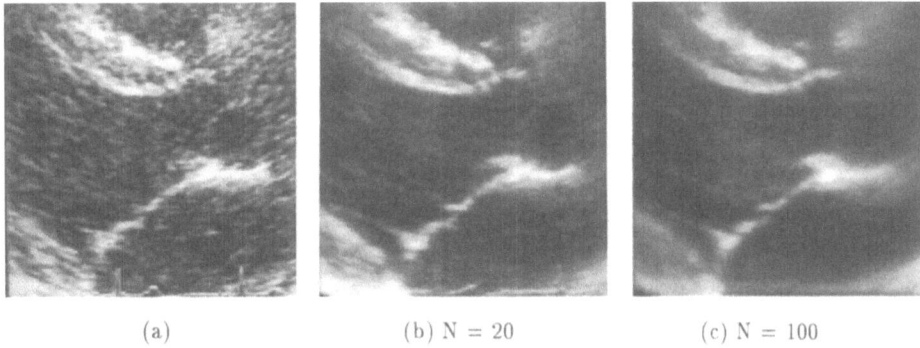

(a)   (b) N = 20   (c) N = 100

Figure 7. (a) is the original image (size:120×120), and (b) and (c) is the resultant images obtained by applying the Temporally Compounding Method.

DISCUSSION

To quantitatively evaluate the effectiveness of the improvement in the quality of images, SNR will be calculated for the result images obtained through this method. SNR of echo images is defined as the average brightness divided by the standard deviation of brightness in the local area of the result images. It is predicted for the SNR to become 1.91 theoretically and to improve in proportion to the square root of the number of overlapping images N. For comparison, SNR will be calculated for local areas of blood as well as for soft-tissues. It is believed that SNR will approach the ratio between the average value $\bar{s}$ of specular components $s(x, z)$ and the standard deviation value σ_s as N increases. Hence, since intensity of $s(x, z)$ on the local area of blood is weak and variation of its brightness distribution, σ_s approaches 0. Consequently, SNR improves in proportion to the root of N as shown in Figure 8. Since $s(x, z)$ in the local area of soft-tissue has variable distribution, when N is sufficiently large, σ_s is converged. Therefore, since $\bar{s}$ is scarcely dependent on N, SNR is converged also. In due consideration, it is believed that in local areas where SNR converges, specular components are preserved regardless of the reduction in speckle components. As shown in Figure 8, it is clear that in the local area of soft tissue, graph inclination is nearly horizontal and that SNR is converged. Hence, it can be said that when this method is applied to reduce speckle noise components, specular components are simultaneously preserved. As shown in Figure 8, when about N = 20, graph inclination of soft-tissue begin converging. Incidentally, as shown in Figure 7, when N = 20, the resultant image is improved image quality enough. Therefore, it is believe that required number of compounded images is approximate 20 for improvement of image quality.

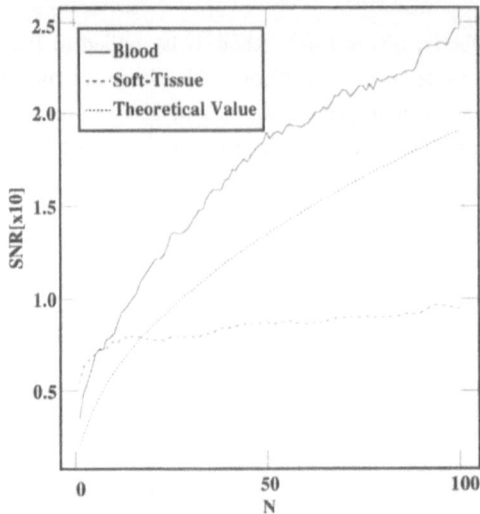

Figure 8. The experimental relation between the number of the compounded images and SNR of the result images. The horizontal axis shown in this figure represents the number of compounded images to be summed. The vertical axis represents the SNR of improved image quality. The plotted values are the average SNR calculated from 10 local are as (size: 6×6 pixel).

CONCLUSION

In htis paper, we examined the Temporal Compounding Method to improve the image quality of echocardiograms. In order to correct fluctuations, a method is proposed whereby any one cycle of echo images is selected as a standard. Then, utilizing M-mode images, partial temporal expansion and/or contraction is performed other B-mode images in the time direction, and each B-mode image is positioned in the depth direction to overlap standard data on same phase in the depth direction. To verify the effectiveness of this method, after applying the method to 100 heartbeats of echocardiograms, it was confirmed that speckle noise components were significantly reduced and quality of images drastically improved for echocardiograms. Furthermore, when SNR was calculated against the result, it was confirmed that reducing speckle noise components was possible without deteriorating specular components by applying this method.

REFERENCES

1. G.E.Trahey, J.W.Allison, S.W.Smith, O.T.von Ramm: "Speckle Reduction Achievable by Spatial Compounding and Frequency Compounding: Experimental Results and Implications for Target Detectability", International Symposium on Pattern Recognition and Acoustical Imaging, SPIE **768**, pp.185-192(1987).

2. M.Unser, L.Dong, G.Pelle, P.Brun, M.Eden: "Restoration of echocardiograms using time warping and periodic averaging on a normalized time scale", Proc. Medical Imaging III; SPIE **1092**, pp.84-93(1989).

APPLICATION OF ORDER STATISTICS FILTERS TO CARDIAC BOUNDARY DETERMINATION IN ULTRASOUND IMAGES

Marek Belohlavek and James F. Greenleaf[1]

Division of Cardiovascular Diseases,
[1]Department of Physiology and Biophysics,
Mayo Clinic and Mayo Foundation, Rochester, MN 55905

INTRODUCTION

Nonlinear order statistics filters are useful for impulsive noise removal and edge preservation. Speckle is a source of impulsive noise in ultrasound images. Edge preservation is important for determination of cardiac boundaries in echocardiographic images. Additionally, order statistics filters also suppress Gaussian noise which is often present in echocardiograms. Pitas and Venetsanopoulos reviewed order statistics filters[2] and described a class of edge detectors[1] based on nonlinear means and medians. They showed that a linear combination of these filters, resulting in a ranked-order filter (that operates with a selected range of ordered maximum and minimum gray level values), will further improve the noise removal capabilities.

In this study, we report initial tests of a ranked-order filter in echocardiographic images of the left ventricle (LV).

METHODS

Both linear and nonlinear filters can be generalized as[3]

$$y_i = \sum_{j=1}^{n} a_j x_{(j)}, \qquad (1)$$

where $x_{(j)}$, $j = 1, ..., n$, are pixels encompassed within an $n = m \times m$ filter window and ordered according to gray scale values. In our implementation, $j = 1$ refers to a pixel with the lowest 8–bit gray scale intensity. Coefficients a_j, $j = 1, ..., n$, define a particular filter.

If a_j in equation (1) is set to $1/n$, the filter becomes an *order mean* filter, which, however, works the same way as a conventional mean filter. Substituting n in (1) by βn, $0 < \beta \le 1$, results in a *ranked-order mean* filter,

$$y_i = \sum_{j=1}^{\beta n} \frac{1}{\beta n} x_{(j)}, \tag{2}$$

where β controls the rank of the ordered input values $x_{(j)}$. If $\beta < 1$, the filter will produce the mean biased towards the low intensities.

If n in (1) is substituted by $n(1-\beta)+1$, i.e.,

$$y_i = \sum_{j=n}^{n(1-\beta)+1} \frac{1}{\beta n} x_{(j)}, \tag{3}$$

the resulting filter will bias the mean towards high intensities because the ordered input values $x_{(j)}$ are summed in the descending direction.

Taking the absolute value of the difference between the filters described by equations (2) and (3) results in a *customized ranked-order mean* (CROM) filter,

$$y_i = \left| \sum_{j=n}^{n(1-\beta)+1} \frac{1}{\beta n} x_{(j)} - \sum_{j=1}^{\beta n} \frac{1}{\beta n} x_{(j)} \right| = \frac{1}{\beta n} \left| \sum_{j=n}^{n(1-\beta)+1} x_{(j)} - \sum_{j=1}^{\beta n} x_{(j)} \right|. \tag{4}$$

This filter generates intensity values proportional to the probability of a presence of the cardiac boundary.

RESULTS

Figure 1 shows an echocardiographic image of the LV and indicates a pathway used for filter testing. The pathway crosses the endocardial boundary in this example. A corresponding function produced by the CROM filter with a 17 x 17 pixel window is demonstrated in Figure 2. A peak of the filter output delimits the location of the LV cavity

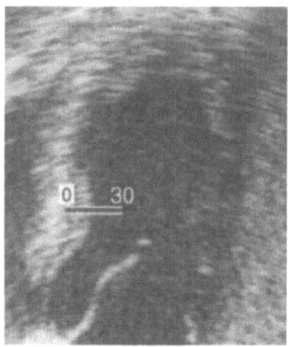

Figure 1. Clinical echocardiographic tomogram of the left ventricle. The customized ranked-order mean filter was tested by translating a 17 x 17 window across the endocardial border along the denoted 31 pixel long pathway.

boundary. This is indicated by the two arrows in the plot, i.e., the boundary region is located approximately in the middle of the filter window pathway which corresponds to what is obvious in Figure 1. In our tests, a mean of ordered values in the first rank has been employed. (Note: Practical implementation of the filter uses discrete ranking, rather than a rank definition by a continuous variable β. The zeroth rank means that only the lowest and highest pixel intensity values were respectively used for calculation of the two subtracted components of the CROM filter; the first rank means that 1 x m lowest and highest pixel intensity values were taken; etc.). In order to optimize the CROM filter for echocardiographic images, with a typical pixel size of 0.4mm in our case, we tested various ranks and sizes of the filter window and found that first rank and 13 x 13 pixel window size represent the best compromise between noise removal and edge preservation. Figure 3 demonstrates application of the CROM filter to different clinical echocardiographic tomograms. The brightness and location of the resulting bands in the filtered images define cardiac boundaries. Normalization of image gray scale intensities to a [0, 1] range would provide probabilities of the presence of the boundary for each pixel location.

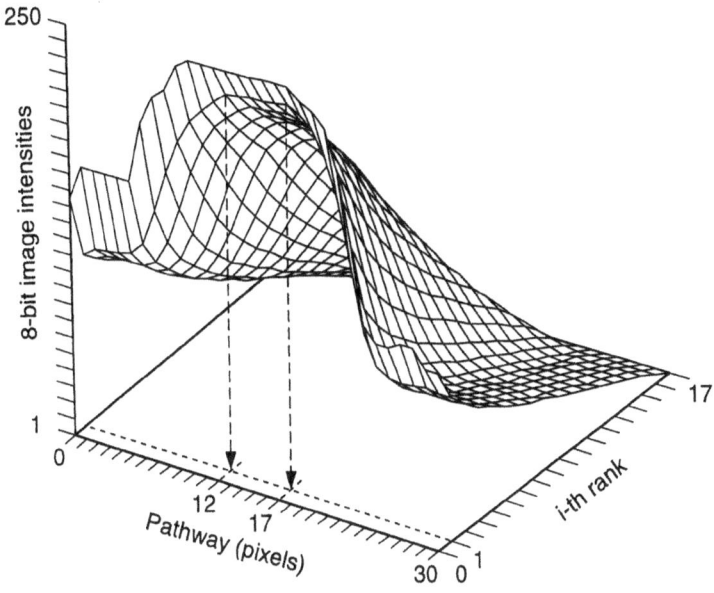

Figure 2. Output function of the customized ranked-order mean (CROM) filter to data across the LV boundary region (shown in Figure 1) using increasing rank values. The two vertical arrows indicate how the output peak of the first-rank CROM filter delimits location of the cardiac boundary.

DISCUSSION AND CONCLUSIONS

The CROM filter effectively converts the original image into a map of probabilities of the presence of boundaries — if its output is normalized to values between 0 and 1. In our case, the CROM filter was intended to provide clues about LV boundaries in echocardiographic images for a neural network algorithm. Noise resistance of the filter facilitated robustness of the algorithm while relatively broad bands (indicating the presence of cardiac boundaries, see Figure 3) provided a certain freedom for optimization of the final delineation of the LV cavity boundary in our particular application. Window size and rank adjustments represent, however, subjective factors in the filter performance. Our initial experience indicates that combinations of appropriate filter window sizes and ranks should allow to customize the filter so that the desired edges are preserved while impulsive and Gaussian noise is suppressed. A thorough study in this regard is yet to be done.

original image   filtered image

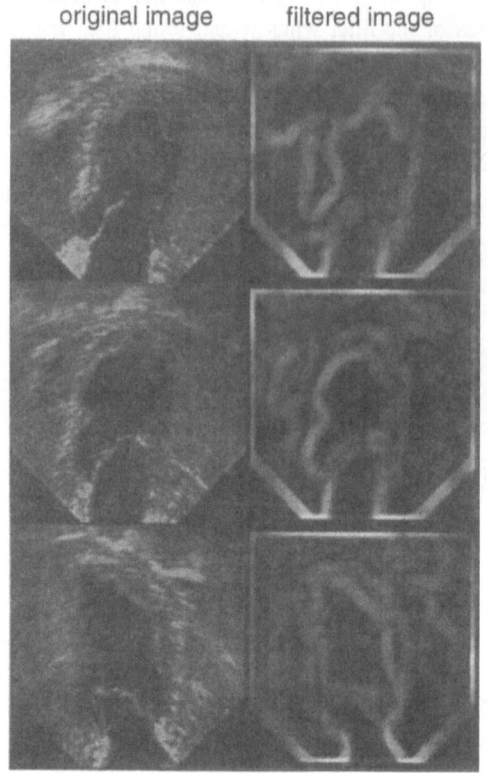

Figure 3. Application of the CROM filter to different clinical echocardiograms.

REFERENCES

1. I. Pitas and A.N. Venetsanopoulos, Edge detectors based on nonlinear filters, *Trans. Patt. Anal. Machine Intell.* 8:1893–1921 (1986).
2. I. Pitas and A.N. Venetsanopoulos. *Nonlinear Digital Filters: Principles and Applications*, Kluwer Academic, Boston (1990).
3. I. Pitas and A.N. Venetsanopoulos, Order statistics in digital image processing, *Proc. IEEE* 80:1893–1921 (1992).

RELATIONSHIP BETWEEN ULTRASONIC ATTENUATION, APPARENT INTEGRATED BACKSCATTER (30 TO 50 MHz) AND THE COMPOSITION OF ATHEROSCLEROTIC PLAQUE

S. Lori Bridal,[1] P. Fornès,[2] and G. Berger[1]

[1]Laboratoire d'Imagerie Paramétrique URA CNRS 1458 75006
[2]Laboratoire d'Anatomie Pathologique Hôpital Broussais 75014
  Paris, FRANCE

INTRODUCTION

Because of the important impact of atherosclerosis on public health, the evaluation of the potential evolution of atherosclerotic lesions and their response to interventional therapy is a subject of much current concern. Studies *in vitro* have demonstrated that the ultimate biological behavior of an atherosclerotic plaque depends, not only on its extent, but also on its biochemical composition and structure.[1] An *in vivo* means for identifying the composition and structure of atherosclerotic plaques would be useful in guiding diagnosis and treatment. Using appropriate signal analysis methods, the acoustic attenuation and the backscatter coefficient may be estimated from echographic signals.[2] These parameters are related to tissue structure and composition and may therefore, offer additional quantitative information useful for the characterization of atherosclerotic plaques.[3-12] We have recently described a method for the construction of quantitative images using local estimations of the attenuation and the apparent integrated backscatter.[11] The current work examines the sensitivity of measurements of the ultrasonic attenuation and the apparent integrated backscatter (30 to 50 MHz) to differences in the local tissue composition in the arterial wall (collagen, lipids, calcifications, and media).

MATERIALS AND METHODS

Backscattered rf from 58 excised segments of human thoracic aorta were acquired using a 50 MHz center frequency acoustic backscatter microscope which has been previously described.[12] For each segment, data were obtained across a total x-y scan plane 1 cm by 0.5 mm with 100 μm between rf lines. Each signal was averaged temporally 256 times, digitized (2000 sample points, 400 MHz) and stored for off-line analysis. After data acquisition was completed, the specimen was removed from the saline bath, and India ink was injected at the limits of the scanned region to provide a marker visible in histologic sections.

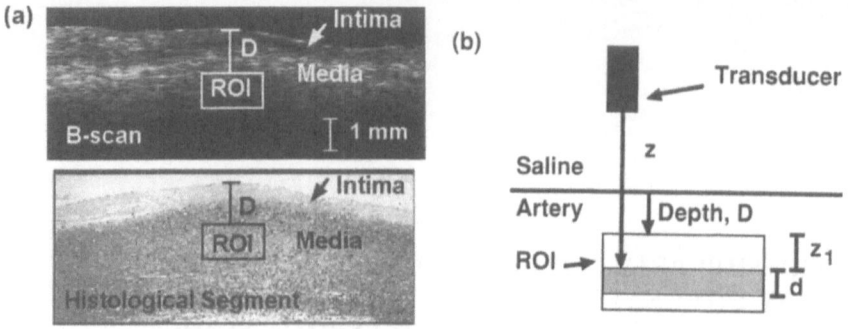

Figure 1. Panel (a): The position of each ROI was noted on the ultrasonic B-scan image constructed from a central scan plane of each arterial segment. The ROI was then relocated on the corresponding histologic section. Panel (b): Schematic diagram of the coordinates of a windowed layer of depth d of radiofrequency data within a ROI used to estimate the integrated attenuation and backscatter.

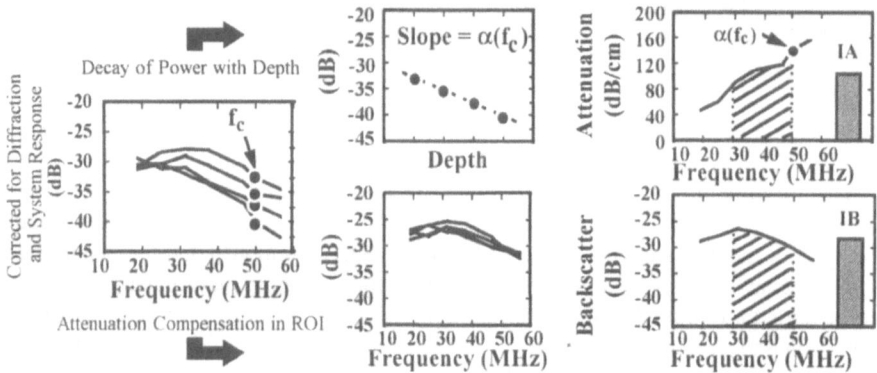

Figure 2. Algorithm for the spectral calculation of the integrated backscatter and attenuation.

Following fixation in 4 % formalin, a 3 mm thick slice of the fixed aorta centered upon the ink-marked region of each scan was embedded in paraffin. Two sections, 5 μm thick, were obtained from each paraffin block and stained with hematoxylin-eosin-saffron for nuclei, cytoplasm and collagen and with orcein for elastic fibers. Regions of interest (ROI), N = 124, were delineated on B-scan images constructed from a central plane of each scanned arterial segment as diagrammed in Figure 1, panel (a). The position of each ROI was noted on the ultrasonic B-scan image, and then each ROI was relocated on the histologic segments obtained from the scanned segment of the artery. The ROI was classified using a semi-quantitative evaluation of the proportion of collagen, calcium and lipids. Measurement regions were identified as: normal media (M), dense collagen (DC), collagen/lipidic (C-L), lipidic (LIP), and containing calcifications (CAL).

Within each ROI, a sliding Hamming window of length d (64 samples, 120 μm) was stepped along each rf line with 50 % overlap (Figure 1, panel (b)). The rf signal in each windowed region was Fourier transformed, squared in magnitude and averaged with the power spectra from the windowed regions of the other rf lines at the same depth z_1 in the ROI to provide a spatially averaged apparent (not compensated for attenuation) backscattered power spectrum $\langle S_m(f,z) \rangle$. The variable z represents the axial distance between the transducer and the center of the 64 sample point windowed region and the brackets denote the spatial averaging. Each power spectrum $\langle S_m(f,z)\rangle$ was corrected for the system response and diffraction effects according to:

$$\langle S_{dif\_corr}(f,z)\rangle = \frac{\langle S_m(f,z)\rangle}{S_p(f,z)} \tag{1}$$

where $S_p(f,z)$ is the spectrum obtained from a reference echo acquired from a polished steel plate normal to the ultrasonic beam with distance z between the plate and the transducer.

As diagrammed in Figure 2, the attenuation in the ROI $\alpha_{ROI}(f)$ was estimated from the decay of these average spectra $\langle S_{dif\_corr}(f,z)\rangle$ as a function of depth using the multinarrow-band method.[2,13] The average value of the attenuation between 30 and 50 MHz was calculated yielding the Integrated Attenuation (IA). Each average spectrum $\langle S_{dif\_corr}(f,z)\rangle$ in a given ROI was then compensated according to equation 2 for the frequency dependent attenuation estimated in the ROI. The third term in equation 2 partially compensates for the frequency dependence of the volume insonified by the transducer where F is the focal length of the transducer, a is the transducer radius, k is the wavenumber, and R_p is the amplitude reflection coefficient of the steel plate (assumed to be one). This compensation term assumes that the directivity of the transducer may be approximated by a gaussian profile and is strictly valid only at the focal length of focused transducers.[14] A compensation for the Hamming gating function and the gate length d would also need to be taken into account to arrive at the fully compensated backscatter coefficient.[2]

The resulting spectra $\langle S_{bs}(f,z)\rangle$ from all depths in each ROI were averaged and used to estimate the Integrated Backscatter (IB) between 30 and 50 MHz as illustrated in Figure 2. The IB is compensated for the attenuation within the ROI itself but remains non compensated for the attenuation between the artery surface and the ROI across the depth D.

$$\langle S_{bs}(f,z)\rangle = \langle S_{dif\_corr}(f,z)\rangle \frac{4\alpha_{ROI}(f)d \times e^{4\alpha_{ROI}(f)\left(z_i-\frac{d}{2}\right)}}{1-e^{-4\alpha_{ROI}(f)d}} \times \frac{R_p^2 k^2 a^2}{8\pi\left[1+\left(\frac{ka^2}{4F}\right)^2\right]} \tag{2}$$

RESULTS

Ultrasonic measurements were grouped according to the histological classifications. The average integrated attenuation ± SD estimated by the multinarrow-band method were media 97 ± 20 dB cm^{-1}, dense collagen 107 ± 33 dB cm^{-1}, collagen/lipidic 142 ± 51 dB cm^{-1}, lipidic 139 ± 53 dB cm^{-1}, and containing calcifications 245 ± 93 dB cm^{-1}.[9] For any given tissue type the estimated value of the integrated backscatter decreased as a function of increasing ROI depth due to the shadowing effect of the overlying tissue. This is demonstrated for regions of media and dense collagen plaque in Figure 3. All regions of media for which the IB is plotted in Figure 3 were free of significant plaque, and thus, the predominate shadowing tissue is normal aortic media. Most regions for IB estimation in collagen plaque were within a more extensive layer of collagen plaque. Thus, for the two cases plotted in Figure 3 the shadowing tissue is primarily normal media or collagen plaque. A linear fit was performed to estimate the rate of decay of the integrated backscatter with depth in the tissue. The slope of these fit lines is on the order of the attenuation estimated for media and dense collagen tissue using the multinarrow-band method.

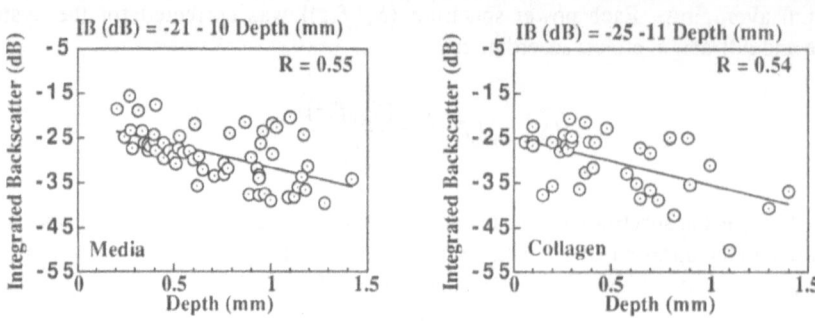

Figure 3. Integrated backscatter of media and collagen as a function of ROI depth beneath the artery surface.

For each tissue type, all values of the integrated backscatter obtained from ROIs at depths between 300 µm and 800 µm were averaged together. The resulting average integrated backscatter values are compared in Figure 4, panel (a). The average integrated backscatter of each tissue group is plotted against the attenuation in Figure 4, panel (b). In these graphs the error bars represent the standard deviation of the mean. The statistical significance of differences was tested between groups using a nonparametric Mann Whitney U test. Neither parameter permits a significant discrimination between normal media and dense collagen plaque. Both parameters provide significant separation between normal media or dense collagen and lipidic regions ($p < 0.05$) and regions containing calcifications ($p < 0.01$). The attenuation values do not permit a separation between lipidic and collagen lipidic tissue, but the values of integrated backscatter for these two groups are significantly different ($p = 0.02$). The attenuation permits a discrimination between collagen-lipidic regions and media ($p = 0.005$) or dense collagen plaque ($p = 0.01$) which is not obtained based upon the integrated backscatter values.

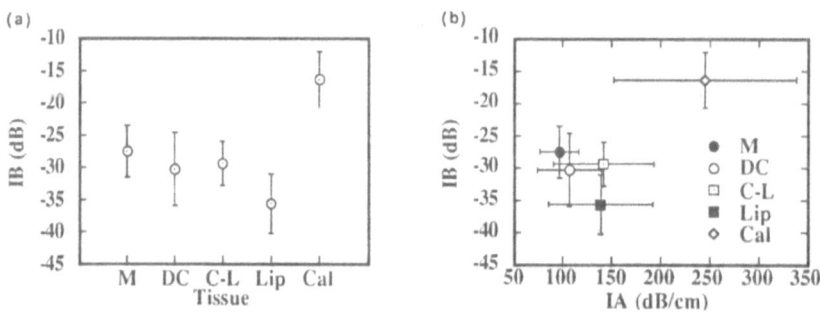

Figure 4. Panel (a): The average integrated backscatter as a function of tissue type. Panel (b): The average value of the attenuation and the integrated backscatter for each tissue type.

As has been shown previously,[11] both attenuation and apparent integrated backscatter can be presented in the form of high resolution parametric images. Parametric images of two segments of atherosclerotic aorta are shown in Figures 5 and 6 with the corresponding histological sections. Parametric images of the integrated backscatter have a resolution of 200 µm lateral by 120 µm axial (depth) by 500 µm slice thickness with 50 % axial overlap between pixels. Because attenuation estimates depend on the measurement of the spectral decay with depth these images have decreased axial resolution (420 µm) where estimates are performed with an 85 % axial overlap in pixels. Several ROIs have been selected in the parametric images of Figures 5 and 6. The comparison between tissue type and parametric values in these regions with the ranges of values associated with different tissues summarized in Figure 4 demonstrates that the two parameters provide complementary information for the identification of local plaque composition.

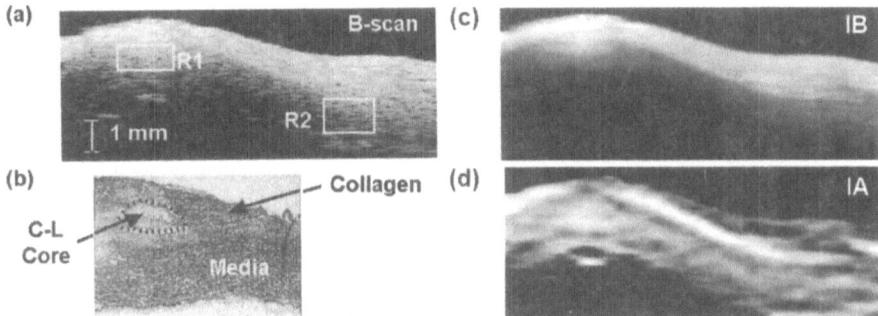

Figure 5. (a) B-Scan image, (b) histologic section (c) IB and (d) IA parametric images showing a plaque containing a collagen-lipidic core (C-L) surrounded by a dense collagen cap. The average IB and IA in R1 (Collagen-lipidic) were - 30 dB and 151 dB/cm and in R2 (normal media) were -33 dB and 98 dB/cm.

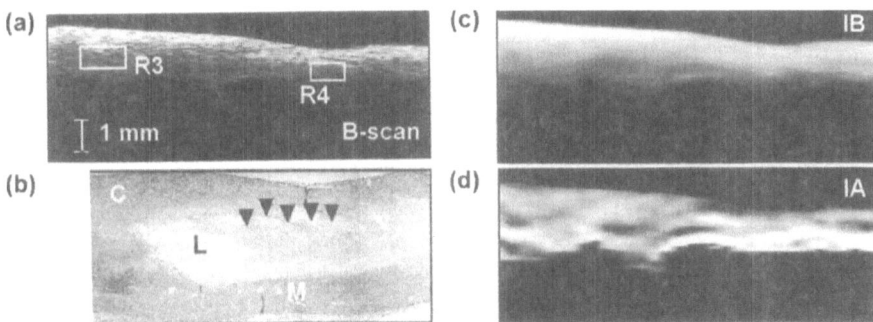

Figure 6. (a) B-Scan image, (b) histologic section (c) IB and (d) IA parametric images of a plaque with a lipidic core (L) containing calcific deposits (denoted by arrowheads) distributed throughout the region of the core towards the right. The average IB and IA in R3 (dense collagen) were - 27 dB and 115 dB/cm and in R4 (lipidic tissue intermixed with calcification) were -18 dB and 292 dB/cm.

DISCUSSION

This work reports high resolution measurements of both IB and IA in human atherosclerotic plaque. The decay of IB measurements with increasing ROI depth was found to be consistent with attenuation estimates obtained from ROIs using a multinarrow-band method. Both attenuation and integrated backscatter demonstrated sensitivity to local atherosclerotic plaque composition. Furthermore, the two parameters provide complementary information for the differentiation of media, collagen-lipid and lipid regions. The uncorrelated nature of these two parameters is of particular interest because it suggests that the attenuation and backscatter are sensitive to different aspects of the tissue composition.

Results at lower frequencies (2 to 10 MHz) reported in the literature[3-6] also provide evidence that ultrasonic backscatter and attenuation contain quantitative information useful for plaque identification. Significant increases in integrated backscatter have been demonstrated between lipidic, fibrofatty and calcificied plaque.[3,4] The global attenuation measured across the arterial wall has been shown not to be significantly different between normal and fibrous arteries but to be significantly elevated across fibrofatty and calcified segments of artery.[5,6]

The results reported here for IA and IB from 30 to 50 MHz are in accord with these lower frequency results. This work represents the first correlation between high resolution estimates of both the IB and IA with local human atherosclerotic plaque composition. Moreover, the estimation of both IA and IB from backscattered rf signals analogous to the signals currently acquired by clinical imaging systems and the display of these estimations in the form of images represent important advances towards the practical application of these parameters for the *in vivo* characterization of the composition and structure of atherosclerotic plaque.

REFERENCES

1. R.T. Lee, S.G. Richardson, H.M. Loree et al. Prediction of mechanical properties of human atherosclerotic tissue by high-frequency intravascular ultrasound imaging: An *in vitro* study, Arteriosclerosis and Thrombosis. 12:1 (1992).

2. V. Roberjot, S.L. Bridal, P. Laugier, and G. Berger, Absolute backscatter coefficient over a wide range of frequencies in a tissue-mimicking phantom containing two populations of scatterers, IEEE Trans UFFC. 43:970 (1996).

3. B. Barzilai, J.E. Saffitz, J.G. Miller, and B.E. Sobel, Quantitative ultrasonic characterization of the nature of atherosclerotic plaques in human aorta, Circ Res. 60:459 (1987).

4. M.P. Urbani, E. Picano, G. Parenti et al, *In vivo* radiofrequency-based ultrasonic tissue characterization of the atherosclerotic plaque, Stroke. 24:1507 (1993).

5. J.F. Greenleaf, F.A. Duck, W.F. Samayoa, and S.A. Johnson, Ultrasonic data acquisition and processing system for atherosclerotic tissue characterization, IEEE Ultrason Symp. 74CH0896-ISU:738 (1974).

6. E. Picano, L. Landini, A. Distante et al, Fibrosis, lipids, and calcium in human atherosclerotic plaque: *In vitro* differentiation from normal aortic walls by ultrasonic attenuation, Circ Res. 56:556 (1985).

7. G.R. Lockwood, L.K. Ryan, J.W. Hunt, and F.S. Foster, Measurement of the ultrasonic properties of vascular tissues and blood from 35-65 MHz, Ultrasound Med Biol. 17:653 (1991).

8. M.G.M. DeKroon, L.F. van der Wal, W.J. Gussenhoven, H. Rijsterborgh, and N. Bom, Backscatter directivity and integrated backscatter power of arterial tissue, Int J of Cardiac Imaging. 6:265 (1991).

9. L.S. Wilson, M.L. Neale, H.E. Talhami, and M. Appleberg, Preliminary results from attenuation-slope mapping of plaque using intravascular ultrasound, Ultrasound Med Biol. 20:529 (1994).

10. S.A. Wickline, J.G. Miller, D. Recchia et al, Beyond intravascular imaging: Quantitative ultrasonic tissue characterization of vascular pathology. IEEE Ultrasonics Symposium. 3:1589 (1994).

11. S.L. Bridal, P. Fornès, P. Bruneval, and G. Berger, Parametric (integrated backscatter and attenuation) images constructed using backscattered radio frequency signals (25 to 56 MHz) from human aortae *in vitro*, Ultrasound Med Biol. 23:215 (1997).

12. S.L. Bridal, P. Fornès, P. Bruneval, and G. Berger, Correlation of Ultrasonic Attenuation (30 to 50 MHz) and Constituents of Atherosclerotic Plaque, Ultrasound Med Biol. in press (1997).

13. M. Fink, F. Hottier, and J.F. Cardoso, Ultrasonic signal processing for *in vivo* attenuation measurement: Short time Fourier analysis, Ultrasonic Imaging. 5:117 (1983).

14. M. Ueda and Y. Ozawa, Spectral analysis of echoes for backscattering coefficient measurement, J. Acoust Soc Amer. 77:38 (1985).

NONINVASIVE EVALUATION OF SPATIAL DISTRIBUTION OF LOCAL INSTANTANEOUS STRAIN ENERGY IN HEART WALL

Hiroshi Kanai,[1] Hideyuki Hasegawa,[1] Noriyoshi Chubachi,[1]
Yoshiro Koiwa,[2] and Motonao Tanaka[3]

[1]Department of Electrical Engineering,
 Tohoku University, Sendai 980-77, Japan
[2]First Department of Internal Medicine, School of Medicine,
 Tohoku University, Sendai 980-77, Japan
[3]Tohoku Welfare Pension Hospital,
 Takasago 10, Fukumuro, Miyagino-ku, Sendai 983, Japan

INTRODUCTION

For the noninvasive diagnosis of heart disease based on the acoustic characteristics of the heart muscle, we have developed a new method[1] for accurately tracking the movement of the heart wall. By this method, a *velocity signal* of the heart wall with a small amplitude of less than 10 μm on the *motion* resulting from a heartbeat with large amplitude of 10 mm can be successfully detected with sufficient reproducibility in the frequency range up to several hundred Hertz continuously for periods of about ten heartbeats. The method has been applied to multiple points preset in the left ventricular (LV) wall along the ultrasonic beam[2] so that the spatial (depth) distributions of the velocity at these points are simultaneously obtained. From the resultant velocity signals, the motion of the heart wall is divided into the following two components: *parallel global motion* of the heart wall and *the change in myocardial layer thickening* at each depth across the LV wall during *myocardial contraction/relaxation*. In this paper, from the local change in thickness, spatial distribution of instantaneous strain energy generated and reserved in myocardium is noninvasively evaluated. This new approach offers potential for research on noninvasive acoustical diagnosis of *myocardial local motility*, that is, the myocardial layer function in the ventricular wall.

MEASUREMENT OF LOCAL THICKENING IN HEART WALL

We herein consider the detection of velocity signals $\{v(x_i; t)\}$ at multiple points $\{i\}$ preset in the LV wall as illustrated in Fig. 1(a) using ultrasound from the chest wall. The instantaneous position in the depth direction of the ith point set in the LV wall is denoted by $x_i(t)$ as shown in Fig. 1(b). The component of the motion *parallel* to the ultrasonic beam is shown in Fig. 1(c). The parallel components at all points along the ultrasonic beam in Fig. 1(c) are the same in phase and magnitude if the beam is perpendicular to the heart wall. The LV wall motion during the cardiac cycle is much larger in magnitude compared with the thickness change. The LV wall shows periodic thickening and thinning along with *myocardial contraction and relaxation* as illustrated in Figs. 1(b) and 1(d). The difference between position $x_i(t)$ of the ith point and position $x_{i+1}(t)$ of the $(i+1)$th point shows the *local thickness* at every instant of time during one cardiac cycle as shown in Fig. 1(d). The thickness change corresponds to the local motility of the myocardial layer across the LV wall.

To establish this noninvasive method for the local characteristics across the LV wall using ultrasound, we have proposed a noninvasive transcutaneous method[1] for detecting small *velocity* signals on the surface of the heart wall. By calculating the constraint cross-correlation function between the sequentially received echoes, the phase change caused by displacement of the ith preset point during the pulse repetition period ΔT is accurately determined and the average velocity $v(x_i; t)$ during the period is obtained. By adding the product of $v(x_i; t)$ and ΔT to the previous object position $x_i(t)$, the next position $x_i(t + \Delta T)$ is estimated. The detected velocity signal shows rapid motion of the heart wall including high frequency components with small amplitudes, which are difficult to recognize by M-mode echocardiography. The validity of the proposed method has been confirmed by experiments[1] using a water tank and has been applied to the *in vivo* detection of small *velocity* signals, with sufficient reproducibility, on the wall of the human heart.

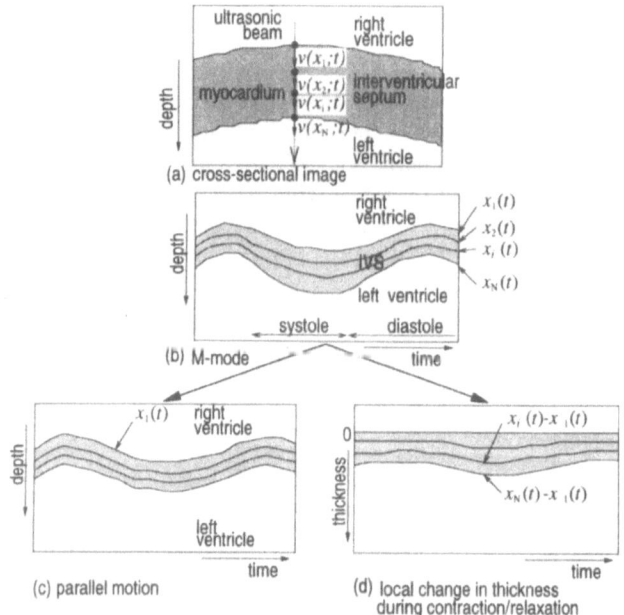

Figure 1. An illustration explaining the process of the evaluation of the local change in thickness during myocardial contraction/relaxation based on noninvasive detection of velocity signals $\{v(x_i; t)\}$ and the instantaneous positions $\{x_i(t)\}$ of multiple N points $\{i\}$ in the heart wall.

PRINCIPLE OF EVALUATION LOCAL STRAIN ENERGY

By applying the above method to each of the multiple N points in the heart wall, the instantaneous object positions $x_i(t)$, and the velocity signals $v(x_i; t)$ are obtained for $i = 1, 2, \cdots, N$ as shown in Fig. 1(b). The initial positions (depth) $\{x_i(t_0)\}$ of the N points $\{i\}$ at a time t_0 are manually preset at even intervals of Δx_S along the ultrasonic beam in the heart wall, where Δx_S is the spatial resolution in the depth direction and $\Delta x_S = 0.75$ mm in the following experiments. We employ the time t_0 at 200 ms before the first R-wave during the A/D conversion, regarding it as near the end-diastole. Thus, the initial positions are given by

$$x_i(t_0) = x_1(t_0) + \Delta x_S \times (i - 1), \tag{1}$$

where $i = 1, 2, \cdots, N$. For example, for the presetting of the initial positions in the interventricular (IVS) with reference to the longitudinal M-mode image, $x_1(t_0)$ and $x_N(t_0)$ correspond to the position on the surface of the right ventricle (RV) and the surface of the left ventricle (LV) of the IVS, respectively. For the setting in the LV posterior wall, $x_1(t_0)$ and $x_N(t_0)$ correspond to the positions on the endocardium and the epicardium in the wall, respectively.

Let us assume that the ultrasonic beam is perpendicular to the heart wall during one cardiac cycle as shown in Fig. 1(a) and that the velocity direction of each point i in the heart wall is parallel to the direction of the ultrasonic beam. Thus, from the difference between the instantaneous object position $x_i(t)$ and the instantaneous position $x_{i+1}(t)$ of the sequentially set point, the *thickness* of the local region between points (i) and $(i + 1)$, which is denoted by $h_i(t)$, is obtained by

$$h_i(t) \quad = \quad x_{i+1}(t) - x_i(t) > 0. \quad [\text{m}] \tag{2}$$

Since the depth axis of Fig. 1(b) is positive in the downward direction, the thickness $h_i(t)$ in Eq. (2) is always positive as shown in Fig. 1(d).

On the other hand, since $x_i(t + \Delta T) = v(x_i; t) \times \Delta T + x_i(t)$ and $x_{i+1}(t + \Delta T) = v(x_{i+1}; t) \times \Delta T + x_{i+1}(t)$, the instantaneous speed $\Delta h_i(t)$ of the local change in thickness is obtained by

$$\Delta h_i(t) \quad = \quad \frac{h_i(t + \Delta T) - h_i(t)}{\Delta T}$$

$$= \quad v(x_{i+1}; t) - v(x_i; t). \quad [\text{m/s}] \tag{3}$$

By dividing the change in thickness, $h_i(t) - \Delta x_S$, from the thickness Δx_S at the end-diastole, by Δx_S, the strain $\varepsilon_i(t)$, which occurs in the assumed homogeneous myocardium between points (i) and $(i + 1)$, is defined by

$$\varepsilon_i(t) \quad = \quad \frac{h_i(t) - \Delta x_S}{\Delta x_S}. \tag{4}$$

By denoting the Young's modulus of the myocardium by E [Pa], the strain energy $U_i(t)$ generated and reserved in ith layer of the myocardium is given by

$$U_i(t) = \frac{1}{2} E \cdot \varepsilon_i^2(t), \quad [\text{Pa}] \tag{5}$$

where we assume the myocardium is isotropic. The increasing rate of the strain energy during the pulse repetition interval ΔT, denoted by $\Delta U_i(t)$, is defined by

$$\Delta U_i(t) = \frac{U_i(t + \Delta T) - U_i(t)}{\Delta T}. \quad [\text{Pa/s}] \tag{6}$$

189

Using Eqs. (5), (4), and (3),

$$\Delta U_i(t) = \frac{1}{2}\frac{E}{\Delta T}\{\varepsilon_i^2(t+\Delta T) - \varepsilon_i^2(t)\}$$

$$= \frac{E}{2\Delta T \cdot \Delta x_S^2}\{(h_i(t+\Delta T) - \Delta x_S)^2 - (h_i(t) - \Delta x_S)^2\}$$

$$= \frac{E}{2\Delta T \cdot \Delta x_S^2}\{(h_i(t) - \Delta x_S + \Delta h_i(t) \cdot \Delta T)^2 - (h_i(t) - \Delta x_S)^2\}$$

$$= \frac{E}{2\Delta T \cdot \Delta x_S^2}\{2(h_i(t) - \Delta x_S)\Delta h_i(t) \cdot \Delta T + (\Delta h_i(t) \cdot \Delta T)^2\}$$

$$\approx \frac{E}{2\Delta T \cdot \Delta x_S^2}2(h_i(t) - \Delta x_S)\Delta h_i(t) \cdot \Delta T$$

$$= E \cdot \frac{x_{i+1}(t) - x_i(t) - \Delta x_S}{\Delta x_S} \cdot \frac{v(x_{i+1};t) - v(x_i;t)}{\Delta x_S} \tag{7}$$

However, the Young's modulus E is unknown and its value highly depends on the strain. In actual *in vivo* measurement, therefore, the strain energy $U_i(t)$ of Eq. (5) and its increasing rate $\Delta U_i(t)$ of Eq. (7) are divided by E and the resultant $U_i(t)/E$ and $\Delta U_i(t)/E$ are evaluated. The spatial distributions $\{U_i(t)/E\}$ and $\{\Delta U_i(t)/E\}$ are color coded and superimposed on the M-mode image. The results quite differs from those obtained in the tissue Doppler imaging.[3,4]

IN VIVO EXPERIMENTAL RESULTS

Firstly, the proposed method is applied to the detection of velocity signals on the IVS of a healthy 26-year-old male volunteer. In the B-mode image of the LV in the short axis, which was obtained by standard ultrasonic diagnostic equipment, 14 points $\{i\}$ are set in the region from point (R) on the RV surface to point (L) near the LV side of the IVS, where $x_R(t) = x_1(t)$ and $x_L(t) = x_{14}(t)$. Since the results obtained by the proposed method depend on the angle between the direction of the velocity vector and the ultrasonic beam, the direction of the ultrasonic beam passing through points (R) and (L) is selected so that the beam is on the LV center in the cross-section and is almost perpendicular to the IVS during the A/D conversion of several cardiac cycles. During the acquisition period, respiration is suspended.

Figures 2(1-a) and 2(1-b) show the ECG and PCG, respectively. The M-mode image which was reconstructed from the magnitude of the digitized signal of the analytic signals. Before applying the above method, by referring to the M-mode image, the positions $\{\hat{x}_i(t_0)\}$ of the 14 points $\{i\}$ are manually preset using the workstation at even intervals of Δx_S=0.75 mm from point (R) near the surface of the RV side of the IV to point (L) near the surface of the LV side. The tracking results $\{\hat{x}_i(t)\}$ of the 14 points $\{i\}$, are superimposed on the M-mode image by the white lines as shown in Fig. 2(1-e).

Figure 2(1-c) shows the superimposed estimates of the velocity signals $\{\hat{v}(x_i;t)\}$ on the tracked points $\{\hat{x}_i(t)\}$, where $i = 1, 2, \cdots, 14$. Its vertical axis is inverted so that the negative value of the velocity, which is shown above the baseline, corresponds to the situation in which the object moves in the direction of the ultrasonic transducer on the chest wall, which is more easily understood. Figure 2(1-d) shows the thickness change, $\{\hat{x}_i(t) - \hat{x}_1(t)\}$, that occurs in the region between the point (R) on surface of the RV side and the ith point in the IVS. For the systolic phase, the IVS becomes about 3 mm thicker than that of the diastolic phase, where the thickness of the IVS is about 10 mm at the end of the diastole.

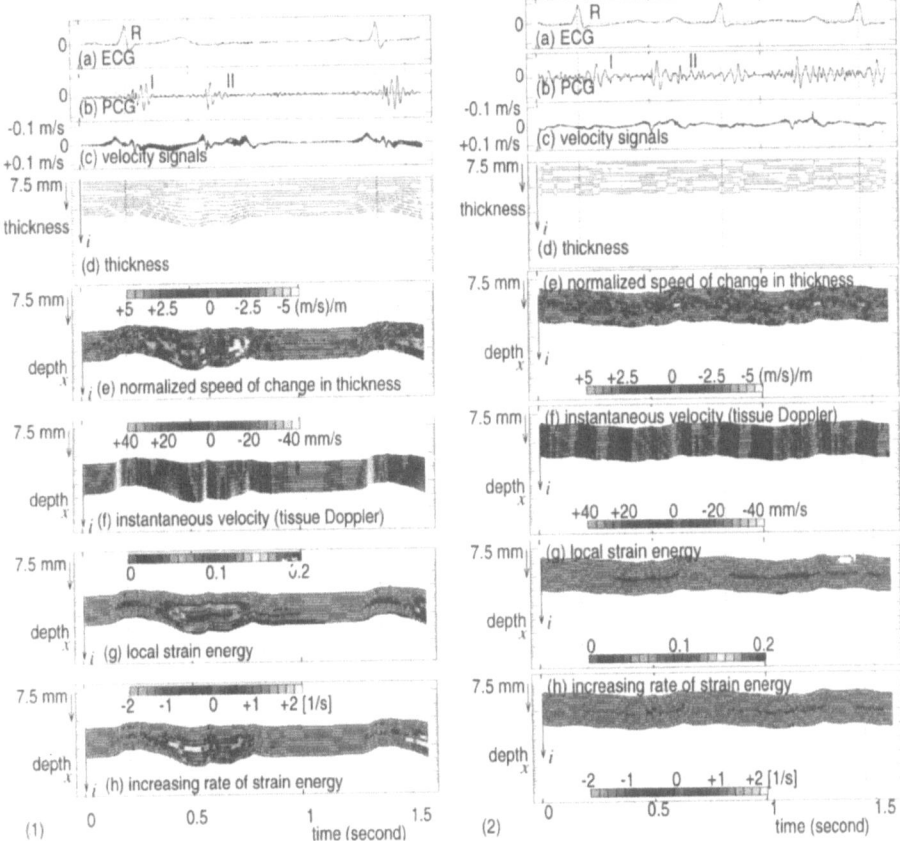

Figure 2. *In vivo* experimental results of the instantaneous object position and the velocity estimated at 14 points $\{i\}$ in the IVS for the first cardiac cycle. (1) a normal subject and (2) a 32-year-old male patient with acute lymphoblastic leukemia and with serious cardiomyopathy, who had been treated with antracenadiones.

Figure 2(1-e) shows the normalized speed of the local change in thickness,[2] $\{v(x_{i+1};t)-v(x_i;t)\}/\Delta x_S$, which occurs in the local region between points $\{i\}$ and $\{i+1\}$. In this figure, their values are color-coded. For the region and the timing where there is a large change in thickness in Fig. 2(1-d) and where there are relatively large differences between the velocity estimates in Fig. 2(1-c), for example, for the beginning and the end of the systole especially at the LV side in the IVS, the instantaneous local change in thickness is found to be large, which correspond to the thickening and thinning, respectively, in the heart wall. For the period around the R-wave and the center of the diastole, however, these velocity estimates coincide with each other, which corresponds to the parallel motion as illustrated in Fig. 1(c).

Figure 2(1-f) shows the results obtained by tissue Doppler imaging applied to the same data set. Its color code is determined from the local instantaneous velocity. Figures 2(1-g) and 2(1-h) show the spatial-time distribution of $U_i(t)/E$ in Eq. (5) and $\Delta U_i(t)/E$ in Eq. (7), respectively.

As an example of noninvasive diagnosis of myocardial damage induced by adriamycin injection, we applied the proposed method to a patient with serious cardiomyopathy. For a 32-year-old male patient with acute lymphoblastic leukemia, who had been treated with antracenadiones (mitoxantrone) and who suffered from Doxorubicincardiotoxicity,[5] the results are shown in Fig. 2(2). As shown in Fig. 2(2), the results are quite different from those of the normal subject represented in Fig. 2(1). For the velocity signals in Fig. 2(2-c), the amplitude of the signals, especially for the higher frequency components, are smaller than that of the normal subject. As can be seen from Fig. 2(2-d), there is little change in thickness of the IVS. The maximum values of the change in thickness of the IVS are about 490 μm in 2(2-d).

There are also clear differences in the spatial-time distributions in Figs. 2(2-e), 2(2-g), and 2(2-h) from those in Figs. 2(1-e), 2(1-g), and 2(1-h), respectively. For the normal subject, thickening and/or thinning occur simultaneously in time and homogeneously across the LV wall. For the serious patient, however, the periodic change in each layer thickening with the cardiac cycle disappears and, moreover, the heterogeneous behavior across the LV wall becomes obvious. These differences are, however, not clearly obtained by the tissue Doppler imaging since the parallel motion is still large in Fig. 2(2-f).

CONCLUSIONS

In this paper we have proposed a new noninvasive method for evaluation of the spatial-time distribution of local thickness change and the strain energy in the heart wall using ultrasound. For a serious patient, the results are quite different from those of the normal subject. Since these preliminary studies in this paper were performed in young patients with clear echocardiograms, much work must be done to find out how the method can be applied in a wider range of subjects and to see whether it can contribute to assessment of myocardial diseases and recovery. It will be necessary to compare the *in vivo* results obtained herein by the proposed method with pathological findings in order to evaluate the range of variability in clinical studies and construct a standard which can be applied to the *in vivo* diagnosis of the heart wall.

REFERENCES

1. H. Kanai, M. Sato, Y. Koiwa, and N. Chubachi, Transcutaneous measurement and spectrum analysis of heart wall vibrations, *IEEE Transactions on Ultrasonics, Ferroelectrics, and Frequency Control*, 43:791(1996).
2. H. Kanai, H. Hasegawa, N. Chubachi, Y. Koiwa, and M. Tanaka, Noninvasive Evaluation of Local Myocardial Thickening and Its Color-Coded Imaging, *IEEE Transactions on Ultrasonics, Ferroelectrics, and Frequency Control*, 44:(1997)(in press).
3. W. N. McDicken, G. R. Sutherland, C. M. Moran, and L. N. Gordon, Color Doppler velocity imaging of the myocardium, *Ultrasound in Med. & Bio.*, 18:651 (1992).
4. K. Miyatake, N. Tanaka, M. Yamagishi, N. Yamazaki, Y. Mine, and M. Hirama, Clinical application of newly developed color coded tissue Doppler echocardiography in detection of abnormal ventricular wall motion, *Journal of the American Society of Echocardiography*, 6:S19 (1993).
5. J. Dunn, Doxorubicin-induced cardiomyopathy, *Journal of Pediatric Oncology Nursing*, 11;152 (1994).

A METHOD FOR INTRAVASCULAR ULTRASOUND FRONTAL VIEWING

Ayumu Matani, Osamu Oshiro, and Kunihiro Chihara

Graduate School of Information and Science
Nara Institute of Science and Technology
8916-5 Takayama, Ikoma, Nara 630-01, Japan

INTRODUCTION

Intravascular frontal viewing is helpful for diagnosis of angiostenosis. Optical measurement methods are not useful in blood vessels filled with blood. Although acoustic measurement methods are useful instead in this case, methods for intravascular ultrasound have been proposed mostly for lateral viewing. This paper describes a newly developed intravascular frontal viewing method with a ring-array probe transmitting spherical pulsed waves. The method employs an optimization algorithm. We have developed an intravascular frontal viewing system with the probe using a conventional synthetic aperture method (Tojo et al., 1994 and 1995). The synthetic aperture method, however, represented elliptically symmetric artifacts. Therefore, even a simple-shaped object was revealed as distorted elliptic planes (lines in 2D) and was difficult to recognize its shape. On the other hand, the optimization method we propose here can eliminate the elliptically symmetric artifacts. To evaluate the method and compare it with our previous method, a computer simulation experiment is performed.

METHOD

Pre-Processing

Figure 1 shows a ring-array probe for the computer simulation experiment. We assume that the cylindrical probe (diameter 2 mm) has 8 ultrasound transducers on the front surface. Assuming one of the transducer S_i transmits an ultrasound spherical burst (central frequency 10 MHz) and a transducer S_j receives the echo reflected by the point reflector R_k, a simple calculation can result in the echo E_{ijk}. By the way, a simple symmetrical relationship between the transmitter and receiver $E_{ijk} = E_{jik}$ brings that the total number of the time sequence echo signal sets will be 28 $(=8(8-1)/2, i \neq j)$. In the next step, the echo is phase detected by cosinusoidal and sinusoidal waves (central frequency 10 MHz) to be envelop signals. The envelop signals are then discretized at sampling frequency 10 MHz to be the time sequence signals. Those time sequence signals are appended for all transmitters and receivers, to be theoretical echo table $\mathbf{T}_k$. This theoretical echo table consists of non-orthogonal bases in the measurement region. On the other hand, the actual echo is also phase detected, discretized, and finally appended to be $\mathbf{E}$.

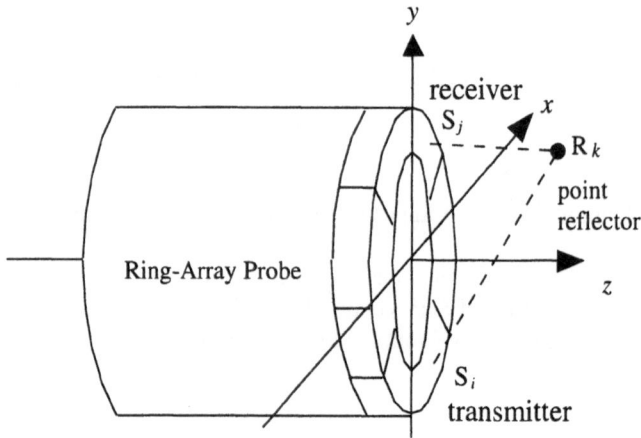

Figure 1. Ring-array probe.

Imaging Method

In conventional synthetic aperture methods, the weight of the kth element, r_k, is simply calculated as $|\mathbf{E} \cdot \mathbf{T}_k|/|\mathbf{E}||\mathbf{T}_k|$. This method is based on a simple expansion algorithm of the measured echo data to the non-orthogonal bases consisting of the theoretical echo table.

Although a optimized weighting method we propose in this paper is basically the recursive calculation of the synthetic aperture method as following steps, this method employs an optimization algorithm.

(1) Register the k where brings cost, $\mathbf{E} \cdot \mathbf{T}_k/|\mathbf{E}||\mathbf{T}_k|$, maximum.
(2) Subtract the theoretical echo $\mathbf{T}_k$ from $\mathbf{E}$.
(3) Plus 1 to r_k.
(4) Back to step (1).

Those steps are repeated until the cost no longer increases. This method is not based on the simple expansion to the non-orthogonal bases but always depends on the already registered point reflectors.

COMPUTER SIMULATION EXPERIMENT

8 point reflectors were assumed in the computer simulation experiment as shown in Figure 2. The 5 reflectors, A through to E, were placed on the same C-mode plane I 4.2 mm apart from the ring-array probe. The other 3 reflectors, F through to H, were aligned on the same C-mode plane II 0.4 mm apart from the C-mode plane I. Those reflectors were spaced so closely that each one could not be distinguished in case of the

estimation results with artifacts.

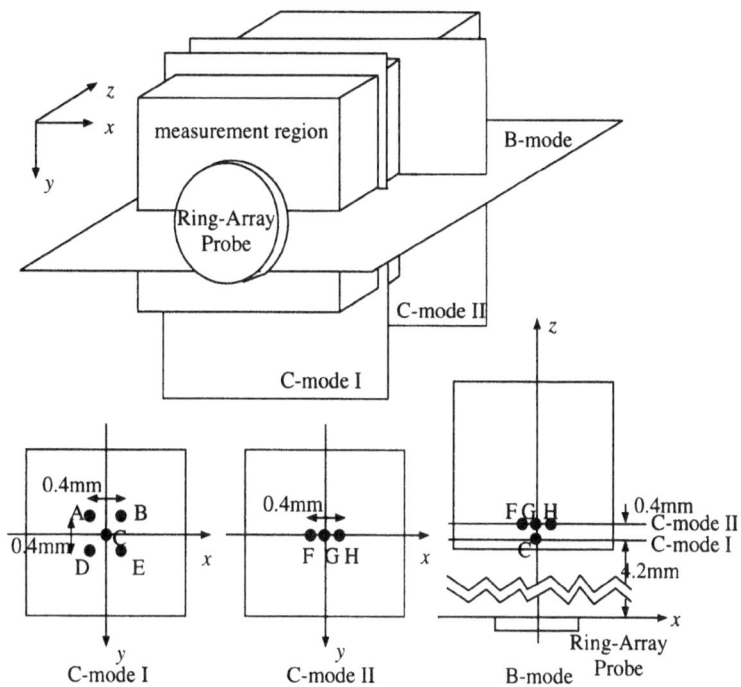

Figure 2. Configuration of point reflectors.

Synthetic Aperture Method

Figure 3(a) shows the estimation results using the conventional synthetic aperture method in 3D. It can be seen clearly that each reflector has elliptically symmetric artifacts (ring shaped in this case). The point reflectors, therefore, could not be distinguished clearly and the configuration of them could not be recognized either. Moreover, reflector C was not weighted conspicuously due to the weighs surrounding reflectors.

The artifacts are caused by the nature of the synthetic aperture method that employs the simple expansion method to the non-orthogonal bases consisting of the theoretical echo table. A weight of a basis is calculated without reflecting the other weights of the bases.

In medical ultrasound measurement, a sort of image processing technique making the final echo image clear using the sensitivity time control. In this case, the observers have knowledge of the shape of the objects (organs) roughly. However, in intravascular ultrasound frontal viewing, the position, shape, and thickness of the angiostenosis are difficult to predict. The image processing technique is not, therefore, appropriate for the intravascular frontal viewing.

Optimized Weighting Method

Figure 3(b) shows the estimation results using the optimized weighting method in 3D. The optimized weighting method represented the 8 closely spaced point reflectors clearly No artifacts could not be seen such that the image processing technique for the compensation was not necessary.

The optimized weighting method always depends on the expansion values of the other bases. The optimization algorithm weights only the basis bringing the maximum expansion value and then the influence of the basis is removed from the measured eco data before the next calculation to find the maximum. In another way, the optimization algorithm searches the next basis as orthogonal as possible to the already registered bases. The method, therefore, dose not weight the points aligned on the elliptic plane with the same expansion values.

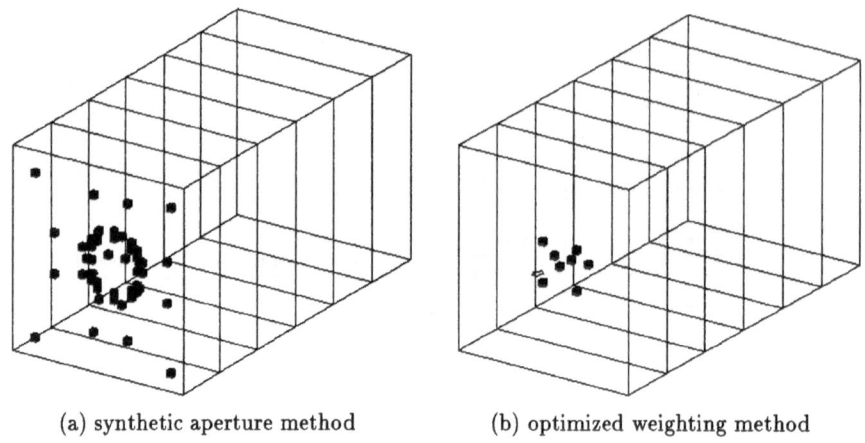

(a) synthetic aperture method   (b) optimized weighting method

Figure 3. Estimation results of the point reflectors.

CONCLUSIONS

We proposed a newly developed intravascular frontal viewing method using an ultrasound ring-array probe as a catheter. This method employs an optimization algorithm for expanding the measured echo data to non-orthogonal bases consisting of theoretical echo data. The computer simulation experiments assuming closely spaced point reflectors was performed to evaluate the method and compare it with a conventional synthetic aperture method. The conventional synthetic aperture method represented the known point reflectors with elliptically symmetric artifacts such that those reflectors were difficult to recognize. On the other hand, this optimized weighting method successfully estimated the location and configuration of the reflectors. This method is expected to be beneficial to intravascular frontal viewing.

REFERENCES

1. H. Tojo, O. Oshiro, K. Chihara, and M. Asao, Three-dimensional echography by a spherical pulsed wave, *Japanese Journal Applied Physics.* 33-1:3162 (1994).
2. H. Tojo, O. Oshiro, A. Matani, K. Chihara, M. Asao, and T. Furukawa, Three-dimensional dynamic imaging by spherical pulsed wave of ultrasound, *Japanese Journal Applied Physics.* 34-1:2857 (1995).

FOUR ULTRASONIC METHODS FOR MEASUREMENT OF VOLUMETRIC FLOW WITH NO ANGLE CORRECTION

Jens Kristian Poulsen

Institute of Experimental Clinical Research, Skejby Sygehus, Aarhus University
8200 Aarhus N, Denmark (email: skejjkp@aau.dk).

ABSTRACT
Four ultrasonic methods are described for the measurement of volumetric flow without using angle correction. The geometrical configurations are spherical, cylindrical, fan and plane scanning. The cylindrical and fan methods are two new innovations. Flow measurement is achieved by numerically integrating the product of velocity and surface area. The flow of interest should be inside the boundary of the surface of integration. Each small area on the surface is chosen to be perpendicular to the direction of the ultrasonic beam. In this way, 3-dimensional (3-D) flow can be measured even though ultrasound Doppler only provides 1-D velocity information. The four methods described offer the possibility of 3-D measurement of flow and allow the use of several different types of ultrasonic transducers.

BACKGROUND
Basically, the Doppler principle only provides velocity information in the direction of the ultrasonic beam. This is a problem when trying to measure flow in-vivo since complicated 3-D flow structures may exist. Doing a simple angle correction of the measured velocities is often not reliable, since it is based on assumptions rather than knowledge of the actual flow patterns encountered. Volumetric flow can be calculated by dividing mass flow [1] by the density of the fluid giving an integration of velocity-area products over the surface of interest, i.e.

$$Flow = \int_{Surface} \vec{V} d\vec{A}, \qquad (1)$$

where $\vec{V}$ is a vector with direction and magnitude equal to the velocity at a given point on the surface and $d\vec{A}$ is the normal vector to the surface at this point with a magnitude equal to the area size. Various methods have been proposed in order to circumvent the 1-D limitation of ultrasound Doppler and measure more than one velocity component, e.g. methods based on multiple transducers: echo correlation [2], diffraction based methods [3] and synthetic aperture methods [4]. There have also been attempts to measure 2-D velocity information using a single transducer, such as speckle tracking [5], correlation based methods [6] or by estimating the angle between the ultrasonic beam and the direction of the fluid [7]. However, these methods have problems such as increased acquisition time (correlation methods), measuring only 2-D information (diffraction based, speckle tracking, single element correlation based) or the requirement for complicated processing and full 2-D ultrasonic arrays (synthetic aperture methods). Angle independent intravascular scanning using a spherically shaped single element transducer has been proposed by Gibson et al. [8]; however, the calculation of area and thus the flow is dependent upon the acoustic attenuation in the fluid.

THEORY

The error encountered when using angle correction can be estimated in the following way (refraction effects and the finite size of the sample volume is not taken into account). The measured velocity when using ultrasound Doppler is the velocity in the direction of the ultrasonic beam:

$$V_m = -\vec{V} \bullet \vec{Beam} = -V\cos\phi, \tag{2}$$

where V_m is the measured velocity, $\vec{Beam}$ is a unit vector in the direction of the ultrasonic beam (pointing from the transducer to the sample volume), $\vec{V}$ is the velocity vector describing the movement of fluid inside the sample volume (if different velocities exist within the sample volume, a spectrum of Doppler shifts will be measured and some statistical measure may be calculated), V is the magnitude of the velocity of the fluid (with sign determined by definition of ϕ) and $\cos\phi$ is defined as the projection of a unit vector in the direction of the beam onto the direction of the fluid. The negative sign occurs because velocities are registered positive when an object is moving towards the transducer. Angle correction can then be used to estimate the real velocities: $V_{est} = \frac{V_m}{\cos\phi_{est}}$, where ϕ_{est} is the estimated angle between the ultrasonic beam and the assumed direction of the flow (which in general is unknown). The estimated velocity is thus found as:

$$V_{est} = \frac{V\cos\phi}{\cos\phi_{est}}. \tag{3}$$

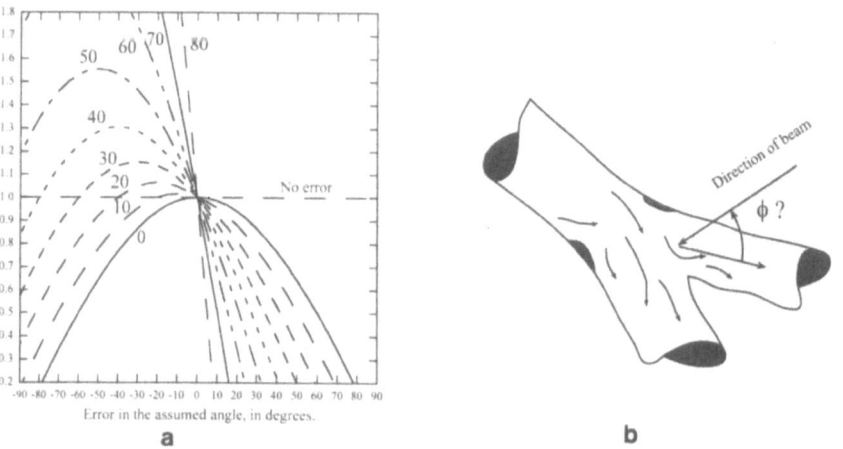

Figure 1. (a) Error plot of angle correction. Angle between assumed direction of flow and ultrasonic beam is shown for 0-80 degrees, in 10 degree increments. (b) Complicated flow patterns may make angle correction difficult.

Figure 1a shows the error encountered when using angle correction based on this equation. As an example, when the assumed angle between the ultrasonic beam and the direction of flow is 55° and the error on this assumption is ±15°, the error in the velocity estimates will reach -40.4% or +33.6% depending on the sign. This is not an unrealistically large error encountered in the clinic since there are errors in the estimation of the angle both in the viewing plane and in a plane perpendicular to this. Effectively, 3-D angle correction may be neccessary, if angle correction is going to be used at all. An example of using 3-D angle correction is shown in [9]. Ferrara [6] gives errors on the estimation of the angle up to 12° for a correlation method in an attempt to measure the lateral velocity component. Thus, reliable in-vivo flow measurements are difficult at best and unreliable at worst. Figure 1b shows an example of this.

However, when using C-mode scanning, that is, collecting all sample points at the same distance from the transducer, there is no need to do angle correction, since the integration of the velocity-area products is performed by choosing a surface of integration perpendicular to the direction of the ultrasonic beam. Figure 2 shows four methods, each using C-mode scanning for collection of data.

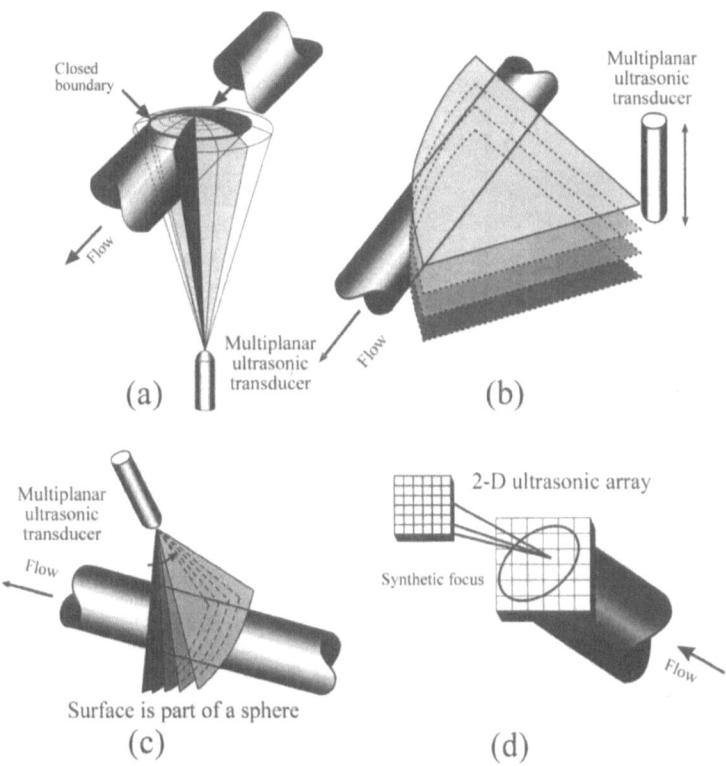

Figure 2. Four ultrasonic methods for measurement of volumetric flow without angle correction: (a) spherical, (b) cylindrical, (c) fan, and (d) planar.

The first method, spherical scanning, uses a rotating multiplanar transducer (see Figure 2a). The transducer is placed at the center of a sphere and sample points are acquired at a fixed distance from the transducer, i.e., on the spherical surface, using B-mode scanning. This enables measurement of 3-D flow using standard ultrasound equipment without angle correction. The transducer is then rotated and more scanplanes are acquired. The flow is calculated by summing the individual products of measured velocities and the corresponding surface areas on the sphere.

The direction of the surface area vector, $d\vec{A}$, is normal to the surface. The normal vector to a small surface, dA, of a sphere at a particular location has the same direction as a vector pointing from the center of the sphere to this particular point on the surface. This means we can write:

$$B\vec{eam}\ dA = d\vec{A}, \tag{4}$$

and thus the flow out of a small area dA is:

$$
\begin{aligned}
\vec{V} \bullet d\vec{A} &= \vec{V} \bullet (B\vec{eam}\ dA) \\
&= (\vec{V} \bullet B\vec{eam})\ dA \\
&= -V_m\ dA. \tag{5}
\end{aligned}
$$

It should be noted that there is no angle correction factor any longer since both V_m and dA are scalars. If we are interested in the flow into a surface, the sign is reversed and by using (1) we get:

$$Flow_{in} = \int_{Surface} Vm\ dA. \tag{6}$$

The area on the surface of a sphere defined by two limiting angles can be found by (see appendix):

$$\Delta A = \Delta \phi R^2 |\cos \theta_1 - \cos \theta_2|. \tag{7}$$

Thus the flow can be calculated as (using a summation over all scanlines):

$$Flow = \Delta \phi R^2 \sum_{Surface} Vm|\cos\theta_i - \cos\theta_{i+1}|. \tag{8}$$

By summation of individual velocity-area products, the total flow can be found. The total cross sectional area of the tube/vessel under investigation is never computed. Rather small individual areas are multiplied by the corresponding velocities. Outside the tube/vessel of interest, the measured velocity is zero, and thus these velocity-area products do not contribute to the total flow.

One drawback of this method is that the scanning pattern results in integrated areas of non-uniform size. This makes it necessary either to place the center of the sphere close to the center of the vessel for minimisation of sampling errors or to resort to the use of a high number of scanplanes in order to accurately describe the flow field of interest. This can be solved by using equidistantly placed sample volumes on a cylindrical, spherical or planar surface to achieve a uniform spatial sampling density.

The second method, cylindrical scanning, uses a translating multiplanar transducer (see Figure 2b). The transducer is placed on the central axis of a cylinder and scanning is performed from this axis. The flow in a vessel is measured by first acquiring velocity data in one scanplane, then repeatedly moving the transducer a fixed distance along the central axis of the cylinder and scanning until adequate data describing the velocity field are taken. For a cylinder, the normal vector to a particular point on the surface has the same direction as a vector starting from the center of the cylinder out to this point. The flow is found by summing the velocity-area products, where the areas are computed using (see appendix):

$$\Delta A = Rh\Delta\theta. \tag{9}$$

Thus, the flow can be calculated with:

$$Flow = \Delta\theta Rh \sum_{Surface} Vm. \tag{10}$$

The third method, fan scanning, uses a tilting multiplanar transducer to scan points placed on a spherical surface (see Figure 2c). The transducer is tilted on an axis perpendicular to the scanplane and all velocities are acquired at a fixed distance from the transducer. This configuration seems well suited for both mechanical scanning and 1.5-D phased arrays. The calculation of flow follows the same equation as the spherical scanning described above; however, the areas show much less variation for a limited field of view. As an example, consider using 60 scanlines and a field of view equal to 45°. Using spherical scanning we get a ratio in the smallest to largest area equal to $\left(\frac{\cos 21.75° - \cos 22.5°}{\cos 0° - \cos 0.75°}\right) = 57.6$. Using the fan method with the same parameters, the ratio is only $\left(\frac{\cos 67.5° - \cos 68.25°}{\cos 89.25° - \cos 90°}\right) = 1.08$. With identical scan parameters, the fan method is statistically more stable and requires fewer ultrasonic samples to obtain the same accuracy. Essentially, the spherical scanning either oversamples in the center region or undersamples at the edges using a given number of scanplanes. This makes it slower compared to fan scanning when the same accuracy is required or less accurate with the same number of scanplanes used.

The fourth method, planar scanning, uses focusing by synthetic aperture radar inspired signal processing (see Figure 2d). The focusing is done by introducing electronically controlled delays to several transducers for each received signal and summing these signals. The delays correspond to different pathlengths travelled by the ultrasonic echoes. It relies on integration over a plane surface and measures the velocity component perpendicular to this surface. These values can thus be used directly in the integration of flow. The formula used to calculate flow is now (ΔX and ΔY are the length of the sides of the rectangles used in the summation):

$$Flow = \Delta X \Delta Y \sum_{Surface} Vm \tag{11}$$

It has the potential advantage of high frame rates but requires large computing power, true 2-D ultrasonic phased arrays and complicated manufacturing technology. However, the current trend with the advent of ultrasonic cameras may soon make this possible. It may be noted that planar scanning is also possible using a single ultrasonic transducer.

DISCUSSION

First studies on C-mode scanning using a 6x6 element 2D array by Moser et. al. [10], showed linearity and small errors with angles of insonation in the range 35° − 55°, but some underestimation in the range 60° − 70° due to incomplete insonation of the flow under investigation. In-vitro studies [11] of spherical scanning showed a good agreement (using 6 scanplanes and color Doppler as the velocity estimator) with the timed collection method, but in-vivo studies [12] (using only 3 scan planes and 4 seconds continous sampling) showed up to 30% error compared to MRI with the largest errors encountered in the smallest patient when measuring cardiac output. Future studies should investigate the relationship between data acquisition time, number of scanplanes and the accuracy of the measurements.

CONCLUSION

Two new methods for measurement of volumetric flow without angle correction has been presented and theoretical performance has been compared to the planar and spherical scanning technique. These new methods have the advantage of uniform or approximately uniform size of the areas of integration. The theoretical foundation for performing angle independent flow measurements using a variety of ultrasonic transducers has been established.

APPENDIX

The areas on the spherical surface can be found by integration in spherical coordinates as shown in [13]:

$$\Delta A = \int_{Surface} dA \tag{12}$$

$$= \int_{\theta=\theta_1}^{\theta_2} \int_{\phi=\phi_1}^{\phi_2} R^2 \sin\theta d\theta d\phi \tag{13}$$

$$= \left[-R^2(\phi_2 - \phi_1)\cos\theta\right]_{\theta_1}^{\theta_2} \tag{14}$$

$$= \Delta\phi R^2 |\cos\theta_1 - \cos\theta_2|. \tag{15}$$

θ is the angle in the viewing plane, ϕ the angle in a plane perpendicular to this. The numerical sign is included for general θ_1, θ_2. In case $\cos\theta = 0$ inside the interval $\theta_1...\theta_2$, the area calculation should be performed in two steps.

Since $\Delta\phi$ measured in radians gives the arc length directly the areas on a cylinder can be found as

$$\Delta A = Rh\Delta\phi. \tag{16}$$

REFERENCES

[1] B. R. Munson, D. F. Young & T. H. Okiishi
     Fundamentals of Fluid Mechanics, Formula 5.4, pp. 225, New York, USA, 1990.
[2] D. Dotti, R. Lombardi R & P. Piazzi
     "Vectorial measurement of blood velocity by means of ultrasound"
     Med. & Biol. Eng. & Comp., vol. 30. pp. 219-225. 1992
[3] D. Vilkomerson, D. Lyons & T. Chilipka
     "Diffractive transducers for Angle-Independent Velocity Measurements"
     IEEE Ultrasonics Symposium, pp. 1677-1682, 1994
[4] U. Moser, M. Anliker & P. M. Schumacher
     "Ultrasonic Synthetic Aperture Imaging used to measure 2-D Velocity Fields in Real Time."
     IEEE International Symposium on Circuits and Systems pp. 738-741. 1991.
[5] L. N. Bohs & G. E. Trahey
     "A Novel Method for Angle Independent Ultrasonic Imaging of Blood Flow and Tissue Motion"
     IEEE Trans. BME vol. 38. No. 3. pp. 280-286. 1991.
[6] K. Ferrara
     "Effect on the Beam-Vessel Angle on the Received Acoustic Signal from Blood."
     IEEE Trans. UFFC, vol. 42, No. 3, pp. 416-428, 1995.
[7] D. Dotti & R. Lombardi
     "Estimation of the Angle Between Ultrasound Beam and Blood Velocity Through Correlation Functions"
     IEEE Trans. UFFC , vol. 43, No. 5, pp. 864-869, Sept. 1996.
[8] W. G. R. Gibson, R. C. S. Cobbold & K. W. Johnston
     "Principles and Design Feasibility of a Doppler Ultrasound Intravascular Volumetric Flowmeter"
     IEEE Trans. Biomed. Eng., vol. 41, No.9, pp. 898-908, Sept. 1994.
[9] Kim WY, Bisgaard T, Nielsen SL, Poulsen JK, Pedersen EM, Hasenkam JM, Yoganathan AP
     Two-dimensional Mitral Flow Velocity Profiles in Pig Models using Epicardial Echo-Doppler-Cardiography.
     24:532-545, J Am Coll Cardiol 1994;
[10] U. Moser, P. M. Schumacher & M. Anliker
     Benefits and Limitations of the C-mode Doppler Procedure
     Acoustical Imaging, Vol. 21, pp. 509-522, 1995.
[11] Poulsen JK & Kim WY
     Measurement of Volumetric Flow With No Angle Correction Using Multiplanar Doppler Ultrasound.
     IEEE Trans. Biomedical Eng., 43;6:589-599, June 1996.
[12] W. Y. Kim, J. K. Poulsen, K. Terp & N.-H. Staalsen
     "A New Method for measurement of Volumetric flow: In-vivo validation."
     J Am Coll. Cardiol., January 1996, pp. 182-192.
[13] G.B. Thomas & R.L. Finney
     Calculus and Analytic Geometry, Addison Wesley, p. 958, 1992.

FORWARD AND INVERSE PROBLEMS IN
ENDOVASCULAR ELASTOGRAPHY

Lahbib Soualmi[1,2], Michel Bertrand[1,2], Rosaire Mongrain[2] and Jean-Claude Tardif MD[2]

[1] Institut de génie biomédical, École Polytechnique C. P. 6079, Succ. Centre-ville, H3C 3A7
Montréal (Québec) Canada
[2] Institut de Cardiologie de Montréal ,500 Bélanger Est, Montréal (Québec) Canada

This paper is about EndoVascular Elastography (**EVE**), a new technique for the diagnostic of arterial disease which gives information about plaque mechanical properties. Given this, **EVE** can provide useful information that could guide therapeutic decisions. Two important problems are considered: a) the **Forward Problem (FP)** of predicting the strain field of a tissue with a known elasticity and subjected to a known surface traction applied as a boundary condition, and b) the **Inverse Problem (IP)** of reconstructing the elasticity distribution from the measured displacement field, the boundary conditions and the governing equations. In the **FP**, arterial elastic properties are disclosed in the tissue displacement induced by a small intraluminal pressure change; the pre- and post-compression ultrasound signals are used to estimate radial (axial) tissue displacement and the corresponding strain. The approach we adopted to solve the **IP** consists in minimizing the least squares error between observed and predicted displacement fields. It uses an iterative procedure where at each iteration a linear inversion scheme based on a perturbation method is implemented. The paper illustrates the **FP** and **IP** implementation we have realized.

INTRODUCTION

EndoVascular Elastography (**EVE**) is a new acoustic imaging technique for arterial tissue characterization. The technique produces images of the elastic properties of arterial tissue. Hence, **EVE** aims at providing an image of the elasticity distribution of compliant arterial wall using intravascular ultrasound. The arterial elasticity distribution is obtained through the estimation of the tissue displacement induced by small intraluminal pressure pushes; the pre- and post-compression ultrasound signals are used to determine the radial (along the r.f. A-lines) tissue displacement and the corresponding strain. The intraluminal pressure pushes are provided by an angioplasty balloon integrated to an ultrasound imaging catheter. This integrated system simultaneously deforms the artery and collects the associated r.f. signals. Because soft material generally exhibits larger strains than hard one, measurements of this quantity can distinguish tissue of differing stiffness. Under the assumption of constant stress-field, the strain-field can be interpreted as a relative measure of elasticity distribution. Since the stress distribution is dependent on the boundary conditions and the elasticity distribution itself, in practice the stress field would not however be constant. This leads to certain limitations of strain imaging which take the form of low contrast-transfer efficiency for soft lesions and strain images artifact for a complex lesions arrangements. Solving the **IP** addresses this issue.

THE FORWARD PROBLEM

The **FP** consists of predicting the pattern of displacement (and strain) field given a distribution of arterial tissue elasticity and a set of boundary conditions. We use the finite element method to solve this problem numerically. The arterial wall tissue including plaque is modeled as an isotropic, incompressible and linearly elastic material. Hence, only the Young's modulus is needed to fully describe the tissue elastic properties. The elasticity equations are solved for a plane-state elasticity problem. The assumption of plane-strain state is justified when the longitudinal dimension of atherosclerotic lesions is in the order of the vessel diameter. This assumption of state plane elasticity reduces the 3D problem to a 2D problem.

The equations relating stress and strain arise from the balance of force in the material medium. For our 2D case we have:

$$\sum_{j=1}^{2} \frac{\partial \sigma_{ij}}{\partial x_j} + f_i = 0 \qquad i = 1,2 \tag{1}$$

where f_i denote the boundary and body forces along x_i and σ_{ij} is one component of the 2nd ranked stress tensor defined as:

$$\sigma_{ij} = 2\mu e_{ij} + \lambda \delta_{ij} e_{nn} \tag{2}$$

The pair of constants λ and μ are called Lamé's constants and μ is referred to as the shear modulus. They are related to the Young's modulus E, and Poisson's ratio υ in a state of plane strain by:

$$\mu = \frac{E}{2(1+\upsilon)} \quad \text{and} \quad \lambda = \frac{\upsilon E}{(1+\upsilon)(1-2\upsilon)} \tag{3}$$

This system of equations must be satisfied at every internal point of the computational domain. The relationship between the strain tensor and the displacement vector (u_1, u_2) is given by:

$$e_{ij} = \frac{1}{2}\left(\frac{\partial u_i}{\partial x_j} + \frac{\partial u_j}{\partial x_i}\right) \tag{4}$$

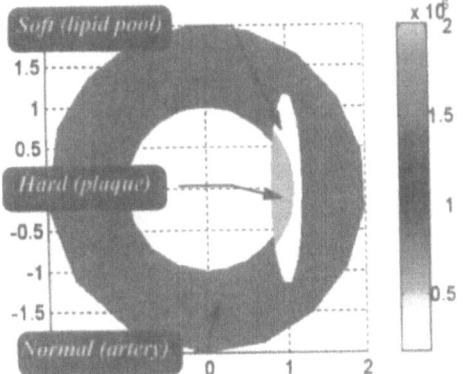

Figure 1 Geometrical model

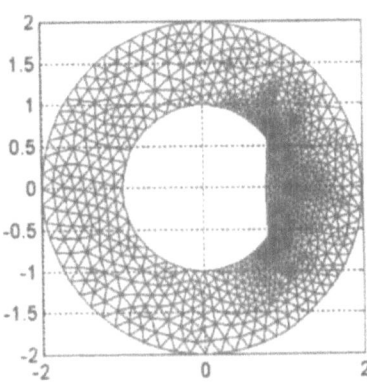

Figure 2 Finite Element Mesh

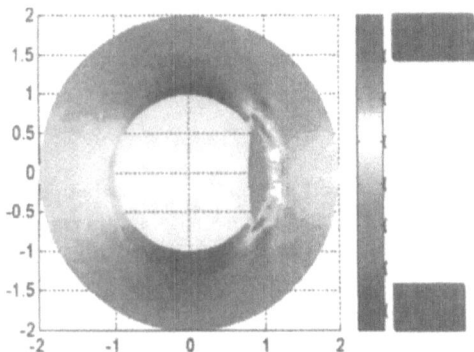

Figure 3 Endovascular Elastogram

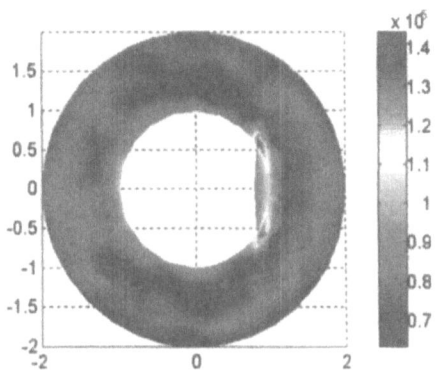

Figure 4 Stress Distribution

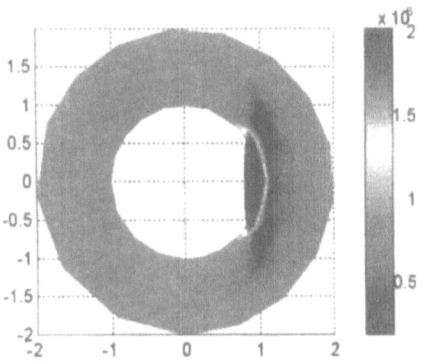

Figure 5 G-N reconstructed Elasticity after
15 iterations

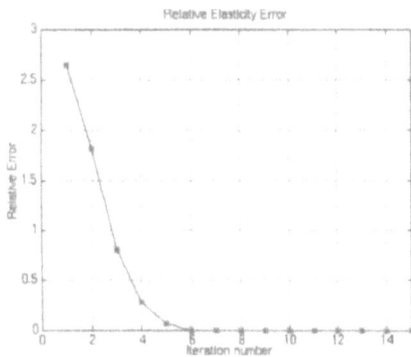

Figure 6 G-N Relative Elasticity Error

From the equations above we get:

$$\frac{\partial}{\partial x_i}\left[\mu\left(\frac{\partial u_i}{\partial x_j}+\frac{\partial u_j}{\partial x_i}\right)+\lambda\delta_{ij}\left(\frac{\partial u_n}{\partial x_n}\right)\right]+f_i=0 \qquad (5)$$

which is a set of elliptic Partial Differential Equations.

The geometrical model chosen for this study represents the cross-section of an atherosclerotic human coronary artery with an embedded intimal plaque. It is illustrated in figure 1, where three regions are defined on the basis of their elastic properties. The hard region represents a fibrous plaque grounded to the inner arterial wall. The soft region represents the subintimal lipid pool often found behind the hard plaque, the remaining part of the material wall is set with "normal elastic properties" with an intermediate value between those of the hard and soft regions. This geometrical model is used to build the unstructured triangular mesh of the computational domain (figure 2). The problem is then solved for a natural (Neumann) boundary condition which means that the pressure inside the vessel drives the inner arterial wall.

THE INVERSE PROBLEM

The **IP** we propose to solve consists of reconstructing the elasticity distribution of the artery given a set of observed displacements and known boundary conditions. The displacement data **u** are related to the elasticity distribution **E** in a non-linear way. The general form of this relationship is:

$$\mathbf{u}=\Im(E_1,E_2,E_3,...,E_p)=\Im(\mathbf{E}) \qquad (6)$$

where $\Im$ is the forward problem function which allows us to calculate the theoretical displacement response to a given set of elasticity distribution and boundary conditions. Given this, the **IP** can be written as:

$$\begin{array}{c} arg\,min\,\|\mathbf{u}-\Im(\mathbf{E})\|^2 \\ \mathbf{E}\in\mathfrak{R}^+ \end{array} \qquad (7)$$

where **u** is the observed displacement field and $\Im(\mathbf{E})$ is the displacement predicted by the **FP**. This problem can be solved in successive approximation using linear least squares method. This involves the conversion into an approximate linear form by expanding the function $\Im(\mathbf{E})$ in Taylor series about an initial guess $\mathbf{E}^0$ of the solution. Retaining only the linear terms of the series, equation 6 can be written as:

$$\Delta\mathbf{u}=\mathbf{S}\,\Delta\mathbf{E}+\mathbf{e} \qquad (8)$$

where $\Delta\mathbf{E}$ is an elasticity perturbation, $\Delta\mathbf{u}$ the resulting displacement perturbation, and **e** represents the residuals associated with observational errors, model errors, and the truncations of the series expansion. **S** is the Jacobian matrix, also called the sensitivity matrix and denote the quantities $\dfrac{\partial\Im_i(\mathbf{E})}{\partial E_j}$ which represent the partial derivative of $\Im(\mathbf{E})$ with respect to each of the model parameter E_j. In practice **S** is constructed using the **FP** formulation. This matrix is analogous to a system impulse response; in such a system the input is a local small elastic perturbation and the output is the displacement field perturbation vector.

In our **IP** we will be searching for the "best" elasticity perturbations or corrections to our initial guess. A solution could be the one that minimizes the least squares error q defined as:

$$minimize \quad q = e^T e = (\Delta u - S \Delta E)^T (\Delta u - S \Delta E) \qquad (9)$$

The minimization is accomplished by setting to zero the derivatives of q with respect to each parameters perturbations ΔE_i. This leads to the solution for the parameter perturbations as:

$$\Delta E = \left[S^T S\right] S^T \Delta u \qquad (10)$$

The perturbation vector ΔE is then added to our initial guess E^0 to yield a better estimate E^1 of the solution to our non-linear problem. We then repeat the calculation of the perturbation (equation 10, using an updated sensitivity matrix) to fit our data using E^1 as the new initial guess. The successive application of this procedure is known as the Gauss-Newton method (**G-N**) for unconstrained iterative least squares minimization.

RESULTS

Figure 3 shows the axial strain image (the EVE) computed with Matlab PDE Toolbox using the model artery of figure 1 and the finite element Mesh of figure 2. In this elastogram, we notice that the soft lipid pool is associated with a complex strain pattern. The pool appears softer at the hard-plaque to lipid pool interface; the central region of the lipid pool, behind the hard plaque, is not discernible as a softer region. On the opposite side of the lesion, in a region with normal morphology, we notice the occurrence of soft artifacts. Finally the stress decay that occurs while getting closer to the outer wall of the vessel, results in an apparent target hardening. **EVE** therefore appears difficult to interpret, a situation that can justify the **IP** solving we propose in order to recover the true elasticity distribution.

Figure 4 shows the principal stress distribution of the arterial model. It reveals high stress concentration at the plaque to normal tissue interface. Such stress concentration may lead to plaque failure, a most important pathological consequence of the arterial lesion. Interestingly, it seems that the elastogram shown in figure 3 can draw attention to this very important phenomenon related to stress concentrations. Indeed, the appearance of a softer region at the hard-plaque to lipid pool interface somewhat coincides with the stress concentration identified in figure 4.

Figure 5 shows a reconstructed elasticity distribution obtained after 15 iterations with the **G-N** method using a noise-free displacement field which consists of both the radial and the lateral components. In this ideal case, we are able to fully retrieve the elasticity distribution. Indeed the relative error for the elasticity reconstruction plotted in figure 6 shows that the method can perfectly recover the elasticity distribution we originally used.

DISCUSSION

We have developed a finite-element based mathematical model to solve the **FP** and **IP** problems for EndoVascular elasticity imaging. In this paper, we solve the **IP** using a **G-N** algorithm , where the convergence scheme is based on the linear perturbation method. To work well, this method must meet several conditions. First, it is necessary that the initial guess be "close" to the solution for the procedure to converge. Second $\left[S^T S \right]$ must be well-conditioned, otherwise the computed elasticity perturbation becomes very sensitive to displacement noise, for example up to a point where the calculated solution becomes physically impossible (negative Young's modulus!). The solution, in this case, is said to *overshoot the linear range.*

To prevent such unbounded solution growth, we propose to use a more robust minimization method, such as for example the Levenberg-Marquart (**L-M**) algorithm were robustness is brought through constrains added to the functional to minimize. Here, when the parameter perturbation ΔE_i is very large, a bound is placed on the size of the perturbations to constrain the step-length of the solutions. Thus the functional to minimize takes the form:

$$\text{minimize } \phi = q_1 + \lambda\, q_2 = e^T e + \lambda(\Delta E^T \Delta E - L_0^2) \tag{11}$$

where we minimize a combination of the prediction error q_1 and solution length q_2 and we have a bound L_0^2 on the energy of solution change. λ is a multiplier that determines the relative importance that will be given to q_1 and q_2. It is referred to as the damping factor. The constrained least squares solution for parameters perturbations for **L-M** method is:

$$\Delta E = \left[S^T S + \lambda I \right]^{-1} S^T \Delta u \tag{12}$$

This is then used in an iterative process to fit the displacement data. The iterative formula is:

$$E^{k+1} = E^k + \left[S^T S + \lambda I \right]^{-1} S^T \Delta u \tag{13}$$

where S is evaluated at E^k and I is the identity matrix. Comparing equations (10) and (12) reveals that one is a regularized form of the other. Indeed the form of equation 12 is identical to the form used in ref. 3. However, the regularizing parameter λ is interpreted here in terms of its damping property for the iterative minimization scheme.

We are presently investigating the for **L-M** method for the case where only the axial displacement is used to solve the **IP.**

ACKNOWLEDGMENTS

A part of the project upon which this work is based was performed pursuant to the University of Texas Grant CA64597-01 with the NIH,PHS. This work was also supported by the National Sciences and Engineering Research Council of Canada, Le ministère de l'éducation du Québec and Le Fond de Recherche de l'Institut de Cardiologie de Montréal.

REFERENCES

[1] Cespedes, E.I; Korte, C.L.; van der Steen, A.F.W.; Norder, B.; te Nijenhuis, K.; "Tissue mimicking material and image artifacts in intravascular elastography", Acoustical imaging, San Antonio, 1996.

[2] Gill, P. E.; Murray, W.; Wright, M.H., "Practical Optimization", Academic Press Inc, 1981

[3] Kallel, F.; Bertrand, M., "Tissue Elasticity Reconstruction Using Linear Perturbation Method", IEEE Trans. on Medical Imaging, Vol.15, No.3, pp. 299-313, 1995.

[4] Gao, L.; Parker, K.J.;Lerner, R.M.; Levinson, S.F.,"Imaging of the Elastic Properties of Tissue- a Review", Ultrasound in Med. & Biol., Vol. 22, No 8, pp. 959-977, 1996.

[5] Shapo, B.M.; Crowe, J.R.; Skovoroda, A.; Eberle, M.J.; Cohen, N.A. and O'Donnell, M., "A new technique for imaging tissue strain imaging of coronary arteries with intraluminal ultrasound", IEEE Trans., Ultras. Ferro. Freq. cont., (43), pp. 234-246, 1996

REFERENCES

APPLICATION OF ANGLE-DEPENDENCE IN THE ULTRASONIC ECHO SIGNAL TO ESTIMATION OF CAROTID PLAQUE CONTENTS

Jens E. Wilhjelm[1], Marie-Louise M. Grønholdt[2] and Henrik Sillesen[3]

Center for Arteriosclerosis Detection with Ultrasound (CADUS). E-mail: wilhjelm@it.dtu.dk. Homepage: http://www.it.dtu.dk/~wilhjelm/cadus.html.
[1]Department of Information Technology, Technical University of Denmark, Building 344, DK-2800 Lyngby, Denmark.
[2]Department of Vascular Surgery, Rigshospitalet, University of Copenhagen, Blegdamsvej 9, DK-2100 Copenhagen Ø, Denmark.
[3]Department of Vascular Surgery, Gentofte Hospital, University of Copen-hagen, DK-2900 Hellerup, Denmark.

INTRODUCTION

Ultrasound imaging of the carotid arteries is today the prevailing method for clinical diagnosis of arteriosclerosis in the carotid arteries. The degree of stenosis can be estimated from the blood velocity image and the appearance of the plaque itself can be assessed from the anatomical B-mode image. Both pieces of information are used to assess the risk of plaque rupture, which - through the production of emboli - can lead to stroke. In particular, recent research has demonstrated that the B-mode image reveals features related to the risk of development of future neurological symptoms and brain infarcts.[1] These image features are related to plaque material and structure, but determination of this relationship is very difficult with the current imaging technique.

The arteriosclerotic plaque mainly consist of five material constituents: thrombus, haemorrhage, lipid materials, fibrous tissues and calcification. The first three constituents probably cannot be distinguished by ultrasound and are typically categorized as "soft materials". The remaining material groups (soft materials, fibrous tissues and calcification) are normally associated with increasing echogenicity with increasing dependence of insonification angle. A clear demonstration of this behavior was made by *Picano et al*:[2] For fatty wall, fibrofatty wall, fibrous wall and calcified wall the maximal backscattering coefficient as a function of angle and the degree of the angular dependence increased.

In contrast to the presumably slap-like shape of the opened aorta wall, the carotid plaques investigated in the present study have a more pronounced three dimensional structure. To fully account for this, three-dimensional ultrasound images were recorded from five different angles, features measuring the angle-dependence in these images were calculated and the results compared to quantitative histological analysis.

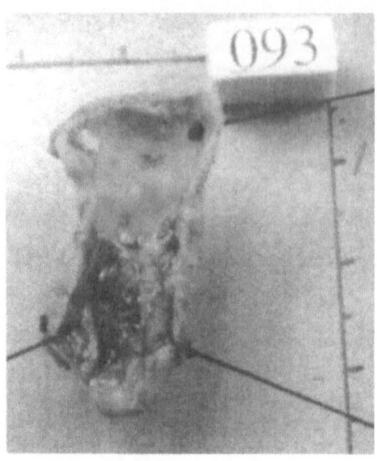

MATERIALS AND METHODS

Plaque Materials

Plaques from thirteen patients with carotid artery disease were removed by pro-phylactic carotid endarterectomy.[3] All patients were referred to the Department of Vas-cular Surgery, Rigshospitalet, Copenhagen, with neurological symptoms from the same side as the stenotic carotid artery. Immediately after removal from the patients, the plaques were fixed in formalin. The subsequent *in vitro* scanning was carried out within 2 weeks. The longitudinally opened plaques were fixed to an acrylic frame in an "as open as possible" condition by means of four sutures, as illustrated in Figure 1. The frame with plaque was next inserted into a thin-walled latex bag with formalin. The plaques had to remain in formalin during scanning, as this could take up to 24 hours.

Histological Analysis

Following ultrasound scanning (to be described subsequently), sutures were removed from the plaques which were subsequently laced back into a shape that matched the *in vivo* form as close as possible. The features analyzed were the volumetric contents of soft materials, fibrous tissues and calcification. The total relative content were 26.9 ±10.1 %, 72.5 ±9.9 % and 0.7 ±0.5 % (mean ± standard deviation), respectively.

Ultrasound System

The experimental ultrasound system consisted of a pulser/receiver driving a 10 MHz, 0.25" diameter, single element spherically focused (at 40 mm) ultrasound transducer. As illustrated in Figure 2, the transducer was mounted to a manual rotational device, such that the ultrasound beam intersected the axis of rotation at a distance of D_{rot} = 58.5 mm. Thus measurements took place in the far field, where the energy of the received signal from a plane reflector - as a function of insonification angle - has a more simple behavior (e.g. Gaussian shaped) than at the near field. The rotational device holding the transducer could be translated in three orthogonal directions. For the purpose of noise reduction, fifty received signals obtained from the same spatial position were averaged in the digital oscilloscope. Also a bandpass filter was applied in the computer to further reduce noise outside the transducer passband.

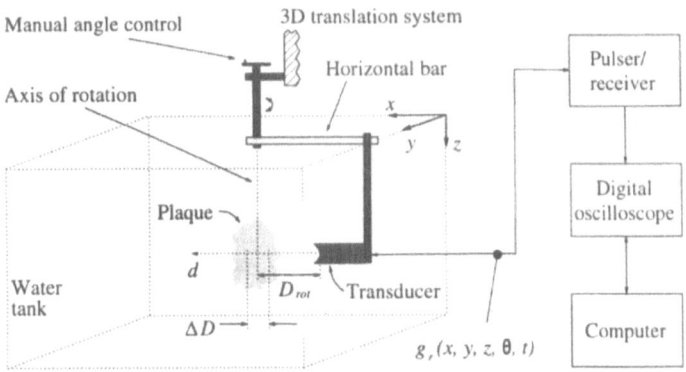

Figure 2 Single element spherically focused ultrasound transducer mounted to rotation axis via a horizontally mounted bar. The rotational system could be translated in space. The acoustic axis intersects the axis of rotation at an angle of 90°.

The beamwidth of the ultrasound transducer was characterized using a 0.1 mm diameter glass sphere moulded into an agar block. At the range D_{rot}, the −3 dB lateral beamwidth was found to 1.4 mm.

To ensure that the ultrasound beam intersected the translation axis as indicated in Figure 2, a special line reflector could be mounted to the physical part of the rotational axis, which was a 13 mm in diameter rod. The line reflector was made from 0.1 mm in diameter stainless steel wire mounted with high precision so that it coincided with the axis of rotation within 0.1 mm. With the line reflector in place, the transducer fixation could be adjusted so that the acoustic axis intersected the axis of rotation.

Recording Procedure

The same part of the ultrasound beam was used for all range cells which required the rotational system with transducer moved in 3D to each specific range cell. The advantage of this time consuming approach was high precision in the spatial definition of the range cell and that all range cells were insonified exactly the same way. The received signal segment, $g_r(x, y, z, \theta, t)$, for a given range cell, (x, y, z) and angle, θ, was extracted by windowing out a small segment from the entire received signal. This segment was located from $T_{rot} − T_w/2$ to $T_{rot} + T_w/2$, where $T_{rot} = 2D_{rot}/c$ and $T_w = \Delta D/c$ is the window length ($\Delta D \sim 4$ mm).

Generation of 3D Energy Images

In order to obtain range cells with an omnidirectional sensitivity function, the axial resolution size had to be made equal to the lateral resolution size. This was done by multiplying the received signal segment, $g_r(x, y, z, \theta, t)$, from each range cell with a Gaussian window. The −3 dB length of this window, T_{-3dB}, roughly corresponded to the −3 dB width of the ultrasound beam at D_{rot} (i.e., 1.4 mm). Thus the axial resolution was degraded by this procedure. The result of the multiplication was then integrated in order to calculate the energy of the received signal as a function of spatial location and angle

$$E(x,y,z,\theta) = \int_{-T_w/2}^{T_w/2} \left| \exp\left[-\frac{2\ln(2)}{T_{-3dB}^2} \left(t - \frac{T_{rot}}{2} \right)^2 \right] g_r(x,y,z,\theta,t) \right|^2 dt \qquad (1)$$

Thus, for a given θ, $E(x, y, z, \theta)$ constitutes a 3D energy image.

Evaluation of Spatial Calibration

The precision of the combined rotation and translation system was investigated with the point target. 3D energy images (15x15x15 range cells) of the point target were recorded from the three angles $\theta = -20°$, $0°$, and $20°$. From these three 3D images, the 3D "point of gravity" (corresponding in 1D to the centroid of *e.g.* a spectrum) was calculated. The three estimated 3D locations should coincide. Comparing them revealed that the maximal spatial difference was below 0.25 mm or less than two wavelengths.

Feature Extraction

Before the 3D energy images of the plaques could be used, the range cells containing sutures had to be masked out manually. Consider a given range cell in a set of five 3D energy images. In Figure 3 it is illustrated how the five associated energy values can be used to calculate six simple features for that cell. In this paper, the "feature volume images" were reduced to single scalars by means of averaging. If N_q is the number of range cells representing plaque, $N_{\theta,q}$ is the number of insonification angles for plaque q, and $\Delta\theta$ is the insonification angle increment, then the three features based on change in energy over angle can be written as:

$$F_{S,\min}(q) = \frac{1}{N_q} \sum_{x,y,z} \min_\theta \left(\frac{|E(x,y,z,\theta) - E(x,y,z,\theta+\Delta\theta)|}{\Delta\theta} \right)$$

$$F_{S,mean}(q) = \frac{1}{N_q} \sum_{x,y,z} \frac{1}{N_{\theta,q}-1} \sum_\theta \left(\frac{|E(x,y,z,\theta) - E(x,y,z,\theta+\Delta\theta)|}{\Delta\theta} \right) \quad (2)$$

$$F_{S,\max}(q) = \frac{1}{N_q} \sum_{x,y,z} \max_\theta \left(\frac{|E(x,y,z,\theta) - E(x,y,z,\theta+\Delta\theta)|}{\Delta\theta} \right)$$

where $\theta = -20°$, $-10°$, $-0°$, $10°$ (for $N_{\theta,q} = 5$). The remaining three features based on energy can be written as:

$$F_{E,\min}(q) = \frac{1}{N_q} \sum_{x,y,z} \min_\theta (E(x,y,z,\theta))$$

$$F_{E,mean}(q) = \frac{1}{N_q} \sum_{x,y,z} \frac{1}{N_{\theta,q}} \sum_\theta E(x,y,z,\theta) \quad (3)$$

$$F_{E,\max}(q) = \frac{1}{N_q} \sum_{x,y,z} \max_\theta (E(x,y,z,\theta))$$

where $\theta = -20°$, $-10°$, $-0°$, $10°$ and $20°$ (for $N_{\theta,q} = 5$).

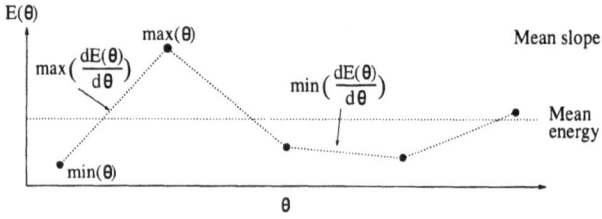

Figure 3 Five energy values from a given range cell, plotted as a function of angle. Six different features can be calculated based on these energy values.

Table 1 Overview of feature performance parameters.

Type	ε_{rms}	p	r
$F_{S,min}$	0.05	0.0002	0.86
$F_{S,mean}$	0.07	0.02	0.63
$F_{S,max}$	0.08	0.09	0.49
$F_{E,min}$	0.06	0.001	0.80
$F_{E,mean}$	0.06	0.004	0.74
$F_{E,max}$	0.07	0.02	0.64

RESULTS

3D energy images were recorded for thirteen plaques. The distance between the center of neighboring range cells was 1.4 mm and signals from approximately 1000 range cells were recorded for each plaque.

The correlation between the histologically determined relative content of fibrous materials, represented in vector $\underline{Y}$, and one of the six ultrasound features, contained in matrix $\underline{X}$, were investigated using the linear model, $\underline{Y} = \underline{X}\,\beta$. The regression parameters, β, were found based on the least-squares solution. The *rms* error (residual) is: $\varepsilon_{rms} = E\{(\underline{Y} - \underline{X}\,\beta)^2\}^{1/2}$, where E{} extracts the mean value.

The *rms* error, the *p*-value and the correlation coefficient, r, for the six features are show in Table 1. The feature values were scaled to best match the relative content of fibrous materials, and a comparison between these for all plaques are show in Figure 4. To ease interpretation, the plaques were sorted in ascending order with respect to relative content of fibrous tissues.

DISCUSSION

The preliminary results in Table 1 and Figure 4 suggest that $F_{S,min}$, $F_{E,min}$ and $F_{E,mean}$ correlated well to the histologically determined relative content of fibrous tissues.

The plaques in the present study mainly consist of soft materials and fibrous tissues. Therefore, based on the results of Picano[2], it should be expected that the energy (averaged over space and angles) as well as the angular dependence increases with relative content of fibrous materials. This was the case for all features.

Despite the good preliminary results presented in this paper, a number of sources of error exist, which should be kept in mind when evaluating the performance of the method:

a) The plaque was only scanned over an angle interval of 40° in one plane. For the 3D-shaped plaques investigated in the present study, perpendicular incidence (to the degree it can be defined) does not happen for all range cells.

b) For the above reason, the rather simple feature extraction scheme might not be optional and other approaches should be investigated as well.

c) The histological classification used might not be completely in agreement with ultrasonic behavior. The tissue class "fibrous tissues" covers a large range of tissues, which might not all behave as "strong" reflectors.

d) Finally, there is a general uncertainty on the histological results. First, sections were cut with a coarse spatial sampling interval of 3 mm. Second, the histological constituents were often mixed, making drawing of exact borders difficult.

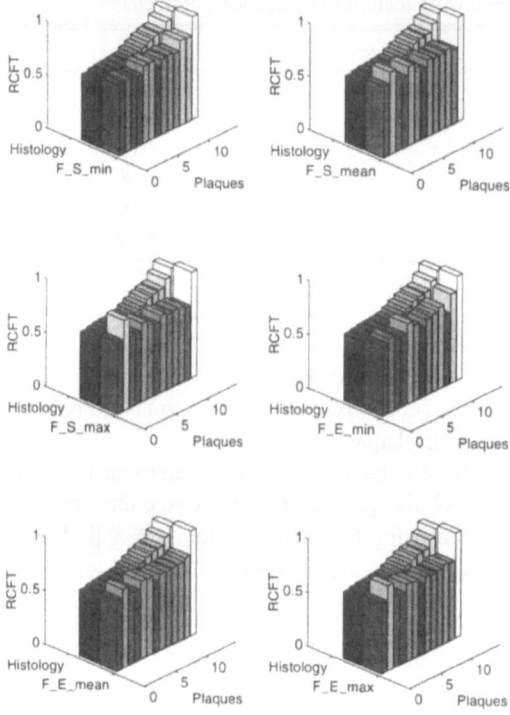

Figure 4 Relative content of fibrous tissues compared to scaled value of ultrasound feature. The plaques are sorted according to the former. RCFT = Relative Content of Fibrous Tissues.

CONCLUSIONS

Thirteen plaques, removed by carotid endarterectomy, have been scanned with ultrasound in 3D from five different insonification angles. The results of the scanning were 3D images representing received ultrasound energy as a function of angle. Six features were extracted from these 3D images and correlated with the histologically determined relative content of fibrous tissue. Good correlation were found for three of these different features.

ACKNOWLEDGEMENTS

CADUS is partly supported by the Danish Technical and Medical Research Councils. The authors gratefully acknowledge the help by M.S. Student Niels Nordmann in carrying out part of the measurements.

REFERENCES

[1] *Cave EM, Pugh ND, Wilson RJ, Sissons GRJ, Woodcock JP:* **Carotid artery duplex scanning: Does plaque echogenicity correlate with patient symptoms?** Eur. J. Vasc. Endovasc. Surg. Vol.10. pp.77-81. 1995.

[2] *Picano E, Landini L, Distante A, Salvadori M, Lattanzi F, Masini M and L'Abbate A:* **Angle dependence of ultrasonic backscatter in arterial tissues: a study in vitro.** Circulation, vol.72, no.3, pp.572-576. 1985.

[3] *Wilhjelm JE, Grønholdt M-LM, Rasmussen S, Martinsen K and Sillesen H:* **Estimation of Plaque Contents With Multi-Angle 3D Compound Imaging.** Proc. of the IEEE International Ultrasonics Symposium, San Antonio, November 3-6, 1996.

FEASIBILITY OF LITTORAL IMAGING WITH A

3 kHz SYNTHETIC APERTURE SONAR

Stephen Celuzza,[1] Philip Abbot,[1] Charles Gedney,[1]
Brad Gillespie,[2] and Kenneth Rolt[2]

[1] Ocean Acoustical Services and Instrumentation Systems, Inc.
5 Militia Drive
Lexington, MA 02173 USA

[2] Sanders, a Lockheed Martin Company
Advanced Systems Directorate
P.O. Box 868
Nashua, NH 03061 USA

INTRODUCTION

The feasibility of using a 3 kHz surface ship bow sonar as a synthetic aperture sonar (SAS) for imaging targets in shallow water is presented. The study consists of modelling the environment with a time-varying ocean surface; a downward refracting sound velocity profile; and bottom roughness, penetration and scattering. Targets placed in the model include both a point target to simulate a mine and confirm imaging resolution, and an array of point targets to simulate the length extent of a submarine. The multipath returns were then processed by standard side-scan sonar (SSS) methods and SAS methods. Replica processing (matched filtering) was employed to improve imaging. Useful SAS imaging is shown.

LIMITATIONS OF SAS IN SHALLOW WATER

In order to focus an synthetic array, the phase errors between the individual elements cannot exceed $\lambda/8$, or about 6 cm at 3 kHz. Thus, during the time period of synthesizing the array, the required phase errors from the combined stability of the ocean and ship motion should be as small as possible, and not exceed $\lambda/8$.

Spatial and Temporal Coherence

The variability of the ocean (both in time and in space) causes phase errors to be introduced along a synthetic array. This variability is discussed by Rolt and Abbot.[1]

Platform Motion

Normal ship motion would exceed the tolerances required for synthetic aperture sonar focussing. However, careful monitoring of the ship's attitude and position can compensate for this motion and allow focussing. Several systems, including an Inertial Navigation System (INS), the Global Positioning System (GPS) and Kalman trackers can be used to monitor these factors.

While standard GPS (or even differential GPS) would not provide the resolution required for aperture synthesis, advanced GPS techniques would provide adequate resolution. In particular, a method originally developed for land surveying, known as carrier phase tracking, is capable of providing resolutions on the order of 1 mm. This has been adapted for moving platforms into a process called real time kinematic (RTK) processing. Commercially available RTK systems currently provide accuracy on the order of 3 cm. Mounting a few of these receivers in distinct locations on the ship and tying them to a Kalman tracker using a ship dynamics model could provide estimates of the bow sonar position.

CHARACTERISTICS OF SITUATION MODELLED

Geometry

The combination source/receiver for this study is a surface ship's hull sonar. The ship is travelling at 5.14 m/s (10 knots) due north in 100 m deep water. This ship transmits a 5 msec ping every 4 sec, or 20.56 m, over a track 2056 m long, for a total of 101 transmit/receive locations.

The targets are east of the ship's track, approximately at the same latitude as the 50th transmit location. They consist of 10 point reflectors, arranged to simulate a mine and a submarine. The mine is simulated by a single point reflector at range 2770 m and azimuth 1058 m, while the submarine is simulated by 9 point reflectors at range 2800 m, spaced 6 m apart, from azimuth 1000 m to 1048 m. All targets are 50 m deep.

Sonar Parameters

The pings emitted by the hull sonar are modelled as 5 msec linear frequency modulated (LFM) slides, sweeping in frequency from 2,250 Hz to 3,750 Hz, resulting in a center frequency of 3,000 Hz and a 1,500 Hz bandwidth. The vertical beamwidth is ±30° and the azimuthal beamwidth is 90°.

Environmental Parameters

The acoustic transmission were modelled by a proprietary ray-based model. A typical downward refracting sound velocity profile was used (sound velocity approximately equal to 1,540 m/s at the surface and 1,500 m/s at the bottom). This resulted in all rays propagating to the bottom within a few hundred meters, and rays propagating to the targets interacting with the bottom several times.

Other model inputs included a time-varying sea surface with waves having a standard deviation of 0.32 m, and a silt-sand bottom with sound velocity 1,700 m/s.

SIGNAL PROCESSING TECHNIQUES AND IMAGES

Side Scan Sonar

In conventional side scan sonar, the received sound pressure levels are stacked, one on top of the other, as in Figure 1. The ping number is related to the azimuthal location by the ping spacing (ping 1 is transmitted at 0 m and received at 20.56 m, while ping 100 is transmitted at 2035.44 m and received at 2056 m). The range is determined from the time delay by an approximated mean sound velocity. The shades of gray in the figure correspond to the magnitude of the received sound pressure level, with black representing the highest pressures and white representing the lowest pressures. (Color graphics illustrate the returns even better.)

Note that these images have a highly distorted X-Y scale. We show a total of 100 m in range, but more than 2000 m in azimuth. This is done to avoid large areas of blank space on the plots, and to enhance the appearance of the range migration hyperbolas. Range migration hyperbolas show the characteristic spread of the echo level into the neighboring source locations, as the receiver moves past a fixed target.

If this figure was the only information about a particular area, one might conclude that there were at least two distance targets in the area. First, a low strength target with a range near 2,770 m, with a closest point of approach (CPA) near ping 52. A second, stronger target occurs at a range of 2,800 m, with a CPA between pings 49 and 52. In fact, these two echoes correspond to the mine and the submarine, respectively.

The other echoes in the figures correspond to multipath propagation of the echoes from the targets. However, without *a priori* knowledge of the number of targets, these could be mistaken for additional targets. One could suppose that the echoes that occur beyond 2,800 m in range, since they occur at the same azimuthal location, are likely to be due to multipath echoes. A more reliable method of determining this is by using matched-field processing, which attempts to predict the location of multipath echoes. We have had success with this technique, but it is beyond the scope of this paper.

The image in Figure 1 is improved upon by replica processing (also called matched filtering), and shown in Figure 2. Replica processing is accomplished by cross correlating the received signal with the transmitted signal. Comparing the two figures shows two improvements: the range resolution is improved (the elevated magnitudes at the mine and the submarine are shorter in the range direction); and the contrast between the targets and the background is improved, as the replica correlation process discriminates against the background noise.

Synthetic Aperture Sonar

Target images can be further enhanced by focusing the synthetic aperture array. During the focusing process, the replica processed data are shifted in time to correct for the spherical wave fronts as they arrive at the individual elements of the synthesized array. Since we do not know the target's position, the array must be focused differently for each assumed target position.

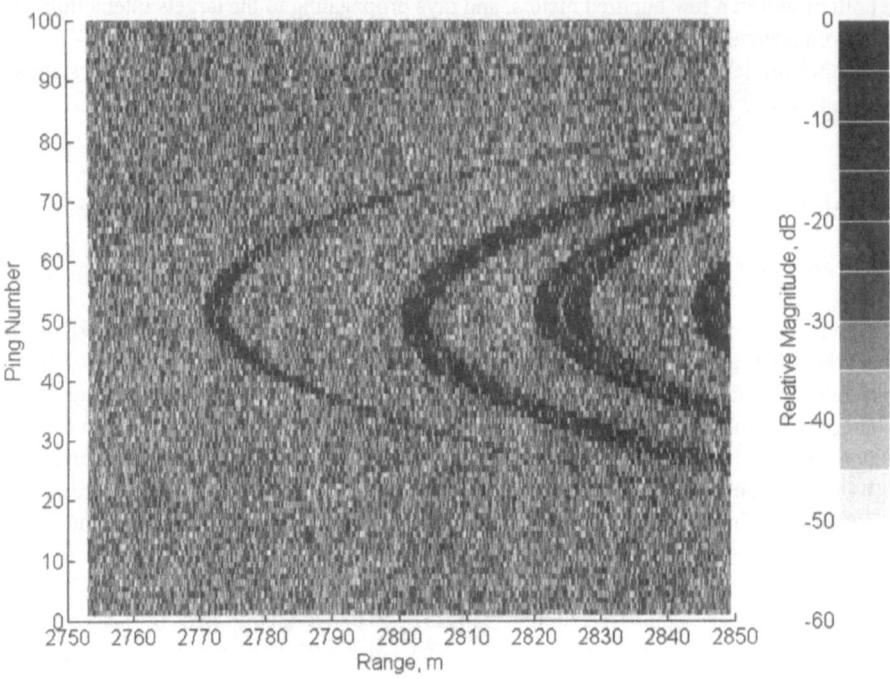

Figure 1. Side Scan Sonar Image with White Noise. The Echo from the Mine can be seen at Range = 2,770 m, and the echo from the submarine can be seen at Range = 2,800 m. Echoes beyond 2,800 m are due to multipath propagation.

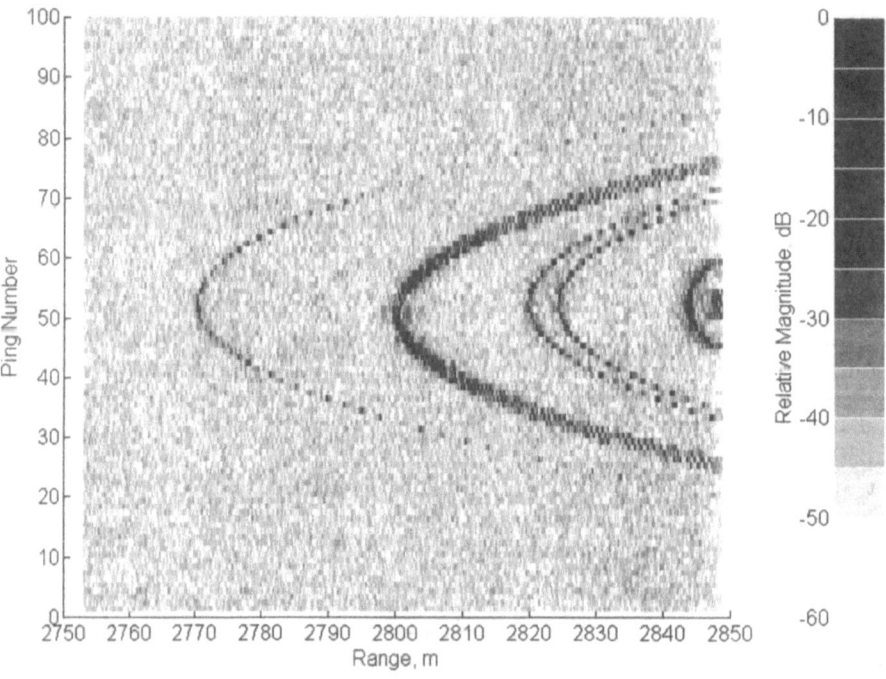

Figure 2. Side Scan Sonar Image with White Noise, Replica Correlated. Note that, compared to Figure 1, the range resolution has improved and the signal-to-noise ratio has increased.

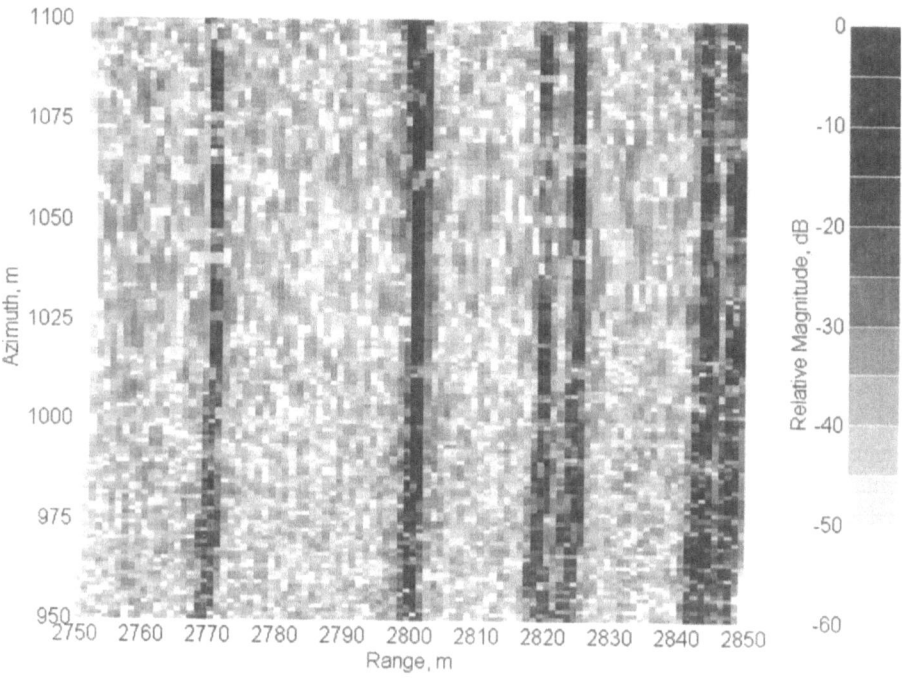

Figure 3. Synthetic Aperture Sonar Image, Focussed with N=4 Elements, Replica Correlated. Notice that the Range-Migration Hyperbolas have been Removed. The Mine is Visible at Range = 2,770 m, Azimuth = 1,058 m, and the Submarine is Visible at Range = 2,800 m, Azimuth = 1,000 m to 1,048 m.

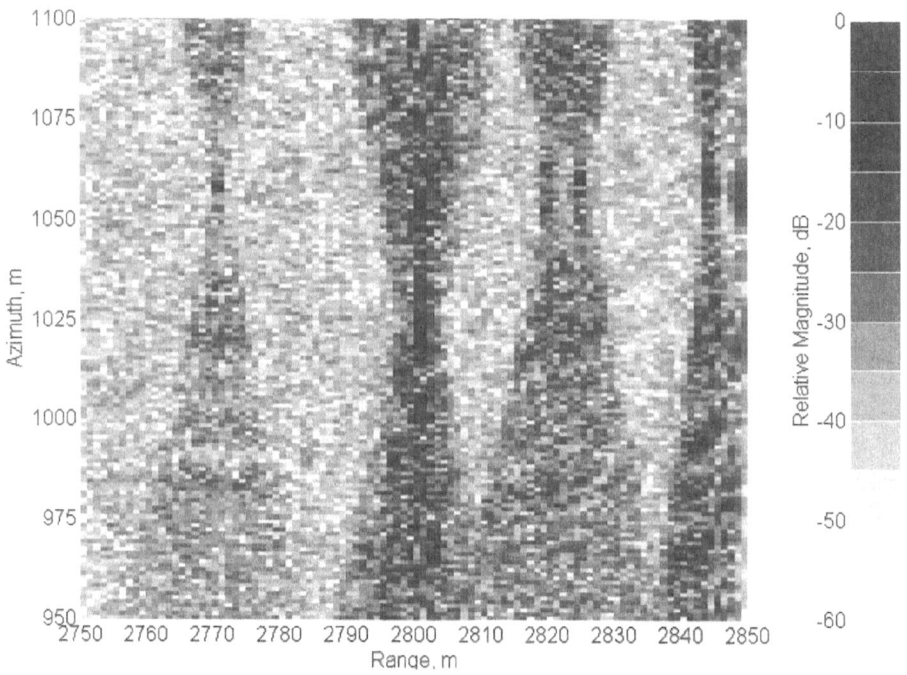

Figure 4. Synthetic Aperture Sonar Image, Focussed with N=32 Elements, Replica Correlated. Compared to Figure 3, the Azimuthal Resolution has been Improved, and the Azimuthal Aliases Reduced.

To illustrate this process, we have selected a 150 m by 100 m horizontal box centered on the submarine and divided it into a grid of 1 m by 1 m cells. Without prior knowledge of the target locations, side scan sonar results must be used to determine the location of the focusing grid. We then selected an array centered on this box and examined the replica processed data for each element of this array. Since the array sources and receivers are located at adjacent locations separated by 20.56 m, the received signal at each element and each instant in time represents a possible target located on an ellipsoidal surface whose foci are located at the source and receiver positions. At each time for which the ellipsoid intersects the focusing grid, the grid function is incremented by the value of the received signal at that time. The focused SAS array output for each grid point is obtained by summing the focused array element outputs over all of the elements in the SAS array.

Figures 3 and 4 show the focused array outputs for the 150 m by 100 m box using $N=4$ and 32 synthetic aperture elements, respectively. Although these figures cover significantly less area and have fewer points than the previous SSS images, we note that the resolution of the mine's position has been enhanced. Further, as the number of elements is increased, the azimuthal resolution is improved. These figures also show a target at the submarine location as well as false targets at larger ranges due to multipath. Aliasing lobes are also apparent in these figures as is expected since the element spacing is much larger than optimal. Note however that the aliasing lobes tend to smear out mostly in azimuth and slightly across range as N is increased.

CONCLUSIONS

Conventional side scan sonar images were shown for a downward refracting shallow water acoustics problem. These images exhibited the range migration hyperbolas which provide a viable way of narrowing the surveillance area in the shallow water environment. They also show multipath hyperbolas, which can be used to correct for platform motion and medium instability, as well as estimate target depth.

Synthetic aperture sonar images demonstrate improved resolution over the side scan sonar images, due to the increased length of the array. Thus, we conclude that it may be possible to use surface ship bow sonars to create synthetic aperture sonars, once the limitations of the environment and ship motion compensation are overcome.

REFERENCES

1. K. Rolt and P. Abbot, Littoral coherence limitations of acoustic arrays, *23rd International Symposium on Acoustical Imaging* (1997).

PARTIALLY-COHERENT ULTRASONIC IMAGING

Richard Y. Chiao, Kai E. Thomenius, and Thomas G. Kincaid[†]

General Electric Corporate R&D Center
P. O. Box 8, Schenectady, NY 12301
[†]Department of Electrical and Computer Engineering
Boston University, Boston, MA 02215

INTRODUCTION

An important problem in ultrasonic imaging is target detection under non-ideal imaging conditions. One example of this occurs in nondestructive testing (NDT) where cracks or inclusions must be found in the presence of normal material inhomogeneities (including surface roughness). The material inhomogeneities may create spatial variations in sound speed [1] as well as multi-path pulse distortion and speckle-like "material noise". Furthermore, the ultrasonic pulse reflected from the target depends on the target shape and composition. Although this work is motivated by problems in NDT, the concept of partially-coherent imaging may also find applications in medical imaging and sonar.

In this paper we approach ultrasonic imaging from a statistical detection framework. Reflections from a single point-like scatterer propagating through a homogeneous medium are considered to be deterministic and known, while those from distributed reflectors and/or propagating through inhomogeneous media are modeled as being random. The optimal detector for the deterministic and known case is shown to be the coherent combination of received data which is the conventional beamformer, while that for the random data results in the proposed partially-coherent beamformer. The partially-coherent beamformer is obtained by breaking the spatial aperture into overlapping segments, computing the coherent delay-and-sum within each segment, and finally computing the incoherent sum over all the segment outputs.

The idea of breaking an aperture into coherent segments and forming the incoherent sum over them is not new and has already been investigated by several researchers mostly for the purpose of speckle reduction [2]-[4]. Mallart et al. proposed partially-coherent imaging using random-phase screens [5]. However, previous approaches for partially-coherent imaging have largely been experimentally-based with little theoretical derivation of the imaging systems presented. Here, we formulate partially-coherent imaging based on detection theory and show that the degree of partial-coherence for a given imaging system should match that of the target signal. We propose a partially-coherent imaging algorithm which adapts to the signal correlation at each spatial position and show experimental results to demonstrate the method.

COHERENT BEAMFORMING

Conventional beamforming assumes the ideal imaging condition of noninteracting point scatterers in a homogeneous isotropic medium [6]. In this section we derive the conventional beamformer as an optimal estimator of scatterer strength, resulting in a beamformer that coherently combines the received data. In the next section we derive the partially-coherent beamformer, which includes the coherent beamformer as a special case, for target detection under non-ideal imaging conditions.

Figure 1 shows a sampled aperture on the surface of an object to be imaged. For simplicity we consider only receive beamforming and assume the subsurface scatterers to be (secondary) sources. The aperture may be sampled by an array of transducers or by a single transducer scanned over the aperture. For a given point source, the received signals at the L sample positions within the aperture are given by

$$u_l(t) = a_0 s(t - \tau_l) + n_l(t), \quad l = 1, 2, \ldots, L \tag{1}$$

where a_0 is the source strength, $s(t)$ is the source pulse shape, $n(t)$ is the thermal noise, and $\tau_l = r_l/c$ is the one-way delay from the source to receiver element l, with r_l as the distance from the source to element 1 and c as the propagation speed of sound.

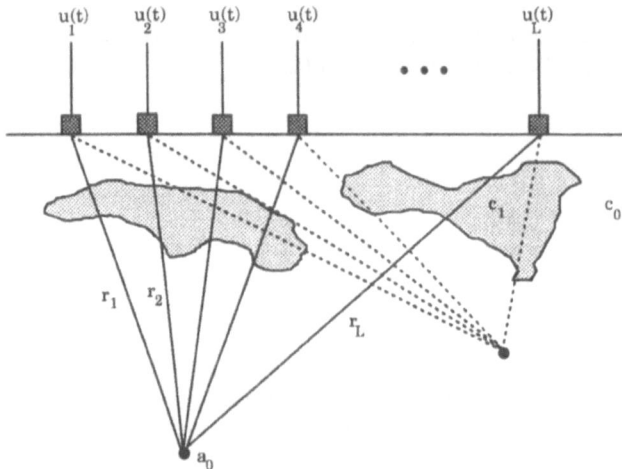

Figure 1. Imaging geometry showing inhomogeneous medium.

The task of the conventional beamformer is to estimate the source amplitude, a_0, at the source location given the data $u_l(t)$, $l = 1, 2, \ldots, L$. To do this, we first Fourier transform the received data to obtain

$$U_l(f) = a_0 S(f) exp(-j2\pi f \tau_l) + N_l(f), \quad l = 1, 2, \ldots, L \tag{2}$$

By discretizing the frequency variable into M frequencies, Eq. 2 can be written in vector form as

$$\overline{U}_l = a_0 \overline{S}_l + \overline{N}_l, \quad l = 1, 2, \ldots, L \tag{3}$$

where

$$\overline{U}_l = [U_l(f_1), U_l(f_2), \ldots U_l(f_M)]^t \tag{4}$$
$$\overline{N}_l = [N_l(f_1), N_l(f_2), \ldots N_l(f_M)]^t \tag{5}$$

and

$$\overline{S}_l = [S(f_1)exp(-j2\pi f_1 \tau_l), S(f_2)exp(-j2\pi f_2 \tau_l), \ldots, S(f_M)exp(-j2\pi f_M \tau_l)]^t \tag{6}$$

We now write the L simultaneous vector equations in Eq. 3 as one vector equation

$$\overline{U} = a_0 \overline{S} + \overline{N} \tag{7}$$

where

$$\overline{U} = [\overline{U}_1^t, \overline{U}_2^t, \ldots, \overline{U}_L^t]^t \tag{8}$$

224

with $\overline{S}$ and $\overline{N}$ similarly defined.

Equation 7 is now in standard form, and the maximum-likelihood estimate of a_0 is given by [7]

$$\hat{a}_0 = (\overline{S}^+\overline{S})^{-1}\overline{S}^+\overline{U} \tag{9}$$

$$= \frac{1}{M}\sum_{m=1}^{M}\frac{1}{S(f_m)}\frac{1}{L}\sum_{l=1}^{L}U_l(f_m)exp(j2\pi f_m\tau_l)$$

where the superscript "+" denotes conjugate transpose. The inner summation is the beam-former at each frequency f_m. The outer summation steps through each frequency which makes the phase delay a time delay. In between the two summations, the transmitted pulse is de-convolved from the received data. In practice, the deconvolution is usually not performed and Eq. 9 reduces to

$$\hat{a}_0 = \frac{1}{L}\sum_{l=1}^{L}u_l(\tau_l) \tag{10}$$

where the received RF data is first delayed to compensate for the propagation delay and then coherently summed across receiver positions.

PARTIALLY-COHERENT IMAGING

Conventional imaging as presented in the previous section produces useful results when the assumptions under which it is derived are not grossly violated. However, in many situations material inhomogeneity causes variations in the propagation speed (see Fig. 1) [1]. In these cases, conventional coherent imaging will be degraded and may not be optimal for target detection. In this section we shift emphasis away from target estimation to detection with the goal of improving target detection under non-ideal imaging conditions.

Starting from Eq. 2, the Fourier transform of the received signal at a given receiver position l is given by

$$U_l(f) = A_l(f)S(f)exp(-j2\pi f\tau_l) + N_l(f), \quad l = 1, 2, \ldots, L \tag{11}$$

where $A_l(f)$ is a complex Gaussian random vector used to model the effects of multiple interacting scatterers and of propagation through inhomogeneous media on the target signal. We also generalize the noise term beyond just thermal noise to include anything that is not considered to be the desired target signal. As in the previous section, we discretize the frequency variable into M samples to obtain the following matrix equation

$$\overline{U} = S\overline{A} + \overline{N} \tag{12}$$

where $\overline{U}$ and $\overline{N}$ are constructed identically as Eq. 8, S is the $LM \times LM$ diagonal matrix given by

$$diag(S) = [S(f_1)exp(-j2\pi f_1\tau_1), S(f_2)exp(-j2\pi f_2\tau_1), \ldots, S(f_M)exp(-j2\pi f_M\tau_L)] \tag{13}$$

and $\overline{A}$ is the LM vector given by

$$\overline{A} = [A_1(f_1), A_1(f_2), \ldots, A_1(f_M), A_2(f_1), \ldots, A_L(f_M)]^t \tag{14}$$

The complex Gaussian random vector $\overline{A}$ with $LM \times LM$ correlation matrix $R_A = E[\overline{A}\overline{A}^+]$ is used to characterize the target signal for detection. The likelihood-ratio detection statistic is given by [7]

$$d = \overline{U}^+ R_N^{-1} R_S (R_N + R_S)^{-1} \overline{U} \tag{15}$$

where

$$R_S = S R_A S^+ \tag{16}$$

is the signal correlation matrix and R_N is the noise correlation matrix. Under the low SNR approximation we obtain

$$d = \overline{U}^+ R_N^{-1} R_S R_N^{-1} \overline{U} \tag{17}$$

225

$$= \overline{V}^{+} \pmb{R_A} \overline{V}$$

where

$$\overline{V} = \pmb{S}^{+} \pmb{R_N}^{-1} \overline{U} \tag{18}$$

is the LM vector of phase-delay compensated, pre-whitened, and match-filtered receiver outputs at each frequency. The processed receiver outputs $\overline{V}$ are combined across frequencies and receivers using the the correlation matrix $\pmb{R_A}$ according to Eq. 17. We use the detection statistic d as the imaging quantity.

Equations 17 and 18 give the general formulation for partially-coherent imaging. We now make some simplifying assumptions about the correlation matrix to obtain a working formulation. First, we assume the frequency correlation to be independent of space, so that the matrix $\pmb{R_A}$ can be written as the Kronecker product of the spatial correlation matrix $\pmb{R_L}$ and the frequency correlation matrix $\pmb{R_M}$

$$\pmb{R_A} = \pmb{R_L} \# \pmb{R_M} \tag{19}$$

where $\#$ denotes the Kronecker product. Physically, we are assuming the pulse distortion at each receiver to be independent of the pulse decorrelation across receivers. Next, we assume the random fluctuations to be stationary in space, such that the spatial correlation matrix is Toeplitz. Finally, we assume the frequencies to be fully correlated (no pulse distortion) such that $\pmb{R_M} = \pmb{1}$ (unity matrix), and Eq. 17 can be written as

$$d = \sum_{l1=1}^{L}\sum_{l2=1}^{L} R_L(l1,l2) \sum_{m1=1}^{M} \frac{S(f_{m1})}{N^*(f_{m1})} U_{l1}^*(f_{m1}) exp(-j2\pi f_{m1}\tau_{l1}) \tag{20}$$

$$\sum_{m2=1}^{M} \frac{S^*(f_{m2})}{N(f_{m2})} U_{l2}(f_{m2}) exp(j2\pi f_{m2}\tau_{l2})$$

$$= \sum_{l1=1}^{L}\sum_{l2=1}^{L} R_L(l1,l2) w_{l1}^*(t+\tau_{l1}) w_{l2}(t+\tau_{l2})$$

$$= \overline{w}^{+} \pmb{R_L} \overline{w}$$

where $\overline{w}$ is the L vector of time-delayed, prewhitened and match-filtered data for a given imaging location.

Figure 2 illustrates the beamformer structure under discussion. As shown, the receive aperture is broken into coherent sub-apertures each of which spans the spatial coherence length. Each sub-aperture is coherently beamformed as in conventional imaging, however the outputs of the sub-apertures are then incoherently summed to form a single pixel value in the reconstructed image. This beamformer structure includes the coherent and incoherent beamformers as special cases: if $\pmb{R_L} = \pmb{1}$ (unity matrix), then there is only one coherent sub-aperture which spans the entire aperture, and all signal components are coherently summed together to form the beamformer output given by $d = |\sum_{i=1}^{L} w_i|^2$. At the other extreme $\pmb{R_L} = \pmb{I}$ (identity matrix), then there are L coherent sub-aperture each containing only a single signal component, and all the signal components are incoherently summed together to form the beamformer output given by $d = \sum_{i=1}^{L} |w_i|^2$. In general, the width of the center diagonal of $\pmb{R_L}$ is proportional to the size of the coherent aperture.

The above partially-coherent imaging algorithm requires knowledge of the signal correlation matrix at each imaging point. The Toeplitz correlation matrix can be obtained by computing the correlation function

$$R_L(n) = \frac{1}{L-n+1} \sum_{l=1}^{L-n+1} w(l)w^*(l+n-1), \quad n = 1,2,\ldots,L \tag{21}$$

which is used to reconfigure the beamformer for optimal detection at each imaging point.

By matching the size of the coherent aperture (degree of partial coherence) with the signal correlation length, we maximize the signal-to-noise ratio (SNR) on the beamformer output given by

$$SNR = \frac{E[d_1]}{\sqrt{E[d_0^2] - E^2[d_0]}} \qquad (22)$$

$$= \frac{\sigma_s^2 \sum_{i=1}^{L} \sum_{j=1}^{L} R_L(i-j)\rho_s(i-j)}{\sigma_n^2 \sqrt{\sum_{i=1}^{L} \sum_{j=1}^{L} \sum_{k=1}^{L} \sum_{l=1}^{L} R_L(i-j)R_L(k-l)[\rho_n(i-k)\rho_n(j-l) + \rho_n(i-l)\rho_n(j-k)]}}$$

where d_1 and d_0 are the beamformer outputs with signal-only and noise-only, respectively; ρ_s and ρ_n are the signal and noise correlation functions, respectively; and σ_s^2/σ_n^2 is the input SNR. Figure 3 plots the SNR given by Eq. 22 for various signal, noise, and beamformer correlation lengths, assuming a flat correlation function, $L = 21$, and unity input SNR.

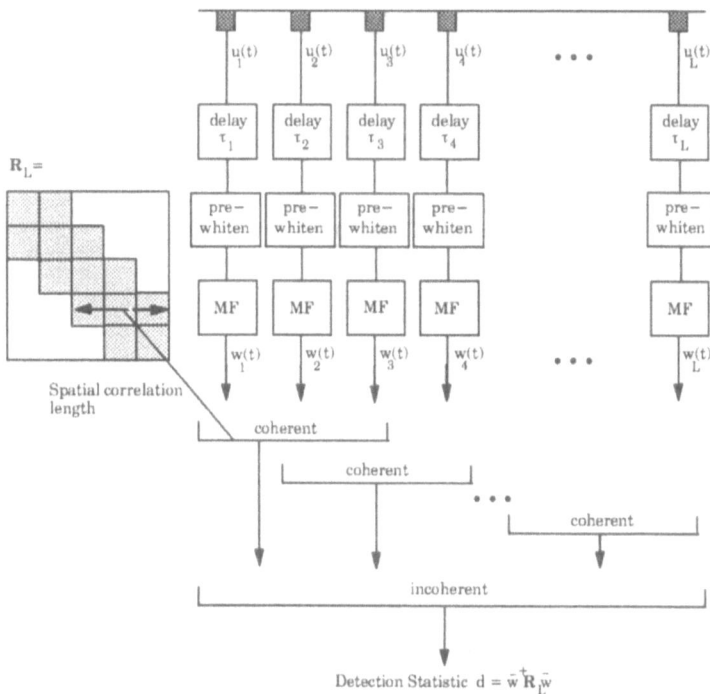

Figure 2. Block diagram of partially-coherent imaging system.

RESULTS

The partially-coherent imaging algorithm described in the previous section has been implemented in software, and in this section we show its application to detecting inclusions in titanium. The particular sample that we image is a rectangular block which has been seeded with weakly-reflecting inclusions at a depth of approximately 45mm. Although the surface of the block is smooth, material noise and propagation speed variations are caused by the anisotropic and randomly oriented grains that make up the metal [1].

A 15MHz spherically-focused transducer with diameter 12.7mm and focal length 76.2mm is used to acquire the synthetic aperture data [6]. The titanium block and transducer are both immersed in water, and the transducer is positioned such that the focal point is at the surface of the block and a diverging wave is transmitted normally into the metal. The transducer is translated by 0.5mm steps over an aperture of 50mm. At each position a broadband pulse is transmitted from the transducer, and the received echos are digitized and stored.

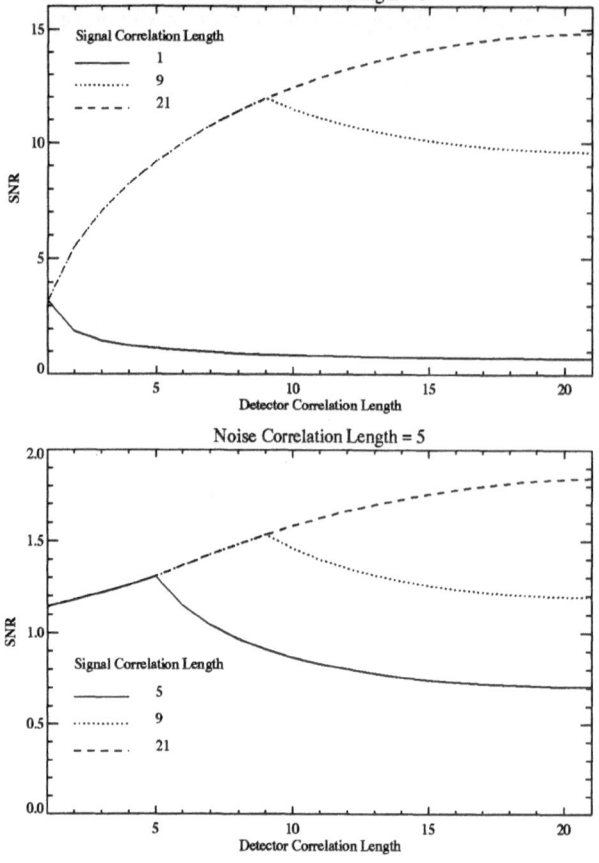

Figure 3. SNR optimization by using appropriate signal correlation length in partially-coherent beamformer.

Figure 4 shows the result of partially-coherent imaging using the experimental data. The top left image shows a raw B-scan where the vertical axis is space along the surface of the titanium block and the horizon axis is depth into the block. The weakly-scattering inclusions are barely visible due to the material noise and defocused beam. The top right image shows the coherently beamformed image where the inclusions are visible amid strong material noise. In the second row of Fig. 4, the left image shows the estimated correlation map and the right image shows the partially-coherent image formed using that correlation map. The two bright horizontal regions in the correlation map indicate high correlation at the inclusions. The partially-coherent image shows reduced material noise fluctuations, but also decreased resolution.

With a-priori knowledge about signal and noise correlation, the correlation map can be used for further clutter rejection. The bottom two rows of Fig. 4 show the results of using two different a-priori target correlation distributions to weight the likelihood-ratio detection statistic.

Figure 5 compares the coherent image with the three partially-coherent images at the top pair of inclusion echos. The solid line shows the coherent reconstruction, while the dotted line shows the partially-coherent reconstructions.

CONCLUSIONS

We have presented a partially-coherent imaging method for target detection under non-ideal imaging conditions. The partially-coherent beamformer has a quadratic form, $\overline{w}^+ R_L \overline{w}$, which includes the coherent and incoherent beamformers as special cases. By estimating the signal correlation on the fly, the beamformer is reconfigured for optimal detection at each imaging point. Experimental results are presented to demonstrate the proposed partially-coherent imaging method.

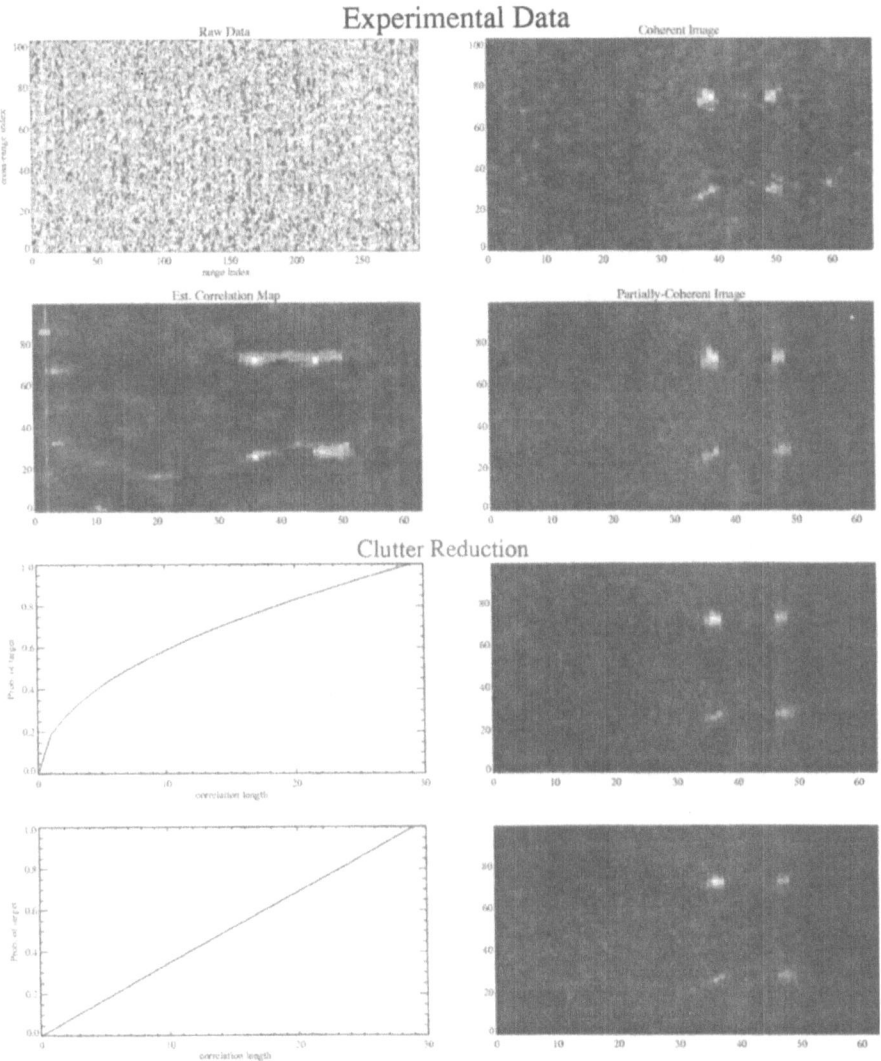

Figure 4. Experimental comparison of partially-coherent and coherent imaging.

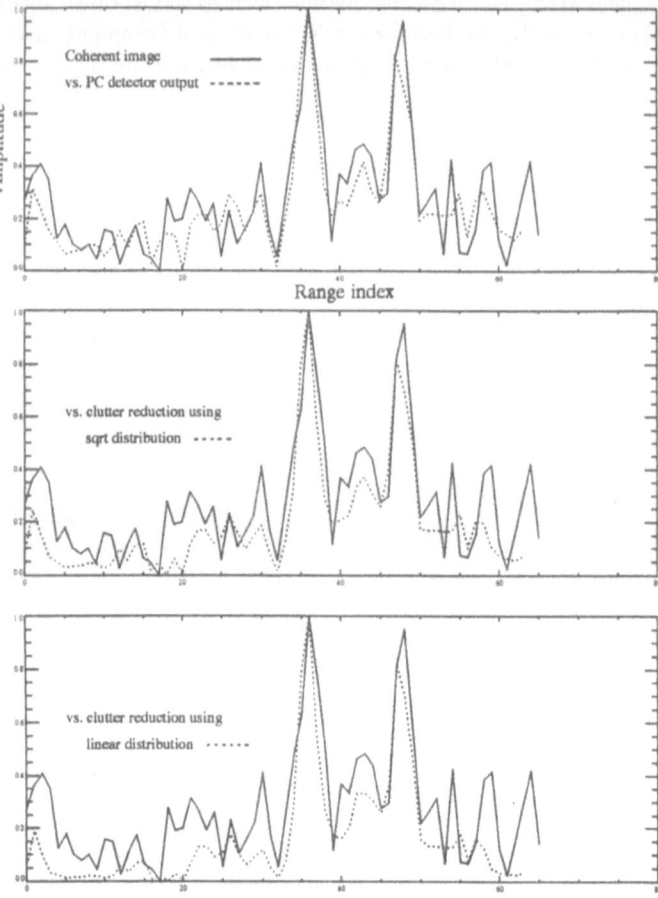

Figure 5. Quantitative comparison of partially-coherent and coherent imaging at one line passing through the top pair of target reflections in Fig. 4.

REFERENCES

[1] R. S. Gilmore, R. A. Hewes, L. J. Thomas III, and J. D. Young, "Broadband Acoustic Microscopy: Scanned Images with Amplitude and Velocity Information," *Acoustical Imaging*, pH. Shimizu, N. Chubachi, and J. Kushibiki, Ed., vol 17, 1989, pp. 97-110.

[2] M. O'Donnell, "Phase-Insensitive Pulse-Echo Imaging", *Ultrasonic Imaging*, vol. 4, 1982, pp. 321-335.

[3] A. T. Kerr, M. S. Paterson, F. S. Foster, and J. W. Hunt, "Speckle Reduction in Pulse Echo Imaging Using Phase Insensitive and Phase Sensitive Signal Processing Techniques," *Ultrasonic Imaging*, vol. 8, 1986, pp. 11-28.

[4] J. J. Giesey, P. L. Carson, D. W. Fitting, and C. R. Meyer, "Speckle Reduction in Pulse-Echo Ultrasonic Imaging Using a Two-Dimensional Receiving Array," *IEEE Trans. Ultrason. Ferroelec. Freq. Contr.*, vol. 39, no. 2, March 1992.

[5] R. Mallart, M. Fink, P. Laugier, and S. Abouelkaram, "Partially Coherent Transducers: The Random Phase Transducer Approach," *Ultrasonic Imaging*, vol. 12, 1990, pp. 205-228.

[6] R. Y. Chiao and L. J. Thomas, "Analytic Evaluation of Sampled Aperture Ultrasonic Imaging Techniques for NDE", *IEEE Trans. Ultrason. Ferroelec. Freq. Contr.*, vol. 41, No. 4, July 1994, pp. 484-493.

[7] A. D. Whalen, *Detection of Signals in Noise*, New York: Academic Press, 1971.

REMOTE ACOUSTIC EMISSION SPECTROGRAPHY (RAES)

Mostafa Fatemi and James F. Greenleaf

Ultrasound Research, Department of Physiology and Biophysics
Mayo Clinic and Foundation
Rochester, MN 55905 U.S.A.

ABSTRACT

The mechanical response of objects to forced vibrations is of interest for nondestructive evaluation of materials. For instance, the resonance frequency of an object has been used as an indicator of possible structural flaws. Here, we present a method to obtain the local frequency response of an object to a remotely controlled force. The vibrating force is generated by the radiation force of two focused ultrasound beams, and is confined to a small region on the object. The force vibrates the object, resulting in an acoustic field (acoustic emission) in the surrounding medium. The resulting responses at a range of vibrational frequencies are recorded and used to form images (spectrograms) representing the object shape and its frequency dependent characteristics at each point. Experiments are performed on a set of three tuning forks with different resonance frequencies. The resulting spectrograms allow discrimination of forks on the basis of their resonance frequencies. This method promises a new material characterization and testing scheme.

INTRODUCTION

The mechanical response of objects to external forces carries information on object structure, and hence, is of interest in nondestructive inspection of materials and material sciences. Conventional methods of ultrasonic nondestructive evaluation of material employ local ultrasonic parameters of the object, such as the reflection and scattering coefficient, to investigate the material. These local parameters can be used to interrogate, or, in some modalities, to image the object with high spatial resolution. However, some characteristics that relate to the global structure of object, such as the mechanical frequency response at low frequencies, can not be obtained from such local parameters. Other methods can be used to look at the global (bulk) parameters of objects. For example, in *"resonant ultrasound spectroscopy"* (Maynard, 1996), a combination of ultrasound source and detector is used to measure the resonance frequencies of a sample with known size and mass. These values are used to calculate the bulk mechanical parameters, including the elastic constants, of the material. In this paper, we present a method to interrogate, or image the object at high spatial resolution, and at the same time, to obtain the low frequency response of the object at each point. The imaging technique used here is based on the radiation force (Torr, 1984; Westervelt, 1951) of two ultrasound beams, similar to the method recently reported by the authors in (Fatemi and Greenleaf, 1996; Fatemi and Greenleaf, 1996a).

In the following, we present the method, simulation and experimental results, and finally the discussion and a summary.

METHOD

The principle of the method to be described here is to apply an oscillatory force at each point on the object, and sweep the frequency of the force over the range of interest. To generate a localized oscillatory force, two intersecting continuous wave (CW) focused ultrasound beams of different frequencies are used. It is only at the intersection region that the ultrasound field energy density is sinusoidally modulated, and hence the field can generate an oscillatory radiation force by interacting locally with the object (Fatemi and Greenleaf, 1996; Fatemi and Greenleaf, 1996a).

We consider two plane wave beams at frequencies ω_1 and ω_2, propagating on the (x,z) plane at angles θ and $-\theta$ with respect to the z-axis, and crossing each other at the origin. The difference frequency is $\Delta\omega = \omega_2 - \omega_1$. Now, consider a target on the $z = 0$ plane. The radiation force exerted normal to this target is found as (Fatemi and Greenleaf, 1996):

$$F_l(t) = Cd_r \cos(\Delta\omega t + \alpha),\qquad(1)$$

where C is a constant proportional to the energy density, d_r is the drag coefficient, which is a function of the scattering and absorbing properties of the object (Westervelt, 1951), and α is a phase constant. This force vibrates the object, resulting an emission of acoustic energy which can be detected by a hydrophone. The amplitude of this field, Φ, can be written as:

$$\Phi = Cd_r|H(\Delta\omega)Q(\Delta\omega)|,\qquad(2)$$

where $Q(\Delta\omega)$ is a complex function representing the mechanical frequency response of the object at the point of excitation, and $H(\Delta\omega)$ represents the combined frequency response of the propagation medium and the hydrophone. $H(\Delta\omega)$ is assumed to be unchanged for any point in the object. By sweeping $\Delta\omega$ across the range of interest, we can obtain the frequency response of the object weighted by $H(\Delta\omega)$. Repeating the same procedure for other points of the object at a given plane (e.g., the z=0 plane), we can collect a set of responses. These data can be mapped into a pictorial format. For this purpose, the received signal, Φ, can be decomposed into three components in three frequency bands using three bandpass filters. The average amplitude in each band is mapped into a monochrome image. By assigning a different color to each image, a multi-color image (spectrograph) is formed by superimposing the three color components. In this case, d_r influences the brightness, while the frequency variations of $|Q(\Delta\omega)|$ are represented by the hue of each image pixel. Since $|H(\Delta\omega)|$ is position invariant, its effect on image color is also position invariant. Hence, variations of image hue versus position qualitatively represent the variations of $|Q(\Delta\omega)|$.

Figure 1 illustrates the imaging system. It includes two RF generators (one sweeping the frequency) driving two transducers. The hydrophone signal is filtered by three bandpass filters, denoted by R, G, and B (representing red, blue, and green image components, respectively). These filters are followed by envelope detectors and the resulting signals are used to form the color image.

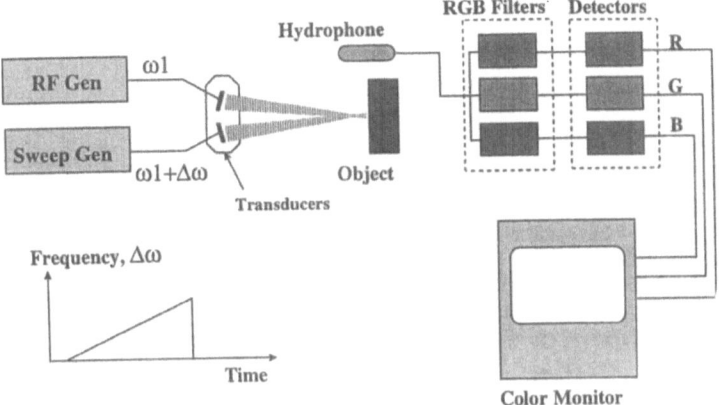

Figure 1. System diagram.

SIMULATION RESULTS

The radiation force of two plane waves is described by Eq. 1. In practice we use two focused transducers, for which the radiation force assumes a more complicated form, and hence we use numerical methods to compute it. In the experimental system, we employed two identical ultrasound transducers. The center frequency, diameter, focal length, and beamwidth were equal to 3.5 MHz, 19 mm, 85 mm, and 2 mm, respectively. The angle between the two beams was set at 20 degrees. Considering these values, the resulting radiation force at the beam crossing region (on the $z = 0$ plane) and at the fixed difference frequency equal to 40 KHz is numerically calculated, and the result is shown in Fig. 2. Because the acoustic emission is proportional to the force, Fig. 2 also represents the point spread function (PSF) of the imaging system. It is seen that the PSF of the system is not radially symmetric. To show this, images of a horizontal and a vertical half planes are simulated. The results, shown in Fig. 3, indicate that the vertical edge is imaged as a series of vertical bars, while the horizontal edge is not detected at all. Hence the system is sensitive only to the vertical edges. Angle dependency of the PSF limits the imaging capability of the method when an object includes sharply defined horizontal edges. To decrease angle sensitivity, one may employ some image processing techniques, or scan the object at two or more beam orientations. We will not explore these options any further in this paper. Here, we apply the method mainly as a tool to obtain the localized frequency response of an object at low frequencies.

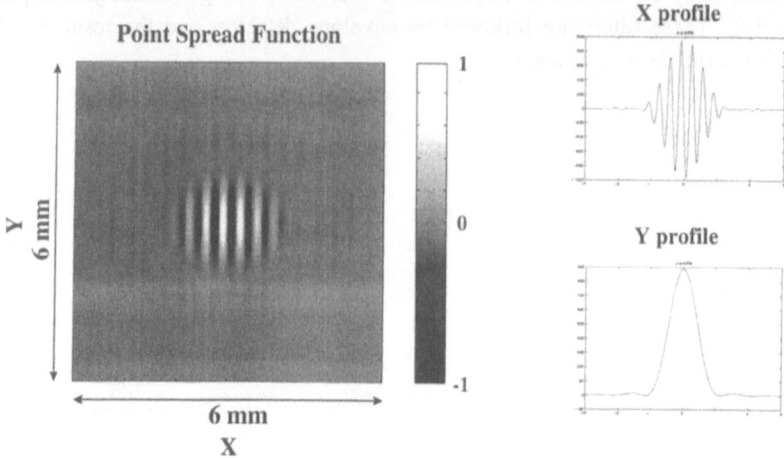

Figure 2. Radiation force distribution in z=0 plane. This also represents the PSF if the system. Horizontal and vertical profiles are shown on the right.

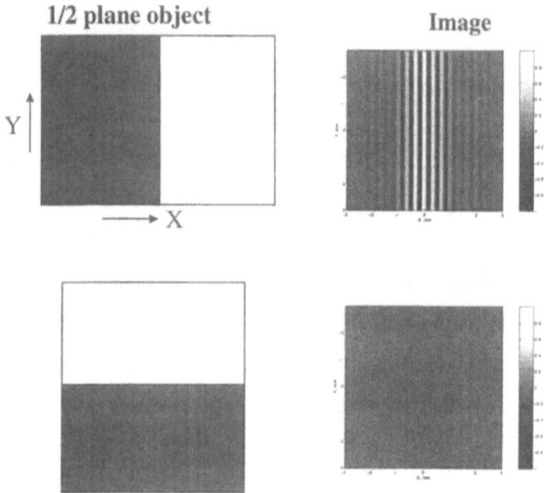

Figure 3. Simulated images of horizontal and vertical half planes.

EXPERIMENTAL RESULTS

The specifications of the transducers used in the experiments are as described in the previous section. The difference frequency was swept from 250 Hz to 2250 Hz. Sweep duration was four seconds.

To show the capability of the method in displaying the frequency responses of an object versus position, a set of three tuning forks were chosen as the test object. The forks and the imaging geometry is shown in Fig. 4. The forks are made from identical material, and have identical finger cross-sections (lengths are different). Resonance frequencies in water are: 407 Hz (right), 809 Hz (middle), and 1709 Hz (left). The forks were scanned in a water tank using the system shown in Fig. 1. The scanning plane covers the front fingers at the bottom part of the forks. The digitized hydrophone signal was filtered by three overlapping bandpass filters each having –6dB bandwidth of 500 Hz. The outputs of the filters with center frequencies 500, 1000, and 1500 kHz were used to produce the red, green, and blue image components, respectively. The three components are shown in Fig. 5 as three gray scale images. To obtain a color image these images can be coded in appropriate colors and superimposed (not shown here).

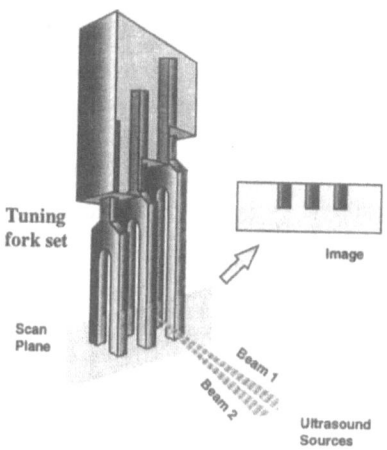

Figure 4. Tuning forks used as the test object. Scan plane covers the front surface at the bottom part of the front fingers.

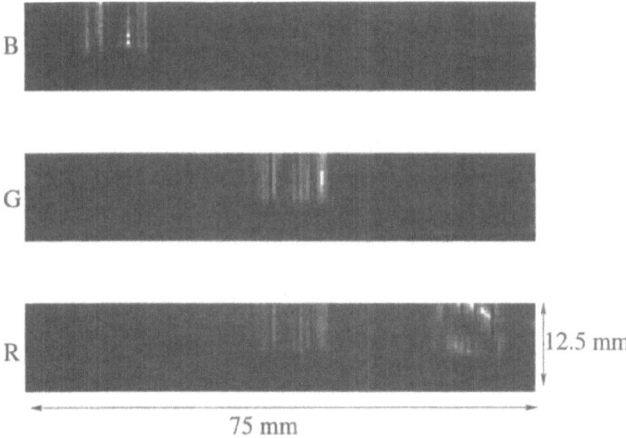

Figure 5. Spectrogram of the tuning forks. The components are shown in gray scale: (R) the red component, (G) the green component, and (B) the blue component.

DISCUSSION

Computer simulation of the PSF of the system, shown in Fig. 2, indicates that the system resolution is about 2 mm, which is about the beam width of the ultrasound system. This is in agreement with the experimental images of the tuning forks. This system can be used as an imaging tool. However one must note that the PSF of the system is angle sensitive. Here, we apply the method mainly as a "spectrography" tool to obtain the localized frequency response of an object at low frequencies.

Figure 5 shows that each fork in the set achieves a higher amplitude in one of the image components, depending on its resonance frequency. The red image component (denoted by R in Fig. 5) displays two forks with center frequencies of 407 and 809 Hz. This is because the resonance frequencies of both forks fall in the passband of the R filter. Figure 5 shows that the system is capable of displaying the frequency response of the object, which is a bulk property of each fork. The sweeping frequency range can be quite wide because the difference frequency is only a fraction of the transducer center frequency, and hence it poses no major technical difficulty. The sweeping range was about one decade in the above experiment.

The method described in this paper provides a combination of high resolution local and bulk parameter of the object. Applications of this method includes nondestructive evaluation of materials and characterization of biological tissues.

SUMMARY

A method for imaging and obtaining the frequency response of an object to a localized force was presented. Simulation results and experimental results were presented. Results indicate that the spatial resolution is about the ultrasound beamwidth. The frequency responses of the object are obtained at low (audio) frequencies in a wide range. Possible applications include nondestructive evaluation of materials and tissue characterization.

REFERENCES

Fatemi, M., and Greenleaf, J.F., 1996, C-scan imaging by radiation force stimulated acoustic emission method, *1996 IEEE Ultrason Symp Proc*, 2:1459–1462.

Fatemi, M., and Greenleaf, J.F., 1996a, Radiation force imaging," in: *Nonlinear Acoustics in Perspective*. R. J. Wei, ed., Nanjing University Press, Nanjing, China, pp 469–474.

Maynard, J., 1996, Resonant ultrasound spectroscopy, *Physics Today*, pp 26–31.

Torr, G. R., 1984, The acoustic radiation force. *Am J Phys*, 52(5):402–408.

Westervelt, P. J., 1951, The theory of steady force caused by sound waves, *J Acoust Soc Am*, 23(4):312–315.

A HAND-CONTROLLED, 3D ULTRASOUND GUIDE AND MEASUREMENT SYSTEM

Ralph C. Fenn, J. Brian Fowlkes, Aaron P. Moskalik
Yong Zhang, Marilyn A. Roubidoux, Paul L. Carson

Department of Radiology
University of Michigan Medical Center
Ann Arbor, Michigan 48109-0553

INTRODUCTION

Summary

Three dimensional (3D) ultrasound images provide superior visualization of anatomical features than two dimensional (2D) power mode or frequency shift color Doppler B-scan ultrasound images. However, 3D data sets require transducer position measurements that can be problematic. One previously investigated approach completely automates transducer positioning and measurements with one dimensional motorized stages. Other system types permit full six degree-of-freedom (DOF) manual transducer manipulation while measuring all six axes. In contrast this work explores the middle ground of measuring the two degrees of freedom most important for free-hand scanning while fixing the remaining axes for minimized complexity.

The device designed, fabricated, and tested allows linear motion perpendicular to the scan plane and along the beam axis, as well as rotation about an axis parallel to the array. The measurement system is designed to minimize size, weight, and cost while maximizing accuracy and ease of operator use. The resulting design weighs 500g in its heaviest configuration and is 150 mm long, 64 mm high, and 70 mm wide. It is capable of 0.1 mm accuracy in the focal plane. Three dimensional data with continually varying translational and rotary speeds was interpolated in 3D by image processing software to minimize distortion in standard Cartesian coordinates. Phantom images show little distortion due to the measuring system or data processing algorithm. This inexpensive, compact, and lightweight system interferes minimally with clinician free-hand scanning and enhances 3D image collection and generation capabilities in the clinical setting.

Three Dimensional Data Acquisition

The goal of three dimensional ultrasound imaging has been pursued by a number of research laboratories. Conventional 2D ultrasound images are inherently limited in their ability to display 3D human anatomy and their diagnostic usefulness is thus reduced. A number of laboratories have investigated various methods of attaining the critical transducer position measurements to create 3D data sets. A common feature of many of these imaging systems is computer controlled, motor-driven linear ultrasound transducer motion (Pretorius et al., 1992, Picot et al., 1993, Moskalik et al., 1995, Tong et al., 1996, and Hernandez et al., 1996). These designs give precise control of the ultrasound transducer position, usually with one degree-of-freedom. However these motorized systems frustrate the clinician's desire for the flexibility of conventional hand-controlled ultrasound examinations. Free-hand control facilitates localizing a specific region of interest (ROI), especially when examining organs of convex or concave external shape such as the shoulder, the uterus, and the liver near the ribs.

In contrast, other studies have provided complete freedom of motion for manual examinations while measuring the positions of each of the 6 axes of motion, both translation and rotation (Detmer et al., 1994, Kelly et al., 1994, Nelson and Pretorius, 1995, and Hughes et al., 1996). Such systems are often less accurate and may be complex, large, and more costly. The approach of the present work uses an intermediate approach with manual transducer control only in the degrees-of-freedom that are most helpful, while avoiding complete 6 DOF freedom of motion that increases complexity and cost. The current design measures 2 DOF and can be easily retrofitted for measurement of the third axis, which is a second translation axis.

METHODS

Instrumentation System Overview

The 3D ultrasound data acquisition system is comprised of three major parts. First is the ultrasound imaging system. The scanning data presented was gathered by a General Electric Logiq 700 MR machine with a 739L transducer but the design is equally applicable to other machine types. Second is the personal computer which handles data acquisition and storage. The current experiments collect position data at the same fixed, 4 Hz rate as the production of ultrasound B-scan images. In future experiments, the personal computer will store B-scans at desired spatial increments through signals to the cardiac gating feature of the ultrasound machine. For color flow studies of maximum sensitivity, cardiac signals will trigger ultrasound image storage and simultaneous position measurements by the personal computer. The third major component, and the focus here, is the mechanical transducer guiding system and its translation and rotation sensors. Postprocessing for 3D image construction is performed on a Digital Equipment Corporation Alpha workstation.

System Design of the Guide and Measurement System

The overall goal of allowing natural, free-hand transducer motion in 2 DOF while measuring translation and rotation drove the design of the measurement system. This goal requires a compact device that allows natural grasp of the transducer grip. The device must not provide too much friction or require movement of significant mass. It should provide range of motion greater than required for screening examinations. The device must have low overall mass to maximize patient comfort. The accuracy and linearity must be high so that these factors contribute little additional distortion to the 3D image. Finally it should be inexpensive. These factors all influenced the design described below.

Figure 1 shows the design developed by this program that provides transducer guidance and measurement. The design uses a "series" design where multiple DOF are created by selecting a base motion type and installing the other DOF on the now moving base stage. In the device described here the base stage provides translation. This is advantageous because of the large stroke required. The alternative of mounting the translation stage on the rotary axis would be very clumsy because of the large radius of the resulting rotating mass. For this reason the rotation stage is mounted on the translation carriage. Finally a second translation stage is mounted on the rotating shaft to provide the third DOF. The first two DOF are measured however the third is not but could be instrumented using straightforward methods. This third DOF can be locked, or left free when positional accuracy is not critical, e.g. while following rough surface contours.

The major components shown in Figure 1 are described below beginning with those components nearest the patient. At the base is a stabilizing platform made of acrylic sheet with a window for transducer array access to the skin. The material is clear for ease of positioning, has low thermal conductivity for patient comfort, and is easily cleanable.

Mounted horizontally on the base plate is a Thompson Series 10 miniature linear

238

bearing. This bearing provides motion of the transducer in the z axis which is defined to be perpendicular to the long dimension of the ultrasound array (x axis).

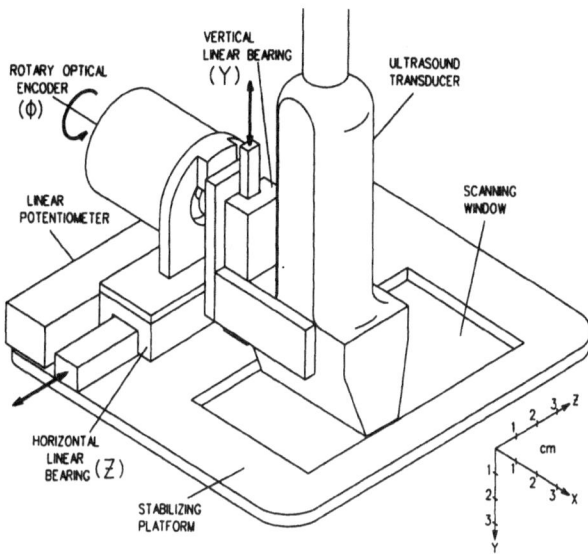

Figure 1. The ultrasound transducer guide and measurement system (drawing by Ben Moskalik).

Low friction is provided by the light rather than heavy preload on the bearings, and by the use of low friction bearing seals. Preload is required to minimize backlash, particularly in the roll axis. Waterproof grease such as petroleum jelly excludes gel and cleaning water from the bearing interior while being nontoxic to the patient.

Parallel to the linear bearing and also mounted on the base plate is a precision linear potentiometer. This off-the-shelf Duncan Electronics 400 Series potentiometer uses a high linearity conductive plastic film and has dual, parallel wipers to reduce noise. This potentiometer measures the translation of the ultrasound array by providing a voltage output when a reference voltage is applied to the resistive element. The friction and damping of the wiper is adjustable by removal of plastic inserts. The wiper access slot has rubber seals that face the side to avoid gel and water intrusion. This sensor has the advantage of providing an analog output which simplifies development of interfaces to various laboratory instruments during system development and experiments. Minimal distortion of the interpolated three dimension model of the tissues requires low position noise and high accuracy. These important potentiometer parameters were measured with results presented at the end of this section.

The capacity to rotate the transducer array around an axis parallel to the array (x axis) is provided by a rotary optical encoder mounted onto the linear bearing carriage. This optical encoder integrates the functions of a rotary bearing and rotary sensor. If separate components were used instead and connected by a belt, for example, backlash and calibration errors would be much larger. The RIS15 encoder from BEI Encoder Systems Division is used and has 2,000 lines on the disk, thus supplying 8,000 counts per turn using quadrature electronics. This disk resolution supplies angle precision of 0.045 degrees or linear precision of 0.075 mm at 50 mm below the skin. The internal encoder bearings provide 22N (5 lbf) load capability which is sufficient to support the transducer and normal operator hand forces. An analog output is provided for simplicity of interface thus avoiding software intensive serial communication.

A major goal of this effort is miniaturization of the measurement system. The resulting design weighs 650g including the ultrasound transducer. This mass is divided into stationary parts weighing 190 g, which includes the base plate, potentiometer, and linear bearing rail. The moving parts weigh 460 g which includes a typical 140 g ultrasound transducer, an optical encoder weighing 170 g, a vertical linear bearing weighing 30 g, and various brackets. The length of the package along the translation (z) axis is 153 mm, and the width (x axis) is 142 mm parallel to the transducer array. The height (y axis) without a ultrasound transducer installed is 64 mm to the top of the optical encoder body. The device fits into a rectangular prism of 1400 cc. In the configuration illustrated the translational range is 70 mm which is easily expandable.

Translation Sensor Linearity and Noise Measurements

Poor linearity and contact noise are traditional weaknesses of potentiometers. The potentiometer was selected for the final design because of the advantages of size and weight over competing linear optical encoders, but only after complete testing of linearity and noise.

Linearity was measured by translating the potentiometer wiper with a lead screw and measuring the potentiometer wiper voltage with a constant voltage applied to the resistive element. The voltage was read by a 12 bit analog-to-digital-converter (ADC) board that was also used for the phantom and clinical data presented in the Results Section. This experimental protocol tests the linearity of the ADC as well as the potentiometer. The error of the digital position measurement was calculated from the best linear model of the data using a least mean square algorithm. The worst cumulative error is 0.47 mm or 0.68 percent. By avoiding the first 2 mm of the 69 mm stroke the nonlinear error can be limited to 0.35 percent. It is important to note that the maximum nonlinear error between successive 0.5 mm slices is limited to 0.16 mm or 32 percent of the slice spacing. This implies that the greatest distortion using this potentiometer would be one third of a space between scan planes. Further improvement in linearity is available in high linearity (0.1%) potentiometers available from the manufacturer.

The electrical noise of the translational sensing system was also measured during a typical slew maneuver of 1 mm per second. Resistance changes as the wipers slid across the conductive film were determined as well as noise from the cabling and ADC circuitry. Figure 2 shows the calculated position during slewing. The primary feature of the plot is the series of steps due to the quantization size of the 12 bit ADC circuit. The resistor output voltage is 10 V for a 100 mm range. This is processed by a 12 bit ADC having 4096 steps and an input range which is also 10 V. The result is quantization steps of 2.4 mV, or 0.024 mm, for the stated position and voltage ranges. Figure 2 shows that this quantization noise dominates the calculated transducer position. Potentiometer noise is small relative to the ADC quantization error and is insignificant for this application.

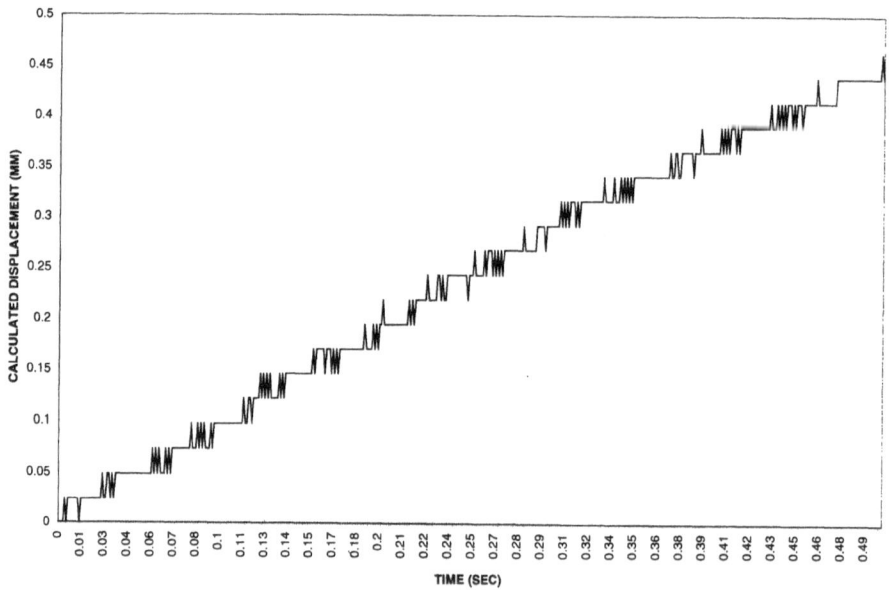

Figure 2. Electrical noise during slewing of the plastic film linear sensor.

RESULTS

The goal of the program is construction of a measurement system that provides linear and rotary position of the ultrasound transducer. The ultimate goal for this data is the construction of 3D data sets where any 2D image section can be presented to the viewer without significant distortion. Real-time measurements avoid the need for evenly spaced B-scans because the 3D image processing software can accomodate the varying slice spacing. Trilinear interpolation is used to assign weighted fractions of the raw image data into the desired 3 D grid. A raw 3D data set was created from scans of a filament phantom. A single B-scan was used as a reference image and a similar section of the interpolated 3D data was evaluated for distortion.

The first step in this evaluation was creation of a reference 2D B-scan image of the phantom. The B-scan was taken perpendicular to the phantom filaments for finest detail and minimal sensitivity to precise transducer alignment. This 2D image is shown in Figure 3.

A 3D data set was created from a set of B-scans taken *parallel* to the phantom filaments, i.e.. perpendicular to the reference 2D scan. Motion was allowed in both translation and

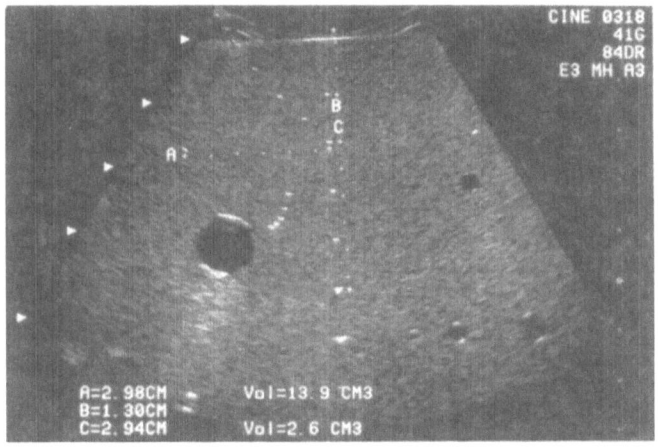

Figure 3. Reference B-scan perpendicular to the phantom filaments.

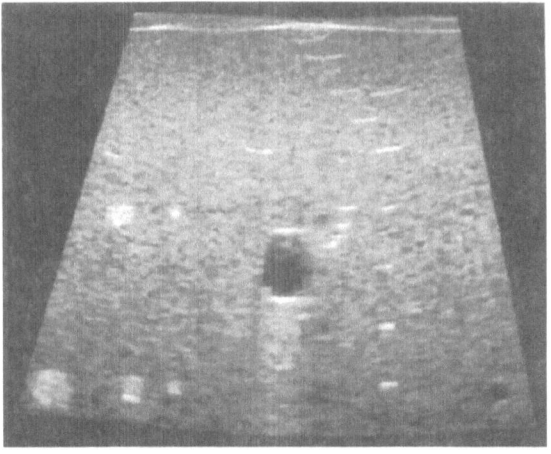

Figure 4. An image from the 3D data set after interpolation for rotation and uneven spacing.

241

rotation. Figure 4 shows an image reconstructed from the 3D data set after interpolation based on the measurement system. Data from each scan plane was placed into the nearest Cartesian grid vertices using trilinear interpolation. The figure shows how the features from Figure 3 are presented with an acceptable level of distortion. Note that the lateral boundaries the image are not parallel and thus accurately reflect the degree of rotation employed during the 3D data collection.

Figure 4 shows some degradation in image quality due to reduced resolution. This effect results from the different precision of beam focusing between the in-plane and out-of-plane directions. Electronic focusing within the B-scan plane provides better focusing, and also high depth of field through changes in focus depth. In the elevational, or out-of-plane direction used by the 3D software to reconstruct objects perpendicular to the constituent B-scans, single array elements provide poorer focusing. The resulting 3D image is smeared in the translation direction because of the large elevational beam width. The 3D image shown is provides the most severe conditions for comparison of B-scan and 3D data.

CONCLUSIONS

This work has developed a small, low cost ultrasound transducer position measuring system that minimally affects the clinician's preferred technique during ultrasound examinations. Manual examination is facilitated by permitting varying speeds of translation while still producing low distortion 3D image sets. Patient comfort is maximized by the device's light weight while its small size permits access to a range of body regions comparable to the ultrasound transducer alone. Measurement of the two axes critical to image acquisition flexibility is provided while motion in other axes is prohibited, thus simplifying measurement, data acquisition, and image processing.

ACKNOWLEDGMENTS

General Electric Medical Systems and Corporate Research Laboratory were particularly helpful with development of rapid acquisition and transfer of digital image data. This work was supported in part by PHS grant R01 CA55076 from the National Cancer Institute and contract No. DAMD 17-96-C-6061 from the Army Medical Research and Development Command.

REFERENCES

Detmer, P.R., Bashein, G., Hodges, T., Beach, K.W., Filer, E.P., Burns, D.H., and Strandness, D.E., 1994, 3D ultrasonic image feature localization based on magnetic scanhead tracking: In vitro calibration and validation, *Ultrasound Med. and Bio.* 20:923.

Hernandez, A., Basset, O., Dautraix, I., and Magnin, I., 1996, Acquisition and stereoscopic visualization of three-dimensional ultrasonic breast data, *IEEE Trans. Ultrasonics,Ferro. and Freq. Ctl.* 43:576.

Hughes, S.W., D'Arcy, T.J., Maxwell, D.J., Chiu, W., Milner, A., Saunders, J.E., and Sheppard, R.J., 1996, Volume estimation from multiplanar 2D ultrasound images using a remote electromagnetic position and orientation sensor, *Ultrasound in Medicine.* 22:561.

Kelly, I.A.G., Gardener, J.E., Brett, A.D., Richards, R., and Lees, W.R., 1994, Three dimensional US of the fetus, *Radiology* 192:253.

Moskalik, P.L., Carson, P.L., Meyer, C.R., Fowlkes, J.B., Rubin, J.M., and Roubidoux, M.A., 1995, Registration of three-dimensional compound ultrasound scans of the breast for refraction and motion correction, *Ultrasound in Med. And Biol.* 21:769.

Nelson, T.R. and Pretorius, D.H., 1995, Visualization of the fetal thoracic skeleton with three-dimensional sonography, *AJR.* 164:1485.

Picot, P.A., Rickey, D.W., Mitchell, R., Rankin, R., and Fenster, A., 1993, Three dimensional colour doppler imaging, *Ultrasound in Med. and Bio.* 19:95.

Pretorius, D.H., Nelson, T.R., and Jaffe, J.S., 1992, 3-dimensional sonographic analysis based on color flow doppler and gray scale image data, *Ultrasound Med.* 11:225.

A LOW POWER IMAGE RECONSTRUCTION TECHNIQUE FOR PHASED ARRAY SCANNERS

J.V. Hatfield[1], P.A. Payne[2] and A.D. Armitage[1]

Departments of [1]Electrical Engineering & Electronics and [2]Instrumentation and Analytical Science
University of Manchester Institute of Science and Technology
PO Box 88, Manchester M60 1QD, UK

INTRODUCTION

Traditionally, medical ultrasound array transducers have used piezoelectric ceramics as their active elements. This technology has been very successfully developed over the years and current medical scanners produce very good images of large structures within the body. For peripheral structures (small parts scanning) higher frequencies can be employed to improve the resolution, but the constructional methods used with piezoelectric ceramic materials limit the width of array elements to about 0.1 mm. This corresponds to a centre frequency of about 7.5 MHz. The decision was therefore taken that the active element for our high frequency array research programme should be one of the piezoelectric polymer materials. Poled polyvinylidene fluoride[1] (PVDF) and its copolymers are readily available from a number of suppliers worldwide, but in addition the research group has a long experience of manufacturing the copolymer of vinylidene fluoride and trifluoroethylene (VDF-TrFE) and using it in a variety of devices[2].

The production of PVDF transducer arrays is non-trivial and has been documented in earlier work by the authors[3]. Unfortunately, when using higher frequency transducers more ultrasonic energy is absorbed by the medium under interrogation, resulting in a highly attenuated ultrasonic echo being received[4]. Using standard ultrasonic imaging architectures, the long wires which carry the analogue signal from the ultrasonic transducer introduce noise into the system. At the 22nd Acoustical Imaging symposium held in Cannes[5] we reported on work being undertaken by the ultrasonics group at UMIST which involved placing both the transmit and receive circuitry as close to the ultrasonic array as possible. We went on to briefly discuss a new approach we were investigating into the development of an integrated ultrasound receiving system. This paper will go on to describe that work in greater detail.

AN UNDER SAMPLING TECHNIQUE

A major problem associated with high frequency ultrasound in phased array scanners is the relatively high sampling rate required to quantize the received echo. For acceptable image reconstruction it has been stated that a sampling rate of at least eight times centre frequency is required[6]. If conventional techniques are employed, the requirements in terms of the receive system can be very demanding. Current digital receivers, for example, sample the data from all the phased array elements simultaneously and store the sampled data into memory. Then the image is built up by reading, and summing, the corresponding points from memory using a focus map. This technique is straightforward for low frequency ultrasound imaging, typically in the region of 5 MHz, as the sampling rate criterion is easily achievable. For a scanner with a centre frequency of 20 MHz, as in the case under discussion, a sampling rate of 160 MHz is required. This makes for difficult high frequency designs which usually result in high power dissipation.

To overcome these problems an under-sampling approach has been devised[7]. The prototype design comprised off-the-shelf hardware with control circuitry implemented as a Field Programmable Gate Array, (FPGA). The version currently under development utilises a custom designed Application Specific Integrated Circuit, (ASIC). The system can also utilise standard PC cards for data acquisition and display. Sampling frequencies in excess of 200 MHz can be readily achieved. Figure 1 illustrates the under sampling technique, whereby the sampling occurs on two separate passes. The second pass is delayed by 1/2 wavelength from the first. The sampling rate has, therefore, effectively been reduced by 50%. The quantised signals must then be combined and re-sequenced to obtain the full digitised signal. Clearly, if more than two passes are made, separated by appropriate delays, very high sampling rates can be "emulated" using the same sampling hardware.

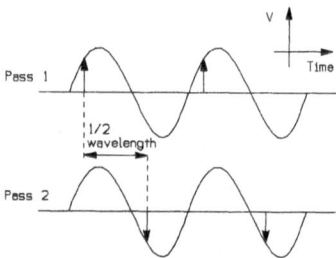

Figure 1. Illustrating the delayed sampling concept.

The implication of Figure 1 is that two signals must be produced in order to build the quantised signal. For ultrasonic imaging the number of pulse-echoes which must be produced doubles each time the emulated sampling rate is doubled. Similarly, there is a halving in data acquisition rate for every doubling of the emulated sampling rate. As we shall later, an ASIC solution provides an answer to this problem.

RECEIVER CIRCUIT OVERVIEW

A block diagram of the prototype scanning system is shown in Figure 2. A controlling Field Programmable Gate Array (FPGA) initiates a sequence of events by triggering the ultrasonic array to produce an ultrasonic pulse. An array element is then connected to the Analogue to Digital Converter (ADC) via a Variable Gain Amplifier (VGA) and analogue multiplexer. At the appropriate instant the Delay Generator instructs the ADC to quantise the returning echo signal. The resulting code is loaded into an ADC, first-in first-out (FIFO), buffer.

As the pulse travels deeper into the scanned tissue the returning echo suffers increased attenuation. This is compensated for by the use of voltage controlled VGAs; the gain controlling voltage being supplied by the Analogue to Digital Converter (DAC). The relevant gains are stored as gain codes which are loaded into a DAC FIFO buffer. The FPGA controls the delay generator and multiplexer operation. The device also clocks the ADC and DAC FIFO buffers at the appropriate time.

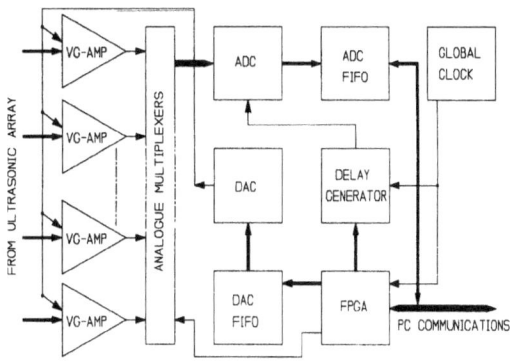

Figure 2. A block diagram of the prototype receiving system.

An array element of the scanner is excited by a pulse that is initiated on the positive edge of a global clock, Φ. The FPGA controller enables Φ to generate a high-voltage pulse by instructing a high voltage pulse generator (not shown) to "fire" the appropriate ultrasonic array element. The same edge of Φ is used to trigger the delay generator which provides programmed delays (Figure 2). The delay element controls the time at which the ADC takes its sample; the delay being referenced to the rising edge of the global clock, Φ. Once the ADC is triggered the controller clocks the First In First Out (FIFO) buffers and resets the delay element.

The *real* sampling rate of the receiver is thus equal to the global clock rate. For a 40 Mhz clock (Φ) a true sampling rate of 40 Mhz is achieved. For an emulated sampling frequency of 200 Mhz, therefore, five ultrasonic pulses must be emitted and hence the delay line is programmed in steps of 5 ns. Once all the echo data has been sampled for that one pulse the delay produced by the delay element is increased by 5 ns and another pulse emitted. This procedure is repeated, by means of the analogue multiplexer block, for each of the phased array elements in turn.

RESULTS

The operation of the scanner is controlled by command codes which are sent to the scanner head unit before each scan commences. These codes control functions such as emulated sampling rate, pulse channel selection, channel receive selection, sampling period, sample delay and signal gain. As a result, scanning can be carried out in more than one mode simply by modifying these software generated codes. The B-mode image of Figure 3 consists of five wires (approximately 0.5 mm in diameter) immersed in a water tank.

The same scanning system can also perform phased scanning. Echoes originating from a one point reflector impinge on different array elements at slightly different times. To obtain the total received signal energy from the reflector point, the various received signals must be re-aligned and summed. The present under-sampling system provides quantised data for all the received signals. It is a simple software task to re-align the signals using pre-calculated time delays stored in a focus map. As an example an NDT test was performed by embedding a metal fragment within a soft plastic medium. The result of performing a phased scan on the plastic sheet is shown in Figure 4. These test results were obtained with a relatively low frequency transducer having a centre frequency of 5 MHz.

A RECEIVE ASIC

Currently both the control logic and the delay generator are being implemented as an Application Specific Integrated Circuit. To save memory the ADC samples are delayed until the

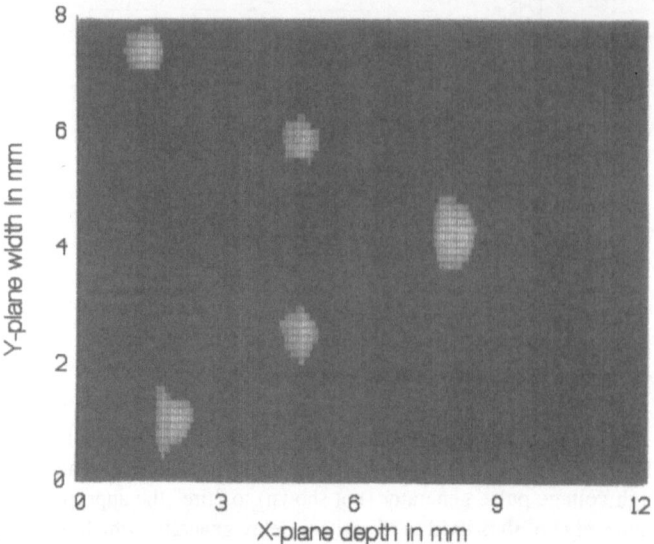

Figure 3. B-scan image of five wires immersed in a water tank.

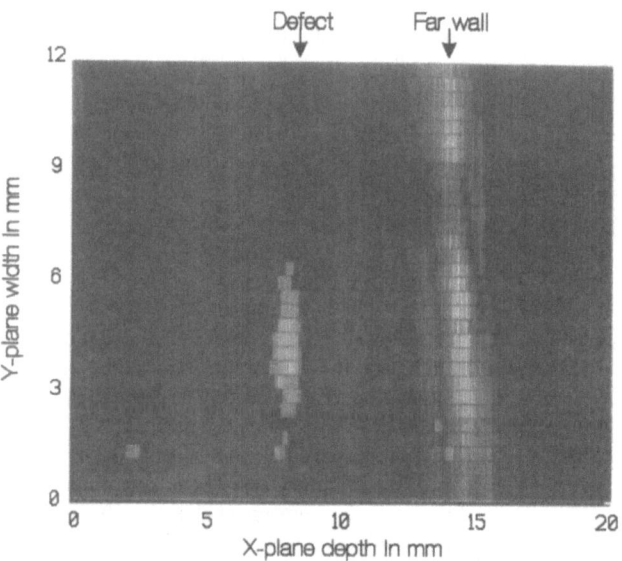

Figure 4 Phased scan of a metal "defect" in a plastic medium.

echoes are incident on the transducer array elements. This is achieved by counters on-board the same controlling ASIC which count at the global clock frequency. Within the ASIC there are two such counters, one delays the time at which sampling is to begin, the second determines the number of samples to be taken.

More importantly, by integrating delay generators onto a single chip, multiple delays can be made accessible at the same time. Figure 5 shows the modified scanner architecture with four scan delays. With this design there need be no data acquisition time penalty when trying to "emulate" high frequency sampling. Multiple sample passes can be avoided as all samples are

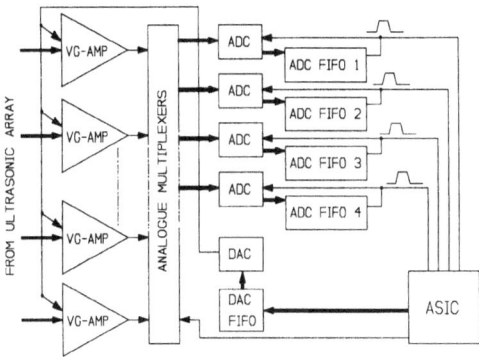

Figure 5. Modified scanner architecture incorporating an Application Specific Integrated Circuit.

taken during the same cycle of the global clock. Hence clocking at 50 MHz with four ADCs, controlled by four separate delay taps, a sampling rate of 200 MHz is obtained. Higher sampling rates can be easily realised by making more phases of the delayed global clock available.

Integrated Delay Generators

The integrated delay generators are implemented as a phase-locked delay line[8]. This has been fabricated and extensively tested. In this scheme a 32 element delay line divides the global clock into 32 equally spaced phases. Any phase can be tapped for echo signal sampling. In the case under discussion the delay line is fed by a 31.25 MHz clock. In conjunction with the feedback circuitry, therefore, the delay line generates clock phase 0 to 31 each of frequency 31.25 MHz. The rising edges of the clock phases are displaced from each other by 1ns, and are both supply voltage and temperature independent.

A delay element is shown in Figure 6. It has been designed to enable simultaneous control of the rise and fall times of both edges of the input pulse. This enables an equal mark space ratio to be preserved and ensures that the 32 ns pulse does not vanish as it traverses the delay line[8]. The control voltage V_C generated by feedback circuitry is increased to reduce the delay through the element and decreased to increase it. Weak transistors T6 and T7 are held on and are designed to flatten the delay versus control voltage characteristic about the intrinsic delay of the delay element.

The basic feedback circuitry is also shown in Figure 6. Upon reset V_C is set, by appropriate potential dividing circuitry, about mid-way between the supply rails. This produces approximate delays in the region of 1ns. The objective of the feedback circuitry is to align the rising edge of Phase N (here N = 31) with the next rising edge of the input, global, clock. This generates 32 clock phases, each separated by 1ns, from a single 31.25 MHz input clock. The rising edge of Phase N will vary with respect to the position of the rising edge of the input clock. If the delay through the delay line is too short, then the rising edge of Phase N will occur before the

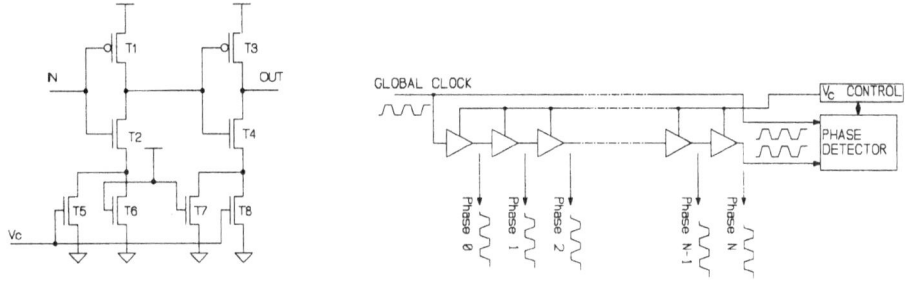

Figure 6. Single delay buffer and circuit schematic of phase locked delay line.

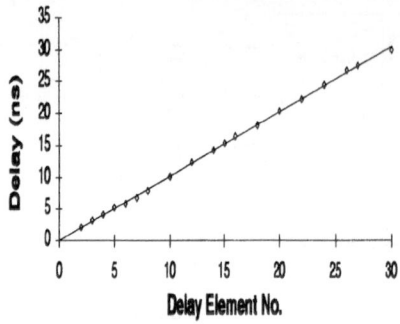

Figure 7. Graph showing relative positions of 20 of the 32 clock phases.

rising edge of the input clock. Transistors T6 and T5 will thus be turned on and the control voltage will fall. A falling V_C increases the delay through the delay elements. If the edge of Phase N rises after the input clock then transistors T1 and T2 will be on and thus reduce the delay through the delay elements. Note that the delay line is extended beyond Phase N in order to provide a WINDOW pulse symmetrically around the rising edge of Phase N. This pulse ensures that V_C can only be adjusted around the rising edge of Phase N and thereby prevents spurious locks to out of phase clock components. One of the primary functions of the prototype IC was to validate the design of the phase locked delay line. Figure 7 shows the externally available clock-phases plotted against time delay. Measurements were taken using a Stanford Research Systems SR620 time interval counter. This particular instrument specifies an absolute error less than 0.5 ns with differential non-linearity typically ±50 ps. For a typical phase output the standard deviation for 5000 samples is approximately 50 ps.

CONCLUSIONS

This paper has presented a report on the development of a new kind of high-frequency ultrasound receiving system utilising an undersampling technique. A prototype version has been developed from off-the-shelf components and tested using basic image reconstruction software routines. The prototype system is of limited application as it requires multiple passes for image reconstruction. These problems will be resolved in an integrated circuit version currently under development. A prototype chip that tests an integrated delay line has been developed and successfully demonstrated. Test results for the system have been presented.

REFERENCES

1. H. Kawai, The piezoelectricity of polyvinylidene fluoride, *Japan J Appl Phys*, 8:975, (1969).
2. Q.X. Chen and P.A. Payne, Industrial applications of piezoelectric polymer transducers, *Meas. Sci. Technol.*, 6:249, (1995).
3. J.V. Hatfield, N.R. Scales, A.D. Armitage, P.J. Hicks, Q.X. Chen and P.A. Payne, An integrated multi-element array transducer for ultrasound imaging, *Sensors & Actuators A: Physical*, 41:167, (1994).
4. P.A. Payne and Q.X. Chen, Chapter 10, in: "Reliability in Non-Destructive Testing, NDT-88C", C. Brook and P.D. Hanstead, eds, Pergamon Press, Oxford, (1989).
5. A.D. Armitage, Q.X. Chen, J.V. Hatfield, P.J.Hicks and P.A.Payne, High frequency integrated ultrasound arrays, in: "Acoustical Imaging, Volume 22", P. Tortoli and L. Masotti, ed., Plenum Press, New York, (1996).
6. G.F. Manes, C. Atzeni and C. Susini, Design of a simplified delay system for ultrasound phased array imaging, *IEEE Transactions on Sonics and Ultrasonics*, SU-30:350, (1984).
7. A.D. Armitage, N.R. Scales, P.J Hicks, P.A. Payne, Q.X. Chen and J.V. Hatfield, An integrated array transducer receiver for ultrasound imaging, *Sensors & Actuators: A*, 47:542, (1995).
8. N.R. Scales, P.J. Hicks, A.D. Armitage, P.A. Payne, Q.X. Chen and J.V. Hatfield, A programmable multi-channel CMOS pulser chip to drive ultrasonic array transducers, *IEEE Journal of Solid State Circuits*, 29:992, (1994).

HIGH-SPEED TRANSMISSION OF IMAGES WITH LIMITED DIFFRACTION BEAMS

Jian-yu Lu

Ultrasound Research, Department of Physiology and Biophysics
Mayo Clinic and Foundation
Rochester, MN 55905 U.S.A.

INTRODUCTION

Limited diffraction beams have a large depth of field.[1] They could have applications in medical imaging,[2] tissue characterization,[3] volumetric imaging,[4] estimation of transverse velocity of blood flow,[5] nondestructive evaluation (NDE) of materials,[6] 2D and 3D high frame rate imaging with simple hardware,[5,7-8] as well as other physics related areas such as electromagnetics[9] and optics.[10]

In this presentation, limited diffraction beams are used as carriers to transfer signals in parallel over a large distance.

PRINCIPLE

Limited diffraction beams are exact solutions of the isotropic-homogeneous scalar wave equation. Because the equation is linear, a linear superposition of these beams of different parameters can form a composite beam that is also a solution.[11] Therefore, according to the Huygens Principle, the composite beam can be produced with a planar wave source such as an acoustic transducer, electromagnetic antenna, or a laser device.[12]

Theoretically, limited diffraction beams can propagate to an infinite distance without changing their shapes if they are produced with an infinite aperture and energy. In practice, the aperture is always finite. In this case, limited diffraction beams have a finite but large depth of field, i.e., they can propagate to a large distance without significant distortion.[13,14] For example, if the diameter of an antenna is 20 m and the Axicon angle of an X wave is 0.005° or smaller, the minimum depth of field is about 115 km or is determined by the Rayleigh distance, whichever is smaller.[13] In addition, limited diffraction beams of different parameters are orthogonal to each other and thus they can be separated when received.[11] This means that limited diffraction beams can be used to carry multiple signals simultaneously over a large distance for wireless telecommunications.

The principle of a telecommunication system using limited diffraction beams is shown in Figure 1. To transfer multiple images in parallel, the images are digitized and then binary encoded. The binary codes are used to modulate (turn on/off) the transmissions of multiple limited diffraction beams of different parameters. The modulated beams are linearly superposed to form a hybrid beam that is transmitted with a physical device. After propagating over a large distance (within the minimum depth of field of the beams), the hybrid beam is received. In the receiver, limited diffraction beams are separated according to their parameters. Multiple binary signals are then detected and decoded to recover the original images simultaneously.

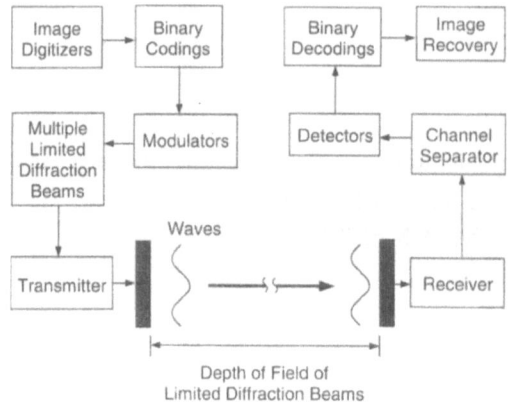

Figure 1. A schematic for parallel transmissions of binary signals with limited diffraction beams.

METHOD

Many types of limited diffraction beams such as Bessel beams,[10] X waves,[13,14] and array beams[4,5,11] can be used to transfer signals. For each type of beam, parameters such as initial phases, orders of the beams, scaling parameters, or Axicon angles, etc. can be used to distinguish beams.[15,16] For simplicity, in the following, zeroth-order acoustic X waves of different Axicon angles are used to demonstrate the principle of the method.[13,17]

To produce a modulated X wave, a ring transducer is placed at the focal plane of a lens (Figure 2). The ring is excited with a signal that is modulated with binary codes that represent a desired image. The acoustic field produced immediately after the lens is approximately the Fourier transform of the wave produced by the ring, which is an X wave. The Axicon angle of the X wave is given by[12,13] $\zeta = \sin^{-1}(a_r/F)$, where a_r and F are the radius of the ring and the focal length of the lens, respectively. The aperture size[13] of the X wave is determined by the diameter of the lens, D. The depth of field[13] of the X wave is given by $XZ_{max} = (D/2)\cot\zeta$.

To produce two X waves of different Axicon angles, two rings must be used. If the two rings are excited simultaneously, a hybrid wave that is a linear superposition of the X waves produced by individual rings is obtained. The hybrid wave can carry multiple signals and propagate to a remote receiver (Figure 3). In the receiver, another lens is used to decompose the hybrid wave back into a ring field at its focal plane corresponding to the Axicon angles of the waves.[12,13] Binary signals are then received with ring detectors and images are recovered simultaneously.

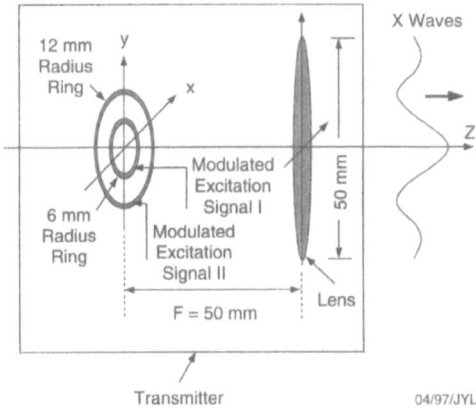

Figure 2. A transmitter system that produces multiple X waves of different parameters (Axicon angles). Both the focal length and the diameter of the lens are 50 mm. Radii of the inner and outer rings are 6 and 12 mm, respectively.

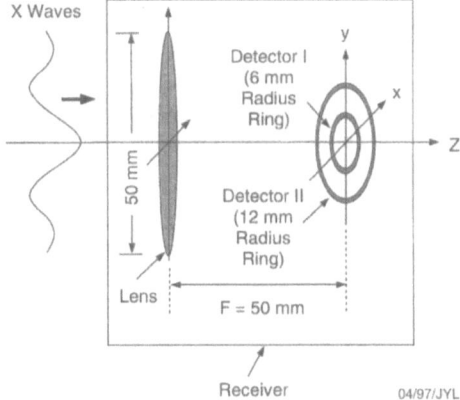

Figure 3. A receiver system that decomposes X waves of different parameters (Axicon angles) back into ring field for detections. For simplicity, the parameters of the lens are chosen the same as those shown in Figure 2.

RESULTS

The X waves produced with the lens of the transmitter in Figure 2 are shown in Figure 4 (lower half of the cross section of the axially symmetric X waves).[13] When a binary code is "0" in a signal channel, no X wave is produced from that channel. If the code is "1", an X wave is produced. The bandwidth of the X wave produced is determined by those of the excitation signal and the ring transducer. In Figure 4, X waves produced with a short pulse (1.5 cycles) and with a long tone burst (10 cycles) are shown in the left and right columns, respectively. In acoustic and microwave cases, short pulses can be produced. In optics, tone bursts that are much longer than 10 cycles are usually used. X waves produced with rings that have radii of 6 and 12 mm are shown in the top and middle rows, respectively. Hybrid X waves produced by simultaneous excitations of both rings are shown in the bottom row of Figure 4. In this figure, the center frequency of the wave is assumed to be 2.5 MHz and the diameter of the transmission lens is 50 mm. With these parameters, the depths of field of the X waves are about 206.86 and 101.12 mm for the rings of 6 and 12 mm, respectively, corresponding to Axicon angles of about 6.892° and 13.89°.

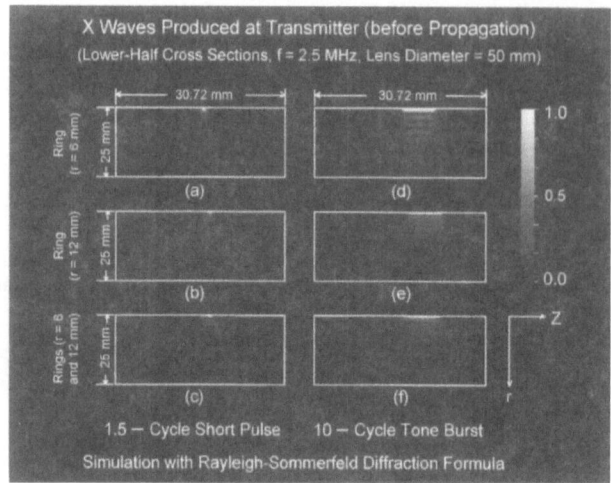

Figure 4. Simulated X waves produced immediately after the transmission lens (see Figure 2). Lower-half of the cross section of the axially symmetric X waves is shown. Panels in the left and right columns are X waves produced with a 1.5–cycle short pulse and a 10–cycle tone burst, respectively. Panels in the top and middle rows are X waves produced with rings of 6 and 12 mm radius, respectively. Panels in the bottom row are a hybrid wave produced when both rings are excited. The Rayleigh-Sommerfeld diffraction formula is used in the simulation. A Blackman window function centered at 2.5 MHz with a relative bandwidth of 81% of the center frequency is assumed. Analytic envelopes of the waves are displayed.

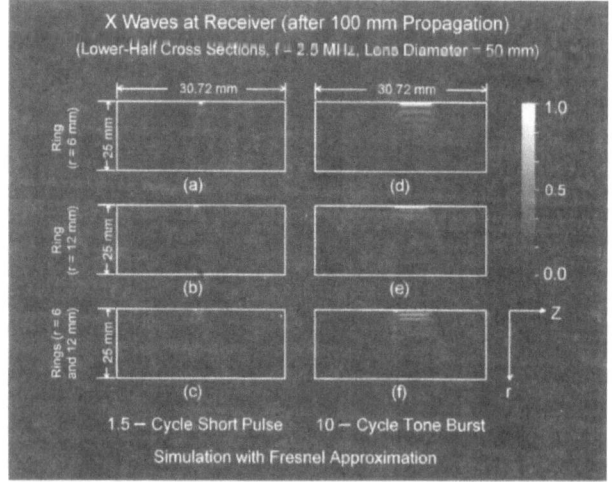

Figure 5. Simulated X waves after propagating over 100 mm. The layout of this figure is the same as that of Figure 4. The Fresnel approximation is used in the simulation.

After propagating over 100 mm, which is approximately the minimum depth of field of the X waves, the results are shown in Figure 5. It is seen that the distortions of X waves are small.

From the X waves shown in Figure 5, ring fields are recovered at the focal plane of another lens (Figure 6). Ring detectors are placed at peaks of the ring field corresponding to the transmit rings in Figure 2 to receive the binary signals and recover images simultaneously. It is seen that the rings are reconstructed very well. From Figure 6, it is clear that more transmit rings and detectors can be added to increase the number of channels of the communication system.

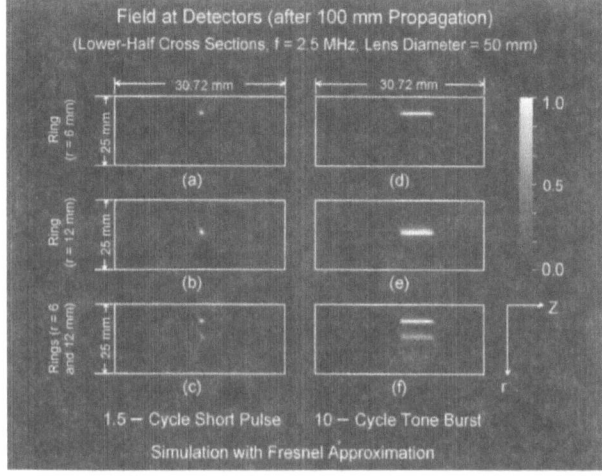

Figure 6. Simulated acoustic fields at the focal plane of the reception lens produced by X waves of different parameters. The layout of this figure is the same as that of Figure 4. The simulation is carried out with the Fresnel approximation.

DISCUSSION

It should be noticed that the method proposed in this paper is only suitable for near-field communication, i.e., within a large depth of field of limited diffraction beams (Figures 4 and 5). Because limited diffraction beams are highly collimated in the near field, this method may be useful for a high-speed wireless private communication. In addition, the collimated beams do not have the problem of multiple reflections from surrounding objects.

In Figure 3, the ring detectors used are phase sensitive. This requires that the transmit rings, lenses, and the ring detectors in Figures 2 and 3 to be coaxially aligned and perpendicular to the beam axis to avoid phase cancellation and thus to increase signal-to-noise ratio. Because of the phase sensitivity, background noise from random radiation sources will be largely cancelled.

From Figure 6, it is seen that the ring fields reconstructed from the X waves have a certain width that increases with the decrease of the diameter of the reception lens. In addition, sidelobes of the ring fields will increase the effective ring width. A larger width will reduce the number of rings that can be placed in the system. The number of rings will also be reduced if the communication distance is increased. This is because a larger distance decreases the largest diameter of the rings.

It is worth noting that array beams[4,6–8,11] can also be used in the communication system. In this case, point sources and detectors are used to replace rings. This will greatly increase the number of channels. However, the signal-to-noise ratio may be reduced because the transmitter and detector sizes will be small and the detectors will be phase insensitive.

CONCLUSION

A method for parallel transmissions of digital binary signals with limited diffraction beams such as X waves has been developed. An acoustic example has been given to demonstrate the method. Because microwave and optic waves obey the same scalar wave equation of the acoustics, the method can also be applied to those areas for high-speed digital wireless telecommunications.

ACKNOWLEDGMENTS

The authors appreciate the secretarial assistance of Elaine C. Quarve. This work was supported in part by grants CA 54212 and CA 43920 from the National Institutes of Health.

REFERENCES

1. J.A. Stratton. *Electromagnetic Theory.* McGraw-Hill Book Company, New York (1941), p. 356.
2. J-y. Lu, H. Zou, and J.F. Greenleaf, Biomedical ultrasound beamforming, *Ultrasound Med. Biol.* 20(5):403–428 (1994).
3. J-y. Lu and J.F. Greenleaf, Evaluation of a nondiffracting transducer for tissue characterization, in *IEEE 1990 Ultrason. Symp. Proc.* 90CH2938–9, 2:795–798 (1990).
4. J-y. Lu, Limited diffraction array beams, *Int. J. Imag. System and Tech.* 8(1):126–136 (1997).
5. J-y. Lu, Improving accuracy of transverse velocity measurement with a new limited diffraction beam, in *IEEE 1996 Ultrason. Symp. Proc.* 96CH35993, 2:1255–1260 (1996).
6. J-y. Lu and J.F. Greenleaf, Producing deep depth of field and depth-independent resolution in NDE with limited diffraction beams, *Ultrason. Imag.* 15(2):134–149 (1993).
7. J-y. Lu, 2D and 3D high frame rate imaging with limited diffraction beams, *IEEE Trans. Ultrason. Ferroelec. Freq. Contr.* (In Press).
8. J-y. Lu, Experimental study of high frame rate imaging with limited diffraction beams, *IEEE Trans. Ultrason. Ferroelec. Freq. Contr.* (Submitted).
9. J. Ojeda-Castaneda and A. Noyola-Iglesias, Nondiffracting wavefields in grin and free-space, *Microwave and Optical Technology Letters* 3(12):430–433 (1990).
10. J. Durnin, J.J. Miceli, Jr., and J.H. Eberly, Diffraction-free beams, *Phys. Rev. Lett.* 58(15):1499–1501 (1987).
11. J-y. Lu, Designing limited diffraction beams, *IEEE Trans. Ultrason. Ferroelec. Freq. Contr.* 44(1):181–193 (1997).
12. J.W. Goodman. *Introduction to Fourier Optics.* McGraw-Hill, New York (1968), chs. 2–4.
13. J-y. Lu and J.F. Greenleaf, Nondiffracting X waves — exact solutions to free-space scalar wave equation and their finite aperture realizations, *IEEE Trans. Ultrason. Ferroelec. Freq. Contr.* 39(1):19–31 (1992).
14. J-y. Lu and J.F. Greenleaf, Experimental verification of nondiffracting X waves, *IEEE Trans. Ultrason. Ferroelec. Freq. Contr.* 39(3):441–446 (1992).
15. J-y. Lu, Bowtie limited diffraction beams for low-sidelobe and large depth of field imaging, *IEEE Trans. Ultrason. Ferroelec. Freq. Contr.* 42(6):1050–1063 (1995).
16. J-y. Lu, Producing bowtie limited diffraction beams with synthetic array experiment, *IEEE Trans. Ultrason. Ferroelec. Freq. Contr.* 43(5):893–900 (1996).
17. J.H. Mcleod, The Axicon: a new type of optical element, *J. Opt. Soc. Am.* 44(8):592–597 (1954).

HIGH -SPEED ULTRASOUND 2D-IMAGING CAMERA

Victor D. Svet, [1] Vladimir I. Sizov, [1] Sergey V. Baykov [1]

[1] Federal Science Center "N. N. Andreyev Acoustics Institute"
Shvernik St. -4, 117036, Moscow, Russia

INTRODUCTION

There are two fundamental approaches to generate acoustic images - acoustic lens and beamforming. Each approach has advantages and disadvantages. Some of them are considered by Houston[1]. For portable and lightweight ultrasonic imaging sonars operating in real time (frame rate 5-25 frames/sec) the main technical problem is the high-speed reading of signals from matrix array and their space-time processing. PZT or PVDF arrays can consists of 100*100 sensors and more to generate the image of high quality. On the other hand the optimal algorithms for space-time processing and acoustical imaging are more complex than in usual optical system, because the last one has the tremendous redundancy in comparison with acoustics imaging system.

Two design approaches are possible. The first is based on microelectronics technology and design of special technology of interconnections in combination with special chips for multiplexing, digitizing and processing. To our knowledge such technology is not ready for use in wide commercial applications.

The second approach is based on beam-switching technology. In Russia such devices have been received the name " unicon- universal converter" which is the modern variant of Sokolov`s tube. Ultrasonic Imaging Camera described below is founded on this converter.

The main application of developed camera is the underwater monitoring of moving targets while operating from small-sized ships or underwater vehicles.

PRINCIPLES OF OPERATION

Ultrasonic Camera is a lens based device where acoustical waves are focused onto a matrix PZT (solid or film) array. Ultrasonic field (image) produces 2D electric potential that is read from PZT by electron beam converting 2D potential to alternative current of unicon. This beam is controlled by line and frame scanning generators. Electric coupling of matrix elements with beam is provided by special design of unicon target. It is a metal-glass fiber plate with very high density of thin wire electrodes (more than 50 wires/mm). The beam scans the inner side of plate and knocks the secondary electrons out. The circular electrode collects these electrons and alternative current is generated. The further transformations of signal are usual: amplification, demodulation, analog-to-digit conversion, digital processing and displaying.

The ultrasound illumination is provided by two projectors with wide crossed beams and wide-band signal radiation. That technical solution in combination with additional space signal digital processing permits to reduce speckle structure in images of targets. Through that the images are getting rather smooth.

General diagram of the Ultrasonic Imaging Camera is shown on Figure 1. Camera consists of two subsystems: underwater block (ultrasound head) and ship-board processing block. Underwater block consists of acoustic lens, unicon, scanning devices, HF amplifier and demodulator. Analog-to-digit converter, digital processor, display and power source are in the ship-board block. The developed software performs all processing from unicon to display and controls the parameters of system.

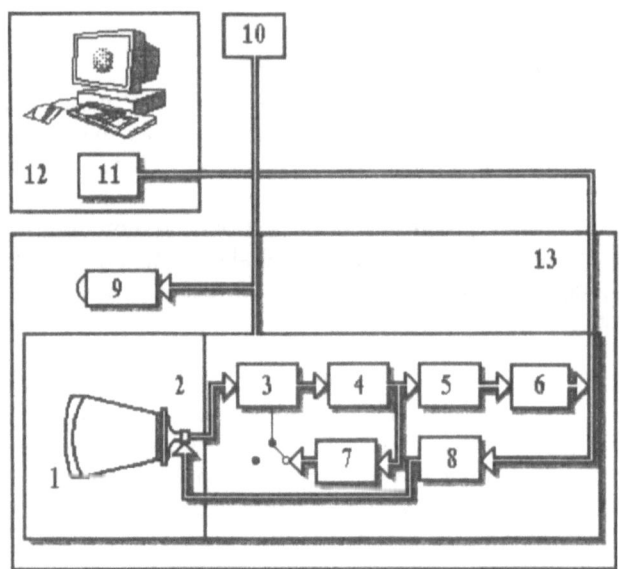

Figure 1. General diagram of the Ultrasound Imaging Camera: 1 - acoustical objective, 2 - unicon, 3 - high frequency band amplifier, 4 - detector, 5 - low frequency filter, 6 - matching amplifier, 7 - automatic gain control unit, 8 - beam deflection unit, 9 - acoustical transmitter, 10 - power supply, 11 - 12-bit analog-to-digit converter and frame processor, 12 - personal computer, 13 - underwater block (head).

The Camera is intended for deploying on the small-sized ships or unattended underwater vehicles. It is especially useful to observe moving targets. The basic technical parameters are listed in Table 1. The attainable parameters are shown in the last column.

Table 1. The features of the Ultrasonic Imaging Camera.

Parameter	Realized	Attainable
Frequency, MHz	1	0,5.... 3,5
Bandwidth, %	20	-
Angle resolution, degree	0,9	0,2
Field of viewing, degree	36	46
Focus distance	constant	variable
Frame rate, Fr./sec	25	-
Pixels in matrix:		
solid PZT	100*100	120* 120
film PZT	120*120	280*280
Illumination	continuous	arbitrary
Weight of underwater block, Kg	11	-
Depth, meters	80	120

EXPERIMENTAL RESULTS

The developed Camera has been tested in the different conditions. Some of acoustical images are presented below.

Figure 2. Acoustic image of a smooth metal cylinder.

Figure 3. Acoustic image of a metal ladder.

Absolutely smooth metal targets.

As known imaging absolutely smooth metal targets (relative to ultrasound wave length) is a very complex problem. The strong mirror reflected components don't permit to extract weak diffraction scattered field from edges of target and receive the image with accessible quality for recognition. The examples are shown on Figures 2, 3. Observation was performed in the hydroacoustical water pool of the Acoustics Institute. The tested objects were a metal cylinder with length -1.2 m and diameter - 0.3 m distanced from the Head on 15 meters and a metal ladder with dimensions 0.8 × 2.5 m on distance of 12 m. In spite of strong mirror reflection it can be seen that the forms of the targets are recognizable.

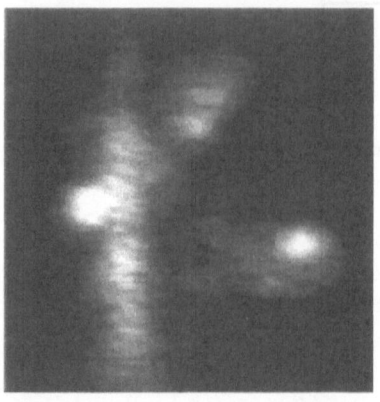

Figure 4. Fragment of a wood pillar with a cross bulk shot in Portsmuth harbor.

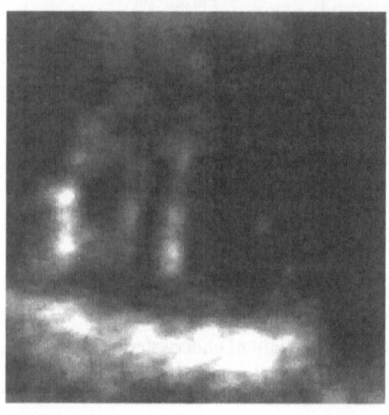

Figure 5. Three steel pillars of an old pier on the Moscow river.

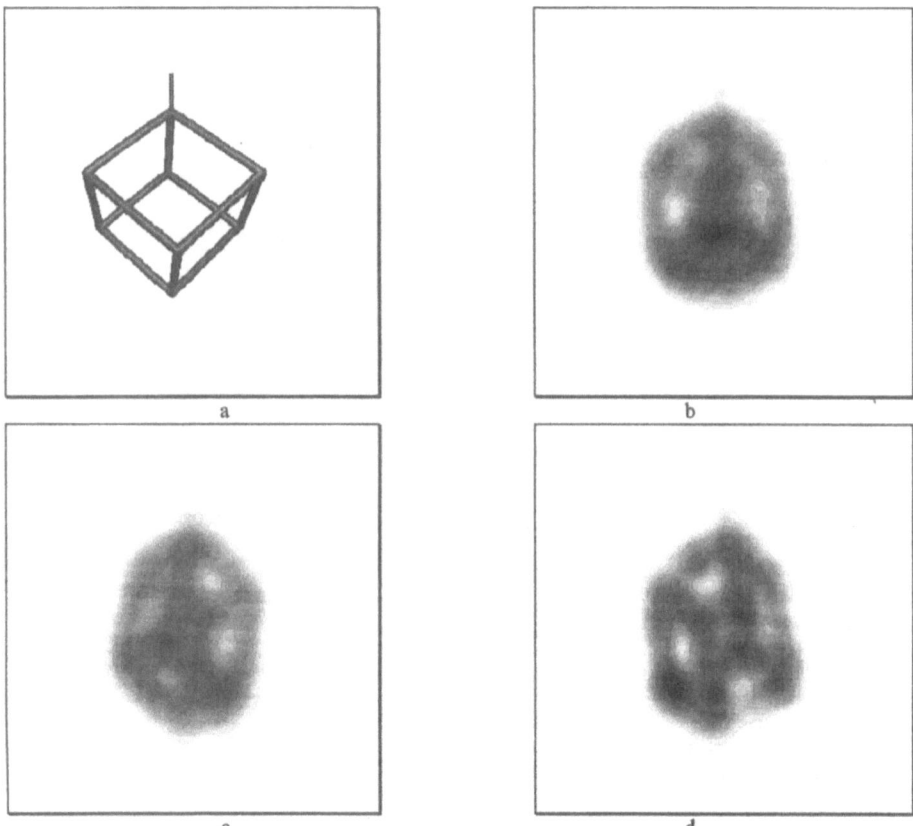

a

b

c

d

Figure 6. The optical prototype (a) and snapshots of the rotating steelworks (b, c, d).

Moving targets with irregular surfaces.

The real targets situated in water for long time have assumed irregular surfaces. It is not very good for long-range detection because of the decreasing of target strength, but it is very good for imaging. The images of such objects are shown on Figures 4, 5, 6. Figures 4 and 5 demonstrates the images of wood and steel pillars of the piers. The fragment of a wood pillar with a cross bulk was observed from distance about 4 meters while testing the Camera in Portsmuth (U. S. A.) but the steel ones were shot in the Moscow river. In last case range was more than 20 meters and deepness of place was about 2 meters so that it is clear seen bottom and surface. Figure 6 presents the optical prototype and the images a steelworks removed from a river. This steelworks was suspended on a rope in the river and twisted. Figure 6 shows the snapshots of the rotating steelworks. One can see that these images are good recognizable.

The last example of acoustical imaging is the detection of egress of the compressed gas in underwater gas pipe.

The moving gas jet has a very high target strength. The high frame rate of the camera gives the possibility not only to detect the jet but also to trace its conception and evolution. It is very important for underwater gas/oil pipe diagnostics while assembling and operating. Some examples are shown on Figure 7. The first image (a) is the moment of jet conception and it can be seen only separate dots (bubbles of gas). Mean while, the next image (b) shows the developed powerful jet.

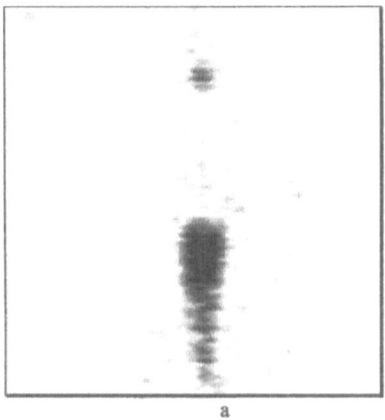 

a   b

Figure 7. The images of gas jet: a - jet conception, b - developed jet.

CONCLUSION

Experience of operating the Ultrasonic Imaging Camera in different conditions with varied targets has shown its wide capabilities in underwater monitoring. The ensuring development of proposed technology can lead to design of Ultrasonic Camera with commercial value for different applications. In spite of some restrictions of beam-switching technology its capability to generate images with high frame rate can be the main factor in practical implementations of acoustical imaging of moving targets.

The prospective directions of further development are, first, using electrostatic scanning instead of electromagnetic to realize the arbitrary rate and law of scanning and, second, impulse mode of illumination. Potentially They will provide passage from 2D-image to 3D-image. But it certainly tends to more sophisticated signal processing and hardware.

ACKNOWLEDGMENTS

The many contributions to V. Kondakov (circuit design), A. Gusev and V. Ryabukhin (nature testing) for their support of camera development.

The authors are grateful to William H. Key (Klein Associates, Inc.) and Thomas P. Kemp, Jr. (USCARR) for their support, kind help and organization of Camera testing in Salem, N.H., U.S.A.

The presentation of this work at the 23rd International Symposium on Acoustical Imaging is supported by ONREUR.

REFERENCES

1. Kenneth M. Houston, 3-D acoustical imaging using micromechanical hydrophones, Sea Technology. pp. 29-34 (September 1996).

A WEIGHTED FRESNEL APPROXIMATION FOR THE DELAYS USED IN FOCUSED BEAMFORMING

Andrea Trucco[1] and Vittorio Murino[2]

[1]Dpt. of Biophysical and Electronic Engineering (DIBE)
University of Genoa.
E-mail: fragola@dibe.unige.it
[2]Dpt. of Mathematics and Computer Science (DIMI)
University of Udine.
E-mail: swan@dimi.uniud.it

INTRODUCTION

Beamforming is a linear technique aimed at processing array signals in order to enhance incoming signals from a selected steering direction and to abate incoming signals from any other direction[1]. Thanks to its flexibility, beamforming can be successfully employed in many application fields with different objectives. In any case, if the array works under far-field conditions, the delays required by the beamforming operation can be easily computed in an exact way, whereas, if it works under near-field conditions, the focalization of beamforming is required to take into account the curvature of waves. In the latter case, a fast computation of the exact delays is often prohibitive, then an approximate version is preferred: generally, the Fresnel approximation (obtained by the expansion of the time-independent free-space Green's function[2,3]) is adopted[1]. Moreover, the Fresnel approximation makes it possible to apply the Fast Fourier Transform (FFT) in the implementation of beamforming even when focalization is necessary[4], thus resulting in a great computational profit. Despite its simplicity and advantages, the Fresnel approximation has a well defined region of validity[2] that forces potential steering directions to be contained inside a narrow scanning region. This constraint is heavy in applications (e.g., acoustic imaging) that require a wide region of view, for both medical and underwater investigations. To avoid this drawback, some imaging techniques that do not need the Fresnel approximation have been devised[5,6] but, unfortunately, they increase the computational load and/or the system complexity.

This paper presents a novel approximation (not too different from the Fresnel one) based on the minimization of the mean square error, so resulting in an acceptable precision inside wide scanning regions. In more detail, the same terms as used for the Fresnel expansion are weighted by coefficients whose values are fixed, given the array geometry, by a least-squares procedure on the basis of the desired scanning region. Thus, the new approximation keeps the low computational load and the low system complexity that characterize the Fresnel expansion and maintains the opportunity to be implemented in an FFT focused beamformer, too.

THE CONVENTIONAL FRESNEL APPROXIMATION

A beam signal (generated by a conventional delay and sum beamformer[1]) steered in the direction of the unit vector $\mathbf{u}$ is defined as:

$$bs_{\mathbf{u}}(t) = \sum_{i=1}^{M} w_i \cdot x_i\big(t - \tau(\mathbf{u},i)\big)$$

(1)

where $x_i(t)$ is the temporal signal received by the i-th sensor, $\tau(\mathbf{u}, i)$ and w_i are the delay and the weight applied to such a signal, respectively, and M is the number of array elements.

Under the far-field hypothesis, the exact delay τ_i can be written as:

$$\tau_{ff}(\mathbf{u},i) = \frac{\big(\mathbf{u}\,\mathbf{v}_i^+\big)}{c}$$

(2)

where $\mathbf{v}_i = [x_i, y_i, z_i]$ is the position vector of the i-th sensor, $+$ indicates the transposition operator (both $\mathbf{u}$ and $\mathbf{v}_i$ are row vectors), and c is the carrier speed. This delay formulation is compatible with the FFT implementation of beamforming. If focalization is necessary, as the far-field hypothesis does not hold any more, then the exact delay is:

$$\tau_{ex}(\mathbf{u},i) = \frac{R - \sqrt{R^2 + \|\mathbf{v}_i\|^2 - 2R\mathbf{u}\,\mathbf{v}_i^+}}{c}$$

(3)

where R is the focalization distance in the steering direction $\mathbf{u}$ and $\|\cdot\|^2$ is the Euclidean norm. These delays are heavy to compute on line (mainly due to the presence of a square root) and inhibit the implementation of focused beamforming by the FFT. For these reasons, one tries to approximate the exact delay by a formulation that allows an easy on line computation and the FFT implementation. This target is generally achieved thanks to the Fresnel expansion, which results in the following approximation[1] to the exact delay:

$$\tau_{Fr}(\mathbf{u},i) = \frac{\big(\mathbf{u}\,\mathbf{v}_i^+\big)}{c} - \frac{\|\mathbf{v}_i\|^2}{2Rc} \approx \tau_{ex}(\mathbf{u},i)$$

(4)

The first term of the addition depends on both $\mathbf{v}_i$ and $\mathbf{u}$, and represents the distance-independent delay equal to that defined in (2). The second term takes into account the wave curvature, depends on $\mathbf{v}_i$ and does not depend on $\mathbf{u}$. Thanks to the latter fact, one can show that the FFT implementation of focused beamforming is feasible[4].

According to Ziomek[2], there are three necessary conditions that define the validity region of the Fresnel approximation. The first imposes a small steering sector and, if one denotes by ϕ the angle between $\mathbf{u}$ and $\mathbf{v}_i$, can be written as:

$$72° \le \phi \le 108°$$

(5)

The second condition stipulates the minimum focalization distance:

$$R > 1.356V \tag{6}$$

where V is the maximum value of $\|\mathbf{v}_i\|$.

The third condition establishes the boundary between near-field and far-field regions:

$$R < \pi V^2 / \lambda \tag{7}$$

where λ is the wavelength of the carrier. These restrictive conditions (in particular, the one that limits the angular extension) are often not satisfied in imaging systems, with potential low performances in the lateral regions of images.

THE WEIGHTED FRESNEL APPROXIMATION

In order to relax the constraint of (5), a possible solution is to weight the terms of the Fresnel approximation by two constants, k_1 and k_2, computed on the basis of the desired steering region and of a fixed focalization distance. The importance of keeping unchanged the two terms of the Fresnel approximation lies in their computational simplicity and in the possibility of implementing focused beamforming by the FFT. One can write the novel delay approximation as follows:

$$\tau_{ls}(\mathbf{u},i) = k_1 \frac{\left(\mathbf{u}\,\mathbf{v}_i^+\right)}{c} + k_2 \frac{\|\mathbf{v}_i\|^2}{Rc} \tag{8}$$

and try to fix the values of the two constants by minimizing the sum of the square differences between the delays provided by the approximation and the exact delays. Square errors can be measured over a two-dimensional grid containing all the possible pairs $(\mathbf{u}, i)$. Denoting by $e(\mathbf{u}, i)$ the error between the approximate and exact delays, and by $\mathbf{u}_1$, $\mathbf{u}_2$, ..., $\mathbf{u}_N$ the steering directions of interest, one can write an overdetermined system of equations by using a matrix formulation:

$$
\begin{bmatrix}
\tau_{ex}(\mathbf{u}_1,1) \\
\vdots \\
\tau_{ex}(\mathbf{u}_1,M) \\
\tau_{ex}(\mathbf{u}_2,1) \\
\vdots \\
\tau_{ex}(\mathbf{u}_2,M) \\
\vdots \\
\vdots \\
\tau_{ex}(\mathbf{u}_N,M)
\end{bmatrix}
=
\begin{bmatrix}
\left(\mathbf{u}_1\,\mathbf{v}_1^+\right)/c & \|\mathbf{v}_1\|^2/(Rc) \\
\vdots & \vdots \\
\left(\mathbf{u}_1\,\mathbf{v}_M^+\right)/c & \|\mathbf{v}_M\|^2/(Rc) \\
\left(\mathbf{u}_2\,\mathbf{v}_1^+\right)/c & \|\mathbf{v}_1\|^2/(Rc) \\
\vdots & \vdots \\
\left(\mathbf{u}_2\,\mathbf{v}_M^+\right)/c & \|\mathbf{v}_M\|^2/(Rc) \\
\vdots & \vdots \\
\vdots & \vdots \\
\left(\mathbf{u}_N\,\mathbf{v}_M^+\right)/c & \|\mathbf{v}_M\|^2/(Rc)
\end{bmatrix}
\begin{bmatrix} k_1 \\ k_2 \end{bmatrix}
+
\begin{bmatrix}
e(\mathbf{u}_1,1) \\
\vdots \\
e(\mathbf{u}_1,M) \\
e(\mathbf{u}_2,1) \\
\vdots \\
e(\mathbf{u}_2,M) \\
\vdots \\
\vdots \\
e(\mathbf{u}_N,M)
\end{bmatrix}
\tag{9}
$$

The system in (9) can be written in a shortened form as:

$$\mathbf{d} = \mathbf{A}\mathbf{k} + \mathbf{e} \tag{10}$$

where **d** is the column vector (*NM* by 1) of the exact delays, **k** is the column vector (2 by 1) of the unknowns, **e** is the column vector (*NM* by 1) of the errors, and **A** is the matrix (*NM* by 2) containing the two terms of the approximation.

By using a least-squares inverse[7], $\mathbf{A}^{ls}$ (2 by *NM*), of the matrix **A**, one can compute a system solution **k*** that minimizes the mean square error:

$$\mathbf{k}^* = \left(\mathbf{A}^{ls}\right)\mathbf{d} = \left[\left(\mathbf{A}^+\mathbf{A}\right)^{-1}\mathbf{A}^+\right]\mathbf{d}.$$
(11)

Once the solution, **k***, has been computed off line, one can begin the focused beamforming operation under the guarantee that the approximate delays τ_{ls} are optimum in the least-squares sense. One can verify that, if one fixes a steering region perfectly overlapped with the validity region of the Fresnel approximation, the differences between the least-squares and Fresnel approximations are negligible, i.e., $k_1 \approx 1$ and $k_2 \approx -0.5$.

RESULTS FOR A LINEAR ARRAY

The effectiveness of the proposed method has been assessed in the case of an equispaced linear array. In this case, the steering faculty is restricted to a plane containing the antenna[1]. Therefore, supposing the array to be placed on the *x* axis, without loss of generality, one can consider the plane $z = 0$ and rewrite the unit vector **u** and the vector $\mathbf{v}_i$ as follows:

$$\mathbf{u} = [\sin\theta, \cos\theta, 0]$$
(12)

$$\mathbf{v}_i = [x_i, 0, 0] = \left[\left(i - \frac{M+1}{2}\right)d, 0, 0\right]$$
(13)

where θ is the steering angle measured between **u** and the *y* axis (i.e., $\theta = 90° - \phi$) and *i* is an integer included between 1 and *M*.

To test the accuracy of the delay approximation, one can use the total mean square error (i.e., $MSE = \|\mathbf{e}\|^2/MN$) plotted versus the focalization distance, the *MSE* as a function of the steering angle (computed at a fixed focalization distance):

$$MSE(\theta) = \frac{1}{M}\sum_{i=1}^{M} e(\theta, i)^2$$
(14)

and the *MSE* as a function of the array element position (computed at a fixed focalization distance):

$$MSE(i) = \frac{1}{N}\sum_{l=1}^{N} e(\theta_l, i)^2$$
(15)

where θ_l is one of the *N* steering angles.

As an example, one can consider an array composed of 120 $\lambda/2$-spaced elements, working at 500 kHz, with a sound speed $c = 1500$ m/s (i.e., $M = 120$, $\lambda = 3$ mm, and $d = 1.5$ mm). A forward-looking sonar system is a potential imaging application of this kind of array. A region defined by $|\theta| \le 32°$ is considered for the computation of (k_1, k_2) and a range domain 0.5 m $\le R \le 10$ m is required. The angular extension of this region is larger than that of the validity region of the Fresnel approximation (i.e., from (5), $|\theta| < 18°$), whereas the range extension is similar (i.e., the Fresnel approximation is valid for 0.12 m < R < 8.34 m). The scanning region is covered by $N = 65$ beams in equispaced steering directions θ_l, from -32° to 32°. Figure 1 shows the total MSE versus the focalization distance R, measured in μs^2, for both the least-squares approximation τ_{ls}, the Fresnel approximation τ_{Fr}, and the far-field delay τ_{ff}. Figure 2 compares the $MSE(\theta)$ of the least-squares approximation with that of the Fresnel approximation. Such errors were computed after fixing $R = 1$ m and measured in μs^2.

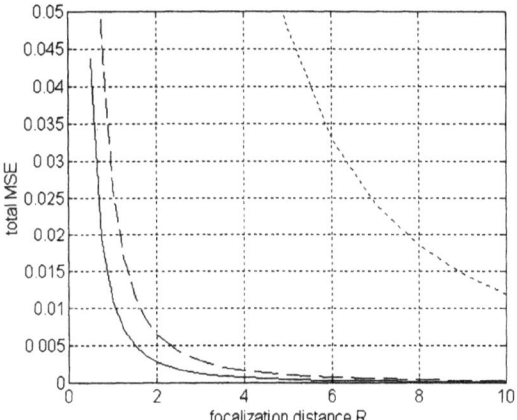

Figure 1. Total *MSE* in μs^2 versus R for the least-squares approximation (solid line), the Fresnel approximation (dashed line), and the far-field hypothesis (dotted line).

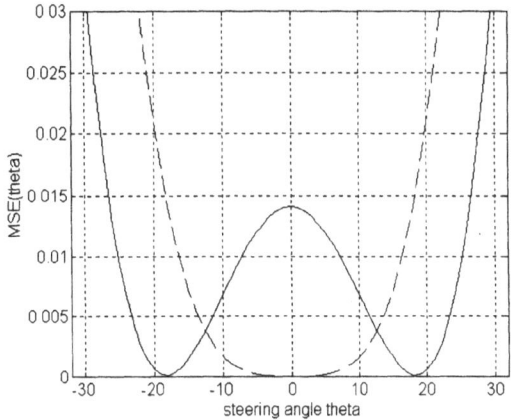

Figure 2. Behaviours of $MSE(\theta)$, computed at $R = 1$ m and measured in μs^2, for the least-squares approximation (solid line) and the Fresnel approximation (dashed line).

DISCUSSION AND CONCLUSIONS

One can notice that, for every delay approximation, the total *MSE* decreases as the focalization distance increases, and that, for a distance out of the near-field region (i.e., $R > 8.34$ m), the far-field hypothesis can be adopted with negligible error differences. Within the near-field region, the proposed approximation has a total *MSE* lower than that of the Fresnel approximation. Moreover, the weights k_1 and k_2 tend to assume constant values for a focalization distance larger than few meters, so excluding the strict necessity for computing them for each R (in particular, k_1 tends to be equal to 1, as in the Fresnel approximation).

Concerning the *MSE* versus the steering angle, one can notice that the error of the Fresnel approximation increases with $|\theta|$ and its value at the border of the validity region is $MSE(18°) = 0.014\ \mu s^2$ (at $R = 1$ m). Despite the error of the least-squares approximation is not null for $\theta = 0°$, its value does not exceed $0.014\ \mu s^2$ over the domain $|\theta| < 26.5°$. Therefore, as within the validity region of the Fresnel approximation the loss of image quality is generally negligible, one can deduce that, moving from the Fresnel approximation to the least-squares approximation, the safe scanning region can be enlarged from $\pm18°$ to $\pm26.5°$, with a gain factor of about 1.5.

By repeating the same reasoning on the errors computed for $R \neq 1$ m, the same conclusion (about the potential enlargement of the scanning region) will be reached. Moreover, if a change of the desired scanning sector is performed, one can verify that the least-squares approximation provides a sort of compromise between the approximation precision and the extension of the desired scanning region.

Finally, concerning the FFT implementation of focused beamforming, one can verify that for a large set of focalization distances, the value of k_1 can be set equal to 1, without a notable loss of precision. Then, the equations that link the spatial frequencies of the FFT to the steering angles can hold their conventional form[4,8]. Instead, if $k_1 \neq 1$, such equations should be slightly updated by an adequate scaling factor.

In conclusion, an approximation for the delays required by focused beamforming has been proposed that minimizes the *MSE* computed over the scanning region of interest. The minimization has been obtained by weighting the terms of the conventional Fresnel approximation through a least-squares solution of an overdetermined equation system. In general, one can observe that, if the desired scanning region is not too extended, the least-squares approximation yields an acceptable precision for many practical operations. At the same time, thanks to the similarity of the proposed approximation to the Fresnel one, the computational load and the system complexity do not notably increase, and the opportunity of implementing focused beamforming via the FFT can be easily kept.

REFERENCES

1. R.O. Nielsen, *Sonar Signal Processing*, Artech House, Boston, 1991.
2. L.J. Ziomek, "Three Necessary Conditions for the Validity of the Fresnel Phase Approximation for the Near-Field Beam Pattern of an Aperture," *IEEE Journ. Ocean. Engin.*, vol. 18, pp. 73-75, January 1993.
3. J.W. Goodman, *Introduction to Fourier Optic*, McGraw-Hill, New York, 1968.
4. V. Murino, A. Trucco, "Dynamic Focusing by FFT Beamforming for Underwater 3D Imaging," *Acoustics Letters*, vol. 17, no. 9, pp. 169-172, March 1994.
5. M. Soumekh, "Array Imaging with Beam-Steered Data," *IEEE Transactions on Image Processing*, vol. 1, pp. 379-390, July 1992.
6. R.K. Hansen, P.A. Andersen, "3D Acoustic Camera for Underwater Imaging," *Acoustical Imaging*, vol. 20, pp. 723-727, 1993.
7. C.R. Rao, S.K. Mitra, *Generalized Inverse of Matrices and its Applications,* Wiley, New York, 1971.
8. V. Murino, A. Trucco, "Underwater 3D Imaging by FFT Dynamic Focused Beamforming," *1st IEEE Intern. Conf. Image Processing*, Austin, Texas (USA), vol. I, pp. 890-894, November 1994.

3D POWER DOPPLER ULTRASOUND IMAGING

OF AN *IN VITRO* ARTERIAL STENOSIS

Louis Allard[1], Guy Cloutier[1], Aaron Fenster[2] and Louis-Gilles Durand[1]

[1] Laboratory of Biomedical Engineering
Institut de recherches cliniques de Montréal
Québec, Canada, H2W 1R7
[2] Imaging Research Laboratories
The John P. Robarts Research Institute
Ontario, Canada, N6A 5K8

INTRODUCTION

Several studies have reported that duplex scanning can be as accurate as angiography to predict and localize arterial stenoses. In spite of this promising performance, recent studies indicated a lower sensitivity for detecting significant stenoses distal to severe or total obstructions (1). The main reason proposed to explain this lower sensitivity is the difficulty to detect the presence of low flow distal to occluded or highly stenotic segments. Power Doppler imaging (PDI) which allows a direct representation of the geometry of the blood flow is a new alternative since it offers several advantages over duplex scanning and color flow imaging: it is characterized by a greater sensitivity to flow, is aliasing free, nearly Doppler angle independent, and little affected by the flow rate (2). PDI has been mainly used for imaging small, low-flow vessels associated with renal, musculoskelatal, fetal and cerebral perfusion (3). This technique has, however, the potential for being applied to larger vessels. Further experimental and clinical studies are required to clearly validate its potential to quantify arterial stenosis. For instance, it is known that the ultrasound backscattering properties strongly depend on the flow characteristics and spatial arrangement of moving scatterers. Quantitative validation of PDI with blood is thus essential since turbulence and red cell aggregation affect the scatterer sizes and the correlation between red cells, and consequently the Doppler backscattered power from blood.

The objectives of the present *in vitro* study were 1) to evaluate the ability of using power Doppler imaging to quantify the flow area and the area reduction of a stenotic vessel, and 2) to measure the variations in echogenicity from blood induced by the presence of the vessel stenosis and red cell aggregation.

METHODS

Two experiments were performed in a horizontal steady flow model at room temperature. A phantom containing a tissue mimic and a wall-less vessel was made to simulate a stenotic vessel with an 80% area reduction. The tissue mimic having an acoustic velocity similar to that of tissue was made of water (89%), agar (3%), and glycerol (8%). The diameter of the nonobstructed vessel lumen was 7.96 mm and the cosine shaped stenosis has a length of 20 mm. The inlet length of the vessel was straight and long enough to ensure a fully developed laminar flow before the stenosis. A roller pump circulated the fluid at constant flow rates of 0.12, 0.25, 0.50, 0.75 and 1.25 l/min, as measured by a cannulating type flow probe (model SF625) coupled to an electromagnetic flowmeter (Cliniflow II, model FM710D, Carolina Medical Electronics).

Ultrasonic measurements were performed with porcine whole blood. Fresh blood was collected from an abattoir and its hematocrit adjusted to 40% by remixing the plasma, white cells, and red cells separated by sedimentation. An ATL Ultramark 9 HDI ultrasound system with a 38 mm aperture and a high resolution linear array probe (L7-4) was used for all measurements. This probe was operated at 5 MHz for B-mode and at 4 MHz for Doppler mode. The selection of the imaging parameters was optimized to provide images with the best visual quality. The B-mode display was used to optimize the power Doppler gain by confining the flow signal to the lumen of the blood vessel. The instrument settings were performed on the proximal section of the flow phantom and were kept constant thereafter. The tissue signal was then shut off to display only the flow field, generating the angiographic image. Since PDI is aliasing free, a pulse repetition frequency (PRF) of 700 Hz was used to increase the imaging sensitivity. The lowest wall filter cutoff frequency (25 Hz) was selected because there was no wall motion. The persistence level of the system, which determines the number of video frames averaged, was set to the maximum (7) to obtain the best vessel contour depiction. Other instrument settings included a color sensitivity of 16, a frame rate of 4.7 Hz, and a color gain of 70%. A gray-scale was used to map the Doppler power. Thus, the background noise (low power) was displayed as almost black and the flow field as white, like a subtraction angiogram.

A 3D imaging system developed at the John P. Robarts Research Institute and produced by Life Imaging System Inc. (London, Canada) was used for 3D image acquisition, reconstruction, and display (4). It includes a Power Macintosh 7500 computer and a motor-driven translation assembly controlled by the computer allowing the scanning of the linear array probe over the tissue mimic. Using discrete displacements (0.5 mm) of the probe with a Doppler angle of 70°, a set of 320 parallel 2D cross-sectional power Doppler images was acquired over a distance of 16 cm (3 cm upstream and 13 cm downstream of the stenosis). These 2D images having a dimension of 256 x 256 pixels were digitized with a precision of 8 bits/pixel. Since a high persistence level was used, the linear scanning was done at a relatively slow speed (0.25 mm/sec) to minimize displacement of the probe during the time required to average seven frames.

After acquisition, the images were reconstructed to a 3D volume following Doppler angle compensation, thus producing 2D slices at 90° with the vessel axis. The 3D volume was rescaled based on the pre-calibrated voxel dimensions in the x, y, and z directions. The voxel dimensions were function of the specific zoom and focus distance selected and, on average, were equal to 0.069 x 0.057 x 0.5 mm/pixel. To quantify the lumen area, the flow and the background noise were differentiated by an automatic segmentation algorithm using a percentage of the maximum intensity of the image as a threshold level. Threshold levels varying between 25% and 75% were tested. The imaged flow area of every 2D slice perpendicular to the lumen's axis was then calculated by automatically counting pixels corresponding to flow. The percentage of area reduction at a given location was also computed as the ratio of the flow area at that position to the maximal flow area found along the vessel (considered as the reference normal area). The calculated values of the flow area (in mm²) and area reduction were plotted as a function of the location along the lumen axis and compared with the true areas and the true area reductions. An absolute error between the area reductions obtained by PDI and the true area reductions was computed and averaged over the different positions along the vessel axis. The maximal area reduction was also determined and compared to the true maximal area reduction (80%). Centerline variations in the echogenicity along the vessel due to the presence of the stenosis (post-stenotic turbulence) and red cell aggregation were quantified by computing mean gray levels in three vessel sections (upstream, within and downstream of the stenosis) as a function of the flow rate.

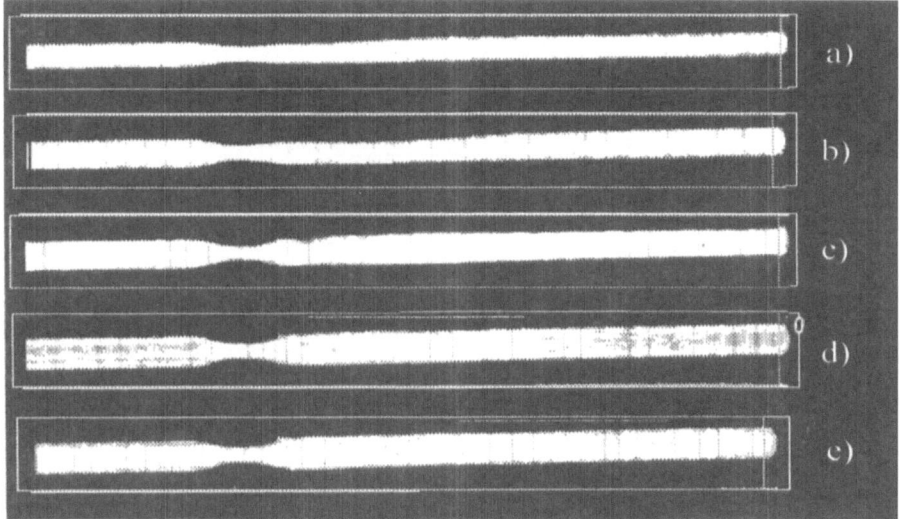

Figure 1. Three-dimensional power Doppler images of the 80% area reduction stenosis under steady flow of 0.12 l/min (a), 0.25 l/min (b), 0.50 l/min (c), 0.75 l/min (d), and 1.25 l/min (e).

RESULTS

Figure 1 is an example of 3D power Doppler images of the 80% area reduction stenosis obtained at different flow rates. It can be observed from this figure that 1) the degree and the length of the stenosis varied as a function of the flow rate, and 2) the area of the nonobstructed section of the vessel was reduced at low flow rates. Figure 2 shows examples of the measured flow areas (panels a and b) and the area reductions (panels c and d) as a function of the position along the vessel. Only the results obtained at the lowest and highest flow rates are presented. At lower flow rates, the flow areas were underestimated because of the effect of the wall filter. The use of a threshold of 25% for image segmentation provided a better estimation of the flow areas. At high flow rates, the flow areas were overestimated with a threshold of 25% but were very accurately estimated with a threshold of 75%.

Because area reduction depends on the ratio of two surfaces, this parameter is not affected by an uniform over- or underestimation of the flow area along the vessel. This is demonstrated in panel d of Fig. 2 where the flow area reductions were little affected by the threshold value. At low flow rates, the area reductions in the poststenotic zone were overestimated due to the effect of the wall filter on the low velocities near the wall. Table 1 presents for two experiments and for different flow rates 1) the absolute error, averaged over the vessel, between the measured and true area reductions, and 2) the maximum area reduction of the stenosis. These results indicate that errors in the estimation of the area reductions (including the maximum area reduction) were less at higher flow rates.

Figure 3 shows the mean power Doppler intensity as a function of the flow rate measured upstream, within and downstream of the stenosis. It can be seen that the power Doppler intensity increased when the flow rate decreased. This relationship between the flow rate and blood echogenicity may be explained by the formation of red cell aggregates at low shear rates which increased the backscattered power from blood. These variations in blood echogenicity were almost identical for the three vessel sections which suggest that similar shear rate conditions in the centerline of the vessel were present along the vessel.

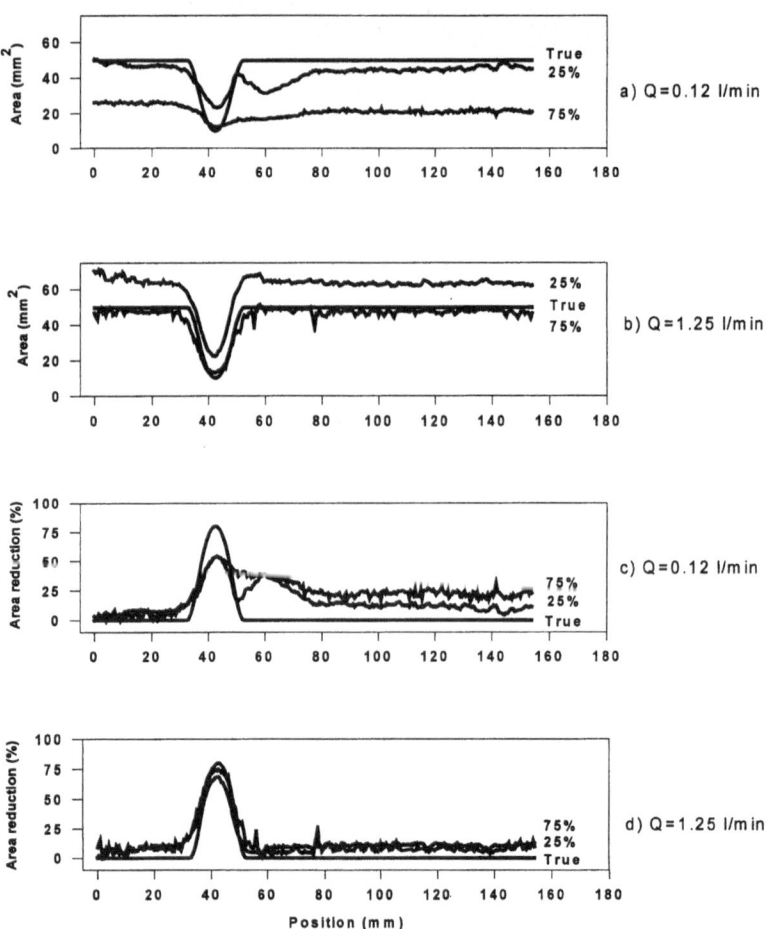

Figure 2. Absolute flow areas (a and b) and percentage of area reductions (c and d) as a function of the position along the vessel for flow rates Q = 0.12 l/min (a and c) and 1.25 l/min (b and d). The segmentation of the power Doppler images was realized with two threshold levels (25% and 75%), as indicated in the figure.

Table 1. Mean absolute errors on the percentage of area reduction (AR) averaged over the vessel length, and maximum percentage of AR of the stenosis as a function of the flow rate. The threshold levels used to perform segmentation of the power Doppler images were those minimizing the mean absolute error for each flow rate. These values are indicated in parentheses.

Flow rates	Mean absolute AR errors (threshold)		Maximum AR of the stenosis	
(l/min)	Experiment 1	Experiment 2	Experiment 1	Experiment 2
0.12	13.3% (40%)	11.7% (25%)	54.3%	59.7%
0.25	14.5% (30%)	12.7% (25%)	64.0%	63.7%
0.50	6.6% (30%)	5.2% (35%)	66.7%	67.4%
0.75	5.7% (30%)	4.2% (35%)	68.4%	67.1%
1.25	6.1% (45%)	4.6% (55%)	70.0%	71.7%

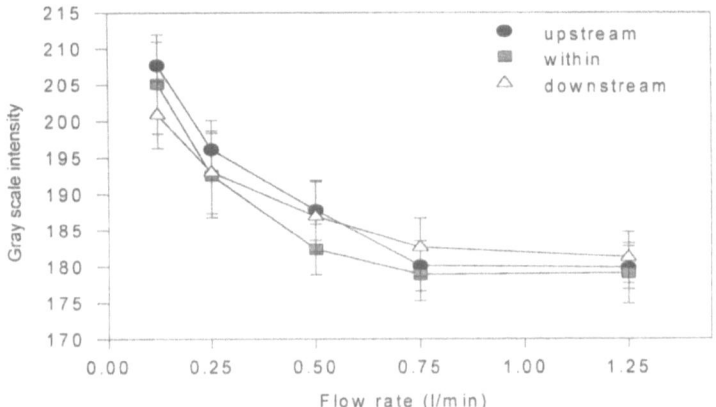

Figure 3. Gray scale intensity as a function of the flow rate in three vessel sections (upstream, within, and downstream of the stenosis). Mean intensities upstream, within, and downstream of the stenosis were computed over 66, 39 and 204 frames, respectively.

DISCUSSION AND CONCLUSION

Recently, results obtained using an *in vitro* stenotic model and blood mimicking fluid indicated that the flow area was generally larger than the true lumen area (5). More specifically, stenotic vessels of 80%, 50% and 30% area reduction were quantified under moderate and high flow rates with an accuracy of 8% and a precision of 7% of the vessel area. Other results obtained using a blood mimicking fluid and a flow model simulating the presence of multiple stenoses indicated that the percentage of the central stenosis was always underestimated: on average, the 80% stenosis was evaluated as a 70% stenosis while the 50% stenosis was evaluated as a 46% stenosis (6).

The accurate estimation of the flow area at low flow rates is clinically important because arterial disease in the lower limbs is often accompanied with the development of collaterals that divert flow around the stenotic segments. In the present study, we demonstrated that the estimation of the flow area with whole blood at moderate and high flow rates is accurate (panels c, d and e of Fig. 1) and comparable to reported results with blood mimic. However, we observed less accuracy at low flow rates due to the effect of the wall filter (panels a and b of Fig. 1). Results also indicated that the flow area was affected by the threshold level used to segment the power Doppler images (panels a and b of Fig. 2). The optimal segmentation of the power Doppler images should be based on a threshold level proportional to the flow velocity. At high flow rates or in vessel sections with high flow velocities (e.g. within the stenosis), a high threshold level should be used. Inversely, at low flow rates or in sections with low flow velocities (e.g. the recirculation zones), a low threshold level should be used to minimize the effect of the wall filter. Fortunately, the estimation of the area reduction is less dependent on the threshold level (panels c and d of Fig. 2). However, as indicated in Table 1, it is also affected by the flow rate.

The variations in power Doppler intensity observed in the present study as a function of the flow rate may be explained by the presence of red cell aggregation. However, no variation was observed between measurements done upstream, within, and downstream of the stenosis for a given flow rate, although it is known that the development of flow turbulence downstream of a severe stenosis is associated with a power Doppler increase (7). Finally, it seems unlikely that the variations in echogenicity observed in the present study had an effect on the estimation of the flow areas.

ACKNOWLEDGMENTS: This work was supported by grants from the Fondation des Maladies du Coeur du Québec, the Medical Research Council of Canada (#MA-12491), and the Whitaker Foundation. The authors gratefully acknowledge Dr. Helen Routh of Advanced Technology Laboratories for loaning an Ultramark 9, Life Imaging System for providing the 3D ultrasound system, and Dr. Zhenyu Guo of the George Washington University for helpful discussions.

REFERENCES

1. Allard L, Cloutier G, Durand LG, Roederer GO, Langlois YE. Limitations of ultrasonic duplex scanning for diagnosing lower limb arterial stenoses in the presence of adjacent segment disease. J Vasc Surg 1994;19(4):650-7.
2. Rubin JM, Bude RO, Carson PL, Bree RL, Adler RS. Power Doppler US: A potentially useful alternative to mean frequency-based color Doppler US. Radiology 1994;190:853-6.
3. Macsweeney JE, Cosgrove DO, Arenson J. Colour Doppler energy (Power) mode ultrasound. Clinical Radiology 1996;51:387-90.
4. Picot PA, Rickey DW, Mitchell R, Rankin RN, Fenster A. Three-dimensional colour Doppler imaging. Ultrasound Med Biol 1993;19(2):95-104.
5. Guo Z, Fenster A. Three-dimensional power Doppler imaging: A phantom study to quantify vessels stenosis. Ultrasound Med Biol 1996;22(8):1059-69.
6. Guo Z, Fenster A, Allard L, Cloutier G, Durand LG. Quantitative evaluation of multiple arterial stenoses using 3D power Doppler imaging. 1996 Canadian Conference on Electrical & Computer Engineering 1996;1:347-50.
7. Cloutier G, Allard L, Durand LG. Characterization of blood flow turbulence with pulsed-wave and power Doppler ultrasound imaging. J Biomech Eng 1996;118:318-25.

REFRACTION ARTIFACTS IN DOPPLER SPECTRA DUE TO DUAL MODE ULTRASOUND PROPAGATION

Paolo Berti[1], Alessandro Gubbini[2] and Piero Tortoli[1]

[1]Electronic Engineering Dept., University of Florence
via S.Marta 3, 50139 Firenze, Italy
[2]Esaote Biomedica, Firenze, Italy

INTRODUCTION

Refraction occurring at the interface between two media having different acoustic impedances can represent a significant source of artefacts in flow imaging systems[1]. In previous studies, only refraction due to *planar longitudinal* waves was considered, and no reference was made to the possible effects of *shear* wave propagation. However, there are cases of interest where sound propagates in both shear and longitudinal modes, e.g. through the rigid tubes used in some flow phantoms.

We have investigated to what extent dual mode ultrasound propagation in plastic materials such as Plexiglas, Nylon and Polyethylene can affect the composition of the Doppler spectrum reflected from moving targets located beyond the material itself. Our attention was concentrated on the behaviour of *focused* ultrasound fields, which in a ray approximation can be decomposed into a set of rays oriented through a corresponding range of angles around the nominal beam-axis-to-plastic-surface angle, θ_i.

Experimental results show that maximum spectrum distortion occurs when θ_i is approximately equal to the first critical angle, θ_c, where the power transmission factor abruptly falls to zero. In this region, that part of the radiation for which $\theta_i > \theta_c$, propagates in the plastic medium as shear waves, whereas that part of the radiation for which $\theta_i < \theta_c$, propagates as both shear *and* longitudinal waves. Since propagation speed and, therefore, refraction angles and propagation paths are different for the two waves, the various contributions can add together in such a way that the Doppler spectrum turns out to be significantly distorted. In particular, we will show that, depending on the actual value of θ_i, this distortion can be interpreted in terms of high-pass ($\theta_i > \theta_c$) or low-pass filtering ($\theta_i < \theta_c$), while for θ_i exactly equal to θ_c, the material acts as a notch filter centered on the mean frequency of the Doppler spectrum. Correspondingly, the error in mean frequency estimates, of interest in flow imaging systems, can be as high as 10%.

DUAL MODE ULTRASOUND PROPAGATION OF FOCUSED BEAMS

It is known that a longitudinal wave incident solid materials may yield the transmission of both *longitudinal* **and** *shear* waves in to the material itself.[2] Their directions of propagation depend on the incident angle through Snell's law (see Fig.1a), while their amplitudes can be numerically estimated on the basis of wave equations represented in terms of potential functions. We have analysed the behaviour of different plastic materials immersed in water, by obtaining power transmission coefficients similar to that calculated for plexiglas, reproduced in Fig.1b.

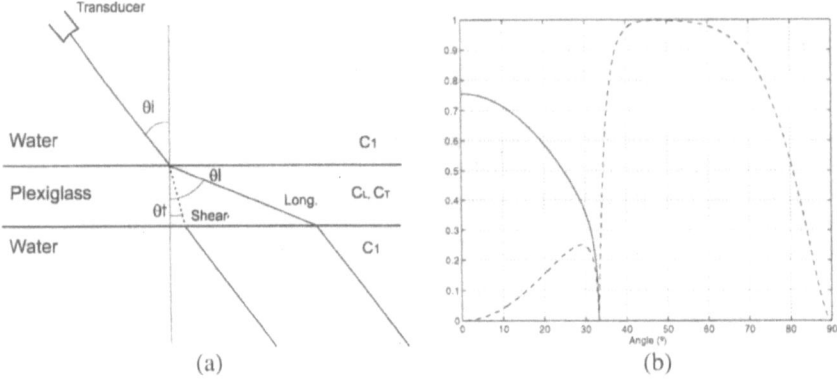

(a)   (b)

Figure1. a) Refraction at the interfaces water-plexiglas and plexiglas-water. b) Power transmission coefficient as a function of incident angle θ_i for a wave in water incident on Plexiglas.

The relevant information from this plot can be summarised as follows:
- For an incident angle $\theta_c = 33°$, which will be referred to as *critical angle*, no power is transmitted to the plexiglas
- For angles higher than θ_c only the shear wave component propagates through the plexiglas, with power comparable to that of the incident wave
- For angles lower than θ_c, the sound wave propagates through both longitudinal and shear wave components, the latter being always of lower amplitude than the former.

Since the velocities in the two propagation modes are different, the two waves are refracted in different directions and they emerge parallel from different points of the material surface. Moreover, the ultrasound beams produced by most transducers used in biomedical applications are typically focused, i.e. they can be decomposed in to a set of plane waves oriented at different angles. Hence, even though the nominal incidence angle is θ_i, an entire set of angles should be considered, covering a range $[\theta_{min}, \theta_{max}]$ centered around θ_i. For example, when $\theta_i = \theta_c$, not all the incident power is lost, since part is transmitted as shear wave (the beam components directed at angles $\theta_c < \theta_i < \theta_{max}$), and part as both shear and longitudinal waves (the beam components directed at angles $\theta_{min} < \theta_i < \theta_c$).

Another point to be outlined is that, according to Snell's law, each "ray" of the ultrasound beam is refracted at a different angle. Depending on the nominal incidence angle as well as on the position of the material relative to the transducer focal depth, a noticeable variation in the width of the ultrasound beam emerging from the material may be yielded.

Fig.2 shows the deformations suffered by an ultrasound beam in the presence of a 3 mm thick plexiglas block interposed between the transducer and a point reflector. The plexiglas was in this case oriented at an angle of approximately 25° to the transducer beam axis. According to Fig.1b, in correspondence of this angle, two waves of comparable power simultaneously propagate through the plastic material. Since the velocity of the longitudinal wave (≈ 2700 m/s) is much higher than the velocity of the shear wave (≈ 1300 m/s), the former is more strongly refracted and the two waves emerge from the plexiglas at a relative distance of about 4 mm. Moreover, the presence of the plexiglas involves a huge beam spread for the longitudinal wave component.

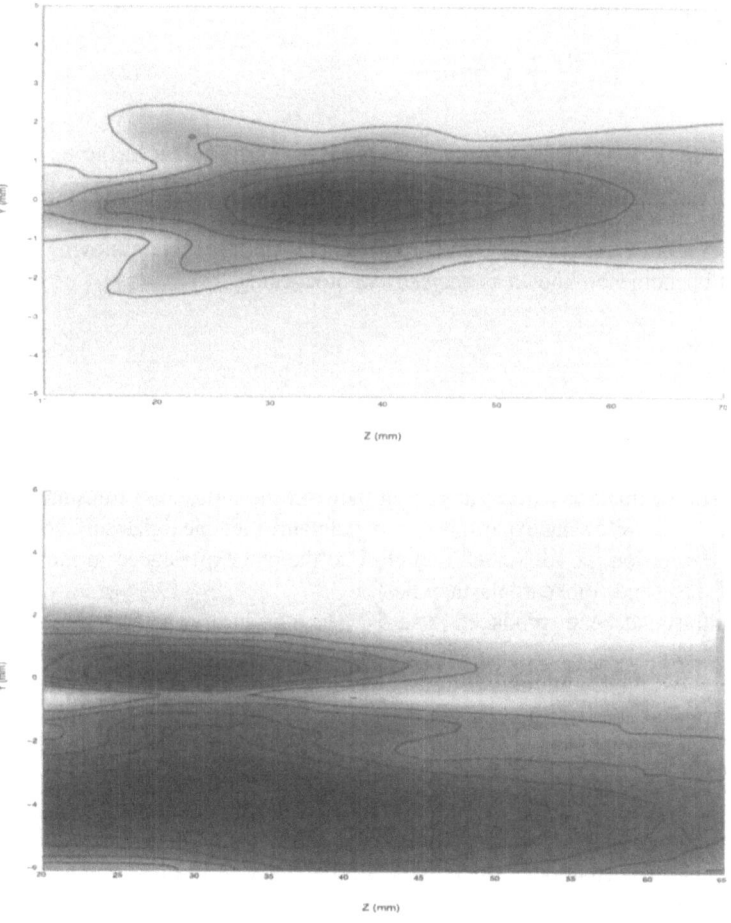

Figure 2: Ultrasound beam plots obtained in normal conditions (a) and with a 3 mm thick plexiglas plate oriented at $\theta_i = 27°$, interposed between the transducer and a point reflector (b). The transducer focal length was 32 mm.

GEOMETRICAL DOPPLER SPECTRUM BROADENING

In ultrasound flowmeters, a flow-line of velocity v insonifyed by an infinite plane wave of frequency f_0 propagating at velocity c, produces a frequency shift described by the well known Doppler equation:

$$f_d = \frac{2f_0}{c} v \cos\phi \qquad (1)$$

where ϕ is the angle between the incident wave and the flow directions. In practical cases, focused ultrasound beams are used. As mentioned above, they can be considered as the combination of a set of plane waves oriented along a corresponding set of directions converging to the focal spot. Each of the "rays" contributing to the focused beam thus generates a different Doppler shift. A full spectrum of frequencies is therefore produced, having a nearly triangular shape and covering a bandwidth represented by:

$$B_d = \frac{2f_0}{c} \frac{W}{F} v \sin\phi \qquad (2)$$

where W is the aperture of the transducer focused at a depth F.[3] It has to be underlined that this bandwidth is centered around a *mean frequency*, f_d, equal to the value given by the eq.(1) with ϕ representing the angle between the flow and the beam *axis*. The production of a full bandwidth of Doppler frequencies from scatterers moving at a single velocity is a phenomenon known as *geometrical* broadening[4].

EXPERIMENTS

In order to test the effects of dual mode propagation on the Doppler spectrum, we have analysed the echoes produced by a thread moving at fixed velocity when plexiglas blocks of variable thickness were interposed between the ultrasound transducer and the thread itself. In the following experiments, we maintained the plexiglas surface parallel to the flow-line direction, i.e. in a condition close to the one experienced in most practical cases (e.g., when fluids move in plastic tubes).

The ultrasound beam produced by a 5 MHz transducer was first directed on the thread maintaining its axis at an incidence angle $\theta_i = 33°$. In normal conditions (i.e., without plexiglas) the Doppler spectrum shown by the dotted line in Fig.3 was obtained. It exhibits a typical nearly-triangular shape with bandwidth B ≈ 1000 Hz. This bandwidth may be considered equivalent to that produced by a set of plane waves oriented at angles in the range [$\theta_{min} = 27°$, $\theta_{max} = 39°$].

The presence of a 2 mm thick plexiglas plate yields a significant change in the spectrum shape, equivalent to that produced by a notch filter. This is consistent with the fact that Doppler contributions (around $f_d \approx 1450$ Hz) due to beam components incident the plexiglas at the critical angle are totally reflected. On the other hand, Doppler frequencies higher and lower than f_d are produced the by shear wave components directed at angles higher and lower than 33°, respectively. Longitudinal waves at angles in the range [27°, 33°] do not contribute significantly, since they are strongly refracted at the water-plexiglas interface, and they suffer a huge attenuation before emerging from the plexiglas.

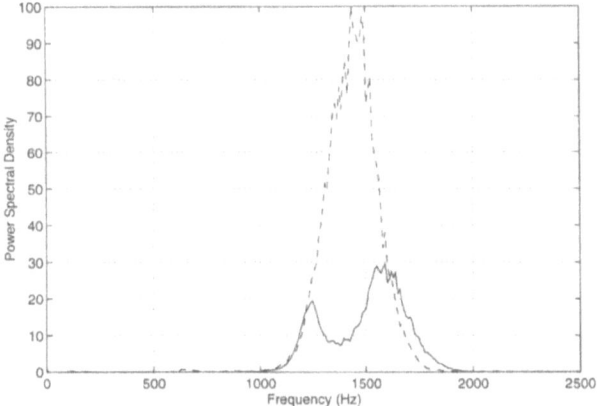

Figure 3. Doppler spectrum obtained when the ultrasound beam axis of a 5 MHz transducer was incident a moving thread parallel to a plexiglas plate at $\theta_i=\theta_c$. The dotted line spectrum obtained without plexiglas is enclosed as a reference.

By slightly decreasing and increasing the incidence angle, the results shown in Fig.4a and in Fig.4b, respectively, were obtained. In both cases the spectra are narrowed because of the strong reflection suffered by wave components oriented at angles close to 33° (see Fig.1b). Because of the presence of the plexiglas block, the original Doppler spectra in Fig.4a and Fig.4b, are low-pass and high-pass filtered, respectively. Note that a further increase of the incidence angle leads to a region where the power transmission coefficient is close to unity, i.e., the incident wave propagates as a shear wave, without reflection, in the plexiglas. In this case, the resultant spectrum remains unmodified by the presence of the plastic plate.

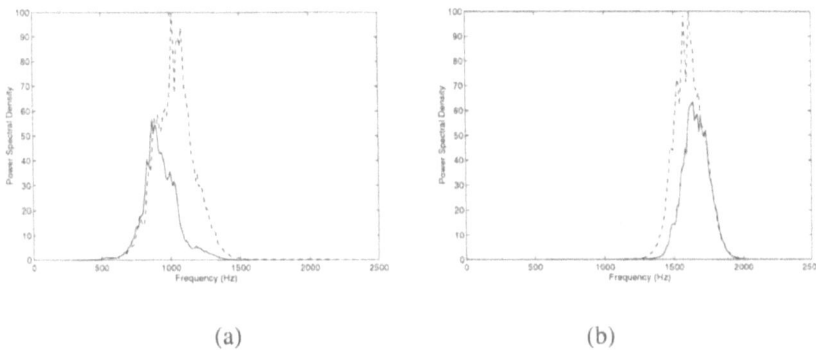

(a)   (b)

Figure 4. Doppler spectra obtained when the ultrasound beam axis of a 5 MHz transducer was incident at 20° (a) and 40° (b) a plexiglas plate parallel to a moving thread. Dotted line spectra obtained without plexiglas are enclosed for reference.

CONCLUSION AND DISCUSSION

The conversion of longitudinal to shear waves in solid materials is a well-known phenomenon, usually of interest in non-destructive testing applications. In this paper we have investigated to what extent it can affect velocity measurements based on the Doppler effect. The practical interest of this study is related to the fact that in many scientific and industrial laboratories flow phantoms based on plastic tubes supporting dual mode propagation are used.

Experimental results have shown that it is not only the *critical* angle which should be avoided in the velocity measurements, but an entire set of angles around it, whose range depend on the transducer focusing. When θ_i is close to θ_c, in fact, the Doppler spectrum is so distorted that, according to our measurements, its mean frequency (the parameter of major interest in flow imaging applications) can deviate as much as 10% from its actual value. These results have been confirmed by a number of experiments performed by using plastic blocks of different materials (Nylon, Polyethylene and Polystyrene) and thickness.

Moreover, there is a further range of angles which may yield significant problems in flow measurements. This corresponds to the condition where both longitudinal and shear waves propagate with comparable amplitude (e.g., for $\theta_i \approx 25°$ in the case of plexiglas). In this case, two parallel ultrasound beams of different width can propagate beyond the plastic plate(see Fig.2). In flow imaging systems using a pulsed emission of ultrasound energy, this means that two spatially distinct *sample volumes*[5] simultaneously contribute to the Doppler signal. In real flow, it is quite probable that these *sample volumes* interrogate completely different flow regions, and their joint contribution can yield misleading results.

Acknowledgements

The authors wish to acknowledge valuable help by Francesco Guidi in obtaining the experimental results, and by Paolo Palchetti in the evaluation of transducer fields.

REFERENCES

1. R.S.Thompson et al., The effect on ultrasound intensity of refraction at a curved surface, *in:* "Acoustical Imaging", Vol.22, P.Tortoli & L.Masotti, eds., Plenum Press, New York, 371:376 (1996).
2. G.S.Kino, "Acoustic Waves: Devices, Imaging & Analog Signal Processing", Prentice-Hall Processing Series, Alan V. Oppenheim Editor, Englewood Cliffs, 100:103, (1987).
3. D. Censor, V.L. Newhouse, T. Vontz, Theory of ultrasound Doppler spectra velocimetry for arbitrary beam and flow configurations, *IEEE Trans. on Biomed. Eng.*, vol.35, 740-746 (1988).
4. S.A. Jones, Fundamental sources of error and spectral broadening in Doppler ultrasound signals, *Critical reviews in Biomedical Engineering*, Vol.21, N.5, 399:483 (1993).
5. Evans D.H., Mc Dicken W.N., Skidmore R., Woodcock J.P., *Doppler Ultrasound, Physics, Instrumentation, and Clinical Application*, John Wiley and Sons, pp. 74-77, 1989.

AIR ULTRASONIC TRANSDUCERS COMBINING HIGH SENSITIVITY, LARGE BANDWIDTH AND WIDE BEAMWIDTH

René Breeuwer,[1] and Jan-Willem Hofstee[2]

[1]TNO Institute of Applied Physics
Delft, The Netherlands
[2]Wageningen Agricultural University
Wageningen, The Netherlands

INTRODUCTION

An ultrasonic Doppler instrument was developed to measure the 3D velocity vector of small particles (diameter 1.5 - 3 mm) traveling in air at high speeds (up to 70 m/s) (Hofstee, 1996). The transducer configuration for this instrument consists of a sensor head with one convex, circular diameter piezoelectric transmitting transducer and three piezoelectric receiving transducers with a flat circular aperture. In passing it, a particle is illuminated by a diverging, monochromatic ultrasonic beam radiated by the transmitting transducer. The Doppler shifted ultrasonic signal backscattered by the particle is recorded simultaneously by all three receiving transducers. The velocity vector is computed from the Doppler shifts of the received signals while the particle diameter is estimated from the amplitude of the signals.

SPECIFICATIONS

The large speed range and the small size of the particles place severe demands on the ultrasonic transducers in the system. The transmit frequency is determined by a compromise between the particle target strength, which increases towards higher frequencies and the sound absorption in air which causes the signal-to-noise ratio to decrease rapidly with frequency. Although not very critical, the optimum lies in the range of 200 - 450 kHz. In this range, the particle diameter is 1.1 - 3.9 wavelengths and the attenuation in air is approximately 10 - 40 dB/m.

In order to cover an adequate measuring volume, both the transmitting and receiving transducers should have a minimum beam width of ±30°. Thus, a large flat aperture is not acceptable.

For the transmitting transducer, the main other requirement is high output. To obtain adequate acoustic power at the required beamwidth, an 18 mm diameter convex element was selected. A narrow bandwidth is permissible and a suitable type of transducer was obtained by special order from a commercial source. It has a center frequency of 200 kHz and achieves a beamwidth of $\approx 35°$.

For the receiving transducers, a large-aperture convex element is not suitable. Particle velocities of 0-70 m/s require a considerable frequency bandwidth (>29 %), somewhat conflicting with the high sensitivity (round trip insertion loss < 50 dB) necessitated by the small signal strength of the scatter signals. Initial efforts to obtain suitable transducers from various commercial sources failed, so eventually these transducers were developed as part of the project. The attention was limited to piezoelectric transducers because the micro-mechanical facilities required to produce high frequency capacitive transducers were not available.

FIGURE-OF-MERIT

The design and construction of efficient wideband piezoelectric air ultrasonic transducers is difficult because of the extreme ratio of the acoustic impedances of air and standard piezoelectric materials. To assess the suitability of various technologies for the present application, available piezoelectric transducers were classified according to relative bandwidth and sensitivity.

These parameters were measured by exciting the transducer under test with a tone-burst from a voltage source, causing it to emit an acoustic signal. A reflector placed in the acoustic path (generally in the nearfield) directs essentially all acoustic energy back to the transducer.

Finally, the transducer response is recorded by a high input impedance amplifier. The ratio of the voltage amplitudes of the excitation and the received signal, the two-way voltage insertion loss, is a measure of the transducer efficiency at that frequency. The center frequency of the exciting toneburst is stepped to determine the efficiency as a function of frequency.

The maximum efficiency as well as the -6 dB relative bandwidth are noted. Together, these figures constitute a 2D figure-of-merit (FOM) for the transducer technology which is independent of the transducer center frequency, diameter, focusing etc. Since, in transducer design, bandwidth to some degree may be traded for efficiency, it gives a good impression of the suitability of a certain type of transducer for a specific application. It may also be readily calculated by standard transducer modeling software.

The reflector used should conform to the shape of the wavefront of the beam: flat in the nearfield, concave in the farfield or for focused transducers. Alternatively, two identical flat transducers placed in each other's nearfield can be used in transmission, thus omitting electrical loading problems by a transmit/receive switch. In all cases, of course, corrections may be made for absorption in the acoustic path or for finite source or input impedances of the electronics.

TRANSDUCER TECHNOLOGY

In Figure 1, the results of a number of measurements of conventional transducers are plotted (round dots), as well as some results of new developments from literature (triangles and hourglass markers). The code letters indicate manufacturers, the code numbers different transducer models. The area enclosed by the heavy lines indicates the specification envelope for the target transducers.

The conventional designs tested cover the range of 200-1000 kHz center frequency and diameters of 10-25 mm. All are assumed to use a single matching layer, usually an RTV rubber or similar material. The trade-off between bandwidth and sensitivity is illustrated by comparing (for instance) S1/S2 and the two H7 samples. The connecting line between these two models can be thought to roughly represent the state of the art for single matching layer piezoelectric

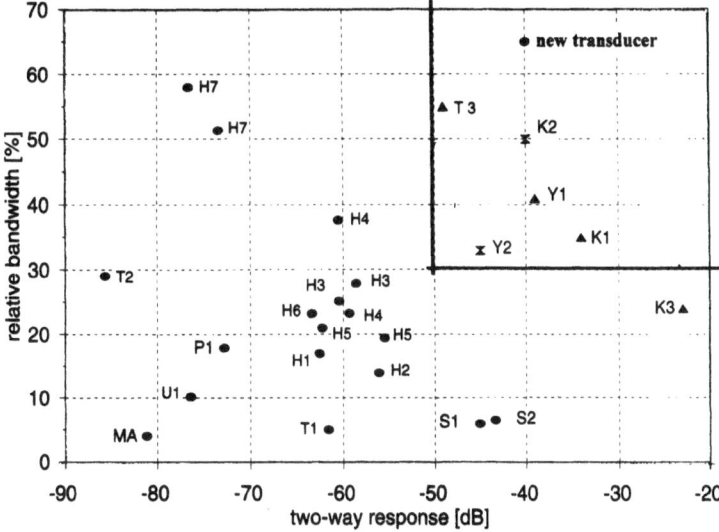

Figure 1: 2D Figure-of-merit of air ultrasonic transducers

air transducers, commercially available at the time of the investigation (1994).

All transducers falling significantly to the left of this line perform rather poorer than necessary. For the present purpose, however, none of the transducers tested fall inside the specification area. The results from literature (Yano, 1987; Khuri-Yakub, 1988) , as indicated by the triangle and hourglass markers, do represent a significant improvement with respect to currently produced commercial transducers, and fall within the performance envelope. Using dual matching layers, they illustrate the importance of the quality of matching the acoustic impedance Z of the piezoelement ($\approx$33 MRayl) to that of air ($\approx$400 Rayl) for efficient wideband operation.

From these results, it was concluded that no single matching layer transducer would yield satisfactory performance but that a transducer constructed with dual matching layers would have this potential. The first (crucial) layer consists of silicone rubber, filled with hollow glass microbubbles, giving an acoustic impedance Z$\approx$0.3 MRayl. For conventional PZT piezoceramics and using this material for the first layer, the second layer requires Z$\approx$5.8 MRayl. This was readily achieved with a filled epoxy.

PIEZO ELEMENT

For the piezomaterial, Ferroperm PZT26 was selected. Considering the position range of the particles and the resulting wave field expected from backscattering, the optimum shape for a receiving transducer aperture is a small disk, with a diameter of 8-10 mm.

The transmit frequency imposed by the transmit transducer already selected demanded a minimum usable frequency range of $\approx$210-280 kHz, corresponding to an element thickness of 4-5 mm for an air-backed design.

Unfortunately, this aspect ratio(diameter/height) of $\approx$9:5 is a very unfavorable: as the aspect ratio is decreased from an infinite diameter plate, the fundamental radial mode frequency increases to approach the fundamental thickness mode frequency. At aspect ratios between $\approx$0.25 and 4, the two modes are strongly coupled (Brissaud, 1987), causing the effective coupling factor of each mode to decrease.

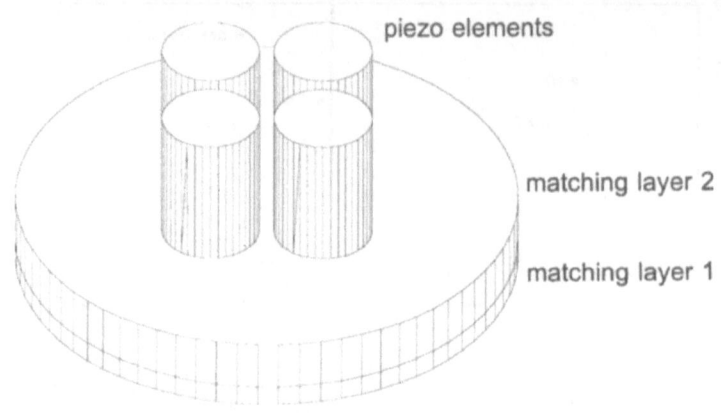

Figure 2: Basic transducer construction

Thus, a transducer built with a single 9 mm diameter disk of the required thickness would perform very poorly. Various ways of cutting the ceramic radially or in a star shape were tested but none performed satisfactorily.

The problem was eventually solved by using 4 separate piezocylinders of 3.7 mm diameter, mounted in a square (see Figure 2) on an oversized second matching layer. The resulting construction could be termed an ceramic-air composite or a small array.

Electrically, the elements were connected in parallel and the transducer was mounted in an aluminium housing of 20 mm outside diameter. To optimize the dynamic range, a special low-noise pre-amplifier incorporating a bandpass filter was integrated with each transducer.

PERFORMANCE

The measured insertion loss of the completed transducers is shown in Figure 3. This figure also shows the result of Piezocad model calculations. It should be noted that at the chosen aspect ratio of 3.7:5 or ≈0.74 the coupling between the radial and thickness modes is still rather strong.

Therefore, the measured insertion loss cannot be expected to agree with the results of 1-D model calculations. Nevertheless, for the lower frequency part of the response spectrum, the Piezocad results fall in the same order of magnitude, with some response fluctuations in the calculations caused by inadequate representation of the transducer damping in the matching layers and the coupling between the modes.

However, the main difference is that in the measured results the bandwidth is considerably larger than in the calculations. This shows the remaining mode coupling used to advantage, similar to stagger-tuning in electrical band-pass filters.

Neglecting the frequency trend in the insertion loss, the usable bandwidth of the transducer is 190 - 360 kHz, a bandwidth of ≈65 %, and its two-way insertion loss is ≈40 dB. These figures are quite satisfactory.

The main disadvantage of the transducer is a large sensitivity to low-frequency vibrations, such as caused by particle impact on the housing. Although electrical filtering may alleviate many problems this causes, an optimization for specifically this aspect would be desirable.

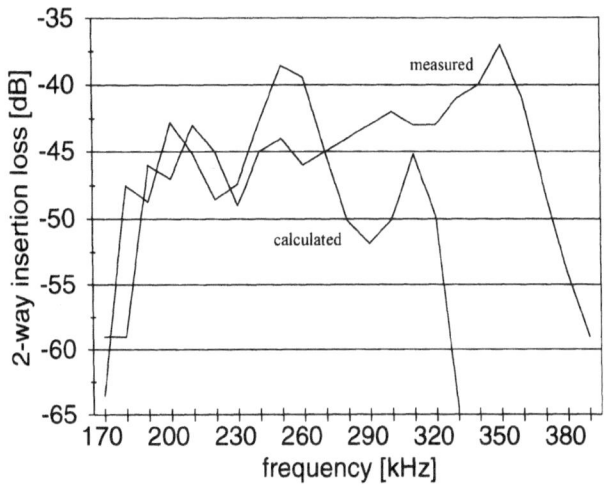

Figure 3: Transducer response

REFERENCES

Hofstee, J.W. and Breeuwer, R., 1996, Evaluation of an alternative method for spread pattern determination of fertilizer spreaders, *AgEng'96*, Paper 96A-019.

Yano, T., Tone, M., and Fukumoto, A., 1987, Range finding and surface characterization using high-frequency air transducers, *IEEE Trans.-UFFC 34 (2) Mar. 1987*:232.

Khuri-Yakub, B.T., 1988, A new design for air transducers, *IEEE 1988 Ultrasonics Symp.*:503.

Brissaud, M., L. Eyraud, and Kleimann, H., 1987, Three dimensional analysis of piezoceramic vibrational modes, *Ultrasonics Int'l '87*:483.

A 3D DOPPLER SCANNING SYSTEM FOR QUANTITATIVE FLOW MEASUREMENTS

M. Calzolai[1], L. Capineri[1], A. Fort[1], L. Masotti[1],
S. Rocchi[1], M. Scabia[1], A. Bertini[2]

[1] Dipartimento Ingegneria Elettronica, Università di Firenze,
50139 Firenze, Italy
[2] Esaote Biomedica S.p.A., 50127, Firenze, Italy

INTRODUCTION

The aim of this work is the development of a three-dimensional (3D) PW Doppler system capable of measuring the velocity vector field (module and direction) without correction for the probe-to-flow angle. The applications of this instrument are in the field medical imaging of blood flow, and for industrial applications. The 3D PW Doppler method was studied theoretically by the authors[1] and it can be used for volumetric scans of flow and quantitative estimation of the velocity vector field in the region of interest. In medical applications where the determination of the probe-to-flow angle is often inaccurate, this technique allows to eliminate the uncertainty due to the angle correction.

In this work the characteristics of a 3D Doppler system based on four annular array transducer are presented and discussed. Finally a 2D Doppler system based on a linear transducer array with electronic scanning and focusing capable to estimate the projection of the velocity vector on the transducer scanning plane is presented.

BASIC PRINCIPLE OF 3D DOPPLER SYSTEM OPERATION

The velocity vector associated to an elementary volume of the investigated flow can be measured by considering that it can be uniquely decomposed along three independent directions. The design of an ultrasonic system capable to measure the Doppler shift along three directions defined by the unity vectors i_1, i_2, i_3 provides a basic solution for the practical implementation of the method. In general, assuming i_0 as the direction of the transmitting element acoustic axis and i_1, i_2, i_3 the directions of the receiving elements, the velocity components (v_x, v_y, v_z) of the unknown velocity $\underline{v}$ are provided by the following relationships:

$$\begin{cases} v_x = \dfrac{c}{f_t} \dfrac{f_{m2} - f_{m3}}{\sqrt{3}\, \sin(\beta)} \\[3mm] v_y = -\dfrac{c}{f_t} \dfrac{f_{m2} + f_{m3} - 2f_{m1}}{3 \sin(\beta)} \\[3mm] v_z = \dfrac{c}{f_t} \dfrac{f_{m1} + f_{m2} + f_{m3}}{3(1 + \cos(\beta))} \end{cases} \qquad (1)$$

where: c is sound velocity, f_{m1}, f_{m2}, f_{m3} are the Doppler shift measured by the three receivers, while β is the angle between the transmitter and each receiver axis, f_t is the transmitted burst central frequency.

This technique can be applied also in two-dimensions (2D PW Doppler) by using a linear array, by considering only the projection of $\underline{v}$ along two directions $\underline{i_1}$, $\underline{i_2}$ along the linear probe scanning plane. In this case the system will provide only the velocity components on the plane containing $\underline{i_1}$, $\underline{i_2}$.

VALIDATION OF THE 3D DOPPLER TECHNIQUE IN VITRO

A prototype system was built to validate the 3D PW Doppler technique with in vitro measurements on a thread phantom[2]. The probe was designed with four confocal transducers: one central transmitter and three lateral receivers. The transducers are focused annular arrays made of piezo-composite material, with central frequency of 2.5 MHz , and -3 dB bandwidth of 1 MHz. The focus of the system was set at a distance of 90 mm in water from the central transmitter.

Several measurements were made with the thread phantom, at various angles and velocities. The overall uncertainty of the velocity module, obtained by varying the probe-to-thread angle in the range from 45° to 75° and with 20 ms acquisition windows, was better than 6%. This is an interesting result if we compare it to the velocity uncertainty in clinical Doppler tests, using monodimensional Doppler systems with angle correction, that can be up to 20-30 %.

One of the main features of the developed prototype system, is the pulse mode operation, which is important for real time scanning along the probe axial direction. Assuming the focus length larger than the equivalent Doppler pulse duration, the multigate method can be applied for a fine flow sampling along the axial direction.

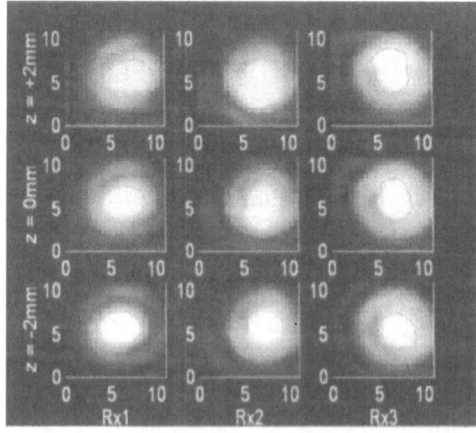

Figure 1 Variation of sensitivity of the probe with three receivers Rx_{1-3} evaluated on focus (z=0mm) and out of focus (z=2 mm above focus, z=-2 mm below focus). Contour level at -3dB. Image size 10x10 mm. Scanning step xy 0.5 mm.

The beam of the four transducers probe was characterized by scanning a pin-head along *xy* planes, perpendicular to the probe acoustic axis *z* at different depths, with *xy* step sizes of 0.5 mm. The central plane was placed on the probe focus corresponding to *z*=0 in Fig. 1. The two planes were placed 2 mm above and below the focus respectively. In Fig. 1 the focal spot size can be estimated for the three receivers by the contour level at -3 dB. It can be observed that the focal spot remains almost of the same shape and size (about 4.5 mm of diameter) also when the probe to target distance is out of focus.

Taking into account these results, the feasibility of the multigate technique applied to out-of-focus measurements was verified. In Fig. 2A and 2B, two scans of the thread phantom related to *xy* planes perpendicular to the probe axis *z* are compared. The angle of the thread phantom with respect to these planes is 55°, the velocity 106 cm/s, the scanning step *(xy)* 1 mm. The two planes are scanned with an analog gate length of 3 mm, and the Doppler frequency estimates are obtained with acquisition windows of 20 ms. In Fig 2A the scan of the plane with axial distance equal to the probe focus (90 mm) is shown, in Fig. 2B the same scan with an axial distance of 93 mm (3 above the probe focus) is presented. In this case the gate position was delayed accordingly.

The 3D velocity map visualization is carried out with a dedicated software[1] which represents the direction of the velocity with an unity vector and the module according to the standard red and blue scale used in color Doppler systems. Both images provide an accurate estimation of the thread velocity (direction and module), hence the feasibility of the multigate technique applied to out of focus measurements is demonstrated. However some differences can be observed between the two velocity maps: the map in Fig. 2B show a larger equivalent thread cross-section because at +3 mm out of focus the probe spot size is enlarged.

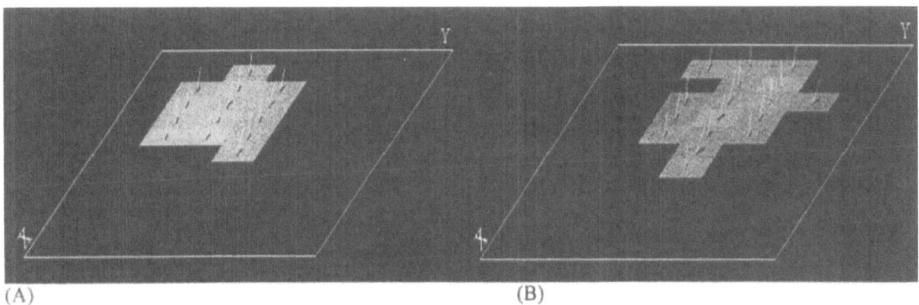

(A) (B)

Figure 2 Comparison of in and out of focus 3D Doppler measurements on the thread phantom: velocity 106 cm/s, probe to velocity angle 55°, acquisition window 20 ms, image pixel size 1x1 mm, probe focus depth 90 mm, gate length 3mm. (A) Thread axial distance 90 mm (In focus), (B) Thread axial distance 93 mm (Out focus)

REAL TIME SCANNING SYSTEM FOR 2D PW DOPPLER MEASUREMENTS

Starting from the in vitro results achieved with the prototype analog 3D PW Doppler system, an implementation for in-vivo scan was developed. The design of a real time 3D Doppler technique requires 2D ultrasonic array transducers with suitable electronic beam-forming circuits. Ideally large matrix of transducers could be employed but they are not yet commercially available. However for this technique they are not essential because 1½ D array of transducers can be used (for example two linear arrays mounted in a T configuration). In this work the design of an electronic scanning system based on a linear array for 2D PW Doppler measurements is presented. In this case the "in vivo" technique was validated by considering the projection of the velocity vector on the scanning plane of the linear array probe.

Scanning method for 2D Doppler measurements with a linear array.

With a linear array provided with an electronic scanning and focusing system, it is possible to perform real time Doppler measurements along two direction over a region of interest. The basic array configuration for the linear scanning scan is shown in Fig 3A and 3B. The central transmitter Tx has a steering angle $\theta_s = 0°$ while the two lateral receivers Rx_1 and Rx_2 have steering angles respectively equal to $+12°$ and $-12°$ and distance d from Tx. The steering angle of the two receivers was selected as the maximum available angle in the considered electronic beam-former. The lateral scan is performed by varying both the Tx and the receivers' position along the array by electronic switching (Fig. 3A). The axial scan is performed by adjusting only the Rx distances from the central transmitter while the steering angle is kept constant. In this way the system focus depth D is changed accordingly to the scanning geometry.

Unfortunately this technique for axial scanning provide only a coarse sampling along the axial direction. Therefore in the designed system the axial scanning is carried out by combining two techniques:

1. The axial scan with array element switching to change the focus depth (Fig 3B)
2. The out of focus scan within an elementary volume (Fig 2)

While the first technique performs coarse axial sampling, the second one performs a fine sampling with the multigate method within an elementary volume. The thickness of the scanned region of interest (ROI) is determined by the characteristics of theacoustic lens placed in front of the array transducer that sets the focus dimensions.

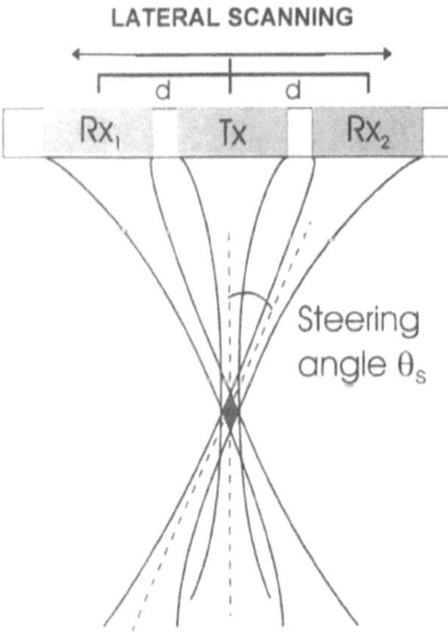

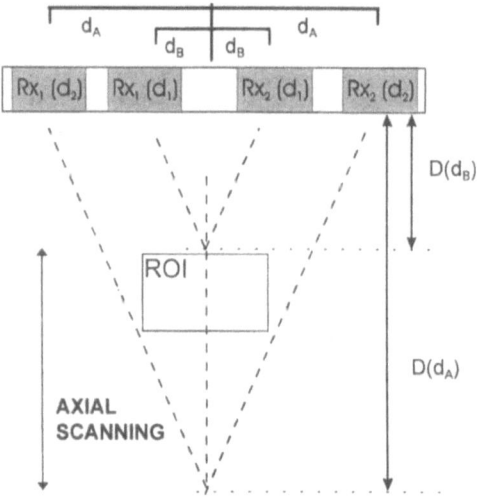

Figure 3A. Lateral scanning method: the active elements for the Rx_1, Rx_2 and Tx are moved along the array by electronic switching. The distance d of Rx_{1-2} from the Tx and steering angle θ_s are fixed.

Figure 3B. Axial scanning method: the focus depth D is varied according to the distance d between Tx and Rx_{1-2}. The steering angle θ_s is set to 12°. ROI is a region of interest.

Scanning system characteristics for 2D PW Doppler measurements and experimental setup.

The prototype system was designed by using a test-bench echographic equipment AU3 Esaote Biomedica S.p.A., Florence, Italy. The basic element is a linear array probe with 128 elements, pitch 0.359 mm and central frequency 5 MHz. The focus size at -6 dB along a plane perpendicular to the transducer beam axis is 1.5 x 2 mm. The Tx and Rx_{1-2} consist of 32 elements each, whose positions can be shifted along the array by electronic switching. Assuming a receivers steering angle of +12° and -12° the maximum focus depth is 81 mm, defining also the maximum triangular region that can be investigated. However a smaller rectangular region of interest (see Fig. 3B) can be obtained by programming the lateral and axial scanning system parameters with a dedicated electronic interface to the echographic equipment. The extension of this area is chosen according to the desired resolution (voxel number per velocity map) and frame rate for a real-time operation mode. The block diagram of the designed electronic system is reported in Fig.4.

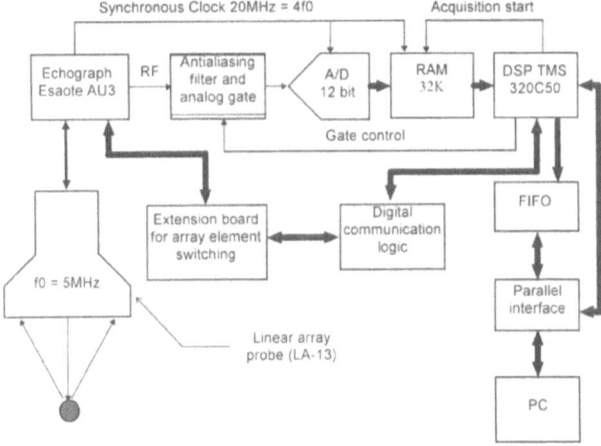

Figure 4. Block diagram of the 2D digital Doppler system

The interface was designed to obtain a complete digital 2D Doppler system and the multigate technique, the phase and quadrature signals are obtained by digital techniques. The radiofrequency signal (RF) is sampled synchronously with the transmitted burst at 4 f_0, f_0 being the central frequency of the transmitted burst, with 12 bit resolution. The numerical calculation of the phase and quadrature Doppler signals is performed by a DSP processor. The samples are then transferred via a high speed parallel port to an external PC, for subsequent off-line processing. The mean Doppler frequencies are extracted by estimating the Power Spectral density with a periodogram algorithm.

EXPERIMENTAL RESULTS

The proposed system was tested with preliminary measurements 'in vitro' on the thread phantom. The linear array probe was first aligned in water on the same plane of the thread consequently the unknown component of the velocity perpendicular to the array scan plane was eliminated. The thread phantom velocity was 74.5 cm/s (± 1%) and the mean Doppler frequency was evaluated with a PRF equal 11.1 kHz, 20 ms acquisition window length, by averaging 20 measurements on a single voxel. The angle between the thread velocity and the probe central axis was set equal to four different values. The estimated velocity module and directions were shown in Table 1.

Table 1. Experimental results with single voxel 2D Doppler in vitro measurements obtained on the thread phantom

Thread phantom angle Θ	Estimated velocity module v	Velocity module relative uncertainty	Estimated angle Θ	Angle Θ absolute uncertainty
50°	72.1cm/s	4.3 %	52.1°	3.4°
55°	74.1 cm/s	4.0 %	55.9°	2.2°
60°	74.1 cm/s	4.2 %	60.9°	2.0°
65°	74.4 cm/s	3.9 %	65.2°	1.6°

The uncertainty on the estimated velocity module was lower than 4.3%, while the absolute uncertainty on the estimated angle was 3-4°. The results obtained with the presented 2D system are even better than those provided by the four discrete transducers' probe. We can explain this result considering that the unknown velocity components are only two in this case having aligned the thread on the linear array plane and so the error sources in the reconstruction formula (1) are reduced accordingly.

CONCLUSIONS

The work presented the characterization of a 3D Doppler technique with in vitro measurements and mechanical volumetric scan and the feasibility of a real-time version of a 2D Doppler system based on a linear array for in vivo applications. The accuracy provided by the developed 2D system are comparable with that found with the 3D system.

ACKNOWLEDGMENTS

The authors wish to acknowledge Paola Lucetti and Andrea Pasqualis for their precious contribution to this work.

REFERENCES

1. G. Bruni, M. Calzolai, L. Capineri, A. Fort, L. Masotti, S. Rocchi, M. Scabia, Measurement and imaging of a velocity vector field based on a three transducers doppler system, *Acoustical Imaging, Vol. 22, Edited by P. Tortoli and L. Masotti*, Plenum Press, New York, 431:437, (1996)

2. M. Calzolai, L. Capineri, A. Fort, L. Masotti, S. Rocchi, M. Scabia, Analysis of factors influencing the accuracy of a 3D PW Doppler technique: simulations and experimental results, *1996 IEEE International Ultrasonics Symposium Proceedings, San Antonio, Texas, USA*, Nov.3-6, (1996), to be printed

SHEAR RATE DEPENDENCE OF NORMAL, HYPO-, AND HYPER-AGGREGATING ERYTHROCYTES STUDIED WITH POWER DOPPLER ULTRASOUND

Guy Cloutier and Zhao Qin

Laboratory of Biomedical Engineering
Institut de recherches cliniques de Montréal
Montreal, Quebec, Canada, H2W 1R7

INTRODUCTION

Sigel *et al.* [1] were the first to demonstrate that the echogenicity of normal human blood at 10 MHz was shear rate dependent. In that study performed in a flow model, the blood echogenicity increased by reducing the shear rate until a maximum was reached at stasis. The increase of the erythrocyte aggregate size at low shear rates was postulated as a mechanism to explain these results. Using a 7.5 MHz pulse-echo ultrasound system, Yuan and Shung [2] showed that the echogenicity of porcine whole blood increased as the shear rate was reduced, whereas that of bovine whole blood and porcine red blood cells (RBCs) suspended in saline were shear rate independent. The absence of rouleau formation for bovine blood and red cell suspensions explained these results. Shehada *et al.* [3] studied porcine whole blood echogenicity at very low shear rates. They showed that shearing around 0.5 s^{-1} provided the maximum backscattered power at 7 MHz because of the higher aggregation level attributed to the increased cell-cell interactions. In a study by our group [4], the Doppler power at 10 MHz backscattered by porcine whole blood showed three specific regions: 1) a rapid reduction of the power between 1 and 5 s^{-1} due to the disruption of large three dimensional aggregates, 2) a transition zone between 5 and 10 s^{-1} related to the dissociation of large rouleaux, and 3) a region above 10 s^{-1} with little variations of the power associated with the separation of small rouleaux. The objective of the present study was to determine the shear rate dependence of normal and pathological levels of RBC aggregation with power Doppler ultrasound.

METHODS

A first series of experiments was performed with porcine whole blood using both the pulse-wave Doppler and the power Doppler imaging techniques. In the second series of measurements, horse blood models characterized by different levels of erythrocyte aggregation were studied with pulse-wave Doppler ultrasound.

Blood sample preparations

An easy way to simulate different aggregation levels is by replacing different proportions of the total volume of plasma with an isotonic saline solution, as described by Weng *et al.* [5]. The following procedure was performed before each experiment with horse blood anticoagulated with EDTA. The plasma was separated from the red and white cells by sedimentation. Several blood samples of 1.5 ml were prepared by replacing 17 to 67% of the total volume of plasma with an isotonic NaCl solution. Using a laser light erythroaggregameter (Regulest, Florange, France), the primary aggregation time (tA), in seconds, was measured for each blood sample at room temperature by adjusting the hematocrit to 40%. According to tA, a dilution level was chosen to obtain the aggregation kinetics desired. Once the correct dilution level was obtained, 1.5 liter of blood was reconstituted using this dilution level and circulated into the flow model for at least half an hour before beginning the experiment. All experiments were performed at 40% hematocrit, as measured by microcentrifugation. For anticoagulated pig blood, no plasma dilution was performed and the hematocrit was also adjusted to 40%.

Pulse-wave Doppler measurements

The steady flow loop model and the methodology used for this portion of the study have previously been described in [4]. Briefly, this model was composed of a peristaltic pump, a Kynar tube with an inside diameter of 12.7 mm, a bottom reservoir of 2 liters, and a top reservoir used to minimize the oscillations produced by the pump. A valve was used to control the flow rate and a cannulating type flow probe was inserted into the flow tubing to measure the flow rate with an electromagnetic blood flowmeter (Carolina Medical Electronics, Cliniflow II, model FM701D). A magnetic stirrer was used to continuously mix the blood in the bottom reservoir. The peristaltic pump circulated blood from the bottom reservoir to the top reservoir.

The 10 MHz Doppler probe was positioned at an angle of 45° with respect to the Kynar tube axis, which was fixed vertically to eliminate the effect of blood sedimentation. The distance between the tip of the probe and the recording sites was 1.9 cm. The pulse-repetition frequency (PRF) was 19.5 kHz, the cut-off frequency of the high-pass wall filter was set at 3 Hz, and the size of the sample volume was 3.7 mm^3 at -3 dB. In order to maintain a constant ultrasound attenuation along the ultrasound path and to allow acoustic coupling, the Doppler probe was immersed in a small tank filled with blood withdrawn from the flow model. The nonfocused Doppler transducer was aligned with the center of the tube and a micrometer was used to radially move the Doppler probe by steps of 0.5 mm. A total of 25 different measurements was performed across the tube. For each position of the sample volume, the Doppler mean velocity and the Doppler backscattered power were computed. No measurement was performed very close to the wall because of the finite dimension of the sample volume. Experiments were performed at flow rates varying between 100 and 1250 ml/min, approximately.

In order to determine the mean shear rate within the Doppler sample volume, the velocity profile across the tube was first fitted to the following power law model:

$$v(r) = v_{max}[1-(r/R)^n]$$
(1)

where v_{max} is the maximum centerline Doppler mean velocity, r is the distance from the center of the tube, R is the radius of the tube, and n is the power law exponent. The 25 velocity measurements and the zero velocity values corresponding to the position of the wall were used to fit the model of Eq. 1. In a second step, a shear rate profile $\gamma(r)$ was obtained by calculating the derivative of $v(r)$:

$$\gamma(r) = nv_{max}r^{(n-1)}/R^n.$$
(2)

The shear rate within the Doppler sample volume $\gamma(r)_{sv}$ was estimated by weighting the shear rate $\gamma(r)$ with a theoretical function describing the radial ultrasonic beam power pattern in the far-field of the transducer. More details about the computation of $\gamma(r)_{sv}$ can be found in Cloutier et al. [4]. For all flow rates tested, the Doppler power was expressed as a function of the shear rate within the Doppler sample volume γ_{sv}.

Power Doppler imaging

A horizontal flow loop model was used in the second part of the study to obtain cross-sectional power Doppler images of moving blood. As described before, the flow was gravity driven from a top to a bottom reservoir. Ultrasound measurements were performed over a Plexiglas box containing tissue mimic composed of 88% water, 9% glycerol, and 3% high-strength agar gel (A-9799, Sigma Chemicals). The 7.9 mm lumen diameter was created by pouring the molten tissue mimic around a rod and then removing the rod after the tissue mimic had set. Blood was circulated in the model at flow rates varying between 62 and 1250 ml/min.

An ATL Ultramark 9 HDI ultrasound system with a 38 mm aperture high resolution linear array probe (L7-4) was used to produce cross-sectional power Doppler images at 4 MHz. Water allowed acoustic coupling between the transducer and the agar gel. The angle between the probe and the axis of the lumen was 50°, the PRF was 600 Hz, and the wall filter was 25 Hz. The transmitted power was constant for all measurements (83 mW/cm², spatial peak time average intensity derated, SPTAd), the Doppler gain was 50%, the image persistence was zero, the image sensitivity was 16, the image filtering was zero (D0), and the system was operating in high resolution mode (HRES). All other settings of the Ultramark 9 were kept constant throughout the study.

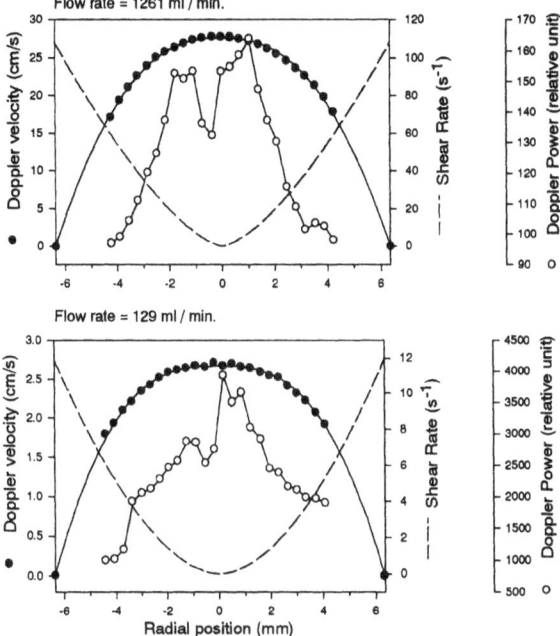

Fig. 1: Doppler mean velocity (black circles), Doppler power (hollow circles), velocity model $v(r)$ (full line), and shear rate model $\gamma(r)$ (dashed line) obtained by circulating porcine whole blood (tA = 1.87 s) at 1261 and 129 ml/min.

A 3D imaging system (Life Imaging Systems, London, Canada) was used for image acquisition, averaging, and display. This system, equipped with a motor-driven translation assembly, was operated in static mode to allow sequential frame grabbing of cross-sectional power Doppler images of the vessel lumen. A linear gray scale power map was used. One hundred consecutive images were digitized and averaged for each flow rate. After manual segmentation of the lumen, the mean and the standard deviation of the gray levels were computed to assess the echogenicity of whole blood as a function of the flow rate.

RESULTS

Shear rate and flow rate dependencies of porcine whole blood

In some experiments, we occasionally noted a reduction of the power backscattered by porcine whole blood at the center of the tube. This power drop, known as the "black hole" phenomenon [3], seemed independent of the flow rate. In Fig. 1, this phenomenon was more pronounced at a flow rate of 1261 ml/min. The shear rate dependence of the blood sample used for Fig. 1 is presented in Fig. 2. An exponentially decaying relationship was found. The power was maximum at low shear rates and reached a plateau around 40 s^{-1}. Using power Doppler imaging, the flow rate dependence of porcine whole blood was also studied. No reduction of the backscattered power was noted at the center of the tube for that experiment, as seen in Fig. 3. Fig. 4 shows the mean gray level as a function of the flow rate for that blood sample. An exponentially decaying relationship similar to that of Fig. 2 was also observed. The mean gray levels dropped as the flow rate was increased and reached a plateau around 500 ml/min.

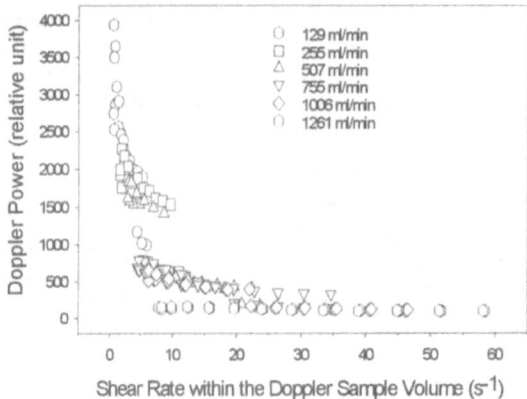

Fig. 2: Doppler power as a function of the mean shear rate within the Doppler sample volume for an experiment performed with porcine whole blood (tA = 1.87 s). The legend gives the flow rate for each measurement.

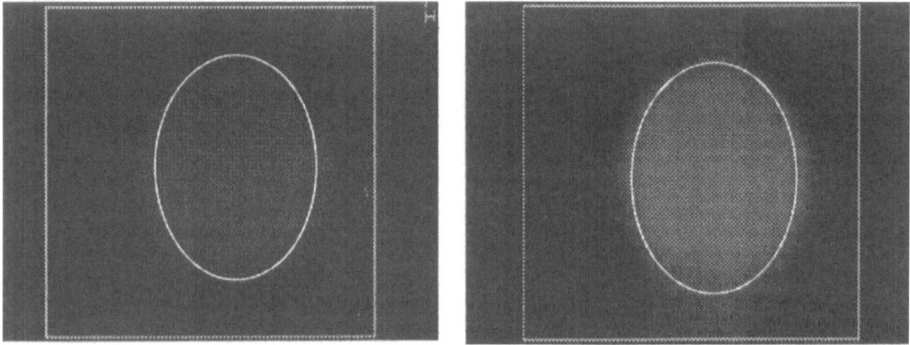

Fig. 3: Cross-sectional representation at 50^0 of the power distribution within the tube for a porcine whole blood sample with tA = 2.38 s. The left panel corresponds to a flow rate of 1250 ml/min whereas the right panel was obtained at a flow rate of 62 ml/min.

Shear rate dependence of horse blood models

Nine experiments performed with tA varying between 1.37 (hyper-aggregating RBCs) and 4.54 s (hypo-aggregating RBCs) were performed. It was observed that 1) for a given tA, the Doppler backscattered power was much higher at 100 ml/min than at 1250 ml/min; 2) for a given flow rate, the Doppler power was higher when tA was low; and 3) the Doppler power was generally low near the wall, maximum between the wall and the center of the tube, and decreased at the center of the tube. The "black hole" phenomenon was observed for almost all experiments with horse blood models. The reduction of the power at the center of the tube was more important for hyper-aggregating RBCs. Fig. 5 shows examples of the shear rate dependence of horse blood models for two experiments with tA = 1.77 and 4.54 s. The power at low shear rates was much stronger for tA = 1.77 s. The plateau of the backscattered power was reached around 30 s^{-1} for tA = 1.77 s, and around 5

s^{-1} for tA = 4.54 s. The large variance of the results for tA = 1.77 s is attributed to the presence of the "black hole", as explained in [6].

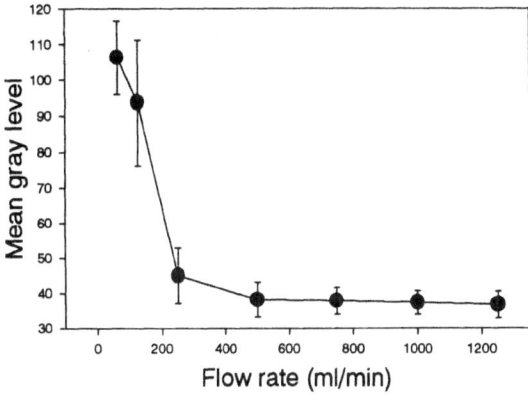

Fig. 4: Mean gray level corresponding to the Doppler backscattered power as a function of the flow rate for an experiment performed with porcine whole blood (tA = 2.38 s). The error bars represent the standard deviation of the gray level across the tube.

DISCUSSION

In the present study, a reduction of the backscattered power at the center of the tube ("black hole") was observed for both porcine and horse blood experiments. In a previous study by our group [4], no "black hole" was noticed with porcine whole blood, whereas Shehada et al. [3] observed it on B-mode images at much lower flow rates than those reported in the current study. The presence of the "black hole" with horse blood has never been reported. It is believed that the structural organization of rouleaux within the tube may contribute to the formation of the "black hole" [6]. The formation of rouleaux instead of clusters for some porcine whole blood experiments may explain the differences between the current study and the results reported in [4].

Direct microscopic observations of normal human blood at shear rates from 5.8 to 230 s^{-1} were performed using a cone-plate viscometer for which the sample cup was remodeled into transparent Plexiglas [7]. At the lowest shear rate tested, clusters of RBCs and no typical rouleaux were seen at physiological hematocrits. By increasing the shear rate, the clusters were disrupted into rouleaux until individual RBCs appeared at shear rates higher than 46 s^{-1}. By repeating similar measurements on patients suffering from acute myocardial infarction [8], the aggregates were larger, more side-to-side attachments leading to clumps were seen, and the aggregates were more resistant to shear. Shear rates up to 230 to 460 s^{-1} were needed to completely disrupt rouleaux.

According to Fig. 5, the formation of large aggregates at low shear rates for hyper-aggregating RBCs resulted in larger backscattered power than that observed for hypo-aggregating RBCs. This observation is consistent with the results of Schmid-Schönbein *et al.* [7,8]. For hyper-aggregating porcine RBCs, the shear rate above which no significant power reduction occurred was around 40 s^{-1} (Fig. 2). For hyper and hypo-aggregating horse RBCs (Fig. 5), the corresponding shear rates were around 30 and 5 s^{-1}, respectively. Since rouleaux were observed at shear rates up to 460 s^{-1} for pathological hyper-aggregating RBCs [8], this suggests that the ultrasound method at 10 MHz may not be sensitive to the formation and disruption of small aggregates of a few RBCs. The use of higher ultrasound frequencies may circumvent this limitation [9].

Acknowledgment: This work was supported by a research scholarship from the Fonds de la Recherche en Santé du Québec, and by grants from the Medical Research Council of Canada (#MA-12491), the Whitaker Foundation, USA, and the Heart and Stroke Foundation of Quebec. The authors gratefully acknowledge Dr. Helen Routh from Advanced Technology Laboratories for giving us access to an ATL Ultramark 9 system, and Mr. Louis Allard and Dr. Louis-Gilles Durand for reviewing the manuscript.

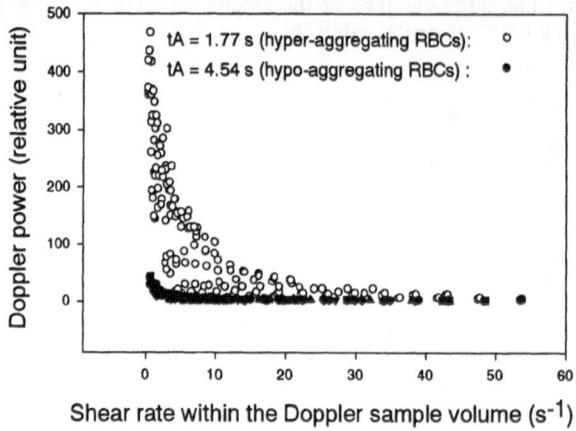

Shear rate within the Doppler sample volume (s^{-1})

Fig. 5: Doppler power as a function of the mean shear rate within the Doppler sample volume for two experiments performed with horse blood models (tA = 1.77 and 4.54 s). The flow rate for those measurements ranged between 100 and 1250 ml/min. Since the characteristics of the transducer were modified, the absolute value of the backscattered power cannot be compared with that of Fig. 2.

REFERENCES

[1] B. Sigel, J. Machi, J. C. Beitler, J. R. Justin, and J. C. U. Coelho, "Variable ultrasound echogenicity in flowing blood", *Science*, vol. 218, pp. 1321-1323, 1982.

[2] Y. W. Yuan and K. K. Shung, "Ultrasonic backscatter from flowing whole blood: I. Dependence on shear rate and hematocrit", *J Acoust Soc Am*, vol. 84(1), pp. 52-58, 1988.

[3] R. E. N. Shehada, R. S. C. Cobbold, and L. Y. L. Mo, "Aggregation effects in whole blood: Influence of time and shear rate measured using ultrasound", *Biorheology*, vol. 31(1), pp. 115-135, 1994.

[4] G. Cloutier, Z. Qin, L. G. Durand, and B. G. Teh, "Power Doppler ultrasound evaluation of the shear rate and shear stress dependences of red blood cell aggregation", *IEEE Trans Biomed Eng*, vol. 43(5), pp. 441-450, 1996.

[5] X. D. Weng, G. Cloutier, P. Pibarot, and L. G. Durand, "Comparison and simulation of different levels of erythrocyte aggregation with pig, horse, sheep, calf, and normal human blood", *Biorheology*, vol. 33(4,5), pp. 365-377, 1996.

[6] Z. Qin, Characterization of red blood cell aggregation dynamics with Doppler ultrasound, M.Sc. dissertation, University of Montreal, pp. 1-149, 1997.

[7] H. Schmid-Schönbein, P. Gaehtgens, and H. Hirsch, "On the shear rate dependence of red cell aggregation in vitro", *J Clin Invest*, vol. 47, pp. 1447-1454, 1968.

[8] J. Goldstone, H. Schmid-Schönbein, and R. Wells, "The rheology of red blood cell aggregates", *Microvascular Res*, vol. 2, pp. 273-286, 1970.

[9] M. S. Van Der Heiden, M. G. M. De Kroon, N. Bom, and C. Borst, "Ultrasound backscatter at 30 MHz from human blood: influence of rouleau size affected by blood modification and shear rate", *Ultrasound Med Biol*, vol. 21(6), pp. 817-826, 1995.

DOPPLER AND CINEMATIC PHASED ARRAY IMAGING VIA DEPTH-FOCUSING

Mehrdad Soumekh

Department of Electrical & Computer Engineering, 201 Bell Hall
State University of New York at Buffalo
Amherst, New York 14260

INTRODUCTION

One of the areas of interest in diagnostic medicine echo imaging is to obtain quantitative as well as qualitative information regarding the condition of the cardio-vascular system using ultrasonic sources. There are two well-known ultrasonic imaging methods for viewing *moving* targets in echocardiography: Doppler-ultrasound, and cinematic-ultrasound. The conventional Doppler-ultrasound systems illuminate the target region repetitively with the focused beam of an aperture [1]-[6]. Motion information is deduced via analyzing the Doppler shift and spreading of the resultant echoed signals, or some form of range correlation processing of the echoes. These motion estimation methods are based on the approximation that the radiated beam is focused on a moving target with a constant velocity distribution in the spatial domain. However, these approximation-based focusing methods are not accurate. As a result, the echoed signal may contain echoes from the the surrounding stationary medium and other moving targets (which may have different velocities) as well as the desired moving target. In the conventional cinematic-ultrasound systems, it is assumed that the heart is frozen (motionless) during data acquisition for a single sweep of the heart [7]-[10]. As a result, the user is forced to utilize larger sample spacing in the beam-steer domain which reduces the signal-to-clutter power ratio.

This paper presents the system model and inversion for an imaging system which utilizes a single phased array to obtain velocity or cinematic (as well as spatial) information on moving targets in an imaging scene. In the system model, we do not approximate the beam-steered radiation pattern of the phased array to be focused in a given angular direction. Moreover, the system model provides a mathematical framework to represent the motion of a moving target in the beam-steering domain which is identified as the *slow-time* domain; this removes the requirement of a motionless target during a single sweep. The inversion provides a reconstruction of the moving targets in the spatial and velocity domains (Doppler imaging), or the spatial and time domains (cinematic imaging). It is shown that a randomized beam steering strategy in the slow-time domain can improve the resolution in the velocity domain. The imaging problem is also formulated for a phased array system which utilizes depth-focusing to improve the target to clutter power ratio.

DOPPLER PHASED ARRAY IMAGING

Consider the two-dimensional spatial domain (x, y). A linear array is located on the line $x = 0$; the array is centered at the origin and its length is equal to $2L$. (While we consider a linear array, however, a similar steps could be used to formulate the problem for a circular array which has *less restrictive* element spacing [12, pp. 343-4].) In the transmit mode, the array's elements are identified by the spatial coordinates $(0, u)$, $u \in [-L, L]$. In the receive mode, the array's elements are identified by the spatial coordinates $(0, v)$, $v \in [-L, L]$.

The transmitted signal is a large bandwidth pulsed function which is denoted with $p(t)$. We call the variable t the *fast-time* domain. The Fourier transform of the transmitted signal with respect to the fast-time is $P(\omega)$ which has a bandpass support region $\omega \in [\omega_c - \omega_0, \omega_c + \omega_0]$ in the fast-time frequency domain. At the beam steer angle θ, the element located at $(0, u)$ transmits the following signal $p(t - \frac{u \sin \theta}{c})$, where c is the wave propagation speed; the target region is irradiated with the integral (sum) of all these pulses in the u domain. In the receive-mode of the beam steer angle θ, the incoming echo at the element $(0, v)$ is delayed by $\frac{v \sin \theta}{c}$; the measured signal is the integral (sum) of all these delayed echoes in the v domain. While we use the term beam steering to identify this data collection strategy, however, we do not assume that the phased array's radiation pattern is focused in the angular direction θ.

Consider a moving target with velocity vector (a, b) in the spatial domain. We identify this target's coordinates (motion trajectory) in the spatial domain via $(x - a\tau, y - b\tau)$, where τ is a variable called the *slow-time* domain, and (x, y) are constants. The distance of this moving target from the transmitting element at $(0, u)$ is $\sqrt{(x - a\tau)^2 + (y - b\tau - u)^2}$ which is a function of the slow-time τ. The radiation experienced by this target due to the illumination by the transmitter at $(0, u)$ at the steer angle θ is $p[t - \frac{u \sin \theta}{c} - \frac{\sqrt{(x-a\tau)^2+(y-b\tau-u)^2}}{c}]$. At the slow-time τ, the distance of the moving target from the receiving element at $(0, v)$ is $\sqrt{(x - a\tau)^2 + (y - b\tau - v)^2}$. Thus, the contribution of this target in the signal in the channel of the receiver $(0, v)$ due to the illumination from the transmitter $(0, u)$ is

$$
p\Big[t - \frac{u \sin \theta}{c} - \frac{\sqrt{(x - a\tau)^2 + (y - b\tau - u)^2}}{c}
$$
$$
- \frac{\sqrt{(x - a\tau)^2 + (y - b\tau - v)^2}}{c} - \frac{v \sin \theta}{c}\Big]
\tag{1}
$$

In practice, the two time variables are the same, i.e., $\tau = t$ (or a shifted version). Moreover, the steer angle θ is also a (sampled) function of time. For instance, in the conventional beam steering, the functional relationship between continuous θ and time τ is linear $\theta = \Omega_0 \tau$, where Ω_0 is a known constant. Provided that the target speed is much smaller than the propagation speed (i.e., $\sqrt{a^2 + b^2} \ll c$), the target's motion effect in a single transmitted pulse for a fixed steer angle is negligible. For instance, when $p(t)$ is a 1 μ-sec rectangular pulse which is amplitude modulating a 3 MHz ultrasonic carrier, the propagation speed is $c = 1.55 \times 10^5$ cm/sec (sound propagation speed in water), and the target's speed is $\sqrt{a^2 + b^2} = 100$ cm/sec, the resultant phase error is 1.44 degrees. Thus, we can assume that the slow-time τ is fixed (the target is stationary) during the illumination at a fixed beam steer angle θ. However, the speed of beam steering could be comparable to the target's speed, i.e., the slow-time τ and the beam steer angle θ are dependent variables. We identify this dependence via the functional form $\theta(\tau)$. The signal in (1) may now be expressed as

$$s_{xy}(\tau, t, u, v) \equiv p\Big[t - \frac{u \sin \theta(\tau)}{c} - \frac{\sqrt{(x - a\tau)^2 + (y - b\tau - u)^2}}{c} - \frac{\sqrt{(x - a\tau)^2 + (y - b\tau - v)^2}}{c} - \frac{v \sin \theta(\tau)}{c}\Big] \tag{2}$$

The contribution of the target in the measured signal is the integral (sum) of the above signal in the transmit-receive (u, v) domains; i.e.,

$$s_{xy}(\tau, t) = \int_u \int_v p\Big[t - \frac{u \sin \theta(\tau)}{c} - \frac{\sqrt{(x - a\tau)^2 + (y - b\tau - u)^2}}{c} - \frac{\sqrt{(x - a\tau)^2 + (y - b\tau - v)^2}}{c} - \frac{v \sin \theta(\tau)}{c}\Big]\, du\, dv \tag{3}$$

Reference [13] oultines an inversion for Doppler phased array system model in (3). The inversion provides a reconstruction of the moving targets in the spatial and velocity domains. It is also shown that a randomized beam steering strategy in the slow-time domain can improve the resolution in the velocity domain. The imaging problem is also formulated for a phased array system which spotlights a target area with its transmitted beam to improve the target to clutter power ratio, and obtains beam-steered data in the receive mode for high-resolution imaging. Matlab programs for these methods can be found at http://www.acsu.buffalo.edu/~msoum.

CINEMATIC PHASED ARRAY IMAGING

One of the challenges in ultrasonic echocardiography is to develop a user-friendly image presentation and image analysis scheme such that the expensive and expert knowledge of the cardiologist is efficiently utilized. While Doppler phased array imaging is a powerful diagnostic tool, however, the type and amount of information (four-dimensional) which it provides is difficult to comprehend and visualize. The other approach for ultrasonic echocardiography is to dynamically display the ultrasonic images of a complete heart cycle to provide a cinematic representation of the heart function (cine mode) [7]-[10]. Cine-ultrasonic images provide a simple cinematic visualization of the target under study which might be sufficient for making a preliminary diagnosis in a doctor's office. However, the cine-ultrasonic images contain similar detailed information as Doppler-ultrasonic images which could be retrieved via more complicated video image processing methods if one desired.

The array signal processing foundation for cinematic phased array imaging system is the same as the one which was outlined in the previous section for Doppler-ultrasonic echocardiography. The main difference is the manner the moving target (heart) is modeled. The target model for spatio-velocity imaging is based on associating a constant velocity to each reflector in the imaging scene resulting in a four-dimensional target function $f(x, y, a, b)$; this model is appropriate to identify the motion (flow) gradient of a moving target such as the blood through arteries. For cine-ultrasonic echocardiography, we model the target via a spatial domain function which varies with the slow-time, i.e., $f(x, y, \tau)$; this is a suitable model for visualizing the evolution of a deformable moving target such as the heart. Both of the above target models have merits in medical diagnosis with ultrasonic echocardiography.

To develop a signal model for phased array echoed data in cine-ultrasonic echocardiography, we consider the phased array system of the previous section. The target function is now identified with a slow-time varying signal $f(x, y, \tau)$. The differential reflector located at (x, y) at the slow-time τ is $f(x, y, \tau)\,dx dy$. The distance of this differential reflector from the transmitting element at $(0, u)$ is $\sqrt{x^2 + (y - u)^2}$ which is invariant of the slow-time τ. The radiation experienced by this target due to the illumination by the transmitter at $(0, u)$ at the steer angle θ is $p\left[t - \frac{u \sin \theta}{c} - \frac{\sqrt{x^2 + (y - u)^2}}{c}\right]$. The distance of this reflector from the receiving element at $(0, v)$ is $\sqrt{x^2 + (y - v)^2}$. Thus, the contribution of this differential reflector in the signal in the channel of the receiver $(0, v)$ due to the illumination from the transmitter $(0, u)$ is

$$s_{xy}(\tau, t, u, v) = f(x, y, \tau)\,dx dy\; p\Big[t - \frac{u \sin \theta(\tau)}{c} - \frac{\sqrt{x^2 + (y - u)^2}}{c}$$
$$- \frac{\sqrt{x^2 + (y - v)^2}}{c} - \frac{v \sin \theta(\tau)}{c}\Big] \tag{4}$$

The contribution of the differential reflector in the measured signal is the integral (sum) of the above signal in the transmit-receive (u, v) domains; i.e.,

$$s_{xy}(\tau, t) = f(x, y, \tau)\,dx dy \int_u \int_v p\Big[t - \frac{u \sin \theta(\tau)}{c} - \frac{\sqrt{x^2 + (y - u)^2}}{c}$$
$$- \frac{\sqrt{x^2 + (y - v)^2}}{c} - \frac{v \sin \theta(\tau)}{c}\Big]\,du\,dv \tag{5}$$

The total mesaured echoed signal is the sum of the echoed signals from all the differential reflectors (see (5)) in the imaging scene

$$s(\tau, t) \equiv \int_x \int_y f(x, y, \tau)\,dx dy \int_u \int_v p\Big[t - \frac{u \sin \theta(\tau)}{c} - \frac{\sqrt{x^2 + (y - u)^2}}{c}$$
$$- \frac{\sqrt{x^2 + (y - v)^2}}{c} - \frac{v \sin \theta(\tau)}{c}\Big]\,du\,dv \tag{6}$$

The system model can be used to develop an inversion for reconstructing $f(x, y, \tau$ from the mcaourements of $s(\tau, t)$.

DEPTH-FOCUSING

As we mentioned earlier, one could utilize a procedure which we refer to as area-spotlighting in transmit-mode to improve signal-to-clutter power ratio. It turns out that one could combine beam-steering and spatial focusing in both transmit and receive mode to improve signal-to-clutter power ratio while collecting the array information base for high-resolution imaging. For this purpose, at the slow-time τ, the transmitting element at $(0, u)$ emits the following signal (a constant delay $-\frac{\sqrt{R^2 + [|Y(\tau)| + L]^2}}{c}$ should be added to ensure causality; this is not shown for notational simplicity):

$$p\Big[t + \frac{\sqrt{R^2 + [Y(\tau) - u]^2}}{c}\Big]$$

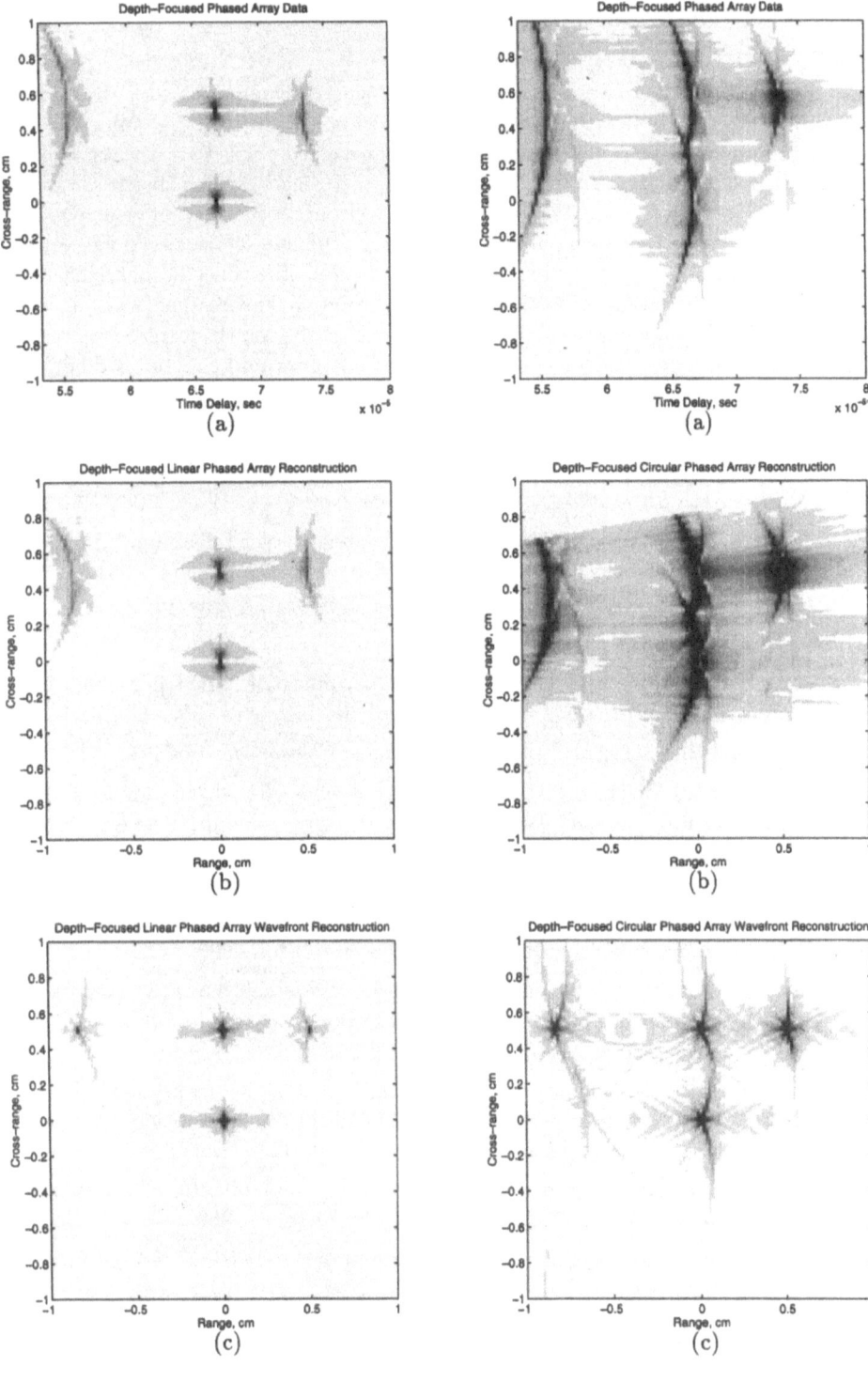

Figure 1   Figure 2

where R is the range of the focused area, and $Y(\tau)$ is the cross-range position of the focused point at the slow-time τ; this corresponds to a *linear* focal contour. A similar delay, i.e., $\frac{\sqrt{R^2+[Y(\tau)-v]^2}}{c}$, is used in the receive-mode.

Figures 1 and 2 show two scenarios for depth-focused beam phased array imaging. In Figure 1, the radiation pattern is focused on the line $R = 5$ cm (range value $R = 0$ in the figure) with a linear array. Figure 1a is the collected phased array data; figure 1b is the target reconstruction by a conventional processing (warping the cross-range as a function of the range); and figure 1c is the wavefront reconstruction principles in [12]. In Figure 2, the radiation pattern is focused on the line $R = 6$ cm (range value $R = 1$ cm in the figure) with a circular array. Figure 2a is the collected phased array data; figure 2b is the target reconstruction by a conventional processing (warping the cross-range as a function of the range); and figure 2c is the wavefront reconstruction principles in [12]. The wavefront reconstruction algorithm could also be modified to process data obtained from an *arbitrary* focal contour.

REFERENCES

1. D. Christensen, *Ultrasonic Bioinstrumentation*, New York: Wiley, 1988.

2. K. Ferrara and R. Algazi, "A new wideband spread target maximum likelihood estimator for blood velocity estimation," *IEEE Trans. UFFC*, Jan. 1991.

3. P. Fish, "Doppler method," in *Physical Principles of Medical Ultrasonics*, C. Hill, Ed., Wiley, 1986, ch. 11.

4. S. Foster, P. Embree and W. O'Brien, Jr., "Volumetric blood flow via time domain correlation," *IEEE Trans. UFFC*, pp. 164-175, May 1990.

5. L. Hatle and B. Angelson, *Doppler Ultrasound in Cardiology*, 1985.

6. V. Newhouse, K. Dickerson, D. Cathignol and J. Chapelon, "Three-dimensional vector flow estimation using two transducers and spectral width," *IEEE Trans. UFFC*, Jan. 1994.

7. O. Petrovic, G. Elsner, R. Wilensky, S. Swanson, and H. Feigenbaum, "Transthoracic echocardiographic detection of coronary artherosclerosis," *American Journal of Cardiology*, vol. 77, no. 88, pp. 569-574, March 15, 1996.

8. B. Bijnens, G. de Paep, M. Herregods, J. Nuyts, P. Suetens, and F. van de Werf, "An open environment for quantitative analysis of left ventricular function using ultrasonic images," *Medical Informatics*, April-June 1995.

9. C. Wolfe, et al., "Assessment of the results of percutaneous transluminal coronary angioplasty using an integrated ultrasound imaging-angioplasty catheter, *Catheterization & Cardiovascular Diagnosis*, pp. 108-112, June 1994.

10. H. Feigenbaum, "Digital echocardiography in myocardial infarction," *Australian & New Zeland Journal of Medicine*, pp. 521-526, October 1992.

11. M. Soumekh, "Array imaging with beam-steered data," *IEEE Trans. on Image Processing*, July 1992.

12. M. Soumekh, *Fourier Array Imaging*, Englewood Cliffs, NJ: Prentice Hall, 1994.

13. M. Soumekh, "Phased array imaging of moving targets with randomized beam steering and area spotlighting," *IEEE Transactions on Image Processing*, May 1997.

A NEW APPROACH TO OBTAIN NON-DIFFRACTION BEAM WITH NEAR-FIELD
RESOLUTION ON LINEAR AND CONVEX ARRAYS

Z. M. Benenson, N. S. Kulberg, T. T. Kasumov

Scientific Council on Cybernetics
Russian Academy of Sciences
117333 Vavilova str. 40, k. 232, Moscow, Russia

INTRODUCTION

The matter of the article is the new approach to the algorithmic synthesis of the «non-diffraction» beam in forming of two- and three-dimension images. Non-diffraction beam (transmitted or received) is supposed to have directional diagram width about several wavelengths, that does not depend on the distance from the aperture. It is known that, using classical method of the synthetic aperture one can obtain «non-diffraction» beam [1], but this method yields very low SNR, especially for the three-dimension variant. Besides that, realization of the synthetic aperture method requires the complicated mechanical driving system of transmitting element which size must be about several wavelengths.

One more algorithm of the non-diffraction beam formation is described in [2] and [3]. This method also uses small (about several wavelengths) transmitting aperture and uses algorithmic synthesis of the directional diagram. For the pulsed signal this method is inexact, and for the wideband signal (bandwidth about the carrier frequency) is inapplicable. Also, SNR of this method is low.

The approach of the present paper allows to use active continuous aperture of any size and wideband pulsed signal and develops results of [4]. New algorithm of the non-diffraction beam synthesis yields high SNR.

SCANNING SCHEME AND INITIAL EXPRESSIONS

Scanning scheme in the method is based on moving of the plane transmitting and receiving apertures subsequently along x and y axes (Fig. 1). We suppose that the active aperture along y axis is continuous and is moved mechanically. Common aperture along x axis consists of a set of small-sized (several wavelengths) elements. Let us suppose that excitation pulse for the transmitting aperture is a wideband signal $V(t)$ and its spectrum is $V(\omega)$. In all expressions we will use the analytic signal:

$$V_a(t) = \frac{1}{2\pi} \int_0^\infty V(\omega) \exp(i\omega t) d\omega \qquad (1)$$

In the signal processing algorithms the received signal is always undergone to Hilbert transform. Since most expressions of the article will use Fourier transform for X and Y coordinates for different z distance to the aperture, (i. e., for different $t = 2z/c$), then let us write principal dependencies in coordinates (Ω_x, Ω_y) and time t..

Spatial spectrum of the transmitted signal $P_t(\Omega_x, \Omega_y, t)$ at the distance $z = ct/2$ is defined by expression:

$$P_i(\Omega_x, \Omega_y, t) = \int F_{tr}(\Omega_x, \Omega_y, \omega) V(\omega) M_1(\omega, z) \exp\left(i\omega\left(\frac{z}{c} - \frac{t}{2}\right)\right) \exp\left(-i\frac{c}{2\omega}(\Omega_x^2 + \Omega_y^2)z\right) d\omega \quad (2)$$

where $F_{tr}(\Omega_x, \Omega_y, \omega) = \iint U_0(x, y, \omega) \exp\left(-i(\Omega_x x + \Omega_y y)\right) dx\, dy$ is the spatial Fourier spectrum of analytic apodization function $U_0(\cdot)$ on the transmitting aperture; $M_1(\omega, z)$ is attenuation function; c is the sound velocity.

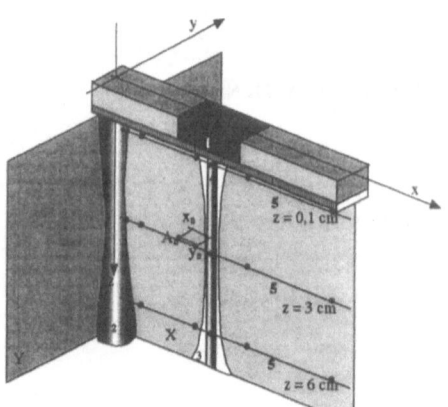

Fig. 1. Scanning scheme. Direction of the electronic scanning is along x axis.
Direction of the mechanical scanning by linear array is along y.
1. Darker elements are active in the current moment. Point A_n coordinates are (x_n, y_n, z).
2. Three-dimension caustic that can be obtained in usual scanning (static focus for transmitting, dynamical focusing for receiving — in X plane, static focus in Y plane);
3. Section of the caustic 2 by X plane (usual scanning);
4. Three-dimension caustic obtained by non-diffraction algorithm;
5. Position of scattering points in digital simulation.

In expression (2) we use the approximation: $\sqrt{(\omega/c)^2 - \Omega_x^2 - \Omega_y^2} \approx \frac{\omega}{c} - \frac{c}{2\omega}(\Omega_x^2 + \Omega_y^2)$ that is valid when $\Omega_x < \omega/3c$, $\Omega_y < \omega/3c$ [5].

Let us suppose that scattering objects $A_n(x_n, y_n, z)$ are static, with complex reflection amplitudes $\gamma_n(z)$. For the spatial Fourier transform of the received signal

$$P_r(\Omega_x, \Omega_y, t) = \sum_n \int d\omega \iiint \gamma_n(z) \exp\left(-i(\Omega_x x + \Omega_y y)\right) P_i(\Omega_x', \Omega_y', t) M_1(\omega, z) \times$$

$$\times F_{re}(\Omega_x - \Omega_x', \Omega_y - \Omega_y', \omega) \exp\left(i\omega\left(\frac{z}{c} - \frac{t}{2}\right)\right) \exp\left(-i\frac{zc}{2\omega}\left((\Omega_x - \Omega_x')^2 + (\Omega_y - \Omega_y')^2\right)\right) d\Omega_x' d\Omega_y' dz \quad (3)$$

where $F_{re}(\cdot)$ is the spectrum of analytic apodization function on the receiving aperture.

Expression (3) is the convolution of (2) with $F_{re}(\cdot)$ multiplied by response function of space. In the spatial coordinates this corresponds to multiplication of transmitting and receiving directional diagrams.

It is evident that function under integral in (3) (let us sign it as $\tilde{S}_r(\omega, \Omega_x, \Omega_y)$) after integration by Ω_x' Ω_y' and z is three-dimension spectrum of received signal $P_r(x, y, t)$.

THEORY OF THE METHOD OF DIFFRACTION-LIMITED (NON-DIFFRACTION) BEAM FORMING

Forming of the non-diffraction beam is based on the choice of the analytic apodization functions of receiving and transmitting apertures, and moving them along x and y coordinates. Apodization function we choose is the product of functions of argument x and y. All expressions analyzed below will be deduced for Gaussian apodization function:

$$u(x, y, \omega) = \exp\left(-x^2\left(\frac{1}{a_{x,v}^2} + i\frac{\omega}{2F_{x,v}}\right)\right) \exp\left(-y^2\left(\frac{1}{a_{y,v}^2} + i\frac{\omega}{2F_{y,v}}\right)\right) \quad (4)$$

where $2a_{x,v}$ $2a_{y,v}$ are corresponding aperture sizes; $F_{x,v}$ and $F_{y,v}$ are focal distances; $v = 1$ for transmitting and $v = 2$ for receiving. With such a choice of the apodization function expression (3) and spectral function $\tilde{S}_r(\omega, \Omega_x, \Omega_y)$ can be represented as product:

$$\tilde{S}_r(\omega, \Omega_x, \Omega_y) = S_{r,x}(\omega, \Omega_x) S_{r,y}(\omega, \Omega_y) V(\omega) \times$$

$$\times \sum_n \int M_1^2(\omega, z) \exp\left(2i\omega\frac{z}{c}\right) \gamma_n(z) \exp\left(-i(\Omega_x x + \Omega_y y)\right) dz \quad (5)$$

Using (5), let us deduce all expressions for the function $S_r(\omega,\Omega)$ (spectrum of directional diagrams), where Ω means any of lateral frequencies Ω_x of Ω_y. For (4) we obtain:

$$S_r(\omega,\Omega) = K\exp\left(-(\alpha_1 - i\zeta_1)\Omega^2\right)\exp\left(\frac{(\alpha - i\zeta_1)^2\Omega^2}{\alpha_1 + \alpha_2 - i(\zeta_1 + \zeta_2)}\right) \tag{6}$$

where K is a scale factor, $\alpha = \dfrac{1}{4\left(\dfrac{1}{a^2} + \dfrac{\omega^2}{4c^2F^2}a^2\right)}$, $\zeta = \dfrac{\dfrac{\omega}{2cz} - \dfrac{\omega}{2cF} - \dfrac{2Fc}{a^4\omega}}{4\left(\dfrac{\omega}{2cz}\left(\dfrac{\omega}{2cF} + \dfrac{2Fc}{a^4\omega}\right)\right)}$ $\tag{6a}$

For computing α_1 and ζ_1 one must substitute $a^{(1)}$ and $F^{(1)}$ instead of a and F; and for computing α_2 and ζ_2 one must substitute $a^{(2)}$ and $F^{(2)}$ with corresponding indexes x and y. It is easy to show that expression (6) can easily be transformed to following form:

$$S_r(\omega,\Omega) = K_1\exp\left(-\left(\tilde{\alpha} - i\tilde{\zeta}\right)\Omega^2\right) \tag{7}$$

where $\tilde{\alpha}$ and $\tilde{\zeta}$ generally are functions of wave number ω/c, aperture sizes $2a^{(1)}$ and $2a^{(2)}$, focal distances F_1 and F_2 and of depth z.

Let us show expressions for $\tilde{\alpha}$ and $\tilde{\zeta}$ in two cases.

I. Transmitting and receiving focal distances, and active apertures are equal.

$$\tilde{\alpha}_1 = \frac{\alpha_1}{2}, \quad \tilde{\zeta}_1 = \frac{\zeta_1}{2}; \quad K_1 = K_I = \frac{\exp(i\theta_1)}{2\sqrt{1/a^4 + \omega^2/4c^2}\sqrt[4]{\alpha_1^2(\alpha_1^2 + \zeta_1^2)}} \tag{8-I}$$

where α_1 and ζ_1 are computed by (6a); and θ_1 is some constant phase.

II. Transmitting aperture is rather greater than wavelength $\lambda = 2\pi c/\omega$, its focus is F. Receiving aperture size is about several λ and is focused at infinity ($F = \infty$).

$$\tilde{\alpha}_{II} = 1\Big/\sqrt{\left[\left(\frac{a\omega}{c}\right)^2\left(\frac{1}{z} - \frac{2}{F}\right)^2 + \frac{16}{a^2}\right]}; \quad \tilde{\zeta}_{II} = \frac{zc}{4\omega} - \left(\frac{2}{F} - \frac{1}{z}\right)\Big/4\left[\left(\frac{\omega}{c}\right)^2\left(\frac{2}{F} - \frac{1}{z}\right)^2 + \frac{16c}{\omega a^4}\right];$$

$$K_1 = K_{II} = \frac{\delta\exp(i\theta_2)\sqrt{\omega/c}}{2\sqrt{2}\left(\sqrt[4]{1/a^4 + \omega^2/4c^2F^2}\right)\sqrt{z - F/2}} \tag{8-II}$$

where δ is a size of single receiving element, and θ_2 is some constant phase.

Expressions (8-II) for $\tilde{\alpha}_{II}$ are valid for $z > 2a$ and $|F| > 2a$. With less z values the influence of limited spatial frequency Ω by wave number ω/c takes place. With less z value $\tilde{\alpha}_{II}$ will increase until about $a_1^2/4$ where $2a_1$ is receiving aperture size.

In the algorithm of focused directional diagram forming for three-dimension imaging there is used value $S_d\left(\omega,\Omega_x,\Omega_y,z_j\right)$ that must be obtained by the formula:

$$S_d\left(\omega,\Omega_x,\Omega_y,z_j\right) = \tilde{S}_r\left(\omega,\Omega_x,\Omega_y\right)\exp\left(-i\tilde{\zeta}_x^2\Omega_x^2 - i\tilde{\zeta}_y^2\Omega_y^2\right) \tag{9}$$

where $z_j = ct_j/2$ is distance to scattering object; $\tilde{\zeta}_x$ and $\tilde{\zeta}_y$ are computed for x and y aperture sizes by formulas (8-I) and (8-II) depending on the construction variants.

One can show, similarly to [4], that signal $S_{I,s}(\omega,x,y,t_j)$, characterized by directional diagram $L(x,y,\omega,z_j) = \exp\left(-x^2/4\tilde{\alpha}_x - y^2/4\tilde{\alpha}_y\right)$, can be determined by the inverse two-dimension Fourier transform of (9) for Ω_x and Ω_y at each value t_j, i. e. receiving time. Here $\tilde{\alpha}_x$ and $\tilde{\alpha}_x$ must be found by formulas (8). Values $2\sqrt{\tilde{\alpha}_x}$ and $2\sqrt{\tilde{\alpha}_x}$ characterize diagram width.

Obvious conclusion from expressions (6) and (8-I) can be made: in the variant (I) direction diagram width is the same for all z, that means non-diffraction beam.

For the variant (II) direction diagram width is reducing with increase of z, when $z > 2a$.

If no apodization is used, formulas (8) will include additional factors that are also functions of ω, Ω_x, Ω_y, z_j, which must be compensated (similarly to [4]).

Making three-dimension Fourier transform for ω, Ω_x, Ω_y for each z_j, one obtain focused non-diffraction signal $S_f(t_j, x, y)$ in spatial and temporal domain. For frequencies Ω_x and Ω_y FFT algorithm can be applied, but for ω it can not be directly applied because function $S_d(\bullet)$ depends on z_j. For this reason let us consider the effective algorithm that takes into account this circumstance.

SIGNAL PROCESSING ALGORITHM IN ORDER TO FORM NON-DIFFRACTION BEAM

Let us consider variant I, when both transmitting and receiving apertures have the same sizes and have the same static focus. In this case $\tilde{\zeta}$ can be represented with satisfactory precision (inaccuracy is about $(c/\omega)^3$)

$$\tilde{\zeta} = \frac{(F-z)c}{4\omega} \tag{12}$$

To compute the target signal $S_f(t_j, x, y)$ let us use the expression:

$$S_f(t_j, x, y) = \iiint \exp\left(-i(\Omega_x x + \Omega_y y)\right) \exp\left(-i\omega t_j\right) \exp\left(-i\tilde{\zeta}(\Omega_x^2 + \Omega_y^2)\right) \tilde{S}_r(\omega, \Omega_x, \Omega_y) d\Omega_x d\Omega_y d\omega \tag{13}$$

Then, let us transform (13), taking into account (12) and substituting $z = ct_j/2$, by entering new variable:

$$\tilde{\omega} = \omega - \frac{c^2}{8\omega}(\Omega_x^2 + \Omega_y^2) \tag{14}$$

As a result we obtain the sub-integral in the expression (13):

$$S_{f,1}(\Omega_x, \Omega_y, t_j) = K_3 \int \exp\left(-i\tilde{\omega}t_j\right) \exp\left(-\frac{iFc}{\psi_1(\tilde{\omega}, \Omega_x, \Omega_y)}\right) \hat{S}_r(\tilde{\omega}, \Omega_x, \Omega_y) \psi_2(\tilde{\omega}, \Omega_x, \Omega_y) d\tilde{\omega} \tag{15}$$

where functions $\psi_1(\bullet)$, $\psi_2(\bullet)$ and $\hat{S}_r(\bullet)$ can be found from the transformation (14). Function $S_{f,1}(\bullet)$ can be computed with FFT algorithm. The result of this must be transformed by FFT algorithm for Ω_x and Ω_y.

Hence, the algorithm for the wideband signal includes three-dimension forward FFT of the received signal for coordinates x, y, t, and, according to (15), inverse three-dimension FFT for the same variables.

For the variant II the algorithm can be applied approximately, because formula (8-II) for $\tilde{\zeta}_{II}$ contains term that depends on $2F/(2z - F)$. Approximately the algorithm can be applied by dividing depth range to short intervals, executing fast Fourier transform for each interval separately.

Another algorithm does not require any Fourier transform for ω. The essence of this method is in executing short convolutions with pre-computed weight functions $W_j(\tau, \Omega_x, \Omega_y)$:

$$S_{f,1}(\Omega_x, \Omega_y, t_j) = \int W_j(\tau, \Omega_x, \Omega_y) P_r(\Omega_x, \Omega_y, t - \tau) d\tau$$

Functions $W_j(\tau, \Omega_x, \Omega_y)$ change with z_j. This function is calculated according to formula (9) for $\tilde{\zeta}$ coefficients. Details of this method were described in [4].

Final choice of the algorithms depends on power of digital signal processors.

SIMULATION, EXPERIMENT AND SOME ESTIMATIONS

Described methods were tested by applying both to simulated signals, and to digital records of signals that were obtained in physical experiment with linear array on the phantom. There were measured both axial and lateral resolution; SNR yielded by described methods was compared to one of the synthetic aperture with the same resolution, element size and signal intensity.

All simulations were done for two-dimension case, because three-dimension direction diagram is a product of two-dimension ones.

Figure 2 presents the direction diagrams of methods I and II, and of synthetic aperture, adding the receiver noise. On the figure there are shown normalized signal intensities. It is obvious that the SNR of method I is better than one of synthetic aperture by about 30 dB. For three-dimension images this ratio increases to 60 dB. Method II yields SNR that differs by 17 dB from the synthetic aperture.

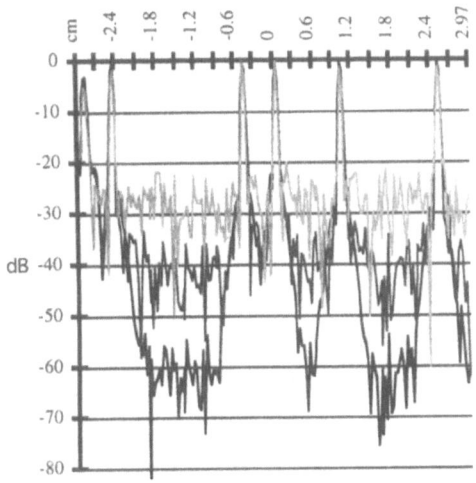

Figure 2. Simulation results for SNR evaluation. Six scattering points were placed at 6 cm from the aperture. First, the graph was built using a synthetic aperture (light-gray line). SNR was 30 dB. For the second graph transmitting was performed by an aperture that was focused at 3 cm. Aperture size was 0.8 cm (24 array elements); signal was received by the single element. After processing by the algorithm SNR has become 40 dB (dark-gray line). In the third case both transmitting and receiving were made by the same aperture (black line). After processing by the algorithm SNR becomes more than 60 dB. In all variants directional diagram width remains the same as for synthetic aperture.

Table 1. SNR expressions for different methods (two-dimension imaging)

Method	SNR	Comments
I	$\dfrac{a^2 \times \omega/c}{2\lvert F - z \rvert}$	$z \neq F$ $(F > 0)$
II	$\dfrac{2(\omega/c)^2 (z - F/2)^2}{z^2 F^2} \delta^2 a^2$	$F < 0$, $\delta \sim \lambda/2 \div 2\lambda$ is aperture element size
Synthetic aperture	$\dfrac{\delta^4 \, \omega/c}{12 \Delta_{lat}^2 \, z}$	Δ_{lat} is lateral array size; $\delta \sim \lambda/2 \div 2\lambda$ is aperture element size

Figure 3 presents the result of the applying of the algorithm I to the signals, received from distance 0.1 cm, with aperture size 0.8 cm. Direction diagram width is the same as for other distances.

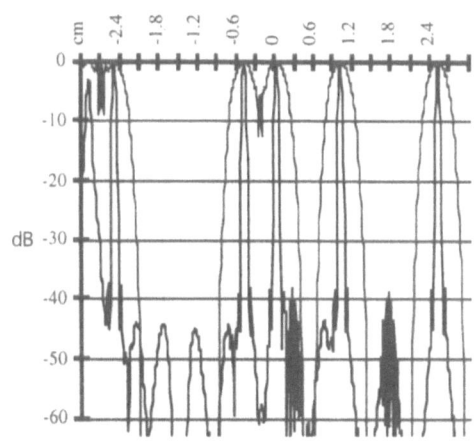

Figure 3. Focusing in the near zone. Six scattering points were placed at 0.1 cm from the aperture. Scanning aperture size is 0.8 cm (24 array elements), with Gauss apodization. Both transmitting and receiving were made with focus at 3 cm. For the received signal (gray line) direction diagram width is the same as the aperture size. After processing diagram width equals to the uniform value that does not depend on distance (compare with diagram at 6 cm on Fig. 2).

In the theory described above, scattering objects were supposed to be stable. For moving objects, synthesized direction diagram changes. These variations are different for axial (v_z) and lateral (v_x and v_y) motion speeds. Axial motion results in Doppler signal phase distortion that yields additional factor to (5), and expressions (7) become the following:

$$S_{r,x} = K_1 \exp\left(-\left(\tilde{\alpha} - i\tilde{\zeta}\right)\left(\Omega_x - \frac{V_z}{V_{s,x}}\frac{\omega}{c}\right)^2\right), \quad S_{r,y} = K_1 \exp\left(-\left(\tilde{\alpha} - i\tilde{\zeta}\right)\left(\Omega_y - \frac{V_z}{V_{s,y}}\frac{\omega}{c}\right)^2\right) \tag{17}$$

where $v_{s,x}, v_{s,y}$ are scanning speeds for corresponding coordinates. Expressions (8) for $\tilde{\alpha}$ and $\tilde{\zeta}$ remain the same. Thus, when $v_z/v_{s,x} < 1$ and $v_z/v_{s,y} < 1$, the diagram width remains the same, but it is shifted by $\delta x = 2\tilde{\zeta}(\omega/c)(v_z/v_{sx})$ and $\delta y = 2\tilde{\zeta}(\omega/c)(v_z/v_{sy})$. For method I this shift is $(v_z/v_{sx}) \times (z - F)/2$. Figure 4 presents simulation results that corroborate these expressions.

For the lateral object motion, expressions (7) become the following: $S_{r,x} = \exp\left(-\left(\tilde{\alpha} - i\tilde{\zeta}\right)\chi_x^2\Omega_x^2\right)$ and $S_{r,y} = \exp\left(-\left(\tilde{\alpha} - i\tilde{\zeta}\right)\chi_y^2\Omega_y^2\right)$, where $\chi_x = 1/(1 - v_x/v_{s,x})$ and $\chi_y = 1/(1 - v_y/v_{s,y})$. This results in diagram widening. For method I the relative diagram widening is by $(1.4(F - z)/F) \times (v/v_s)$ times. It is four times less than for synthetic aperture.

Also algorithm was successfully tested on digital records of physical signals from phantom.

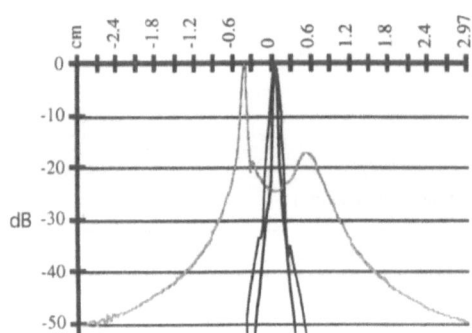

Figure 4. Imaging of moving objects in Y plane for three-dimension case. Black line shows directional diagram for stable point. Dark-gray line shows directional diagram for the point that is moving along y axis. Light-gray line shows directional diagram for the object moving along z axis. Motion speed is about 0.2 of scanning speed.

CONCLUSION

It is shown that it is possible to form algorithmically non-diffraction beam for all depths for any active aperture size and any signal bandwidth.

Results of the present paper are valid for any active aperture sizes and any signal bandwidths. Yielded SNR's are by 60 dB higher than ones for synthetic aperture. It is shown that for objects, moving in axial direction, the directional diagram is shifted; for the lateral motion the widening of diagram takes place. To reduce these defects, there are needed maximal scanning speeds along x and y axes. Supposed approach is flexible enough to increase sufficiently scanning speed for both axes.

Proposed algorithm, with some modifications, easily can be applied to convex arrays.

Principal operations of the algorithm can be reduced to two- and three-dimensional fast Fourier transforms, that allows to perform processing in real-time mode.

ACKNOWLEDGMENTS

This work has been supported by Russian Foundation of Fundamental Research (grant No 96-01-00107)

REFERENCES

1. P. O. Corel and G. S. Kino, A Real-Time Synthetic Aperture Imaging System, Acoustical Imaging, vol. 9, K. Y. Wang ed., Plenum Press, New York, pp. 341-355, 1980.

2. J. Ylitalo et al, Ultrasound Holographic B-Scan Imaging, IEEE Trans. Ultrason., Ferroelectric and Freq-Contr., vol 36, pp. 376-383, 1989.

3. J. Ylitalo and H. Ermet, Ultrasound Synthetic Aperture Imaging: monostatic approach, IEEE Trans. Ultrason., Ferroelectric and Freq-Contr., vol. 41, pp. 333-339, 1994.

4. Z. M. Benenson. N. S. Kulberg, Dynamical Focusing of the both Transmitted and Received Beams via Digital Processing of Pulsed Acoustical Signals on a Single-Element Scanning Aperture. Acoustical Imaging, vol. 22, P. Tortoli and L. Massotti ed., Plenum Press, NY, pp. 531-536, 1996.

5. J. W. Goodman, Introduction to Fourier Optics, Moclaw Hill Book Company, 1968.

6. J. Ylitalo, An Efficient Algorithm for computed ultrasound imaging using curved synthetic aperture, Acoustical Imaging, vol. 22. P. Tortoli, L. Massotti ed., Plenum Press, New York, pp. 525-530, 1996.

STATISTICAL ESTIMATIONS IN THERMOACOUSTICAL INTROSCOPY

V.A.Burov, E.E. Kasatkina

Moscow State University, Faculty of Physics, Department of Acoustics,
Moscow, 119899, Russia

A heat chaotic motion of atoms and molecules is a physical cause of noise radio-frequence and acoustical body radiation. Proportionality of the radiation intensity to absolute temperature allows to estimate inner temperature of the object by means of passive thermolocation methods. Infrared and superhigh radio-frequency thermovision methods are the most developed ones at the present time. However, these methods have insufficient depth of penetration in absorption medium or low resolution. Therefore in the number of cases, they can yield to methods based on registration of thermal acoustical radiation [1].

An algorithm of uncoherent wave tomography for an optimum estimation of inner temperature distribution of heated and absorbing medium is considered in this paper. Data are obtained by an observation of acoustical noise radiation. The solution of the wave radiation inverse problem has been obtained by A.J. Devaney [2] for uncorrelated sources:

$$\Gamma_{ff}(\mathbf{r}_1, \mathbf{r}_2) = I(\mathbf{r}_1)\delta(\mathbf{r}_2 - \mathbf{r}_1), \tag{1}$$

here $I(\mathbf{r}_1)$ is a continuous, nonnegative function called the "intensity profile" of the source and $\delta(\mathbf{r}_2 - \mathbf{r}_1)$ is the three-dimensional Dirac delta function.

This model is acceptable for the solution of problem of biological medium acoustothermography, where spatial coherence length of radiation is less than the linear size of spatial resolution element [3]. Some more general model representation is possible in the case of structural media. Cells of such medium or structural elements of mechanical system performing stochastic oscillations have the set of eigenvectors and corresponding eigenvalues. In the case of the three-dimensional problem the dimensional redundancy of experimental information allows to generalize the solution obtained in the paper [2] to complicated configurations of sources. For example, let suppose that the coherency function of the form $\Gamma_{ff}(\frac{\mathbf{r}_1 + \mathbf{r}_2}{2}, \mathbf{r}_1 - \mathbf{r}_2)$ describing the sources (here the first argument characterizes the slow change of the profile of the function Γ_{ff} which oscillates fast depending on difference argument) belongs to some functional space with basis $\Gamma_i(\mathbf{r}_1 - \mathbf{r}_2)$. For example, this basis may correspond to free oscillations of elastic subsystem or to harmonic basis in a "cube" with linear size determined by maximum radius of correlation. Then the source coherency function may be presented as follows:

$$\Gamma_{ff}(\mathbf{r}_1, \mathbf{r}_2) = \sum_i A_i(\mathbf{r}_1)\Gamma_i(\mathbf{r}_1 - \mathbf{r}_2)A_i^T(\mathbf{r}_2), \tag{2}$$

where $A(r) = I^{1/2}(r)$ and index "T" means a transposition operation. If a character of correlation connections changes slightly along the correlative radius of sources, then the description through variables $r_- = r_1 - r_2$ and $r_+ = (r_1 + r_2)/2$ is convenient. The coherency function of the field of the sources in terms of $\Gamma_i(r_1 - r_2)$ can be expressed as

$$\Gamma_{uu}(y_1, y_2) = \iint_R G(y_1, y_2, r_-, r_+) \sum_i I_i(r_+) \Gamma_i(r_-) dr_+ dr_- . \tag{3}$$

Here R is the region of source localization and Y is the region of receiving. The function $\hat{G}(y_1, y_2, r_-, r_+)$ in eq. (3) is the kernel of integral operator $\hat{G}_o(y_1, r_1) \hat{G}_o^+(r_2, y_2)$, where $G_o(y, r)$ is the Green's function of homogeneous medium and $I_i(r_+)$ is the product $A_i(r_1) A_i^T(r_2)$. Integration of expression (3) over variable r_- leads to equation:

$$\Gamma_{uu}(y_1, y_2) = \int_R \sum_i \Theta_i(y_1, y_2, r_+) I_i(r_+) dr_+ . \tag{4}$$

Function $\int_R G(y_1, y_2, r_-, r_+) \Gamma_i(r_-) dr_- = \Theta_i(y_1, y_2, r_+)$ is coherency of signals received in the points y_1 and y_2, if distributed source with space coherency $\Gamma_i(r_1 - r_2)$ is located in the point r_+.

Estimation of combination of the intensity profiles $I_i(r_+)$ describes spatial-correlative structure of the sources and can be obtained by the maximum likelihood method, i. e. by maximization of the corresponding functional. Logarithm of this functional (the terms which don't take part in the variation are omitted) has the following form:

$$\ln \omega \{U(y)\} \sim -\ln \det K_{uu}(y_1, y_2) - \iint_Y W(y_1, y_2) K_{uu}^{-1}(y_2, y_1) dy_1 dy_2 . \tag{5}$$

Coherency function $K_{uu}(y_1, y_2)$ includes both the coherency function $\Gamma_{uu}(y_1, y_2)$, and the coherency of measurement errors:

$$K_{uu}(y_1, y_2) = I_{uu}(y_1, y_2) + nN(y_1, y_2). \tag{6}$$

Here N is interference coherency matrix and n is interference power.
In eq. (5) $W(y_1, y_2)$ is a coherency matrix of data. With accumulation of M independent measurements:

$$W(y_1, y_2) = \frac{1}{M} \sum_{m=1}^{M} U_m(y_1) U_m^+(y_2). \tag{7}$$

The variation of the functional (5) with regard to unknown distributions of the intensities $I_j(r_+)$ leads to the system of integral relations:

$$\iiiint_Y K_{uu}^{-1}(y_1, y_2) \Theta_j(y_2, y_3, r_+) K_{uu}^{-1}(y_3, y_4) W(y_4, y_1) dy_1 \ldots dy_4 =$$

$$= \iint_Y K_{uu}^{-1}(y_1, y_2) \Theta_j(y_2, y_1, r_+) dy_1 dy_2, \tag{8}$$

$j = \overline{1,J}$, where J is the dimension of expansion basis $\{\Gamma_j(r_-)\}$ in the parametric model (3), which is taken to be finite for the solution of practical problems.

The variation of eq.(5) relative to the unknown interference power n leads to the equation:

$$\int\int\int\int_Y K_{uu}^{-1}(\mathbf{y}_1,\mathbf{y}_2)N(\mathbf{y}_2,\mathbf{y}_3)K_{uu}^{-1}(\mathbf{y}_3,\mathbf{y}_4)W(\mathbf{y}_4,\mathbf{y}_1)d\mathbf{y}_1\ldots d\mathbf{y}_4 =$$
$$= \int\int_Y K_{uu}^{-1}(\mathbf{y}_1,\mathbf{y}_2)N(\mathbf{y}_2,\mathbf{y}_1)d\mathbf{y}_1 d\mathbf{y}_2. \tag{9}$$

The system of equations (8),(9) is nonlinear with respect to $I_i(r_+)$ and n (because these functions are contained in $K_{uu}(\mathbf{y}_1,\mathbf{y}_2)$) and its solution seems to be a very difficult problem. However, the using some simplified assumptions permits to advance in solution of this problem. For instance, the signal/noise ratio is very low for most acoustothermography problems. Because of this, it is possible to replace the coherency matrix K_{uu}^{-1} by $(n_0N_0)^{-1}$, without any significant loss in noise-resistance, where n_0N_0 is a known matrix describing the coherency properties of all object radiation and noise. Such a replacing leads to linearization of the system (8),(9) relative to the estimations $I_i(r_+)$ looked for:

$$\int\int\int\int_Y N_0^{-1}(\mathbf{y}_1,\mathbf{y}_2)\Theta_j(\mathbf{y}_2,\mathbf{y}_3,\mathbf{r}_+)N_0^{-1}(\mathbf{y}_3,\mathbf{y}_4)W(\mathbf{y}_4,\mathbf{y}_1)d\mathbf{y}_1\ldots d\mathbf{y}_4 =$$
$$= \int\int\int\int_Y N_0^{-1}(\mathbf{y}_1,\mathbf{y}_2)\Theta_j(\mathbf{y}_2,\mathbf{y}_3,\mathbf{r}_+)N_0^{-1}(\mathbf{y}_3,\mathbf{y}_4) \times$$
$$\times \left[\int_R \sum_i \Theta_i(\mathbf{y}_4,\mathbf{y}_1,\mathbf{r}'_+)I_i(\mathbf{r}'_+)d\mathbf{r}'_+ + nN(\mathbf{y}_4,\mathbf{y}_1)\right]d\mathbf{y}_1\ldots d\mathbf{y}_4, \tag{10}$$

$$\int\int\int\int_Y N_0^{-1}(\mathbf{y}_1,\mathbf{y}_2)N(\mathbf{y}_2,\mathbf{y}_3)N_0^{-1}(\mathbf{y}_3,\mathbf{y}_4)W(\mathbf{y}_4,\mathbf{y}_1)d\mathbf{y}_1\ldots d\mathbf{y}_4 =$$

$$= \int\int\int\int_Y N_0^{-1}(\mathbf{y}_1,\mathbf{y}_2)N(\mathbf{y}_2,\mathbf{y}_3)N_0^{-1}(\mathbf{y}_3,\mathbf{y}_4)\left[\int_R \sum_i \Theta_i(\mathbf{y}_4,\mathbf{y}_1,\mathbf{r}'_+)I_i(\mathbf{r}'_+)d\mathbf{r}'_+ +\right.$$
$$\left. +nN(\mathbf{y}_4,\mathbf{y}_1)\right]d\mathbf{y}_1\ldots d\mathbf{y}_4 \tag{11}$$

Here the identity transformations of the right parts of the equations (8), (9) were previously performed:

$$\int\int_Y K_{uu}^{-1}(\mathbf{y}_1,\mathbf{y}_2)\Theta_j(\mathbf{y}_2,\mathbf{y}_1,\mathbf{r}_+)d\mathbf{y}_1 d\mathbf{y}_2 \equiv ,$$
$$\equiv \int\int\int\int_Y K_{uu}^{-1}(\mathbf{y}_1,\mathbf{y}_2)\Theta_j(\mathbf{y}_2,\mathbf{y}_3,\mathbf{r}_+)K_{uu}^{-1}(\mathbf{y}_3,\mathbf{y}_4)K_{uu}(\mathbf{y}_4,\mathbf{y}_1)d\mathbf{y}_1\ldots d\mathbf{y}_4,$$

$$\iint_Y K_{uu}^{-1}(\mathbf{y}_1,\mathbf{y}_2)N(\mathbf{y}_2,\mathbf{y}_1)d\mathbf{y}_1 d\mathbf{y}_2 \equiv$$

$$\equiv \iiiint_Y K_{uu}^{-1}(\mathbf{y}_1,\mathbf{y}_2)N(\mathbf{y}_2,\mathbf{y}_3)K_{uu}^{-1}(\mathbf{y}_3,\mathbf{y}_4)K_{uu}(\mathbf{y}_4,\mathbf{y}_1)d\mathbf{y}_1...d\mathbf{y}_4 , \tag{12}$$

and thereafter the products of the form $K_{uu}^{-1}\Theta_j K_{uu}^{-1}$, $K_{uu}^{-1}NK_{uu}^{-1}$ in the left and right parts of eq. (8), (9) were simplified by the same way. In some instances one can suppose that N_0 is a unit matrix. Then the system of the equations (10), (11) is reduced to especially simple form after integration over variables $\mathbf{y}_1...\mathbf{y}_4$:

$$\int_R \sum_i A_{ji}(\mathbf{r}_+,\mathbf{r}'_+)I_i(\mathbf{r}'_+)d\mathbf{r}'_+ + nB_j(\mathbf{r}_+) = C_j(\mathbf{r}_+), \tag{13}$$

$$\int_R \sum_i A'_i(\mathbf{r}'_+)I_i(\mathbf{r}'_+)d\mathbf{r}'_+ + nB' = C'. \tag{14}$$

The situations when the interference n is weak and accuracy of measurements of the observed correlation matrix $W(\mathbf{y}_1,\mathbf{y}_2)$ is high, are also very interesting. In this case the replacing $K_{uu}^{-1} \to W^{-1}$ and $W \leftrightarrow K_{uu}$ is possible. As a result, the symmetrized system of eq. (8),(9) becomes nonlinear relative to the observed coherency matrix, but it becomes linear relative to the unknown values $I_i(\mathbf{r}_+)$:

$$\iiiint_Y W^{-1}(\mathbf{y}_1,\mathbf{y}_2)\Theta_j(\mathbf{y}_2,\mathbf{y}_3,\mathbf{r}_+)W^{-1}(\mathbf{y}_3,\mathbf{y}_4)\left[\int_R \sum_i \Theta_i(\mathbf{y}_4,\mathbf{y}_1,\mathbf{r}'_+)I_i(\mathbf{r}'_+)d\mathbf{r}'_+ + \right.$$

$$\left. + nN(\mathbf{y}_4,\mathbf{y}_1)\right]d\mathbf{y}_1...d\mathbf{y}_4 = \iint_Y W^{-1}(\mathbf{y}_1,\mathbf{y}_2)\Theta_j(\mathbf{y}_2,\mathbf{y}_1,\mathbf{r}_+)d\mathbf{y}_1 d\mathbf{y}_2 , \tag{15}$$

$$\iiiint_Y W^{-1}(\mathbf{y}_1,\mathbf{y}_2)N(\mathbf{y}_2,\mathbf{y}_3)W^{-1}(\mathbf{y}_3,\mathbf{y}_4)\left[\int_R \sum_i \Theta_i(\mathbf{y}_4,\mathbf{y}_1,\mathbf{r}'_+)I_i(\mathbf{r}'_+)d\mathbf{r}'_+ + \right.$$

$$\left. + nN(\mathbf{y}_4,\mathbf{y}_1)\right]d\mathbf{y}_1...d\mathbf{y}_4 = \iint_Y W^{-1}(\mathbf{y}_1,\mathbf{y}_2)N(\mathbf{y}_2,\mathbf{y}_1)d\mathbf{y}_1 d\mathbf{y}_2 . \tag{16}$$

The approximate solution of the system (15), (16) is the known Capon's estimation of source power distribution, if contributions to the sums in left parts of the equations of all the terms except the term looked for are neglected. However, the last estimation does not reproduce all the correlative characteristics of the field observed.

After discretization of the systems of eq. (10), (11) and (15), (16) uniqueness of the solution is not provided, in general case. Nonuniqueness will take place if the number of independent values of the observed matrix $W(\mathbf{y}_1,\mathbf{y}_2)$ is less than the number of resolution elements. If nonuniqueness takes place for problems with high accuracy of measurements (as (15), (16), for example), then optimal quasi-solution with the help of principles of the minimal norm or the maximum entropy can be used. However, for the most thermotomography problems the signal/ noise ratio is so low that "superresolution" is not possible. Then to analyze the picture in detailes is useless.

The wide-band signals require not only space but also time compensation. Then the correlation approach requires very large calculation volume if antenna containing many transducers is used. However, the fast progress of the systems of processing on the base of matrix chips and programmed architecture of parallel conveyors allows to consider in perspective the following scheme of thermotomograph (Fig.1).

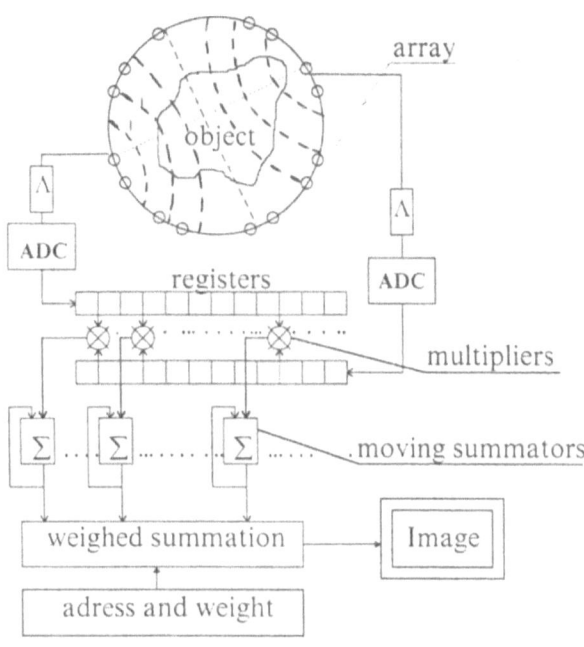

Figure1. Structure of thermotomograph.

Signals (in digital form) from the circular array of transducers are delayed by the shift registers, in which the maximum time of "propagation" is equal to the time of sound propagation in the object. Then each pair of array signals is multiplied and averaged for different time delays. Each of summators accumulates the signals from the set of object pixels lying on hyperbola corresponding to the fixed difference of propagation times. The set of signals from all pairs of receivers, which are placed at the same angular distance gives the values of the coherency function on a family of hyperbolaes displaced on the certain angular step from each other. Set of data for tomography in "hyperbolic beams" will be formed by this manner. Other data set corresponding to other angular distance between receivers gives identical information about the object, but with different noise background. Simultaneous use of all the sets of hyperbolaes can be used for additional accumulation of the information and increase of the algorithm sensitivity. A total volume calculations for reconstruction of the $N \times N$ pixels image by means of the data from K receivers, M temporal shifts and at averaging T time samples is as follows: about K^2MT operations of multiplication and summation for estimation of a coherency matrix and N^2K^2 operations for tomographic reconstruction. Let's assume that $K \approx N \approx M \approx 3 \times 10^2$ and $T \approx 10^7$ then the total number of operations at the stage of "accumulation" makes $10^{13} \div 10^{14}$ multiplications, which are carried out by hardware way with the help of matrix of K^2M multipliers. The operations of time delay and multiplication are simplified to a considerable extent if clipping amplifiers are used, since the shift registers become one-bit, the multiplication is realized by operation "excluding or" and accumulation - by counters with an addressed access. The realization about $10^9 \div 10^{10}$ operations in order to form an image for tens of seconds seems to be a simpler operation. The use of clipping amplifiers reduces the sensitivity of system insignificantly (about 20 %), but requires calibration of average absolute temperature of the image through average temperature of object measured independently (that is relatively easy).

The use of processing systems, built on the base of matrix chips, therefore, allows to produce the processing of the signals from arrays containing about a hundred or more receiving elements. However, the use of signals from the large number of receivers for multiplicative processing is connected with large constructive and schemotechnical difficulties. Therefore the consideration of various simplified variants of correlative algorithm realization is expendient at the given stage. For instance, a possibility of the use of rather small number of transducers with large surfaces for arbitrary spatial-temporal scanning of object is estimated. The receivers can move in space and be phased by means of various time delays. The common architecture of the system is the same, as above considered. An another regarded scheme consists of the system of large linear subarrays the signals from which are multiplied (in pairs) with different time delays. These delays are such that the sources of correlated signals lie on a direct line parallel to bisector of an angle between subarrays. In the elementary case two such subarrays are parallel to each other and are located on the opposite sides of object tested. It is possible to have a parallel shift of this line by changing the time delay. A rotation of these subarrays permits to cover the area researched by such direct lines under different angles. A classical tomographic scheme of final reconstruction of the object takes place, thus. Simplifications of the construction and reduction of calculation volume in these schemes are achieved for account of reduction of sensitivity and resolution or increase of the total time of information accumulation during angular scanning. For instanse, the obtained estimations of potential sensitivity and resolution of the regarded thermotomograph block-schemes have shown that the detection of domains with linear sizes about of 1 mm and temperature contrast about of 0.1 absolute degree within analyzed layer is possible with the help of complete schemes containing the large number of receiving elements. The linear size of resolution element must be larger or it is necessary to increase the averaging time, in order to determine such the temperature contrast by schemes containing small number of transducers with $d^2 / \lambda \geq D$ (where d is linear size of transducer aperture, D is linear size of tested region and λ is average received wavelength).

It should be noted that one of the peculiarities of the systems considered is the fact that in contrast to usual passive detection-location, the increase of resolution capability leads to decreasing the signal/noise ratio. It takes place because the whole researched volume represents the distributed source of signals of the same type, as signal from a chosen local heated volume and is the interference, thus. Moreover, when the acoustothermography problems are considered, it is necessary to take into account that a contribution to the acoustic radiation power from a volume element dV is proportional to local values of temperature $T(\mathbf{r})$ and of absorption coefficient $\alpha(\mathbf{r})$. Because of this fact the quantitative estimation $T(\mathbf{r})$ with absolute accuracy within 0.1 absolute degree requires the knowledge of distribution map of the absorption coefficient $\alpha(\mathbf{r})$ with the same relative accuracy (about 10^{-4}). The independent measurements of the absorption coefficient and phase velocities can be carried out by means of methods of the active acoustical tomography [4] (however, the problem consists in achievement of necessary very high accuracy!). Data obtained as a result of active mode of tomography can be used to make program of image formation more informative and precise, when the degree of medium inhomogeneity essentially distorts a form of the Green's function. The thermoacoustical introscopy in a combination with the active acoustical tomography can give the important diagnostic information (even as the distribution $\alpha(\mathbf{r})T(\mathbf{r})$) in addition to the tomographic image of the acoustical parameters of medium.

REFERENCES

1.   T. Bowen, Acoustic passive temperature sensing, in: *Acoustical Imaging-12*, E. A. Ash and C.R. Hill ed., Plenum Press, New York (1982).
2.   A.J. Devaney, The inverse problem for random sources, *J. Math.Phys.* 20:1687 (1979).
3.   V.I. Mirgorodsky, V.V.Gerasimov, and S.V. Peshin, Three-dimensional ultrasonic imaging of temperature distributions, in: *Acoustical Imaging-22*, P. Tortoli and L. Masotti ed., Plenum Press, New York (1996).
4.   V.A. Burov, M.N. Rychagov, and A.V. Sascovets, Account of multiple scattering in acoustic inverse problems of tomographic type, in: *Acoustical Imaging-20*, Y. Wei and B. Gu, ed., Plenum Press, New York (1993).

EXPERIMENTAL OBSERVATION OF SPECKLE MOTION ARTIFACT WITH ROTATION

S. Dupont[1], M. Bertrand[1,2], T. Hall[3], M. Cyr[1], and F. Kallel[4]

[1]Institut de génie biomédical, École Polytechnique de Montréal
[2]Institut de Cardiologie de Montréal
[3]University of Kansas Medical Center
[4]University of Texas Medical Center

INTRODUCTION

Ultrasound speckles may move in a way that bears no simple relationship to the motion of the corresponding tissue. In some instances, the speckle motion does not replicate the underlying tissue motion and has a strong artifactual component. Kallel et al.[1] proposed an image formation model to explain the motion artifact under tissue rotation. We are now validating this model by imaging a rotating phantom with a linear array. For a rotating tissue, the model predicts an apparent movement composed of the expected rotation plus a strong horizontal translation. The model explains this translation by the non-linear phase characteristics which originate from the curvature of the system point spread function (PSF). In the far field, the translation artifact is proportional to the scan depth and the rotation angle. Using a correlation method to compute the displacement field, we can determine the amplitude of the motion artifact. This paper reports on an experimental validation of the motion artefact model predictions.

EXPERIMENTAL METHOD

A cylindrical phantom made of an agar-based gel with graphite powder for absorption and glass beads for scattering was imaged using a 5 MHz linear array transducer. For each of the five angular position (0 to 4 degrees), one hundred scan planes were taken parallel to the longitudinal axis of the rotating cylinder, thus providing 3-D r.f. speckle data. Ninety-eight cross sectional r.f. images of the cylinder were then assembled. This scanning technique was used in order to acquire r.f. images under wide beam conditions; according to the model, this leads to the largest motion artefact.

Acoustical Imaging. Vol 23
Edited by Lees and Ferrari, Plenum Press, New York, 1997

315

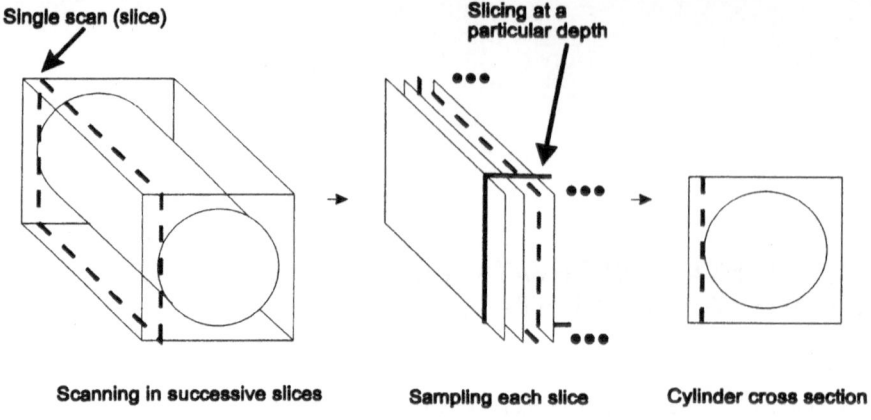

Single scan (slice)

Slicing at a
particular depth

Scanning in successive slices   Sampling each slice   Cylinder cross section

Figure 1: Technique used for volume reconstruction of the phantom.

MODEL PREDICTIONS

In a transverse cut of the cylinder, the speckle motion between successive rotations of one degree should be, as stated earlier, the result of the expected rotation field plus a translation component proportional to the scan depth. To better isolate the artifact we compensate for the movement of the cylinder by appropriately counter-rotating the post-motion image, i.e. the counter-rotation removes the rotational component of the movement and leaves only the motion artifact. Following this, the motion artefact is estimated using a correlation based method.

For a gaussian PSF, the model predicts the lateral artefact t_x as:

$$t_x = a \left(v_0 + \frac{1}{2\sigma_x^2 v_0} \right) \theta \quad \text{with: } a = \frac{v_0}{\dfrac{d}{\sigma_x^4} + \dfrac{v_0^2}{d}} \quad \text{and } v_0 = \frac{f_{tr}}{c} \qquad (1)$$

where 'd' is the radius of curvature of the PSF and in the far field it is equal to the distance between the transducer and the region of interest, 'θ' is the angle of rotation between the two images, 'v_0' the spacial frequency of the PSF in the axial direction, 'f_{tr}' is the transducer frequency and 'c' is the speed of sound in the target. Notice that the lateral artefact is proportional to θ.

A very small axial motion artifact t_y is also predicted:

$$t_y = \left[a \left(\frac{3}{2} v_0 + \frac{1}{4 v_0 \sigma_x^2} \right) - a^2 \left(\frac{v_0^2}{d} + \frac{1}{d\sigma_x^2} \right) \right] \theta^2 \qquad (2)$$

where 'a' is the same as in the lateral artefact expression. In practice the axial motion artifact can be neglected, being small for correlated speckle motion. Indeed for a rotation of one degree, t_y is in the order of a few tens of microns in our region of interest.

316

Equations (1) and (2) can be simplified for a PSF in the far field. For a large 'd' the expression for the lateral artefact can reduced to:

$$t_x = d\theta \qquad (3)$$

while the axial artefact is reduced to:

$$t_y = 1.5d\theta^2. \qquad (4)$$

Equation (3) shows that in the far field conditions the lateral artefact is proportional to the scan depth (and thus to the width of the PSF since the beam is diffracting). Such motion is that of a lateral shear. The equation above suggest that the best conditions under which the artefact can be observed between a pair of images would be that of a large rotation angle. This however is not so, because speckle decorrelation increases with rotation angle, and this up to the point where the two images become so dissimilar that it is no longer possible to perceive a "structured" motion. Hence the requirement for the small angular increment (1 degree) we used.

RESULTS AND DISCUSSION

When animating the sequence of cross-section images from 0 to 4 degrees we clearly see the motion artifact. The impression is that the cylinder does not rotate along its longitudinal axis but rather rolls around a much lower point under the image. A lateral motion vector field added to a rotation field produces such an effect.

From equation (1) we expect the lateral motion artifact to be about 0.2 to 0.4mm per degree depending on the distance between the ROI (Region Of Interest) and the transducer. Because the r.f. line of the images are 0.40mm apart, a subpixel motion estimation algorithm is required to assess the fine translation at different scan depths.

Here, to compute the displacement field between the pre-motion and motion compensated images we used an image block processing cross-correlation technique. To improve the resolution for localizing the correlation peak we used 2-D parabolic interpolation.

Our model validation requires that we first estimate the PSF parameters, i.e. axial and lateral dimensions and radius of curvature. Parameters σ_x and σ_y were obtained from manual measurements of the full-width-half-maximum of the 2-D autocorrelation function of the r.f. image of the phantom. This simple method in effect assumes that the radius of curvature is large. Parameter 'd' of the PSF was approximated as the distance between the ROI and the transducer.

Results of the computed motion artefact averaged for all the slices are shown in Figure 2.

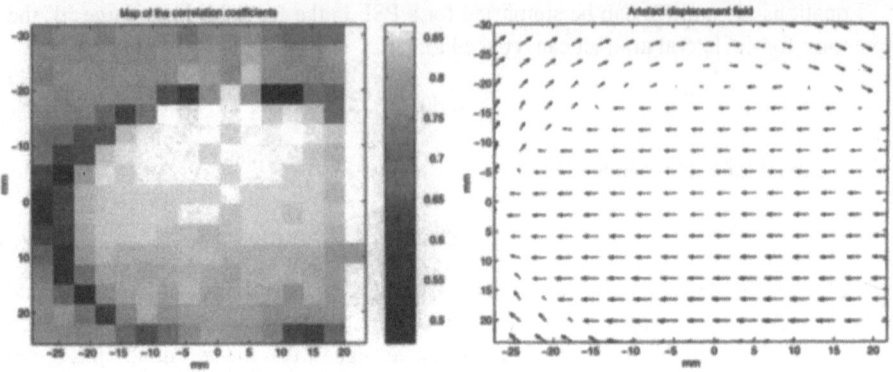

Figure 2: Map of the correlation coefficients (left) and vector field showing the motion artefact (right)

The left-hand part of Figure 2 is a map of the local correlation coefficients. Each pixels represents the maximum correlation between two blocks of pixels, one in the pre-motion image and the other one at the same position in the compensated post-motion image. It provides a data reliability map useful for interpreting the displacement field shown on the right. On the rotation axis in the middle of the cylinder (the cylinder is the circular region with highest correlation) the displacement vectors show a clear leftward motion. This lateral motion is the artefact and varies from 0.3 mm in the upper region to about 0.5 mm in the lower region.

The apparent motion around the cylinder is due to the motion compensation applied on the post-motion image. The fluid in which the cylinder was immersed did not move except at the interface with the cylinder when it was rotated. The low correlation at the cylinder-fluid interface is due to the high local shearing; the cylinder-fluid interface movements cannot be properly measured with the simple correlation technique we used.

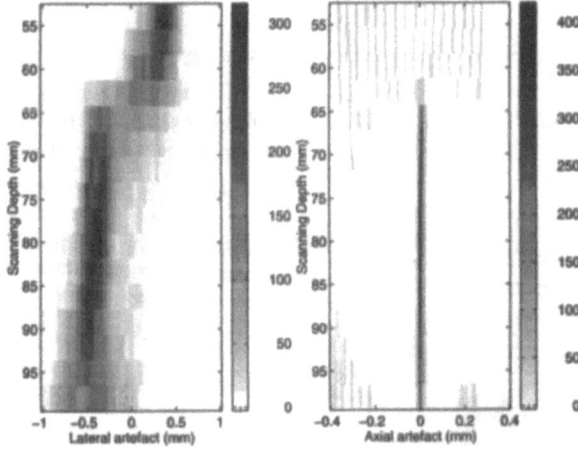

Figure 3: Histograms at multiples depths for a 1 degree rotation. Each line of these images is an histogram computed with all the measured displacements on all the transverse slices of the cylinder at the specified depth.

The two images in Figure 3 illustrate the lateral and axial motion artefacts as a function of the scan depth. These two images are the artefact amplitude histograms shown in gray level. The lateral artefact is more than 0.3 mm and increases with depth. The axial artefact appears negligible as expected.

Figure 4 compares the observed lateral motion artefact with the one predicted with equation (1). It is shown that the experimental and the theoretical curves display the same trend with depth but the artefact measured from the experimental data (—) is larger than the one the model predicts (- -). However if we assume a PSF width (σ_x) 17% larger, then we can get a very good match between model (- • -) and the experiment.

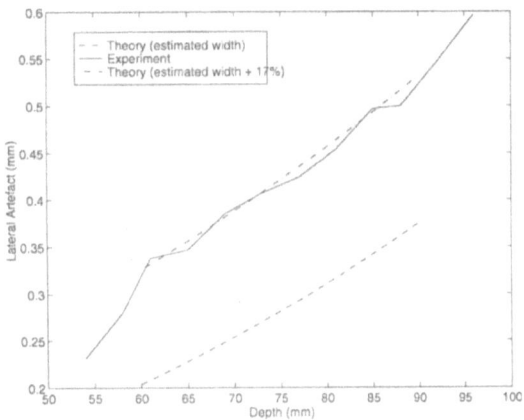

Figure 4: Observed and predicted lateral motion artefact. The theoretical curve (- -) was computed from eq. (1) using the estimated PSF parameters. The theoretical curve (-•-) was computed with a 17% larger σ_x.

It therefore appears that the width of the PSF would have been underestimated by 15-20%. This underestimation could be caused by the observed phase jitter between adjacent r.f. lines and/or the simple PSF model we used for estimating the curvature. The r.f. jitter is due to the A/D clock not perfectly synchronized to the pulse generation logic of the scanner. The 50 MHz A/D conversion rate we used leads to a small phase error on the r.f. signals but this is enough to reduce significantly the lateral correlation and thus the estimated width of the PSF.

The other source of error we are currently investigating has to do with the simple model we used for estimating the size of the PSF. For a motion artefact to be present, the PSF has to be curved and its autocorrelation is not strictly separable; however to determine the PSF size parameters using simple manual measurements, we implicitly assumed separability. This method certainly underestimates the PSF size parameters.

We are also examining the effects of several noise sources which could have a significant impact on the measurements. One of these sources is the interpolation noise that results from the required geometrical transformation for motion compensation.

The effects of all these sources of error are being investigated using a computer model of the image formation.

CONCLUSION

Understanding the motion artifact and compensating for it will make way for better motion analysis in a sequence of images.

In certain circumstances, speckle motion artefact can dominate the observed lateral motion, thus it could be used as a motion amplifier. Designing a transducer aperture that induces enhanced motion artefact could result in an even more sensitive motion detection.

1. F. Kallel, M. Bertrand, J. Meunier, *Speckle motion artifact under tissue rotation*, IEEE Transaction on ultrasonics, ferroelectrics and frequency control, vol. 41., no.1, pp. 105-122, 1994.

* Supported by NIH grant no. CA64957-01 with the NIH, PHS.

A NEW FORM OF ACOUSTICAL IMAGING, THE SOLITON IMAGING

Woon S Gan & Cheong K Gan

Acoustical Services Pte Ltd
29 Telok Ayer Street
Singapore 048429
Republic of Singapore

INTRODUCTION

In this paper, a new form of acoustical imaging, the soliton imaging is proposed. This type of image formation is based on using the solitary wave or soliton as the incident sound beam. The reason for using them is due to their several interesting properties: (1) they can propagate to an infinite distance without changing their shapes, i.e. having an infinite depth of field. This will improve greatly on the image resolution, (2) the solitons have almost limitless capacity in information transmission and processing over infinite distances, (3) the solitons are stable over long propagation distances, (4) soliton is localized so that it decays or approaches a constant at infinity, (5) soliton can interact strongly with other solitons without losing its identity. This will enable linear summation methods to be used for nonlinear scattering problems, (6) solitons have non-dispersiveness characteristics.

APPROACH TO THE PROBLEM

Solitary waves or solitons are a type of solution of nonlinear differential equation. There are three types of nonlinear differential equations which will produce solitary wave solutions: Korteweg-de Vries (KdV) equation, nonlinear Schrodinger equation and Sine-Gordon equation. Originally solitons only occur in water[1]. Later it is found also in air[2], in plasma physics[3], in solid state physics[3], in elastic strings[4] and in solids[3]. Our imaging system is based on the diffraction tomography system, and the inverse problem is used.

We first consider the propagation of sound waves in inhomogeneous medium. The medium is dispersive and there is no assumption of weak inhomogeneities. The propagation of waves will be described by the nonlinear KdV equation:

$$u_t - c_0 u_x + \beta u_{xxx} + \alpha u_x u = 0 \tag{1}$$

where the subindex denotes a partial derivative with respect to a coordinate $\beta, \alpha =$ constants, u=acoustic pressure and c_0=sound velocity and the medium exhibits nonlinearity of hydrodynamical type, namely, of the type of uu_x.

Proceeding to a reference frame moving with the velocity c_0, and inserting dimensionless variables we can write down the KdV equation in the canonical from

$$u_t - 6uu_x + u_{xxx} = 0 \tag{2}$$

The factor 6 has been incorporated into the nonlinear term to simplify the shape of numerical coefficients in the final solutions. First of all we will analyze general properties of the steady state KdV solutions-solitary stationary waves (solitons and multisoliton states) as well as nonlinear periodic solutions. We will try steady state solutions to this equation as travelling waves, $u(x,t) = u(\xi) = u(x - vt)$;

$u(\xi)$ is inserted into (2) to give

$$-vu_\xi + u_{\xi\xi\xi} - 6uu_\xi = 0$$

Single integration yields

$$-vu - 3u^2 + u_{\xi\xi} = C_1 \tag{3}$$

where C_1 is an integration constant, ξ=displacement and v=particle velocity. This equation is consistent in form with that for nonlinear oscillator vibrations. Both sides of (3) are multiplied by u_ξ, to be again integrated to yield

$$-\frac{v}{2}u^2 - u^3 + \frac{1}{2}(u_\xi)^2 = C_1 u + C_2$$

where C_2 is the second integration constant. From this we derive

$$(u_\xi)^2 = 2C_1 u + 2C_2 + 2u^3 + vu^2$$

This solution for the solitons can be written as

$$u = -\frac{v}{2} \operatorname{sec} h^2 [\sqrt{\frac{v}{2}}(x - vt - x_0)] \tag{5}$$

FORWARD PROBLEM

The purpose here is to determine the scattered field. From the previous section, the scattered field is given by (5).

FRACTAL STRUCTURE AS A DIFFRACTION MEDIUM

This paper will be applied to medical imaging and the human tissue used will be brain. The human brain has fractal characteristics as several researchers have shown that the EEG

322

data from the brain are chaotic in nature. Here the random walk model[5] and the Sierpinski gasket cells are used to represent the fractal structure of the human brain.

Let $P(r,t)$ =probability of finding the random walker on sites at a fixed distance r from the starting point. The probability $P(r,t)$ to find the walker at l at the time t is a Gaussian.

$$P(l,t) = P(o,t)\exp(-l^2/4Dt) \tag{6}$$

where D=fractal dimension for Sierpinski cells. The moments of the probability density $<P^q(r,t)>$ can be written as a convolution integral:

$$<P^q(r,t)> = \int_0^\infty Q(r/l)P^q(l/t)dr \tag{7}$$

where $Q(r/l)$==probability of finding the sites separated by a chemical distance l and Endidean distance r. The chemical distance is the shortest path between two sites on the cluster. In the general case, the qth moment, $<P^q(r,t)>$ can be written as

$$<P^q(r,t)> = \frac{1}{N_r}\sum_{i=1}^{N_r}P_i^q(r,t) \tag{8}$$

where the sum is over all N_r sites located at a distance r from the origin (N_r may include) many configurations or a single configuration with a very large number of cluster sites. The sum equation(8) can be separated into sums over different l values (N_m values of l_m):

$$<P^q(r,t)> = \frac{1}{N_r}\{\sum_{i=1}^{N_1}P_i^q(l_1,t) + \sum_{i=1}^{N_2}P_i^q(l_2,t)+\cdots\}$$

$$= \frac{1}{N_r}\sum_m N_m \times \frac{1}{N_m}\sum_{i=1}^{N_m}P_i^q(l_m,t)$$

$$= \frac{1}{N_r}\sum_m N_m <P^q(l_m,t)> \tag{9}$$

This covers all the scattering points within the fractal medium. In this problem it is assumed that the random walker starts at the origin D and after t time steps can be found at $r[x]$ with very different probabilities at different sites.

For the scattering of sound by a fractal medium one needs to treat all sities of the fractal as starting points and the various parameters like sound velocity, attenuation coefficients etc have to be modified for fractals. Fractal media are characterized by not having a very characteristic length scale and they have a very inhomogeneous density distribution. One can therefore expect to find very different physical properties in materials with fractal structure compared to the ordinary solids. Furthermore, real fractals are disordered and highly irregular. In some sense they can be regarded as ideally disordered materials. In conventional diffraction tomography theory, one considers only scattering by one point by ignoring the object size. This is known as Born approximation. Here the object size is taken into account as consisting of several scattering points and all sites of the fractal are considered as scattering points. We call this type of diffraction :"fractal diffraction".

WAVE SCATTERING MODIFIED BY THE FRACTAL MEDIUM

The expression for the scattered acoustic pressure wavefield is modified by the correlation coefficient which contains the fractal dimension of the medium. We have obtained scattered wavefield amplitude fluctuation from eq.(5) as

$$u(x,t) = -\frac{v}{2}\sec h^2[\sqrt{\frac{v}{2}}(x - vt - x_0)]$$
(10)

For diffraction of sound wave by a fractal medium, one needs to consider all sites of the fractal as scattering points. For the reason, the correlation coefficient is chosen as eq.(9). By modifying eq.(10) by eq.(9), then the autocorrelation function for the amplitude fluctuation is given by the following formula:

$$\overline{u_1(t)u_2(t)} = \int_0^{R_1}\int_0^{R_2}\int\int\int\int_{-\infty}^{+\infty} u_1(v_1,x_1,t_1)u_2(v_2,x_2,t_2) < P^q(r,t) >$$
(11)

$$dv_1 dv_2 dx_1 dx_2 dt_1 dt_2$$

where the coordinates of the receivers are $(R_1,0,0)$ and $(R_2,0,0)$. The power spectral density (PSD) of the scattered field=Fourier transform of autocorrelation function

$$= \int_{-\infty}^{+\infty} \overline{u_1(t)u_2(t)}e^{-2j\pi ft} dt$$
(12)

where f=frequency. The overall amplitude of the acoustic pressure of the scattered field is proportional to the square root of the PSD.

INVERSE PROBLEM

The purpose here is to obtain sound velocity field in the medium from the scattered sound pressure field. The method of nonlinear iteration will be used. The aim is to obtain velocity images under diffraction tomography format. The nonlinearity at tomographic inversion here is related to heterogeneity of the human tissue. The vital difficulties in the inverse scattering problems are the typical nonlinearity caused by the strong disturbances which cannot be solved by direct employment of Born or Rytov approximations in order to study the characteristics of nonlinear inversion, one needs instructions from the theory of nonlinear systems. The scattered wavefield (acoustic pressure) during the forward problem will be needed in the inverse problem. This will be the square root of the P.S.D. given by eqs(11) and (12).

RECONSTRUCTION ALGORITHM

The successive linearization methods is used for inversion of scattered traces, the system characteristics of the inversion iteration, and the features of the output sequences of acoustic impedance. The method of iteration will be used.

For reconstruction, the planar wave equation is used:

$$[\Delta - \frac{1}{c^2(x)}\frac{\partial^2}{\partial t^2}]u(x,t) = \alpha$$
(13)

where c is the wave velocity, u is the acoustic wavefield and $\Delta = \nabla^2 = \dfrac{\partial^2}{\partial x^2}$

The following iteration formulae are needed:

$$c_k^{-2}(x) = c_{k-1}^{-2}(x) + \gamma_k(x) \tag{14}$$

$$u_k(x,t) = u_{k-1}(x,t) + v_k(x,t) \tag{15}$$

with $\lim\limits_{k \to \infty} c_k(x) = c(x)$

From eqn(13), the following equations are obtained:

$$[\Delta - \dfrac{1}{c_k^2(x)} \dfrac{\partial^2}{\partial t^2}]u_K(x,t) = \alpha \tag{16}$$

$$[\Delta - \dfrac{1}{c_{k-1}^2(x)} \dfrac{\partial^2}{\partial t^2}]u_{k-1}(x,t) = \alpha \tag{17}$$

From Eqn(16) and (17), one obtains:

$$v_k(x,t) = \int\int G(x,x',t,t')u_k(x',t')\gamma_k$$

$$u_k(x,t) = u_{k-1}(x,t) + \int\int G(x,x',t,t')u_k(x',t')\gamma_k(x')dx'dt' \tag{18}$$

where the Green's function satisfies

$$[\Delta - \dfrac{1}{c_{k-1}^2(x)} \dfrac{\partial^2}{\partial t^2}]G(x,x',t,t') = -\dfrac{\partial^2}{\partial t^2}\delta(t-t')\delta(x-x')$$

From eqn (18), since u_k in the right hand side is not known,

$$u_k(x,t) = u_{k-1}(x,t) + \int\int G(x,x',t,t')[u_{k-1}(x,t) + v_{k-1}(x,t)]\gamma_k(x')dx'dt' \tag{19}$$

Set $v_k = \mu v_{k-1}$ and $0 < \mu < 1$ for iteration to improve the approximation. As a result, eqn (19) becomes :

$$u_k(x,t) = u_{k-1}(x,t) + \int\int G(x,x',t,t')[u_{k-1}(x,t) + \mu v_{k-1}(x,t)]\gamma_k(x')dx'dt' \tag{20}$$

γ_1 can be found by listing two initial scattered field values from the output data of forward problem , $u_0(x,t)$ and $u_1(x,t)$, and by setting $v_0 = 0$.

From eqn (15), $v_1(x,t) = u_1(x,t) - u_0(x,t)$
Hence,

$$v_1(x,t) = \int\int G(x,x',t,t')u_0(x,t)\gamma_1(x')dx'dt'$$

Solve for γ_1 as a constant by using Composite Simpson Rule .

By given the initial velocity propagating in tissue be $c_0(x) = 1750ms^{-1}$, $c_1(x)$ can be calculated by eqn(14) as below:

$$\frac{1}{c_k^2(x)} = \frac{1}{c_{k-1}^2(x)} + \gamma_k(x)$$

$$c_k(x) = \frac{c_{k-1}(x)}{\sqrt{1 + \gamma_k(x)c_{k-1}^2(x)}}$$

This is then followed by iteration to calculate new values of u_k, and v_k and solve for γ_k. The iteration produces a sequence of $c_k(x)$, where $k = 1,2,\cdots\cdots$.
In this project , since the value of scattered field values are given in terms of x only , so all the variables are confined to one variable only , ie x. The Green's function of the differential equation is given by $G(x,x') = \dfrac{e^{jk_0|x-x'|}}{4\pi|x-x'|}$.

The Green's function represents the solution of the wave equation for a single delta function, because the left hand side of the wave equation is linear. The solution can be written by summing up the scattered field due to each individual point scatterer . After hundreds of iterations , from the results obtained , the sequence of $c_k(x)$ converges . The last value of the sequence is taken . The above procedures are repeated for the next two values of scattered field until the last pair are used. Finally , all the scattered field values are converted into velocity values.

In this project , a total of 10 projections are used in order to reconstruct the image . So , the scattered field values for each projection are converted to velocity values which corresponding to the line integral of that projection . After all the line integrals of the projections are obtained , an image recontruction method called filtered backprojection algorithm is used.

The filtered backprojection algorithm can be implemented as :
Sum for each of the K angles , θ, between $0°$ and $180°$
Measure the projection
Fourier transform it to find $S_\theta(w)$
Multiply it by the weighting function
Sum over the image plane the inverse Fourier transform of the filtered projection

VERIFICATION OF WRITTEN PROGRAMS

In this paper , 4 programs have been written to implement the inverse problem .
They are FIELD .C, INVERSE .C , BACK .C ,and PLOTIMAG .C . Their respective function[5] are described in Table 1.

Table 1 .   Functions of respective programs

Programs	Functions
FIELD.C	To calculate the scattered field by using data obtained from the output of forward problem

INVERSE.C	To implement the proposed inverse algorithm
BACK.C	To implement filtered backprojection algorithm
PLOTIMAG.C	To display the image on screen

Both FIELD.C and PLOTIMA.C programs are straightforward , hence further proof is not necessary. The other two programs are rather complicated which implement the above mentioned algorithms. Thus , verification of both programs needs input data and expected output data .

For verification of INVERSE .C , the expected output of the program must be tally with a known input. Since the algorithm is applied to medical imaging , the corresponding input and output values should be approximately the same with the range that obtained experimentally in medical field. For the input data , the range of scattered field value used are in the range of 0.01 to 1.0 W/m^2. Running the program with the specified range of input data , the output data is expected to converge and fall into the specified range which in this case is the sound velocity in the tissue (1500 m/s to 1750 m/s). Since the output data are in the expected range , the program can be proved to be valid until this initial stage. Of course , further verification of the program can only be realized when the physical system is built. By using the experimental input data and obtained measured output data , a further improvement of the program can be achieved.

For the program (BACK.C) , the verification of the reconstruction algorithm can be done by firstly writing a program that will simulate the projection data

$$g(s,\theta) = \{ \begin{array}{l} \dfrac{2ab\sqrt{sm^2 - s^2}}{sm^2} f_0, |s| \leq sm \\ 0, |s| > sm \end{array}$$

where $sm^2 = a^2 \cos^2\theta + b^2 \sin^2\theta$, sm=maximum distance of ray from the origin.

In this program , values of a and b used are 16 and 9 respectively for a horizontal ellipse. The data generated is used as input to BACK.C and the output data obtained is the reconstructed image data. PLOTIMAG .C is then used to display the image in grey-scale form.

From the display of image , any ellipse is observed. This proves that the reconstruction algorithm and the display program are correctly implemented. It is also observed that the resolution of image improves when the number of projection increases and that the algorithm is still valid for the other orientation of the shape of ellipse.

We are still in the process of the numerical evaluation of the scattered solitary wave in the forward problem. This is needed before the final simulated image can be obtained.

CONCLUSIONS

In order to be able to use solitary wave in ultrasonic medical imaging, one has to be able to produce a solitary wave source in the air. The ultrahigh capacity of the soliton and that solitary wave has no change in shape over the propagation distance , will manifest in the acoustical image that it will have much better resolution and that the sound wave will have deep penetration into the solids. But these remain to be verified in an actual medical imaging system.

REFERENCES

1. W. Ni , R.Wei , X.Wang. Q. Xu ,G.Wei and Y.Xu , The periodical passing through of a pair of solitary waves , Proceedings of 14th International Congress on Acoustics , Beijing , China , 1992 , Vol 1, pp A3-2

2. N.Sugimoto , Acoustic solitary waves in a tunnel with an array of Helmholtz resonators , J.A.S.A.99 :1971 (1996)

3. F.Abdullaev, Theory of Solitons in Inhomogeneous Media , John Wiley & Sons , Chichester (1994).

4. M.S.El Naschie , Stress , Stability and Chaos in Structural Engineering : an Energy Approach , McGraw-Hill , London (1992).

5. H.E Stanley , Fractals and multifractals : the interplay of physics and chemistry, in : Fractals and Disordered Systems , A.Bunde and S. Havlin , eds., Springer-Verlag , Heidelberg (1991).

OBJECT IMAGING SYSTEM
USING ACOUSTIC FOURIER TRANSFORM

Tadashi Honda, Yoshimasa Honda, Masaomi Ikeda,
and Masato Nakajima

Department of Electrical Engineering, Faculty of Science
and Technology, Keio University, Japan

INTRODUCTION

In the field of image processing, Forier Transform is often used in the process that the information of the object is dealt with or the image is reconstructed. Though Fast Fourier Transform (FFT) by the computer is often used for the way of Forier Transform of the image, the optical Forier Transform by using a lens has been known well since the old days, too.

We researched about Forier Transform not optically but acoustically since now, and proposed a new method which was called Acoustic Fourier Transform [1]. The acoustical distribution obtained by Acoustic Fourier Transform is perfect Forier Transform of the object. Therefore, it is thought that we can reconstruct the image of the object easily and quickly, if this distribution is taken in the computer and Forier Transform (FFT) is performed with this distribution. So, by using the principle of this Acoustic Fourier Transform, we try to produce the real-time ultrasound camera that reconstruct the object in the flame, the mud, the mist and so on (we call these conditions, "the sight-limited environment"). Furthermore, we decide to add the reference wave electrically for a substitute to irradiate the reference wave from the reference source. Moreover, we measure the distance from the object to the transmitter-receiver array using the pulse echo method, and reconstruct the image of the object at any distance clearly.

PRINCIPLE

COLLECTION OF OBJECT INFORMATION
BY ACOUSTIC FOURIER TRANSFORM

This imaging system is shown in Figure 1. The figure shows that the object and the reference source are placed in the front of the transmitter-receiver array which one transmitter and many receivers are installed in. The transmitter source, the receiver array, the object and the reference source must be placed on the about same perimeter as Figure 2.

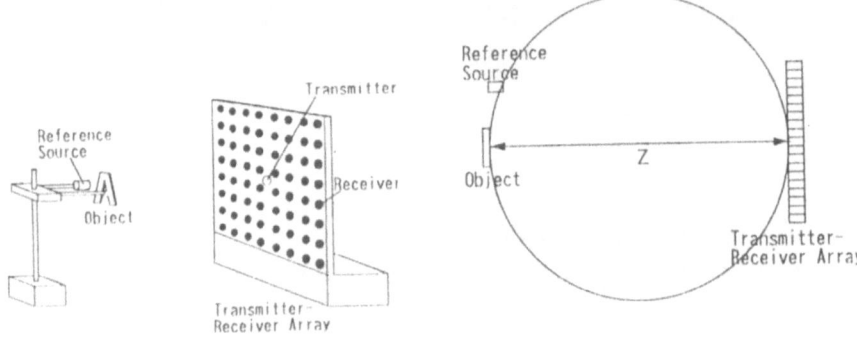

Figure 1. System of Acoustic Fourier Transform.

Figure 2. Arrangement of transmitter-receiver array, object and reference source.

First, the spherical wave is irradiated to the object from the transmitter arranged in the center of the transmitter-receiver array as Figure 3.

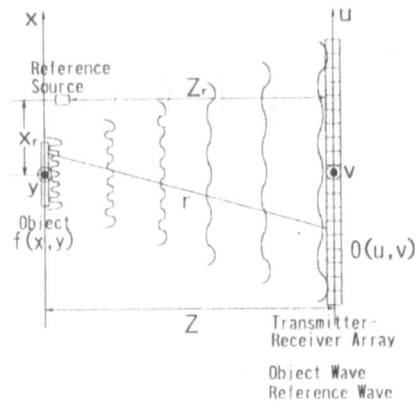

Figure 3. Principle of Acoustic Fourier Transform.

Then, the wave (object wave), which reflected at the surface of the object, is received at the receiver elements arranged in the receiver array. The acoustical distribution $O(u, v)$ of the object wave on the receiver array becomes

$$O(u, v) = \int \int f(x, y) \frac{A}{r} e^{jkr} dx dy \qquad (1)$$

Where, $f(x, y)$ is the distribution of the object surface, A is a constant amplitude $k = \frac{2\pi}{\lambda}$, λ is the wavelength, r is the distance from a point (x, y) on the object surface to a point (u, v) on the receiver array. Next, the following condition is thought about, and the equation (1) is approximated.

$$Z \gg x, Z \gg y$$
$$x, y \text{ are very small.}$$

Where, Z is the distance from the object to the receiver. the acoustical distribution $O(u, v)$ becomes approximately

$$O(u, v) \approx \frac{Ae^{jkz}}{z} e^{jk(\frac{u^2+v^2}{2z})} F(u, v) \tag{2}$$

Where, $F(u, v)$ is Forier Transform of the object $f(x, y)$. As the equation (2), the acoustical distribution $O(u, v)$ is shown to be the product of Fourier Transform and a phase.

Next, when the ultrasound irradiated from the transmitter source is reflected at the surface of the object, a spherical wave is irradiated to the receiver array from the reference source arranged in the neighborhood of the object as Figure 3. The acoustical distribution $R(u, v)$ of the reference wave on the receiver array becomes

$$R(u, v) = \frac{A_r}{r_r} e^{jkr_r} \tag{3}$$

Where, A_r is a constant amplitude, r_r is the distance between the position (x_r, y_r) of the reference source and a point (u, v) on the object surface. Furthermore, when the above-mentioned condition is given to the equation (3), it becomes due to the approximation

$$R(u, v) \approx \frac{A_r e^{jkz}}{z} e^{jk\frac{u^2+v^2}{2z}} e^{-jk\frac{ux_r+vy_r}{z}} \tag{4}$$

The object wave and the reference wave, which arrive at the receiver array, interfere with each other, and the interference fringe is formed. The power distribution $I(u, v)$ of this interference fringe is shown with

$$\begin{aligned} I(u, v) &= |O(u, v) + R(u, v)|^2 \\ &= |O(u, v)|^2 + |R(u, v)|^2 \\ &\quad + O(u, v)R^*(u, v) + O^*(u, v)R(u, v) \end{aligned} \tag{5}$$

When we substitute the equation (2) and the equation (4) to the equation (5), and put in order, the power distribution $I(u, v)$ becomes

$$\begin{aligned} I(u, v) &= |O(u, v)|^2 + |R(u, v)|^2 \\ &\quad + \frac{AA_r}{z^2} e^{jk\frac{ux_r+vy_r}{z}} F(u, v) \\ &\quad + \frac{AA_r}{z^2} e^{-jk\frac{ux_r+vy_r}{z}} F^*(u, v) \end{aligned} \tag{6}$$

As shown in the third term of the equation (6), we obtain Forier Transform of the object with the phase that is decided by the position of the reference source. It is interesting that the power distribution $I(u, v)$ at the receiver array has all the information of the object which should be expressed as the complex in itself. In other words, it is not necessary to measure the wave shape itself at each receiver element arranged in the receiver array. And we can obtain the all object information which are necessary for the reconstruction by getting only amplitude into the computer.

RECONSTRUCTION OF OBJECT IMAGE

The reconstruction of the object image is attained easily by performing Inverse Fourier Transform with the power distribution $I(u, v)$ in the computer. $i(x, y)$, that is Inverse Fourier Transform of $I(u, v)$, becomes

$$
\begin{aligned}
i(x, y) \;=\;& o(x, y) + r(x, y) \\
&+ \frac{A A_r}{z^2} f\!\left(x + \frac{k x_r}{z}, y + \frac{k y_r}{z}\right) \\
&+ \frac{A A_r}{z^2} f\!\left(-x - \frac{k x_r}{z}, -y - \frac{k y_r}{z}\right) \qquad (7)
\end{aligned}
$$

Where, $o(x, y)$ and $r(x, y)$ are Inverse Fourier Transform of $|O(u, v)|^2$ and $|R(u, v)|^2$ respectively. Figure 4 shows the sketch of the reconstructed image after we perform Inverse Fourier Transform with the distribution, that is obtained by Acoustic Fourier Transform, in the computer (the alphabet character "A" as the object).

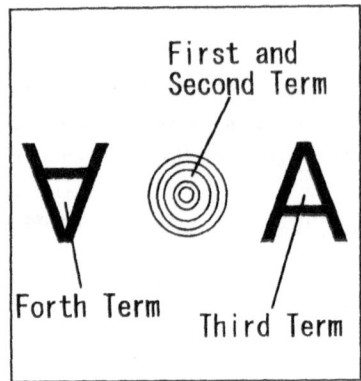

Figure 4. Sketch of reconstructuin object.

A meaningless circular-shaped pattern appears in the center of the image, and two reconstruction images of the object appear on both sides. In other words, the first term and the second term of the upper equation are the pattern of the center, the third term and the fourth term show the reconstruction images.

ELECTRICAL ADDITION OF THE REFERENCE WAVE
AND AUTOMATIC FOCUS

It isn't practical to put a reference source by the side of the object as Figure 1. In this method, the reference wave is made in the computer and is added with the object wave electricity on the circuit for the substitute of irradiation of the reference wave from the reference source. By the electrical addition of the reference wave, we can easily make the reference wave irradiated from any positions in the computer. So, we make that we can reconstruct the object in any position to raise the practical use of this method further. First, the distance to the object is measured by using the pulse echo method. Next, the position of the reference source corresponding to the distance of the object is demanded in accordance with the circle of Figure 2. Then, the reference wave, that is irradiated from the obtained position of the reference source, is made in the computer. As the above, it is thought that we can reconstruct the object at any positions.

EXPERIMENT

In this research, we decided to make the experimental system and to reconstruct the object actually for the verification of this method.

First, we simulated the experiment on the computer to determine the specification of the experimental system to make. As shown in Figure 5, the system parameters are three as follows, the number and the sampling interval of the elements of the receiver array, the irradiation angle of the reference wave.

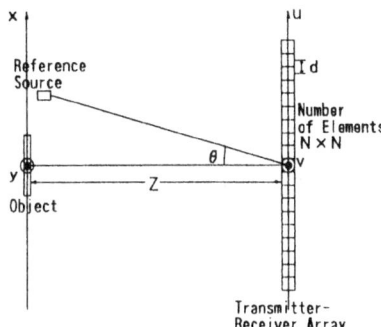

Figure 5. System parameters.

In the simulation on the computer, we changed the value of the system parameters variously and reconstructed some simple objects (alphabet characters, a geometrical figures). Then, we determined the specification of the experimental system by the subjective evaluation of the reconstruction images which were obtained with the simulation. The determined system specification is shown in Table 1.

Table 1. System parameters determined by computer simulation.

Number of the elements of the receiver array	22 × 22
Sampling interval of the elements of the receiver array	20 mm
Irradiation angle of the reference wave	9 deg.

We made the experimental system like Figure 6 by using these system parameters.

Figure 6. Experiment system.

The reference source doesn't exist in the system as shown in Figure 6, because the reference wave interferes with the object wave electrically as the above-mentioned. Using the made experimental system, we tried to reconstruct some alphabet characters and some geometrical

333

figures made of the acrylic fiber as shown in Figure 7. The size of the object is 100mm × 100mm with both as well.

Figure 7.   Objects.

And, the object reconstruction by this experimental system could be done in about 20 seconds using the personal computer which had the CPU of Pentium 133MHz. Figure 8 shows the obtained result in the order as follws, the object in the left, the computer simulation result (reconstruction image) in the center, the experiment result (reconstruction image) in the right. Moreover, we show the actual distance Z and the measured distance Z' in the experiment result to prove the validity of the automatic focus. As Figure 8, we could confirm the reconstruction image both of the computer simulation and the experiment.

And, the object in the smoke tried to be reconstructed as an example of the object reconstruction under the sight-limited environment. The result is shown in Figure 9. As this figure, it could be confirmed that the object in the smoke was reconstructed sufficiently.

But, the quality of the experiment result deteriorates much more than the quality of the result by the computer simulation. One of the factor is the influence of the electric noise which an experimental machine raises, and the other is the imperfection of interference of the object wave and the reference wave at the receiver array. There are few influences of the first factor, and we think that the influence of the second factor is big. In other words, it is thought that the timing of the interference of the object wave and the reference wave made in the computer deviates a little, and it is the interference fringe which we require. So, it is a future subject to solve this problem and make the reconstruction image of the object clearly.

CONCLUSION

The object reconstruction system by Acoustic Fourier Transform was developed as one of the imaging systems of the object under the sight-limited environment.

In this method, the information of the object is acquired by Acoustic Fourier Transform, and an object image is reconstructed by Inverse Fast Fourier Transform in the computer. Moreover, the practical use is raised by the electrical addition of the reference wave and the removal of the limit of the object position using the automatic focus.

An experimental system was made by using the system parameter decided on the computer simulation, and the validity of this method was verified by the reconstruction of the simple object. However, the quality of the reconstructed image obtained by the experiment was not so sufficient. From now on, it will be necessary to consider about the cause of the quality deterioration of the reconstructed image.

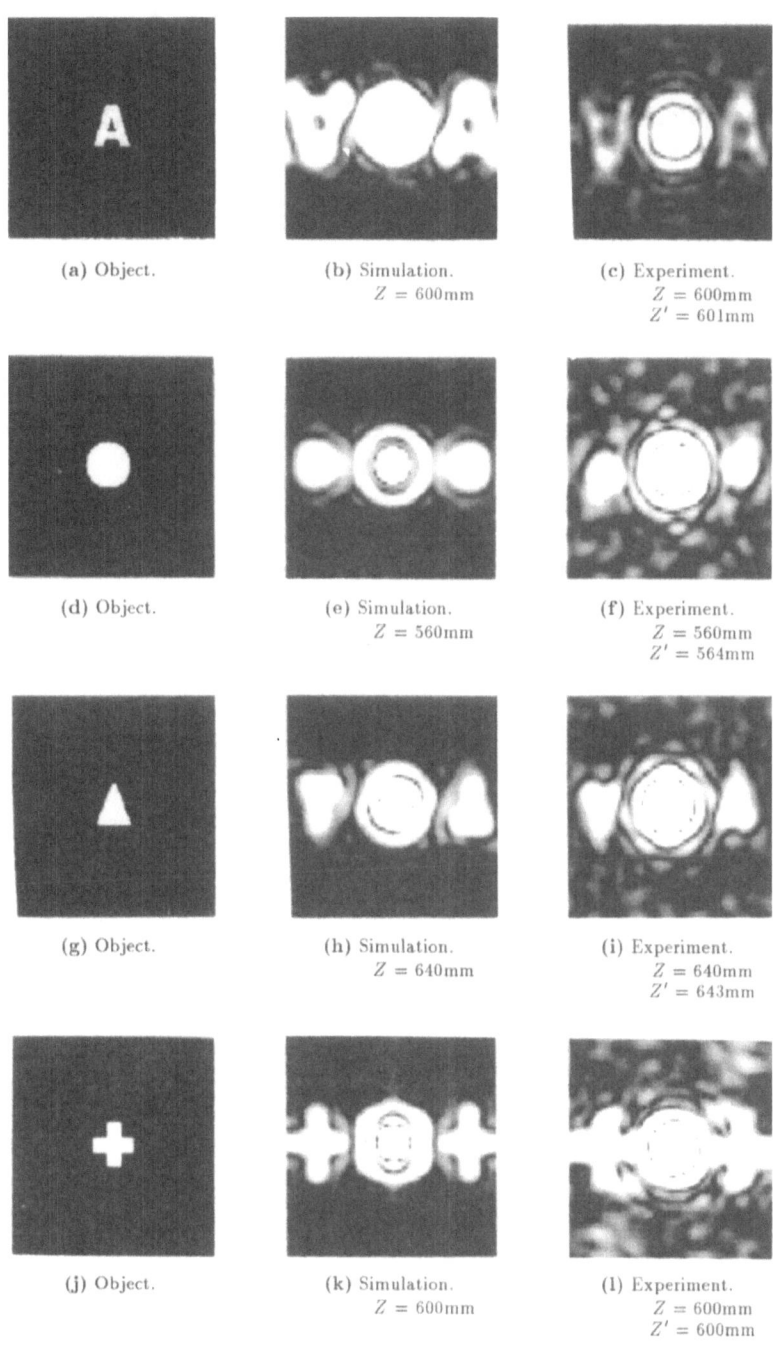

(a) Object.

(b) Simulation.
$Z = 600$mm

(c) Experiment.
$Z = 600$mm
$Z' = 601$mm

(d) Object.

(e) Simulation.
$Z = 560$mm

(f) Experiment.
$Z = 560$mm
$Z' = 564$mm

(g) Object.

(h) Simulation.
$Z = 640$mm

(i) Experiment.
$Z = 640$mm
$Z' = 643$mm

(j) Object.

(k) Simulation.
$Z = 600$mm

(l) Experiment.
$Z = 600$mm
$Z' = 600$mm

Figure 8. Results of the reconstructed object.

(a) Object.   (b) Not in the smoke.   (c) In the smoke.

Figure 9. Result of the reconstructed object in the smoke.

REFERENCES

1. T.Honda and M.Nakajima: "Invisible Object Recognition using Acoustic Fourier Transform", 6th International Conference on Signal Processing Applications and Technology, pp. 1146-1150, Oct. 1995.

ESTIMATION OF MEAN SCATTERER SPACING

BASED ON RF ECHO DATA OBTAINED FROM

DIFFERENT TRANSDUCER SYSTEMS

Klaus V. Jenderka,[1] Tilo Gaertner,[1] Hans Heynemann,[2] Frank Heinicke[1]

[1]Institute for Medical Physics and Biophysics
[2]Urological Clinic
Martin-Luther-University Halle-Wittenberg
Medical Faculty
D-06097 Halle, Germany

INTRODUCTION

Next to Ultrasound Attenuation and Backscatter Parameters measurements the estimation of the Mean Scatterer Spacing is useful for the description of the pathological state of tissue. The backscattered rf signal is influenced from regular and diffuse scattering within the tissue. In liver tissue for example, the hepatic lobule represents a quasi-periodic structure

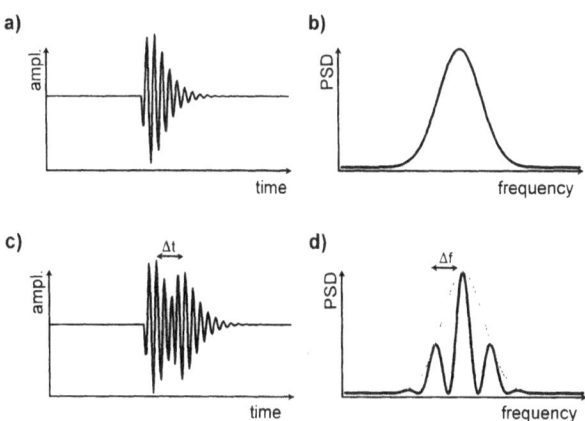

Figure 1. Time signals and spectra of a simulated single (a,b) and double (c,d) puls.

which effects the regular scattering. This structure is disturbed by diffuse scattering of cellular structures.

A number of different techniques for the Mean Scatterer Spacing estimation using the amplitude [1,2,3,4,5,6] or amplitude and phase [7,8] information contained in the spectra are described in the literature. Additional to the well known Fast Fourier Transformation (FFT) different methods like correlation, cepstral and auto regressive techniques are applied to the Power Spectral Density (PSD) estimation.

This paper examine the influence of transducer type (single or multi element), focusing, centre frequency, effective bandwidth and windowing of data and compare different PSD estimation methods. The investigations were performed on simulated data, phantom measurements and rf data of ischemic rat liver (single transducer) and human testis and prostate (multi element transducer) investigations.

THEORY

The spectra of the backscattered rf signals shows some characteristics depending on the spacing between the regular scatterers.

Using the shift property of the Fourier Transformation:

$$\mathcal{F}[p(t-T)] = e^{-i\omega T} P(\omega) \quad , \tag{1}$$

we find for two pulses p_1 and p_2 shifted by 2T and p_2 attenuated by $K(\omega)$:

$$\mathcal{F}[p_1(t+T) + p_2(t-T)] = ((1 - K(\omega))^2 + 4 K(\omega) \cos^2(\omega T))^{-1/2} P(\omega) \ . \tag{2}$$

The spectrum of the echo from a single scatterer is superposed with a $\cos^2$ function. The distance between the maxima Δf is related to $2T = \Delta t$, the time of flight between the two scatterers (Figure 1):

$$\Delta t = \frac{1}{\Delta f} \ . \tag{3}$$

For investigation of real tissue spectra we have to analyse a mixture of different peak spacings in the spectra.

MEAN SCATTERER SPACING (MSS) ESTIMATION METHODS

The estimation of the MSS is based on digitized rf signals and the power spectrum density estimation (Figure 2.). The Fast Fourier Transformation is a good choice for most of the applications. The resolution depends on the sample rate, but a step by step changing is possible by zero padding. An advantage is the simple back transformation in the time domain.

Other types of PSD estimators such as the Autoregressive Spectral Analysis, the Correlogram Method or the Maximum Entropy Method are able to change the resolution and the smoothing of the spectra by adjustment of the parameters.

The MSS estimation is further influenced by the signal bandwidth and the type of the applied window function. The application of broadband transducers improve the resolution of small spacings and the stability of the estimation. Window functions with strong side lobes in the frequency domain will generate additional spacings, which are not related with specimen.

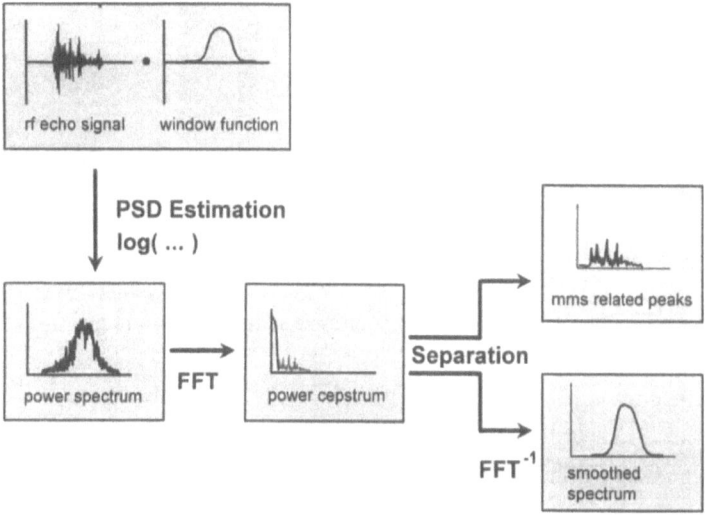

Figure 2. Flow chart for the MSS estimation.

The detection of the mean time spacings between the single echos from the scatterers is possible by processing the Autocorrelation Function or the Power Cepstrum. A problem is the selection of the relevant peaks by elimination of subharmonics, higher order and mixed harmonics (Figure 3.). A very simple technique is to look for the first maximum peak.

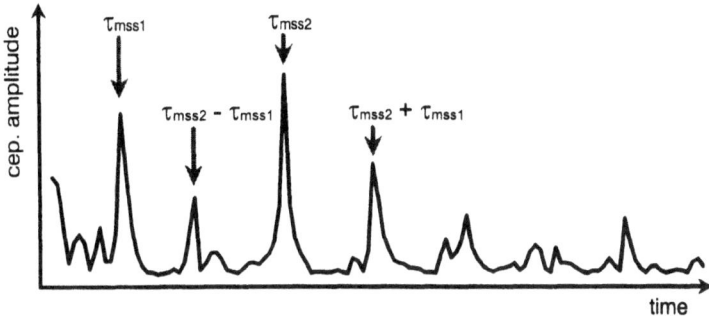

Figure 3. Power Cepstrum with mixed harmonics of a simulated signal containing three pulses spaced by different times (t_{mss1} and t_{mss2}).

EFFECTS OF FOCUSING AND LAYERED MEDIA

The time delay error between the time of flight of pulses from different elements of a multi element transducers for example, is only small in the synthetic or natural focus area. Outside of this areas the timing is not exact and time delays occur.

Applying an unsymmetrical layer with a different speed of sound causes that the focus area is moving. Differences in the time delay for large aperture transducers up to 1,3 μs occurs (Figure 4.). This time differences corresponds to a scatterer spacing of 1 mm, which is typical for liver tissue.

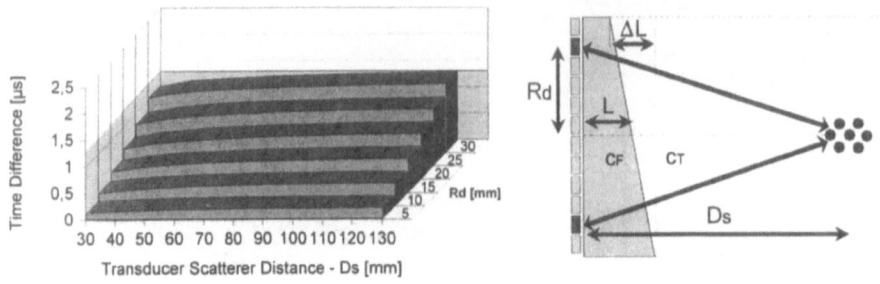

Figure 4. Difference in time delay influenced by layered media (fat layer: L=15 mm, ΔL=10 mm, speed of sound in fat c_F=1420 m/s, speed of sound in tissue c_T=1580 m/s).

EFFECTS OF SCATTERER ORIENTATION

The measured MSS is not direct comparable with the mean histological distance between the scattering structures[9]. Assuming the scatterers arranged in a regular hexagonal lattice (like in liver), the detected spacings depend on the orientation of the lattice relative to the transducer.

That means, if 'd' is the histological distance, the transducer will receive spacings between d and 1/2 d (Figure 5.). In the most cases 1/2 d will not be resolvable for the measurement system and a orientation error of 13,4 % can be expected.

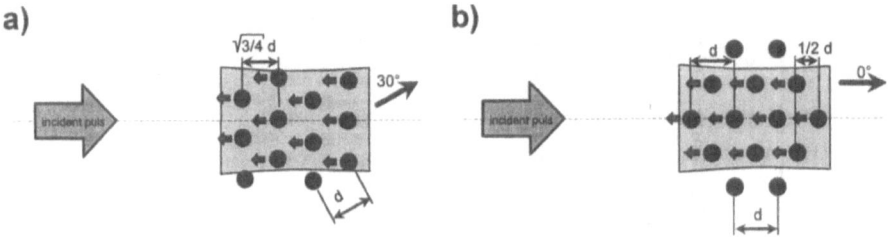

Figure 5. Possible spacings of scatterers arranged in a regular hexagonal lattice.

MEASUREMENTS AND RESULTS

Additional to the simulation of echo signals measurements with different specimens (Agar/Graphite phantom, rat liver, prostate) and transducer systems (linear array with 4-10 MHz, single transducer with 5-13 MHz and curved array (TR probe) with 4-7 MHz) were executed.

Figure 6. shows power spectra of the phantom echo signals processed with the estimation methods as above. The smoothing effect of the Autoregressive method or the Maximum Entropy method is clear to identify.

If the range of possible scatterer spacings is well known the advantage of higher resolution can used. For optimal detection of the Mean Scatterer Spacings the parameters of the AR model can adapted to the expected spacings[10].

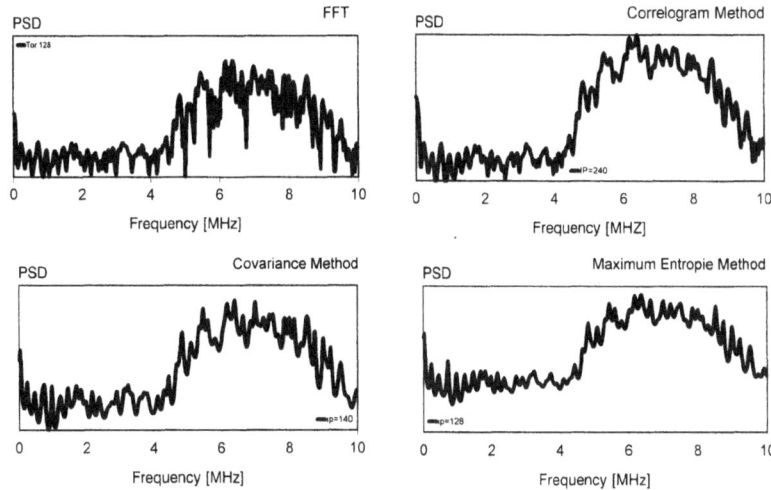

Figure 6. Power spectra of a scan line segment (Agar/Graphite phantom, ATL U9 HDI) estimated by different methods.

The power cepstra estimations of all examples bases on the inverse Fourier Transformation of the logarithm power spectra (Figure 7.).

The cepstra of the phantom echos signals shows for all methods a dominant peak at 2 µs and a smaller one at 1 µs. In the liver example the peaks differ between the estimation methods. The reason for this could be the higher resolution of the cepstra, because the sample rate of the single transducer system was fixed at 100 MS/s. The peaks in the prostate

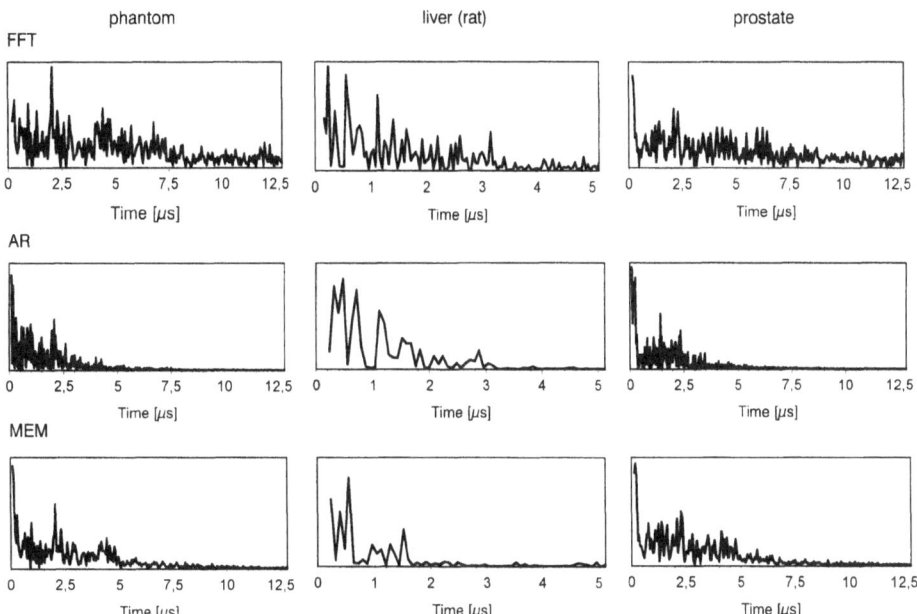

Figure 7. Power cepstra of echo signals from different specimen and different PSD estimators (first column: Agar/Graphite phantom, second column: rat liver (single transducer), third column: prostate, first row: FFT, second row: Covariance method (AR model), third row: Maximum Entropie method).

cepstra are located approximately at 2.5 µs. In the cepstrum from the autoregressive method the first maximum peak appears at 1.5 µs.

CONCLUSION

Next to the system properties like transducer and sampling parameters the power spectrum estimation method and the technique of mean scatterer spacing estimation effects the results strongly. The basis for an successful determination of the spacings is a good quality of the data without timing errors generated by large apertures and layered media. The center frequency and bandwidth of the signals have to be large enough. A gate function without side lopes in the spectrum must used for the windowing of the sampled data.

By choosing the PSD estimator parameters adjusted to the range of expected scatterer spacings, the quality of the mean scatter spacing can improved.

ACKNOWLEDGMENT

This work was supported by the 'Deutsche Forschungsgemeinschaft' He 2156/2-2.

REFERENCES

1. L. Fellingham and F. Sommer, Ultrasonic characterization of tissue structure in the in vivo human liver and spleen, *IEEE Transactions on Sonics and Ultrasonics*, SU-31: 418-428 (1984).
2. R. Kuc, K. Haghkerder and M. O'Donnel, Presence of cepstral peaks in random reflected ultrasound signals, *Ultrasonic Imaging*, 8: 196-212 (1986).
3. K.V. Jenderka, Measurement of attenuation and mean scatterer spacing by ultrasound spectroscopy, *The Journal of nuclear medicine and allied sciences*, 32: 196-198 (1988).
4. L. Landini and L. Verrazzani, Spectral characterization of tissues microstructure by ultrasounds: A stochastic approach, *IEEE Transactions on Ultrasonics, Ferroelec. and Freq. Control*, 32: 448-456 (1990).
5. K. Wear, R. Wagner, M. Insana and T. Hall, Application of autoregressive spectral analysis to cepstral estimation of mean scatterer spacing, *IEEE Transactions on Ultrasonics, Ferroelec. and Freq. Control*, 40: 50-58 (1993).
6. C. Simon, R. Seip and E. Ebbini, Estimation of mean scatterer spacing based on autoregressive spectral analysis of prefiltered echo data, *Proceedings of the 1995 IEEEUltrasonics Symposium*, 2: 1153-1156 (1996).
7. T. Varghese and K. Donohue, Mean scatterer spacing estimates with spectral correlation, *Jour. Acoust. Soc. Am.*, 96: 3504-3515 (1994).
8. U. Abeyratne, A. Petropulu and J. Reid, On modeling the tissue responses from ultrasonic B-scan images, *IEEE Transactions on Medical Imaging*, 15: 479-490 (1996).
9. E.G.M.P. Jacobs, J.M. Thijssen and B.J. Oosterveld, A simulation study of structural scattering in ultrasonic B-mode imaging, in: *Acoustical Imaging vol. 19*, H. Ermert and H.P. Harjes, ed., Plenum Press, New York (1992).
10. T. Varghese, K.V. Donohue, V.I. Genis and E.J. Halpern, Order selection criteria for detecting mean scatterer spacings with the AR model, *IEEE Transactions on Ultrasonics, Ferroelec. and Freq. Control*, 43: 979-984 (1996).

TRANSIENT WAVE FIELD CALCULATION FOR ULTRASONIC TRANSDUCERS USING INTEGRAL TRANSFORM METHODS

Elfgard Kuehnicke[1], Uwe Voelz[2]

[1]Dresden University of Technology, Institute of Technical Acoustics
D-01062 Dresden, Germany
[2]Federal Institute for Materials Research and Testing, 12200 Berlin, Germany

INTRODUCTION

For many applications of ultrasound, the description of the propagation of elastic waves emitted by a transducer into a solid is of fundamental importance. In many cases, the sound wave generated by the transducer passes layers of different impedances until it reaches the object to be tested. Thus the sound field in the specimen depends on the parameters of the probe and the parameters of all layers passed by the sound field.

Integral transform methods have been proposed to calculate the transducer generated three-dimensional, transient field [1,2,3]. In [1,2] the transient field radiated by a circular or rectangular transducer into a solid half-space and in [3] the transient field radiated by a linear array into an immersed solid half-space is calculated. These approaches, however, are restricted to special problems with plane interfaces.

For the calculation of transducer generated three-dimensional, transient wave fields in layered media with curved interfaces a separation approach in combination with integral transform methods was proposed[4]. The separation method has been applied for transducer design programs based on time-harmonic waves[5,6]. To extend the separation method to the calculation of transient wave fields it is necessary to develop an efficient approach. That is why an exact method will be compared with an optimized approximation method.

APPROACH

A Green's function defines the response of the rigidly supported body to an impulse point load. Such Green's functions are available for a half-space and for a horizontally layered medium (two-dimensional geometry). Basing on Green's functions the field generated by a wide source can be obtained a spatial convolution[1] or by adding the effects due to all elementary portions of the source, as it will be shown in this paper. Due to the point source synthesis, the calculation time results from the product of the time to calculate the transient signal for a point source and the number of point sources. Therefore, the development of an efficient algorithm for simulation calculations in transducer design is always connected with the search for a fast calculation of the Green's functions. There are two possible ways to get the transient Green's functions:

    (1) by a direct calculation or
    (2) by a superposition of harmonic waves.

In order to find out a method with a low computation time both methods will be compared.

Calculation of Transient Green's Functions by the direct method

The propagation of transient waves in a linear elastic isotropic medium may be analyzed by integral transform methods. A transform in time and space yields a set of coupled ordinary differential equations for the displacement potentials. In the Laplace transformed domain the response of the specimen to a vertical point source is expressed in terms of the infinitesimal displacement

$$\bar{u}_\alpha(r,z,s) = as^2 \bar{G}_{\alpha z}(r,z,s) = as^3 \overline{I}_{\alpha z}(r,z,s) \quad \text{with} \quad a = \frac{f(s)}{4\pi\rho s^2}. \tag{1}$$

where α stand for the cylindrical coordinates z and r, s is the parameter of transform, $f(s)$ is the transformed time function of the source and ρ the density of the layer. $\bar{G}(r,z,s)$ are the transformed Green's functions for a layer with parallel interfaces. Multiple reflections are neglected.

$$\bar{G}^m_{zz}(r,z,s) = s \int_0^\infty S^m W^m_z e^{-sp_m z} J_0(swr) w \, dw \quad, \quad \bar{G}^m_{rz}(r,z,s) = s \int_0^\infty S^m W^m_r e^{-sp_m z} J_1(swr) w \, dw \quad, \tag{2}$$

I is the answer to a step function. The index m has to be replaced by L for the longitudinal wave and by T for the transverse wave. For the L-wave p_m is $p_L = \sqrt{w^2 + c_L^{-2}}$ and for the T-wave yields $p_T = \sqrt{w^2 + c_T^{-2}}$. w is the parameter of a Hankel-transform with respect to the coordinate r and c_L and c_T represent the wave speeds of the longitudinal and transverse waves, respectively. J_0 and J_1 are the Besselfunctions. Different source functions S^m are given in [4]. For instance, the source functions of a free surface S^* are:

$$S^{L*}(w) = 2(w^2 + p_T^2)c_T^{-2} N^{-1} \quad S^{T*}(w) = -4p_L w \, c_T^{-2} N^{-1} \quad N = \left[4p_T p_L w^2 - (w^2 + p_T^2)^2 \right] \tag{3}$$

W is the receiver function and it depends on whether the displacement is calculated inside the layer or at the interface.

$$W^L_z = -p_L \quad W^T_z = W^L_r = -w \quad W^T_r = -p_T \quad W^L_z = \frac{2p_L\left(p_T^2 + w^2\right)}{c_T^2 N} \quad W^T_z = W^L_r = \frac{4p_L p_T w}{c_T^2 N} \quad W^T_r = \frac{2p_T\left(p_T^2 + w^2\right)}{c_T^2 N} \tag{4}$$

For the calculations of the transient wave propagation, the double transform is inverted by the Cagniard-deHoop method, that means by an analytical inverse transform followed by a numerical inverse transform of the remaining integral. Details are given in [7]. The inverse Laplace transform yields

$$I^k_{zz}(t) = H(t - t_A)\frac{2}{\pi} Im \int_0^{w(t)} SW_z \frac{w}{K} dw \quad I^k_{rz}(t) = H(t - t_A)\frac{2}{\pi} Im \int_0^{w(t)} SW_r \frac{w}{K} \frac{t - p_k}{wr} dw \quad K = \sqrt{w^2 r^2 + (t - zp_k)^2} \tag{5}$$

where $H(t)$ is the Heaviside step function. Given a source time function $f(t)$, the answers for $u_\alpha(t)$ are determined by a convolution between the transient Green's function and $f(t)$

$$u_\alpha(r,z,t) = \frac{1}{4\pi\rho} f'(t) * I = \frac{H(t - t_A)}{4\pi\rho} \int f'(t - \tau) I(r,z,\tau) d\tau \tag{6}$$

where the condition $f(0) = f'(0) = 0$ has been assumed.

The field due to a distribution of sources can be obtained by adding the effects due to each elementary portion of source. Then the displacement for an extended source results from an integral over the source distribution

$$U_\alpha = \int dA \, u_\alpha(r,z,t). \tag{7}$$

Transient Green's Functions by Superposition of Time-harmonic Green's Functions

When the time function of the source is harmonic in the form $e^{-i\omega t}$ only one transform is necessary to get a set of ordinary differential equations for the displacement potentials. Equations (2) are also the Green's functions for a steady state response when s is replaced by $i\omega$. For instance, in a half-space with the source function for a free surface the displacement u_z is

$$u_z(r,z) = \frac{1}{4\pi\rho\omega^2} \int_0^\infty \frac{1}{N} \left[\underbrace{2(2w^2 - k_T^2)\sqrt{w^2 - k_L^2} \, e^{-i\omega\sqrt{w^2 - k_L^2}\, z}}_{S^L \cdot W^L} + \underbrace{4\sqrt{w^2 - k_L^2}\, w^3 \, e^{-i\omega\sqrt{w^2 - k_T^2}\, z}}_{S^T \cdot W^T} \right] J_0(\omega wr) w \, dw \tag{8}$$

$$\text{with } N = \omega^{-4} c_T^2 \left[4w^2 \sqrt{(w^2 - k_L^2)(w^2 - k_T^2)} - (2w^2 - k_T^2)^2 \right] \tag{9}$$

344

where $k = \omega/c$ is the wave number. No numerical evaluation of integrals is necessary, if asymptotic expressions for large r, and a deformation of the path of the w-integration are used according to [8]. This results in a spherical wave and a directivity pattern of the point source. The same results have been obtained by an application of a Fraunhofer approximation to the Helmholtz-Huyghens' integral representation. In this approximation Equ.(8) results in

$$u_z(r,z,k) \approx \frac{\cos \gamma}{4\pi\rho R}[S^L(w) \cdot W_z^L(w) \cdot e^{-ik_L R} + S^T(w) \cdot W_z^T(w) \cdot e^{-ik_T R}] \tag{10}$$

The directivity pattern for the point sources S^L and the receiver functions W are obtained from Eqs.(3) and (4) replacing w by $\sin\gamma/c_L$ or $\sin\gamma/c_T$. A discussion of different point source models is given in [9]. As an example Equ.(11) gives the directivity pattern for a normal load on a free surface

$$S^L = \frac{\cos(\gamma)\left(1-2\left(\frac{c_T}{c_L}\right)^2\sin^2(\gamma)\right)}{\left(1-2\left(\frac{c_T}{c_L}\right)^2\sin^2(\gamma)\right)^2 + 4\left(\frac{c_T}{c_L}\right)^4\sin^2(\gamma)\,\cos(\gamma)\,\sqrt{\left(\frac{c_L}{c_T}\right)^2-\sin^2(\gamma)}} \tag{11}$$

R is the distance between the point under consideration and the source point. γ is the angle between the local vector R and the normal vector in the source point.

The transient wave motion generated by a load of arbitrary time dependance can be expressed as a superposition integral over the response to a corresponding time harmonic load.

$$u_z^t(r,z,t) = \frac{1}{4\pi\rho}\int_{-\infty}^{\infty} d\omega\, u_z(r,z,k)f(\omega)\, e^{i\omega t} \tag{12}$$

For a real excitation signal the transformed time function $f(\omega)$ has a limited bandwidth and the integrand in Equ. (12) becomes zero for higher frequencies. Thus the number of time harmonic fields which have to be calculated is small and the method becomes very efficient.

The field for an extended source results from Equ.(7). Due to the application of the Fraunhofer approximation to the field of a point source the approximated Equations (10,11) are valid only in a distance of some wavelengths from the surface. This fact however, gives only minimal restrictions to the calculation of the nearfield of the transducer.

CALCULATION EXAMPLES

First results are available for a solid half-space (ρ=7,77g/cm^3 ,c_T=5840ms^{-1} ,c_L=3170ms^{-1}). The exciting element is positioned in x-y-plane at z=0 and centers in the origin of the coordinates. In Fig.1 the normal displacement u_z for a circular element (diameter d=6mm) is calculated by means of the direct calculation method for selected points on the z-axis (columns in Fig.1) and for different time excitations (lines in the figure). Line 1 represents the normal impulse response. The transient displacement waveform is obtained by a convolution of the impulse response with the time function of the source. Lines 2,3 show the transient signal for an excitation by a sine burst of n cycles at 5MHz according to

$$f(t) = \sin(\omega t)\,\sin\left(\frac{\omega}{2n}t\right) \qquad \text{for} \qquad 0 \le t \le nT \qquad \text{with} \qquad \omega = \frac{2\pi}{T} \tag{13}$$

The displacements are normalized to the same value for all impulse excitations in line 1 and also for all transient displacement waveforms from line 1 to line 3, so that all graphs in line 2 and 3 can be compared with each other. The time is normalized to the arrival time of the L-wave t_L at the observation point z, i.e. the beginning of the signal corresponds to the arrival time of the direct L-wave from the middle of the element. The arrival times of the L-wave from the edge t_{LK} and from the T-wave from the edge t_{TK} are also indicated in the figures. At a point of the z-axis, for instance at the point P_1 =(0,0,z=3.3 mm) in line 1, the beginning of the impulse response can be described by a step function with a front edge caused by the arrival of the direct L-wave from the middle of the element. The rear edge is caused by the arrival of the compression edge wave. The width of the rectangle is determined by the difference between the arrival times, and consequently, it depends on the distance z. After the rectangle the signal increases slowly and forms a ramp function until the arrival of the head wave forms a peak. The signal decreases sharply with the arrival of the shear edge wave. Similar results are described in [2].

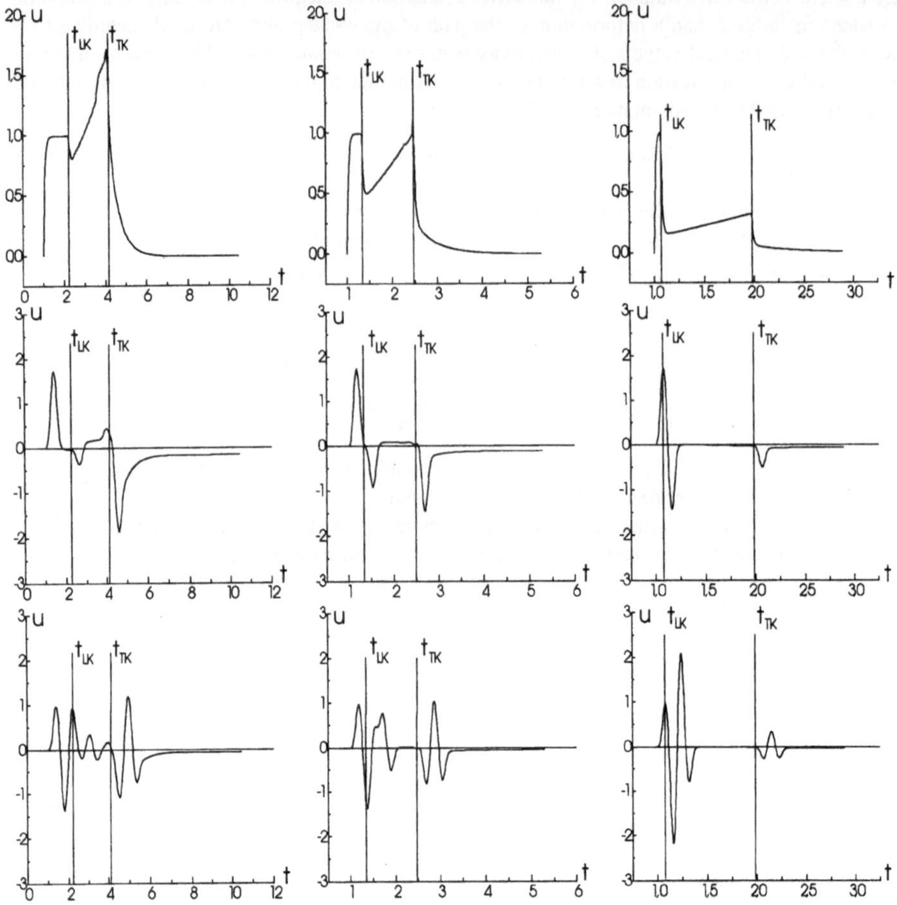

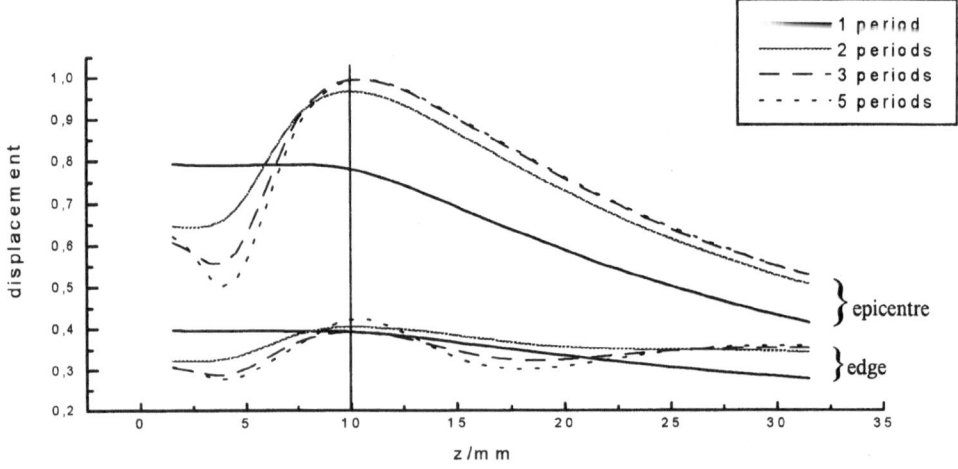

Fig.1: Normalized displacement $u_z(t)$ for different z (columns) and for different excitation functions
columns: 1- for z=1.5 mm, 2- for z=3.3 mm, 3- for z=7.5 mm,
lines: 1-impulse response, 2- sine burst of 1 period, 3- sine burst of 2 periods

Fig.2: Magnitude of the L-wave for a rectangular element

346

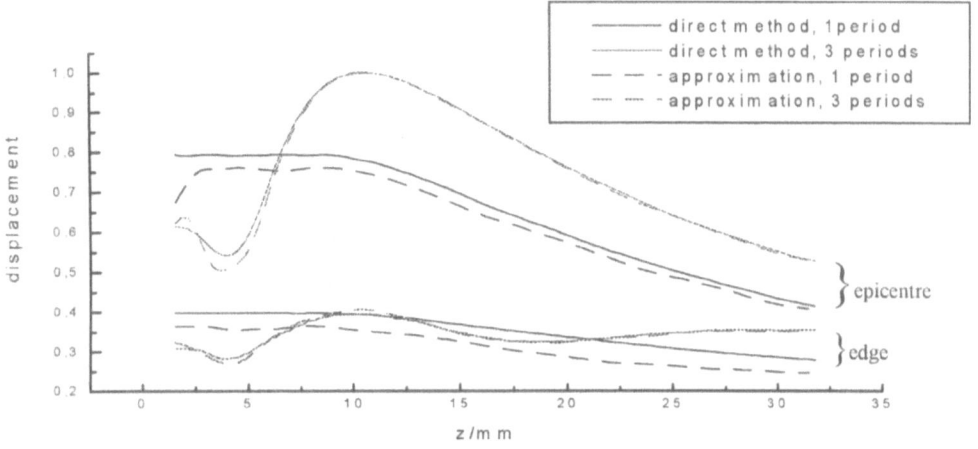

Fig.3: Comparison of the L-wave magnitude for the rectangular transducer element which results from
direct calculation and harmonic synthesis approximation

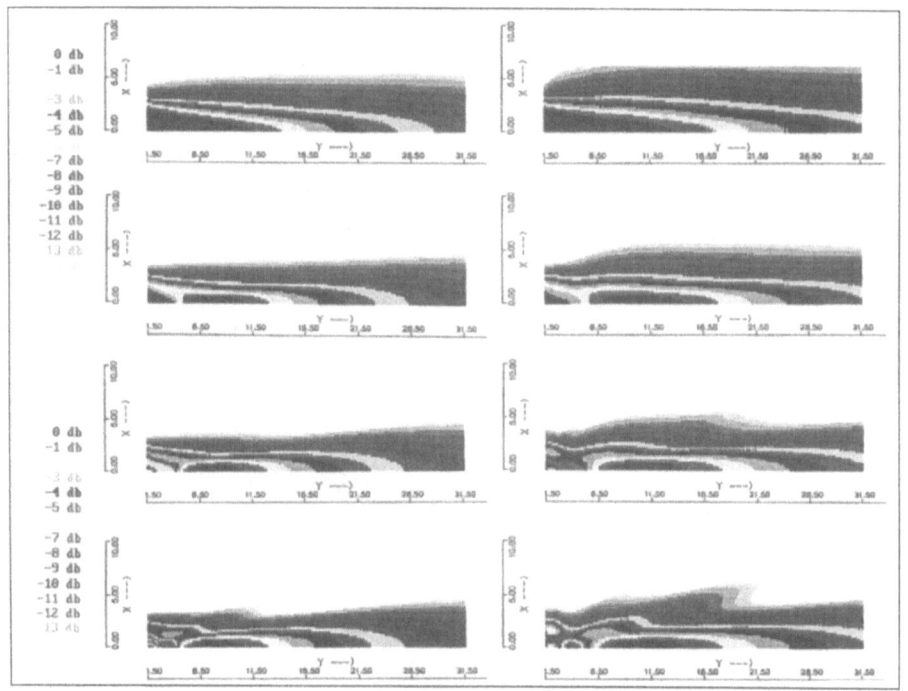

Fig.4: Sound field for a circular (left) and rectangular (right) transducer element (d=6mm, f=5Mhz)
Excitation functions: sine burst of 1 period, 2 periods, 3 periods and 5 periods (from top to bottom)

Because the transient signal (line 2,3) is a convolution of the impulse response and the integrated excitation function, each jump in the impulse response (jump at the arrival time of the direct compressional wave, the compressional edge wave and the shear edge wave) produces a signal with a number of oszillations which depends on the excitation signal. The shape of the signal after the convolution for one jump can be seen at the time t_{TK} because here is no overlapping.

Fig.2 shows the peak magnitude of the compression mode normal displacement along the z-axis and a parallel axis to the z-axis at y=3mm on the edge of an rectangular element (length a=6mm). There is a difference in the peak magnitudes in dependance of the time excitation of the element. For a time excitation with a sine function of one cycle the peak magnitude function contains no maxima and minima in the near field. In contrary, for a longer excitation of up to 1,5 to 2 cycles the development of the typical near field features starts. Due to the longer duration of the signal, the effects from the two edges in the impulse response superpose.

For comparison, in Fig.3 the sensitivity functions are calculated by the direct method and by the optimized field calculation method according to Eqs.(10,11). The results show a high degree of consistence between the two approaches. Fig.4 shows sound fields calculated by means of the time-optimized method..

CONCLUSIONS

A comparison of both presented methods shows a high coincidence of the results. Using a superposition of time-harmonic waves, for the examples shown above only a small number of 5 to 7 harmonic wave fields has to be calculated to get the near field of a real broadband transducer. But even in such a case, the calculation of one time-harmonic wave field for the center frequency of the excitation signal is a good approximation in the far field. Especially the near field length is the same for transient and time-harmonic excitation.

In this paper the formulae are written for the displacement and partly for the most simple cases (half-space, free surface) only to give the general idea of the two approaches. Both methods, however, can be used in connection with the separation method[4] to calculate the transient wave field in a layered media with curved interfaces.

Because in the direct method for each point source an integral has to be calculated numerically, the computation of the exact impulse response to an extended source is very time expensive. Using a superposition of time-harmonic waves in combination with a Fraunhofer approximation no numerical integral calculation is necessary and in dependance on the length of the excitation signal and the field point, only a small number of harmonic wave fields has to be calculated to get the field for a broadband transducer. This gives the chance for the development of a fast tool for transducer design for layered test objects with curved interfaces.

REFERENCES

1. L.F.Bresse, D.A.Hutchins, Transient generation of elastic waves by a disk-shaped normal force source, J.Acoust.Soc.Am., 86, 1989, pp.810-817
2. H.Djeloua, J.C.Baboux, Transient ultrasonic field radiated by a circular transducer in a solid medium, J.Acoust. Soc.Am., 92, 1992, pp.2932-2941
3. Wu Ping, R.Kazys, T.Stepinski, Calculation of transient fields in immersed solids radiated by linear focusing arrays, 1995 IEEE Ultrasonics Symposium, pp.993-996
4. E.Kühnicke, Three-dimensional waves in layered media with non-parallel and curved interfaces - A theoretical approach, J. Acoust. Soc. Am. 100(2), 1996, 709-716
5. E.Kühnicke: Simulation calculations for monofrequent sound fields in layered media, 21st International Symposium on Acoustical Imaging Laguna Beach, USA, March 28-30, 1994, ed. by J.P.Jones, Plenum Press New York, vol.21, pp.47-53
6. E.Kühnicke; Calculation of three-dimensional harmonic waves in layered media, 1995 IEEE International Ultrasonics Symposium, Seattle, November 7-10, 1995
7. Y.-H.Pao, R.R.Gajewski: The generalized ray theory and transient response of layered elastic solids, in Physical Acoustics, edited by W.P.Mason, G.H.Thurston, Academic New York 1977,vol.13,chap. 6
8. G.F.Miller, H.Pursey: The field and radiation impedance of mechanical radiators on the free surface of a semi-infinite isotropic solid, Proc. R. Soc.London Ser.A 223, 521-541, (1954)
9. E.Kühnicke, Directional field of a point source for calculation of three-dimensional harmonic waves in layered media, 22nd International Symposium on Acoustical Imaging, Florence, Italy, 1995, ed. by P.Tortoli and L.Masotti, Plenum Press New York, vol.22, pp.9-14

A NEW THREE-DIMENSIONAL MODEL OF PIEZOELECTRIC ELEMENTS FOR COMPOSITE MATERIALS

N. Lamberti[1], A. Iula[2], M. Pappalardo[2]

[1] Dip. di Ingegneria dell'Informazione ed Ingegneria Elettrica
   Università di Salerno
   Via Ponte Don Melillo, I-84084 Fisciano (SA), ITALY.
[2] Dipartimento di Ingegneria Elettronica - Università di Roma III
   Via della Vasca Navale, 84 - 00146 Roma - ITALY

INTRODUCTION

Composite piezoelectric materials are widely used in ultrasonic transducers for acoustical imaging and non destructive testing applications, because of their high electromechanical conversion efficiency and their good matching to water. A piezoelectric composite is usually made by combining piezoceramic and inert elements in several connections. The first step to obtain a model able to describe a piezocomposite of any kind of connection, is to develop a three–dimensional model of the single piezoelectric element, which takes its interaction with the surrounding media into account. A three–dimensional model of piezoceramic plates loaded on each face was recently proposed by Brissaud [1], but it is not clear the way in which the boundary conditions are satisfied. In this work a new three–dimensional model of the rectangular piezoceramic is proposed. Three general orthogonal wave functions, each depending only on one Cartesian axis corresponding to the propagation direction, are taken as solutions of the differential wave equation system which describes the vibration of the element. As a consequence of this choice, mechanical and electrical boundary conditions cannot be satisfied in a punctual form; however, an integral form can be adopted for mechanical conditions. As far as the electrical condition is concerned two solutions are explored: in the first even this condition is applied in an integral form, as it was done by the authors for the 3–D model of the piezoceramic disk [2]; in the second, in analogy with the 2–D model of the single array element proposed by some of the authors [3], the piezoelectric coupling constant in the transverse direction is neglected. In both cases the proposed model is able to compute the electrical input impedance and the transfer functions (in transmission and reception) of the piezoelectric element. In order to compare the two approximations, the electrical input impedance of a sample was computed in the two cases and compared with experimental results; the comparison shows that the integral condition also for the electrical boundary is more adequate. The proposed 3–D model of piezoceramic elements is able to take into account the interaction with the surrounding medium and the coupling between the thickness and the lateral modes; for these reasons it is well suited to be used to describe complex composite structures of any connection.

THE THREE–DIMENSIONAL MODEL

Fig. 1 shows a piezoelectric element of comparable geometric dimensions; the polarization axis is parallel to the z axis and the metallized surfaces are orthogonal to the same axis. On these surfaces $E_1 = E_2 = D_1 = D_2 = 0$ and, because of the permittivity of the piezoelectric material, greater than that of the surroundings, the fringing electric field and displacement are prevented and these quantities can be neglected everywhere [4]. Finally, for the piezoelectric ceramics $e_{34} = e_{35} = e_{36} = 0$ and therefore the stress and strain shear components are uncoupled with the others. With these hypotheses the set of the more useful constitutive equations is:

$$T_1 = c_{11}^D S_1 + c_{12}^D S_2 + c_{13}^D S_3 - h_{31} D_3$$

$$T_2 = c_{12}^D S_1 + c_{11}^D S_2 + c_{13}^D S_3 - h_{31} D_3$$

$$T_3 = c_{13}^D S_1 + c_{13}^D S_2 + c_{33}^D S_3 - h_{33} D_3 \qquad (1)$$

$$E_3 = - h_{31} S_1 - h_{31} S_2 - h_{33} S_3 + \beta_{33}^S D_3 .$$

Due to the above mentioned high permittivity of the piezoelectric material, the electric flux lines are assumed to be parallel to the z axis, and therefore $\partial D_3/\partial x = \partial D_3/\partial y = \partial D_3/\partial z = 0$; the wave differential equations describing the element vibration are:

$$\rho \frac{\partial^2 \xi_1}{\partial t^2} = c_{11}^D \frac{\partial^2 \xi_1}{\partial x^2} + c_{12}^D \frac{\partial^2 \xi_2}{\partial x \partial y} + c_{13}^D \frac{\partial^2 \xi_3}{\partial x \partial z}$$

$$\rho \frac{\partial^2 \xi_2}{\partial t^2} = c_{12}^D \frac{\partial^2 \xi_1}{\partial x \partial y} + c_{11}^D \frac{\partial^2 \xi_2}{\partial y^2} + c_{13}^D \frac{\partial^2 \xi_3}{\partial y \partial z} \qquad (2)$$

$$\rho \frac{\partial^2 \xi_3}{\partial t^2} = c_{13}^D \frac{\partial^2 \xi_1}{\partial x \partial z} + c_{13}^D \frac{\partial^2 \xi_2}{\partial y \partial z} + c_{33}^D \frac{\partial^2 \xi_3}{\partial z^2} .$$

where ξ_1, ξ_2 and ξ_3 are the displacements in the x, y and z directions respectively. As solutions of this equation system we choose the three orthogonal wave functions:

$$\xi_1 = \left[A_1 \sin\left(\frac{\omega x}{v_1}\right) + B_1 \cos\left(\frac{\omega x}{v_1}\right) \right] e^{j\omega t}$$

$$\xi_2 = \left[A_2 \sin\left(\frac{\omega y}{v_1}\right) + B_2 \cos\left(\frac{\omega y}{v_1}\right) \right] e^{j\omega t} \qquad (3)$$

$$\xi_3 = \left[A_3 \sin\left(\frac{\omega z}{v_3}\right) + B_3 \cos\left(\frac{\omega z}{v_3}\right) \right] e^{j(\omega t + \pi)},$$

where the phase velocities of the three uncoupled waves are computed by substituting the relations (3) in the (2):

$$v_1 = \sqrt{\frac{c_{11}^D}{\rho}} \; ; \; v_3 = \sqrt{\frac{c_{33}^D}{\rho}} . \qquad (4)$$

The constants A_1, B_1, A_2, B_2, A_3 and B_3 must be computed by satisfying the mechanical boundary conditions: in order to take into account the external media around the element, we assume the continuity of the velocities on the element external surfaces:

$$\dot{\xi}_1(0)=u_1 \ ; \ \dot{\xi}_1(w)=-u_2 \ ; \ \dot{\xi}_2(0)=u_3 \ ; \ \dot{\xi}_2(L)=-u_4 \ ; \ \dot{\xi}_3(0)=u_5 \ ; \ \dot{\xi}_3(l)=-u_6 \quad (5)$$

With these boundary conditions the wave functions become:

$$\xi_1 = \frac{1}{j\omega}\left[u_1 \cos\left(\frac{\omega x}{v_1}\right) - \frac{u_1 \cos\vartheta_1 + u_2}{\sin\vartheta_1}\sin\left(\frac{\omega x}{v_1}\right)\right]$$

$$\xi_2 = \frac{1}{j\omega}\left[u_3 \cos\left(\frac{\omega y}{v_1}\right) - \frac{u_3 \cos\vartheta_2 + u_4}{\sin\vartheta_2}\sin\left(\frac{\omega y}{v_1}\right)\right] \quad (6)$$

$$\xi_3 = \frac{1}{j\omega}\left[u_5 \cos\left(\frac{\omega z}{v_3}\right) - \frac{u_5 \cos\vartheta_3 + u_6}{\sin\vartheta_3}\sin\left(\frac{\omega z}{v_3}\right)\right].$$

The external behavior of the element is computed by imposing the continuity between the stresses and the forces on its surfaces, but the functions (6) do not satisfy these conditions; to this end we can use a weaker form: we assume that on every external surface of the element only the integral of the stress is equilibrated by the external force:

$$\int_{\sigma_1} T_1(0)\,d\sigma = -F_1 \ ; \ \int_{\sigma_2} T_1(w)\,d\sigma = -F_2 \ ;$$

$$\int_{\sigma_3} T_2(0)\,d\sigma = -F_3 \ ; \ \int_{\sigma_4} T_2(L)\,d\sigma = -F_4 \ ; \quad (7)$$

$$\int_{\sigma_5} T_3(0)\,d\sigma = -F_5 \ ; \ \int_{\sigma_6} T_3(l)\,d\sigma = -F_6 \ ;$$

where σ_1 and σ_2 are the external surfaces of the element orthogonal to the x axis in $x = 0$ and $x = w$ respectively, σ_3 and σ_4 are the surfaces orthogonal to the y axis in $y = 0$ and $y = L$, while σ_5 and σ_6 are the surfaces orthogonal to the z axis in $z = 0$ and $z = l$. By substituting the constitutive eqs. (1) in the boundary eqs. (7), and by using the relation between the current and the electrical displacement ($D_3 = I/(j\omega\sigma)$), we obtain:

$$F_1 = \frac{Z_1}{j}\left(\frac{u_1}{\tan\vartheta_1}+\frac{u_2}{\sin\vartheta_1}\right)+\frac{c_{12}^D\,l}{j\omega}(u_3+u_4)+\frac{c_{13}^D\,L}{j\omega}(u_5+u_6)+\frac{h_{31}\,l}{j\omega\,w}I$$

$$F_2 = \frac{Z_1}{j}\left(\frac{u_1}{\sin\vartheta_1}+\frac{u_2}{\tan\vartheta_1}\right)+\frac{c_{12}^D\,l}{j\omega}(u_3+u_4)+\frac{c_{13}^D\,L}{j\omega}(u_5+u_6)+\frac{h_{31}\,l}{j\omega\,w}I$$

$$F_3 = \frac{c_{12}^D\,l}{j\omega}(u_1+u_2)+\frac{Z_2}{j}\left(\frac{u_3}{\tan\vartheta_2}+\frac{u_4}{\sin\vartheta_2}\right)+\frac{c_{13}^D\,w}{j\omega}(u_5+u_6)+\frac{h_{31}\,l}{j\omega\,L}I$$

$$F_4 = \frac{c_{12}^D\,l}{j\omega}(u_1+u_2)+\frac{Z_2}{j}\left(\frac{u_3}{\sin\vartheta_2}+\frac{u_4}{\tan\vartheta_2}\right)+\frac{c_{13}^D\,w}{j\omega}(u_5+u_6)+\frac{h_{31}\,l}{j\omega\,L}I \quad (8)$$

$$F_5 = \frac{c_{13}^D\,L}{j\omega}(u_1+u_2)+\frac{c_{13}^D\,w}{j\omega}(u_3+u_4)+\frac{Z_3}{j}\left(\frac{u_5}{\tan\vartheta_3}+\frac{u_6}{\sin\vartheta_3}\right)+\frac{h_{33}}{j\omega}I$$

$$F_6 = \frac{c_{13}^D\,L}{j\omega}(u_1+u_2)+\frac{c_{13}^D\,w}{j\omega}(u_3+u_4)+\frac{Z_3}{j}\left(\frac{u_5}{\sin\vartheta_3}+\frac{u_6}{\tan\vartheta_3}\right)+\frac{h_{33}}{j\omega}I,$$

where:

$$\vartheta_1 = \frac{\omega w}{v_1} \quad ; \quad \vartheta_2 = \frac{\omega L}{v_1} \quad ; \quad \vartheta_3 = \frac{\omega l}{v_3} . \tag{9}$$

$Z_1 = \rho\, v_1\, L\, l$, $Z_2 = \rho\, v_1\, w\, l$, $Z_3 = \rho\, v_3\, w\, L$ can be seen, respectively, as the piezoceramic acoustic impedances of the element along the x, y and z directions. As far as the electric field is concerned, by substituting eqs. (6) in eqs. (1), we have:

$$E_3 = \beta_{33}^S D_3 - j \left\{ \frac{h_{31}}{v_1} \left[u_1 \sin\left(\frac{\omega x}{v_1}\right) + \frac{u_1 \cos \vartheta_1 + u_2}{\sin \vartheta_2} \cos\left(\frac{\omega x}{v_1}\right) \right] + \frac{h_{31}}{v_1} \left[u_3 \sin\left(\frac{\omega y}{v_1}\right) + \right. \right.$$

$$\tag{10}$$

$$\left. \left. + \frac{u_3 \cos \vartheta_2 + u_4}{\sin \vartheta_2} \cos\left(\frac{\omega y}{v_1}\right) \right] + \frac{h_{33}}{v_3} \left[u_5 \sin\left(\frac{\omega z}{v_3}\right) + \frac{u_5 \cos \vartheta_3 + u_6}{\sin \vartheta_3} \cos\left(\frac{\omega z}{v_3}\right) \right] \right\}.$$

The voltage V across the electrodes is computed by integrating E_3 along the z axis. The obtained result is a function of x and y, in contrast with the assumption that the surfaces orthogonal to the z axis are metallized and therefore equipotential. This is a direct consequence of the choice of three orthogonal wave functions as solutions of the differential wave equations. This discrepancy can be overcome by integrating relation (10) also along x and y [2, 5], obtaining:

$$V = \frac{h_{31}\, l}{j\omega w} (u_1 + u_2) + \frac{h_{31}\, l}{j\omega L} (u_3 + u_4) + \frac{h_{33}}{j\omega} (u_5 + u_6) + \frac{1}{j\omega C_0} I, \tag{11}$$

where $C_0 = (w\, L)/(\beta_{33}^S\, l)$ is the so called piezoceramic "clamped capacity".

An alternative approach to make the voltage V independent of x and y is to impose $h_{31} = 0$ [3]. With this assumption, a new equivalent material is obtained, slightly less piezoelectrically active; the external behavior of the element, in this second case, is described by 7 equations obtained from eqs. (8) and eq. (11) simply by imposing $h_{31} = 0$. With both approximations, eqs. (8) and eq. (11) can be written in a matrix form: $F = A\, u$ where F is the vector of the forces F_i and of the voltage V, u is the vector of the velocities u_i and of the current I and A is the 7x7 matrix representing the element model in the frequency domain. By loading the mechanical ports with the acoustical impedances of the surrounding media and by applying an a. c. voltage V to the electric port, it is possible to compute the electrical input impedance ($Z_i = V/I$), the transmission ($TTF = F_i/V$) and the receiving ($RTF = V/F_i$) transfer functions.

RESULTS OF THE THREE–DIMENSIONAL MODEL

In order to compare the results obtained with the two previously described alternative approximations we computed, in both cases, the electrical input impedance of a piezoelectric element with $w = 20\ mm$, $L = 10\ mm$ and $l = 4\ mm$ and we compared the obtained results with a measurement made by means of an HP 4194–A impedance analyzer. As it can be seen in fig. 2 the best approximation of the measured result is obtained by integrating the electric field also along y and z (Fig. 2a).

In order to ascertain the capabilities of the model to describe the coupling between the three fundamental (thickness, length and width) modes and their harmonics, we computed the frequency spectrum of an element with $L = 0.5\ mm$, $l = 1\ mm$, by varying the ratio $G = w/l$ from 0.1 to 10 and by assuming resonance at the frequency of maximum electrical input impedance. Fig. 3 shows the results obtained; the radius of the circles is proportional to the effective coupling factor which, as it is well known, for each mode is defined as:

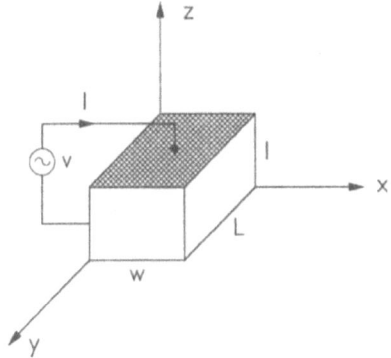

Figure 1. Geometry of the piezoceramic element.

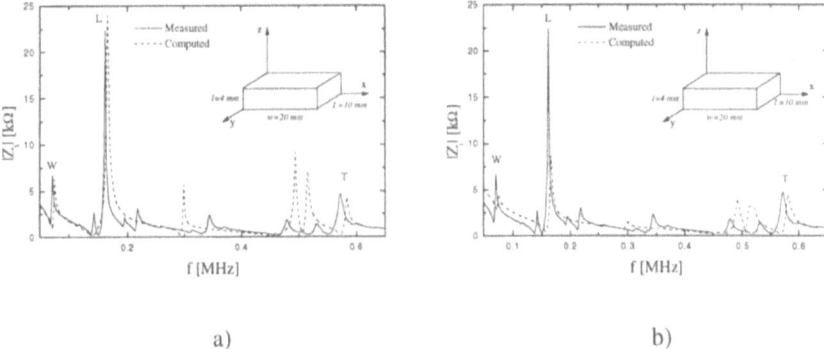

a)   b)

Figure 2. Comparison between the measured electrical input impedance and those computed by integrating also the electric field along x and y (a) and by imposing $h_{31} = 0$ (b).

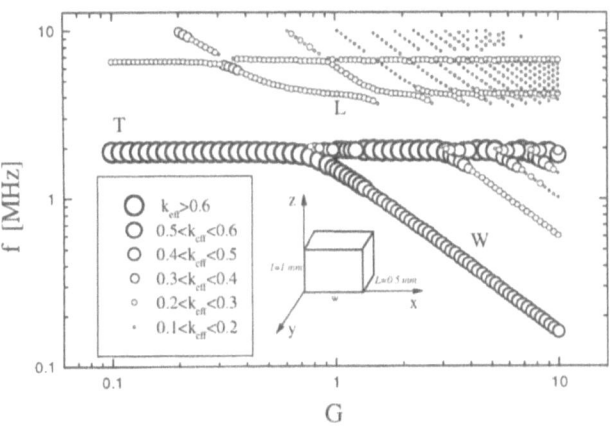

Figure 3. Frequency spectrum of the piezoceramic element shown in figure. T, W, and L indicate the fundamental thickness, width and length modes respectively. The radius of the circles is proportional to the effective coupling factor.

$$k_{eff}^2 = \frac{f_p^2}{f_p^2 - f_s^2} \qquad (12)$$

where f_p and f_s are the frequency of maximum and minimum impedance respectively. In the spectrum it is easy to recognize the three fundamental modes (T, W, L) and their harmonics; the coupling between them is pointed out by a shift of the resonance frequencies. As expected the higher k_{eff} value is obtained for the thickness mode T and, when the width mode W is strongly coupled with it, i.e. it has a resonance frequency close to that of the thickness, ($G \approx 1$) an increase of its k_{eff} can be observed.

CONCLUSIONS

In this work a new three–dimensional model of the rectangular piezoceramic element, able to take into account the interaction with the surrounding media, is described. Three general orthogonal plane wave functions, each depending only on one Cartesian axis corresponding to the propagation direction, are taken as solutions of the differential wave equation system which describes the vibration of the element. As a consequence of this choice, mechanical and electrical boundary conditions are satisfied only approximately, in an integral form. With this approach the external behavior of the piezoelectric element is described in the frequency domain by a system with 6 mechanical ports (one for each external face) and 1 electrical port. The electrical input impedance of a sample was computed by the model and compared with experimental results, obtaining a good agreement. The proposed model is able to preview the coupling of the thickness mode with the width (x direction) and the length (y direction) modes. It can be used to design the best geometry of the element in order to maximize the conversion energy of the desired mode. Further, it can be considered as a first step in order to develop a model of a piezoelectric composite of any connection.

REFERENCES

1. M. Brissaud: "Rectangular piezoelectric plate loaded on each face: a three–dimensional approach", *Ultrasonics*, vol. 34, pp. 87–90, 1996.

2. A. Iula, N. Lamberti, G. Caliano, M. Pappalardo: "A new three–dimensional model for circular piezoelectric transducers", *Acoustical Imaging*, Vol. 21, J. P. Jones Ed., Plenum Press, New York, pp. 139–144, 1995.

3. N. Lamberti and M. Pappalardo: "A general approximated two–dimensional model for piezoelectric array elements", *IEEE Trans. on Ultras., Ferroelec. and Freq. Control*, vol. 42, no. 2, pp. 243-252, March 1995.

4. D. A. Berlincourt, D. R. Curran and H. Jaffe, "Piezoelectric and piezomagnetic materials and their function in transducers," in *Physical Acoustic*, vol. 1, W. P. Mason Ed., New York, Academic Press, pp. 169-270, 1964.

5. G. Hayward and D. Gillies, "Block diagram modeling of tall thin parallelepiped piezoelectric structures," *J. Acoust. Soc. Am.*, vol. 86(5), pp. 1643-1653, Nov. 1989.

SCANNING AND IMAGING USING LAMB WAVES

Ali C. Lassal and Peter A. Payne

Department of Instrumentation and Analytical Science
UMIST
P O Box 88
Manchester M60 1QD, UK

INTRODUCTION

Lamb waves propagate readily along thin plates and have been exploited for non-destructive testing applications and for use as a means of remote sensing in which the active transducer can be sited a considerable distance away from the measurand, which could be of a hazardous nature.

Here we report on some experimental studies using Lamb waves at somewhat higher frequencies than have been employed previously. The original impetus for these studies was the wish to investigate possibilities for ultrasonic measurements of hard and soft tissue within the mouth. The eventual aim of this would be to offer diagnostic techniques that avoid some of the hazards of dental X-rays. A Lamb wave-based waveguide can be constructed from thin stainless steel material and our experiments at a centre frequency of 20 MHz indicate that a zero order asymmetrical Lamb wave will propagate several centimetres along such a waveguide without significant loss or beam spread. We wish to employ waveguides since it proves difficult, if not impossible, to introduce ultrasound transducer arrays into the mouth.

In order to make use of such a device, we would expect to bring the waveguide into contact with soft gingival tissue in the mouth which, due to its liquid-like properties, will mode convert the Lamb wave into a compressional wave. Echoes from targets will cause return compressional waves which will mode convert at the tissue-waveguide junction giving a return Lamb wave for subsequent processing. If the device were to be used for imaging hard tissue junctions in the teeth, such as those between the enamel, the dentine and the pulp, then the provision of a water coupling jet would be fairly simple. This would enable the same mode conversion process to occur, enabling B-scan images to be obtained.

ULTRASOUND WAVEGUIDES IN MEDICINE

Reports on studies of ultrasonic waveguides go back as far as 1943[1] and there are many subsequent reports on work for industrial non-destructive testing and evaluation purposes.

Among the first reports of the use of waveguides in a medical context dates back to 1974[2] with follow-up work reported in 1983[3]. A series of papers has been published by Nicholson and McDicken and colleagues over the years from 1984 to 1996[4-9]. These papers reported on the use of wire and tubular waveguides at frequencies of up to 5 MHz. They looked at a number of solid wire waveguides between 0.5 and 1.0 mm in diameter and concluded that the best material for such a guide is stainless steel.

The work that we report in this paper has been carried out as part of doctoral programme aimed at better understanding some of the practical problems and solutions in connection with scanning and imaging for dentistry applications.

WAVEGUIDE MATERIALS STUDIES

For reasons of resolution requirements for dental investigations, we elected to work at a centre frequency of 20 MHz, although even higher frequencies would be preferable[10]. We also decided to work with only the A_0 Lamb wave which means that attention must be paid to the frequency-thickness product for the material to be used. Using equations given by Victorov[11] for both symmetrical and antisymmetrical modes, we have computed the group and phase velocities for both the A_0 and S_0 modes over a frequency-thickness product range of 0 to 6 MHz-mm for materials with longitudinal velocities of 5,700 m s^{-1} and shear velocities of 3,150 m s^{-1}. These velocities correspond to stainless steel.

Since we are also interested in the mode conversion mechanisms at the interface between water or liquid-like materials and the waveguide, the angle of incidence against frequency-thickness product for 0 to 6 MHz-mm was computed. The results of both computations were then used to guide us in the experiments reported below.

Experimental Work

Initially, we set out to determine experimentally whether the various types of thin stainless sheet available gave rise to differences in performance as waveguide materials. From the work done previously, the thickness of stainless steel required is of the order of 150 μm and we were able to acquire samples of eleven different compositions of stainless steel for our measurements. Some of these were described as hard stainless steels and others as annealed. The thickness ranged from 0.10 mm up to 0.20 mm. These materials were each fabricated into a waveguide suitable for inserting into our test rig which is shown in diagrammatic form in Figure 1 (transmit end only). This is similar to the method described by Avioli[12], but we found the use of a small aperture at the base of the water tank sealed by an O-ring far superior to that of the use of a thin polymer window which Avioli employed. For the pitch-catch experiments we used a function generator, power amplifier and digital oscilloscope all under computer control. In Figure 2 we show typical received signals when the spacing between the launching and receiving water tanks was 10, 40 and 70 mm. The driving signal was a tone burst of five sinusoids at 20 MHz modulated by a weighting function in order to produce a reasonably smooth frequency spectrum[13].

In addition to pitch-catch experiments, we also conducted pulse echo measurements in which the echo was generated from the edge of the waveguide, as illustrated in Figure 3. Figure 4 shows a typical echo signal, together with its frequency spectrum for a waveguide shaped as shown in Figure 3 and where the narrow part of the guide was 5 mm wide. In this case, the waveguide was 0.15 mm thick. Most of the waveguide materials tested exhibited surface characteristics which easily identified the direction of rolling, therefore, our measurements were also conducted along the rolling direction and at right angles to it.

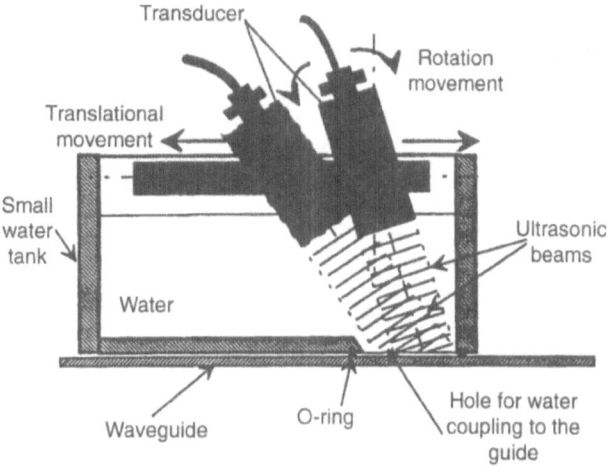

Figure 1. Diagram illustrating the method used to adjust the incident angle for excitation of A_0 waves. For pitch-catch experiments a second transducer and water tank were positioned at the far end of the waveguide.

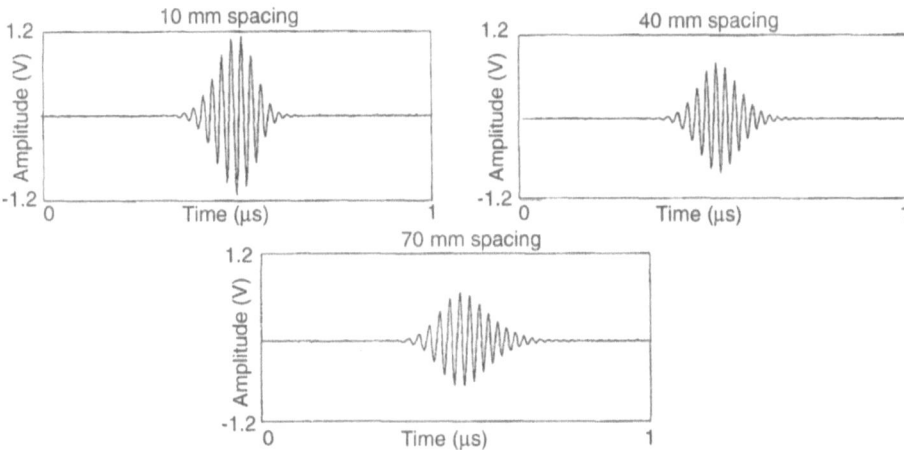

Figure 2. Received signals in the pitch-catch mode at 10, 40 and 70 mm water tank spacing.

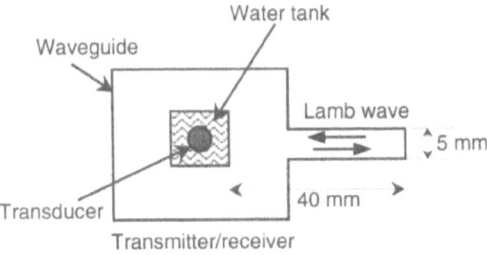

Figure 3. Pulse-echo experimental arrangement.

357

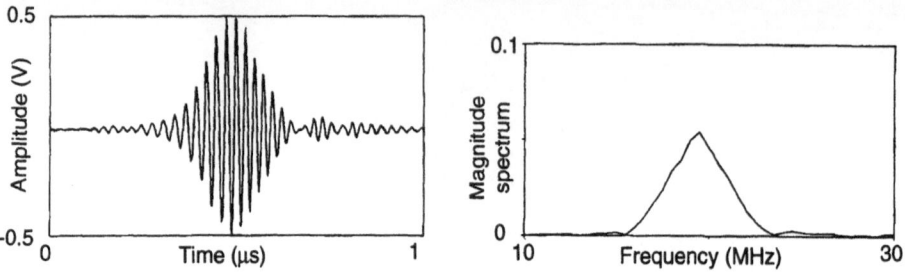

Figure 4. Echo signal obtained from the waveguide end and its frequency spectrum.

Figure 5 shows the results of measurements of angle of incidence which we found were typical for the materials tested. For annealed materials, the angle of incidence for the rolling direction and the transverse direction were almost identical, whereas for the hard stainless steels, quite considerable differences in angle of incidence were measured. Further work is required to fully explain these differences.

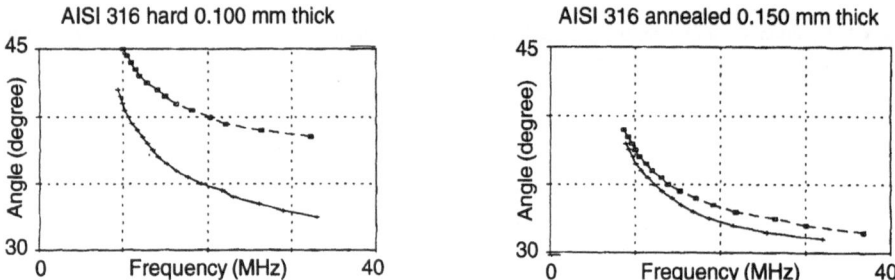

Figure 5. Measured angles of incidence at a range of spot frequencies for hard and annealed stainless steel waveguides.

By using the type of result illustrated in Figure 2 we were able to accurately measure the attenuation characteristics of the various types of stainless steel available. Table 1 shows the results obtained. The various gaps in the table simply indicate unavailability of material at the

Table 1. Measured attenuation (dB/mm) for different types of stainless steel waveguide materials.

Type of stainless steel	Thickness (mm)			
	0.100	0.125	0.150	0.200
AISI 302 (hard)	1.50	1.54	2.72	3.05
AISI 302 (annealed)	-	1.80	-	4.72
AISI 304 (annealed)	1.25	1.25	-	4.27
AISI 316 (hard)	1.26	-	-	-
AISI 316 (annealed)	-	3.12	1.15	-
AISI 321 (annealed)	0.70	0.97	1.55	4.18
AISI 347 (annealed)	0.94			1.22
15-7 PH (annealed)	-	0.82	-	-
17-7 PH (annealed)	-	0.80	-	-
Dunham (hard)	-	2.75	-	2.03
Stubs (hard)	-	-	3.79	-

thickness shown. All results were obtained using a 20 MHz centre frequency signal, as described previously. Clearly, the attenuation coefficients are smaller in the thinner waveguides, 0.1 and 0.125 mm thick. We know that the ratio of the vertical to longitudinal displacement components increases monotonically with frequency-thickness product and these results confirm that as this ratio increases, so the attenuation coefficient increases. This is in line with the work of Meitzler[14].

BEAM PLOTTING EXPERIMENTS

One of the aims of the work reported here was to discover the type of compressional wave field established following mode conversion at a water interface. To this end, our experimental rig was altered to enable a miniature underwater hydrophone to be used to probe the fields produced by a number of different waveguides. The polyvinylidene fluoride (PVDF) hydrophone was attached to a X-Y-Z scanning system and, in most cases, the field plotting was conducted under computer control over a period of several hours (usually overnight). The transducer was a 6 mm diameter flat device used in the near-field. Figure 6 shows hydrophone responses as it is traversed across the beam in front of the waveguide. As can seen, little beam spread has occurred when we compare results at distances of 10, 25 and 40 mm from the guide which was 45 mm long. These results are quite typical of many obtained.

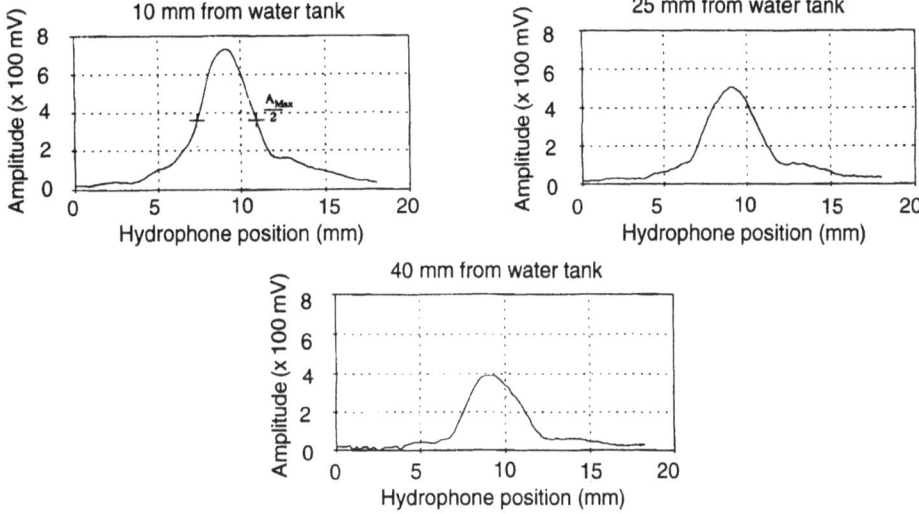

Figure 6. Scans across the mode converted field for a waveguide length of 45 mm and the hydrophone places at 10, 25 and 40 mm from the guide.

The last set of experiments reported here are connected with the concept of producing a focused field. Figure 7 shows the three-dimensional plot of the ultrasonic field distribution for a simple flat waveguide when the hydrophone is positioned at 25 mm from the guide. For the same set-up, we then introduced a focused area into the waveguide by the use of a steel ball bearing to produce a depression in the plate (see Figure 8). If we then scan in the same position at about 25 mm from the waveguide, the improved field plot is as indicated in Figure 9. A considerable degree of focusing is clearly obtained. In fact, by scanning at a distance of 17 mm from the waveguide, an even tighter focused field is obtained.

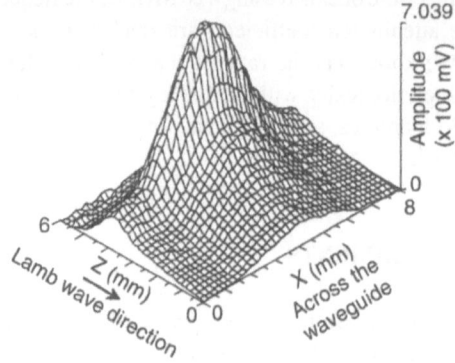

Figure 7. Three-dimensional plot for the ultrasonic field distribution from a flat waveguide with a hydrophone 25 mm from the guide.

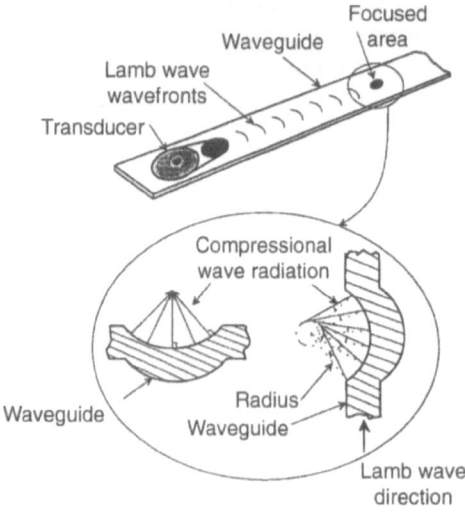

Figure 8. Diagram illustrating method used for focusing (radius of depression = 10 mm).

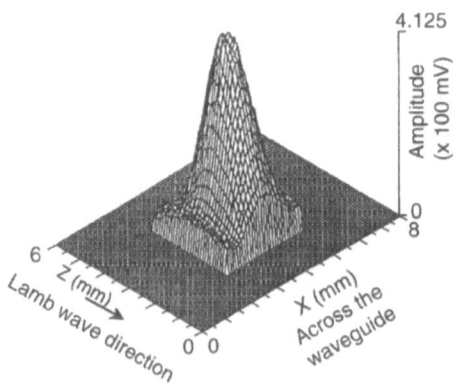

Figure 9. Three-dimensional plot for the ultrasonic field distribution from a focused waveguide with a hydrophone 25 mm from the guide.

CONCLUSIONS

We have conducted a series of experiments aimed at better understanding of the influence of various types of stainless steel employed as high frequency ultrasonic waveguides. Our results indicate that considerable differences exist between hard and annealed stainless steels in terms of incident angles, but that there is little difference between these materials in terms of attenuation coefficients. Attenuation coefficient is mainly determined by the frequency-thickness product which needs to be less than 2 MHz-mm. Our beam plotting experiments show that Lamb waves will travel considerable distances in the waveguide without significant spreading. They also indicate that simple techniques can be readily employed to obtain a significant degree of beam focusing.

REFERENCES

1. G.E. Hudson, Dispersion of elastic waves in solid circular cylinders, *J. Acoust. Soc. Am.* 63:46 (1943)
2. B. Woodward and A.J. Allen, Importance of calibration in medical ultrasonics, *Brit. J. Radiol.* 47:707 (1974)
3. B. Woodward, A waveguide pulse-echo system for ultrasonic absorption measurements in biological media, *J. Biomed. Eng.* 5:343 (1983)
4. N.C. Nicholson and W.N. McDicken, Waveguides in medical ultrasonics, *Ultrasonics* 26:27 (1988)
5. N.C. Nicholson, W.N. McDicken and T. Anderson, Waveguides in medical ultrasonics: an experimental study of mode propagation, *Ultrasonics* 27:101 (1989)
6. N.C. Nicholson and W.N. McDicken, Mode propagation in waveguides used in medical ultrasonics, *Ultrasonics* 29:133 (1991)
7. N.C. Nicholson and W.N. McDicken, Mode propagation of ultrasound in hollow waveguides, *Ultrasonics* 29:411 (1991)
8. N.C. Nicholson, W.N. McDicken and T. Anderson, Waveguides in medical ultrasonics: effect of waveguides medium upon mode amplitude, *Ultrasonics* 30:82 (1992)
9. N.C. Nicholson, W.N. McDicken and T. Anderson, A comparison of coupling horns for waveguides used in medical ultrasonics, *Ultrasonics* 34:747 (1996)
10. R. Wichard, J. Schlegel, R. Haak, J.F. Roulet and R.M. Schmitt, Dental diagnosis by high frequency ultrasound, in: *Acoustical Imaging, Vol. 22*, P. Tortoli and L. Masotti, ed., Plenum Press, New York (1996)
11. I.A. Viktorov. *Rayleigh and Lamb Waves*, Plenum Press, New York (1967)
12. M.J. Avioli, Lamb wave inspection for large cracks in centrifugally cast stainless steel, *EPRI Research Project 2405-23*, Georgetown University (1988)
13. D. Alleyne and P. Cawley, A two-dimensional Fourier transform method for the measurement of propagating multimode signals, *J. Acoust. Soc. Am.* 89:1159 (1991)
14. A.H. Meitzler, Selective attenuation of elastic wave motions in strips of polycrystalline metals, *J. Acoust. Soc. Am.* 34:444 (1962)

FIELD PROPAGATION VIA THE
ANGULAR SPECTRUM METHOD

Sidney Leeman and Andrew J. Healey

Department of Medical Engineering and Physics
King's College School of Medicine and Dentistry
King's College Hospital (Dulwich)
London SE22 8PT, U.K.

INTRODUCTION

The angular spectrum technique is the standard method for extrapolating ultrasound fields from a data set consisting of point measurements of the pressure field in a single plane[1]. Since the field distribution on any other plane may then, in principle, be predicted via relatively straightforward computation, the method may be seen as an exact technique for imaging the entire acoustic field. In practice, however, the approach suffers from two significant disadvantages:

(a) Backward propagation is necessary when the field is desired over a region which lies closer to the transducer than the measurement region. This is an important case, since field extrapolation back to the transducer face is commonly seen as an important first step towards characterising the *transducer* (rather than the field). However, backward propagation is accurate over only relatively short distances because of the occurrence, in the formalism, of problematic evanescent waves.

(b) The method is essentially devised for harmonic fields, and present algorithms for imaging transient fields are cumbersome and computationally intensive.

This paper addresses both of these problems: it presents a novel approach towards handling the evanescent waves, and also points the way towards a new, more efficient, approach for moderately rapid prediction of transient fields over an extended volume, from a data set obtained only in a region of relatively limited spatial extent. However, the latter is not fully proven here: some results are presented. When such a method is coupled with an appropriate data visualisation technique, it may be described as a 'digital acoustic camera'. Since the (measured) data set contains essentially all the information for establishing the field at all space and time points, it is the obvious starting point for seeking an effective means for characterising not only the field, but also, to a more limited extent, the transducer itself.

THE ANGULAR SPECTRUM TECHNIQUE

At the outset, it should be borne in mind that the angular spectrum approach was originally formulated in the context of harmonic fields with frequency ω, whose (spatial) part can be written as:

$$p_\omega(x,y,z) \doteq \int\limits_{-\infty}^{\infty} dk_x dk_y \, P(k_x, k_y, z) \, \exp(-i[k_x x + k_y y]) \tag{1}$$

If the field $p_\omega(x,y,z)$ is considered to be a solution of the Helmholtz equation describing an ideal lossless medium with acoustic velocity C, then it is easily shown that

$$\frac{\partial^2}{\partial z^2} P(k_x, k_y, z) + k_z^2 P(k_x, k_y, z) = 0, \quad \text{with} \quad k_z^2 = (\omega/C)^2 - \left(k_x^2 + k_y^2\right) \tag{2}$$

The above equation has as its forward propagating (into the z-direction) solution:

$$P(k_x, k_y, z) = P(k_x, k_y, 0)e^{ik_z z} \equiv A_\omega(k_x, k_y)e^{ik_z z} \tag{3}$$

A_ω denotes the angular spectrum, for the frequency ω, and is essentially the Fourier transform of the known values of the (harmonic) field, as measured on some plane, designated by $z = 0$. Thus the field may be expressed in terms of its angular spectrum via

$$p_\omega(x,y,z) = \int\limits_{-\infty}^{\infty} dk_x dk_y \, A_\omega(k_x, k_y) \, \exp(-i[k_x x + k_y y]) \, \exp(ik_z z) \tag{4}$$

This last equation allows the field on any plane, z, to be expressed in terms of the angular spectrum, obtained from measurements carried out over the plane $z = 0$. However, there is the possibility that k_z may become a negative, purely imaginary, quantity: for forward extrapolation ($z > 0$) there is no problem, but, in the reverse direction, such contributions grow exponentially with associated numerical problems. This part of the contribution to the angular spectrum decomposition of the field is referred to as the 'evanescent wave' contribution, and consists of spectral components that are exponentially damped (growing) in the positive (negative) z-direction, even for a lossless medium. Clearly, the evanescent waves are 'unphysical', and arise only because of the particular field decomposition that is posited.

In practice, the harmonic pressure field over a plane is measured with a point hydrophone, and the angular spectrum computed from the data by Fourier transformation. The field distribution on some other z-plane is then computed by applying Equation (4). The complicating effects of the evanescent waves appear when propagating backwards, and (if measurement- and noise- considerations are disregarded) are a consequence of the *numerical* errors that arise when dealing with the potentially unlimited growth of the associated exponential factors. In fact, the evanescent components are essential for successful back propagation, and good results can be achieved provided that such short distances are involved that the exponential growth of the evanescent terms remains computationally manageable. This is explicitly demonstrated in Figure 1, which utilises the computer simulation of the field from a circular aperture over which $p_\omega(x,y,z_0)$ is uniform for all time. The field is calculated (via the exact angular spectrum technique) on a plane located at 1/16 of the last axial maximum distance. This is then taken as the input to a backpropagation algorithm based on the angular spectrum technique. The field at the transducer face

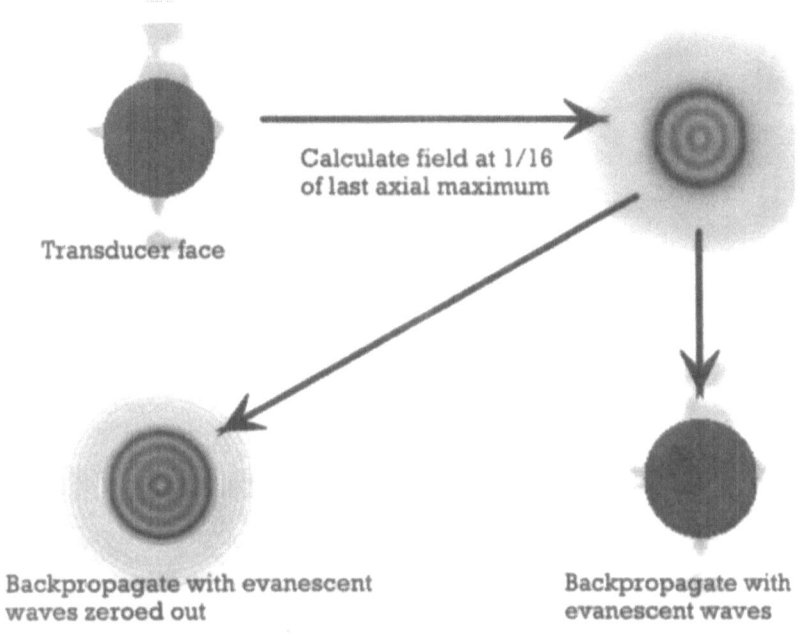

Calculate field at 1/16
of last axial maximum

Transducer face

Backpropagate with evanescent
waves zeroed out

Backpropagate with
evanescent waves

Figure 1. Demonstration that evanescent waves are essential for accurate backpropagation. The distance involved is small enough for numerical problems to be contained.

is recovered perfectly if the evanescent waves are included. Note that, for such a short distance, exponentially increasing terms arising from the evanescent waves remain manageable. If the evanescent waves are zeroed out before backpropagation, somewhat less accurate results are obtained: clearly, such a procedure should be adopted with caution when the retention of the evanescent wave component results in numerical difficulties. Poor results (not shown) are obtained when this computer experiment is repeated for a simulated field on a plane located at distance equal to 1/8 of the last axial maximum: in this case the exponential terms already become numerically unmanageable (IEEE 8 byte floating point precision), and the neglect of the evanescent wave component -- although somewhat improving matters -- continues to give an unsatisfactory result.

PROJECTING OUT THE EVANESCENT WAVES

The problems posed by the evanescent waves may be overcome by considering some remarkable properties of the projections of the field. Consider the particular case of the projection, $P_{PROJ}(z)$, formed by integrating over the x-y plane located at distance z:

$$P_{PROJ}(z) \equiv \iint_{-\infty}^{\infty} dxdy\, p_\omega(x,y,z) = \iint_{-\infty}^{\infty} dxdy \iint_{-\infty}^{\infty} dk_x dk_y\, P(k_x,k_y,z)\, \exp\left(-i[k_x x + k_y y]\right)$$

$$= \iint_{-\infty}^{\infty} dxdy \iint_{-\infty}^{\infty} dk_x dk_y\, A_\omega(k_x,k_y)e^{ik_z z}\exp\left(-i[k_x x + k_y y]\right)$$

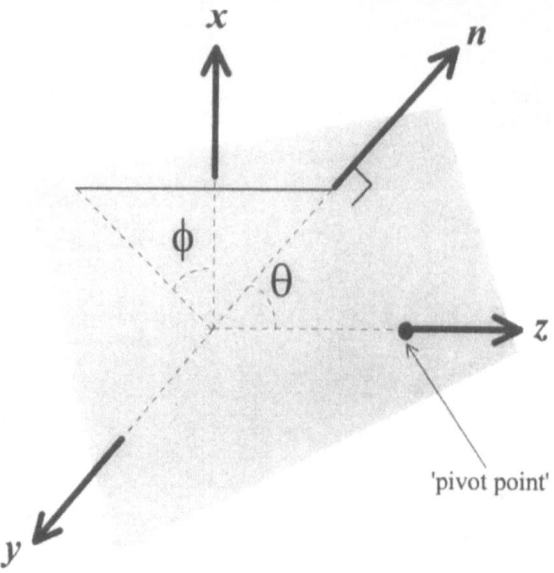

Figure 2: Orientation of plane over which the 'pivot point' projections are taken.

$$= \iint\limits_{-\infty}^{\infty} dk_x dk_y \, A_\omega(k_x, k_y) e^{ik_z z} \delta(k_x) \delta(k_y) = A_\omega(0, 0) e^{i\omega z/C} \tag{5}$$

where δ denotes the Dirac delta function.

Note that the above result proves that evanescent waves are not involved in the 'propagation' of this particular projection of the field, which obeys a (spatial) propagation law similar to that of a plane wave of frequency ω. Thus, the projections of the field may be backpropagated, over even relatively large distances, without any fear of exponentially increasing factors! This result may be extended to the general case, with the projection plane taken as indicated in Fig. 2. In this case, the value of the projection relates (apart from some geometrical factor) to the angular spectrum value $A_\omega(\omega n_x, \omega n_y)$, where $\boldsymbol{n}$ is the unit vector normal to the projection plane. This projection 'propagates' along the $\boldsymbol{n}$-direction as a plane wave of frequency ω. Once again, evanescent waves are not involved. In the case illustrated here, the pivot point (*i.e.*, the unique point which all projection planes of a given data set share) is assumed to be located at the same z-value as the plane in which the field measurements for the evaluation of the angular spectrum are made. However, this is not an essential assumption, and the theory is readily accommodated to the situation that this requirement is not satisfied. The important feature is that, at no stage, are angular spectrum values corresponding to evanescent waves utilised.

Thus the angular spectrum may be measured in the usual way, and its (non-evanescent) values related to the values of the appropriate pivoted angle projections. The latter may be straightforwardly propagated -- either forwards or backwards -- with no complications, to give the values of the angle projections corresponding to some other pivot point. It remains only to show how the field in a plane located at a z-value corresponding to the new pivot point may be reconstructed from the pivoted angle projections, in order to establish this new method for bi-directional field propagation using measured angular spectrum values.

A FIELD RECONSTRUCTION METHOD

The basis of the reconstruction technique is the directivity spectrum, $D(\mathbf{k})$, representation of the space-time dependence of the field [2]

$$p(\mathbf{r},t) = \int\limits_{-\infty}^{\infty} d^3\mathbf{k}\, D(\mathbf{k}) \exp{(i[\mathbf{k}\bullet\mathbf{r} - Ckt])} \qquad (6)$$

which represents the general solution of the canonical wave equation for lossless media. The important feature of this formalism is that there is no need for evanescent waves. For the case of interest here (harmonic waves), this field description reduces to:

$$p_{\omega}(x,y,z) = \int\limits_{-\infty}^{\infty} dk_x dk_y dk_z\, D(k_x, k_y, k_z).\exp{(i[k_x x + k_y y + k_z z])}.\delta\left(k - \frac{\omega}{C}\right)$$

$$= \left(\frac{\omega}{C}\right)^2 \int\limits_0^{\pi} \sin\vartheta d\vartheta \int\limits_0^{2\pi} d\varphi.\, D\left(\vartheta, \varphi, \frac{\omega}{C}\right).e^{i\frac{\omega}{C}z\cos\vartheta}.e^{i\frac{\omega}{C}\sin\vartheta(x\cos\varphi + y\sin\varphi)} \qquad (7)$$

where the directivity spectrum has been expressed in terms of the spherical polar coordinate representation of the $\mathbf{k}$-variable, (k, ϑ, φ).

The directivity spectrum components appearing in Equation (7) are, apart from an inessential factor, the values of the pivoted angle projections of the field, as may be verified with moderate effort. Thus, in this way, the field may be reconstructed from the appropriate projections. The following algorithm for field propagation from angular spectrum data, without evanescent waves, is thus proposed:

- Measure the field and calculate the angular spectrum in the usual way.
- Multiply the non-evanescent wave components of the angular spectrum by the appropriate k-space magnitude and phase factors, to give the pivoted angle projection values.
- Propagate the pivoted angle projections to the desired location (only a phase factor is required).
- Insert the modified pivoted angle projection values into an integral of form Equation (7) to reconstruct the field on a new (x,y)-plane.

The procedure works equally well for both forward and backward propagation. At no stage are evanescent waves encountered.

Although not specifically dealt with here, the procedure works equally well with wide-band fields. An example of the good results that can be obtained with this technique, when backpropagating empirically measured transient fields over significant distances, is shown in Figure 3.

CONCLUSIONS

It has been demonstrated that unphysical evanescent waves are not an essential feature of field propagation by the angular spectrum technique. Problems associated with back-

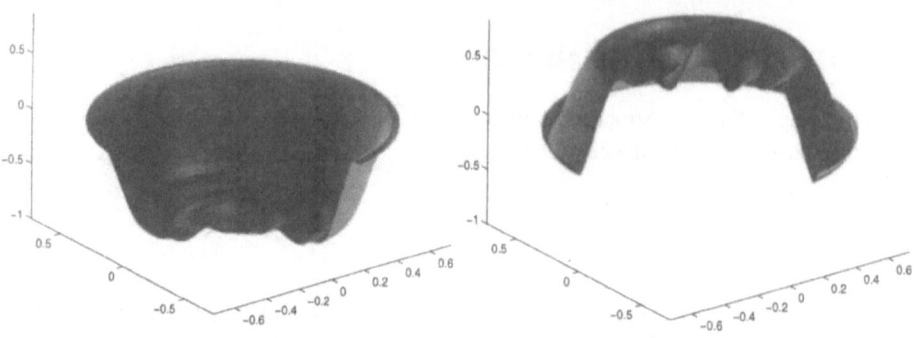

Figure 3: Pressure field back propagated to plane of transducer face, shown at two different time instants. The field was measured at 3.2 cm from the transducer face. The transducer was a Panametrics V309 5 MHz planar 19 mm diameter aperture, with impulse excitation.

propagation can be removed by transforming the problem to a new variable space in which only 'physical' (non-evanescent) waves appear. However, only data from the original angular spectrum are used. Transient fields may also be handled by the proposed method, but only some results, rather than the full theory, have been shown.

ACKNOWLEDGEMENTS

The Royal Society and The Wellcome Trust are thanked for their support.

REFERENCES

1. C.J. -Vecchio, M.E. Shafer, and P.A. Lewin, Prediction of ultrasonic field propagation through layered media using the extended angular spectrum method, *Ultrasound Med. Biol.* 20:611 (1994).
2. S. Leeman and A.J. Healey, Description and measurement of transient ultrasound fields, in: *Advanced Mathematical Tools in Metrology*, F. Pavese, ed., World Scientific, Singapore, 1995.

APPROXIMATIONS FOR LEAKY WAVE AMPLITUDES IN ACOUSTIC IMAGING: APPLICATIONS TO HIGH FREQUENCY SONAR

Philip L. Marston

Department of Physics
Washington State University
Pullman, WA 99164-2814

INTRODUCTION

Leaky waves that have been excited on elastic objects in water by ultrasonic waves can be a major contributor to the "ultrasonic images" of the elastic object. To facilitate the interpretation of such images, a quick and simple method of approximating the spatial dependence of leaky wave contributions to the radiated amplitude is needed which allows the object to be curved and/or the incident wavefronts which illuminate the object to be curved. The purpose of this paper is to summarize the development and testing of such an approximation. While the analytical details are being published elsewhere[1,2] some novel numerical tests of the method are included in the present paper. In addition, an example is shown of how to estimate the range of conditions for strong leaky wave excitation by combining spatial considerations involving the shape of the surface with information about the attenuation of the excited leaky wave and dephasing relative to the incident wave.

Figure 1(a) illustrates an example from the imaging experiments of Kaduchak et al.[3] where leaky waves produce the dominant contrast mechanism in an acoustical imaging system. In the experiments a leaky wave is excited on a tilted cylindrical shell illuminated by a 600 kHz ultrasonic beam from a rectangular transducer. The source and receiver were approximately 1.1 m from the target which was 62 cm long. The width of the beam at the target was much less than the length of the target while the illumination across the diameter of the target at a given point along its length was nearly uniform on the illuminated side of the target. The backscattering was detected with a rectangular transducer having a directional sensitivity corresponding to the footprint of the beam. The source and receiver were simultaneous scanned perpendicular to the direction of propagation of the illumination such that the illuminated and viewed footprints could scan the length of the cylinder and beyond. To summarize the experimental results it is necessary to introduce the tilt angle γ which, as shown in Fig. 1, is an angle defined by the dominant incident wave vector and the axis of the cylinder such that $\gamma = 0$ corresponds to broadside illumination. The experiments were concerned with tilted cylinders such that the specular reflection was directed away from the receiver. It was found that for certain tilt angles the most distant end of the cylinder would appear to be very much brighter than any of the other contributors to the acoustical image. The tilt angles γ for this enhancement were such that $\gamma \approx \theta_l$ where $\theta_l = \arcsin(c/c_l)$, $c_l > c$, c_l is the phase velocity of the leaky wave on the scatterer and c is the velocity of sound in the surrounding fluid (water); θ_l is the usual coupling angle for the synchronous excitation of the lth class of leaky wave.[4,5] In the experiments of Kaduchak et al.,[3] the condition that $\gamma \approx \theta_l$ leads to the geometric requirement that the leaky wave excited by the incident acoustic wave travels down the meridian of the cylinder which lies in the plane containing the cylinder's axis and the incident wave vector. Reflection of this meridional ray from the cylinder's end, which is cut perpendicular to the cylinder's axis, produces a reflected leaky meridional ray. Acoustic radiation from that ray produces acoustic wavefronts directed back towards the source of the sound. The signal received is especially strong since it may be shown[1] using the analytical methods summarized below that the Gaussian curvature of the radiated wavefront vanishes. It can be shown geometrically that points on the outgoing wavefronts having a vanishing Gaussian curvature produce farfield caustics.[6]

The formulation summarized here for quantitative analysis of radiated acoustic amplitudes for this type of situation is based on a surface convolution of the local pressure amplitude of the illuminating wave with an approximate leaky wave response function, which may be thought of as an amplitude "point spread function." The analytical and numerical results summarized here pertain to the excitation of meridional waves on cylinders. For the case of a truncated circular cylinder, helical leaky rays may be excited which reflect off the end of the cylinder and radiate sound back to the source. That class of rays, which occurs for $\gamma < \theta_l$, while evidently not as important as meridional rays in the experiment of Kaduchak et al.[3], has been demonstrated to occur in backscattering by solid truncated cylinders[7] and in hollow shells (see e.g. Ref. 8). Other observations of large amplitude meridional ray contributions to scattering have been carried out for shells[8] and solid cylinders.[9] In addition to the backscattering geometry shown in Fig. 1(a), meridional rays are also relevant to the understanding of the scattering of sound computed for the direction in the meridional plane of an infinite cylinder coincident with the direction of the specular reflection from the cylinder shown in Fig. 1(b). This is relevant to the interpretation of exact partial-wave series calculations for a solid cylinder tilted at the Rayleigh wave coupling angle.[10]

REVIEW OF APPROXIMATIONS FOR LEAKY WAVE AMPLITUDES

Bertoni and Tamir[11] introduced a method of approximating leaky Rayleigh wave contributions to the reflection of acoustic beams from flat surfaces which was subsequently put in the form of a one-dimensional convolution integral.[4] For the case of curved surfaces the initial approaches were based on separable geometries such as spheres and infinite cylinders. For example, Williams and Marston[12] numerically evaluated a residue associated with the Rayleigh wave pole in the Watson-transform formulation of backscattering by a solid sphere and obtained good agreement with measurements. The method was subsequently used to synthesize the exact backscattering amplitude as a function of frequency. The method of numerically evaluating residues tends to mask what are the physically relevant properties in determining the leaky wave amplitude. This motivated Marston[13,14] to derive an approximation for the relevant residues in the form of a "coupling coefficient" for the cases of spheres and circular cylinders at normal incidence which is found to be proportional to the leaky wave attenuation rate from radiation damping. The derivation was based on using the ray formulation to approximate resonances and then matching the relevant residues in the complex frequency plane found from resonant scattering theory. Examples and comparisons with experimental results are summarized in Refs. 6, 15, and 16. The coupling coefficient has also been used to interpret backscattering experiments with shells having leaky waves having group velocities in the opposite direction of the phase velocity[17] and subsonic or "creeping" flexural waves on the shell.[18] A convolution formulation for curved two-dimensional scattering problems[5] was tested by recovering previous results[13] for circular cylinders. Several other approximations for leaky wave scattering amplitudes have been given for thin elastic shells.[19,20]

CONVOLUTION FOR THREE-DIMENSIONAL SCATTERING

The present formulation[1,2] sums the leaky waves which travel along all paths on the scatterer from illuminated regions to a given surface point S of interest. For simplicity we restrict attention to the situation where the leaky wave phase velocity c_l and attenuation rate and may be approximated as independent of direction. Thus for each frequency $\omega = kc$ of interest there is a complex leaky wavenumber $k_p = k_l + i\alpha$ where $k_l = kc/c_l$ for each leaky wave of interest and the time dependence is $\exp(-i\omega t)$. For the situation of interest $\alpha \ll k_l$. For large k and weakly curved surfaces, it may be satisfactory to approximate k_p from poles of the flat-surface reflection coefficient though other approaches may be useful.[1] The outgoing leaky wave amplitude at a surface point S is approximated as

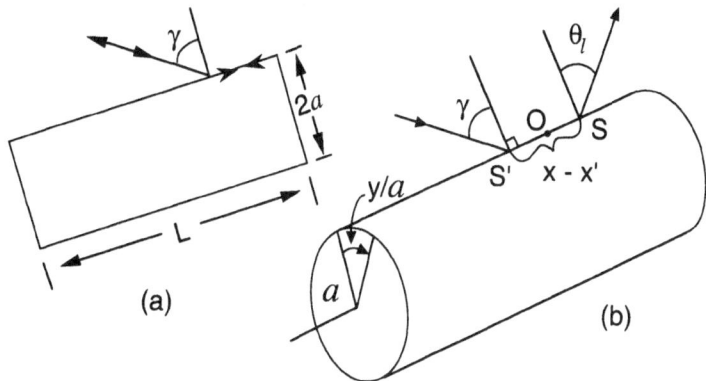

Figure 1. (a) Meridional ray backscattering enhancement mechanism demonstrated in Ref. 3. (b) Coordinates for approximating the outgoing meridional ray amplitude from Eq. (1) for an infinite cylinder.

$$p_l(S) \approx \int_{\mathcal{D}} p_i(S')[\kappa H_0^{(1)}(k_p s)] d\mathcal{A}' , \tag{1}$$

where $H_0^{(1)}$ is the Hankel function having an argument proportional to the geodesic distance s between S and an illuminated point at S' where the incident wave complex amplitude is $p_i(S')$ and $d\mathcal{A}'$ is the differential of area of the contributing surface patch. The domain $\mathcal{D}$ is restricted as summarized below depending on the directional properties of the outgoing wave of interest and[1,2] $\kappa \approx -\alpha k_l \exp(i\phi_{bl})$, where ϕ_{bl} is a background phase. Propagation of p_l gives the farfield amplitude. The restriction of $\mathcal{D}$ may be illustrated for the axially propagating leaky wave in the positive x direction in Fig. 1(b): $\mathcal{D}$ is restricted to x' < x. For the situation illustrated, $H_0^{(1)}(k_l s)$ may be approximated by its large argument form giving[1]

$$p_l(x,y) \approx \kappa(2/\pi)^{1/2} e^{-i\pi/4} \int_{-\infty}^{x} dx' \int_{-\infty}^{\infty} dy' \, p_i(x',y')(k_p s)^{-1/2} e^{-\alpha s} e^{ik_l s} , \tag{2}$$

where the orthogonal surface coordinates y and y' are as shown in Fig. 1(b) where y = 0 and y' = 0 for points on the meridian previously defined. For this choice of coordinates S' = (x',y'), S = (x,y), and s = $[(x - x')^2 + (y - y')^2]^{1/2}$. The limits of integration are shown for an infinitely long cylinder. Since the illuminated region is restricted to $y_L < y' < y_U$ where $y_{U,L} = \pm \pi a/2$ (see Fig. 1(b)), p_i vanishes outside that strip. An alternative form for Eqs. (1) and (2) is to express $p_l(S)$ in terms of a convolution over an infinite domain of p_i with a point spread function given by $\kappa H_0^{(1)}(k_p s)$ weighted by a unit-step directional factor which vanishes outside the physically relevant sector. For the case of a surface curved only along x' where the wavefield is made two-dimensional by taking p_i to be independent of y', the integrations over y' in Eqs. (1) and (2) may be performed and the result from Ref. 5 is recovered. Another way of deriving the coefficient κ is from a spatial spectral decomposition of the reflected wavefield in the flat surface case.[2]

MERIDIONAL RAY AMPLITUDE FOR AN INFINITE CYLINDER

To test this method, Eq. (2) was approximated[1] for the case of an infinite cylinder where exact partial-wave series (PWS) solutions are known for the solid cylinder[10] and uniform shell cases for tilted plane wave illumination. To construct the farfield scattering amplitude it was necessary to obtain the local principal radius of curvature for the wavefront radiated by the leaky wave which is curved in only one direction. This was achieved by considering $p_l(x,y)$ not only in the meridional plane where y = 0 but also for small values of y. The PWS result is usually expressed as a complex form function f which is related in the meridional plane to the far-field amplitude by

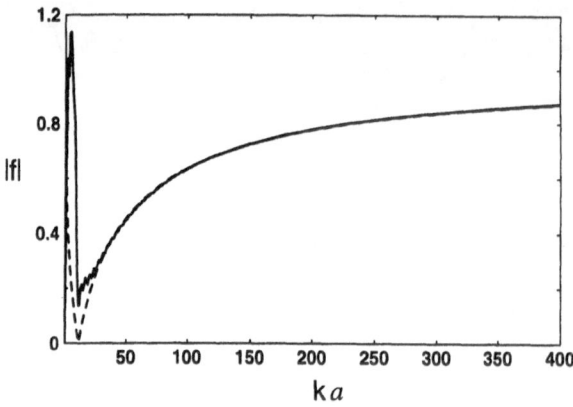

Figure 2. The solid curve is the exact form function for an infinite aluminum cylinder in water tilted at the Rayleigh angle of $\theta_l = 30.3°$. The dashed curve is the approximate ray result, Eq. (5), based on Eq. (1).

$$P_{sca} = P_{io} \, (a/2r)^{1/2} \, f \exp(ik_x x + ik_r r) , \qquad (3)$$

where p_{io} is the reference amplitude of the incident wave, $k_x = k \sin\gamma$, $k_r = k \cos\gamma$, γ is the tilt angle as in Fig. 1(b), r is the radius from the cylinder's axis, and a is the radius of the cylinder. The ray approximation for the meridional leaky wave contribution to f for $\gamma \approx \theta_l$ is[1]

$$|f_l(\gamma)| \approx 2^{3/2}|(\alpha/v)F(\mu")|\left\{ 1 + 2\mu[(1/F(\mu)) - 1] \right\}^{-1/2} , \qquad (4)$$

where $v = \alpha + i(k \sin\gamma - k_l)$, $\mu = \alpha a \, \tan\theta_l$, $F(\mu) = (\pi\mu)^{1/2} \, e^{\mu} \, \mathrm{erfc}(\mu^{1/2})$, and $\mu" = \mu(\cos\theta_l/\cos\gamma)[1 + i(k \sin\gamma - k_l)/\alpha]$. To approximate the total $|f|$ for $\gamma = \theta_l$ it is necessary to superpose f_l with a specular contribution, taking into account the phase of both which gives[1]

$$|f_{ray}| \approx |[1 - |f_l| \exp(i\varphi_{bl})]| . \qquad (5)$$

Equations (4) and (5) were previously tested by comparison with PWS results for a solid stainless steel cylinder in water, where as in the example below, the water-loaded flat surface values of α and k_l are used. An approximate method of isolating the meridional ray contribution in the PWS result for f is to compute $|f - f^{(r)}|$, where $f^{(r)}$ is the PWS result for an infinite tilted rigid cylinder.[1] This procedure subtracts off a background contribution to f. When $|f - f^{(r)}|$ is evaluated as a function of γ a broad peak is found near $\gamma \approx \theta_l$ due to the meridional ray which dominates all other contributions except near γ of 90°.

The new results shown here in Figs. 2 and 3 are for the case of Rayleigh waves on a solid aluminum cylinder in water and for comparison Fig. 3 also includes the previous results for the stainless steel cylinder. The new results are of interest because α/k_l is somewhat larger for the aluminum case where $\alpha/k_l \approx 0.0288$ in comparison to $\alpha/k_l \approx 0.0116$ for a stainless-steel water interface, though the Rayleigh angles are similar ($\theta_l \approx 30.3°$ and 30.7°, respectively). Figure 2 demonstrates that both the PWS and ray results for $|f|$ approach unity and are nearly indistinguishable for large ka at $\gamma = \theta_l$, which is the appropriate limit since $\theta_l \approx 30.3°$ exceeds the shear critical angle for aluminum so that the flat-surface reflection coefficient is unimodular and the normalization in Eq. (3) is such that $|f| \to 1$ in the meridional plane as ka $\to \infty$. The approach to unity is slow because of the asymptotic behavior of the function $\mathrm{erfc}(\mu)$ within $F(\mu)$ in Eq. (4) where μ is proportional to ka. The dashed and solid curves in Fig. 3 show the PWS results for the broad meridional ray peaks in $|f - f^{(r)}|$ for aluminum and stainless steel cylinders in water, respectively, both for ka = 100. The adjacent dotted curves are the respective results for $|f_l|$ from Eq. (4) where the outgoing wavefront curvature was approximated to be the value found for $\gamma = \theta_l$ even when $\gamma \neq \theta_l$. Other sources of error in the comparison may be the use of flat surface values for k_l and α and the approximation of the background as rigid. The ray model is seen to give the approximate magnitude, location, and width of the meridional ray peaks.

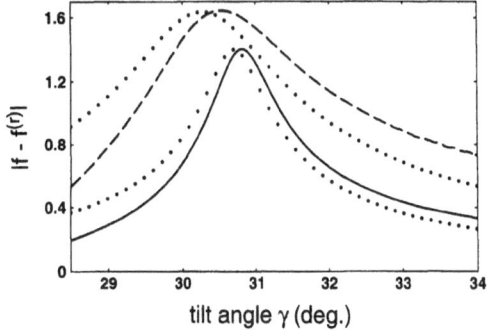

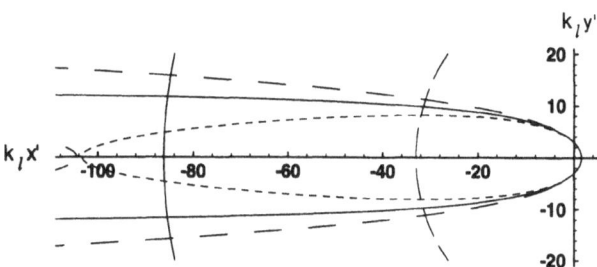

Figure 3. The dashed and solid curves give the background-subtracted form functions from the partial wave series for aluminum and stainless steel cylinders, respectively, at ka = 100. The dotted curves give the meridional ray contribution from Eq. (4) near the coupling maximum.

ANGULAR WIDTH OF THE COUPLING REGION

Inspection of Fig. 3 shows that the angular width of the coupling region is somewhat greater for aluminum than for steel. This may be understood from the dephasing and attenuation of the locally excited leaky waves relative to the incident acoustic wave. Attention is restricted to S in the meridional plane with $(x,y) = (0,0)$ and a plane wave incidence having a pressure $p_{io} = p_o \exp(i\varphi_i)$ at the surface where $\varphi_i(x',y') = ka[1 - \cos(y'/a)] \cos\gamma + kx'\sin\gamma$. The form of Eq. (2) indicates the dephasing by an amount $\Delta\varphi$ is given by the locus in (x',y') of $k_l s + \varphi_i(x',y') - \varphi_i(0,0) = \Delta\varphi$. For the case of $\gamma = \theta_l$ all of the paths along the meridian with $x' < 0$ have the same phase and correspond to the set of Fermat acoustic and Rayleigh wave paths to S. The locus of (x',y') for which $\Delta\varphi = \pi$ gives the boundary of a Fresnel coupling patch in which paths, though defective in that they may deviate from a Fermat path, have a cumulative dephasing of π radians or less. The Fresnel patch boundary is the solid curve that opens to the left in Fig. 4 for the case where $\theta_l = 30°$ and ka = 100. Attenuation as well as dephasing also affects the size of the coupling region and two circular arcs are shown[1] corresponding to radii of $k_l s = k_l/\alpha$ where the attenuation is exp(-1) corresponding to aluminum (dashed arc) and stainless steel (solid arc). Tilting the cylinder away from $\gamma = \theta_l$ causes an additional dephasing. The curves with long dashes opening to the left are for $\gamma = \theta_l + 1°$ and $\Delta\varphi = \pm \pi$ and the oval with short dashes is for $\gamma = \theta_l - 1°$ and $\Delta\varphi = \pi$. Inspection of Fig. 4 shows that while for aluminum taking $|\gamma - \theta_l| = 1°$ only weakly affects the size of the attenuation-limited coupling region, for stainless-steel there is a much greater effect. This explains why the stainless-steel peak in Fig. 3 is narrower.

MERIDIONAL RAY BACKSCATTERING ENHANCEMENT FROM ENDS

For a finite cylinder of length $L \gg 1/\alpha$ the outgoing leaky wave amplitude with $\gamma = \theta_l$ in Fig. 1(a) has been analyzed for reflection from the cylinder's end.[1] The resulting

Figure 4. Coupling regions in normalized unwrapped coordinates for a surface point S at the origin which is on the meridian. Fresnel patches for ka = 100 are shown for different tilts (see text).

farfield scattering amplitude must be normalized with a spherical (rather than cylindrical) spreading factor with the result that the farfield backscattering magnitude is predicted to generally be as large as the signal reflected by a rigid sphere having the same radius as the cylinder. The brightness of the localized image enhancements reported in Ref. 3 are thereby explained. The contributions can be shown to be brighter and more localized in γ than the ordinary GTD end contributions from sound diffracted in water by the ends of the cylinder.[1]

Acknowledgment: This work was supported by the Office of Naval Research.

References

1. P.L. Marston, Approximate meridional leaky ray amplitudes for tilted cylinders: end-backscattering enhancements and comparisons with exact theory for infinite solid cylinders, *J. Acoust. Soc. Am.* (accepted for publication).
2. P.L. Marston, Spatial approximation of leaky wave surface amplitudes for three-dimensional high-frequency scattering: Fresnel patches and application to edge-excited and regular helical waves on cylinders, *J. Acoustic Soc. Am.* (submitted).
3. G. Kaduchak, C.M. Wassmuth, and C.M. Loeffler, Elastic wave contributions in high resolution acoustic images of fluid-filled finite cylindrical shells in water, *J. Acoust. Soc. Am.* 100:64 (1996).
4. H.L. Bertoni, Ray-Optical Evaluation of V(z) in the Reflection Acoustic Microscope, *IEEE Trans. Sonics and Ultrasonics* SU-31:105 (1984).
5. P.L. Marston, Leaky waves on weakly curved scatterers. II. Convolution formulation for two-dimensional high-frequency scattering, *J. Acoust. Soc. Am.* 97:34 (1995).
6. P.L. Marston, Geometrical and catastrophe optics methods in scattering, in: *Physical Acoustics,* R.N. Thurston and A.D. Pierce, eds., Academic, Boston, (1992), 21:1.
7. X.L. Bao, Echoes and helical surface waves on a finite elastic cylinder excited by sound pulses in water, *J. Acoust. Soc. Am.* 94:1461 (1993).
8. S.F. Morse and P.L. Marston, High frequency backscattering enhancements any thick finite cylindrical shells in water at oblique incidence: experiments, interpretation and calculations, *J. Acoust. Soc. Am.* (submitted).
9. K. Gipson, (Washington State University, unpublished communication, 1997).
10. L. Flax, V.K. Varadan, and V.V. Varadan, Scattering of an obliquely incident acoustic wave by an infinite cylinder, *J. Acoust. Soc. Am.* 68:1823 (1980).
11. H.L. Bertoni and T. Tamir, Unified theory of Rayleigh-Angle phenomena for acoustic beams at liquid-solid interfaces, *Appl. Phys.* 2:157 (1973).
12. K.L. Williams and P.L. Marston, Backscattering from an elastic sphere: Sommerfeld-Watson transformation and experimental confirmation, *J. Acoust. Soc. Am.* 78:1093 (1985); 79:2091 (1986).
13. P.L. Marston, GTD for backscattering from elastic spheres and cylinders in water and the coupling of surface elastic waves with the acoustic field, *J. Acoust. Soc. Am.* 83:25 (1988).
14. P.L. Marston, Variable phase coupling coefficient for leaky waves on spheres and cylinders from resonance scattering theory, *Wave Motion* 22:65 (1995).
15. S.G. Kargl and P.L. Marston, Background contributions and coupling coefficients for backscattering by thick shells, *J. Acoust. Soc. Am.* (accepted for publication).
16. P.L. Marston, S.G. Kargl, and N.H. Sun, Elastic resonance amplitudes described by generalized GTD and by product expansions of the S-matrix, in: *Acoustic Resonance Scattering,* H. Überall, ed., Gordon and Breach Science, Philadelphia (1992).
17. G. Kaduchak, D.H. Hughes, and P.L. Marston, Enhancement of the backscattering of high-frequency tone bursts by thin spherical shells associated with a backwards wave: Observations and ray approximation, *J. Acoust. Soc. Am.* 96:3704 (1994).
18. G. Kaduchak and P.L. Marston, Observation of the midfrequency enhancement of tone bursts backscattered by a thin spherical shell in water near the coincidence frequency, *J. Acoust. Soc. Am.* 93:224 (1993).
19. L.B. Felsen, J.M. Ho, and I.T. Lu, Three-dimensional Green's function for fluid-loaded thin elastic cylindrical shell: Alternative representations and ray acoustic forms, *J. Acoust. Soc. Am.* 87:554 (1990); 89:1463 (1991).
20. A.N. Norris and D.A. Rebinsky, Acoustic coupling to membrane waves on elastic shells, *J. Acoust. Soc. Am.* 95:1809 (1994).

NUMERICAL TECHNIQUES FOR MODELING ULTRASOUND SYSTEMS

Peder C. Pedersen
Dept. of Electrical and Computer Engineering
Worcester Polytechnic Institute, Worcester, MA 01609

INTRODUCTION

The relationship between the electrical output signal from an ultrasound pulse-echo system and the features of the reflecting structure or interface is complex, yet the ability to model the complete electro-acoustic interaction between an ultrasound system and a reflecting structure is essential for the development of quantitative ultrasound measurement techniques. A number of variables affect the electrical output signal: i) parameters of the transmitting and receiving transducers (geometry, aperture size, frequency response and excitation signal), ii) properties of the medium (density, speed of sound, absorption and scattering), and iii) features of the reflecting structure (size, shape, surface characteristics, orientation, and location).

For an ultrasound pulse-echo system with specified transmitting and receiving transducers (which can be different when using transducer arrays), the modeling can reveal the influence of a given reflector parameter, *e.g.* orientation or geometry, on the received signal. Alternatively, modeling can help determine which transducer geometry and thus which type of acoustic field is best suited for estimating a given parameter of the reflecting structure, and which feature(s) of the received signal (or spectrum) should be extracted over which frequency range. Modeling studies can potentially enhance the performance of ultrasound systems, for applications such as characterization of cracks and flaws, localization of corrosion, classification of arterial plaque types, and recognition of complex objects; modeling may also aid in the design of ultrasound systems which are optimized for discrimination between different reflecting structures.

Approaches to modeling the output voltage from pulse-echo systems have been investigated in recent years. An analytical solution was developed by Chen et al[1] for a flat plate insonified under normal incidence by a focused transducer, while McLaren and Weight[2] used the *radiation–scattering by point target–reception* formulation to model planar, normally aligned targets. This formulation was extended to targets of arbitrary shape by Lhemery[3,4]. Finite element modeling of the received signal from planar surfaces and wires has been presented by Lerch[5].

In this paper, three different modeling techniques will be reviewed and contrasted. One technique uses complex integration of the pressure field from a transmitting image transducer over the surface of the receiving transducer; this technique is applicable only to planar reflectors large, enough to intercept the whole beam. The second technique is based on the angular spectrum approach whereby the *insonifying* field is decomposed into monochromatic plane waves. Each plane wave in the insonifying field is diffracted into a new distribution of plane waves when impinging on a reflecting structure. The output from the receiving transducer is found as the sum of the contributions from each of these diffracted plane waves. This technique is applicable to planar reflectors of arbitrary size, location and orientation. The last technique is named the Diffraction Response Interpolation Method (DRIM) and can be used for reflectors of arbitrary geometry. Here, the received signal from reflectors of arbitrary geometry is modeled by tessellating the surface into modest-sized triangular or rectangular tiles and then calculating the response to each tile by an efficient numerical integration of the pulse-echo diffraction response over the surface of the tile.

It should be emphasized that the modeling techniques described in this paper incorporate the fully diffracted field from realistic transducer types, *e.g.*, planar and focused single element transducers as well as linear and annular array transducers, and take into consideration all diffraction effects of the reflector. No assumption regarding placing the transducer in the far field is required. Only two modeling assumptions must be fulfilled: i) the medium between the transducer and the reflecting surface is assumed to be homogeneous (attenuation effects can be incorporated); ii) the modeling is treated as a linear problem which precludes incorporation of multiple reflections and shadowing effects.

INTEGRATED PRESSURE FIELD FROM IMAGE TRANSDUCER

Planar, smooth reflectors, with dimensions greater than the lateral beam width, may be formed from delaminations or may exists naturally as surfaces or interfaces in layered structures. For measurements on such structures, the received signal is dependent on the orientation of the surface relative to the transducer axis and on the transducer type[6,7].

The physical situation is shown in Figure 1(a) where angle α denotes the angular deviation from normal incidence or the misalignment angle. A computational model is obtained by replacing the interface with an image transducer, located as shown in Figure 1(b). This results in a transmission configuration with the image transducer as the transmitter and the actual transducer as the receiver. The magnitude of the transmitted field is scaled by the reflection coefficient of the interface. The output from the pulse echo transducer is found by a complex, numerical integration of the pressure field from the image transducer over the surface of the actual pulse-echo transducer.

The most general way of representing the effect of the misalignment angle is to observe the output spectrum , $V_{out}(\omega,\alpha)$, as a function of angle α for a given broadband transducer excited with an impulse function. This function will be referred to as IP-ASD, for *Infinite Planar - Angle-dependent Spectral Distortion*, and may be normalized with the spectrum at $\alpha = 0°$, as shown in (1).

$$IP - ASD = \frac{V_{out}(\omega,\alpha)}{V_{out}(\omega,\alpha=0)} \qquad (1)$$

From the complex IP-ASD function, the output voltage for a given excitation voltage and transducer transfer function can be calculated. To reduce computation time, an efficient method of calculating the complex pressure field over a wide frequency band is required. Such a technique was developed based on the velocity potential impulse response, using an optimized decimation/filtering procedure (zoom FFT)[8].

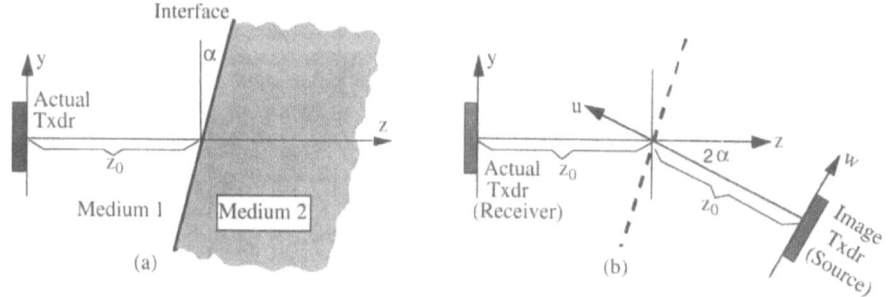

Figure 1. (a). The physical model with a transducer operating in pulse-echo mode and insonifying a large planar interface; (b). The computational model consisting of an image transducer, replacing the interface.

Figure 2 and 3 present IP-ASD results for a planar and a focused transducer, respectively, with specifications given in the figure captions. These IP-ASD results have not been normalized with the response at $\alpha = 0°$. The IP-ASD curves for the planar piston transducer, shown in Figure 2, reveal that the response at $\alpha = 0°$ is nearly independent of frequency, but that for $\alpha \neq 0$ the level drops off very rapidly, even for misalignment angles of a fraction of a degree. Figure 3 shows IP-ASD curves for a focused transducer. Note that the IP-ASD spectrum drops off rapidly with frequency for $\alpha = 0°$, and in fact a higher signal level is obtained for $\alpha = 8°$. The signal from the focused transducer exhibits much less angle dependence than the signal from the planar piston transducer; in fact, the angle dependence is still further reduced near the focal point where the beam width is at a minimum. While not evident in Figures 2 and 3, the IP-ASD results for a focused transducer will exhibit significant *range* dependence, whereas the output from a planar piston transducer will have very little range dependence.

IP-ASD curves have been also obtained experimentally[3], and a good agreement with the modeling results observed.

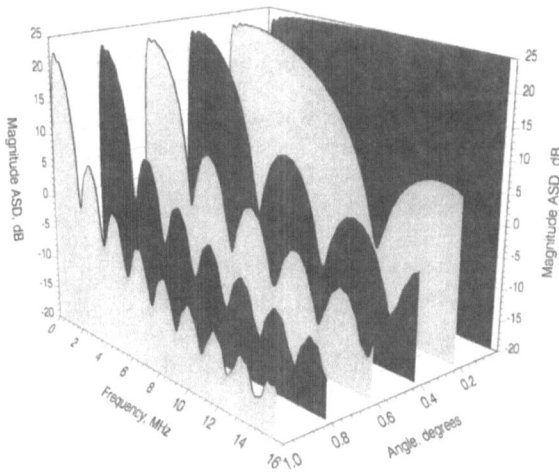

Figure 2. Un-normalized IP-ASD curves for planar piston transducer with radius 1.27 cm, and with reflecting surface placed 5 cm from the transducer. The misalignment angles cover the range from 0° to 1°. Reprinted with permission from JOURNAL OF THE ACOUSTICAL SOCIETY OF AMERICA, vol. 92, #5, Nov. 1992, pp. 2883-2889. Copyright 1992 Acoustical Society of America.

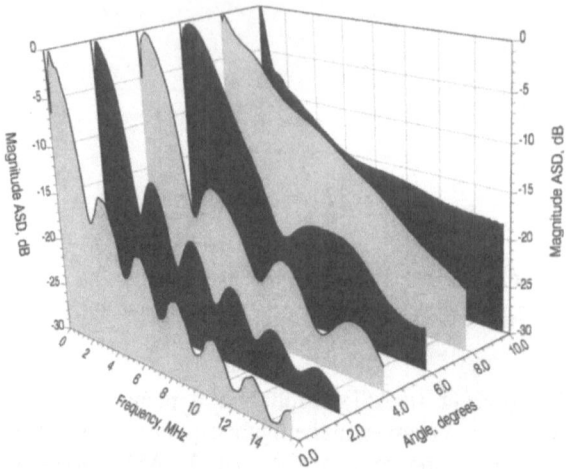

Figure 3. Un-normalized IP-ASD curves for focused transducer with radius 1.27 cm and focal distance equal to 5 times radius or 6.35 cm. The reflecting surface is placed 5 cm from the transducer or 80% of the focal distance. The misalignment angles cover the range from 0° to 10°. Reprinted with permission from JOURNAL OF THE ACOUSTICAL SOCIETY OF AMERICA, vol. 92, #5, Nov. 1992, pp. 2883-2889. Copyright 1992 Acoustical Society of America.

ANGULAR SPECTRUM BASED MODELING

The received signals from a pulse-echo ultrasound system due to planar, smooth reflectors of arbitrary shape can be modeled using an angular spectrum approach. Here the transmitted and diffracted signals are decomposed into their monochromatic, plane wave components[9,10]. The received signal from non-planar objects can be computed by tessellating the surface into planar tiles and summing the individual responses. In this paper, only a single reflector will be considered. Given that, the technique can be used to determine the pulse-echo output as a function of: i) reflector orientation with respect to the transducer axis; ii) shape and size of the reflector; and iii) the transducer geometry or insonifying acoustic field; iv) distance between the transducer and reflector; v) lateral position of the reflector.

Figure 4(a) depicts the physical model where a planar smooth reflector of arbitrary shape is insonified by a pulse-echo transducer. The location and orientation of the reflector is specified by the position vector, $\vec{r}$, and the surface normal unit vector, $\hat{n}$. For modeling purposes, the reflector is replaced by a complementary aperture placed in a perfect absorber of infinite dimensions and insonified by an image transducer located as shown in Figure 4(b). The location and dimensions of the complementary aperture are identical to those of the actual reflector. Likewise, the transmission coefficient of the complementary aperture and its angle and frequency dependence must match the reflection coefficient of the actual reflector.

When parameters ii) to v) above have been specified, the output spectrum for a broad-band transducer excited with an impulse function may be modeled as a function of reflector orientation, $V_{out}(\omega,\alpha)$ where angle α is the angle between $\hat{n}$ and the transducer axis = z-axis. The function $V_{out}(\omega,\alpha)$ is referred to as the FP-ASD, or *Finite Planar - Angle-dependent Spectral Distortion*, and may be normalized with the spectrum at $\alpha = 0°$, as shown in (2).

$$FP - ASD = \frac{V_{out}(\omega,\alpha)}{V_{out}(\omega,\alpha=0)} \qquad (2)$$

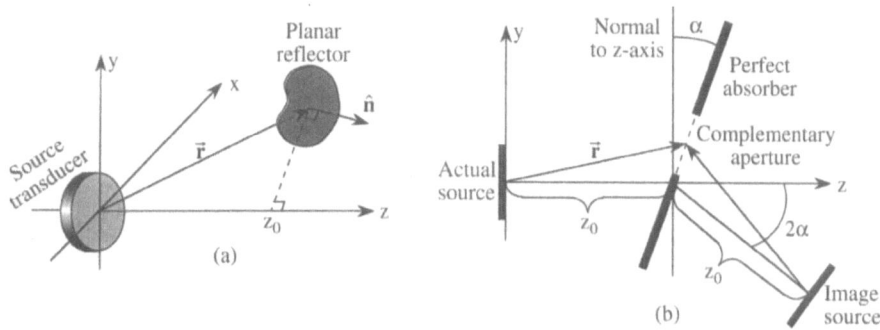

Figure 4. (a). The physical model consists of a transducer in pulse-echo mode, insonifying a planar reflector of arbitrary dimensions; (b). The computational model consists of a complementary aperture in a perfect absorber replacing the reflector and insonified by an image transducer.

To carry out angular spectrum based modeling, the (complex) normal velocity distribution, $u(x,y,\omega)$, over the surface of the transmitting transducer must be specified for all frequencies of interest. This is depicted in Figure 5. The normal velocity distribution reflects the transducer's size, apodization, focusing, excitation, etc. As shown in (3), a scaled spatial Fourier transform of $u(x,y,\omega)$, for a given ω, yields the k_x and k_y components of the plane waves which constitute the pressure field from the transmitting transducer.

$$\tilde{P}_i(k_x,k_y,\omega) = \frac{1}{Z} F\{u(x,y,\omega)\} \tag{3}$$

A plane wave is specified by its propagation vector, $\mathbf{k} = k_x\hat{\mathbf{x}} + k_y\hat{\mathbf{y}} + k_z\hat{\mathbf{z}}$. Thus, for a given (temporal) frequency, k_x and k_y uniquely specify the direction of a given plane in the angular spectrum, since the k_z components is found as:

$$k_z = \sqrt{\left(\frac{\omega}{c}\right) - (k_x)^2 - (k_y)^2} \; ; \; |\mathbf{k}| = \frac{\omega}{c} \tag{4}$$

The distribution of plane waves which make up a given field is generally referred to as the angular spectrum or the k-space distribution. The magnitude and phase of a given plane wave, $\tilde{P}_i(k_x,k_y,\omega)$, in the angular spectrum are specified by the Fourier transform in (3), with the k_z component given by (4).

In Figure 5, $\mathbf{k_q}$ represents the propagation vector for an arbitrary plane wave in the insonifying field from the transmitting transducer. To calculate the aperture diffracted field due to the plane wave represented by $\mathbf{k_q}$, it is noted that in the spatial (or physical) domain, the field immediately to the right of the perfect absorber in Figure 5 is the *product* of the free field (field in the absence of the absorber) and the aperture transmission function. However, multiplication in the spatial domain corresponds to convolution in the k-space domain in which a plane wave is represented as a spatial delta function.

An aperture coordinate system (u,v,w) is defined so that the aperture is located in the u-v plane of this coordinate system and is given as $a(u,v)$, as seen in Figure 5. The spatial Fourier transform of the aperture function yields $A(k_u,k_v)$. To obtain the angular spectrum of the diffracted field due to $\mathbf{k_q}$, $\mathbf{k_q}$ is interpreted in the aperture coordinate system as

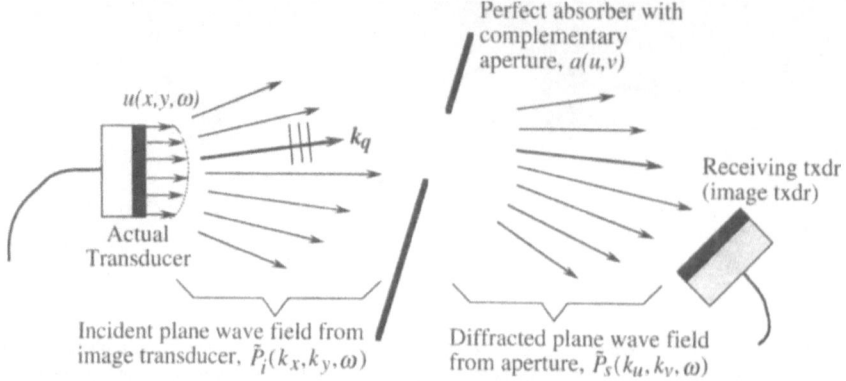

Figure 5. Angular spectrum components of the incident and diffracted field, used in the modeling formulation

$$\mathbf{k_q} = k_{q,u}\hat{\mathbf{u}} + k_{q,v}\hat{\mathbf{v}} + k_{q,w}\hat{\mathbf{w}} \tag{5}$$

The result of the k-space convolution of the the plane wave represented by $\mathbf{k_q}$ and the aperture function, $A(k_u, k_v)$, gives:

$$\tilde{P}_s(k_u, k_v, \omega) = \tilde{P}_q \, A(k_u - k_{q,u}, k_v - k_{q,v}, \omega) \tag{6}$$

In the final step, all of the plane waves in the diffracted field, $\tilde{P}_s(k_u, k_v, \omega)$, must be propagated to the receiving transducer (image transducer), interpreted in the coordinate system of the receiving transducer, and the output due to each plane wave calculated. Closed form expressions exist for several transducer geometries.

To obtain a given FP-ASD function, one must sum over all temporal frequencies of interest, and for each temporal frequency, over all plane waves which make up the insonifying field. Further, for each plane wave in the insonifying field, the sum of the output voltages due to the diffracted plane wave components must be computed. The formulation of an un-normalized FP-ASD is given in (7) where $V_{rec}(\omega)$ is the output from the receiving transducer due to a given plane wave.

$$FP - ASD = V_{out}(\omega, \alpha) = \sum_{\substack{\text{all } \omega}} \quad \sum_{\substack{\text{all plane waves} \\ \text{from transm. txdr.}}} \quad \sum_{\substack{\text{all plane waves} \\ \text{in diffr. field}}} V_{rec}(\omega) \tag{7}$$

The computational aspects of angular spectrum decomposition are described in papers by Orofino and Pedersen[11,12]. As seen from (7), the angular spectrum based modeling is computationally demanding, and specific strategies have been employed to accelerate the computation, as described in a previous publication[9]. A number of FP-ASD results for rectangular planar reflectors of various sizes have been obtained with this modeling technique. The accuracy of the method has been verified by comparison to other modeling techniques, and a good agreement between modeling and experimental results has also been observed [9].

Figure 6 shows a single set of results for a 1.5 cm square target at various orientations relative to the transducer axis, and with the specific modeling parameters given in the caption. By comparing with the IP-ASD results in Figure 2 for the same type of transducer, a similar curveform can be observed, albeit with a different null pattern.

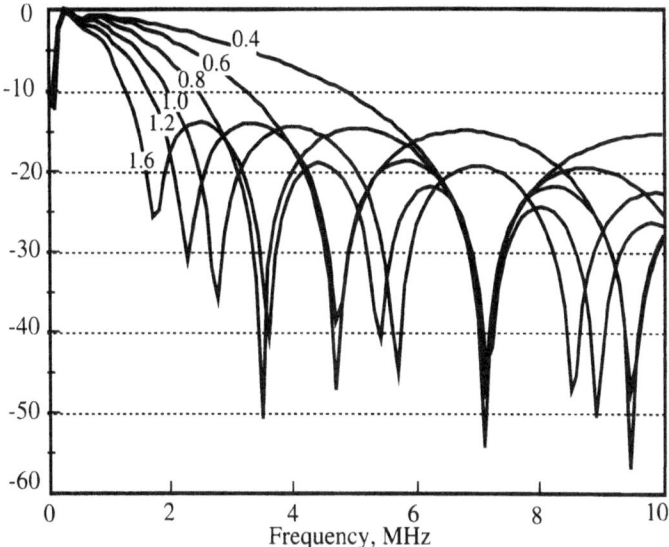

Figure 6. Magnitude FP-ASD curves for a 1.5 cm x 1.5 cm square target, placed on the transducer axis at a distance of 5 cm from a 1.27 cm radius planar piston transducer. Curves are obtained for angular deviation from normal incidence of 0.4° to 1.6°. Reprinted with permission from IEEE. IEEE/UFFC 1996, vol. 43, pp. 303 - 311. Copyright 1996 IEEE.

DIFFRACTION RESPONSE INTERPOLATION BASED MODELING

The Diffraction Response Interpolation Method (DRIM)[13-15] permits modeling of the pulse-echo output voltage due to smooth surfaces of arbitrary geometry and size, and this technique is thus the most general of those described in this paper. Whereas the angular spectrum based modeling is based on decomposing the acoustic field into its plane wave components, the DRIM operates, in principle, by decomposing the given surface into a large number of planar reflecting elements whose dimensions must be much smaller than the wavelength at the highest frequency of interest.

A pulse-echo system insonifying a single such reflecting element with area dA is shown in Figure 7(a) and can be described by the linear systems model in Figure 7(b). The output voltage can be formulated analytically, based on the velocity potential impulse response of the pulse-echo transducer, $h(\vec{r},t)$. Let the pulse-echo diffraction impulse response, or diffraction response for short, be defined as follows:

$$D(\vec{r},t) = \frac{\partial^2}{\partial t^2}\left(h(\vec{r},t) \otimes h(\vec{r},t)\right) \tag{8}$$

Based on $D(\vec{r},t)$, the output voltage from the transducer due to the small reflecting element with area dA can be found to be

$$dv(\vec{r},t) = A_1 \cos[\psi(\vec{r})]\, dA\, \{u(t) \otimes D(\vec{r},t)\} \tag{9}$$

where $\cos[\psi(\vec{r})]$ is an obliquity factor, $u(t)$ is the normal velocity on the surface of the transmitting transducer, and A_1 is a constant which among other things incorporates the reflection coefficient.

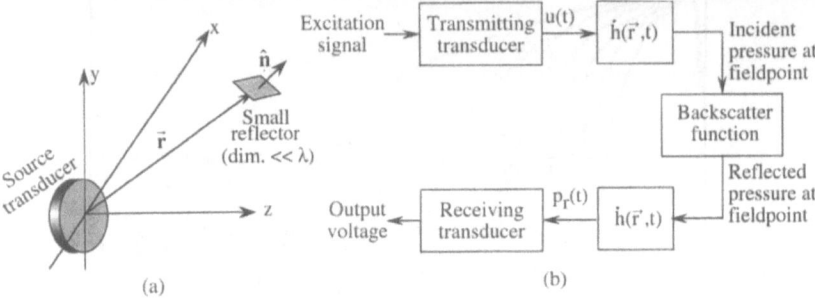

(a)   (b)

Figure 7. (a). Physical measurement situation where a pulse-echo transducer insonifies a small reflecting element. (b). Linear systems model of the measurement situation in (a).

The formulation in (9) is basically a form of Huygens's principle applied to pulse-echo systems. The output voltage due to the actual surface is then found as the summation over all the reflecting elements which make up this surface. Unfortunately, this modeling approach is computationally very demanding, and the DRIM is a numerical method which will significantly speed up the computation.

The basic principle of the DRIM is as follows: The actual reflector is tessellated into *planar* rectangular or square tiles of moderate size, with typical dimensions of 5 - 10 λ at the highest frequency, as illustrated in Figure 8. The appropriate tile size is both a function of the curvature of the surface and of the spatial rate of change of the velocity potential impulse response.

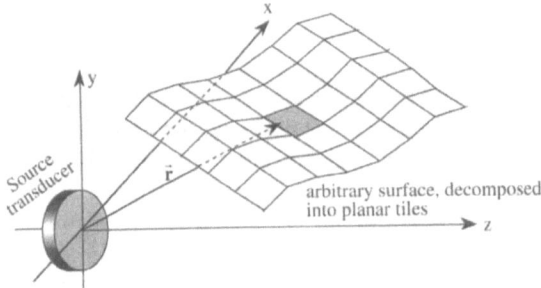

arbitrary surface, decomposed into planar tiles

Figure 8. The basic concept in modeling with the Diffraction Response Interpolation Method whereby the given surface is tessellated into planar tiles of moderate size. Reprinted with permission from IEEE. Submitted to IEEE/UFFC 1996. Copyright 1996 IEEE.

Considering an arbitrary tile with area A, as the one highlighted in Figure 8, the output voltage from the tile is formally an integration of $D(\bar{r},t)$ over the surface, as expressed in the first line in (10). The spatial variation of $D(\bar{r},t)$ can be found to consist mainly of a spatially varying delay and much less of a spatially varying change in shape. The key aspect of the DRIM is the finding that integrating $D(\bar{r},t)$ over the tile surface is equivalent to lowpass filtering $D(\bar{r},t)$ with a filter whose response is determined by the variation in delay over the tile surface. The filtering operation is computationally much faster that the integration.

On this basis, the output voltage is computed by lowpass filtering and scaling the calculated diffraction response from a single, small reflecting element, typically chosen to be an element in the middle of the tile, as shown in the second line of (10).

$$v(\vec{r},t) = A_1 \cos[\psi(\vec{r})] \left\{ u(t) \otimes \iint_A D(\vec{r},t) \, dA \right\}$$

(10)

$$= A_1 \cos[\psi(\vec{r})] \left\{ u(t) \otimes D(\vec{r}_{mid},t) \otimes \text{lowpass filter impulse response} \right\}$$

The computational and implementational aspects of the DRIM have been described in the literature[13-15]. An evaluation of the accuracy and speed of the DRIM is carried out below by computing the spectrum of the received pulse-echo signal for the modeling situation, shown in Figure 9(a) and described in the caption. The output spectrum was calculated by the following techniques: i) The DRIM using 900 tiles; ii) angular spectrum method; iii) Huygens methods, as formulated in (8) and (9), using 22,500 tiles.

As seen from the results in Figure 9(c), the three methods produce virtually indistinguishable results, while the computational time were dramatically different, 2.3 min, 93 min, and 70 min for the three methods, respectively.

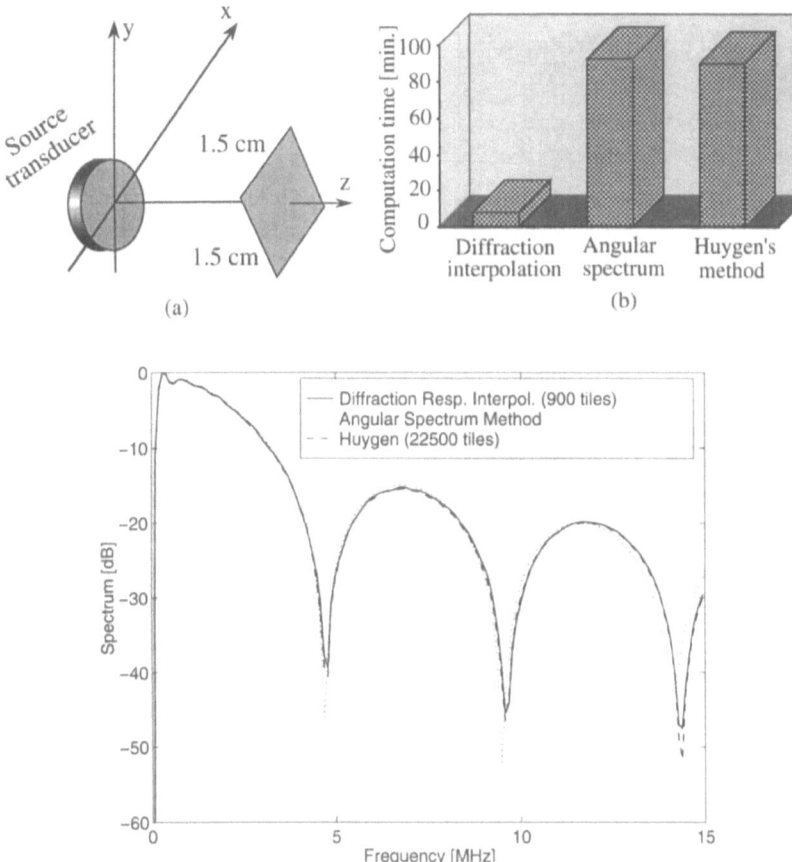

Figure 9. (a). Modeling set-up with a planar piston transducer, radius = 1.27 cm, and planar square reflector with sidelength = 1.5cm, located 5 cm from the transducer. The reflector is rotated 0.6° with respect to the z-axis. (b). Computational time for calculating the spectrum of the output voltage of the modeling situation in (a), over the frequency range 0 - 15 MHz, using the Diffraction Response Interpolation Method, the Angular Spectrum, and Huygens's Method. The computations were performed on an HP 735 workstation. (c). Output spectrum for the modeling situation in (a), based on the three modeling techniques. Reprinted with permission from IEEE. Submitted to IEEE/UFFC 1996. Copyright 1996 IEEE.

CONCLUSIONS

Three modeling techniques have been presented together with results. The *Integrated Pressure Field* determines the output voltage by complex integration of the pressure field from the transmitting transducer over the surface of the receiving transducer. This technique is fairly fast, but only applicable to large planar reflectors. The *Angular Spectrum* modeling technique is able to handle planar reflectors of arbitrary size by decomposing the incident and diffracted fields into their plane wave constituents, but is computationally demanding. Finally, the *Diffraction Response Interpolation Method* is applicable to smooth surfaces of arbitrary shape and integrates the diffraction response over the reflector surface in a computationally efficient manner .

REFERENCES

1. X. Chen, K.Q. Schwarz and K. Parker, Acoustic coupling from a focused transducer to a flat plate and back to the transducer, *J. Acoust. Soc. Am.*, 95:3049 (1994).

2. S. McLaren and J.P. Weight, Transmit-receive mode responses from finite-sized targets in fluid media, *J. Acoust. Soc. Am.*, 82:2102 (1987).

3. A. Lhemery, Impulse-response method to predict echo responses from targets of complex geometry. Part I: Theory, *J. Acoust. Soc. Am.*, 90:2799 (1991).

4. A. Lhemery and R. Raillon, Impulse-response method to predict echo responses from targets of complex geometry. Part II: Computer implement. and exper. verification, *J. Acoust. Soc. Am.*, 95:1790 (1994).

5. R. Lerch, H. Landes, and H.T. Karman, Finite element modeling of the pulse-echo behavior of ultrasound transducers, *Ultrasonics Symp Proc.*, Cannes, France, Nov. 1994, 1021.

6. I. Lifshitz, P.C. Pedersen, and P.A. Lewin, The reconstruction of the acoustic impedance profile of a multilayer medium, *Ultrasound Imag*, 14:40 (1992).

7. D.P. Orofino and P.C. Pedersen, Angle-dependent spectral distortion for an infinite planar fluid-fluid interface, *J. Acoust. Soc. Am.*, 92:2883 (1992).

8. D.P. Orofino and P.C. Pedersen, Multirate digital signal processing algorithm to calculate complex acoustic pressure fields, *J. Acoust. Soc. Am.*, 92:563 (1992).

9. P.C. Pedersen and D.P. Orofino, "Modeling of received ultrasound signals from finite planar targets," *IEEE Trans UFFC*, 43:303 (1996).

10. P.C. Pedersen and D. Orofino, Modeling of received signals from finite reflectors in pulse-echo ultrasound, *1994 IEEE Ultrasonics Symp Proceedings*, Cannes, France, Nov. 1994, 1177.

11. D.P. Orofino and P.C. Pedersen, Efficient angular spectrum decomposition for acoustic sources. Part I: Theory, *IEEE Trans UFFC*, 40:238 (1993).

12. D.P. Orofino and P.C. Pedersen, Efficient angular spectrum decomposition for acoustic sources. Part II: Results, *IEEE Trans UFFC*, 40:250 (1993).

13. S.K. Jespersen, P.C. Pedersen, and J.E. Wilhjelm, "Modeling of received signals from interfaces of arbitrary geometry," *1995 IEEE Ultrasonics Symp Proc*, Seattle, WA, Nov. 1995, 1561.

14. P.C. Pedersen and S.K. Jespersen, "The diffraction response interpolation method. Part I: Theoretical foundation," submitted to *IEEE Trans UFFC*, 1996.

15. S.K. Jespersen, P.C. Pedersen and J.E. Wilhjelm, "The diffraction response interpolation method. Part II: Implementation and results," submitted to *IEEE Trans UFFC*, August, 1996.

OPTICAL VISUALIZATION OF WIDEBAND ULTRASOUND FIELDS

Todd A. Pitts, Randall R. Kinnick, James F. Greenleaf

Ultrasound Research
Mayo Clinic and Foundation
Rochester, MN 55905

INTRODUCTION

Acoustooptic interaction has been extensively studied for use in visualization and mensuration of ultrasound fields. Typical methods model narrow band ultrasound fields as optical phase gratings and produce cross-sectional estimates of power, amplitude, or squared amplitude in the beam. In this paper the authors model interaction of a coherent optical (scalar) field with a wideband ultrasound pulse as weak scattering by an arbitrary refractive index perturbation.

Figure 1 describes an experiment in which a succession of identical wideband ultrasound pulses from a typical medical transducer (center frequency 2.25 MHz) is created in a water tank. The ultrasound bursts are stroboscopically synchronized with pulses from a collimated diode laser ($\lambda = 810$ nm). As the light passes through the region of water supporting the acoustic pulse it interacts with a spatially varying refractive index. At sufficiently low pressures the change in local refractive index is approximately linearly proportional to the difference between local and ambient pressure. On exit from the sound field portions of the optical wavefront have been delayed. The amount of delay is proportional to a line integral (taken in the direction of propagation) of the refractive index and thus the *instantaneous* pressure in the acoustic pulse. The exiting optical field then passes through two successive lenses and impinges on a CCD array. Adjustment of the lens system allows measurement of optical field intensity in arbitrary planes subsequent to passage through the sound field. Estimates of the integrated phase delay suitable for tomographic reconstruction of instantaneous pressure are obtained via iterated projection onto constrained sets.

PHYSICAL MODEL

Piezooptic Coefficient

To first order we may model the interaction of optical and acoustical fields via a simple acoustically induced refractive index variation. We say that the local pertur-

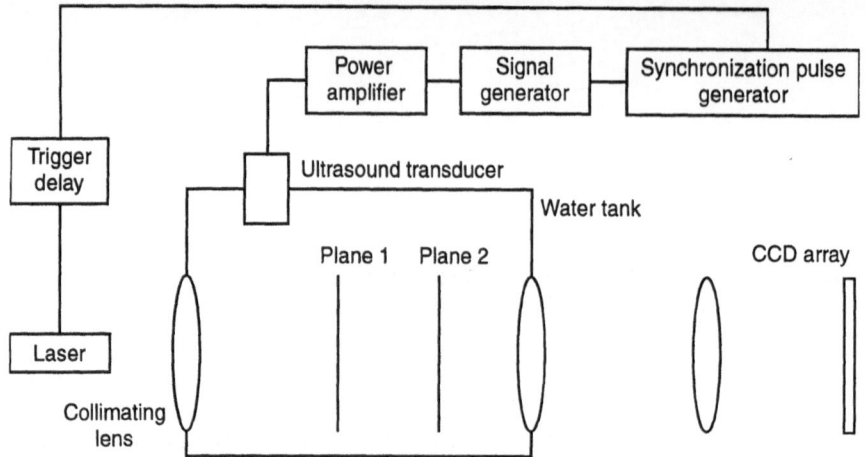

Figure 1. The experiment is conducted so that the laser ($\lambda = 810$nm) pulse and ultrasound pulse interact in the water tank just prior to plane 1 (object plane). Timing circuitry allows this interaction to take place at the same point in the temporal evolution of the sound field each time data are taken. After exiting the sound field (plane 1) the optical wave has phase given by equation 3. Its magnitude has not been significantly altered due to the weak scattering properties of the sound field (see equation 1). The optical field is allowed to propagate to plane 2 (diffraction plane) where it is imaged onto the CCD array, digitized, and stored on disk.

bation in refractive index is directly proportional to local increases in instantaneous pressure and write

$$n(\mathbf{r}, t) = n_o + p(\mathbf{r}, t) \left(\frac{\partial n}{\partial p} \right)_s. \tag{1}$$

The piezooptic coefficient $(\partial n/\partial p)_s$ has been measured at 5461 Angstroms by Riley and Klein[1] as

$$\left(\frac{\partial n}{\partial p} \right)_s = 1.52963 \times 10^{-5} \left[\text{atm}^{-1} \right]. \tag{2}$$

Optical Phase Delay

If the induced refractive index variation is sufficiently small optical scattering may be modelled as a simple accumulative phase delay. We refer to the sound field as a *phase object* denoting the fact that the *magnitude* of the optical wave remains essentially unchanged. In such a case the scattered optical field immediately after exiting the sound field has phase given by

$$\theta = k \int \delta n(\mathbf{r}) \, dl. \tag{3}$$

The phase of the optical field therefore represents a line integral of the instantanous pressure in the ultrasound pulse. A complete set of line integrals for 180 degree rotation of the ultrasound field about an axis normal to the direction of optical wave propagation constitutes the Radon transform of the pulse.

Fresnel Diffraction

If the angular spectrum of the optical field is sufficiently narrow we may use the Fresnel diffraction model for arbitrarily short distances[2]. We may thus calculate the

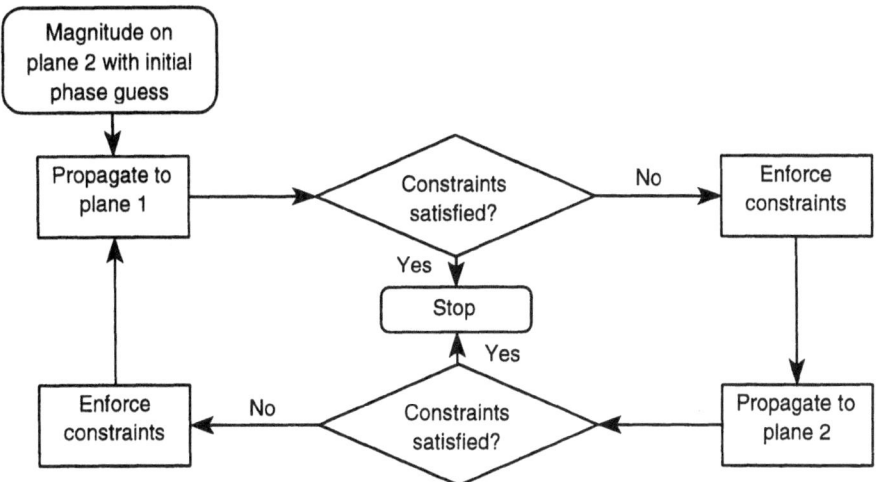

Figure 2. The above algorithm may be viewed as reconstruction via projection onto constrained sets. We begin with the magnitude data measured in plane 2 of Figure 1 and an initial phase guess. We then propagate this field to plane 1 and apply any known constraints such as region of support for the phase function or a magnitude constraint. After enforcing the constraints we propagate the resulting field back to plane 2 in Figure 1 and apply any known constraints on values in that plane. In our case the principle constraint on plane 2 is the measured magnitude. We continue this process until the constraints are satisfied (within some tolerance) without explicit enforcement.

optical field a few centimeters after exiting the sound field via a fast Fourier transform method.

SIGNAL PROCESSING

If we are to infer the desired pressure information we need to know the *phase* of the optical field in the plane immediately after the ultrasound pulse. Much work has been done on the problem of *phase recovery*[3, 4]. Questions of uniqueness and convergence for the various proposed algorithms are topics of current research. In our work we use a version of the Gerchberg-Saxton algorithm applied to Fresnel diffraction. We begin with the magnitude of the field in plane 1 and an estimate for the phase. We then propagate this estimated field to plane 2 and apply any known constraints to the field. In this case we have measured the intensity there and thus know the magnitude. The propagated phase is retained. We then propagate the field back to plane 1 and apply the magnitude constraint again. In our model the acoustic pulse is a phase object and therefore the magnitude in plane 1 is always unity. The algorithm stops when the corrected (measured) magnitude and the predicted magnitude in one of the planes are not significantly different.

CONCLUSIONS

An optical technique for measurement of integrated instantaneous pressure in an ultrasound pulse has been developed. The model used to describe the physics of the exper-

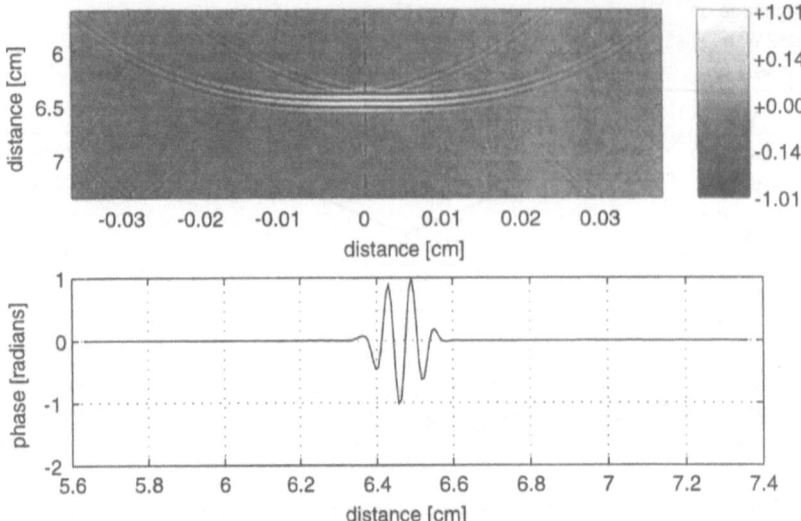

Figure 3. (a) A uniformly shaded circular ultrasound transducer with center frequency 2.25 MHz and an 80 percent bandwidth was simulated via a Fresnel model. The transducer was driven with two cycles at the center frequency and the pulse allowed to propagate approximately 6.8 cm before being imaged. The image represents instantaneous pressure integrated normal to the principle direction of acoustic propagation (e.g. in the direction of optical wave propagation). It also constitutes the phase of the scattered optical wave in plane 1 of Figure 1. (b) Line profile of integrated instantaneous pressure along the dashed line in the image.

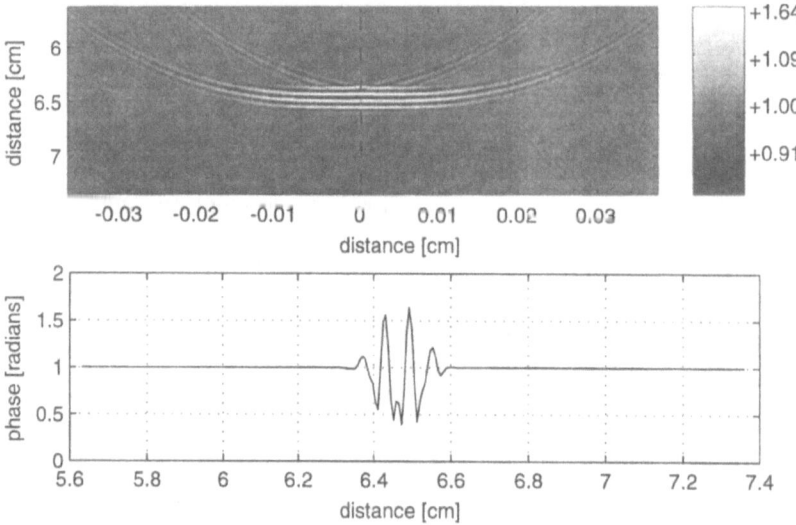

Figure 4. (a) The image represents the magnitude of the optical field in plane 2 of Figure 1 as computed via a Fresnel model. The optical field in plane 1 is assumed to have unit amplitude and phase as shown in the part (a) of Figure 3. The separation between object and diffraction plane is 13 cm. (b) Line profile of the computed optical field magnitude along the dashed line in the image of part(a).

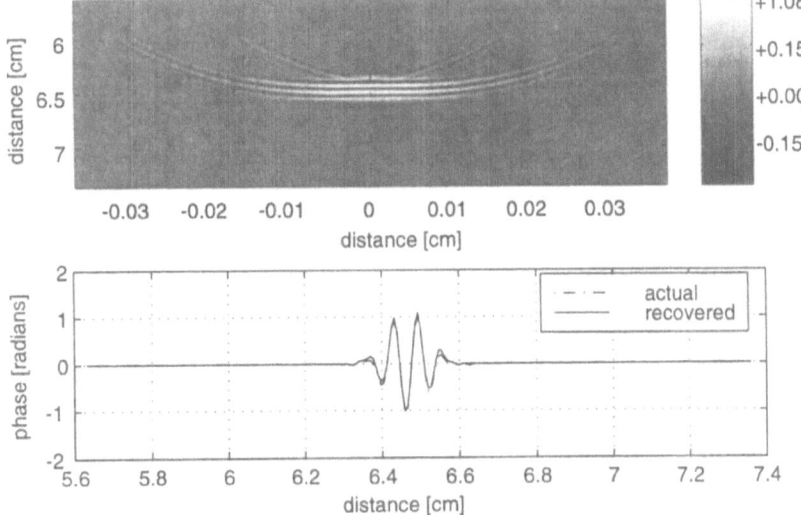

Figure 5. (a) Using the algorithm described in Figure 2 and the computed magnitude of the optical field in the diffraction plane (see Figures 1 and 4) the phase of the optical field in the object plane was estimated as shown in the image. After 15 iterations the phase changed only slowly and the algorithm was terminated. (b) Line profile comparing the estimated optical phase along the dashed line in the image of part (a) with the actual value (see Figure 4).

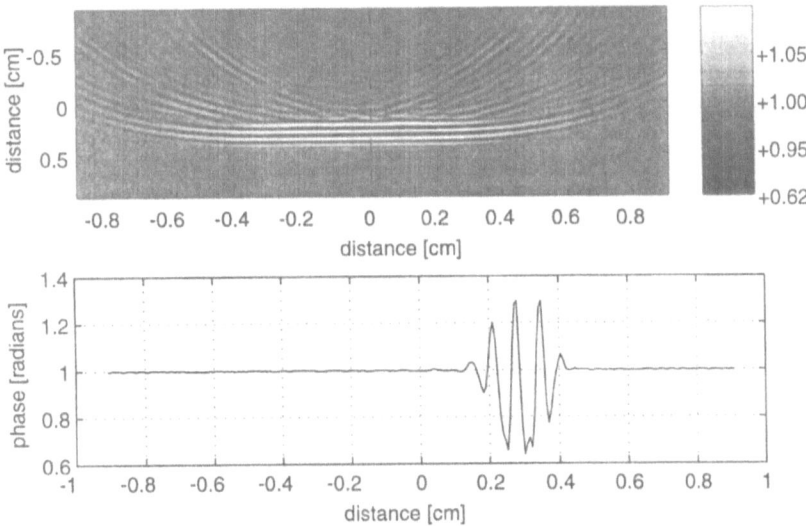

Figure 6. (a) Experimental data from an unfocused 2.25 MHz PZT transducer with a diameter of 1 cm were taken for a diffraction plane distance of approximately 13 cm. The sound field has propagated approximately 2 cm. The image shows the magnitude of the optical field in the diffraction plane. (b) A line profile of the measured optical field magnitude along the dashed line shown in part (a).

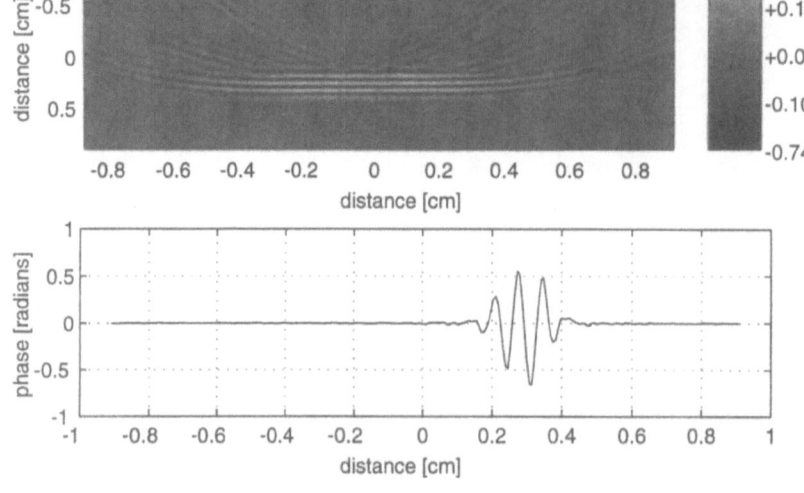

Figure 7. (a) Using the algorithm describe in Figure 2 and the measured magnitude of the optical field in the diffraction plane (see Figures 1 and 6) the phase of the optical field in the object plane was retrieved. After 10 iterations the phase changed only slowly and the algorithm was terminated. (b) Line profile showing the recovered phase along the dashed line in the image.

iment assumes weak optical scattering and represents the ultrasound pulse as a stationary refractive index perturbation. The required measurement is the two-dimensional distribution of optical intensity in a diffraction plane of the imaging system. An iterative phase retrieval algorithm is used to recover the phase of the optical field. This phase is closely related to the local integral (in the direction of optical propagation) of the pressure field. The method appears to have very good sensitivity and resolution and a high signal to noise ratio. As can be seen from the images a good intuitive feel for the overall shape of the field can be obtained quickly.

REFERENCES

[1] W. A. Riley and W. R. Klein, "Piezo-optic coefficients of liquids", *Journal of the Acoustical Society of America*, vol. 42, no. 6, pp. 1258–1261, 1967.

[2] Salvatore Sollmeno, Bruno Croosignani, and Paolo DiPorto, *Guiding, Diffraction, and Confinement of Optical Radiation*, Harcourt Brace Jovanovich, New York, 1986.

[3] Mohammad H. Maleki and Anthony J. Devaney, "Phase-retrieval and intensity-only reconstruction algorithms for optical diffraction tomography.", *J. Opt. Soc. Am. A, Opt. Image Sci.*, vol. 10, no. 5, pp. 1086–92, May 1993.

[4] J.N. Cederquist, J.R. Fienup, C.C. Wackerman, S.R. Robinson, and D. Kryskowski, "Wave-front phase estimation from Fourier intensity measurements.", *J. Opt. Soc. Am. A, Opt. Image Sci.*, vol. 6, no. 7, pp. 1020–6, July 1989.

ON UNDERSTANDING THE RELATIONSHIP BETWEEN ULTRASOUND SPECKLE AND THE SCATTERING MICROSTRUCTURE

N.A.H.K. Rao[1], S. Venkataraman[1], M. Helguera[1,2]

[1] Center for Imaging Science
Rochester Institute of Technology
Rochester, NY, 14623
[2] CENIDET, Cuernavaca, México

ABSTRACT

Recent studies on the statistics of the envelope of the ultrasound echo signal from a random scattering medium suggest that the statistical moments of the signal may carry quantitative information about the scattering microstructure.

The concept of effective cell volume and its relationship to the system's point spread function is established. The influence of the imaging system's point spread function on the statistical moments is considered.

A microstructure is probed experimentally with multiple bandwidth pulses with center frequency matched to the transducer center frequency. Variation of the second normalized intensity moment with the cell volume is considered and exploited experimentally for structure characterization.

To estimate the effective cell volume, independent point spread function measurements at different bandwidths are performed.

Two parameters, the slope and the intercept of the normalized second intensity moments versus the inverse of the effective cell volume prove to be useful features of the scattering microstructure.

INTRODUCTION

Figures 1 (a), (b), and (c) show the surface image of the three different sponge structures used in our experiments. Figures 2 (a), (b), and (c) show their B-scan images taken after they are immersed in water. Even qualitatively it is difficult to differentiate them from their speckle images. This paper describes the data acquisition and analysis strategy that brings out quantitative descriptors for structure characterization.

The theoretical framework for studying the non-Gaussian behavior of scattered waves was laid down by Jakeman [1]. We present here the results for the second normalized intensity moments derived by applying random walk concepts under the narrow bandwidth assumption [2] with emphasis on the effective cell volume definition. Consider a narrow bandwidth signal driving the transducer as the real part of $\mathbf{p(t)} = \overline{\mathbf{A}}(t)e^{j2\pi f_0 t}$ where f_0 is the center frequency and $\overline{A}(t)$ is the pulse envelope (e.g. a Hanning window). The transmit/receive frequency response of the transducer is h(f). h(f) is typically a Gaussian shaped function centered at some resonant frequency f_0. It is important to note that in our experiments the center frequency of the drive signal is matched to the center frequency of the transducer, but its bandwidth Δf is gradually increased.

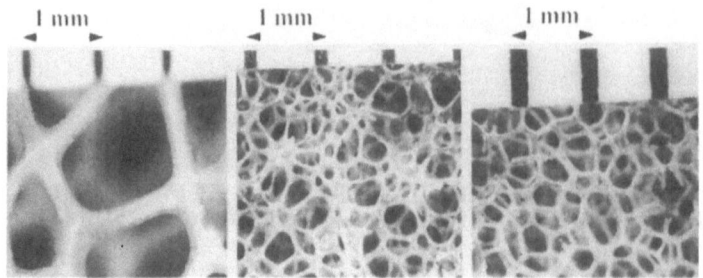

Fig. 1. Photograph of the surface of sponges. From left to right: sample 1, sample 2, and sample 3. Distance between bars is 1mm

If $\overline{A}'(f)$ and $\overline{A}(t)$ are Fourier transform pairs, the received pulse $A(t)e^{j2\pi f_0 t}$ has the same center frequency f_0 but its envelope $A(t)$ will now be different from $\overline{A}(t)$ due to the bandpass filtering by the finite bandwidth frequency response of the transducer. If we assume that the drive signal bandwidth Δf is very small compared to the bandwidth of the transducer frequency response $h(f)$, then $A(t) \approx \overline{A}(t)$.

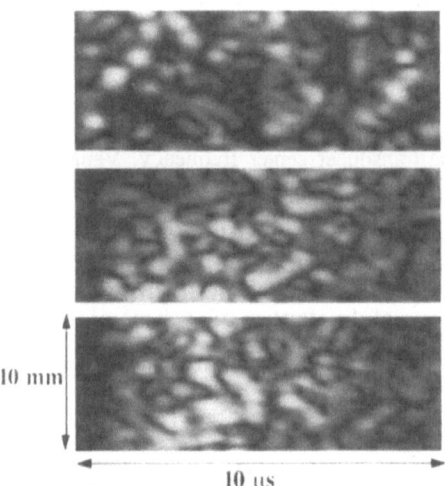

10 mm

10 µs

Fig. 2. B-scan images. From top to bottom: sample 1, sample 2, sample 3

The combined transmit/receive transducer beam profile $B(r)$ is assumed to be circularly symmetric, where r is the perpendicular distance from the transducer beam axis which is assumed to be along the z axis (Fig. 3). The medium is considered to be nonattenuating and uniform except for small size scatterers (impedance discontinuities) distributed randomly in a three dimensional space. The location of the nth scatterer in the beam can be represented by (r_n, z_n) and the two way travel time is $t_n = 2\sqrt{r_n^2 + z_n^2}/c$ where c is the speed of ultrasound in the medium.

The echo signal from the nth scatterer is given by:

$$s_n(t) = F^{-1}\left[\overline{A}'(f-f_0)h'(f)a'_n(f)b'(r_n,z_n,f)e^{-j2\pi t_n f}\right] \approx a_n B(r_n,z_0)A(t-t_n)e^{j2\pi(t-t_n)f_0} \quad (1)$$

where $a'_n(f)$ is the frequency dependent backscatter coefficient of the nth scatterer and $b'(r_n,z_n,f)$ is the two way propagation transfer function of the transducer (diffraction filter) [3]. The approximation in Eq. (1) follows from the assumption that the bandwidth Δf of the drive signal, i.e. $\overline{A}'(f-f_0)$ is "small enough" ($\Delta f \leq 1$ MHz). The validity of this approximation was examined by us through simulations but is not presented here [3,5].

Eq. (1) essentially states that the echo signal is a time shifted and bandpass filtered version of the drive signal. It's amplitude is modified by two factors, $a_n \approx a_n'(f_0)$ the backscatter coefficient at f_0 and $B(r_n,z_n)$ the transducer beam profile value at the scatterer location.

The echo signal at any time t from a volume distribution of scatterers (M being the number of scatterers in a volume V_T that is formed by considering a cylinder of diameter equal to 20 dB beamwidth and length equal to 20 dB pulse width) is given by:

$$s(t) = e^{j2\pi f_0 t}\sum_{n=1}^{M} a_n B(r_n,z_0)A(t-t_n)e^{-j2\pi f_0 t_n} = e^{j2\pi f_0 t}\sum_{l=1}^{M} E_n e^{\phi_n} \quad (2)$$

where

$$t_n = 2\sqrt{r_n^2 + z_n^2}/c; \quad E_n = a_n B(r_n,z_0)A(t-t_n); \quad \phi_n = -2\pi f_0 t_n$$

The conditions imposed in defining the volume V_T ensure that any scatterer outside this volume will make a negligible contribution to the echo signal at t.

The problem can now be stated in the context of random walk as described by Jakeman[1]. Eq.(2) can be considered as a vector sum of M phasors in the complex plane[2]. The M phasors that make significant contributions to the signal at time t will generally have random amplitudes E_n and random phase ϕ_n. A given scatterer with some r_n and z_n will make a significant contribution to the signal at time t with phasor amplitude E_n only if both $B(r_n, z_0)$ and $A(t-t_n)$ are not significantly small, i.e. the scatterer must be located well within the volume V_T centered at some depth $z=c/2t$ as shown in Fig. 3.

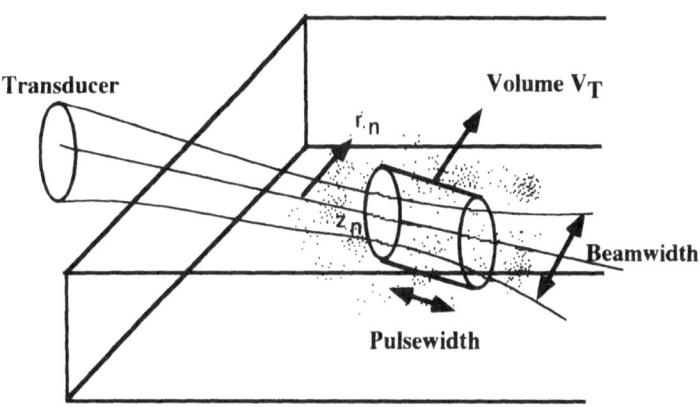

Figure 3. Transducer scanning geometry and the concept of resolution cell volume

The phasor amplitudes E_n also depend on the scattering microstructure through the scattering coefficient terms a_n which we assume to be statistically independent. If the interscatterer spacing in the microstructure is such that the phases ϕ_n due to time delays t_n are independent and uniformly distributed between 0 and 2π, i.e. the impedance discontinuities exist on a scale small compared to a wavelength, then following Jakeman we can write the second normalized moment of the intensity distribution $I(t)=|s(t)|^2$ as

$$\frac{\langle I^2 \rangle}{\langle I \rangle^2} = \frac{2\langle M(M-1) \rangle}{\langle M \rangle^2} + \frac{\langle E^4 \rangle}{\langle M \rangle \langle E^2 \rangle^2} \tag{3}$$

where $\langle ... \rangle$ stands for ensemble average. If M, in a given sonicated volume V_T is large and is Poisson distributed [7] then $\langle M(M-1) \rangle = \langle M^2 \rangle = \langle M \rangle^2$. If we assume that a_n, $B(r_n,z_0)$ and $A(t-t_n)$ are statistically independent we have

$$2\frac{\langle M^2 \rangle}{\langle M \rangle^2} + \frac{\langle a^4 \rangle \langle B^4 \rangle \langle A^4 \rangle}{\langle M \rangle \langle a^2 \rangle^2 \langle B^2 \rangle^2 \langle A^2 \rangle^2} \tag{4}$$

Replacing now the ensemble average over the terms A and B by its spatial average over the volume V_T as:

$$\frac{\langle B^4 \rangle \langle A^4 \rangle}{\langle B^2 \rangle^2 \langle A^2 \rangle^2} = \frac{\frac{1}{V_T}\langle\langle B^4 \rangle\rangle\langle\langle A^4 \rangle\rangle}{\left[\frac{1}{V_T}\langle\langle B^2 \rangle\rangle\langle\langle A^2 \rangle\rangle\right]^2} = \frac{V_T}{V_E} \tag{5}$$

where $\langle\langle ... \rangle\rangle$ stands for integration over the volume V_T. The effective volume V_E results from the volume integration [4]

$$V_E = \frac{c\left[\int 2\pi r B^2(r)dr \int A^2(t)dt\right]^2}{2\left[\int 2\pi r B^4(r)dr \int A^4(t)dt\right]} \equiv \frac{c\left[\int\int g(r,t)2\pi r dr dt\right]^2}{2\left[\int\int g^2(r,t)2\pi r dr dt\right]} \tag{6}$$

Where $g(r,t)$ is the experimental intensity point spread function to be defined in the next section. If we use $\langle M \rangle / V_T = \langle N \rangle$ as the volume scatterer number density, then Eq. (3) becomes:

$$\frac{\langle I^2 \rangle}{\langle I \rangle^2} = 2 + \frac{\langle a^4 \rangle}{\langle a^2 \rangle^2}\frac{1}{\langle N \rangle V_E} = 2 + \frac{1}{M_{eff}} \tag{7}$$

This equation is similar to the one derived by other researchers.[7,8]

A. Resolution Cell Volume and System Point Spread Function

Consider a wire target at depth z_0 in Eq. (1). By scanning in the direction perpendicular to the wire and taking the square of the echo envelope, we obtain the intensity point spread function $g(r,t)$ as

$$g(r,t) = \left|F^{-1}\left[\overline{A}'(f-f_0)h'(f)b'(r,z_0,f)\right]\right|^2 \approx \left|B(r,z_0)A(t)\right|^2 \tag{8}$$

where $t = 2\sqrt{r^2 + z_0^2}/c$. The approximation follows the narrow bandwidth arguments presented earlier. Strictly speaking, what we measure with the wire targets is the projection of the radially symmetric three dimensional PSF. We will take it to represent a two dimensional radial slice of the PSF due to small differences in practice.

Mathematically, the extreme right-hand side of Eq. (6) is a functional of the positive function $g(r,t)$, which we will consider to be the experimental intensity PSF of the imaging system. This functional possesses the properties required to define a three dimensional volume. We will refer to it as the "effective volume" V_E. Note that the term V_E that appears in Eq. (7) and is defined by Eq. (6) results from expressing $g(r,t)$ as a product of two separable functions $B(r)$ and $A(t)$. $A(t)$ is the filtered version of the drive signal envelope $\overline{A}(t)$ and hence takes into account the frequency response of the transducer. We assume that $B(r)$ depends on the center frequency f_0 of the pulse but not on Δf. Eq. (7) then predicts that the normalized second moment should increase linearly with the drive pulse bandwidth Δf, when f_0 is held fixed.

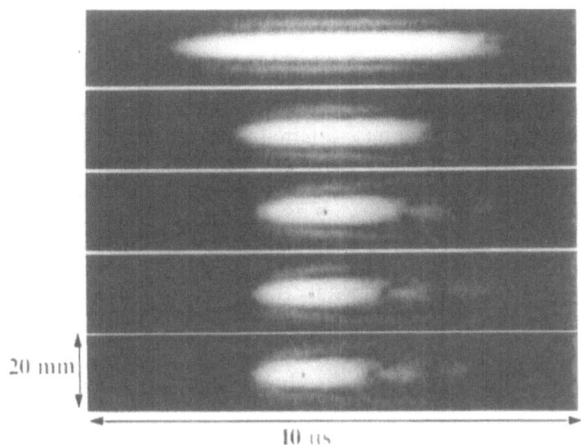

Fig. 4. Experimental two dimensional system PSF on a logarithmic scale over 60 dB at five different bandwidths. From top to bottom: Δf= 0.3 MHz, 0.6 MHz, 0.9 MHz, 1.2 MHz, and 1.5 MHz respectively. f_0= 2.2 MHz in all cases.

The experiments involve probing the microstructure with different bandwidth pulses. In each case the normalized intensity moments (left-hand side of Eq. (7)) are estimated from the echo signal. V_E is also determined from a separate calibration experiment. Eq. (7) predicts that the slope of the $\langle I^2 \rangle / \langle I \rangle^2$ vs V_E^{-1} plot should depend only on the scattering microstructure (i.e. $\langle a^4 \rangle / \langle a^2 \rangle \langle N \rangle$) and hence can serve as a useful parameter. The intercept is the well known high density limiting value 2, when the speckle becomes a fully developed speckle[1] (Gaussian limit). The slope estimate can be roughly seen as the inverse of the "effective scatterer number density" evaluated at the frequency f_0.

B. Effects of Subresolution Periodicity

From Eq. (7) we can also see that in the limit $V_E \to \infty$ or $\Delta f \to 0$, the expected value of the second moment is 2. This number can go below 2 if substantial unresolved coherent component is present in the backscattered signal [2,6]. In Eq. (2) it was assumed that the random positions of the scatterer introduce path differences that exceed the dominant wavelength $\lambda_0 = c/f_0$ of the pulse so that $\phi_n = 2\pi f_0 t_n$ can be regarded as being uniformly distributed between 0 and 2π. However, when there is subresolution periodicity in the interscatterer spacing, constructive interference effects [6] can ensue whenever the periodic spacing becomes half-integer multiple of the dominant wavelength λ_0. In the random walk context, this amounts to adding a constant phasor to the complex signal $s(t)$. The PDF of the envelope signal follows a Rice distribution and the normalized second moment is given by [2]

$$\frac{\langle I^2 \rangle}{\langle I \rangle^2} = \frac{2 + 2\gamma + \gamma^2}{(1+\gamma)^2} \tag{9}$$

where γ is the ratio of the constant phasor intensity I_s to the variance of the incoherent random Gaussian distribution. If the constructive interference effects at f_0 contribute to a significant build up of I_s, then we expect $\gamma > 0$ and in the limit of large cell volume, i.e. $V_E \rightarrow \infty$, the value of the intensity moment given by Eq. (9) can be less than 2. This limiting behavior has been considered separately in order to explain some of out experimental observations. This consideration at present is incomplete in that its effect on the derivation of E. (7) is not clear yet.

EXPERIMENTAL VERIFICATION

The previous theory was verified by recording the echo signals from three 14 cm × 7 cm × 6 cm blocks of sponge structures immersed in water making sure there were no trapped air bubbles in the medium. One sponge, referred to as sample 1, had larger mean pore size (d= 2 mm). The other two sponge structures had much smaller mean pore size (d=.5 mm). However, the microstructure of sample 3 appears to be much more ordered than sample 2.

The 13 mm diameter circular disk, 2.25 MHz medium focus transducer has a focal length of 5.5 cm. Its 6 dB bandwidth was approximately 1.2 MHz. The sponge surface was positioned at 5 cm from the transducer and held perpendicular to the beam axis. The transducer was excited with programmable signals generated by a waveform generator (Analogic Corporation, Model 2020) as follows:

$$p(t) = \overline{A}(t)\cos(2\pi f_0 t) = 0.5\left[1 - \cos\left(\frac{2\pi t}{T}\right)\right] \cdot \cos(2\pi f_0 t); \ 0 \leq t \leq T \tag{10}$$

f_0 was set at 2.2 MHz. The 6 dB pulse bandwidth Δf, measured on the power spectrum of the signals was varied between 0.3 MHz and 1.5 MHz in steps of 0.1 MHz by changing T from 6.6 μs to 1.32 μs. 10 μs of echo signal, centered in the focal zone are recorded in each case at a sampling interval of 0.02 μs with an 8 bit digitizer (Analogic Corporation DATA 6500). For each of the 13 bandwidths and each sponge sample echo signals were recorded at a 100 different locations on the sponge. The two dimensional PSF was also measured in separate experiments by placing and scanning a 0.25 mm nylon wire at 5.5 cm from the transducer. Five of the 13 pulses, those with $\Delta f = 0.3$ MHz, 0.6 MHz, 0.9 MHz, 1.2 MHz, and 1.5 MHz were used in the PSF measurement. All the measured PSF's are shown in Fig. 4 on a logarithmic scale over a 60 dB range. This entire two dimensional function represents g(r,t) in Eq. (8).

Fig. 5 shows a plot of $1/V_E$, the inverse of the "effective volume" defined in Eq. (6), numerically calculated from the experimental PSF g(r,t) for the five different bandwidths. The solid line is a 3rd order polynomial fit to the data.

RESULTS

Envelope detection was performed on the RF echo signal using Hilbert Transform [4]

The normalized second intensity moments as a function of interrogating pulse bandwidth Δf, for the three different sponge structures were obtained by performing both time averaging over a 10 μs window and ensemble averaging over 100 different A-lines. Fig. 6 shows the normalized second intensity moment as a function of $1/V_E$. The normalized moments increase approximately linearly with $1/V_E$, at least for bandwidths below 1 MHz. The slope of the linear section is much higher for the large pore size sponge (sample 1) compared to samples 2 and 3. The intercept at $\Delta f=0$ is close to 2 for samples 1 and 2 but is below the Rayleigh limit value of 2 for sample 3.. The value for $1/V_E$ at each Δf was obtained from Fig. 5. The slope and intercept on this graph only depend on the scattering microstructure. For the sake of clarity, the error bars have been omitted and data for all three samples are shown in this graph. The three straight lines are the least square fit to the six data points with smallest $1/V_E$ values for the three respective samples. The intercept for samples 1, 2, and 3 are found to be 1.95, 2.08, and 1.76 respectively. Similarly, the slope for samples 1, 2, and 3 are 25.6 mm³, 4.03 mm³ and 4.6 mm³ respectively.

The intercept value for sample 3 was found to be significantly below the Gaussian limiting value of 2. The rationale for this was presented in the theoretical section. In order to obtain an indirect confirmation that this phenomenon is related to the fact that we may have periodic spacing of the order of the dominant wavelength $\lambda_0/2$, separate experiments were performed at $f_0 = 1.8$, 2.0, 2.4, and 2.6 MHz. In each case the intercept increased to a value of 2 with no significant effect on the slope.

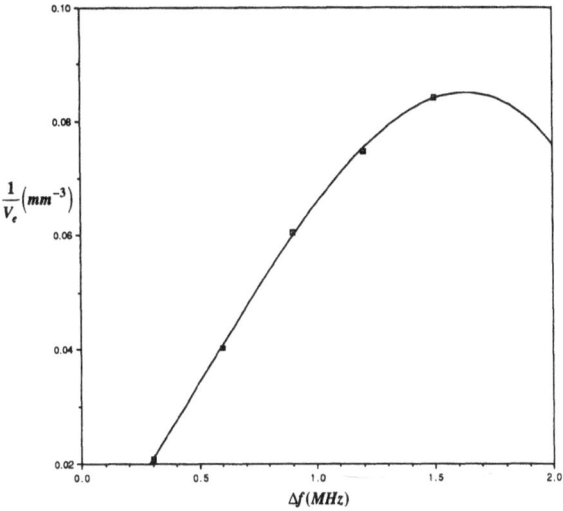

Fig. 5. $1/V_E$ values calculated from PSF measurements at five different bandwidths Δf

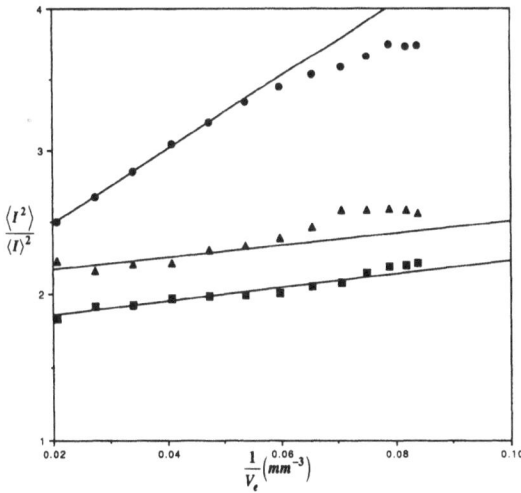

Fig. 6. Normalized second intensity moments vs $1/V_E$ for the three sponge structures.

▼ sample 1, ◆ sample 2, ● sample 3

CONCLUSIONS

We conclude with the following observations. The theory that was presented to extract scattering microstructure parameters was based on small, but variable bandwidth probe pulse assumption. The prediction that second normalized intensity moments should vary linearly with $1/V_E$ was verified for the three sponge structures. It is reasonable to conclude that the assumptions are valid, at least for the sponge structures under study. These scattering structures were studied because they have some relevance to the lobular scattering microstructure of liver tissue (spacing $\approx$ 1 mm) and grain boundary structure in metal nondestructing testing. The slope , according to the theory depends on $\langle a^4 \rangle / N \langle a^2 \rangle^2$, a term that involves scatterer number density and the probability distribution function of scattering coefficient values at f_0. We were able to determine statistically significant differences in this parameter for the three samples. Our method however is not able to separate the two component terms, but the parameter itself can be useful for differential diagnosis. Finally, the intercept (specially the departure form the limiting value of 2) may also serve a useful diagnostic purpose. It is an indicator that an underlying unresolved periodic structure with spacing $= \lambda_o/2 = c/2f_0$ is present in the sample. This behavior is of course a function of f_0 and the degree of periodic order (or randomness) of the structure, and is not fully understood at present. Nevertheless, it appears that this intercept value may also serve a useful role in microstructure characterization and deserves further study.

Acknowledgments

This work was supported in part by a grant No. R15CA71004-01 from NIH.

REFERENCES

1. E. Jakeman and R. J. A. Jough, Non-Gaussian models for the statistics of scattered waves, *Advances Phy.* 37:471 (1988).
2. J. W. Goodman, Statistical properties of laser speckle patterns, in: *Laser Speckle and Related Phenomena*, J. C. Dainty, ed., Topics in Applied Physics, 9:9-74, Springer-Verlag (1984).
3. M. A. Fink and J. Cardoso, Diffraction effects in pulse-echo measurements, *IEEE Trans. Sonic. Ultrason.* SU-31:313 (1984).
4. N. A. H. K. Rao and H. Zhu, Simulation study of changes in ultrasound speckle statistics with the system point spread function, *J. Acoust. Soc. Am.* 95:1161-1164 (1994).
5. W.H. Round and R. H.T. Bates, Modification of spectra pulses from ultrasonic transducers by scatterers in non-attenuating and in attenuating media, *Ultrasonic Imaging* , 9:92-105 (1987).
6. N. A. H. K. Rao and W. M. Aubry, Evaluation of a pulse coding technique for spatial structure characterization, *IEEE Trans. Ultrason. Ferroel. Freq. Control* . 41:660 (1994).
7. J. F. Chen, J. A. Zagzebski and E. L. Madsen, Non-Gaussian versus non- Rayleigh statistical properties of ultrasound echo signals, *IEEE Trans. Ultrason. Ferroel. Freq. Control*, 41:435-440 (1994).
8. V. M. Narayanan, P. M. Shanker and J. M. Reid, Non-Rayleigh statistics of ultrasonic backscattered signals, *IEEE Trans. Ultrason. Ferroel. Freq. Control*, 41:845-852 (1994).

MODELING OF ULTRASONIC BEAM INTERACTION
WITH AN ANISOTROPIC MULTILAYERED MEDIUM,
USING ANGULAR SPECTRUM METHOD

Aziz U. REHMAN,[1] C. POTEL,[1,2] J. F. de BELLEVAL[1]

[1]Univ. Tech. Compiègne. LG2mS, UPRES ass. CNRS, Compiègne. France.
[2]Univ. Picardie Jules Verne. IUT Amiens. Dept. OGP Soissons, France.

ABSTRACT

In many domains of acoustic field propagation, such as ultrasonic non-destructive testing, a realistic calculation of ultrasonic beam patterns requires treatment of the effects of refraction and reflection from finite and infinite media. Application of the Angular Spectrum Analysis (ASA) to calculate the acoustic fields has become an increasingly interesting topic, because the ASA is easy to be numerically implemented upon the basis of the Fast Fourier Transform and can be used effectively to solve a variety of complex problems. especially those involved with boundaries and multilayered anisotropic media.

Here we describe the development of a computationally efficient technique capable of calculating the intensity distribution in reflected fields from a multilayered anisotropic finite medium submerged in water. The technique is based on presenting the given acoustic field using angular spectrum decomposition, and involves a systematic extension to account for various properties of medium, modifying the phase factor to account for refraction, reflection, anisotropic effects and to provide three/two dimensional prediction of reflected beams, whatever the form of the incident profile may be.

Key words: nonspecular reflection. multilayered anisotropic structure.
ultrasonic beam. angular spectrum analysis. anisotropy

INTRODUCTION

Ultrasonic nonspecular reflection effects were originally discussed and studied experimentally by Schoch[1] using Schlieren photography. Schoch calculated that the reflected beam was displaced laterally from the position expected from geometrical acoustics, while the reflected beam profile retained the same general shape as the incident beam profile. Although the basic framework of Schoch's analysis has been used by others, experiments have shown that, in contrast to Schoch's simple picture, the reflected field is much more complicated than simply a shifted beam.

The discrepancies between Schoch's theory and experiments were studied by Bertoni and Tamir[2] in several parametric regimes for the fluid-loaded half-spaces. They have developed an analytical model to describe the nonspecular reflection of ultrasonic beams from a fluid-solid interface for incidence at the Rayleigh angle. They assume that the incident field is given by a Gaussian distribution of the normal velocity along the interface and that the characteristic width of this distribution is large compared to the wavelength in the fluid. The normal velocity distribution of the reflected field is then obtained along the interface by means of a Fourier representation. In their results, the specular reflection and the reradiation of Rayleigh waves appear clearly. Their method was employed in later studies dealing with propagation away from the interface[3] and extension to fluid-loaded plates[4]. Rousseau and Gatignol[5] employed Bertoni and Tamir representation for the incident beam and approximated the reflected field integrals, for single interfaces and plates, by paraxial high-frequency asymptotics. Ngoc and Mayer[6] used the numerical integration approach to extend the Bertoni and Tamir results to arbitrary incidence on fluid-solid interfaces and submerged plates. Afterwards, Claeys and Leroy[7] have provided a new model for the beam as a finite superposition of inhomogeneous plane waves and they were led to similar results.

The present article is intended to apply the angular spectrum decomposition method to investigate the nonspecular reflection effects for anisotropic multilayered medium.

THEORETICAL DESCRIPTION

Spectral Representation of an Acoustic Beam

Fig. 1. shows an ultrasonic beam of width $2a$ incident on a solid (multilayered anisotropic) plate of thickness h. The beam of an angular frequency $\omega = 2\pi f$ is seen to be incident at an angle θ_i. The coordinate system to be used is shown in the diagram,

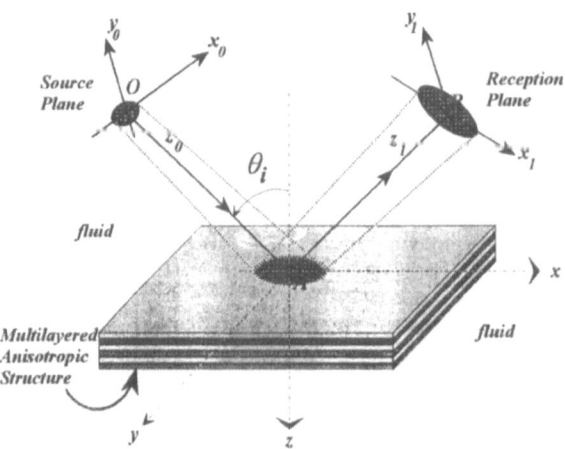

Figure 1. Representation of the situation to be studied showing the choice of coordinate system.

and it will be assumed that the plate's incidence surface forms the xy plane. In the context of angular spectrum analysis, an acoustic beam can be represented as a superposition of an infinite number of plane waves which have different amplitudes and wave vectors. It is shown[4-7] by using Fourier analysis that the field of an incident beam can be uniquely determined at any point, if its field distribution is known in any plane. Thus a spectral representation of a well-defined incident beam can be expressed by the following Fourier integral transform pair.

$$\tilde{G}\left(k_{x_0},k_{y_0},\omega\right) = \int\int_{-\infty}^{+\infty} \tilde{g}\left(x_0,y_0,\omega\right) \times \exp\left[-j\left(k_{x_0}x_0 + k_{y_0}y_0\right)\right] dx_0 dy_0 \tag{1}$$

$$\tilde{g}\left(x_0,y_0,\omega\right) = \frac{1}{(2\pi)^2} \int\int_{-\infty}^{+\infty} \tilde{G}\left(k_{x_0},k_{y_0},\omega\right) \times \exp\left[j\left(k_{x_0}x_0 + k_{y_0}y_0\right)\right] dk_{x_0} dk_{y_0} \tag{2}$$

where k_{x_0} and k_{y_0} are the wave numbers in the x_0- and y_0-directions, whereas the z-component of the wave number vector $\vec{k}$ is defined by $k_{z_0} = \sqrt{k^2 - k_{x_0}^2 - k_{y_0}^2}$. $\tilde{g}$ is specified over the source plane and is here taken to be the normal velocity field $\tilde{u}$.

It is implied, with no loss of generality, that the normal vector to the source plane lies parallel to the z_0-axis. The amplitude $\tilde{G}$ is specifically replaced by $\tilde{U}$ in these situations. A tilde (~) is used here to denote complex quantities, and capitals denote the spatial Fourier transform of the corresponding lowercase quantities. In this article, the time dependence, $exp(-j\omega t)$, in the Fourier transform pair is suppressed, and the incident beam is taken to have a *Gaussian* profile.

The physical interpretation of *(eq. 2.)* is that the incident beam is composed of an infinite number of plane waves having the same wavelength but incident at different angles. The principle of spectral representation can be extended to describe an acoustic beam reflected from multilayered anisotropic medium. If $\tilde{R}\left(k_{x_1},k_{y_1}\right)$ denotes the plane wave reflection coefficient for a particular layered structure, then the associated ultrasonic field at a distance $\vec{r} = \left(\overline{OA} + \overline{AB}\right)$, can be represented by

$$\tilde{u}\left(x_1,y_1,z_1;\omega\right) = \frac{1}{(2\pi)^2} \int\int_{-\infty}^{+\infty} \tilde{U}\left(k_x,k_y,\omega;\vec{r}\right) \times \tilde{R}\left(k_x,k_y\right) \times \exp\left[j\left(k_{x_1}x_1 + k_{y_1}y_1\right)\right] dk_x dk_y \tag{3}$$

where

$$\tilde{U}\left(k_{x_1},k_{y_1},\omega;\vec{r}\right) = \tilde{U}\left(k_{x_0},k_{y_0},\omega;\vec{0}\right) \cdot \exp\left(j k_{z_1}\vec{r}\right)$$

Plane Wave Reflection Coefficients

It is evident from *(eq. 3.)* that the relevant plane wave reflection coefficient $\tilde{R}\left(k_{x_1},k_{y_1}\right)$, which is the function of the incident angle, number and orientation of layers and the product fh, strongly influences the beam profile. Indeed, *(eq. 3.)* indicates that the resulting profile is constructed by the interference of the individual plane waves, which have amplitude $\tilde{U}\left(k_{x_1},k_{y_1},\omega;\vec{r}\right)$ and undergo a phase shift upon reflection.

The amplitude plane wave reflection coefficients for the cases of interest, i.e. for a medium consisting 6 periods of *0.47mm {0°/45°/90°/135°}₀°* *Carbon/Epoxy* plate at an incidence frequency of *f =3.0 MHz*,[8] as shown in *fig. 2*. It was described in [9] that by writing the boundary conditions at each interface separating two successive layers, the transfer matrix of one superlayer can be found, which allows the displacement

401

amplitudes of the plane waves in the first layer of a superlayer to be expressed as a function of those in the first layer of the next superlayer. Due to the fact that, locally, the acoustical state is characterised by six quantities, this matrix is of the 6th order, and the waves which correspond to the eigen vectors of this matrix are Floquet waves. If the general solution is decomposed on the Floquet wave basis, the transfer matrix becomes a diagonal matrix. These six Floquet waves are the propagation modes of an infinite periodically multilayered medium. It was shown in [8] that for certain incident angles, the *multilayered Rayleigh modes* exist in the multilayered anisotropic structures when all the Floquet waves are inhomogeneous and the phase changes from π to -π (see *fig. 2.*).

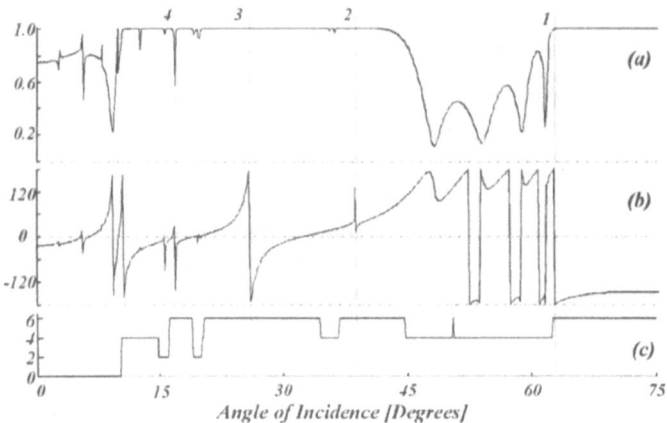

Figure 2. *a)* Magnitude and *b)*Phase of reflection coefficients with the *c)* Number of inhomogeneous Floquet waves for a Carbon/Epoxy medium consisting of 6-periods, at an incidence frequency of *f=3.0 MHz*. (medium characteristics and other acoustic parameters are taken from *reference 8.*)

REFLECTED PROFILES FOR INCIDENT GAUSSIAN BEAM.

The angular spectrum decomposition method is then applied to study the reflected beam profiles from a multilayered anisotropic structure immersed in a fluid. The integral to be evaluated is *(eq. 3.)*, which describes the profiles of the reflected beam. The incident frequency is fixed at $f = 3.0$ *MHz* and the diameter of the incident gaussian beam is taken to be *12.7 mm*, the profile of which is shown in *fig. 3a*. It has been established[4-7] that for a fluid-solid-fluid system, the nonspecular phenomena are prominent at such an angle of incidence that either or both of the modulus and phase of reflection coefficients vary significantly. Therefore, it is interesting to investigate the profiles for different incident angles, corresponding to the *leaky modes*. The profiles shown in *fig. 3.* are calculated arround the *62.6°* — modulus and phase of reflection coefficients vary significantly arround that angle — to investigate the phenomena of nonspecular reflection.

Fig. 3. shows the reflected profiles in three dimensions for the incidences arround *62.6°*. It should be noted that the reflections show the general features like lateral beam displacement, null intensity region, and a trailing ultrasonic field. It is also important to be mentioned that the presence of plate modes in the vicinity of the selected incident angles enhances significantly the presence of null intensity regions in the calculations of reflected beam profiles.

Reflected profile representation in two dimensions is also made (*Fig. 4.*) to facilitate the comparison in the resultant profiles for different incident angles arround the *leaky mode.*

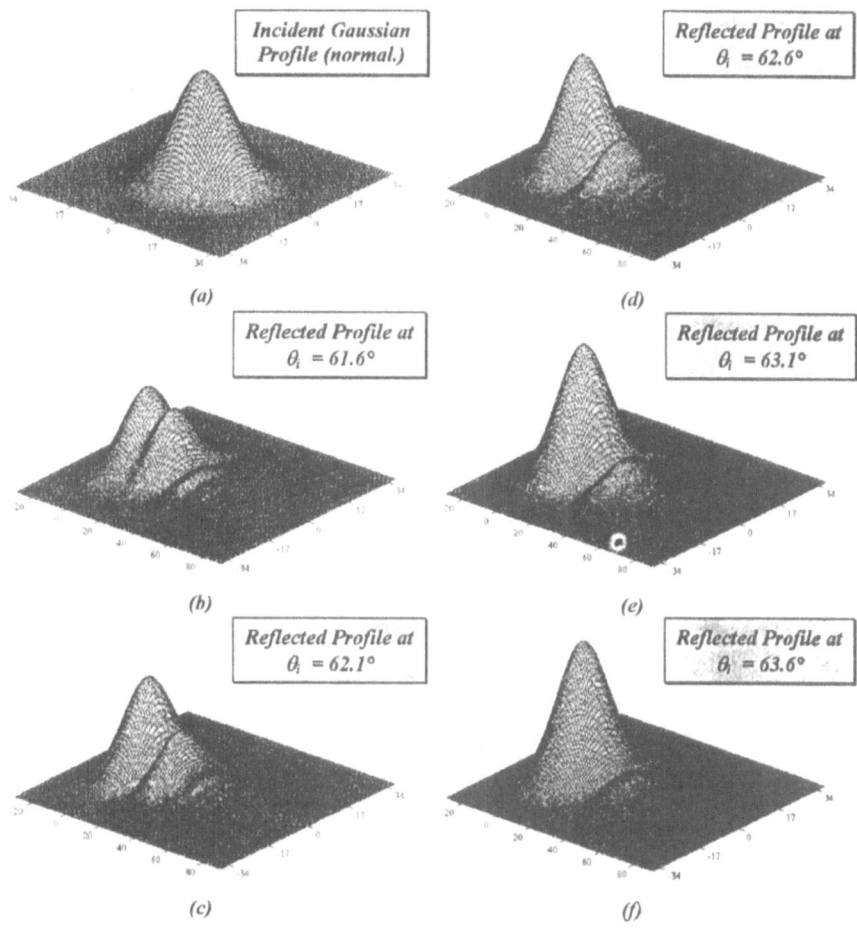

Figure 3. *a)* Incident Gaussian beam profile (*normalized at 400mm*). *b-f)* Reflected beam profiles for the incident angles mentioned for each profile at $OA=AB=200mm$ from the interface fluid-solid.

CONCLUSION

The numerical method described above, when applied to a fluid-solid (multilayered anisotropic)-fluid system yields informations about nonspecular reflectivity. It can be concluded from the results presented that the nonspecular reflection effects for an ultrasonic beam incident on a solid plate immersed in a fluid occur at or arround the *phase matched leaky mode incidences.* Further, it has been shown that in order to predict the reflected beam profiles, it is necessary to have a knowledge of the reflection coefficients. The analysis presented indicates that the reflected acoustic field is constructed by the interference of the reflected infinite plane waves which make up the incident beam. Each plane wave is generally reflected with a different phase change. Since the nonspecular reflections take place at the leaky mode incident angles, it is tempting to interpret the reflection mechanism by postulating the generation of a leaky vibrational mode which manifests itself in the form of what has been called a trailing acoustic field, which is clearly visible in *fig. 4.*

Maximas higher than unity are observed for some values of incident angles. The reason behind this is the concentration of energy flux in certain directions for anisotropic materials, which causes the concentration of the acoustic waves to occur, for the incident angles coinciding with these directions.

Taken together, these conclusions suggest that the proposed method furnishes numerically efficient, and accurate algorithm for addressing beam-structure interaction for broad range of fluid-immersed elastic configurations. It also quantifies the phenomena observed when the incident acoustic beam interacts with anisotropic structures, especially with respect to nonspecular reflection effects near phase-matched leaky mode incidences.

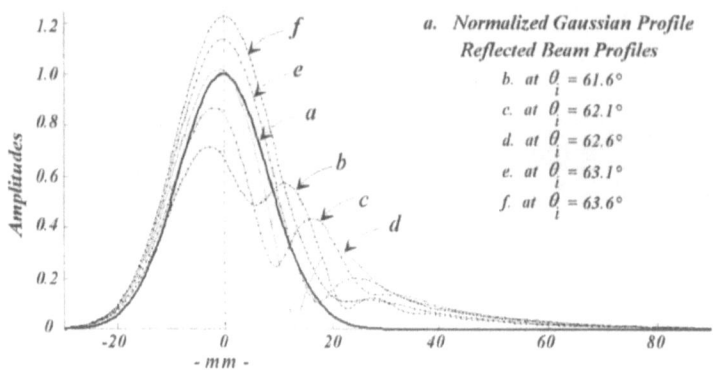

Figure 4. Two dimensional representation of amplitude distributions for reflected profiles compared with the incident Gaussian beam profile normalized at *(OA+OB) 400mm.*

REFERENCES

1. A. Schoch: *"Der Schalldruchgang durch Platten* [Sound Transmission Through Plates]." Acustica. 2. p. 1-18. (1952).
2. H.L. Bertoni. T. Tamir: *"Unified theory of Rayleigh-Angle phenomena for acoustic beams at liquid-solid interfaces."* J. of Appl Phys. 2, p, 157-172. (1952).
3. M. A. Breazeale. L. Adler. G.W. Scott: " *Interaction of ultrasonic waves incident at the Rayleigh angle onto a liquid-solid interface."* J. of Appl. Phys.. 48. p. 530-537. (1973).
4. L. E. Pitts. T. J. Plona. W. G. Mayer: *"Theory of nonspecular reflection effects for an ultrasonic beam incident on a solid plate in a liquid."* IEEE Trans. Sonics Ultrason., SU-24. p. 101-109. (1977).
5. M. Rousseau. Ph. Gatignol: *"Asymptotic analysis of nonspecular effects for the reflection and transmission of Gaussian beam incident on a solid plate."* J. of Acoust. Soc. Am.. 80. p. 325-332. (1985).
6. T. D. K. Ngoc. W. G. Mayer: *"General description of ultrasonic nonspecular reflection and transmission effects for layered media."* IEEE Trans. Sonics Ultrason.. SU-27. p. 229-236. (1980).
7. J. M. Claeys. O. Leroy: *"Reflection and transmission of bounded beams on halfspaces and through plates."* J. of Acoust. Soc. Am.. 72. p. 585-590. (1982).
8. C. Potel. J. F. de Belleval. E. Genay. Ph. Gatignol: *"Behavior of Lamb waves and multilayered Rayleigh waves in an anisotropic periodically multilayered medium: Application to the long-wave length domain."* Acta Acustica. 82. p. 738-748. (1996).
9. C. Potel. J. F. de Belleval Y. Gargouri: *"Floquet waves and classical plane waves in an anisotropic periodically multilayered medium: Application to the validity domain of homogenization."* J. of Acoust. Soc. Am.. 97. p. 2815-2825. (1995).

SHEAR RELAXATION TIMES OF SOME PERCHLORATES
EVALUATED FROM ULTRASONIC VELOCITY MEASUREMENTS

Dip Singh Gill,* Parvinder Singh,[†] and Jasbir Singh

Department of Chemistry
Panjab University
Chandigarh -160014 India

Ultrasonic velocities (u), densities (ρ), and viscosities (η) of the compounds $LiClO_4$, $NaClO_4$, $Mg(ClO_4)_2$, and $Al(ClO_4)_3$ have been measured in water in the concentration range 0.02–5.0 mol kg^{-1} at 298 K. The isentropic compressibilities (K_s) of various salts have been calculated and used to evaluate shear relaxation times (τ_s). The relaxation time for all electrolytes shows a strong dependence on salt concentration. For $Al(ClO_4)_3$, $Mg(ClO_4)_2$, and $LiClO_4$, the relaxation time increases strongly with increase of salt concentration while for $NaClO_4$ it decreases with increase of salt concentration. At a particular concentration of the perchlorates investigated, the relaxation time has the order of magnitude

$$Al(ClO_4)_3 > Mg(ClO_4)_2 > LiClO_4 > NaClO_4$$

INTRODUCTION

The dissipative response of liquids to time dependent stresses has been exhaustively studied theoretically and experimentally.[1–3] When the solution is subjected to ultrasonic wave, the viscous properties of the medium not only produce attenuation in amplitude but also dispersion[4] in the velocity of propagation of the wave. The dispersion of the ultrasonic wave produced by the shear coefficient of viscosity has a characteristic relaxation time (τ_s). A relationship between shear relaxation time, τ_s and isentropic compressibility, K_s has been well known.[4] The measurement of this relaxation time provides sufficient information about the viscoelastic behavior and structure of the solution. In this paper, therefore, we have made ultrasonic velocity, density and viscosity measurements of $LiClO_4$, $NaClO_4$, $Mg(ClO_4)_2$, and $Al(ClO_4)_3$ in water at high concentrations to evaluate their shear relaxation times for examining their solvation behavior in a more rigorous way.

---

*Main author.
[†]Presented paper.

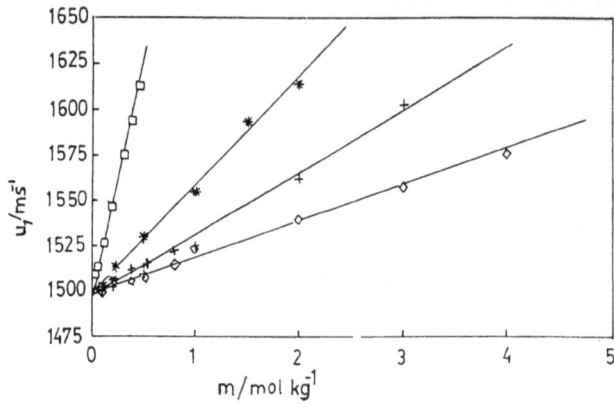

Figure 1. Plots of ultrasonic intensity (u) vs. molality (m) of some perchlorates in water at 298 K. ◇ LiClO$_4$; + NaClO$_4$: * Mg(ClO$_4$)$_2$; □ Al(ClO$_4$)$_3$.

EXPERIMENTAL

Doubly distilled water with a specific conductance $< 1 \times 10^{-6}$ S cm^{-1} was used for all measurements. LiClO$_4 \cdot$ 3H2O, NaClO$_4 \cdot$ H$_2$O, Mg(ClO$_4$)$_2 \cdot$ 6H$_2$O, and Al(ClO$_4$)$_3 \cdot$ 6H$_2$O were prepared from the corresponding carbonates by the standard methods. Ultrasonic velocities were measured as before[5-8] with an accuracy of 2 parts in 10^4 at 2 MHz using an ultrasonic time intervalometer model UTI-101 from Innovative Instruments, Hyderabad, using a pulse-echo overlap method. Viscosities were measured using an Ubbelohde suspended bulb viscometer. The overall accuracy of the viscosity measurements was better than ±0.1%. The densities of the salt solutions were measured using an Anton Paar Digital Densimeter model 60 and a calibrated cell Type 602 with a reproducibility of ±1 × 10^{-5} g cm^{-3}.

RESULTS AND DISCUSSION

Ultrasonic velocities (u), densities (ρ), and viscosities (η) of the compounds LiClO$_4$, NaClO$_4$, Mg(ClO$_4$)$_2$, and Al(ClO$_4$)$_3$ have been measured at several concentrations of the salts in the concentration range 0.02–5.0 mol kg^{-1} in water at 298 K. In case of Al(ClO$_4$)$_3$ the solubility was not very high, therefore, the measurements were restricted to a maximum concentration of 0.5 mol kg^{-1}. The plots of u versus molality (m) of the salts are linear in all cases while the plots of ρ and η versus m are nonlinear. For illustration the plots of u versus m are shown in Fig. 1. The plots of isentropic compressibility (K_s) calculated from the relation

$$K_s = \frac{1}{u^2 \rho} \qquad (1)$$

versus m for all the electrolytes are also nonlinear. The nonlinearity of the plots of K_s versus m in the present studies arises from ion-solvation and solvent–solvent interactions.

The shear relaxation time, τ_s is a better and more informative parameter than isentropic compressibility as it incorporates the shear viscosity effects also. An equation mathematically derived by Kinsler and Frey[4] for the evaluation of τ_s from K_s values can be written as:

$$\tau_s = \frac{4}{3}\eta K_s \qquad (2)$$

The applicability of equation (2) has been tested for pure solvents and for binary mixtures in our previous work.[9] By using equation (2), the shear relaxation times for the solution

Table 1. Shear Relaxation Time (τ_s) of $LiClO_4$, $NaClO_4$, $Mg(ClO_4)_2$, and $Al(ClO_4)_3$ at Different Salt Concentrations in Water at 298 K

$LiClO_4$		$NaClO_4$		$Mg(ClO_4)_2$		$Al(ClO_4)_3$	
m/mol kg^{-1}	$\tau_s/10^{-13}$ s	m/mol kg^{-1}	$\tau_s/10^{-13}$ s	m/mol kg^{-1}	$\tau_s/10^{-13}$ s	m/mol kg^{-1}	$\tau_s/10^{-13}$ s
0.0	5.31	0.0	5.31	0.0	5.31	0.0	5.31
0.1	5.76	0.1	5.29	0.05	5.42	0.0267	5.36
0.2	5.84	0.2	5.28	0.1	5.79	0.052	5.43
0.5	5.95	0.5	5.28	0.2	5.90	0.12	5.87
1.0	6.12	1.0	5.20	0.5	6.23	0.19	5.94
2.0	6.35	2.0	5.07	1.0	6.54	0.31	6.62
3.0	6.70	3.0	5.11	1.5	7.07	0.38	7.23
4.0	7.00			2.0	8.09	0.46	7.83

of $LiClO_4$, $NaClO_4$, $Mg(ClO_4)_2$, and $Al(ClO_4)_3$ in water has been calculated from their compressibility data and the results are reported in Table 1.

The results of Table 1 show that the relaxation time decreases with increase of molality in case of $NaClO_4$ and significantly increase in cases of $LiClO_4$, $Mg(ClO_4)_2$, and $Al(ClO_4)_3$. At a fixed concentration of the perchlorates the relaxation time decreases in the order:

$$Al(ClO_4)_3 > Mg(ClO_4)_2 > LiClO_4 > NaClO_4$$

These results indicate that $NaClO_4$ shows overall less structure in water and the structural effects decrease with increase of salt concentration. On the other hand, $Al(ClO_4)_3$, $Mg(ClO_4)_2$, and $LiClO_4$ show strong structural effects. Since ClO_4^- is poorly solvated in water, therefore, the changes in structural effects are mainly due to the solvation effects of cations only. As Al^{3+}, Mg^{2+}, and Li^+ because of their high charge density have very strong solvation effects in water as compared to those for Na^+, therefore, strong structural effects in case of $Mg(ClO_4)_2$, $Al(ClO_4)_3$, and $LiClO_4$ are according to the expectations and in accordance with the previously well established fact that Al^{3+} is more strongly solvated than Mg^{2+} and Mg^{2+} more strongly solvated than any monovalent cation in water. In conclusion it can be found that shear relaxation times can well explain the solvation behavior of ions in solution.

ACKNOWLEDGMENTS

Research grant from CSIR, New Delhi under the research scheme 1 (1412)/96-EMR-II is gratefully acknowledged.

REFERENCES

1. T. A. Litovitz and C. M. Davis, in: *Physical Acoustics*, edited by W. P. Mason, Academic Press, New York, Vol II A, p. 281 (1965).
2. D. A. Pinnow, S. Landau, J. L. Macchia, and T. A. Litovitz, *J. Acoust. Soc. Am.* 43:131 (1967).
3. R. D. Mountain, *J. Res. Natl. Bur. Stand., Sect. A.* 72: 95 (1968).
4. L. E. Kinsler and A. R. Frey, *Fundamentals of Acoustics*, Wiley Eastern, Ltd., New Delhi, p. 224 (1978).
5. D. S. Gill, T. Kaur, H. Kaur, I. M. Joshi, and J. Singh, *J. Chem. Soc. Faraday Trans.* 89: 1737 (1993).
6. J. Singh, T. Kaur, V. Ali, and D. S. Gill, *J. Chem. Soc. Faraday Trans.* 90: 579 (1994).
7. D. S. Gill, R. Singh, V. Ali, J. Singh, and S. K. Rehani, *J. Chem. Soc. Faraday Trans.* 90:583 (1994).
8. D. S. Gill, P. Singh, J Singh, P. Singh, G. Senanayake, and G. T. Hefter, *J. Chem. Soc. Faraday Trans.* 91 : 2789 (1995).
9. D. S. Gill, P. Singh, J. Singh, S. K. Rehani, and R. Khajuria, *Indian J. Chem. Section A* (Communicated).

MULTIPARAMETER INVERSION OF ELASTIC WAVE
WITH A SINGLE IMPULSE SOURCE

Xiuping Tao, Jianchun Cheng, and Shuyi Zhang

Institute of Acoustics and Laboratory of Modern Acoustics
Nanjing University, Nanjing 210093, P.R.China

INTRODUCTION

We present a single impulse source method to reconstruct two-dimensional variations in elastic materials with inhomogeneous density and elastic Lame parameters. For the case of seismic exploration, i.e., the scattering object is embedded in an elastic medium at some depth under the surface, the source and receivers are located on the surface above the object. For single impulse source, the receiver array will detect elastic wave traces scattering from the object at a series of receiving positions. The advantage of this data collection strategy is that we can obtain much more data from one pulse and there is no need to rotate the system.

From the elastic wave equation in an inhomogeneous isotropic elastic solid, the reconstruction formulas are developed for this reflective imaging configuration under the far field condition and Born approximation. The scattering of the incident wave can be characterized by the four mode conversion processes, i.e., P-to-P, P-to-SV, SV-to-P and SV-to-SV. Because the P-to-SV, SV-to-P and SV-to-SV scatterings are independent of the variation of Lame parameter λ, we only consider the P-to-P scattering in order to inverse variations of all the three parameters.

In the process of P-to-P scattering, we reduce this problem to solve the generalized inversion of linear algebraic system. The transform of the detected signal in the time domain is related to a linear integral of a function of the elastic parameters along a straight line. The line depends on time t and the angle θ between the incident and the scattering wave directions. For a series of time of the detected signal, we can obtain the integrals along the parallel straight lines, and for different receivers, the integrals of different angle θ can also be given. The elastic parameters can then be determined by solving an over-determined linear equations.

LINEARIZED INVERSION OF ELASTIC WAVES

The elastic wave equation in an inhomogeneous isotropic elastic solid is,

$$\rho\frac{\partial^2 U_\alpha}{\partial t^2} = \frac{\partial}{\partial x_\alpha}(\lambda\frac{\partial U_\beta}{\partial x_\beta}) + \frac{\partial}{\partial x_\beta}[\mu(\frac{\partial U_\alpha}{\partial x_\beta} + \frac{\partial U_\beta}{\partial x_\alpha})] + \rho f_\alpha \qquad (1)$$

where U_α (α=1,2) is the displacement components, λ and μ are the Lame constants, ρ is the density, and f_α is the applied force density. As shown in Fig. 1, the source and receiver are located at $\mathbf{X}_s=(X_s, D)$ and $\mathbf{X}_r=(X_r, D)$, respectively, and the vector $\mathbf{X}'=(x', y')$ represents the scatter point in the scatter region σ. The displacement field can be expressed by the incident wave U_α^i and scattered wave U_α^s,

$$U_\alpha = U_\alpha^i + U_\alpha^s \qquad (2)$$

The incident wave U_α^i is produced in homogeneous medium by an impulse point source with force vector $f_\alpha(\mathbf{X},t)=f_\alpha(t)\delta(\mathbf{X}_s-\mathbf{X})$ located at $\mathbf{X}_s$, which can be expressed by, in frequency domain,

$$U_\alpha^i(\mathbf{X}_s,\mathbf{X},\omega) = \int_\alpha f_\beta(\omega)\delta(\mathbf{X}_s - \mathbf{X}')g_{\alpha\beta}(|\mathbf{X} - \mathbf{X}'|)d\mathbf{X}' = f_\beta(\omega)g_{\alpha\beta}(|\mathbf{X} - \mathbf{X}_s|) \qquad (3)$$

where $g_{\alpha\beta}$ is the dyadic Green function. The elastic wave traces detected by the receiver located at $\mathbf{X}_r$ can be expressed by

$$U_\alpha^s(\mathbf{X}_s,\mathbf{X}_r,\omega) = \int_\sigma f_\beta^s(\mathbf{X}',\omega)g_{\alpha\beta}(|\mathbf{X}_r - \mathbf{X}'|)d\mathbf{X}' \qquad (4)$$

where f_β^s comes from the scattering of inhomogeneous region,

$$f_\beta^s(\mathbf{X}',\omega) = \rho_1(\mathbf{X}')\omega^2 U_\beta + \frac{\partial}{\partial x_\beta'}[\lambda_1(\mathbf{X}')(\nabla' \cdot \mathbf{U})] + \frac{\partial}{\partial x_s'}[\mu_1(\mathbf{X}')(\frac{\partial U_\beta}{\partial x_s'} + \frac{\partial U_s}{\partial x_\beta'})] \qquad (5)$$

where $\rho_1=\rho-\rho_0$, $\lambda_1=\lambda-\lambda_0$ and $\mu_1=\mu-\mu_0$ are the variations of the density and Lame parameters, respectively, ρ_0, λ_0 and μ_0 are the homogeneous background parameters. In order to inverse ρ_1, λ_1 and μ_1 from the integral equation [Eq.(4)], two approximations are employed,

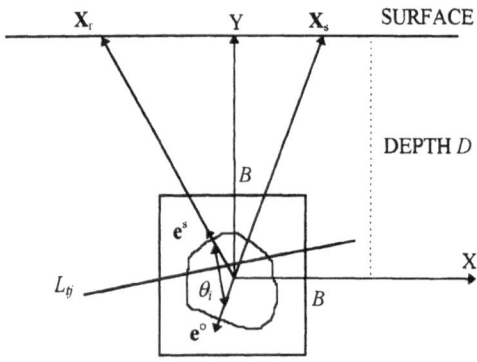

Figure 1. The configuration of the data collection system. The single impulse source and the receiver array are located at the surface.

(1) the linearized Born approximation: in the right side of the Eq.(5), U_α is replaced by U_α^I;
(2) the far field approximation, i.e. $|\mathbf{X}_s|$ and $|\mathbf{X}_r| >> |\mathbf{X}'|$, thus we have

$$|\mathbf{X}' - \mathbf{X}_s| \approx |\mathbf{X}_s| + \mathbf{e}^o \cdot \mathbf{X}'; \quad |\mathbf{X}' - \mathbf{X}_r| \approx |\mathbf{X}_r| + \mathbf{e}^s \cdot \mathbf{X}' \tag{6}$$

where the $\mathbf{e}^o$ and $\mathbf{e}^s$ are unit vectors in the incident and scattering directions respectively. The two-dimensional dyadic Green function can be represented by the Hankel function,

$$g_{\alpha\beta}(|\mathbf{X} - \mathbf{X}'|) = \frac{i}{4\rho_0\omega^2} \langle \frac{\partial^2}{\partial x_\alpha \partial x_\beta} [H_0^{(1)}(K_S|\mathbf{X} - \mathbf{X}'|) - H_0^{(1)}(K_P|\mathbf{X} - \mathbf{X}'|)]$$
$$+ K_S^2 \delta_{\alpha\beta} H_0^{(1)}(K_S|\mathbf{X} - \mathbf{X}'|) \rangle \tag{7}$$

where $K_P = \omega/C_P$ and $K_S = \omega/C_S$ are the longitudinal and transverse wavenumbers, C_P and C_S are the longitudinal and transverse wave velocities, respectively. Now, the scatterings of the incident wave can be characterized by the four mode conversion processes:
   (1) P-to-P

$$(U_\alpha^s)^{P-P}(\mathbf{X}_s, \mathbf{X}_r, \omega) = e_\alpha^s \frac{i\omega(\mathbf{f} \cdot \mathbf{e}^o)\exp[iK_P(|\mathbf{X}_s| + |\mathbf{X}_r|)]}{8\pi\rho_0 C_P^3(|\mathbf{X}_s||\mathbf{X}_r|)^{1/2}} F^{PP}(\mathbf{K}_P, \theta) \tag{8}$$

where $\mathbf{K}_P = K_P(\mathbf{e}^s - \mathbf{e}^o)$ and $F^{PP}(\mathbf{K}_P, \theta)$ are defined as

$$F^{PP}(\mathbf{K}_P, \theta) = [\rho_1(\mathbf{K}_P)/\rho_0]\cos\theta - [\lambda_1(\mathbf{K}_P) + 2\mu_1(\mathbf{K}_P)\cos^2\theta]/[\lambda_0 + 2\mu_0] \tag{9}$$

and θ is the angle between $\mathbf{e}^o$ and $\mathbf{e}^s$;
   (2) P-to-SV

$$(U_\alpha^s)^{P-SV}(\mathbf{X}_s, \mathbf{X}_r, \omega) = (e_-^s)_\alpha \frac{i\omega(\mathbf{f} \cdot \mathbf{e}^o)\exp[i(K_S|\mathbf{X}_r| + K_P|\mathbf{X}_s|)]}{8\pi\rho_0(C_PC_S)^{3/2}(|\mathbf{X}_s||\mathbf{X}_r|)^{1/2}} F^{PS}(\mathbf{K}_{PS}, \theta) \tag{10}$$

where $\mathbf{K}_{PS} = -K_P\mathbf{e}^o + K_S\mathbf{e}^s$ and $F^{PS}(\mathbf{K}_{PS}, \theta)$ are defined as

$$F^{PS}(\mathbf{K}_{PS}, \theta) = [\rho_1(\mathbf{K}_{PS})/\rho_0]\sin\theta - [\mu_1(\mathbf{K}_{PS})/\mu_0]C_SC_P^{-1}\sin 2\theta; \tag{11}$$

and e_-^s is the unit vector in the transverse direction of e^s
   (3) SV-to-P

$$(U_\alpha^s)^{SV-P}(\mathbf{X}_s, \mathbf{X}_r, \omega) = e_\alpha^s \frac{i\omega[f^2 - (\mathbf{f} \cdot \mathbf{e}^o)^2]^{1/2}\exp[i(K_S|\mathbf{X}_s| + K_P|\mathbf{X}_r|)]}{8\pi\rho_0(C_PC_S)^{3/2}(|\mathbf{X}_s||\mathbf{X}_r|)^{1/2}} F^{SP}(\mathbf{K}_{SP}, \theta) \tag{12}$$

where $\mathbf{K}_{SP} = -K_S\mathbf{e}^o + K_P\mathbf{e}^s$ and $F^{SP}(\mathbf{K}_{SP}, \theta)$ are defined as

$$F^{SP}(\mathbf{K}_{SP}, \theta) = [\rho_1(\mathbf{K}_{SP})/\rho_0]\sin\theta - [\mu_1(\mathbf{K}_{SP})/\mu_0]C_SC_P^{-1}\sin 2\theta; \tag{13}$$

   (4) SV-to-SV

$$(U_\alpha^s)^{SV-SV}(\mathbf{X}_s, \mathbf{X}_r) = (e_-^s)_\alpha \frac{i\omega[f^2 - (\mathbf{f} \cdot \mathbf{e}^o)^2]^{1/2}\exp[iK_S(|\mathbf{X}_s| + |\mathbf{X}_r|)]}{8\pi\rho_0 C_S^3(|\mathbf{X}_s||\mathbf{X}_r|)^{1/2}} F^{SS}(\mathbf{K}_S, \theta) \tag{14}$$

where $\mathbf{K}_S = K_S(\mathbf{e}^s - \mathbf{e}^o)$ and $F^{SS}(\mathbf{K}_S, \theta)$ are defined as

$$F^{SS}(\mathbf{K}_s, \theta) = [\rho_1(\mathbf{K}_p) / \rho_0] \cos\theta - [\mu_1(\mathbf{K}_s) / \mu_0] \cos 2\theta; \qquad (15)$$

It is apparent that the *P-to-SV*, *SV-to-P* and *SV-to-SV* scatterings are independent of the variation of λ. Therefore, in order to inverse three parameters ρ_1, λ_1 and μ_1, we only consider the *P-to-P* scattering. In the process of *P-to-P* scattering, the detected signal of the longitudinal wave can be represented by

$$S(\mathbf{X}_s, \mathbf{X}_r, \omega) = \sum_{\alpha=1}^{2} (U_\alpha^s)^{P-P} e_\alpha^s \qquad (16)$$

We define a "signal" $Q(\mathbf{X}_s, \mathbf{X}_r, t)$ in the time domain,

$$Q(\mathbf{X}_s, \mathbf{X}_r, t) \equiv F^{-1} \left\langle \frac{F\{H[S(\mathbf{X}_s, \mathbf{X}_r, t)]\}}{\omega} \right\rangle \qquad (17)$$

where F, F^1 and H represent the Fourier, inverse Fourier and Hilbert transforms, respectively, and the Hilbert transform is introduced for causality. From the Eq.(17), we can obtain the relation between $Q(\mathbf{X}_s, \mathbf{X}_r, t)$ and the elastic parameters with a generalized Radon integral,

$$Q(\mathbf{X}_s, \mathbf{X}_r, t) = \frac{(\mathbf{f} \cdot \mathbf{e}^o)}{4\rho_0 C_P^3 (|\mathbf{X}_s||\mathbf{X}_r|)^{1/2}} \int_\sigma F^{PP}(\mathbf{X}', \theta) \delta\{[t - \frac{1}{C_P}[|\mathbf{X}_s| + |\mathbf{X}_r| + (\mathbf{e}^o - \mathbf{e}') \cdot \mathbf{X}']\} d\mathbf{X}' \qquad (18)$$

In previous researches[1,2,3], because of using the constant θ gathers, one can directly inverse the function $F^{PP}(\mathbf{X}, \theta)$ by the inverse Radon transform. In such procedure, the source array and receiver array must be employed in order to get enough information to inverse the $F^{PP}(\mathbf{X}, \theta)$. Here, we present an algorithm based on the theory of generalized linear inversion for over-determined system to inverse directly the elastic parameters ρ_1, λ_1 and μ_1. In this algorithm, we need only single impulse source rather than a source array. The Eq.(18) represents a linear integral along straight line L, which is decided by the following equation

$$C_P t = |\mathbf{X}_s| + |\mathbf{X}_r| + (\mathbf{e}^o - \mathbf{e}') \cdot \mathbf{X}' \qquad (19)$$

The straight L depends on the angle θ and time t, and for a series of time $\{t_1, t_2, \ldots\}$ of the detected signal, we can obtain the integrals along the parallel straight lines $\{L_{t1}, L_{t2}, \ldots\}$, and for different receivers, the integrals along the different angle θ are given.

COMPUTER SIMULATIONS

As shown in Fig. 1, the scatterer locates at the rectangular region [2B, 2B] in depth D, and the impulse source locates at $\mathbf{X}_s = (0, D)$, the receiver is at $\mathbf{X}_r = (-D \tan(\theta), D)$. The time t and angle θ are sampled in the interval $2D/C_P < t < 2(D+B)/C_P$ and $135° < \theta < 225°$ with the sampling number N respectively. The inhomogeneous region is divided into $M \times M$ pixels with pixel width $\Delta = 2B/M$. Then the line integral [Eq.(18)] can be discretized into,

$$Q(i, j) = \frac{(\mathbf{f} \cdot \mathbf{e}^o)}{4\rho_0 C_P^3 D} |\cos\theta_i|^{1/2} \sum_{k,l=1}^{M \times M} [\gamma_\rho(k, l) \cos\theta_i - \gamma_\lambda(k, l) - 2\gamma_\mu(k, l) \cos^2\theta_i] G(i, j, k, l)$$

$$i, j = 1, 2, \ldots, N \qquad (20)$$

where $G(i,j,k,l)$ is the cutting length at the pixel $[k,l]$ by the straight line L_{ij}, and

$$\gamma_\rho(k,l) = \rho_1(k,l) / \rho_0, \quad \gamma_\mu(k,l) = \mu_1(k,l) / (2\mu_0 + \lambda_0), \quad \gamma_\lambda(k,l) = \lambda_1(k,l) / (2\mu_0 + \lambda_0) \qquad (21)$$

We have employed the theory of generalized linear inversion to solve Eq.(20) to obtain the elastic parameters. Because there exist data errors in both the measured signals and the discretized process in Eq.(20), we seek a pseudo inversion of Eq.(20) by the Total Least Squares (TLS) method [4]. The Eq.(20) can be expressed by a matrix formula,

$$AX = Q \qquad (22)$$

where A is an $N^2 \times M^2 (N^2 > M^2)$ matrix, X is the $M^2 \times 1$ pixel vector, and Q is the $N^2 \times 1$ "observation" vector. The TLS method is to find the generalized solution to make

$$\text{Minimize} \left\| V[E|e]T \right\|_F$$
$$\text{Subject to } Q + e \in \text{Range}(A + E) \qquad (23)$$

where $\|.\|_F$ denotes the Frobenius norm [4], V and T are the nonsingular weighting matrices, E and e are the error matrices corresponding to A and Q, respectively. In practical problems, the noise is unavoidable. We add the "noise" $N(t)$ to the signal $S(t)$, as follow

$$S'(t) = S(t) + N(t) \qquad (24)$$

where $S(t)$ represents the detected signal of the longitudinal wave, $S(\mathbf{X}_s, \mathbf{X}_r, t)$. The $N(t)$ is the random number series in the range: $-8 \times 10^{-4} \sim 8 \times 10^{-4}$.

The simulation has been implemented at Computer CD4680. Fig.2(a) shows the model used to simulate. The density variation is 0 in the white area, 0.1 in the shaded area, and 0.15 in the black area. Fig.2(b) is the result of numerical simulation. Fig.3 shows the cross-sections of the model and the simulation along line x=0 and y=0 respectively. In the simulating procedure, the parameters are $B=100$m and $D=2000$m, and the longitudinal velocity $C_p=5000$m/s, $N=64$ and $M=32$.

(a)

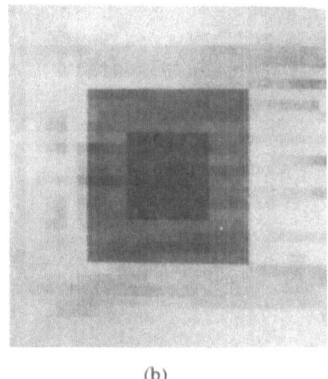
(b)

Figure 2(a) The model: three regions of the density variation, 0 in the white area, 0.1 in the shaded area and 0.15 in the black area; **(b)** The numerical simulation.

413

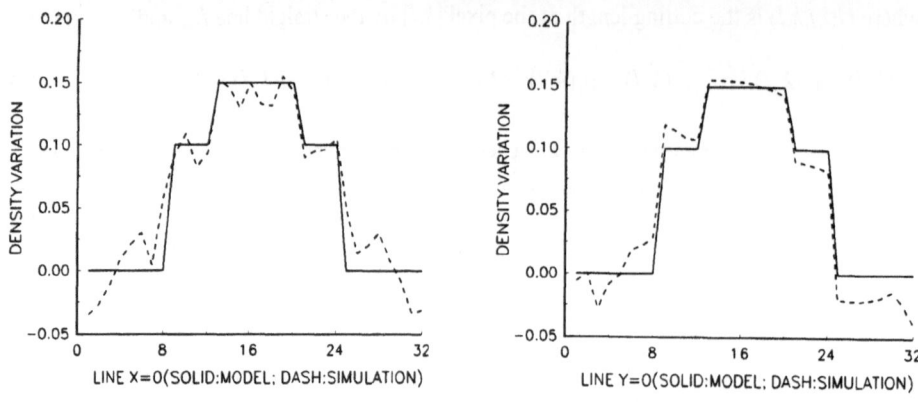

Figure 3. The cross-sections of the model and the numerical simulation.

CONCLUSION

We have developed an algorithm to inverse the two dimensional elastic parameters for the seismic exploration under the Born approximation and the far field condition. The algorithm is different from previous works, and only single impulse source is employed. The numerical simulations show that this algorithm is effective and available. Although the Born approximation is limited in practical application, it has been shown to be an appropriate means of characterizing the property of elastic medium [1,2,3] On the other hand, it is worth to point out that the algorithm can be easily extend to the case with spatially varying background parameters.

ACKNOWLEDGMENT

This work is supported by the Natural Science Foundation of China.

REFERENCES

1. G.Beylkin and R.Burridge. Linearized inverse scattering problems in acoustics and elasticity, *Wave Motion.* 12:15 (1990).
2. J.M.Blackledge, R.E.Burge. K.I.Hopcraft and R.J.Wombell, Quantitative diffraction tomography: II. Pulsed elastic waves *J.Phys. D.* 20:11(1987).
3. F.Ying and Y.Wei, A generalized radon transform method for reconstructing density and lame parameters of elastic medium using field data. in: *Acoustical Imaging,* Y.Wei and B.Gu. ed.. Plenum Press, New York (1993). Vol.20. P.611.
4. G.H.Golub and C.F.Van Loan. An analysis of the total least squares problem. *SIAM J.Numer.Anal.* 17:883 (1980).

MODULAR SYSTEM FOR DETECTION OF FLAWS IN METAL BLOCKS

Pedro Acevedo-Contla, Martín Fuentes-Cruz, and Felipe Rauda-García

Departamento de Electrónica y Automatización (DEA).
Instituto de Investigaciones en Matemáticas Aplicadas y en Sistemas
(IIMAS).
Universidad Nacional Autónoma de México (UNAM).
Apdo. Postal 20-726, Admon. No. 20, Del. Alvaro Obregón
C.P. 01000 México D, F., México.

INTRODUCTION

Ultrasound is widely used in Nondestructive Testing (NDT), it can be used to detect flaws in materials under inspection. NDT is a technique in which tests are made to materials and structures with the aim of evaluating their quality and consistency without destroying them. NDT began at the Second World War mainly used in the quality control field. Since then, NDT has been developed as a technique oriented to industry to locate defects in already manufacturated parts. Also, many advances have been made in materials researching, this simultaneously has provided new tools to develop new material technologies. NDT is very useful, it is extensively and systematically applied not only to examine finished products, but also to periodically check parts that are in continuous use.

This paper describes the design and construction of an ultrasonic imaging modular system, based on an IBM compatible Personal Computer (PC) for the nondestructive evaluation of internal defects in metal blocks. The surfaces of the block are roughly scanned by a dual probe ultrasonic transducer to record data. The system mainly consists of an ultrasonic transducer (Tx/Rx), an x-y positioning arrangement, an Excitation and Reception module, an Industrial Process Controller (IPC) and the necessary software.

GENERAL DESCRIPTION

The system uses a Transmitter/Receiver Transducer (Tx/Rx) and three main Modules. These are described as follows:

Transducer (Transmitter/Receiver).

The transducer is a central component in transmission and reception of ultrasound. By enabling us to produce and "hear" ultrasound, the transducer acts as an interface between the material under test and the electronics that provide an image [2]. On the material side are all the events connected with ultrasound as a mechanical wave: propagation, refraction, reflection, scattering, and attenuation. On the electronics side are regulated the events needed to convert the ultrasound information into an image. The aim in designing a system for ultrasonic image acquisition is to make the display represent the external and internal features of the material under test with a maximum resolution. The Transducer is a Krautkramer-Branson dual element probe. This is connected to the receiving and excitation module by means of a coaxial cable.

Excitation and Reception Module.

This module interfaces the transducer with the other two modules. In this stage the transducer is excited with a narrow pulse to produce a signal that bounces back wherever an acoustic impedance occurs within the material under test, the bounced signal is amplified, demodulated and filtered, after all these steps the signal can be captured using the acquisition system.
This module is divided in four sections:

Clock section.- All the clock signals used by the system are generated in this section, using a 20 MHz master clock and a series of counters which generate the frequencies needed.

Excitation section.- This section generates the excitation pulses, a signal from the clock section is taken as basic signal and the pulse width is adjusted using a monostable, controlling in this way the duty-cycle. The circuit has a transformer which amplifies the excitation voltage from 5 V dc to 120V dc, this 120 V dc pulse is abruptly cut using a SCR, this is made with the aim of exciting the piezoelectric element and to absorb the necessary energy to reduce the ringing and in this way to be able to delimit simply the starting point of the pulse to be measured.

Reception section.- This section has four main stages:
a) The first stage enables and disables the receiver's Input using a high speed analogue switch. The receiver's input is disabled while the transmitter is excited. This stage allows the system to work with a single transducer as transmitter and receiver.
b) The second stage is a High gain RF amplifying circuit used to amplify the echo-pulse signal to a detectable level.
c) In the third stage the echo-pulse signal is demodulated and filtered.
d) In the fourth and last stage the AF signal is amplified within the 0 to 5 V range. Also, in this stage the signal is finally captured using the acquisition system to be then processed.

Digital Display section.- This section has the objective of generating a digital count that gives a lecture of the depth or distance at which the echo-pulse is produced. Both, the signal from the AF amplifier and the excitation pulse signal are input to a flip-flop to obtain a pulse which width is proportional to the distance at which the echo-pulse is produced, this pulse is used as a window to perform a number of counts using the master clock, this produces a digital count which is measured by a counter/driver display device. Finally, the count is displayed using a crystal liquid display.

Industrial Process Control Module.

The heart of this module is a general purpose board (IPC), this board has the following features:

a) A 12 bit A/D converter with a maximum conversion time of 25 μseconds and four analogue inputs. This stage is used to digitise the signal from the Reception and Excitation module.

b) Digital outputs to activate the relays which give the sequence to the step motors to set the Tx/Rx Transducer in a bidimensional plane.

c) Digital inputs controlled by optic sensors to detect the position of the Rx/Tx Transducer in the bidimensional plane.

d) Control signals to control the Reception and Excitation module.

e) A D/A converter to reproduce the analogue signal, using this signal is possible to compare different sample data.

Also, this board has the necessary hardware to fit in one of the PC internal expansion slots using the reserved addresses for the prototype board (300H-31FH)[3].

Application Software. A programme written in C is used to control automatically the whole system: position, transmitter activation, digitising and storage, signal processing and display of results. The processing algorithm is shown in figure 1.

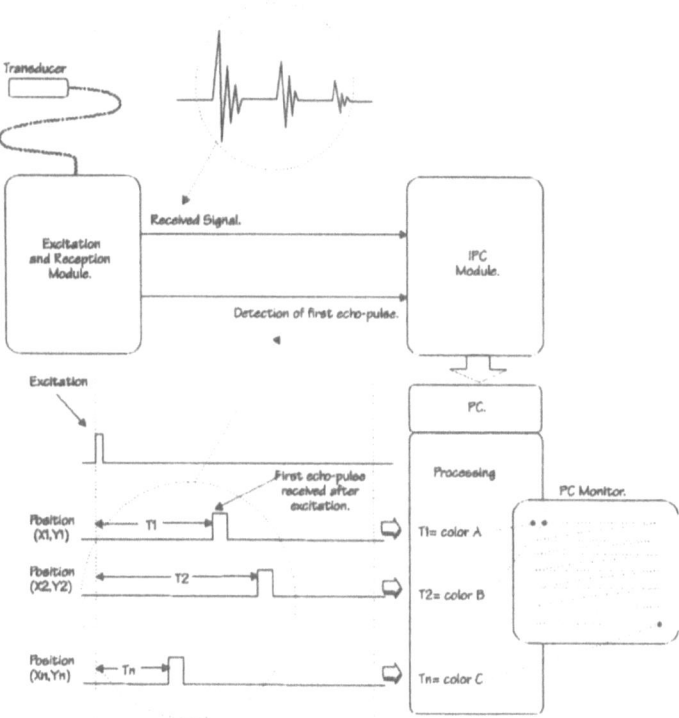

Figure 1. Processing algorithm.

Figure 2 shows a general diagram of the system. The diagram shows the IPC board which is interconnected to one of the PC expansion slots, the Reception and Excitation module board, the Rx/Tx ultrasonic transducer using an acoustic matching and the power stage to drive the step motors.

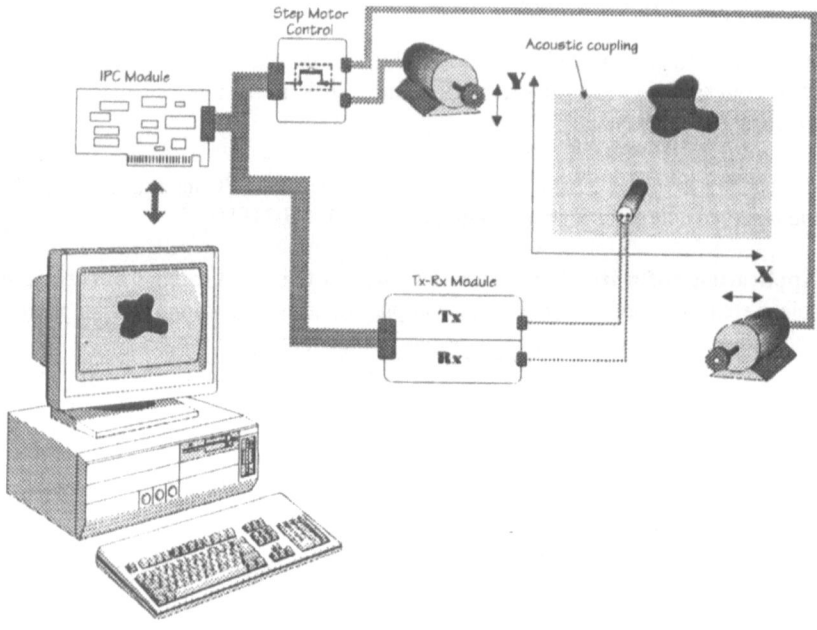

Figure 2. General diagram of the system.

SYSTEM OPERATION

The system's Tx/Rx Transducer moves in a bidimensional plane scanning the material under test with the aid of the step motors which are controlled from the PC by the IPC board. The Tx/Rx Transducer is placed on a fixed coordinate (x,y) and it is excited by means of a short pulse, then using an instrumentation amplifier and an analogue filter the echo-pulses generated by the transmitter are captured. These echo-pulses are digitised by the 12 bit A/D converter and then stored in memory. This operation is repeated until the material under test is fully scanned. Stored data is processed and then displayed on a super VGA monitor.

RESULTS

To examine the system's performance steel test blocks, some with different shapes and some with drilled holes at different locations were used. Figure 3 shows the image of a curved steel test block and figure 4 shows the image of steel block with two drilled holes. In both figures, shape and flaws are displayed with acceptable clarity.

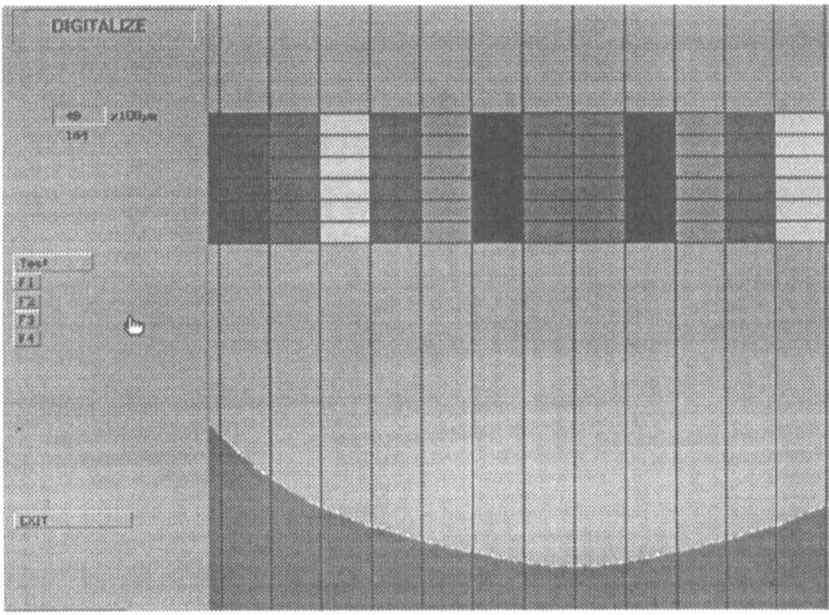

Figure 3. A Steel block showing a curved shape.

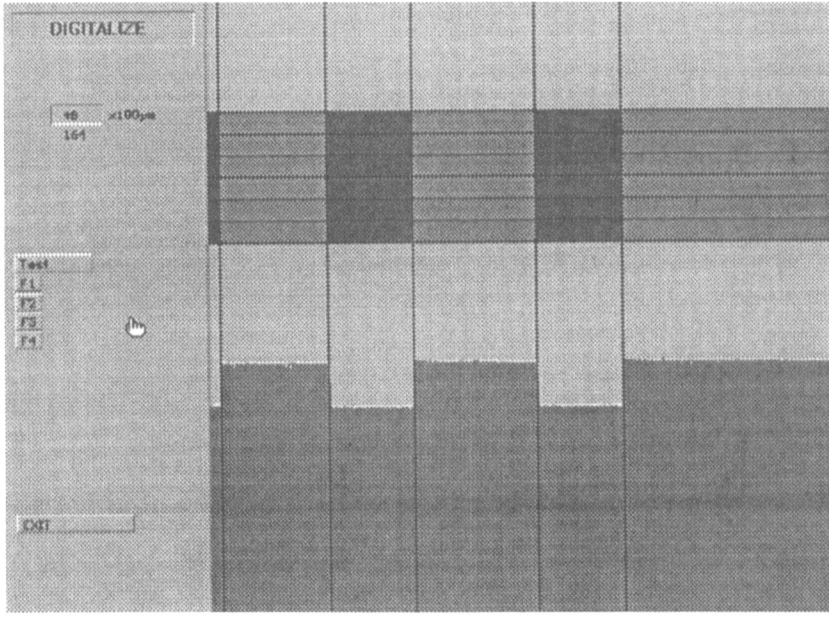

Figure 4. A Steel block showing two drilled holes.

CONCLUSIONS

The designed and constructed modular system for detection of flaws in metal blocks, it is an efficient alternative to inspect metal blocks to confirm their quality. Flaws are displayed in an easy way using a PC. Although the system does not work on real-time, it is intended in a near future to increase the capacity of the system incorporating an ultrasonic transducer array together with the use of parallel processing techniques to display images on real-time. It is also intended to use this system as an auxiliary tool in the development of larger ultrasonic flaw detection systems.

REFERENCES

1. M.G.Silk, "Ultrasonic Transducers for Nondestructive Testing", Adam Hilger Ltd, Bristol, pp.63-72, 1984.
2. R. L. Powis and W. J. Powis, "A Thinker's Guide to Ultrasonic Imaging", Urban & Schwarzenberg, Baltimore-Munich, pp. 51-85, 1984.
3. N. Zuech, "Handbook of Intelligent Sensors for Industrial Automation", Addison-Wesley, 1988.

ADVANCED IMAGE PROCESSING TECHNIQUES FOR AUTOMATIC INTERPRETATION OF TIME-OF-FLIGHT DIFFRACTION IMAGES

L. Capineri[1], P. Grande[1], L. Masotti[1], C.G. Windsor[2] and J.A.G. Temple[3]

[1] Dipartimento Ingegneria Elettronica, Università di Firenze,
   50139 Firenze, Italy
[2] UKAEA Fusion, D3, Culham Laboratory, Culham, OX14 3DB, UK
[3] AEA Technology, Harwell Laboratory, Didcot, OX11 0RA, UK

INTRODUCTION

Time-of-flight diffraction techniques are widely used in NDT and commercial equipment which can produce good quality B-scan images is widely available. Their present application is concentrated on the inspection of metals using frequencies from 1 to 5 MHz and a pulse length of a few cycles. In this range the ultrasonic wavelengths are comparable to the characteristic size of the defects and the geometrical theory of diffraction accurately describes the propagation phenomena involved. In many instances a defect can be considered as a point-like reflector.

This means that there is a hyperbolic relationship between the time-of-flight of the back-reflected signal and the position of the surface probe. Thus the B-scan images are often characterized by hyperbolic arcs, each associated with a point like-defect. Unfortunately the interpretation of these hyperbolic patterns is not straightforward and requires expertise and time.

The work presented here concerns the development of a new technique for the automatic interpretation of time-of-flight images aimed to reduce the operator intervention and so increase the reliability and productivity of the method.

AUTOMATIC INTERPRETATION BASED ON THE HOUGH TRANSFORM OF FEATURES EXTRACTED FROM B-SCAN IMAGES.

Basic theory of Randomized Hough Transform (RHT) for hyperbolae.

The technique is carried out by using a modified version of the Hough Transform[1] for hyperbolic arcs. The Hough Transform works on a pre-filtered set of pixels selected at random among the bright pixels of the image. Each selected triplet of pixel is Hough transformed and it is accumulated in the space of defect positions (depth and horizontal coordinates). In this way the highest peaks in the accumulator are considered as possible positions for a defects. Finally for each defect the most likely position is determined by correlating the hyperbolic curves corresponding to the highest peaks in the accumulator to the image. This method is efficient because the coefficients of the hyperbolic curves are calculated only on a limited number of triplets. We found the accuracy of the method better than 5% on synthetic images and good results were obtained also with experimental images.

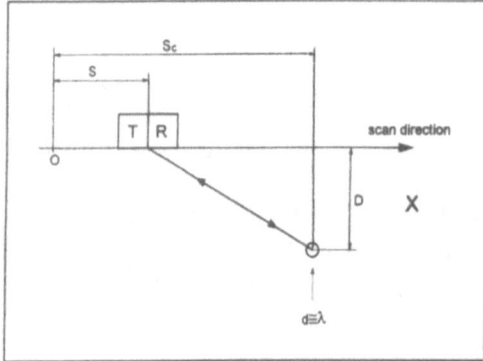 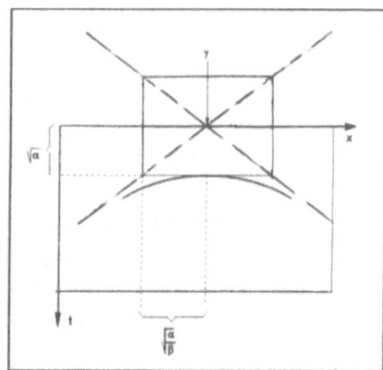

Figure 1A Scanning geometry with a monostatic probe   Figure 1B Geometrical meaning of α, β, γ.

The hypothesis for the application of these methods to a gray scale image produced with the scanning geometry shown in Figure 1A are: object size d comparable to wavelength λ; homogeneous medium and constant velocity V. The object position is in depth D and horizontal S_c along direction X while S denotes the position for the monostatic probe. Then we obtain the following relationship for the time of flight:

(1)

$$t_{TOF} = 2 \left(D^2 + (S - S_c)^2 \right)^{1/2} / V$$

By discretizing the time and the horizontal probe positions with sampling steps D_T and D_x respectively, we can replace the continuous variables in (1) with discrete indexes according to the relationships $t_{TOF} = j\, D_T$, $S = i\, D_x$, $S_c = i_c\, D_x$, providing a parametric equation of an hyperbola:

(2)

$$j^2 = \alpha + \beta\, (i\text{-}\gamma)^2$$

The geometrical meaning of parameters α, β and γ are shown in Fig.1B, and they are used in the Hough Transform to identify an hyperbola that is related to the unknown object position. The calculation of the Hough Transform for an hyperbola is a time consuming process if all the pixels of the image are involved. A dramatic improvement can be achieved by using a random selection of bright pixels (or dark depending on the image) which leads to the method call Randomized Hough Transform[2].

The Randomized Hough Transform algorithm. By picking three distinct bright points ($P_L=(i_L, j_L)$, $P_M=(i_M, j_M)$ and $P_N=(i_N, j_N)$) on a hyperbola, we can solve the following three equations to obtain the values of α, β and γ, given by:

$$(3)$$

$$\gamma=[(i_N^2-i_L^2)(j_M^2-j_N^2)-(i_M^2-i_N^2)(j_N^2-j_L^2)]/[2(i_N-i_L)(j_M^2-j_N^2)-(i_M-i_N)(j_N^2-j_L^2)]$$

$$\beta=(j_M^2-j_N^2)/[(i_M-i_N)(i_M+i_N-2\gamma)]$$

$$\alpha=j_L^2-\beta(i_L-\gamma)^2$$

The accumulation of votes of the triplets α, β and γ, corresponding to possible hyperbolic arcs, occurs in a three dimensional array by scanning over the bright pixels of the B-scan image. In this way we build up a distribution in this three dimensional parameters space which has peaks and ridges. These high spots provide the parameters that identify the most likely hyperbolic arcs appearing in the image.

The performances achievable with the HT are largely dependent from the sampling step of the coefficients α, β and γ. In fact with high resolution (large size of the accumulator) leads to the phenomena of spreading of votes that has been already pointed out in other works and reviewed by Davies[1]. The spreading of votes leads to several spurious peaks in the accumulator array around the true positions leading to a loss of detection capability. We can overcome this limit by using the process of "re-accumulation" of votes[3].

METHODS

The description of the developed program for automatic interpretation of B-scan images is reported below. The only input parameters required from the user are the maximum number of significant lines or hyperbolae sought in the defined region of interest of B-scan image and the threshold selection method. The program can also analyze straight line patterns and hyperbolae separately in the same image. The output of the identification program is a file, containing the coefficients of the identified curves in the form indicated above, which is used as input for a program developed in MATLAB which has the following tasks:
- Compute a correlation index between the curves found and the input image;
- Select the curve with the highest correlation index and display it, superposed to the original gray scale B-scan image, in color.

RESULTS AND FINAL REMARKS

The application of the Randomized Hough Transform method to ultrasonic images was carried out with two samples. The first one is an isolated defect in a test object and the results are comapared to those obatined with a standard SAFT method. The second one shows the possibility to analyse images containing multiple defects.

The first example considers a B-scan image obtained with a data acquisition from a test object consisting of a side drilled hole ($\varnothing$=2 mm) in a steel block (Vl_{ong}=5900 m/s). The synthetic aperture length was 70 mm, step 2.41 mm$\approx$2λ, realized by sweeping a KB

Aerotech probe at 5MHz, diameter 0.25 inch, bandwidth -3dB = 1.75 MHz. In Figure 2A is shown the SAFT reconstruction and in Figure 2B the original B-scan image

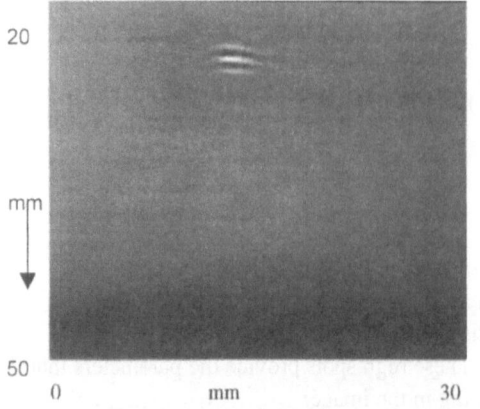 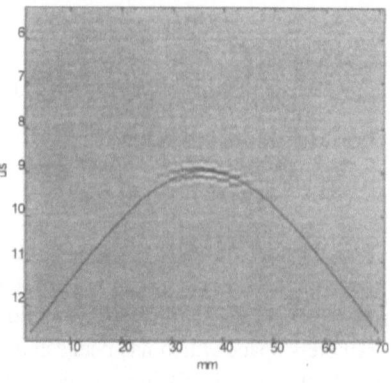

Figure 2A. SAFT image reconstructed with the original data of an isolated defect. Image 256x256 pixel, grey levels.

Figure 2B. Detected hyperbola superposed to the original B-scan image of an isolated defect.

In a second example we applied the method to an ultrasonic image obtained with a "pitch and catch" probe configuration, using two contact probes, 5 MHz, 45° angle , diameter 6.35 mm. The scan lenght was 196 mm, 71 steps on a test object consisting of four side drilled holes in a steel block (V_{long}=6240 m/s). The defect positions (horizontal and vertical) are listed in Table 2. The hyperbolae found with the automatic method are displayed in Figure 3. Besides the image is rather difficult because the presence of strong interferences between the arcs and the reduced number of pixels, the method can locate accurately three defects and the fourth is identified with correct depth but not with wrong lateral position. The estimated defect coordinates are also reported in Table 2 where TOF is converted in equivalent depth.

Table 1. Results for test image with a side drilled hole in steel with SAFT and automatic method based on Random Hough Transform (see Fig. 2A-B)

Defect position					
SAFT			RHT		
X	Depth	TOF	X	Depth	TOF
mm	mm	µs	mm	mm	µs
33	24.7	8.37	33.8	26.43	8.96

Table 2. Results for test image with 4 side drilled holes in steel (see Figure 3).

Defect positions					
Real			Estimated		
X	Depth	TOF	X	Depth	TOF
mm	mm	µs	mm	mm	µs
68	48.5	15.54	68.7	48.47	15.53
88	48.0	15.38	87.5	48.04	15.4
108	47.5	15.22	101.6	47.6	15.25
128	47.5	15.22	129.7	47.6	15.25

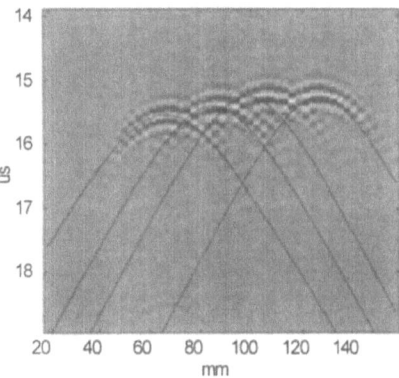

Figure 3. Automatic defect identification by hyperbolae detection.
Test object is a steel block containing four side drilled holes.

SAFT images can determine the object position with the resolution of the transmitted pulse but the reconstruction process is slow (about 300s on a 100 MHz) compared to the Random Hough Transform method and they need expertise for interpretation or further processing. In order to improve the robustness of the automatic interpreation method we developed another method which extracts a list of points following the hyperbolic arc based on the time domain cross-correlation between adjacent ultrasonic traces of the image and followed by the RHT of this list of points. This method is slightly faster than RHT but has the advantage that are not required any input parameters. However at present this method works well only with isolated defects. The future research work will try the analysis of more difficult images with multiple arcs and will consider the extension of the methods with synthetic and experimental three-dimensional data acquisitions.

ACKNOWLEDGMENTS

The authors wish to acknowledge the support to this work from the Cooperation Initiative between The British Council and the Italian Ministry for Technological Research and the NATO Collaborative Research Grant , project CRG 940333.

REFERENCES

1. Davies *Machine Vision:Theory, Algorithm, Practicalities,* Academic Press Ltd, London, (1990)
2. Xu L., Oja E., Kultanen P., A new curve detection method: randomized Hough transform (RHT), Journal of Pattern Recognition Letters, (1989)
3. Gerig G, F. Klein, Fast contour identification through Hough transform and simplified interpretation strategy, Proc. 8th Int. Conf. on Pattern Recognition, vol.1,Paris, France October 27-31, pp. 498-500, (1986)

APPLICATION OF ULTRASONIC FORCE MICROSCOPE TO FRACTURE

Woon.S.Gan

Acoustical Services Pte Ltd
29 Telok Ayer Street, Singapore 048429
Republic of Singapore

INTRODUCTION

Ultrasonic force microscope (UFM) has been applied to the detection of subsurface defects in HOPG samples[1]. The advantages of UFM over the conventional scanning acoustical microscope (SAM) is the capability of nanoresolution for subsurface images, without the wavelength limitation of the SAM. It can visualize the quantum mechanical microscopic processes. In this paper we propose the application of the ultrasonic force microscope (UFM) to study fracture in material. Our paper will concentrate on the interpretation of images, leading to the understanding of the physical process in fracture. This will have application in non-destructive evaluation such as to know how long it takes the turbines of a jet to deteriorate to such an extent that they are dangerous and to design materials that resist mechanical ageing.

ENERGY AT GRAIN BOUNDARY SLIDING

Grain boundary sliding (ie., the sliding of one grain over another parallel to the boundary interface that occurs in response to applied forces) plays an important role in the deformation and fracture processes of polycrystalline materials, particularly at high temperature[2] G.Molteni et al[3] uses quantum mechanics to investigate how quickly does a given material distort under stress, how is this rate affected by temperature and what processes occur as the material goes under repeated distortions that eventually lead to its fracture and failure. Although almost every material can be studied using this approach, G.Molteni et al[3] considered germanium because it is one of easiest materials to simulate. The primary phenomena they studied is the sliding of two tiny crystals or grains, past each other in the material. The sliding is essential in considering the strength and reliability of a material. Their study presents a framework for extracting important mechanical properties from quantum mechanical calculations.

They took two critical steps. First, they used the fact that materials science theory shows how the mechanical behaviour of a material is controlled by the very thin but critical regions that depart from perfect crystalline order. These regions are known as extended defects, and they can be studied in terms of the collective behaviour of as few as 100 atoms. Second, they used density functional theory to calculate the average behaviour of about 10^{25} electrons. Such works foster great insight into the atomic phenomena that underly mechanical behaviour.

The computational experiments performed by G.Molteni et al[3] involved forming two grains of germanium, rotating the grains with respect to one another at a predetermined angle and then bringing them together. They studied the boundary region between the grains as they dragged the grains past each other. The results reveal that the system goes through "stick-slip" cycles as the grains move. As the grains are forced past each other, energy is build up and then released in a series of sudden, discontinuous events. The evolution of the grain boundary energy during the sliding is shown in Fig. 1.

The calculations of G.Molteni[3] et al provide new insights into the mechanical ageing process and show the way to quantitative predictions of grain boundary slip rates. Interestingly, the events in the study involved only a few atoms in each slip, which afterwards were able to form new bonds and thus maintain the mechanical strength of the grain boundary. With continued sliding, they observed that imperfections in rebonding spread away from the interface, penetrating into the bulk of the material, where they cannot be corrected by rebonding during further slips. Concomitant with the growth of this flaw, the grains begin to slide more easily past one another, eventually leading to mechanical failure of the interface. This gave us our first insight into the mechanical ageing process of grain boundaries.

Their result relates to the question of how quickly an object, such as a support beam in a skyscraper, bends when subjected to a force, such as the weight of the building. One of the factors that discourages the beam from bending is that the grains in the material slide past one another extremely slowly. If the beam were heated to a sufficiently high temperature, the grains would slide past one another much more quickly and sagging of the beam would become noticeable.

MICROELASTICITY IMAGING

The purpose of this paper is to use UFM to study the mechanism of the fracture process. On very small length scales ($\leq 10^{-6} m$) fracture is a topic of materials science. From the electronic level to the level of dislocations or grain boundaries the mechanism of fracture are highly material dependent. In order to describe fracture, one needs to know first how an unfractured solid responds to an applied force. For this purpose, let us consider Young's experiment, ie. a homogeneous block of size $l \times w^2$ subjected to a uniaxal force F. For small forces one expects a linear response of the type

$$\sigma = \frac{F}{w^2} = E \frac{\Delta l}{l} = \frac{E}{v} \frac{\Delta w}{w} \tag{1}$$

where E=Young's modulus, Δ =Poisson's ratio, σ =the stress, l=length, w=width

Usually very small forces are not enough to fracture the solid. The behaviour for an arbitrary force is given by the so-called constitutive law. The linear regime, which is called elastic, is valid up to a force F_n beyond which one will find nonlinear deviations.. As long as the force is less than the yield-point F_y the behaviour is reversible, ie. the original shape is recovered if the force is reset to zero. Beyond F_y the behaviour is plastic, ie. the system deforms irreversibly so that when the force is reset to zero a finite elongation d called "dilatancy" remains. If, in the plastic regime, the material can flow without increasing the force, the behaviour is called ideal plasticity, but most metals behave rather that after an elongation S, the material again has a finite toughness called strain hardening. Plasticity, which in metals is due to dislocations and in polymers to chain reorientation, is described by a rather involved, nonlinear formalism.

In our work, the purpose is to use the UFM's nanoresolution capability to see the grains of the material at the boundary region of material slipped past each other with continued sliding. As the crack grows, the grains begin to slid more easily past each other and eventually leads to fracture. This is imaging beyond the elastic limit and investigating the atomic nature of the physical mechanism of fracture.

This is an extension of the elasticity imaging of J.Ophir et al[4] to the nonlinear region with the capability of imaging the microelastic properties of a fracture process. We call this "microelasticity imaging". Microelasticity imaging is performed by obtaining a set of vibration amplitudes from the UFM on a target subjecting to a large deformation causing fracture. Time shift estimation along the direction of the applied loads are computed by performing piecewise cross-correlation on congruent pairs of vibration amplitude segments. The time-shift estimations are then converted to grain boundary energy information that is displayed in the form of a 2-D longitudinal energy imaging.

DERIVATION OF THE ENERGY EQUATION

One of the powerful conclusions from the theory of UFM response is that it can be described simply, by the introduction of a new force-versus-separation dependence $F_m(z)$, derived from the original dependence F(z) by its averaging over a vibration period. It reduces the problem of UFM response to the solving of the standard for Atomic Force Microscope (AFM) force balance equation, $F_m(z) = k(z - z_s)$, when z_s is the displacement of the object surface out of the area of contact, K-static rigidity of cantilever and $z_c = z - z_s$, is the cantilever displacement, the value measured in the experiment. The $F_m(z)$, therefore, is defined by the original F(z) as well as by vibration amplitude a,

$$F_m(z, a) = \frac{1}{2\pi} \int_0^{2\pi} F(z - a \cos \chi) d\chi \qquad (2)$$

The increase of vibration amplitude a modifies the $F_m(z)$, and hence, changes the cantilever deflection z_c, thus plotting the deflection-versus-ultrasonic-amplitude dependence-z(a) curve.

In the UFM, the sample is vibrated at ultrasonic frequencies and the tip is elastically indented into the sample. Subsurface features of rigid objects are observed using a soft cantilever. When the direction of the tip sample vibration force is controlled and the cantilever torsion as well as the deflection are monitored, subsurface features sensitive to compression and shear are selectively imaged.

In this paper, we are interested in the fracture process. So the tip has to be indented into the sample and high excitation voltages are required. In this case, the cantilever vibration spectrum will contain many signals and it is nonlinear in nature.

We first look at the cantilever deflection as well as the tip sample force and distance during the UFM operation. The cantilever is deflected by z_c from its free position by applying a tip sample static force F_c. Due to this force, the tip is indented in the sample, and the tip sample distance is determined from the force curves $d_c = F^{-1}(F_c)$.

When the ultrasonic frequency vibration (UFV) is applied, the cantilever cannot follow the sample vibration due to inertia, since the frequency of a UFV above 1 MHz is much higher than the cantilever resonant frequency. However due to the additional repulsive force caused by contact with the vibrated sample, cantilever deflection is increased from z_c to z_0. This increase is bought about by the nonlinear rectifying effect[5], which is determined by solving an integral equation[6]

$$kz_0 = \frac{1}{T}\int_0^T F(z_s + a\cos\omega t - z_0)dt \qquad (3)$$

where z_s is the sample stage displacement required to lift the cantilever by z_c, determined by

$$z_s = z_c + F^{-1}(kz_c) \qquad (4)$$

Once z_c is assumed, z_0 is obtained by solving eqs.(3) and (4), and the additional cantilever deflection $z_a = z_0 - z_c$ due to the UFV is calculated as a function of a.

EXTRACTION OF INFORMATION USING HIGHER ORDER STATISTICS

It was shown by O.Kolosov et al[7] that measurement of z(a), or Ultrasonic Force Spectroscopy (UFS), allows to determine the local elastic properties of objects, namely the effective elastic stiffness $K_{eff} = dF / dz$ of the contact. Now we need to extract information on the fracture process from the reflected ultrasonic vibration signal by using higher order statistics (HOS) since the vibration spectrum is nonlinear in nature. The experimental setup is the same as that of O.Kolosov et al[7] only that our process of image analysis is different.

The slip energy is given by the intensity of the reflected ultrasonic vibration signal as

$$intensity = \left|vibration\ amplitude\right|^2 \qquad (5)$$

During the experimental process, we would expect the reflected ultrasonic vibration signal from the slip boundary to be of higher peaks than other echoes. In the linear vibration process, the intensity is given by the power spectral density which is the Fourier transform of the autocorrelation function. For the nonlinear vibration process, we will limit to the bispectrum case and the trispectrum case.

If $\{X(k)\}, k = 0,\pm1,\pm2,\pm3,\cdots$ is a real stationary random process and its moments up to order n exist, then

$$Mom = [X(k), X(k+\tau_1),\cdots X(k+\tau_{n-1})]$$
$$= E\{X(k)\bullet X(k+\tau_1)\cdots X(k+\tau_{n-1})]\}$$

will depend only on the time differences $\tau_1,\tau_2,\cdots,\tau_{n-1},\tau_i = 0,\pm1,\pm2,\cdots$ for all i. We now write the moments of a stationary random process as:

$$m_n^x(\tau_1, \tau_2, \cdots, \tau_{n-1}) \stackrel{\Delta}{=} E\{X(k), X(k+\tau_1) \cdots X(k+\tau_{n-1})\} \tag{6}$$

For the bispectrum case, the autocorrelation function becomes the third-order moment sequence.

Consider the narrow-band process

$$Z(k) = X(k)\cos(\omega_0 k) + Y(k)\sin(\omega_0 k) \tag{7}$$

where $X(k)$, $Y(k)$ are independent stationary random processes with $E\{X(k)\} = E\{Y(k)\} = 0$, $m_2^x x(\tau) =$ autocorrelation sequence $= E\{X(k)X(k+\tau)\} = m_2^y(\tau)$ and $m_3^x(\tau_1, \tau_2) = E\{X(k)X(k+\tau_1)X(k+\tau_2)\} = m_3^y(\tau_1, \tau_2)$.

$\{Z(k)\}$ is a wide sense stationary random process and the third-order moments are

$$Mom[Z(k), Z(k+\tau_1), Z(k+\tau_2)]$$
$$= m_3^x(\tau_1, \tau_2)[\cos(\omega_0 k)\cos(\omega_0(k+\tau_1))\cos(\cos(k+\tau_2))$$
$$+ \sin(\omega_0 k)\sin(\omega_0(k+\tau_1))\sin(\omega_0(k+\tau_2))] \tag{8}$$

and the quantities in square brackets are independent on k for τ_1, τ_2. Hence $\{Z(k)\}$ is nonstationary in its third-order statistics.

For the trispectrum case, the 4th order moments are

$$Mom[Z(k), Z(k+\tau_1), Z(k+\tau_2), Z(k+\tau_3)]$$
$$= m_4^x(\tau_1, \tau_2, \tau_3)[\cos(\omega_0 k)\cos(\omega_0(k+\tau_1))\cos(\omega_0(k+\tau_2))\cos(\omega_0(k+\tau_3))$$
$$+ \sin(\omega_0 k)\sin(\omega_0(k+\tau_1))\sin(\omega_0(k+\tau_2))\sin(\omega_0(k+\tau_3))] \tag{9}$$

From the moments sequence, one obtains the cumulants sequences as :
3rd-order cumulants:

$$c_3^x(\tau_1, \tau_2) = m_3^x(\tau_1, \tau_2) - m_1^x[m_2^x(\tau_1) + m_2^x(\tau_2)$$
$$+ m_2^x(\tau_2 - \tau_1)] + 2(m_1^x)^3 \tag{10}$$

where $m_3^x(\tau_1, \tau_2)$ is the 3rd-order moment sequence.
4th-order cumulants:

$$c_4^x(\tau_1, \tau_2, \tau_3) = m_4^x(\tau_1, \tau_2, \tau_3) - m_2^x(\tau_1) \bullet m_2^x(\tau_3 - \tau_2)$$
$$- m_2^x(\tau_2) \bullet m_2^x(\tau_3 - \tau_1) - m_2^x(\tau_3) \bullet m_2^x(\tau_2 - \tau_1)$$
$$- m_1^x[m_3^x(\tau_2 - \tau_1, \tau_3 - \tau_1) + m_3^x(\tau_2, \tau_3) + m_3^x(\tau_2, \tau_4)$$
$$+ m_3^x(\tau_1, \tau_2)] + (m_1^x)^2[m_2^x(\tau_1) + m_2^x(\tau_2) + m_2^x(\tau_3)$$
$$+ m_2^x(\tau_3 - \tau_1) + m_2^x(\tau_3 - \tau_2) + m_2^x(\tau_2 - \tau_1)] - 6(m_1^x)^4 \tag{11}$$

The bispectrum is defined as the Fourier transform of the third-order cumulant sequence and the trispectrum is defined as the Fourier transform of the fourth-order cumulant sequence of a stationary random process.

It is also possible to estimate the grain displacement from the vibration amplitude. From the time shift between two echoes from the grain boundary and multiplied by the ultrasound wave velocity in the specimen, one could obtain the grain distance at the grain boundary. The data can then be applied to the grain boundary energy versus grain displacement plot given in Fig. 1.

431

CONCLUSIONS

We have given a framework for the image analysis of the application of the ultrasonic force microscope to study fracture. The fracture process is considered from a quantum mechanical point of view by the estimation of the grain boundary energy during the grain sliding process during fracture. This would increase our understanding of structural failure of aerospace materials and other composite materials and will be a valuable tool in nondestructive evaluations.

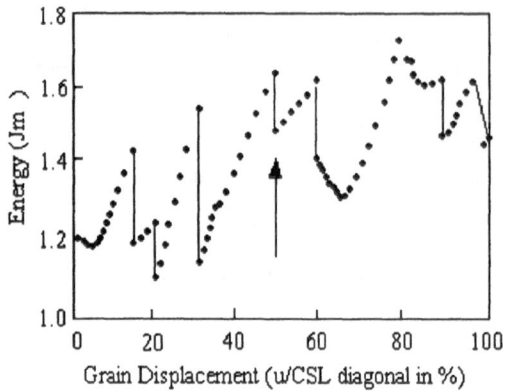

Fig.1. The grain boundary energy during the sliding process. The arrow shows where disorder starts to migrate away from the boundary surface. (after C.Molteni et al [3])

REFERENCES

1. K.Yamanaka, New approaches for noncontact measurement, in: Advances in Acoustic Microscopy, A.Briggs, ed., Academic Press, London (1995)
2. A.P.Sutton and R.W.Balluffi, Interfaces in crystalline materials, Oxford University Press, New York (1995)
3. C.Molteni, G.P.Francis, M.C.Payne and V.Heine, First principles simulation of grain boundary sliding, P.Rev Letter 76: 1284 (1996)
4. J.Ophir, I.Cespedes, H.Ponnekanti, Y.Yazdi and X.Li, Elastography: a method for imaging the elasticity in biological tissues, Ultrasonic Imaging, 13:111 (1991)
5. W.Rohrbeck and E.Chilla., Detection of surface acoustic waves by scanning force microscopy, Phys. Stat. Sol. A131: 69 (1992)
6. O.Kolosov and K.Yamanaka , Nonlinear detection of ultrasonic vibrations in atomic force microscope.Jpn.J.Appl.Phys.32:1095 (1993).
7. O.Kolosov , H.Ogiso and K.Yamanaka , Proceedings of 3rd Japan SAMPE Symposium , Dec 8-10 , 1993 , Tokyo , Japan , pp.2196.

Ultrasonic Inspection of Tendon Ducts in Concrete Slabs Using 3D-SAFT

XXIII Int. Symposium on Acoustical Imaging
Boston, April 1997

M. Krause, W. Müller[*], H. Wiggenhauser

Federal Institute of Materials Research and Testing (BAM)
12200 Berlin, Germany

[*] Fraunhofer Institute for Nondestructive Testing
University, Building 37, 66123 Saarbrücken, Germany

Introduction

The main purpose to use ultrasonic pulse echo techniques for concrete are the following testing problems:

- Injections faults in tendon ducts because they lead to a loss of the basic protection of the tendon steel and can result in corrosion damage
- Compaction faults or honeycombing because they reduce the concrete strength. They influence the static stability beginning from a diameter of about 50 mm.

Lately, there were many activities in developing non-destructive testing methods for these tasks, so that we can state an enormous progress[1,2,3,4,5].

In order to compare NDT methods applicable to the above mentioned tasks, a research project funded by BASt (Federal Highway Research Institute) was carried out by BAM and several other institutes[6]. For that purpose specimens made from concrete, which is typically used for prestressed concrete constructions were fabricated at BAST containing several intentionally positioned defects. Blind tests were performed by 9 working groups using the methods:

- Impulse Radar

- Ultrasonic Methods
  - Monostatic and bistatic measured A-Scans
  - Direct B-Scan
  - Array Technique
  - Reconstruction by means of L-SAFT
  - 2 D synthetic aperture measuring and 3D-SAFT-reconstruction
- Impact-Echo

Additionally simulation calculations for the propagation of acoustic and microwaves were carried out to point out possible improvements in the experiments.

This paper is focused on the results obtained by means of 2D synthetic aperture and 3D-SAFT reconstruction (SAFT: Synthetic Aperture Focusing Technique) [7,8].

The biggest obstacle in using ultrasonic pulse echo methods for concrete is the limitation of the usable frequency to the range of 100 kHz corresponding to a wavelength in concrete of about 40 mm. The reason is, that concrete is a compound material made from cement paste and aggregates with a typical maximum size of 16 or 32 mm, and the pores and entrained air attenuate enormously the acoustic waves.

Additionally there is the effect of frequency dependent attenuation[9]. That means e. g. that even a broadband transmitting pulse with a centre frequency of 200 kHz will result in a pulse having a centre frequency of 100 kHz after a through transmission of 0.6 m concrete. For that frequency range the diameter of the transducers and the wavelength are in the same order of magnitude so that longitudinal and shear waves are transmitted in all directions inside the half space.

In recent years many efforts were undertaken to overcome the difficulties resulting in applications of ultrasonic echo methods for concrete which were not possible before:
- Broad-band low frequency transducers are available now which allow direct A-scan or B-scan analysis, respectively[1,10,11]
- Transducer array combined with cross correlation analysis are used[12]
- The application of synthetic aperture techniques with different transmitting and receiving positions can be evaluated using 2D or 3D reconstruction methods [1]

The last mentioned technique has been developed in an joint project by Fraunhofer IZFP and BAM. It was evident, that in order to apply 2- or 3D-SAFT reconstruction, either the constant coupling of moving transducers on concrete surfaces must be ensured or other detection techniques must be applied. The use of a scanning doppler laservibrometer as ultrasonic detector solved the problem of constant coupling as well as the question how the large data sets can be collected by an automated inspection system.

Experiments

The experimental set up is shown in figure 1. The broad-band transducer is in a fixed position and the aperture scanned by the laser vibrometer.

The vibration of the surface in the direction of the laser beam is registered by the laser vibrometer which scannes the surface by means of a mirror scanner. Mechanical and ultrasonic vibrations are separated using a bandpass filter and the signal to noise ratio is improved by time averaging. The data are stored in a PC for further evaluation.

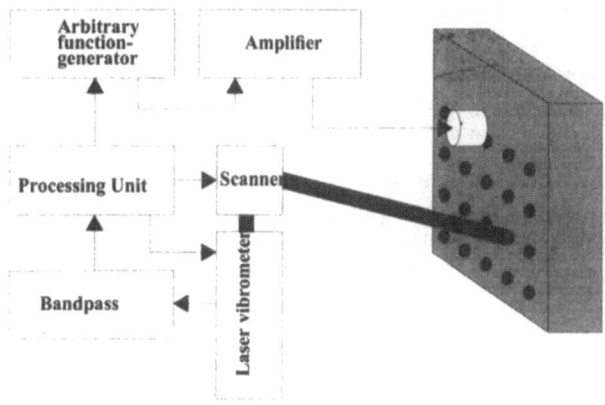

Figure 1 Outline of the experimental set up for scanning a synthetic aperture with a laser vibrometer as ultrasonic receiver.

A typical size of a scanned area is 0.6 x 0.6 m^2, with an increment width of 10 mm. A typical distance between the surface and the laser vibrometer is 2 m. Up to now retroreflecting colour must be applied to the surface to ensure that sufficient laserlight is reflected to be detectable.

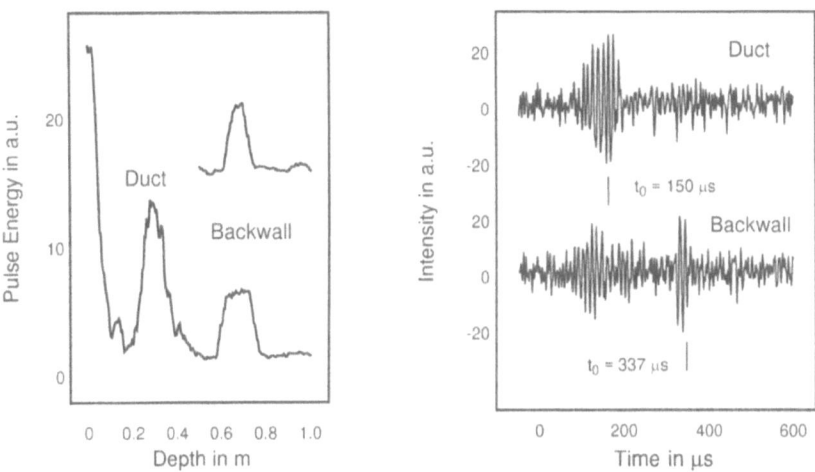

Figure 2 Evaluation by means of time corrected superposition of a data set. **Left side**: Integration of the pulse energy vs. depth; integration interval 40 µs (upper curve) and 70 µs (lower curve). From this the depth of the duct and the backwall can be measured. **Right side**: Superposition for the depth measured for the duct and the backwall of the specimen, respectively (t_0: time of flight corresponding to the depth).

435

The specimens for the blind test were fabricated at BASt. They were made from a concrete with a maximum aggregate size of 16 mm in the dimensions 2 x1.5 x 0.7 m³. There are two areas, one without and one with non-prestressed reinforcement. For the tests described here the specimen contains two metallic ducts with several unknown faults: injection faults in and compaction faults around the duct.

First the ducts were localized with impulse radar. Then twelve positions for the transmitting transducer were chosen along the duct location and for each of these positions the ultrasound amplitude was measured point by point in a corresponding aperture with the scanning laser vibrometer.

Results

An example of the evaluation by means of time corrected superposition is shown in figure 2. The pulse velocity may be measured in through transmission at one point of the specimen to get the necessary value for the analysis. Shown is the integral of the measured ultrasonic energy versus the depth of the specimen. In the superposition shown in figure 2 the echo of the duct and the backwall of the specimen clearly appear when the reconstruction is made for the corresponding depth. Thus

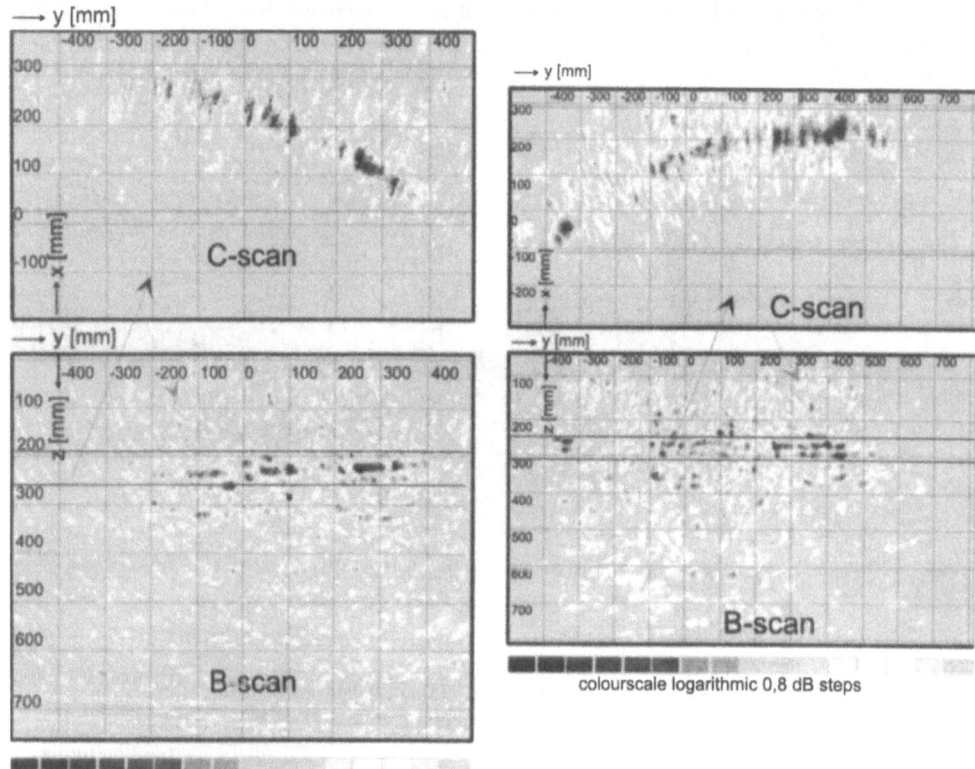

Figure 3 Result of 3D-SAFT reconstruction of the part without reinforcement rebars

Figure 4 Result of 3D-SAFT reconstruction of the part containing reinforcement rebars

the concrete coverage of the duct can be measured. But to really analyse the inside of the specimen a three dimensional reconstruction is necessary.

These calculations were carried out by means of a three dimensional version of the SAFT method [7,8]. Briefly described is it an integration of the time of flight surface in the x-y-z-t data field for each voxel of the specimen. It results in a three dimensional representation of the backscatter intensity from the inside of the specimen. To interpret such data field, projection planes have to be chosen which are the well known C-scans (parallel to the surface) and the B-scan (perpendicular to the surface).

These results for that part of the specimen without reinforcement bars are shown in figure 3. In the C-scan the location of the duct parallel to the surface appears clearly whereas from the B-scan the depth below the surface can be read.

There is an important detail in the B-Scan: The near as well as the far side of the duct can be identified in the reconstruction. That means: the ultrasonic waves pass the duct and the cement paste

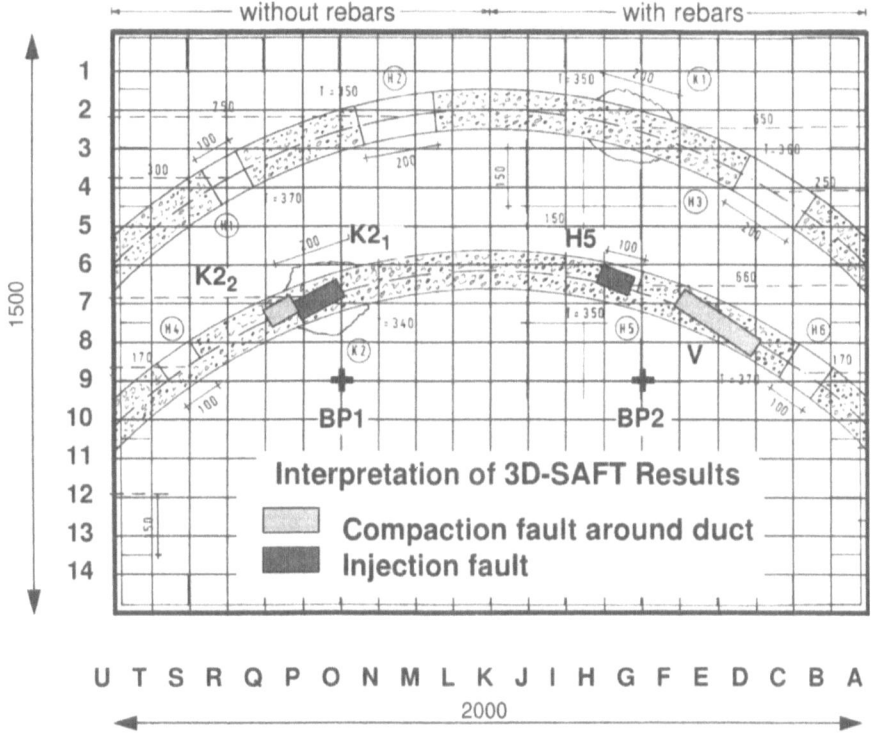

Figure 5 Comparison of the location of the defects interpreted from the 3D-SAFT reconstruction to the construction plan of the specimen (Dimensions in mm). BP*: Orientation points for the scanning laser vibrometer; K*, H*: Indication of compaction and injection faults, resp.

inside and are reflected from the far side. This observation is also predicted by simulation calculations of ultrasonic waves in concrete[1,13]. It leads to the following interpretation when defects have to be identified:

- When only the near side of the duct is detectable and not the far side the ultrasonic waves cannot pass the duct. This is certainly caused by reflection at air voids or delaminations in the duct. Such an observation is identified as a defect inside the duct.
- When neither the near side nor the far side of the duct are detectable, the ultrasonic energy is attenuated earlier which can be interpreted as multiple reflections in a compaction fault. Such an observation is identified as a defect around the duct.

The corresponding results for that part containing reinforcement rebars is shown in figure 4. The signal to noise ratio is a little bit low but the duct is clearly detectable in both, the C-scan and B-scan representation.

All these data were analysed carefully looking at the reflection from the upper and lower side of the duct. The interpretation concluded is shown in figure 5. Two injection faults and two compaction faults around the duct were identified inside the measured area.

After this analysis and interpretation the true positions and kind of defects were revealed: The injection H5 corresponds very well with the experiments as well as the compaction fault K2, when we consider that a part of this compaction fault was misinterpreted as an injection fault.

There is a very clear compaction fault which was not intended (V), which must be confirmed by opening the specimen. The injection faults H6 and H4 were not identified because of their unfavourable location relatively to the synthetic aperture.

Conclusions

An ultrasonic pulse echo technique with a low frequency broadband transmitter and a scanning laser vibrometer as receiver was applied to a concrete specimen (2 x 1.5 x 0.7 m, maximum aggregate size 16 mm). In the specimen a tendon duct and one layer of non prestressed reinforcement were placed. In and around the duct some compaction and injection defects were intentionally placed.

In a blind test two of the defects could be localized by applying 3D reconstruction calculations (3D-SAFT), a third one has to be confirmed by destructive testing. These results clearly indicate, that non-destructive testing of prestressed concrete elements with ultrasound can identify defects in and around tendon ducts.

Acknowledgement

We gratefully acknowledge financial support from the Federal Highway Research Institute (Bundesanstalt für Straßenwesen, BASt). The software for the scanning laser vibrometer was written by S. Thiele and G. Wilsch, the experiments were assisted by F. Mielentz. The graphics were edited by K. Borchardt.

References

1 Krause, M., Bärmann, R., Frielinghaus, R., Kretzschmar, F., Kroggel, O., Langenberg, K., Maierhofer, C., Müller, W., Neisecke, J., Schickert, M., Schmitz, V., Wiggenhauser, H. und F. Wollbold: *Comparison of Pulse Echo Methods for Testing Concrete.* In: Schickert, G., Wiggenhauser, H. (eds.),

Proceedings of the International Symposium Non-Destructive Testing in Civil Engineering (NDT-CE), Vol. 1, Berlin: DGZfP 1995, pp. 281-296

2 Maierhofer, Ch. und J. Wöstmann: *Investigation of Dielectric Properties of Brick Materials as a Function of Moisture and Salt Content Using a Microwave Impulse Technique at Very High Frequencies.* In: Bungey, J.H. (ed), Conference Proceedings of the British Institute of NDT, Non-Destructive Testing in Civil Engineering, International Conference Liverpool NDT-CE'97, Vol. 2, 1997, pp. 743 - 754

3 Krause, M. und H. Wiggenhauser: *Ultrasonic Pulse Echo Technique for Concrete Elements Using Synthetic Aperture.* In: Bungey, J.H., (ed), Conference Proceedings of the British Institute of NDT, Non-Destructive Testing in Civil Engineering, International Conference Liverpool NDT-CE '97, Vol. 1, 1997, pp. 135 - 142

4 Kroggel, O., Jansohn, R. und M. Ratmann: *Novel Ultrasound System to Detect Voids Ducts in Posttensioned Bridges.* In: Proceedings of the 6th International Conference on Structural Faults and Repair, Vol. 1, Ed. by M. C. Forde London, UK, July 1995, Engineering Technics Press, pp. 203-208

5 Krause, M., Maierhofer, C. und H. Wiggenhauser: *Thickness measurement of concrete elements using Radar and Ultrasonic Impulse Echo Techniques.* In: Proceedings of the 6th International Conference on Structural Faults and Repairs, Vol. 2, Ed. by M. C. Forde, London, UK, July 1995, Engineering Technics Press, pp. 17-24,

6 Krieger, J., Krause, M. und H. Wiggenhauser: *Anwendung von zerstörungsfreien Prüfmethoden bei Betonbrücken.* Abschlußbericht zum Forschungsvorhaben 9.94241 Fl, to be published

7 Schmitz, V., Kröning, M. and K. J. Langenberg: *Quantitative NDT by 3D Image Reconstruction.* Acoustical Imaging, Vol 22, Edited by P. Tortoli and L. Masotti, Plenum Press, New York, 1996, pp. 735 - 744

8 Schmitz, V. and W. Müller: *Evaluation and Interpretation of SAFT Images,* theese proceedings

9 Schickert, M.: *Einfluß der frequenzabhängigen Schallschwächung auf die Ultraschall-Laufzeitmessung an mineralischen Stoffen.* In: Berichtsband zur DGZfP-Jahrestagung, Timmendorfer Strand, 9.-11. Mai 1994, Berlin: DGZfP e.V. 1995, Nr. 43, Teil 2, S. 479-485

10 Wollbold, F. und J. Neisecke: *Ultrasonic-impulse-echo-technique - advantages of an online-imaging technique for the inspection of concrete.* In: Schickert, G., Wiggenhauser, H. (eds.), Proceedings of the International Symposium Non-Destructive Testing in Civil Engineering (NDT-CE), Vol. 2, Berlin: DGZfP 1995, pp. 1135 - 1143

11 Schickert, M.: *Towards SAFT-Imaging in Ultrasonic Inspection of Concrete.* In: Schickert, G., Wiggenhauser, H. (eds.), Proceedings of the International Symposium Non-Destructive Testing in Civil Engineering (NDT-CE), Vol. 1, Berlin: DGZfP 1995, pp. 411 - 418

12 Jansohn, R., Kroggel, O. und M. Ratmann: *Detection of Thickness, Voids, Honeycombs and Tendon Ducts Utilising Ultrasonic Impulse-Echo-Technique.* In: Schickert, G., Wiggenhauser, H. (eds.), Proceedings of the International Symposium Non-Destructive Testing in Civil Engineering (NDT-CE), Vol. 1, Berlin: DGZfP 1995, pp. 419 - 427

13 Marklein, R., Langenberg, K., Bärmann, R., Mayer, K. und S. Klaholz: *Nondestructive Testing with Ultrasound: Numerical Modeling and Imaging.* In: Tortoli, P., ed., "Acoustical Imaging", Vol. 22, Edited by P. Tortoli and L. Masotti, Plenum Press, New York, 1996

BISTATIC CIRCULAR ARRAY IMAGING
WITH GATED ULTRASONIC SIGNALS

S. Arnfred Nielsen[1] and L. Bjørnø[2]

[1] Materials Research Department, Risø National Laboratory,
DK-4000 Roskilde, Denmark
[2] Department of Industrial Acoustics, Technical University of Denmark,
DK-2800 Lyngby, Denmark

INTRODUCTION

Ultrasound computed tomography (UCT) is a relatively new imaging technique in nondestructive evaluation (NDE) of solid materials that offers an improved characterization of inhomogeneties compared with conventional A-, B- and C-scan. Specimens under investigation are insonified by ultrasound, and transmitted or reflected signals are used for reconstruction of a display of cross-sectional information of the specimen in form of an image. However, compared to classical tomography with x-rays UCT suffers from several artifacts in the projection data which degrade the final image. The most important of these artifacts are; diffraction[1], reflection and refraction[2] of the sound pulse. Common assumptions are, therefore, often either diffraction, reflection, or refraction free measurements along with straight-line propagation of ultrasound signals[3].

An ultrasonic imaging technique has been developed for generating two-dimensional cross-sections, or tomograms, of cylindrically shaped solid materials. The current work shows how the effect of ultrasonic beam refraction may be reduced compared to using a linear array of receivers[4]. A circular aperture array is used to avoid mode conversion of the incident beam. This ensures that the mode inside the sample is longitudinal with little or no transversal wave present. Diverging ultrasonic beams are used to insonify the cylindrical object, and the resultant transmitted and reflected signals are recorded by only considering received signals measured in a time gate. The divergent nature of the ultrasonic beam reduces the total scanning time and back-projection with digital filtering is applied in order to reduce reconstruction time. The circular aperture array is realized by object rotation and ultrasonic signals are acquired via a single mobile transmitter and receiver, respectively. Initial experimental results obtained on artificial defects in Plexiglas® phantoms are presented. The method described here may be applicable to non-destructive evaluation of many industrial, cylindrically shaped components where a relatively fast inspection is required, e.g. axles, pipes, rods, cranks, cylinders etc.

IMAGING CYLINDRICAL OBJECTS

Ultrasound inspection of cylindrical objects is attractive due to the high penetration depth compared to x-ray inspection but, as mentioned, it is limited by refraction of the sound path when the beam is incident at angles different from normal.

The main purpose of this paper is to explore an experimental method to acquire refraction limited ultrasonic signals in a circular aperture array, as shown in Fig. 1. The imaging system considered here have N transducer elements of the array placed uniformly along a circular aperture with radius R. The flexibility of the experimental system allows the circular aperture array to obtain ultrasonic signals in a monostatic or bistatic mode, respectively. In the monostatic mode one transducer element act both as transmitter and receiver. An ultrasonic pulse is transmitted into the specimen and the amplitude of the back scattered sound pulse is measured by the same transducer element. In the bistatic mode the forward transmitted ultrasonic sound field is received over a circular arc of receivers. The monostatic mode has been used to reconstruct a cross-section of the object's reflection coefficient by various authors[5], while the bistatic mode has been used to reconstruct an image of the object's attenuation coefficient or the index-of-refraction[6].

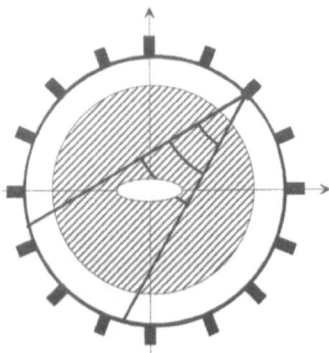

Figure 1. Schematic representation of a circular aperture array with uniformly distributed transducer elements along a radius R. One transducer transmits and receives, respectively, a short pulse (monostatic mode) while a sector of transducers receives the transmitted pulse (bistatic mode). The process is repeated for N transducer elements.

In this work the two modes of the circular array is used to obtain qualitative images of the reflection and attenuation coefficient, respectively. It is assumed that the ultrasonic beam is wide in the horizontal plane of the circular array, and the central line of the beam pattern is directed to the center of the circular plane. It is also assumed that the beam is collimated in the plane perpendicular to the circular plane in order to insonify an infinitesimal thin slice of the specimen. These assumptions will, obviously, only be partly fulfilled in a realistic imaging system, where either relative long transducer elements or cylindrically focused transducer elements are applied.

Reconstruction of the image in the monostatic mode is performed under a set of idealized assumptions, such as ideal point scatters and constant speed of sound. The back-scattered echoes represent line integrals of the acoustic reflectivity over concentric arcs centered at the transducer position[5]. A schematic representation of these arcs is also shown in Fig. 1. To reconstruct the image, the amplitude of the A-scan echoes evaluated at time $2\rho/c$, where c is the speed of sound, is backprojected over an arc of radius ρ in the image space. The transducer is moved along the circular aperture and the process repeated. The summation of the backprojected A-scans for every transducer element position forms the image of the object reflectivity function.

Let $f(r,\theta)$ be the two-dimensional reflectivity function of the object in the plane of the circular array, where the origin of the polar coordinate system is the center of the array. The backpropagation over circular arcs can be found as

$$f(r,\theta) = \tfrac{1}{2\pi} \int_0^{2\pi} g(\rho(\phi,r,\theta))d\phi \tag{1}$$

where a distance from the transducer element (R, ϕ) to the reconstruction point at (r,θ) is given by the relation

$$\rho(\phi,r,\theta) = \sqrt{R^2 + r^2 - 2Rr\cos(\theta - \phi)} \tag{2}$$

In analogy with x-ray tomography the simple backprojection of the A-scans leads to blurring of the image. The shape of the blurring function is the $1/r$ encountered in x-ray tomography for backprojection along straight lines, plus a smaller second order term resulting from the curvature of the circles (the path along the chord is not equal to the diameter of the circular array). To avoid this blurring the A-scans can be convolved with a deconvolution filter before backprojection as performed in x-ray tomography.

The image in the bistatic mode is reconstructed using the same assumption as mentioned above for the monostatic mode. A standard filtered-backprojection[7,8] algorithm is used in this work because it is relatively simple to implement, gives reasonably short processing time, and allows some flexibility in the reconstruction via the choice of the projection filter. Here the process of backprojection simply involves taking an "amplitude" projection (or fan-beam projection[9]) and extending each value across image space back in the same direction along which the projection was obtained. By repeating this operation for every projection an image of the attenuation distribution is found.

EXPERIMENTAL METHOD

A setup of the experimental system is shown in Fig. 2. Here single-element transducers have been used in preference to fixed circular arrays due to the flexibility of the system. This allows us for example to change transducer configuration, for example center frequency, focal length, piezoelectric element diameter etc., but also to image large specimens, which are only restricted by the scanning tank dimensions. Because only two mobile transducers are used in this experiment, the setup and the signal processing is much simpler than those needed for ultrasonic tomography based upon fixed transducer array systems.

The scanning facilities used in this experiment are the following: The tank dimensions are 70 x 107 x 47 cm^3 and all measurements are performed in water at 20.0 ± 1 °C. Two different broadband transducers, one acting as a transmitter and the other as a receiver, are mounted normal to the cylinder axis on each side of the specimen with a well-defined distance to the specimen. The transmitter and the receiver are high-frequency ceramic transducers with a center frequency of 2.25 MHz (wavelength 0.7 mm in water). The cylindrical specimen can be rotated about a center of rotation by a stepping motor controlled by a computer. The receiving transducer can also be rotated around the target by a second stepping motor controlled by the same computer. This allows high flexibility in the experimental procedure. Data acquisition is performed by an ultrasonic inspection system HFUS 2000 with a peak detection gate. Transmitted and reflected ultrasonic signals are digitized with a high-speed digitizer at 6.25 MHz and data is stored in a special DIM-file format which contains all scanning information such as number of receivers, number of projections etc. Approximately 0.5 Mbytes data are required per DIM-file.

The scanning procedure in the bistatic case is the following: The transmitted signal is received with 0.57 °/rec. resolution along a sector of receivers, amplified and displayed on a digital oscilloscope. A time gate is set to measure the peak amplitude for each signal. The specimen is then rotated 0.38 °/proj. and the sector is scanned again until the required number of projections have been measured. The positioning accuracy of the system is estimated to be less than 0.2 mm. The total scanning time is about 8 minutes for 128 receiver positions and 360° rotation of the target.

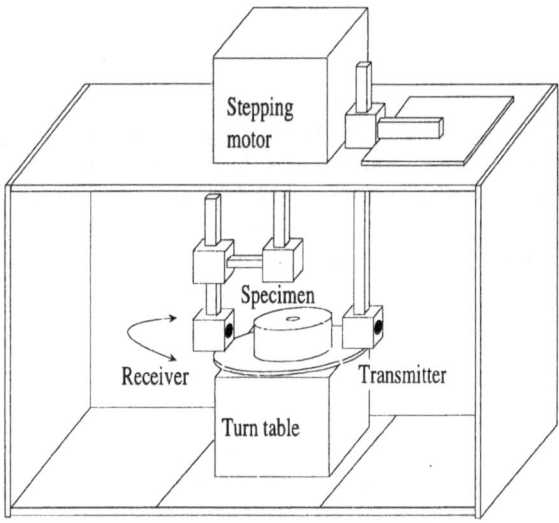

Figure 2. Experimental configuration of the tomographical system. One stepping motor rotates and controls the receiver while another stepping motor rotates the specimen.

RESULTS AND DISCUSSIONS

Ultrasonic signals obtained from simple homogeneous, isotropic Plexiglas® specimens (diameter 100 mm) with a machined water filled elliptical hole (20 mm (l) by 8 mm (w)), was used to evaluate the two modes of the circular aperture array.

In the monostatic mode a peak detection gate was set to cover only the reflected signals from the interfaces of the ellipse. The gate threshold was set above the average level of material noise to avoid detection of multiple reflections. A total of 1469 transducer positions were used over 360°, each signal containing 256 samples.

Fig. 3a shows the experimental reflection data obtained as a function of receiver position and relative delay time. The minimum time delay, A, and the maximum time delay, B, between two reflections in Fig. 3a corresponds to the major and minor axis of the ellipse. The ratio A/B = 0.56 is in agreement with the ratio between the minor and major axis of the ellipse, as shown in Fig. 3b, where a typical UCT tomogram of the ellipse based on the reflection coefficient is presented. The dark regions in the figure stand for high reflection, representing the interface of the ellipse. This is in accordance with the fact that large acoustical impedance mismatch between water and Plexiglas reflect the majority of incident sound energy. The tomogram shows the ellipse in the Plexiglas specimen on a matrix size of 1024x1024 pixels. The image gives a good qualitative impression of the size of the ellipse. Fig. 4a shows a typical data profile of the transmitted amplitude data as a function of relative receiver position. The solid line is a polynomial fit to the experimental values shown as black dots. The profile is confined within a sector of two polar angles defined by the minimum signals that can be measured by the receiver. The profile is very dependable of the signal amplification and forms, therefore, a practical system limitation.

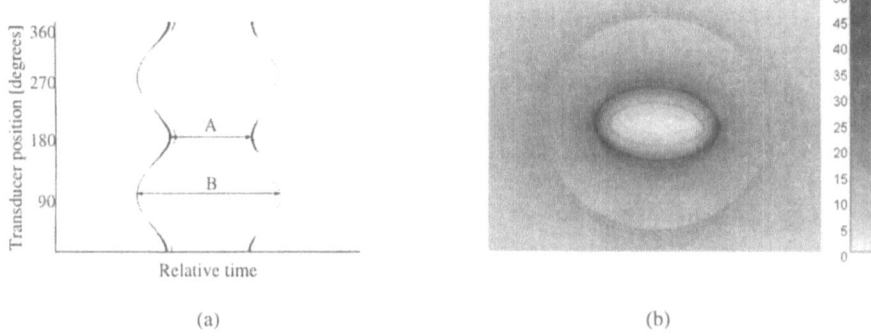

(a)   (b)

Figure 3. (a) Back-scattered data as a function of transducer position and relative propagation time. (b) Reconstructed scattering data of an ellipse in a Plexiglas specimen. The time delay A and B corresponds to the minor and major axis of the ellipse in (b), respectively.

Fig. 4b shows a typical UCT tomogram of attenuation in a cross section of the ellipse obtained from 128 receiver positions and 1469 amplitude profiles. In the image the dark central region represents the water filled ellipse with low attenuation and the bright region represents the Plexiglas specimen with high attenuation. The amplitude data from another cylindrical Plexiglas specimen, but without hole is subtracted from the amplitude data. This method is used due to difficulties in calculating realistic sound fields off-axis. A good qualitative image of the ellipse is found. The bright spots inside the ellipse, which degrade the quantitative interpretation of the image, may be due to diffraction limitations of the ultrasonic beam and due to experimental difficulty in aligning the specimen on the rotational axis. Other practical artifacts such as multiple reflections and positioning of the target is not compensated for in these initial measurements. Another limitation is due to the characteristic impedance difference between water and Plexiglas, which makes a transmitted signal almost impenetrable.

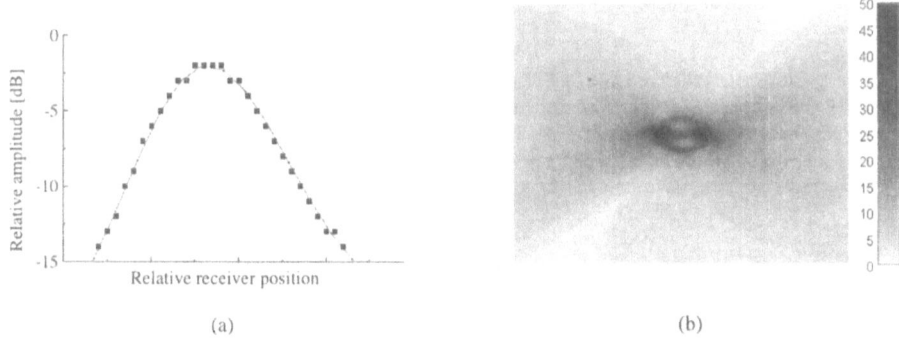

(a)   (b)

Figure 4. (a) Relative amplitude profile as a function of receiver position. (b) Reconstructed attenuation tomogram relative to the reference specimen.

445

CONCLUSIONS

In this study, we have demonstrated an experimental method to acquire refraction limited ultrasonic signals within cylindrical specimens. The monostatic and bistatic mode of a circular aperture array is used to generate UCT tomograms of inhomogeneties based on reflectivity and attenuation, respectively. Diverging ultrasonic beams are used to insonify the cylindrical specimen, and the resultant transmitted and reflected signals are recorded by only considering received signals measured in a time gate. Experimental data are obtained from cylindrical Plexiglas specimens containing defects with strong reflection boundaries. These data are reconstructed using backprojection techniques and good qualitative agreement is obtained compared with real dimensions.

There is a great potential for this type of imaging methods in Non-Destructive Evaluation (NDE) of isotropic, homogeneous materials, where each scan is processed separately and the final image is formed by superimposing the two images. These images give more information of the specimen than either of the individual images would do on their own.

The experimental method described here may be applicable to many industrial cylindrically shaped materials, such as axles, pipes, rods, cranks, cylinders etc. However, an extension to advanced materials, e.g. fiber composites require more advanced reconstruction methods. More studies are needed to better understand advanced materials and to extract more information from divergent ultrasonic signals.

ACKNOWLEDGMENTS

The authors would like to thank Mr. Lillegaard for his practical assistance in the construction of our experimental system. We are grateful to The Danish Research Academy for funding this study and to Risø National Laboratory for financial support.

REFERENCES

1   C. F. Schueler, H. Lee and G. Wade, Fundamentals of digital ultrasonic imaging, *IEEE Trans. on Sonics and Ultrasonic*, 31:195 (1984).

2   J. L. Rose and J. J. Ditri, Ultrasonic computed tomography considerations in the NDE of solid materials, *IEEE Ultrasonic Symposium*, 1: 991 (1990).

3   R. A. Kline and Y. Q. Wang, Application of tomographic imaging principles to the ultrasonic characterization of polymers, *Materials Evaluation*, 4:1385 (1990)

4   S. A. Nielsen, K. K. Borum and H. E. Gundtoft, Verifying an ultrasonic reconstruction algorithm for non-destructive tomography, in: *Proc. 1st World Congress on Ultrasonics*, J. Herbertz, ed., Gefau, Germany, 1:447 (1995).

5   S. J. Norton and M. Linzer, Ultrasonic reflectivity tomography: Reconstruction with circular transducer arrays, *Ultrasonic Imaging*, 1:154 (1979).

6   G. H. Glover and J. C. Sharp, Reconstruction of ultrasound propagation speed distributions in soft tissue: Time-of-flight tomography, *IEEE Trans. on Sonics and Ultrasonic*, 24:229 (1977).

7   N. Sponheim and I. Johansen, Experimental results in ultrasonic tomography using a filtered back-projection algorithm, *Ultrasonic Imaging*, 13:56 (1991).

8   A. J. Devaney, A fast filtered backpropagation algorithm for ultrasound tomography, *IEEE Trans. Ultrason. Ferroelect. Freq. Control*, 34:330 (1987).

9   P. Defranould, Acoustical fan-beam measurements by transducer arrays for tomography reconstruction, *IEEE Trans. on Sonics and Ultrasonic*, 28:418 (1981).

EVALUATION AND INTERPRETATION OF SAFT IMAGES

V. Schmitz, W. Müller

Fraunhofer Institut Zerstörungsfreie Prüfverfahren
Saarbrücken - Germany

INTRODUCTION

One of the classical tasks of ultrasonic NDT is the detection, sizing and characterization of material damages like cracks of welded metallic structures. The characterization implies the definition of types of defects. The defect may be a surface or an internal one, one-dimensional, two-dimensional, planar or volumetric. Therefore it is important to expand the 2D-imaging technique of LineSAFT as shown by Schmitz,[1] to 3D imaging with two-dimensional scanning.

In a second step the reliability of the image spots is investigated. Indications which are detected, but not correlated to defects are called artefacts. They are caused by different phenomena, e.g. propagation of the transmitted ultrasonic pulses on different paths with different velocities.

A software tool has been developed which allows to identify the different pulses received by a contact technique probe. The benefit is that most of the artefacts can be explained to avoid unnecessary repairs.

Finally a defect characterization strategy will be explained. For feature extraction it is necessary to decide whether an objects shape looks more like a point, a line or an area. It is recommended to compare the three orthogonal views of the images and to derive the classification by the combination of the evaluation. In an example a pipe segment with a coarse grain structure of an austenitic weld material including an interface of a ferritic pipe to an Inconell buttering has been investigated with 3 MHz at 60° insonification angle.

2D/3D - IMAGING CAPABILITIES

Ultrasonic Imaging is primarily based on wave propagation and interaction effects with the defect. At present SAFT is a strictly scalar procedure and is applied for longitudinal and for shear wave imaging. If ideal data are supplied, e.g. time domain data of "infinite" band width and a synthetic aperture surrounding the object completely, the Generalized Diffraction Tomography is in its SAFT version the "best" available imaging procedure.

In automatic ultrasonic NDT, the Synthetic Aperture is realized by a manipulator which moves a probe across the surface of a component. The processed image is a B-scan image in the plane of the movement and the insonification direction. Perpendicular to this plane no image processing will be performed. Pseudo 3D-SAFT images can be obtained by scanning many parallel lines in x-direction, by processing each line individually and by normalizing all images to their common maximum. All images are arranged in a 3D-data field and can be visualized as a projection of subgroups in a vertical plane or in arbitrarily slices parallel to the surface. The full imaging quality can be obtained only in those cases, where all signals of all lines are stored and processed both in x- and in y- direction.

To allow a full check of the performance of a PC-based 3D-SAFT system, a resolution test block has been manufactured which contains three rows of flat bottom holes with constant sizes of 1 mm, 3 mm and 5 mm and with border to border distances between 1 mm and 10 mm - fig.1.

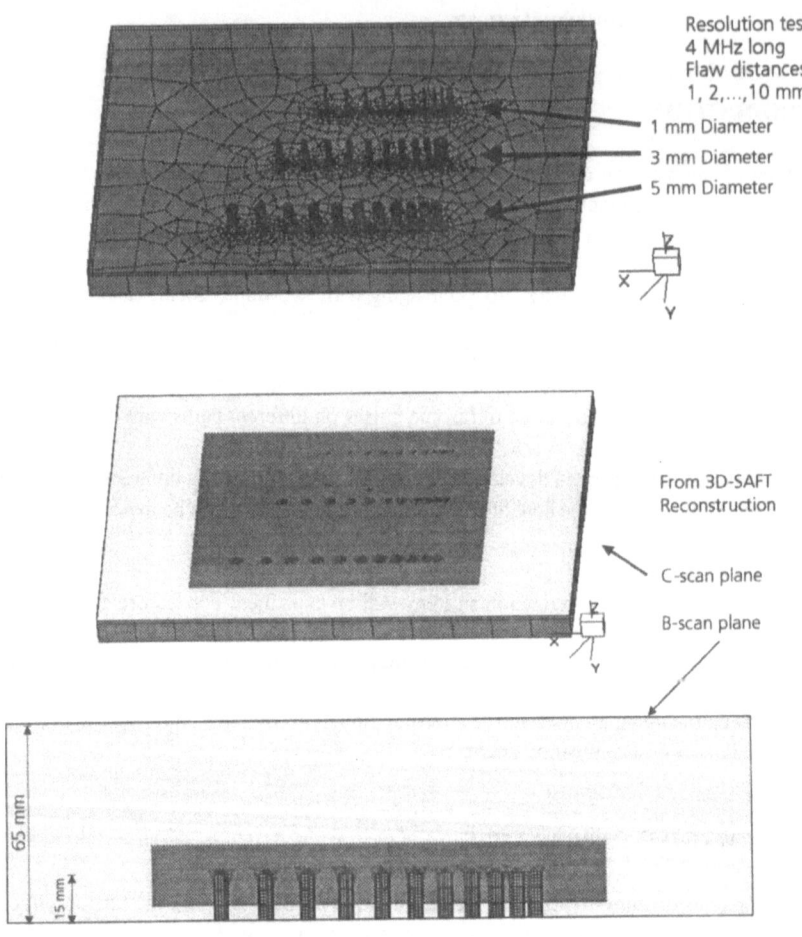

Fig. 1 Implementation of a 3D-SAFT Reconstruction into a 3D-CAD-Drawing

The block has a thickness of 65 mm and the holes are drilled from the back side 15 mm into the material. Fig. 1 shows a cut in the plane of the holes from a 3D-CAD drawing, a C-scan view presented in a perspective view and a side view. These confirm the ability to separate defects with distances of 1 mm or more if a wavelength of 1.5 mm (4 MHz longitudinal wave) is used.

FLAW CHARACTERIZATION PROCESS

Besides localization and sizing accuracy, NDE has the task to characterize defects. From Eriksson,[2] and other researcher, time domain and frequency spectral data have been recorded from several thousand data points to develop algorithms and train networks to provide an enhanced means for automatic data interpretation. It has been anticipated that classification of the defects into planar or spherical voids or cracks can be performed.

Another way to characterize defects is based upon a comparison with mathematical-numerical derived images from simulated A-scan data. The basic mathematical tool which has been used is the Elastodynamic Finite Integration Technique - "EFIT", derived by Fellinger[3] . The complexity of the task to compare acoustic images obtained by experimental data of unknown defects with acoustic images obtained from simulated data of known defects will be explained using a typical example taken from an industrial application.

In the following we will assume a metal plate with a thickness of 35 mm. This plate has a surface connected crack with a depth extension of 18 mm, 9 mm and 5 mm - fig. 2.

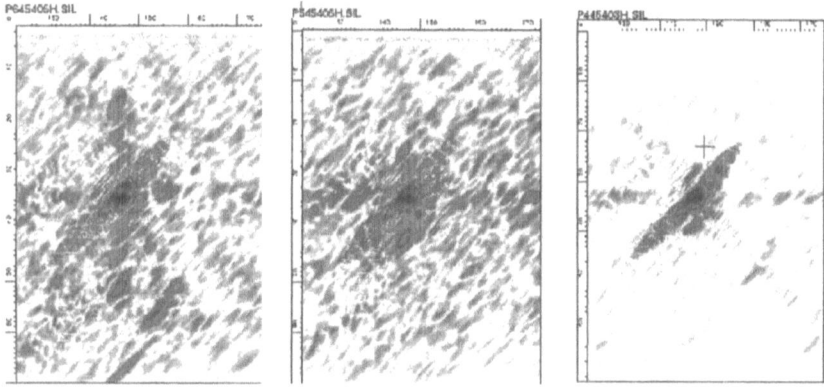

Fig. 2 Appearance of surface connected cracks with different depths

The essential information about the defect is the crack tip, the imaged spots at the crack surface and the corner echo which is imaged with high intensity due to the corner trap effect. An additional information is the mirrored crack tip which occurs due to the sound path via the back wall. A 4 MHz shear wave probe has been used for data acquisition. The time needed for the SAFT reconstruction process is less than 1 minute on a Pentium PC.

If the decision is not clear that the imaged defect is the image of a crack, one has to simulate a crack and a pore. In Fig. 3 a 10 mm crack and a pore which has been placed at the position of the crack tip has been modelled by EFIT.

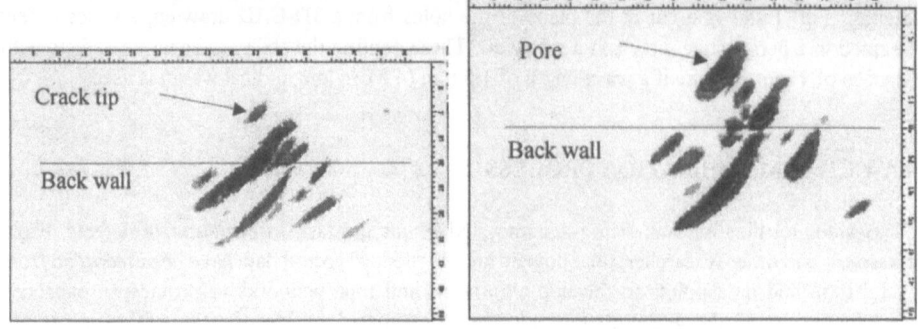

Fig. 3 Synthesized SAFT image of a crack and a pore
using the EFIT simulation program

The two last figures demonstrate the difficulties one encounters in comparing two different images. The EFIT simulated SAFT-reconstruction shows a pattern of image spots, but only one of them corresponds to the real defect. Other indications are due to mode conversion effects like creeping wave around the pore or to different wave paths. The SAFT-reconstruction with experimental data are still more complex, and the three different cracks do not have a characteristic structure. Therefore it is recommended not to follow immediately ways to solve the problem, but to identify image spots first, to check if they are relevant and to reduce the complexity of the image.

Software Module for the Identification of Image Spots Generated by Different Sound Paths

A simple computer model has been generated for a plate with an inner surface connected crack. In the model the position of the probe can be scanned and the TOF-values calculated for the nine different models sketched in fig. 4:

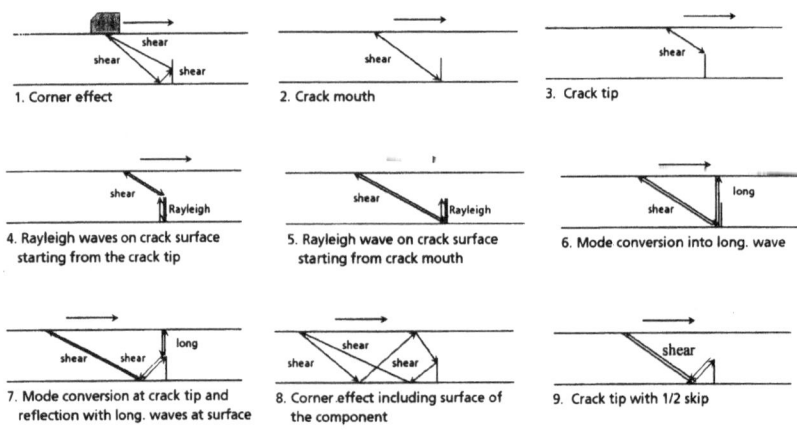

Fig. 4 Models for the prediction of ultrasonic pulses

After application of the models, the program allows to compare either the original TOF-data or the SAFT-reconstructions. First the experimental and the modelled high-frequency ultrasonic data are loaded. The region of interest is selected and arranged and zoomed to be identically - fig. 5.

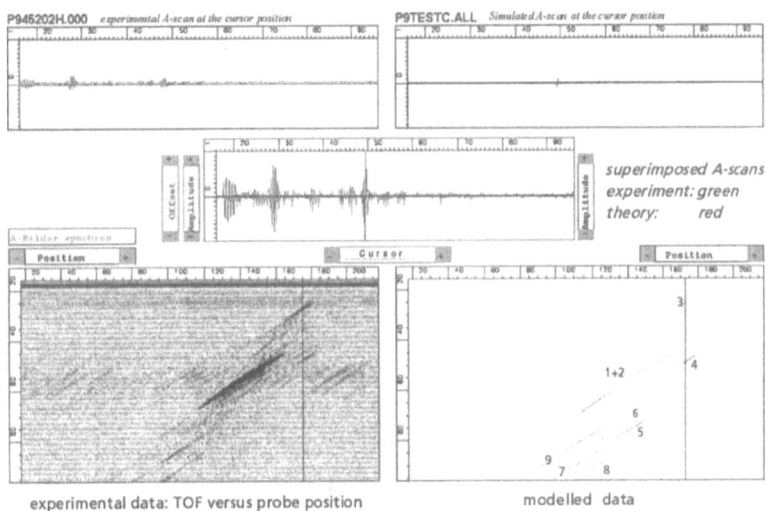

experimental data: TOF versus probe position   modelled data

Fig. 5 Comparison of high-frequency ultrasonic data in a scan-line

The modelled TOF-data are shown in fig. 5 in the lower right window and assigned to the different models. If the cursor is moved it will move simultaneously in the left window where the experimentally obtained TOF data are depicted as a function of the probe position. The A-scans belonging to these lines are shown separately in the upper part of fig. 5 and as an overlay in the middle. To allow a careful comparison, possible DC components of the experimental A-scans can be compensated by the "offset"-button and the amplitudes can be adapted by the "amplitude"-button.

With the identification of TOF-curves or the correspondent image spots, those which do not contribute to the image of the defect are removed by applying a commercially available image process-ing tool (IMAGE PRO+ from Media Cybernetics). Some of the features used are: masking out disturbing parts of the image, defining an amplitude level for contour estimation, counting and measuring objects, thinning and splitting operations.

Classification Scheme using an Image Processing Tool and 3D SAFT

The strategy is based on a 3D-SAFT reconstruction of the area which has been inspected with ultrasound. From this three possible main orientations which have to be selected are:

* parallel to the inspection surface and parallel to the scanning direction (C-scan image)
* perpendicular to the inspection surface and parallel to the scanning direction (B-scan image)
* perpendicular to the inspection surface and perpendicular to the scanning direction (D-scan)

Each plane is allowed to have a "thickness" within a predefined depth zone of the third coordi-nate. The decision "planar" or "non-planar" is made by the following classification scheme - table 1.

Table 1. Classification scheme for pores/inclusions or crack like defects

Table 1	B-scan	C-scan	D-scan
pore/inclusion	point	point	point
slag inclusion	point	line	line
crack-like	line	line	plane
inclined crack-like	line	plane	plane

In the example of an inclined crack in the buffering of a weld the C- and D- scans look more like a plane, the B-scan like a line; therefore the classification was performed as „inclined crack-like". The dimensions of the defect are 20 mm in depth and 40 mm in circumferential extension.

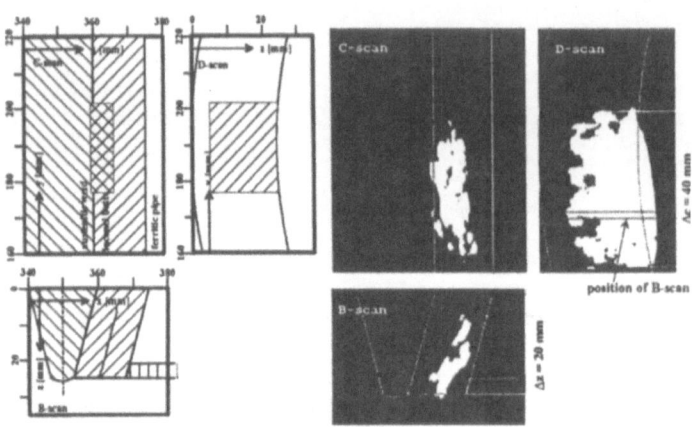

Fig. 6 Classification of a crack in a weld using
the identification module, the classification scheme and 3D-SAFT

Summary

A case study has been performed to get insight in different defect interpretation strategies. It was felt that, before comparing experimental results of unknown defects with simulated results of known defects, it is important to "clean" ultrasonic images from non-relevant indications. In the final evaluation, knowledge based systems or neural network classifyer can be used to classify the image or features derived from the image. A different way would be to use geometrical features for different planes which can be created by a 3D-reconstruction program.

References

1. V. Schmitz, W. Müller, G. Schäfer, Synthetic Aperture Focussing Technique; State of the Art, in *Acoustic Imaging, Vol.19, H.Ermert and Hans-Peter Harjes, ed., Plenum Press, New York and London*
2. Bo Eriksson, Object Characterization using Radar and using Ultrasound, *UPTEC 95 118 R; ISSN 0346-8887*, Box534, 75121 Uppsala, Sweden, Sept. 95
3. P. Fellinger, R. Marklein,K.J. Langenberg, S. Klaholz, Numerical Modeling of Elastic Wave Propagation and Scattering with EFIT - Elastodynamic Finite Integration Technique, *Special Issue of WAVE MOTI ON on behalf of the Miklowith Memorial Symposium, University of Virginia, VA, USA, June 1993,*

HOLOGRAPHIC RECONSTRUCTION OF ACOUSTIC FIELDS
BY DISCRETE CONVOLUTION

Makoto Tabei and Mitsuhiro Ueda

Department of International Development Engineering,
Tokyo Institute of Technology,
2-12-1 O-okayama, Meguro-ku, Tokyo, 152 Japan

INTRODUCTION

The conventional holographic reconstruction produces large error when it is applied to the long range propagation of the acoustic wave.[1, 2] This is because the propagator that is defined by the sampled frequency response causes strong artifacts by overlaying wrapped images on the true reconstruction. In this paper, a robust reconstruction technique is proposed based on the convolution of observed data and the propagator rather than the direct manipulation of data in the frequency domain. It is shown that when the observed data are backpropagated to the plane that includes the finite sound source, the sampling interval of the data can be taken to be significantly larger than the half-wavelength. On the other hand, by taking the pixel interval of the reconstructed source image as less than or equal to the half-wave length, the acoustic field at an arbitrary point can be evaluated by the successive forward propagation of the source image.

PROPAGATOR IN SPATIAL DOMAIN

The propagation of sound pressure $p(x, y, z)$ between two planes $z = z_1$ and $z = z_2$ can be evaluated by approximating the diffraction integral with discrete summation.[1]

$$p(x, y, z_2) = \int_{-\infty}^{+\infty} \int_{-\infty}^{+\infty} p(\xi, \eta, z_1) \cdot h(x - \xi, y - \eta, z_2 - z_1) \, d\xi \, d\eta \qquad (1)$$

$$\simeq \Delta_x \Delta_y \sum_{i_\eta = -\infty}^{\infty} \sum_{i_\xi = -\infty}^{\infty} p(\Delta_x i_\xi, \Delta_y i_\eta, z_1) \cdot h(x - \Delta_x i_\xi, y - \Delta_y i_\eta, z_2 - z_1) \qquad (2)$$

where Δ_x and Δ_y are sampling intervals in x and y coordinates. $h()$ is the propagator in spatial domain[1] which is expressed using temporal frequency f and the sound velocity c.

$$h(x, y, z) = \frac{z \left(1 + j2\pi(f/c)\, r\right)}{2\pi r^3} \exp\left(-j2\pi(f/c) r\right), \quad r = \sqrt{x^2 + y^2 + z^2}. \qquad (3)$$

If the following condtions hold; the spatial frequency of $p(\xi, \eta, z_1)$ is bandlimited below f/c, $z_2 - z_1$ is larger than a few wave length, and Δ_x and Δ_y are less than $c/(2f)$, then the difference between (1) and (2) becomes negligible.

SAMPLING CONDITION FOR BACKPROPAGATION

In the following we will show that the general sampling condition ($< c/(2f)$) is alleviated in the special case that the sound field from the finite sound source is backpropagated to the source plane. Because of simplicity, we start the discussion with the case of point source. The unit amplitude point source on $z = 0$ is expressed as $\delta(\xi, \eta)$. The sound pressure caused by this point source is obtained on plane $z = z_1$ as

$$
\begin{aligned}
p(x, y, z_1) &= \int_{-\infty}^{+\infty} \int_{-\infty}^{+\infty} \delta(\xi, \eta) \cdot h(x - \xi, y - \eta, z_1) \, d\xi \, d\eta \\
&= h(x, y, z_1).
\end{aligned}
\tag{4}
$$

This can be backpropagated to the source plane ($z = 0$) using complex conjugate propagator $h^*(x, y, z)$ instead of $h(x, y, z)$.

$$
\begin{aligned}
p(x, y, 0) &= \int_{-\infty}^{+\infty} \int_{-\infty}^{+\infty} h(\xi, \eta, z_1) \cdot h^*(x - \xi, y - \eta, z_1) \, d\xi \, d\eta \tag{5} \\
&\simeq \Delta_x \Delta_y \sum_{i_\eta=-\infty}^{\infty} \sum_{i_\xi=-\infty}^{\infty} h(\Delta_x i_\xi, \Delta_y i_\eta, z_1) \cdot h^*(x - \Delta_x i_\xi, y - \Delta_y i_\eta, z_1) \tag{6}
\end{aligned}
$$

The condition that (6) coincides with (5) is that the highest frequency of the integrand $h(\xi, \eta, z_1) \cdot h^*(x - \xi, y - \eta, z_1)$ with ξ and η to be less than $1/(\Delta_x)$ and $1/(\Delta_y)$. (Sampling theorem) Because of the phase cancellation between $h(\xi, \eta, z_1)$ and $h^*(x - \xi, y - \eta, z_1)$, the highest frequencies are dependent on x and y, and given approximately by

$$
\frac{f/c}{\left(\left(\frac{1}{2} \right)^2 + \left(\frac{z_1}{x} \right)^2 \right)^{\frac{1}{2}}}, \qquad \frac{f/c}{\left(\left(\frac{1}{2} \right)^2 + \left(\frac{z_1}{y} \right)^2 \right)^{\frac{1}{2}}}.
\tag{7}
$$

As a result, by limiting the range of reconstruction in $|x| \leq x_{\max}$ and $|y| \leq y_{\max}$, the sampling conditions are given by the following,

$$
\Delta_x < \frac{c}{f} \cdot \left(\left(\frac{1}{2} \right)^2 + \left(\frac{z_1}{x_{\max}} \right)^2 \right)^{\frac{1}{2}}, \qquad \Delta_y < \frac{c}{f} \cdot \left(\left(\frac{1}{2} \right)^2 + \left(\frac{z_1}{y_{\max}} \right)^2 \right)^{\frac{1}{2}}.
\tag{8}
$$

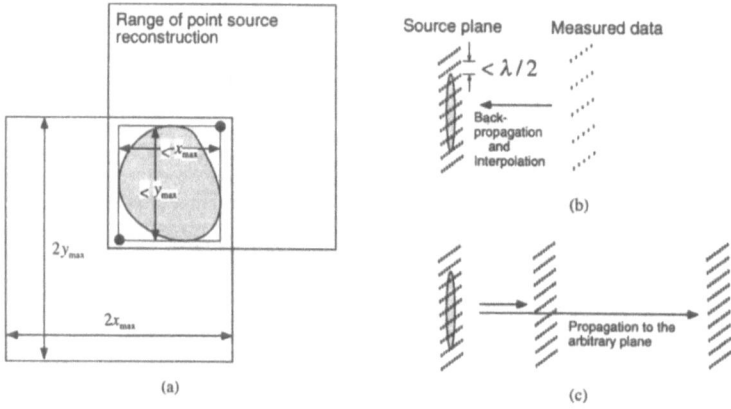

Figure 1. Accurate range of finite sound source reconstruction (a); and two step reconstruction of the field (b), (c).

The finite sound source can be interpreted as the superposition of point sources and the accurate reconstruction is obtained in the union of all the accurate range with each point source, that is, $(2x_{max} - x\_source\_size) \times (2y_{max} - y\_source\_size)$. As a result, maximum reconstructible source size is $x_{max} \times y_{max}$. (Fig.1(a))

On the other hand, (2) suggests that the sound pressure can be obtained at an arbitrary coordinate on the reconstruction plane, i.e., the sound pressure can be interpolated. By combining this interpolation characteristics and the special sampling condition, an useful procedure can be formed for the reconstruction of acoustic field. The sound pressure from the finite sound source is measured on the observing plane with relatively large sampling interval that satisfies (8). Then it is backpropagated to the source plane and reconstructed with the bin distance less than $c/(2f)$ (Fig.1(b)). The reconstructed data can be used for the reconstruction of acoustic pressure at an arbitrary point by successive forward propagation because it satisfies the general sampling condition. (Fig.1(c))

IMPLEMENTATION

If we assume the input pressure field has finite extent to accommodate actual finite aperture, (2) can be rewritten using finite sum.

$$p(\Delta_x i_x + x_0, \Delta_y i_y + y_0, z_2) \simeq (S_x \Delta_x)(S_y \Delta_y) \sum_{i_\xi=0}^{N_\xi-1} \sum_{i_\eta=0}^{N_\eta-1} \delta_{S_x,S_y}(i_\xi, i_\eta) p(\Delta_x i_\xi, \Delta_y i_\eta, z_1)$$

$$h(\Delta_x(i_x - i_\xi) + x_0, \Delta_y(i_y - i_\eta) + y_0, z_2 - z_1) \qquad \begin{pmatrix} 0 \le i_x \le N_x - 1, \\ 0 \le i_y \le N_y - 1 \end{pmatrix} \qquad (9)$$

where, Δ_x and Δ_y are bin distance, N_ξ, N_η and N_x, N_y are observation and reconstruction data sizes, respectively. $\delta_{S_x,S_y}(i_\xi, i_\eta)$ is a sparsing operator that is useful for interpolation.

$$\delta_{S_x,S_y}(i_\xi, i_\eta) = \begin{cases} 1 & \text{if } i_\xi \bmod S_x = 0 \text{ and } i_\eta \bmod S_y = 0, \\ 0 & \text{otherwise.} \end{cases} \qquad (10)$$

By storing the input sound pressure p sparsely in every S_x and S_y bins, the actual input distances become $S_x \Delta_x$ and $S_y \Delta_y$ while keeping the output bin distances as Δ_x and Δ_y. x_0 and y_0 are the offsets of the output relative to the input, and it is often convenient to choose $x_0 = \frac{1}{2}\Delta_x(N_\xi - N_x)$ and $y_0 = \frac{1}{2}\Delta_y(N_\eta - N_y)$ because the centers of input and output data coincide.

Since (9) has common bin distance for input and output, it can be implemented using DFT-based fast convolution. In (9), $i_x - i_\xi$ and $i_y - i_\eta$ ranges in $-(N_\xi - 1) \le i_x - i_\xi \le (N_x - 1)$ and $-(N_\eta - 1) \le i_y - i_\eta \le (N_y - 1)$, therefore, $(N_\xi + N_x - 1) \times (N_\eta + N_y - 1)$ samples of h is required. The result of linear convolution of $N_\xi \times N_\eta$ and $(N_\xi + N_x - 1) \times (N_\eta + N_y - 1)$ extends to $(2N_\xi + N_x - 2) \times (2N_\eta + N_y - 2)$. The storage for computation can be saved by overlapping "partially convolved part" at the fringe of the result that does not affect desired $N_x \times N_y$ output. The technique is known as "overlap-save method".[3] The realization of convolution by 2-D version of overlap-save method is depicted in Fig.2(a). Both input pressure field and the propagator should be stored in $(N_\xi + N_x - 1) \times (N_\eta + N_y - 1)$ 2-D array. The input pressure field is put in the lower left corner of the array and the rest is filled with zeroes. The DFTs of these two arrays are multiplied in frequency domain, and the output pressure field is obtained in the upper-right corner of its IDFT. (It originates at $(N_\xi - 1, N_\eta - 1)$.) Upper and right ends of the convolution that exceeds the DFT size is wraparound and overlapped on the lower and left ends. Surprisingly, these DFT operations may be done by either $(N_\xi + N_x - 1)$ column $(N_\eta + N_y - 1)$ row 2-D DFT or $(N_\xi + N_x - 1) \times (N_\eta + N_y - 1)$ (very long) 1-D DFT. The difference of these methods appears in the wrapped portion of the result and $N_x \times N_y$ output is identical.

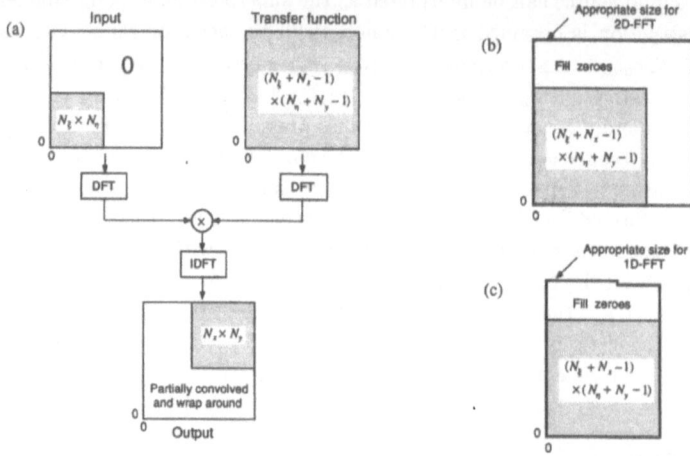

Figure 2. DFT based convolution using "*Overlap-save method*" (a); and zero filling for FFT (b),(c).

When calculating the DFT using typical FFT, the data size needs to be a power of 2 by further filling with zeroes. Fig.2(b) and (c) explain the zero filling for 2-D FFT and 1-D FFT implementations. In the case of 2-D FFT, the data should be filled with zeroes to the power of 2 in each dimension, whereas the 1-D FFT requires only the total number of data to be the power of 2. As a result, the 2-D FFT based implementation requires double storage and computation of the 1-D FFT's in the worst case.

EXAMPLES OF RECONSTRUCTION

First, to confirm the validity of sampling condition (8), the field of point source is back-propagated. The distance between the source and the observation is set to 100 wave length and the sizes of observation is chosen as 960 × 960 (wave length) in fig.3(a) and 96 × 96 in fig.3(b). The observation bin interval is set to 8 wave length (16 times the Nyquist interval) that makes accurate reconstruction within the range of $x_{max} = y_{max} = 12.5$ wave length. Note that in both figures, the faulty responses at $\pm x_{max}$ and $\pm y_{max}$ that are produced by the grating beams do not have sidelobes toward the true images of the point source. It is also confirmed that the change in the size of observing aperture does not affect the sampling condition.

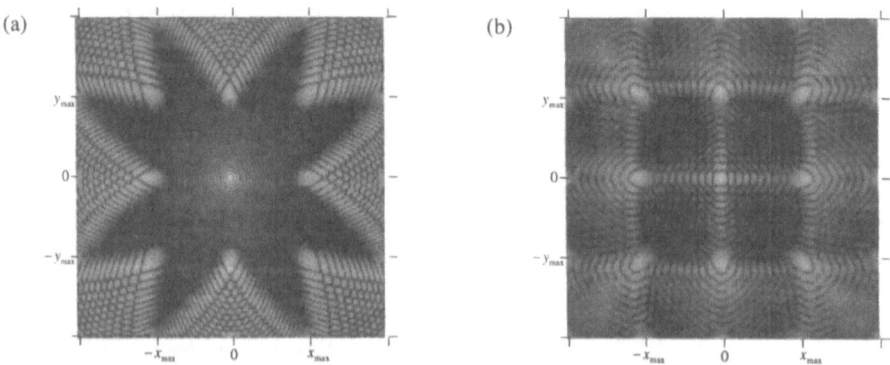

Figure 3. Back-propagated image of point source. $\Delta_x = \Delta_y = 8c/f$. (DR: 60dB)

Next, we have reconstructed the source images and the acoustic fields of the actual transducers. Throughout the following reconstructions, the data are obtained by the experiments in water, and the transducers are driven by the 2.25MHz burst sinwave (length of wavelet is 30 cycles). The data are obtained by scanning a 1mmϕ PVDF hydrophon mechanically. The electrical signal is A/D converted and complex amplitude is extracted by numerically Fourier transforming the sequence at 2.25MHz. For the reconstruction shown in fig.4, a 50×50 obser-

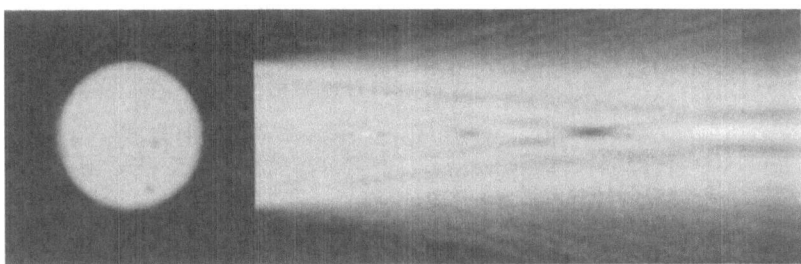

Figure 4. Reconstructed source image of 25.4mmϕ flat transducer and acoustic field obtained by the forward propagation. (DR:30dB)

vation is made in the range of 50mm from the 25.4mm circular plane source. The bin distance for observation is 0.9mm that is nearly 3 times larger than $c/(2f)$. (The wave length c/f is 0.667mm in water.) The observed data is inserted two zeroes in between each datum to form a 148×148 input with 0.3mm intervals. (This corresponds to choosing the sparsing parameter $S_x = S_y = 3$ in (9)). Then observed data is backpropagated to the source plane, using (9). We have chosen the size of reconstruction as 100×100. The minimum DFT size is 247×247 in 2D or 61009 in 1D. We used 65536-point 1D FFT by pad some zeroes at the end of the sequence. The reconstructed source image is automatically interpolated as seen on the left of fig.4. Because of the coarse observation, the faulty images that are produced by the grating beams are supposed to appear in some distance from the true image, however, they are not included in the 100×100 reconstruction. (The absolute sampling condition given by (8) is around 1.35mm, therefore, 0.9mm gives some margin.) Finally, The reconstructed source image is used for the successive calculation of the acoustic field by the forward propagation. On the right of fig.4, the cross sectional acoustic field along the transducer axis is displayed. It is obtained by repeating the reconstructions by changing the distance from the source plane. The size of each reconstruction is 150×1 and it was repeated up to 100mm from the source with 0.3mm increments. Since it satisfies absolute half-wavelength sampling condition, the clean image is obtained in the whole range. We can use smaller size FFT for this because it requires only one line reconstruction crossing the transducer axis. The minimum DFT size is 247×100 in 2D or 24700 in 1D and we used 32768-point 1D FFT.

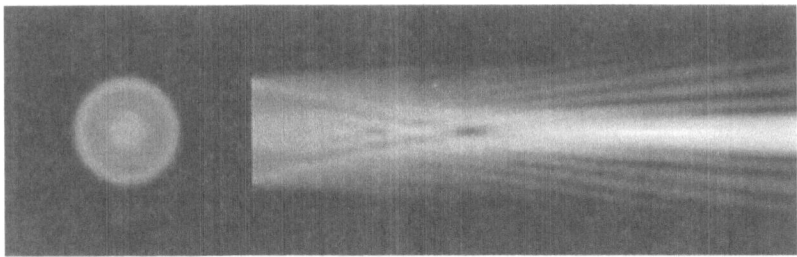

Figure 5. Reconstructed source image of 19mmϕ focal transducer (focal length=100mm) and its acoustic field. (DR:30dB)

In fig.5, the source image and the acoustic field of a 19mmϕ focal type transducer are reconstructed. The observation is made at 80mm from the transducer surface with 25×25 and 1.8mm intervals. The data are inserted five zeroes in between each datum to make 145×145 data. A 100×100 source image is reconstructed with 0.3mm bin intervals.

Fig.6 (a) and (b) show the result of direct reconstruction of an acoustic field of the same observation as fig.5. The observed data are expanded to 145×145 and the field is evaluated directly from this data. In fig.6 (a), because of the undersampling of the field, several grating beams are emitted together with the true acoustic beam from the observing plane. However, it can be seen that the true beam is completely separated from the grating beams near the source plane.

 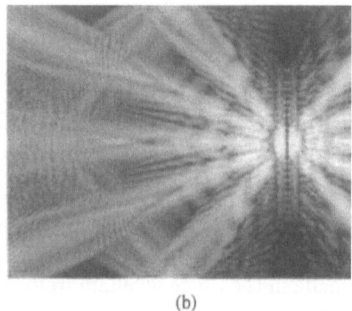

(a)   (b)

Figure 6. Direct reconstruction of 19mm focal transducer from the measurement at 80mm by discrete convolution (a), and by conventional method (b). (DR:30dB)

In fig.6 (b), the conventional frequency domain implementation of holographic reconstruction is tested for comparison. The same 145×145 expanded data is directly propagated using frequency domain representation of the propagator and 256×256 FFT. At near observing plane, the reconstructed field looks similar to that of fig.6 (a), however, on the source plane, the source image is degraded by the overlaid wrapped beams because the sampled frequency response gives wrapped transfer function in spatial domain. The error due to the wrapping may be reduced by using larger size FFT and/or introducing frequency domain filtering for restricting the angular direction of the propagation.[1,2] However, the control of frequency characteristics of the transfer function without reducing the resolving power is not easy task because it is dependent on the propagation distance and the DFT span.

SUMMARY

In this paper, a technique for the reconstruction of the acoustic fields using discrete convolution is proposed. The backpropagation to the source plane is used as the intermediate process to put the constraints on the undersampled measurement. It enables the robust reconstruction of the full acoustic field.

REFERENCES

1. R.C.Waag, J.A.Campbell, J.Ridder and P.R.Mesdag, "Cross-sectional measurements and extrapolations of ultrasonic fields," *IEEE Trans. Sonics Ultrason.* **SU-32**, 26-35 (1985)
2. P.T.Christopher and K.Parker, "New approaches to the linear propagation of acoustic fields" *J. Acoust. Soc. Am.* **90**(1), 507-521 (1991)
3. J.S.Lim, "*Two dimensional signal and image processing*," pp.145-149, Prentice-Hall (1990)

SHAPE RECONSTRUCTION OF A PENETRABLE SCATTERING BODY VIA DIFFRACTED WAVES AND CANONICAL SOLUTIONS

A. Wirgin and T. Scotti

Laboratoire de Mécanique et d'Acoustique
31 chemin Joseph Aiguier 13402 Marseille cedex 20, France

INTRODUCTION

This work is concerned with the shape reconstruction of a penetrable body from measurements of the scattered field when the body is exposed to a plane acoustic wave. The resolution of the forward problem, during the inversion, is bypassed by employing exact solutions of a canonical problem involving a body of simpler shape. This is the basis of the the Intersecting Canonical Body Approximation (ICBA). The inverse problem then reduces to the search, in each scattering direction, for the radius of a circular cylinder having the same known composition as that of the real body which gives the same scattered field as the measured scattered field and to the identification of this radius with the local radius of the unknown body. The same method has already been employed for shape reconstruction of acoustically soft and hard bodies (Scotti and Wirgin, 1995) and has been compared, as concerns the forward problem, to other algorithms (Wirgin and Scotti, 1996).

PROBLEM INGREDIENTS

The exterior medium Ω_0 is homogeneous and unbounded. The unknown cylinder (occcuped by medium Ω_1), filled with an homogeneous fluid, bounded by a surface whose trace in the xy plane is Γ, is assumed to be representable by the parametric equation $\tau(\varphi)$: $0 < \varphi \leq 2\pi$. Its axis is the z axis of the $Oxyz$ cartesian coordinate system where O is assumed, for convenience, to be located within Γ. The incident wave vector lies in the Oxy plane so that the wavefields do not depend on z (2D problem). The $\epsilon^{-i\omega t}$ time dependence is omitted. u^i will represent the incident plane wave field, u_0 and u_1 the total fields respectively in Ω_0 and Ω_1. These fields do not depend on z. u_0 and u_1 are square integrable in Ω_0 and Ω_1, governed by the Helmholtz equations ($(\Delta + k_j^2)u_j = 0$ in Ω_j with k_j the wave number in Ω_j $j = 0, 1$) , satisfy the outgoing wave

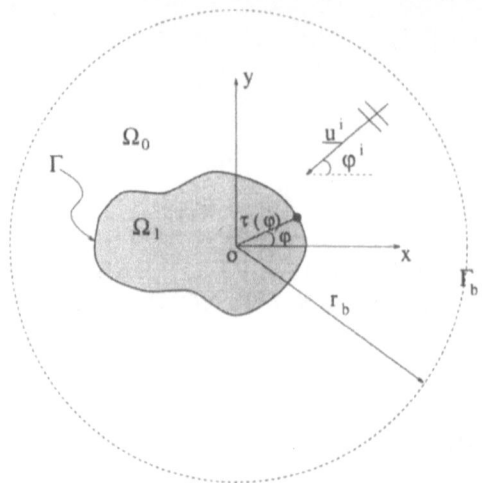

Figure 1: Cross section of scattering configuration.

condition at infinity (as concerns u_0) and obey the transmission boundary conditions on Γ:

$$\begin{cases} \alpha_0 u_0 /_\Gamma = \alpha_1 u_1 /_\Gamma \\ \beta_0 \partial_n u_0 /_\Gamma = \beta_1 \partial_n u_1 /_\Gamma \end{cases} \quad (1)$$

with $\alpha_1 \; \alpha_0 \; \beta_0 \; \beta_1$ known constants. For the inverse problem, the measured field u_0 is supposed to be known at given points on a circular surface Γ_b (whose radius is $r_b \geq max\{\tau(\varphi)\}$). This data is actually computed by the Rayleigh theory (Bolomey and Wirgin, 1974). Thus, given u_0 on Γ_b, φ^i, k_0 and k_1, the objective is to fully or partially, determine the bounding curve Γ of the body.

FIELD REPRESENTATION

From Green's theorem it is found, regardless of the shape of the scattering cylinder, that the total wave field can be represented in $\Omega_0^+ = \{r \geq \bar{r} = Max[\tau(\varphi) ; 0 \leq \varphi < 2\pi]\}$ by (Bolomey and Wirgin, 1974) ·

$$u_0(r,\varphi) = u^i(r,\varphi) + \sum_{n=-\infty}^{+\infty} A_n \; H_n(k_0 r) \; e^{in\varphi}, \quad (2)$$

with

$$u^i(r,\varphi) = \sum_{n=-\infty}^{+\infty} J_n(k_0 r) \; e^{-in(\varphi^i + \frac{\pi}{2})} = e^{-ikr\cos(\varphi - \varphi^i)}. \quad (3)$$

Similarly, in $\Omega_1^- = \{r \leq \underline{r} = Min[\tau(\varphi) ; 0 \leq \varphi < 2\pi]\}$,

$$u_1(r,\varphi) = \sum_{n=-\infty}^{+\infty} C_n \; J_n(k_1 r) \; e^{in\varphi}, \quad (4)$$

wherein J_n and H_n are the n-th order Bessel and Hankel functions of the first kind respectively.

460

DETERMINATION OF THE SHAPE OF A NON CIRCULAR PENE-TRABLE CYLINDER

Preliminary forward problem via ICBA

The problem is to determine the field on Γ_b for known $\Gamma, \varphi^i, k_1, k_0$. The ICBA method consist in the following approximation (Scotti and Wirgin, 1995) : if the body is not much different from a circular cylinder, we assume that, for each particular scattering direction φ^l, the field is given by Eq.(2) wherein the A_n are those of the intersecting circular cylinder of radius η^l where :

$$\underbrace{\eta^l}_{\text{circular cylinder}} = \underbrace{\tau(\varphi^l)}_{\text{local radius of body}} ,$$

Eq.(2) is now replaced by the approximation :

$$u_0(r = r_b, \varphi^l) \simeq u^i(r_b, \varphi^l) + \sum_{n=-N}^{+N} A_n^l(\eta^l, \varphi^i) \, H_n(kr_b) \, e^{in\varphi^l} , \; l = 1, 2, ...L, \qquad (5)$$

wherein we have reduced the infinite series in Eq.(2) to a finite series for computational purposes and the $A_n^l(\eta^l, \varphi^l)$ are given (via the introduction of Eqs.(2)-(4) into Eq.(1)) by :

$$A_n^l(\eta^l, \varphi^i) = \frac{\alpha_0 \, \beta_1 \, k_1 \, J_n(k_0\eta^l) \, \dot{J}_n(k_0\eta^l) - \beta_0 \, \alpha_1 \, k_0 \, \dot{J}_n(k_0\eta^l) \, J_n(k_1\eta^l)}{-\alpha_0 \, \beta_1 \, k_1 \, H_n(k_0\eta^l) \, \dot{J}_n(k_1\eta^l) + \beta_0 \, \alpha_1 \, k_0 \, \dot{H}_n(k_0\eta^l) \, J_n(k_1\eta^l)} \, e^{-in(\varphi^l + \frac{\pi}{2})},$$
$$(6)$$

Wherein $\dot{H}_n(\xi)$ and $\dot{J}_n(\xi)$ are the derivatives of J_n and H_n with respect to ξ.

Inverse (shape reconstruction) problem

The problem is to determine Γ, i.e., $\tau(\varphi)$, given k_1, k_0 and u_0 on Γ_b. For a particular scattering direction φ^l, we match the expression of u_0, given by Eq.(5), with the given data u_0:

$$\underbrace{u_0(r_b, \varphi^l)}_{\text{given data}} - \underbrace{\left[u^i(r_b, \varphi^l) + \sum_{n=-N}^{+N} A_n^l(\eta^l, \varphi^i) \, H_n(kr_b) \, e^{in\varphi^l} \right]}_{\text{theoretical values}} \approx 0 \; ; \; l = 1, 2...L. \qquad (7)$$

Since the A_n^l are known, analytically speaking (see Eq.(6)) to within the single parameter η^l, which is the radius of the circular cylinder that gives, for this particular scattering direction, the same diffracted field as the data, Eq.(7) enables one to determine η^l. The inverse problem reduces to : 1. searching, for each scattering direction φ^l, the radius η^l of a circular cylinder which gives the same diffracted field as that given by the measurements and 2. associating η^l with the local radius of the body. If this is done for a set of measurements in an angular sector (or all around the body) then the shape function $\tau(\varphi)$ is thereby partially (or totally) reconstructed.

Remarks

1) if we take L measured samples of the diffracted field at angles φ^l, we have to solve a system of L non-linear equations (one for each sample) in L unknowns (one "radius"

η^l for each scattered direction φ^l); 2) these equations are not coupled in terms of η^l, so that the system can be solved equation-by-equation instead of globally; 3) η^l should be real, but, because errors always exist in numerical computations, the solution η^l of Eq.(7) is, in fact, complex; 4) we keep only the real part of η^l to test our results; 5) for each equation, the solution is not unique and we have developed an algorithm to eliminate spurious roots; 6) k_1 could be determined in the same way provided a, k_0, φ^i, Γ and the field on Γ_b were known.

COMPUTATIONAL PROCEDURES

The Bessel and Hankel functions with (real) arguments $k_0 b$ were computed by means of the IMSL (IMSL, 1991) subroutines DBSJS and DBSYS. The Bessel and Hankel functions having (complex) arguments k_1 were computed by means of the IMSL (IMSL, 1991) subroutines DCBJS and DCBYS . The nonlinear system Eq.(7) was solved, equation by equation, by means of the (IMSL, 1991) subroutine DZANLY. The latter computes the complex zeros of a complex function by the so-called Müller method which is a variant of the Newton-Raphson scheme. It is important to point out that each equation possesses an infinite number of roots. Practically, the number is finite and user-specified for each call to DZANLY. This number must be chosen large enough to make sure that one has not left out the sought-for root.

POST PROCESSING TO CHOOSE THE 'RIGHT' PROFILE

We first eliminated profiles for which the real part of $\tau(\varphi^l)$ is negative, unreasonably large (or larger than Γ_b in the near field measurements), then we chose the profile corresponding to the smallest imaginary part of $\int \tau(\varphi) d\varphi / \int d\varphi$. For backscattering measurements in the the far field, we used the scattering diagram to get an idea of the larger dimension of the body (by measurements of the maximal amplitude of the main lobe), thus eliminating larger bodies, and then chose profiles corresponding to the smallest imaginary part of $\int \tau(\varphi) d\varphi / \int d\varphi$.

APPLICATION TO A HIDDEN OBJECT CONTAINED WITHIN AN-OTHER OBJECT

The unknown (penetrable) object (e.g., organ : medium Ω_3 bounded by a surface whose trace in the $x - y$ plane is Γ_3 assumed to be representable by the parametric equation $r = \tau_3(\varphi)$) is now hidden within two other penetrable objects (e.g., body composed of media Ω_1 and Ω_2 bounded by two surfaces whose trace in the $x - y$ plane are Γ_1 and Γ_2, assumed to be representable by the parametric equation $r = \tau_1(\varphi)$ and $r = \tau_2(\varphi)$ respectively). A variant of the ICBA explained above is employed for solving this inverse problem : in a particular scattering direction (φ^l), exact solutions to the canonical problem of diffraction of a plane wave by the three intersecting circular (concentric) cylinders (of radii $a = \tau_3(\varphi^l), b = \tau_2(\varphi^l), c = \tau_1(\varphi^l)$) are used to obtain an approximation of the diffracted field by the real composite body.
Assuming that the geometrical properties of the body are known, the inverse problem

then reduces to the search, in each scattered direction, for the radius of an inner circular cylinder, having the same known composition as that of the real hidden object, which gives the same diffracted field as the "measured" diffracted field. We then identify this radius with the local radius of the hidden inner object.

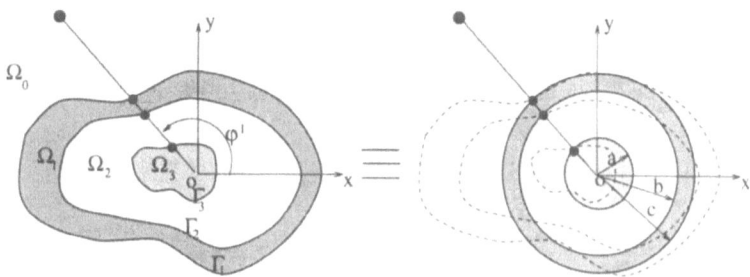

Figure 2: Scattering configuration and intersecting cylinders.

More specifically, the boundary conditions are, at each interface Γ_j,

$$\left|\begin{array}{l} \alpha_{j-1}u_{j-1} - \alpha_j u_j = 0 \\ \beta_{j-1}\partial u_{j-1} - \beta_j \partial u_j = 0 \end{array}\right. \quad j = 1, 2, 3$$

and the expression of the total field in $\Omega_0, \Omega_1, \Omega_2, \Omega_3$, are

$$u_0(r, \varphi) = u^i(r, \varphi) + \sum_{n=-\infty}^{+\infty} a_n H_n^{(1)}(k_0 r) e^{in\varphi},$$

$$u_1(r, \varphi) = \sum_{n=-\infty}^{+\infty} \left\{ b_n H_n^{(1)}(k_1 r) + c_n J_n(k_1 r) \right\} e^{in\varphi},$$

$$u_2(r, \varphi) = \sum_{n=-\infty}^{+\infty} \left\{ d_n H_n^{(1)}(k_2 r) + e_n J_n(k_2 r) \right\} e^{in\varphi},$$

$$u_3(r, \varphi) = \sum_{n=-\infty}^{+\infty} f_n J_n(k_3 r) e^{in\varphi}.$$

From here one proceeds in the same way as that which follows Eq.(4) to solve both the forward and inverse problems via the ICBA.

CONCLUSIONS AND RESULTS

We employed "measurements" in either the near ($r_b \ll \infty$) or far field ($r_b \to \infty$), $\varphi^l = \varphi^i$ (backscattering), the incident angles being equally spaced all around Γ_b. Fig.3 illustrates the fact that reconstructions using near or far field measurements are of comparable accuracy and quite acceptable, even for an ellipse of rather high excentricity. We chose for Fig.3 : $L = 40$, $N = 6$, $k_0 = 1.5$ and $k_1 = 1 + 1.1i$, $r_b = 2$ and $\alpha_0 = 1$, $\alpha_1 = 1.5$ $\beta_0 = 1$, $\beta_1 = 1.5$, and for Fig.4, $r_b = 3$, $k_0 = 3.5 + i0$ (water), $k_1 = 3.6 + i0$, $k_2 = 3.3 + i0$ (fat), $k_3 = 3.2 + i0$ (glandular medium), $\alpha_0 = \alpha_1 = \alpha_2 = \alpha_3 = 1.0$, $\beta_0 = \beta_1 = 1.0$, $\beta_2 = 0.94$, $\beta_3 = 0.83$

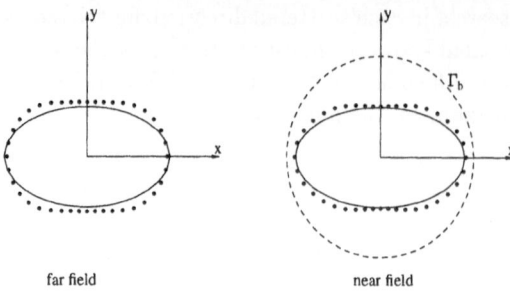

far field   near field

Figure 3: Reconstruction using backscattering measurements for an ellipse with horizontal and vertical semi-axes $a_h = 1.8$, $a_v = 1$. The actual boundary is the full line curve, the reconstructed boundary is the set of points.

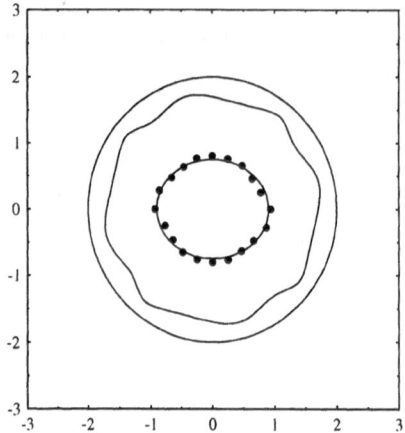

Figure 4: Cross section of a female breast (treated as cylinder) immersed in water. The dots are the reconstruction (using backscattering measurements) of a hypothetical tumor within the glandular medium. The data was taken on a cylinder (in the near field zone) of radius $r_b = 3$.

References

Angell, T.S., Kleinman, R.E., and Roach, G.F., 1986, An inverse transmission problem for the Helmholtz equation, *Inverse Probs.*, (**3**):1987, p 149-180

Bolomey J.C. and Wirgin A., 1974, Numerical comparaison of the Green's function and the Waterman and Rayleigh theories of scattering from a cylinder with arbitrary cross section, *Proc.IEE*, (**121**):794, p 794-804

IMSL, 1991 User's manual Fortran subroutines for mathematical applications: MATH/ LIBRARY, special functions, Version 2.0 *Ref.SFLB and MALB-USM-PERFECT-EN9104-2.0*, *IMSL, Houston.*

Scotti T. and Wirgin A., 1995, Shape reconstruction using diffracted waves and canonical solutions, *Inverse Prob.*, (**11**), p 1097-1111

Wirgin A. and Scotti T., 1996, Wide band approximation of the sound field scattered by an impenetrable body of arbitrary shape, *Journ.Sound.Vibr.* (**194**), p 537,572

RECONSTRUCTION OF FLOW AND REFRACTIVE OCEAN PARAMETERS BY TOMOGRAPHY METHODS

Sergey V. Baykov, Valentin A. Burov, and Sergei N. Sergeev

Physics Faculty, Acoustics Department
Moscow State University
Moscow 117234, Russia

This paper dwells upon the development of the practical scheme of the experiment and the algorithm of data processing in the tomography investigation of big regions of the inhomogeneous stratified and moving ocean. In such studies different experimental schemes and sound field descriptions are available.

In the authors previous works devoted to the moving ocean tomography [1,2] common approach to the ocean tomography as an inverse problem was developed. The base of the approach is the wave equation for inhomogeneous and slowly $(|\bar{v}(\bar{r})| \ll c_0)$ moving medium:

$$(\nabla^2 + k_0^2)p_s(\mathbf{r}) = \left(\frac{\omega^2}{c_0^2} - \frac{\omega^2}{c^2(\mathbf{r})}\right)p(\mathbf{r}) - \frac{2i\omega}{c_0^2}\mathbf{v}(\mathbf{r})\nabla p(\mathbf{r}), \tag{1}$$

where $\mathbf{v}(\mathbf{r})$ is the speed of water in the point with coordinates $\mathbf{r}$, $p(\mathbf{r}) = p_0 + p_s(\mathbf{r})$ is the full field that is sum of incident p_0 and scattered p_s fields. Several methods of description can be used for the tasks of sound fields representation for inhomogeneous stratified moving ocean. In this work the following representation is used. The solution is represent as an expansion in terms of modes $\psi_n^0(z)$ of some "reference" stationary ocean medium with some "average" sound profile $c_0(z)$. So it is assumed that in this case $\mathbf{v}(\mathbf{r})=0$ and the phase velocity $c_0(\mathbf{r})$ dependence on the vertical coordinate z only. In such approach the declination of $c(\mathbf{r}) \equiv c(x,y,z)$ from the average profile and the medium motion $\mathbf{v}(\mathbf{r}) = \mathbf{v}(x,y,z)$ are the perturbations and they cause the scattering: the declination of observing fields from expecting ones in reference direct problem.

This perturbation results in the expansion terms φ_n dependence upon the coordinates:

$$p(\mathbf{r}) = \sum_n \varphi_n(x,y)\psi_n(z) \tag{2}$$

which can be described by the equation system

$$\Delta_r \varphi_n(x,y) + p_{0_n}^2 \varphi_n(x,y) + S_{nm} \varphi_m(x,y) + \frac{2i\omega}{c_n^2} V_{nm} \nabla_{xy} \varphi_m(x,y) = f_n(x,y)\psi_n(z) \qquad (3)$$

(There is a sum on repeating indexes). Here c_n is the phase velocity of n-th mode in nonperturbated medium, $f(\mathbf{r}) = f_n(x,y)\psi_n(z)$ - the space distribution of sources.

The scalar and vector operators are

$$S_{nm}(x,y) = \int_{-H}^{0} \omega^2 \left(\frac{1}{c^2(\mathbf{r})} - \frac{1}{c_0^2(z)} \right) \psi_n(z)\psi_m(z)dz$$

$$(4)$$

$$V_{nm}(x,y) = c_n^2 \int_{-H}^{0} \frac{\mathbf{v}(\mathbf{r})}{c_0^2(z)} \psi_n(z)\psi_m(z)dz.$$

The infinite system of concerned wave equations for multi-component field $\varphi_n(x,y)$ decomposes to the independent equations in the case of adiabatic approximation for which S_{nm} and V_{nm} become diagonal operators.

The perturbations being introduced by the operators S_{nm}, V_{nm} become strong because in real conditions the additional phases may be greater then 2π on the distances in tens and hundreds kilometers. It means that the scattered field becomes equal or more then incident field and mode focusing due to the horizontal refraction also amplifies the scattering effects. This circumstance makes the Born approximation unusable for the solution of the inverse problem of ocean inhomogeneities reconstruction upon the perturbed field observations using set of antennae surrounding the investigated region. The uniqueness of the inverse problem solution can be achieved by the bigger redundancy in scattered data [3]. It can be done by several ways:
1. A number of estimated parameters can be reduced by the *a priori* information.
2. A number of space points of receivers and sources can be increased.
3. A number of frequencies can be increased.

Reduction of a number of hydrophones in vertical arrays can be (may be partially) compensated by the increase of the frequency range but a number of source and receiver locations in horizontal plane is defined by the quantity of independent parameters which characterize the investigated region in this plane.

The flow velocity can be determined by the violation of the reciprocity theorem. This violence is present in observed data by some way (directly or indirectly).

Because known regorous functional-analytical methods for solution of inverse problem haven't had practical generalizations on the case of multichannel scattering the method we use is the gradiant-iterational method. Redundancy of the data makes the solution to be unique and *a priory* information (i.e. reduction of a number of estimating parameters) allows to increase this redundancy and stability of the solution.

The influence of antenna declination from the vertical profile due to the water flows is neutralized by use of "four-frequency" algorithm [4]. In this algorithm data set on four frequencies is present in the kind of product

$$M_4 = U_{\omega_1} U_{\omega_2} U_{\omega_3}^* U_{\omega_4}^* \qquad (5)$$

with the condition $\omega_1 + \omega_2 = \omega_3 + \omega_4$. As a result the demands on the accuracy of antenna positioning and the value of possible unknown declinations of their profile from vertical line are reduced. If these declinations are less then the smallest period of space beating for used frequency range and propagating modes (it is determined by the channel parameters) then the influence of these declinations on the M_4 is negligible. The reconstruction of the kinetic parameters of the medium V_{nm} also can be done by means of the reciprocity violation in the products M_4 if one or two cofactors are replaced by the contrary data from opposite direction. The estimation process is automatic in the algorithm if two ways data are included in the functional of residual.

The four frequency algorithm for reconstruction of the sound speed and flow velocity inhomogeneities has been tested on the following model. The region under investigation (see Figure 1) is surrounded by 40 vertical arrays deployed from surface to bottom. Each array consists of 20 equal spaced receivers. The distance between the nearest arrays is 10 km. The linear flow crosses the region. Its cross size is 50 km. The flow consists of 5 horizontal layers (each - 10 km in cross). Velocity distribution is shown on Figure 2. Sound speed profile (Figure 3 b) in these layers is different from the profile of the upper and down parts of the region (Figure 3 a). Corresponding sound speed ingomogenety is shown on Figure 4. The estimated parameters are the values of sound speed and flow velocity in each horizontal layer of the flow on horizons of 400, 800, 1200, 1600, 2000 m. So that the whole number of unknown parameters is 50.

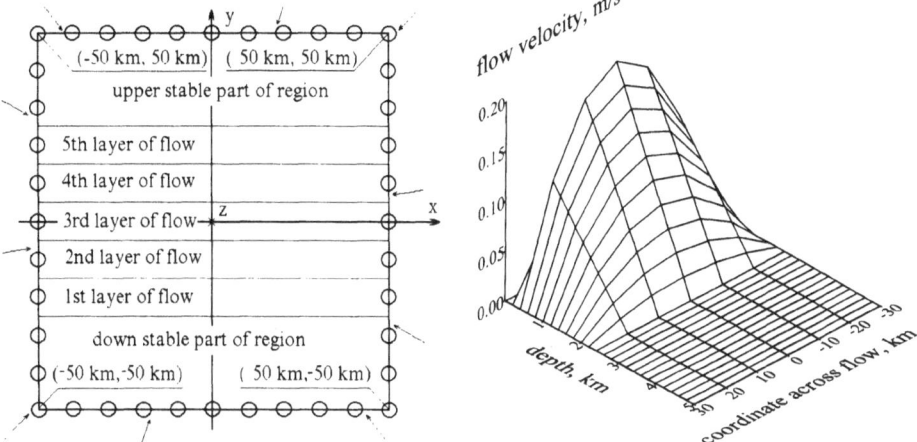

Figure 1. The model of region for computer simulation. Outside lines indicate directions of the probe waves.

Figure 2. Flow velocity distribution.

The probe wave is plane in horizontal and has mode structure in vertical direction. Such model approximately describes the situation when the investigated region is irradiated by sources which are located near remote shores. There is the set of 10 probe waves different by its direction of propagation as shown on Figure 1. To provide minimum data for the four frequency algorithm the frequencies of 8, 9, 10, 11 Hz are used to generate acoustic field. It is calculated according to adiabatic approximation of mode propagation by summing first 10 modes.

As initial approximation the hydrology of the upper (down) part of the region is used to start the iterative process of the hydrology's reconstruction.

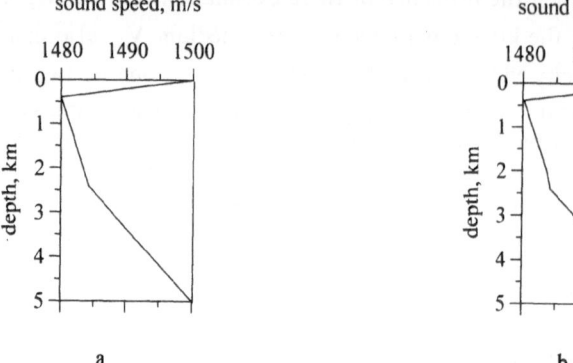

sound speed, m/s   sound speed, m/s

a   b

Figure 3. Sound speed profiles used in computer simulation: a - for upper and down parts of region, b - for flow layers.

Assuming above configuration it was sufficient to use the iterative process on base of the Newton algorithm to obtain unknown parameters. The solution after 6 iterations is equal to the original sound speed profiles. As to the flow velocity's distribution the deviation from the actual one is shown on Figure 5. Evidently, it is nearly equal to the original inhomogeneity.

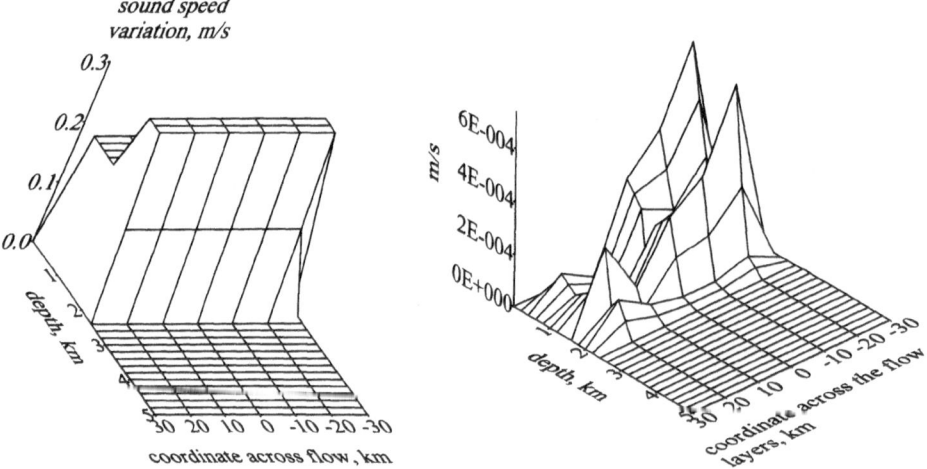

Figure 4. Sound speed inhomogeneity for computer simulation.

Figure 5. The absolute of the difference between the calculated flow velocity and the actual one.

Convergence of the iterative process is illustrated by Figure 6. It presents the norms of, first, difference between actual field and the field calculated for i-th approximation of the inhomogeneities (i is the iteration number) - the scattering field's norm, second, difference between the calculated inhomogeneity and the actual one - the residual inhomogeneity's norm and, third, the calculated inhomogeneity itself - the solution norm. The scattering field's norm is normalized by the actual field's norm, meanwhile, the others are normalized by the actual inhomogeneity's norm. It is clearly seen that the scattering field's norm decreases steadily while the residual inhomogeneity's norm "explodes" sharply and only then goes to zero. This behavior is because of the low-conditionality of the problem for such small set of frequencies.

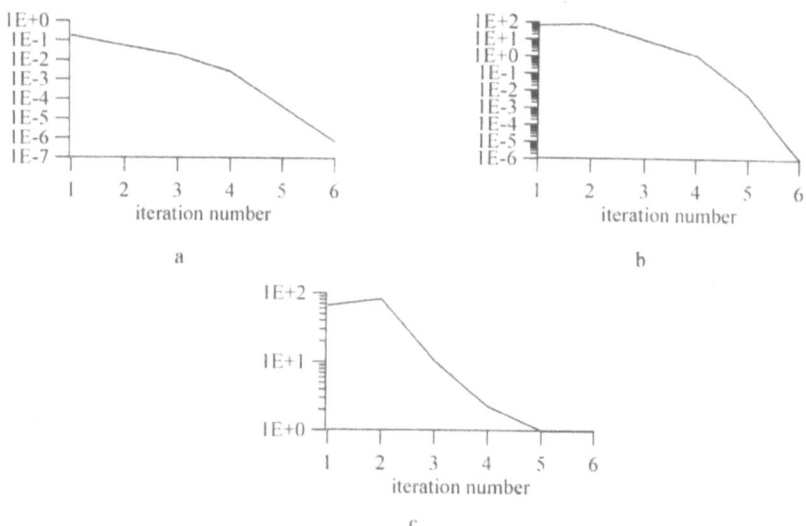

Figure 6. Convergence of the iteration process: a - scattering field's norm, b - residual solution's norm, c - solution norm.

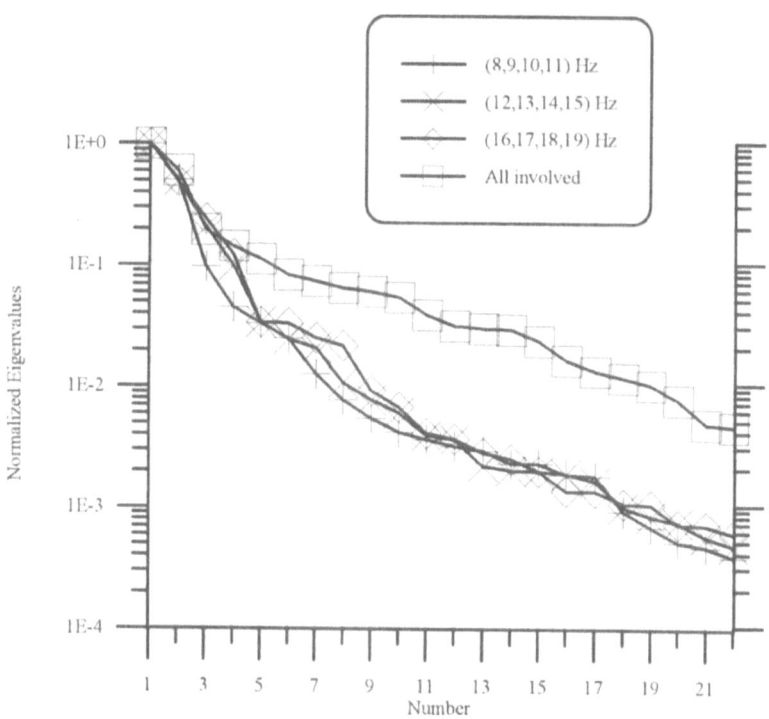

Figure 7. Eigenvalues of the matrix for first iteration and different sets of data.

It is supposed that the larger spectral band of the probe signals the more stable solution and faster convergence. An evidence of that is presented on Figure 7. It shows first 22 eigenvalues of the reduced matrixes for solving first iteration. Three of curves correspond to using one set of four frequencies' data. The fourth one corresponds to involving all three sets of data in the iterative process. From this example it is seen the conditionality of the problem with multifrequencies' data is better than with only four frequencies data.

REFERENCES

1. V.A. Burov and S.N. Sergeev, The ocean tomography as an inverse problem, in: *The formation of acoustical fields in oceanic waveguides*, IAP RAS, N.Novgorod (1995).
2. S.V. Baykov, V.A. Burov, and S.N. Sergeev, Mode tomography of moving ocean, *Proc. 3rd European Conf. on Underwater Acoust*. Heraclion-Crete. 845 (1996).
3. V.A. Burov and O.D. Rumiantseva, Influence of the scattering data redundancy on uniqueness and stability in reconstruction of strong and complicated scattering, *Acoustical Imaging*, P. Tortoli Ed., Plenum Press, New York 22:107 (1996).
4. V.A. Burov and S.N. Sergeev, Ocean tomography by vertical arrays, *Acoustical Imaging*. P. Tortoli Ed., Plenum Press, New York 22:119 (1996).

ACOUSTIC SOURCE SPECTRAL ESTIMATION
FOR A SHALLOW WATER CHANNEL

A. A. (Louis) Beex and John F. Tilki

Digital Signal Processing Research Laboratory
The Bradley Department of Electrical Engineering
VIRGINIA TECH
Blacksburg, VA 24061-0111

INTRODUCTION

We aim to find an unbiased estimate of the spectral content of an acoustic source in a shallow water channel, from measurements on an array of acoustic sensors. The acoustic source spectrum is assumed to be a mixture of discrete and continuous spectral components. Due to multipath arrival at the sensors the spectral distribution at any individual sensor is no longer simply a scaled version of the source spectrum. Consequently, spectra associated with individual sensor signals produce biased estimates of the source spectrum.

Information about the spectral content of the source is present in the sensor array signals. The signals arriving at the sensors have, however, been modified by the channel characteristics from source to sensor. We aim to equalize the channel characteristics that produced each sensor signal, to generate multiple estimates of the source spectrum. The "volatility" of the channel characteristics for a particular sensor in the array is taken into account when fusing the source spectral estimates produced by the individual sensors.

An example integrated throughout the paper shows the efficacy of this procedure, predicated on being able to obtain accurate source-sensor transfer characteristics.

PROBLEM SCENARIO

Let us focus on the problem first. Assume that we have a shallow water channel with a depth of 104 m. Assume a vertical sensor array of 34 sensors, spaced 3 m apart, and thus spanning the depth of the channel. We emphasize that in our procedure there is no need for a vertical array, equal spacing, or spanning the depth of the channel; indeed, the procedure applies for a completely arbitrary array configuration. The acoustic source is assumed to be located at a range of 1000 m and a depth of 40 m.

A Matlab version of KRAKEN[1] is used to model the littoral channel, implementing the normal mode approach. For a given sound velocity profile, a realistic one measured in the Gulf of Mexico, bottom absorption, attenuation due to surface and bottom scatter, and the given channel depth, array sensor locations, and source position the transfer functions from the source to the individual sensors can be evaluated numerically with this water - sediment - rock layer model. This is done for a range of frequencies of interest, we used 15-150 Hz, and the results stored. Given the transfer function information we then generate the appropriate complex exponential received at a given sensor, and sample this signal at an appropriate rate (our sampling frequency was 300 Hz). Thus we use the normal mode approach to generate a sensor data array for a particular discrete frequency. Assuming the channel to be linear, the generation process can be repeated for several frequencies (we used the set f=[75, 100, 125] (in Hz), with corresponding amplitudes A=[1, 1/3, 1/5] and phase angles ϕ=[0, 0, 0], to produce a periodic signal as might be expected from rotating machinery), and combined into the discrete sensor data array Xd. A sensor data array Xc, due to a continuous source spectral component, was generated using the KRAKEN source-sensor impulse responses with an autoregressive input; AR(2) with poles at 90 and 115 Hz, and pole radii at 0.95. With Xn representing the wideband sensor noise, the 100x34 sensor array data matrix X is thus given by

$$X = Xd + Xc + Xn \tag{1}$$

The discrete-signal-to-wideband-sensor-noise ratio is set to approximately 30 dB, while the continuous source component is visible above the wideband noise floor at the sensors. We used 100 snapshots of sensor array signals.

SOURCE SPECTRAL ESTIMATION PROCEDURE

To find out which discrete frequencies are present in the source, we perform column-wise spectral estimation on the data matrix X. Several methods are available to do this, ranging from the classical periodogram (DFT-based) to the modern, model-based approaches. Assuming that the relevant frequency components are spaced apart (here, 25 Hz) by much more than the Rayleigh limit of resolution (300/100 = 3 Hz here), the periodogram-based methods are nearly optimal frequency estimators for the white noise case.[2] To derive accurate frequency estimates one can use lots of zero-padding. With a quadratic least squares fit in the neighborhood of spectral peaks, some immunity to the presence of noise results.

Discrete Source Component

Note that the periodogram exhibits processing gain, thereby increasing the emphasis on discrete spectral components over wideband noise as more data (before zero-padding) is used. Each of the column-wise spectral density (or power) estimates reflects the discrete source components, as modified by the source-sensor transfer functions. Consequently, the relative amplitudes and phase angles of the discrete spectral components have changed. Indeed, due to a null, at a particular frequency, in a particular source-sensor transfer function, one or more of the discrete spectral components may not be present at all. However, it is unlikely that such a null occurs for all of the sensors in the array. Information from all column-wise spectral estimates must be fused, to detect that set of discrete frequencies most likely to have occurred at all sensors. To illustrate, we simply use the sum of all the column-wise spectral estimates and then find the peak location frequencies. This results for our example, using a Blackman window and zero-padding to 8192 points, in the frequency estimates f=[75.008, 99.977, 124.88]. Note that the estimation error is generally larger for the weaker discrete frequency components.

Source Localization

Knowing a discrete frequency estimate, we can use KRAKEN to generate a database of complex (amplitude/phase) sensor information, $\underline{D}(d,r)$, for varying source depth and range, given the sound velocity profile and sensor array configuration. Narrowband filtering of the array data at the individual discrete frequency estimates, generates a number of filtered array data matrices (one for each discrete frequency) or, after Fourier transformation or spectral modeling, the equivalent row-wise spectral sensor measurements, $\underline{Y}(f)$.

A multitude of matched field processors (MFP) is available to compare the measurement $\underline{Y}(f)$ with the database $\underline{D}(d,r)$ to yield source depth and range estimates which correspond best with the received sensor data[3]. The results for each of the discrete frequencies can be fused, to produce an improved estimate.

Our present aim is to show that bias-free estimation of source spectra is feasible. Therefore, we assume that the source can be localized accurately via MFP, meaning availability of a source depth estimate $d=40$ m, and a source range estimate $r=1000$ m.

Discrete Source Amplitude/Phase Estimation

If we let $\underline{s}$ represent the sensor array geometry, we can now use KRAKEN to generate the array $H(f,\underline{s},d,r)$ of sensor-measured responses to a unit amplitude complex exponential source. The latter allows us to apply its inverse to the spectral sensor measurements $\underline{Y}(f)$, thereby equalizing the effects of the source-sensor transfer functions. This yields, for each of the sensors, an estimated source amplitude and phase angle for each of the discrete frequency components, as shown in Figure 1. The variability of these estimates comes from 3 sources: sensor measurement noise, differences in source-sensor transfer characteristics, and the presence of the continuous spectral component in the source. For the sake of simplicity, we fuse all estimates into their weighted average, where the weighting is proportional to the absolute value of the source-sensor transfer function value at the given frequency. The latter is aimed at countering the noise amplification effects from inverting small transfer function values. This fusing process results in source amplitude and phase angle estimates of $\underline{A}=[1.024, 0.3391, 0.1952]$ and $\underline{\phi}=[-0.006499, -0.0006275, -0.001556]$ (in π radians) respectively.

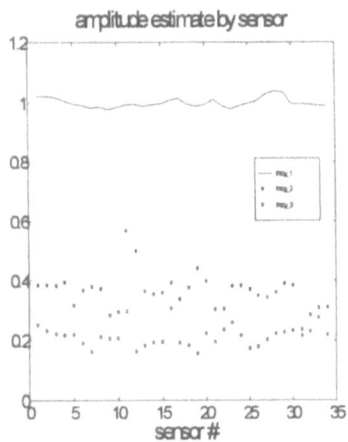 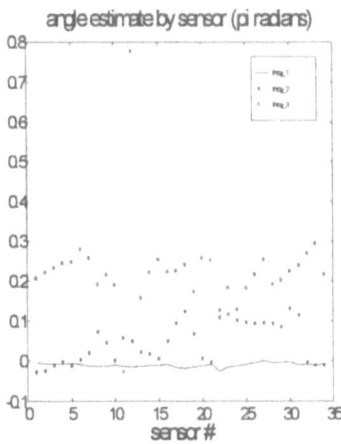

Figure 1 Discrete Amplitude Estimates (L) and Discrete Phase Angle Estimates (R).

Source Inference and Quality Control

On the basis of the source amplitude and phase angle estimates we now reconstruct or estimate the discrete source component. Inference on the source can then be obtained, for example by template matching, audio identification, or other pattern recognition procedures.

The source estimates, together with the earlier obtained source-sensor transfer function information $H(f,s,d,r)$, can be used to reconstruct or estimate Xd. Now X-Xd ($=Xc+Xn$), the continuous sensor residual, can provide an idea of how well the discrete source component was captured. When localization is in error, or when discrete frequency estimates are off, be it in frequency, amplitude, or phase angle, the continuous sensor residual is less in tune with sensor noise statistics. This quality control process can also be done spectrally, taking advantage of the processing gain for discrete components. In any case, reconstruction helps to ascertain the integrity of the many processing steps.

Continuous Source Component

So far we have seen that accurate results can be obtained for discrete source components, given conducive processing and environment. We therefore continue on the basis of the continuous sensor residual estimate.

As samples from an arbitrary signal can be represented as a linear combination of discrete frequency components, via the DFT frequencies, one might be tempted to apply the above procedure for discrete frequencies to the continuous spectral component. However, this method does not readily extend to the case of a continuous spectral component. This is seen when we examine what happens when we have a single discrete frequency component which cannot be captured by a single DFT frequency. As a result of spectral leakage the power in this single frequency component will be distributed over all, or most, of the uniformly spaced DFT frequencies. As the source-sensor transfer functions are different for each of the DFT frequencies, the DFT frequency components are multiplied by different complex constants corresponding to the transfer function values at those frequencies. The actual single frequency component, however, would undergo a change by a single complex constant, corresponding to its source-sensor transfer function value. The decomposition corresponding to the representation components that were subject to the different transfer functions no longer corresponds to the representation for the original discrete component that was subject to a single transfer function value! Consequently, a decomposition type representation can not be used here.

Assuming that the sensor signals are bandpass filtered, so that no DC component is measured and aliasing is small enough after sampling, the continuous sensor signal satisfies:

$$y(t) = \int H(\omega) X(\omega) e^{j\omega t} d\omega \qquad (2)$$

where $H(\omega)$ represents the source-sensor transfer function, and $X(\omega)$ represents the Fourier transform of the source signal. While in its Riemann sum approximation the integral can be thought of as a linear combination of discrete sinusoids, this will only approximate the signal $y(t)$ well if the discretization in the frequency variable ω is fine enough. The latter means that the product $H(\omega)X(\omega)$ is nearly constant over the discretization interval. While $X(\omega)$ is fairly smooth over frequency, $H(\omega)$ for the underwater acoustic problem is not. This can be seen by finding $H(\omega)$ over the sensor bandpass filter's frequency range of interest. Figure 2 shows the source-sensor magnitude response for sensor #17, as evaluated by KRAKEN at the 2048 DFT frequency samples. Note the resonant volatility of $H(\omega)$, which makes it more difficult than in the discrete source frequency case to apply the corresponding inverse, or equalizer, to the sampled measured sensor signals. In principle, $H(\omega)$, inclusive of our bandpass filter, corresponds to an impulse response. The latter can be approximated by discretization, resulting in the unit pulse response $h(n)$. The sampled version of (2) then expresses the received sensor signal samples $y(n)$ as the convolution of the unit pulse response and the source signal samples $x(n)$. An inverse DFT of the 2048 DFT frequency samples for $H(\omega)$ yields the approximate (aliased) impulse response shown in Figure 2.

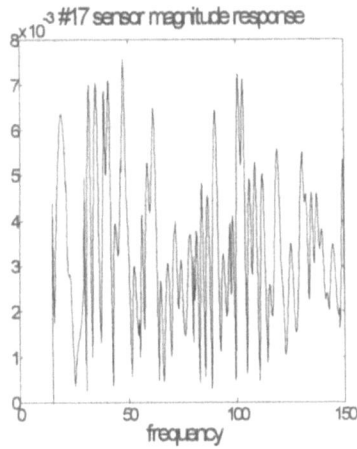 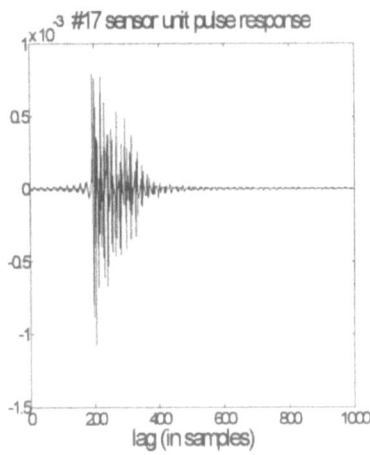

Figure 2 Source-Sensor#17 Characteristics: Magnitude Response (L) and Unit Pulse Response (R).

For a sound velocity of 1500 m/s, it takes 2/3 of a second for sound to travel 1000 meters. At a sampling frequency of 300 Hz it takes 200 samples to travel 1000 meters. The source-sensor unit pulse response must reflect this 200 sample interval, as seen in Figure 2. If too few DFT frequency samples are taken severe aliasing results, reflected in the lack of such necessary physical attributes. In the sequel 1024 point DFT results were used.

As before, the application of spectral estimation techniques to the sensor signals does not reflect the source spectrum directly, because the source signal has been subject to the source-sensor transfer functions. Knowing the accuracy of the unit pulse response, the corresponding DFT can simply be "inverted" on a frequency sample-by-sample basis. As the measured sensor signals are of a bandpass type, the DFT values in the stopbands are considered zero, and so are their "inverses." An IDFT of the "inverse" DFT frequency samples then yields a filter that, when cascaded with the unit pulse response filter, yields an equalized response on the order of 1 over the range of passband frequencies.

Improvements in the inverse filter design process, over the frequency sampling design approach taken here, are certainly possible. However, while the result is by no means perfect on a frequency-by-frequency basis, it goes a long way towards equalizing the response for a stochastic process that is broadband relative to the local perturbations in such an equalizer.

The modified covariance method of linear prediction[2] is used on each of the equalized sensor signals to yield a corresponding AR power spectral density estimate. The gain of the AR model is normalized by matching its autocorrelation at lag zero to the variance of the observed sequence. The order of the AR model used was 3, i.e. one higher than the underlying AR component, so that the extra pole could model the (equalized) wideband noise, leaving the other two poles to model the spectral energy at 90 and 115 Hz. A weighted average, according to the energy in the corresponding unit pulse response, fuses the AR power spectral densities, the AR parameters, or the reflection coefficients from all sensors. Figure 3 shows the source power spectral density (heavy line), differently averaged AR estimators (thin smooth lines), and the averaged periodogram (noisy line), all applied to the equalized estimated continuous sensor residual. Comparing the actual AR source spectral density and the continuous residual spectral estimates indicates that our procedure estimates the major continuous spectral features reasonably well in the present noisy spectrally mixed scenario, while the effects of the "equalized" measurement noise on the continuous residual spectral estimates are not generally negligible.

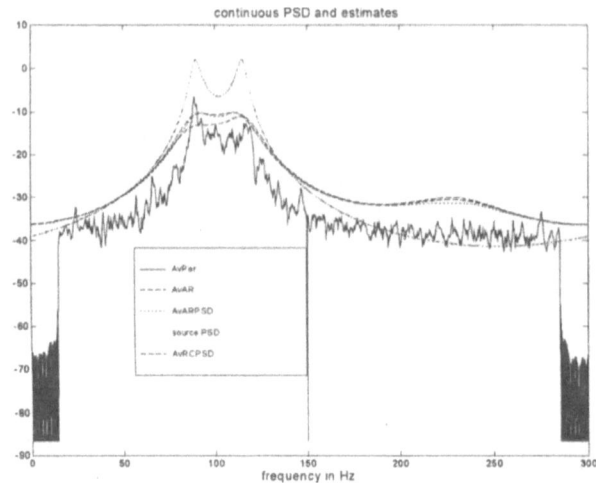

Figure 3 Continuous Sensor Residual Spectral Estimates (100 snapshots).

Without measurement noise, the overall AR spectral density estimate improves noticeably around the spectral peaks. When the number of snapshots is increased the spectral estimates behave as expected, and move closer to the actual source spectral density. Therefore, no systematic bias is apparent.

CONCLUSION

A sensor-equalization approach was demonstrated, based on arbitrary geometry sensor array measurements, for estimating the discrete and continuous spectral components assumed to exist in a source spectrum. The accuracy of the estimates depends on the sensitivity of the frequency functions being sampled, over which control can be exercised, the number of snapshots of available data, the relative strengths of the continuous and discrete source components, and the strength of the measurement noise. Fusing of the results associated with individual sensors reduces estimation variance, because each sensor signal reflects the same source spectrum, while being corrupted by different sensor noise.

REFERENCES

1. J. Ianniello, A Matlab version of the KRAKEN normal mode code, TM 94-1096, NUWC/NL, 10/3/94.
2. S. M. Kay, *Modern Spectral Estimation: theory & application*, Prentice-Hall, 1988.
3. A. Tolstoy, *Matched Field Processing for Underwater Acoustics*, World Scientific, 1993.

THE DEVELOPMENT OF A VERY LARGE, DENSE PACKED, TWO-DIMENSIONAL, ULTRASONIC IMAGING ARRAY

Kim C. Benjamin

Underwater Sound Reference Detachment
1176 Howell Street Building 1171A
Newport, Rhode Island 02841

ABSTRACT

The design, construction, and measured results for a prototype module of probably the world's largest dense packed, two dimensional, ultrasonic (1.5 MHz), receive array is described. Containing over a quarter million diced elements, and a centrally located, constant beam width, spherical cap projector, the complete array was to consist of 25 modules, of which the central module is described. The individual receive element dimensions were (.91 x .91 x .89) mm, with a center spacing of 1.07 mm. The ultrasonic spherical cap projector element had a 29 mm radius of curvature and an active angular aperture of 40 degrees. Both projector and receive elements were designed to achieve a minimum 3dB beam width of 30 degrees at 1.5 MHz. In both cases the active material was piezoceramic. Various material trade-offs associated with the selection of the: 1) active layer, 2) backing layer, 3) mechanical, and electrical connections are discussed. Also the issues of dicing and backfilling of the inter-element interstices will be addressed. Measured acoustic calibration results for a small subset of receive elements and the projector are presented.

SONAR SYSTEM OVERVIEW

Figure 1 provides an overview of the full sonar imaging system which relies on single-ping 3-D volumetric image processing. The receive elements' outputs are combined electronically to form 12,830 individual beams. The entire receive aperture of 60 cm (24 in), contains 280,900 individual diced PZT elements, of which 1728 elements are actually used in a sparse distribution. Figure 2 shows some of the sparse element distributions which were considered during the systems development. Although the entire system was not built, two of the 25 modules were constructed and calibrated in order to demonstrate the feasibility of the array fabrication technology. Figure 3 shows the center module of the final array design, which was modified to simulate a lower resolution system configuration. This module is 12 cm (4.7 in) square and contains the centrally located 29 mm (1.14 in) diameter ultrasonic projector. The receiver portion of this module consists of 128 active hydrophone elements arranged in 16 rings with 8 elements per ring.

The 128 active elements were selected from among 11,236 PZT-5H ceramic elements which remained after dicing a 12 x 12 cm (4.7 x 4.7 in) ceramic plate. The element dimensions were .91 mm (0.036 in) square x .89 mm (0.035 in) thick and were mounted to a common tungsten/epoxy substrate. Laser drilled holes filled with silver loaded epoxy provided the electrical pathways for the positive connections.

The 128 active elements and 128 additional unused spares can be seen in Figure 3. The locations

- SINGLE- PING 3-D VOLUMETRIC IMAGE PROCESSING
- RECEIVE ELEMENTS' OUTPUTS COMBINED ELECTRONICALLY TO FORM 12,830 INDIVIDUAL BEAMS
- FREQUENCY: 1.5 MHz
- CROSS-RANGE RESOLUTION: (0.75 - 2.00) cm
- DOWN RANGE RESOLUTION: 0.5 cm
- RECEIVE ARRAY APERTURE: 60. cm
- 280,900 INDIVIDUAL DICED PZT ELEMENTS
- 1,728 ELEMENTS USED IN SPARSE DISTRIBUTION

Figure 1   Sonar System Overview

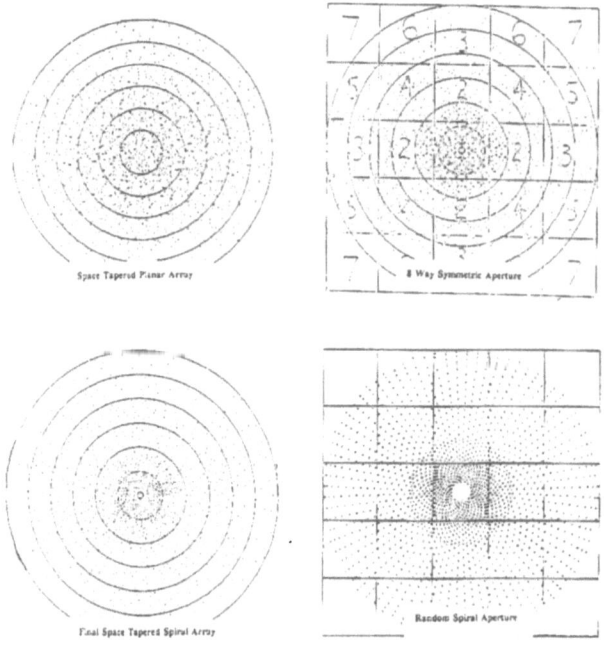

Figure 2   Comparison of Various Sparse Element Distribution

of these elements have been highlighted in the figure by a film overlay positioned over the array surface.

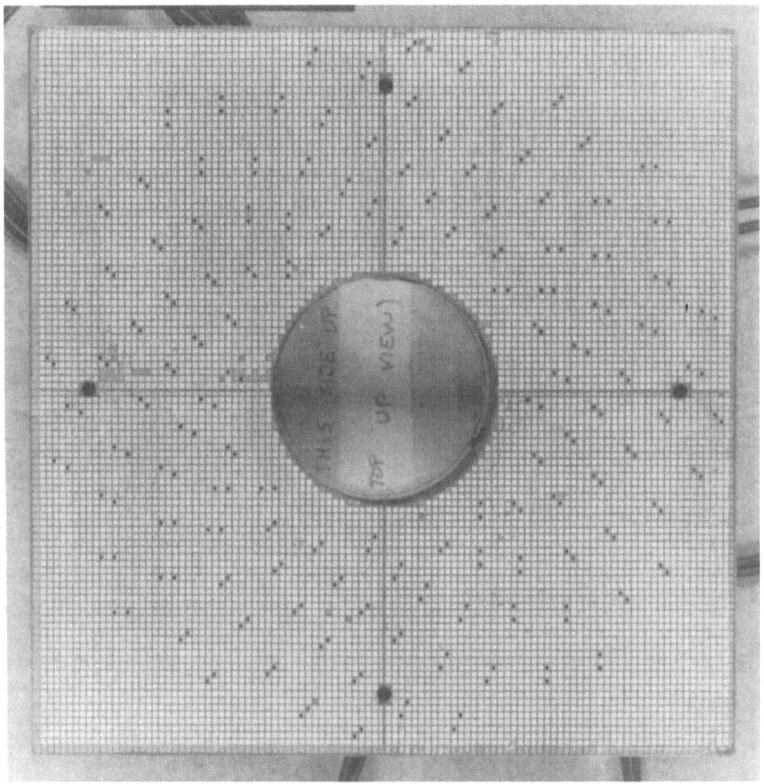

Figure 3   Active Elements and Unused Spares

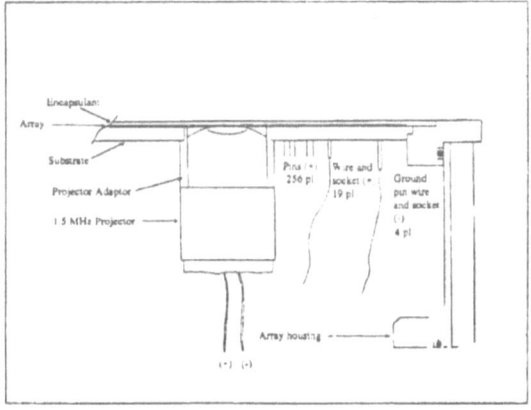

Figure 4   Module Assembly

The ultrasonic projector consisted of a 29 mm (1.14 in) diameter PZT-8 ceramic spherical cap mounted to the same backing substrate material used for the hydrophone array. It is supported within the module with the projector adapter as show in Figure 4.

To test the receiver portion of the array module, 16 of the 128 active hydrophone elements were connected to preamplifiers located in an attached housing positioned in close proximity to the module elements. Approximately 10 inches of very low capacitance wire was used to electrically connect each hydrophone/preamplifier assembly. The low cable capacitance was necessary to minimize the insertion loss associated with the low capacitance (28 pF) hydrophone elements. The gain of each preamplifier was 46 dB.

The mechanical dimensions and measured acoustic performance of the individual hydrophone elements and the ultrasonic projector used in the central array module are summarized in Table I. A photograph of the completed center module attached to the test housing is show in Figure 5.

Table I. Performance Parameters

Receive Array

Element dimensions	(.91 x .91 x .89)mm	(0.036 x 0.036 x 0.035)"
Element sensitivity	-215 dB re uPa.	
Element capacitance	28 pF.	
Nominal beam width	30 degrees	
Operating frequency	1.5 MHz.	

Projector

Spherical cap radius	29 mm	(1.142")
Element thickness	1.55 mm	(0.061")
Element diameter	20 mm	(0.787")
Element angular aperture	40 degrees	
Element beam width	24 degrees	
Element TVR	153 dB re uPa//m.	
Maximum SPL at 5 V/mil.	202.5 dB re uPa//m.	
Operating Frequency	1.5 MHz	
Mechanical Q	<3.0	

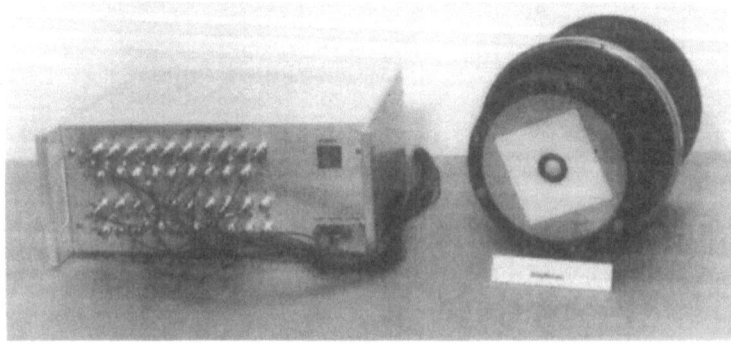

Figure 5   Assembled Central Array Module

Figure 6 shows measured hydrophone beam patterns at 1.5 MHz for 16 randomly selected array module elements. Measured receiving voltage sensitivity curves over the frequency band 800 kHz to 1.8 MHz for these same elements are also shown in Figure 6. The uniformity of the individual element performance shown by these measurements demonstrates that good process control was maintained during the module assembly.

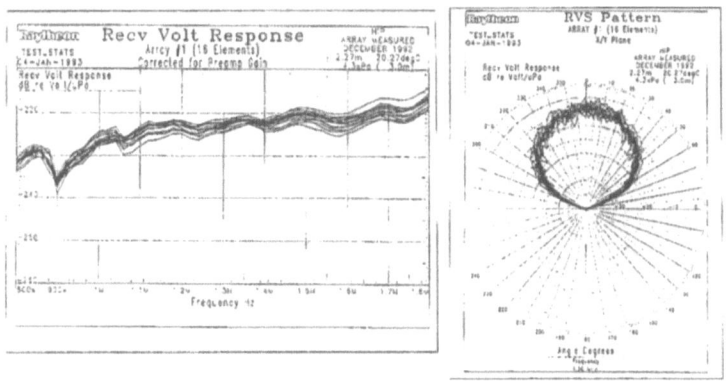

Figure 6   Measured Hydrophone Beam Patterns and Measured Receiving Voltage Sensitivity for 16 Elements

TRANSDUCER ARRAY FABRICATION

Several prototype array configurations were initially considered and resulted in various design trade-offs. The selected approach, shown in Figure 7 consisted of a thin layer of ceramic material cemented to a tungsten/epoxy backing substrate. The ceramic was then diced to form an array of elements of the desired geometry. The remaining interelement grooves were then back filled with a non-conducting soft elastomeric material to electrically and mechanically isolate the individual array elements. A common ground plane was then deposited over the top surface of the array. Silver epoxy filled, laser drilled holes in the backing substrate provided the individual positive connections for each active array element.

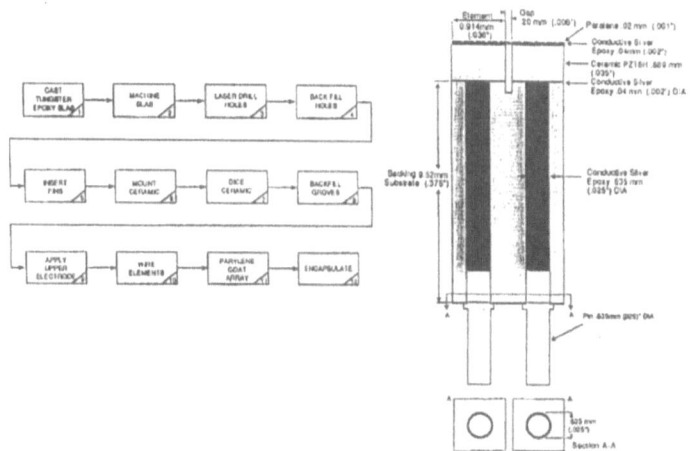

Figure 7   Final Prototype Fabrication Process

Considerable effort was spent developing the backing substrate configuration as it represented a critical array component. The backing had to provide a rigid support as well as mechanical and electrical isolation between individual elements. The tungsten/epoxy composite was moldable, machinable afforded high attenuation at ultrasonic frequencies, and was structurally rugged. The backing consisted of a tungsten powder, 70 um diameter (+325 mesh) and epoxy mixture. A mixture of Armstrong A2 structural adhesive and tungsten powder (50% by volume) was selected. A set of 256 .63 mm (.025 in) diameter laser drilled holes were machined in the central module backing.

Each hydrophone element within the array must have 2 electrical connections. The positive connection on the backside of each individual element had to be unique whereas the negative connection could be common to all element. Since the holes drilled in the back layer (used for making electrical connections to each element) represented a major portion of the area behind each element, the material used to back fill the holes was selected to have mechanical properties very similar to, the surrounding backing material. Each hole was back filled with conductive silver epoxy (TRACON Tra-Duct 2902), and a pin was inserted to the rear side of the substrate before the epoxy cured. Excess silver epoxy residue between the pins is removed using a micro-sand blaster. The front face of the back filled backing is machined flat before the ceramic is mounted. Four thin PZT plates were cemented to the backing using the same conductive silver epoxy used to back fill the holes in the backing layer. This adhesive was silk screened onto the tungsten/epoxy backing as well as to the ceramic plate. A 6.9 to 13.8 kPa (1-2 psi) compressional pressure was used to bond the plates. After a 72-hour room temperature cure the assembly was ready to dice. Four ceramic plates were bonded simultaneously to the backing. Figure 8 shows a completely diced module. Dicing of the ceramic was accomplished using a 152 micron (0.006 in) thick diamond wheel. The 8 tabs around the outside are used for alignment and are machined off in a non-critical machining process. The back filling of the inter-element grooves with a lossy polyurethane was accomplished under vacuum in order to avoid the random dispersion of tiny air bubbles.

Figure 8   Completely Diced Module

PROJECTOR FABRICATION

The projector configuration was selected for spatial response, power handling considerations, and ease of fabrication. The required 30 degree wide beam at the frequency of interest would require a very thin ceramic having a very small aperture, and the associated volume could not support the input power densities required for the system. Using a curved, spherical PZT cap, a constant beam width projector element was simply constructed as shown in Figure 9. A ceramic spherical cap was cut to size for the required beam width, wire leads were attached, and the element was placed into the simple mold shown. A mixture of tungsten powder and epoxy was poured into the mold and cured. The epoxy phase of the mixture forms an intimate bond with the ceramic element. The larger curved element with its greater volume allows sufficient input power densities and provides a wide constant beam width above a nominal cut-off frequency. The transmit voltage response, and transmit spatial response, are also shown in Figure 9. The ripple in the main lobe of the beam pattern is a result of the edge discontinuity in the velocity distribution and may be minimized using area amplitude shading.

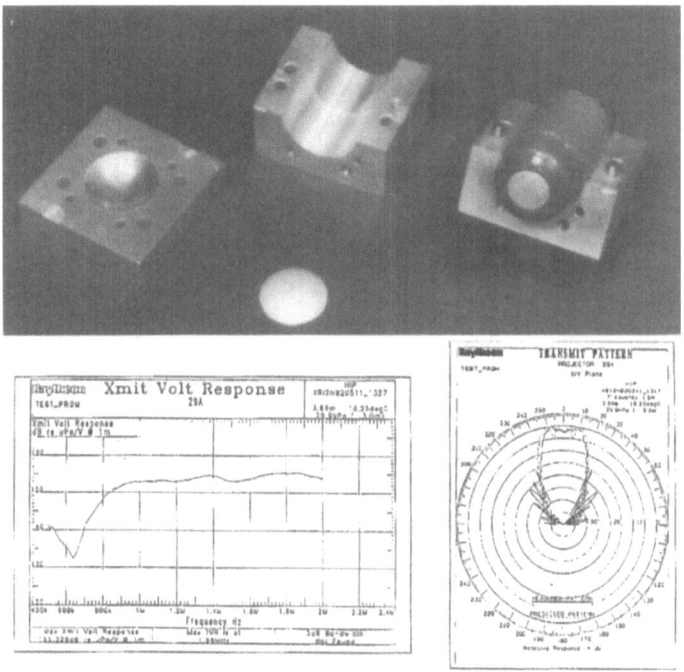

Figure 9   Projector Fabrication Tooling and Measured Projector Performance

ACKNOWLEDGEMENTS

The author would like to acknowledge the valuable assistance of the following individuals as well as the Raytheon Company for the support of this project.

Felix Petine
Pat Brogan
Frank Liotine
Marty Ring
Mike Janik
Fred Dutton

A 128 X 128 (16k) ULTRASONIC TRANSDUCER HYBRID ARRAY

Ken Erikson, Allen Hairston, Anthony Nicoli, Jason Stockwell & Tim White

Lockheed Martin IR Imaging Systems
Lexington, MA 02173

INTRODUCTION

Ultrasonic imaging in the low MHz frequency range with large two dimensional arrays presents many challenges in design and fabrication. In this paper, a 128 x 128 (16,384 total) element receiver array, consisting of a 1-3 composite piezoelectric transducer array bonded directly to large custom integrated circuits is described. This Transducer Hybrid Array (THA) is intended for use in a real-time 3D imaging system or acoustical camera (Fig. 1) for medical and underwater applications.

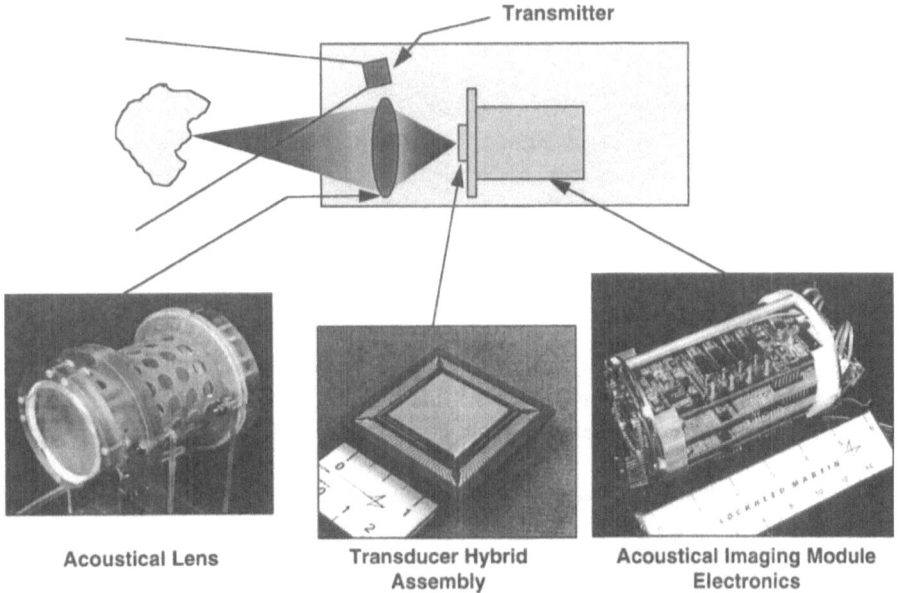

Acoustical Lens | Transducer Hybrid Assembly | Acoustical Imaging Module Electronics

Figure 1. Real-time 3D Acoustical Camera using the Transducer Hybrid Array.

Traditional ultrasonic imaging systems, whether for medical or underwater use, are also operated in a monostatic mode, i.e., the same array element is used for both transmission and reception. This poses practical difficulties in fabricating an integrated circuit with both the high density required for massively parallel on-chip signal processing and the high voltage capabilities required for adequate transmitted power. The bistatic mode, where the transmitter is separated from the receiver was selected for the array discussed here. Although the THA was first demonstrated[1] over 20 years ago, advances in microelectronics technology have now made such a device practical.

Conventional ultrasound systems use micro-coaxial cable to connect the array to the front end electronics. Although micro-coax technology has improved dramatically in the past decade, interconnecting 16,384 array elements with separate wires remains a formidable challenge. In addition to this practical fabrication issue, the capacitance of a long coaxial cable (typically 40 pF/m) is much larger than that of a typical 2D array element (< 1pF). This creates a voltage divider that severely reduces the signal-to-noise ratio of the channel. The direct connection method used in the THA (Fig. 2) reduces the interconnect length to less than 0.1 mm, reducing interconnection capacitance to the level where it is no longer a dominant factor in the channel signal-to-noise ratio.

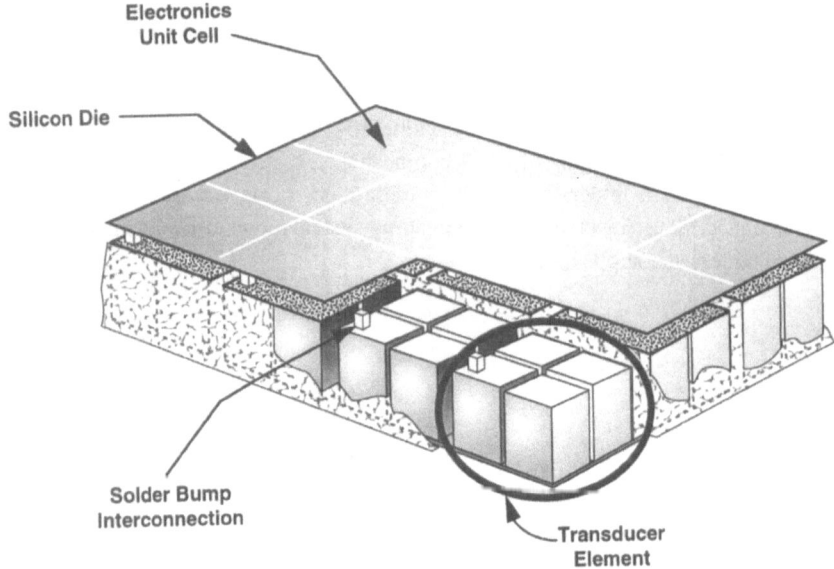

Figure 2. Cross-section of Transducer Hybrid Array showing the interconnection method.

Finally, real-time 3D imaging imposes additional severe requirements on an acoustic array and signal processing system. Due to the finite velocity of sound, ranges greater than a few mm in tissue or water require either multiple parallel beam forming in a B-scan system or C-scan parallel data acquisition.

Conventional B-scan systems acquire data one line at a time, typically waiting until well after information from the furthest range has been received before transmitting the next pulse. With single plane imaging, this is not a serious limitation. Indeed, real-time imaging is one of the key strengths of present day medical ultrasound.

3D Ultrasound (3DUS) has been gaining increasing acceptance in medical and underwater imaging applications due to the improved image understanding it provides and to the additional applications it enables. 3DUS systems have been developed using conventional B-scan probes mechanically scanned in the third dimension. The B-scan images are subsequently assembled into 3D volume images off-line. To scan out a volume with comparable resolution in the third dimension, however, results in very low frame rates. For example, to scan an 80 mm cubic volume centered at 120 mm with 1 mm^3 resolution requires 6,400 separate pulses, which in turn requires well over one second for data acquisition under the most favorable assumptions.

Various strategies[2] have been devised to improve the frame rate of B-scan ultrasound systems. Improvements of a factor of 3 or 4 may be achieved; however, this is still far from real-time. Another successful approach[3] uses a wide insonification beam together with multiple parallel receive beamformers to present a somewhat lower resolution image at real-time rates sufficient for cardiac imaging.

A C-scan format (Fig. 3) was chosen for the present system to enable real-time 3D imaging.

figure 3

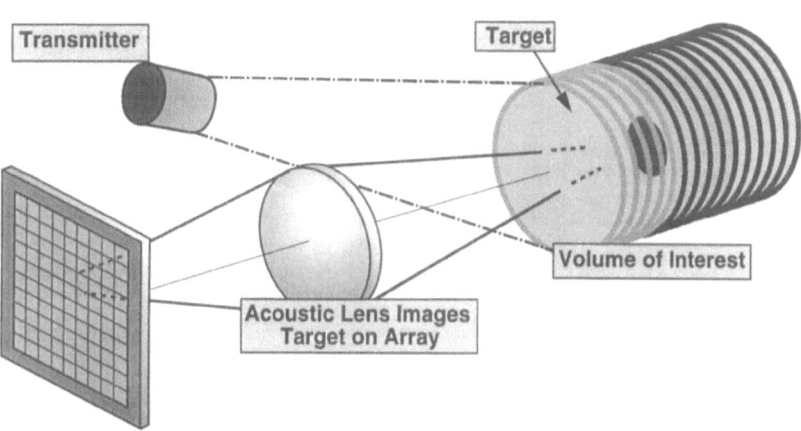

Figure 3. C-Scan "Plane at a Time" imaging enables real-time 3D operation.

With this C-scan, bistatic system a separate pulsed transmitter insonifies the entire volume of the target. An acoustical lens images successive planes in the target onto the acoustical array. As shown in Fig. 3, five range planes can be recorded from a single pulse of sound. By simply delaying the range gate timing on the next pulse, the next five planes in the target are captured, subject to depth of field limitations of the acoustical lens.

The acoustical frame rates enabled by this system are very high. For example, to record data from an 80 mm cubic volume centered at 120 mm range with a 1 mm^3 resolution results in an acoustic frame rate of approximately 300 Hz. This is more than enough to support a 30 Hz real-time display, so the extra frame rate may be applied to other signal processing techniques such as frame averaging for noise reduction.

TRANSDUCER HYBRID ARRAY (THA) REQUIREMENTS

The THA is the heart of the real-time 3D system in development. Table 1 lists the performance requirements for the device and are derived from the system performance goals.

Table 1. Performance Requirements for the Transducer Hybrid Array (THA)

Number of elements:	Medical - 128 x 128 (rectangular matrix)
	Underwater - 64 x 64
Element size:	$< \lambda \times \lambda$ square (λ = wavelength in water)
Aperture:	25 mm x 25 mm minimum
Ultrasound frame rate:	≥ 200 Hz maximum
(Supports frame averaging for a 15-30 Hz display update rate)	
Acoustic performance:	
Center frequency:	Medical - 5 MHz
	Underwater - 2.5 - 3.5 MHz
Fractional bandwidth:	> 40%
Sensitivity:	> -200 dB re 1V/μPa
Crosstalk:	< - 30 dB
On-chip signal processing:	
a. Amplifier overvoltage protection	
b. Gain: x2, x20, x200	
c. Maximum Sampling frequency: ≥ 20 Megasample/sec	
d. Type of sampling:	
(1). Sample and hold (20 data points)	
(2). Quadrature detection:	
Supports sequential acquisition of 5 planes	
(range gates) per acoustic pulse.	
Electronic Performance:	
Frequency response:	Bandpass, 150 kHz to > 8 MHz.
Instantaneous dynamic range:	> 60 dB
Noise level	< 50 nV/Hz$^{1/2}$ (referred to input)
Crosstalk:	< - 50 dB
Fixed pattern noise:	< 0.5 dB standard deviation re average level
(Due to pixel to pixel electronic gain and offset variations)	
Total harmonic distortion:	-60 dB @ 2 V output swing
Power:	< 1 watt total
Size (including package and leads):	< 50 mm x 50 mm x 10 mm

ARRAY DESIGN

A single square of 1-3 composite piezoelectric material is used for the ultrasound array[4]. For the 5 MHz medical imaging application, a 0.2 x 0.2 mm center-to-center spacing is used together with a single matching layer. The underwater imaging application uses 2.5 to 3.5 MHz elements on 0.4 mm centers. The elements were delineated by sawing through one electrode layer. A common ground layer was used on the matching layer side. Figure 4 is a photograph of a typical 64 x 64 element array and Fig. 5, a single element. The elements are typically 45% piezoceramic volume fraction and have a capacitance of several hundred femtofarads (fF).

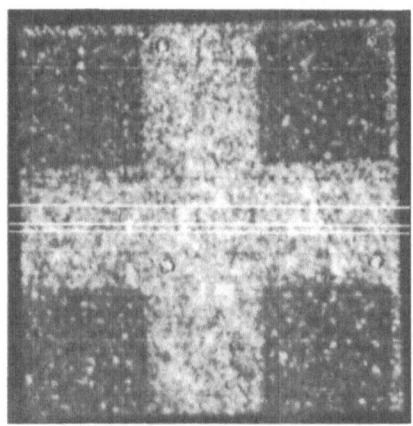

Figure 4. 2D Transducer Array[2]: 25.6x25.6 mm.

Figure 5. Typical 2.5 MHz array element: 45% volume fraction on 0.4 mm centers.

INTERCONNECTION

The interconnection method, a variation of flip-chip bonding is shown in Fig. 2. Solder bumps are deposited on both the array and the silicon integrated circuit. During hybridization the bumps are optically aligned and brought into contact to make an electrical connection. The contact area of the bumps on the array is approximately 20 x 20 microns, which provides adequate mechanical integrity, good electrical contact and has a minimal (but not negligible) effect on the acoustics of this essentially air-backed transducer design.

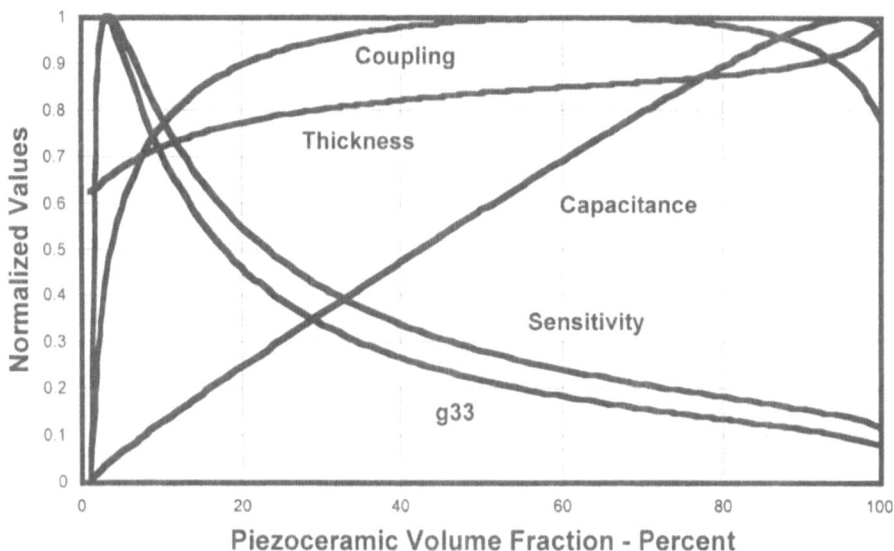

Figure 6. Normalized 1-3 Piezoelectric Composite Parameters *at Constant Resonance Frequency.*

With this interconnection method, the input capacitance of the electronics is less than 100 fF. This provides a unique opportunity to optimize the composite piezoelectric material that is generally not available to conventional ultrasound systems.

Figure 6 plots piezoelectric coupling coefficient, capacitance and sensitivity at *constant resonance* frequency as a function of piezoceramic volume fraction[5]. By reducing the volume fraction of ceramic, sensitivity can be increased at the expense of lower capacitance while maintaining high coupling.

Figure 7 shows the frequency response of a typical element of a 16x16 test array. Sensitivity is excellent and the bandwidth acceptable.

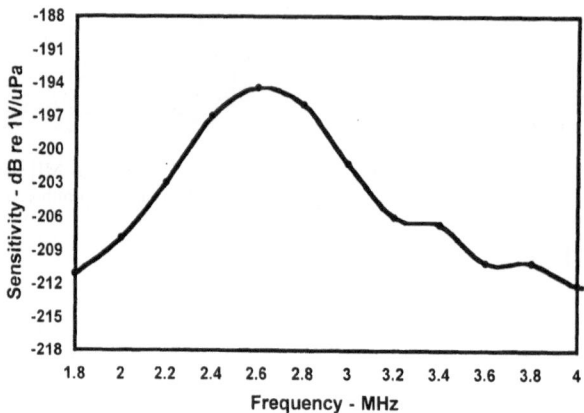

Figure 7. Frequency response of typical array element without an optimized matching layer. The sensitivity was measured with a calibrated PVDF membrane hydrophone. The measured -6 dB bandwidth is 700 kHz.

READOUT INTEGRATED CIRCUIT (ROIC) DESIGN

From a silicon foundry perspective, matching the silicon unit cell to the array dimensions would result in very large die. A four-chip hybrid was selected to keep the IC size reasonable. To avoid a row of dead pixels in the center of the array, the silicon is designed to be butted along two adjacent edges, with all bond pads on the other two edges. The resulting die is 15.5 mm x 15.5 mm and contains 64 x 64 (4,096 total) unit cells together with multiplexing circuits for readout. There are about 1 million active devices per die.

Figure 8 is a schematic diagram of the ROIC. In each unit cell, amplification, filtering, analog sampling and storage of the output of all the transducer elements in the array occurs simultaneously (in parallel) during data acquisition periods (range gates) which are typically a few microseconds long. After data acquisition, the stored data are read out through four 10 MHz pixel rate outputs.

Two modes of operation are possible: sample-and-hold and quadrature sampling. In the sample-and-hold mode, the received waveform is simply sampled at a 20 Megasample/sec maximum rate when the range gate is activated. In the quadrature sampling mode, the waveform is sampled at a rate 4 times the acoustic center frequency (20 Megasamples/s maximum) and stored. Through this well known technique[6], the in-phase and quadrature components of all 16,384 waveforms are simultaneously recorded at 5 separate range gate times. Following data acquisition, the stored data are read out to the system through four outputs per ROIC for a total of 16 readout channels.

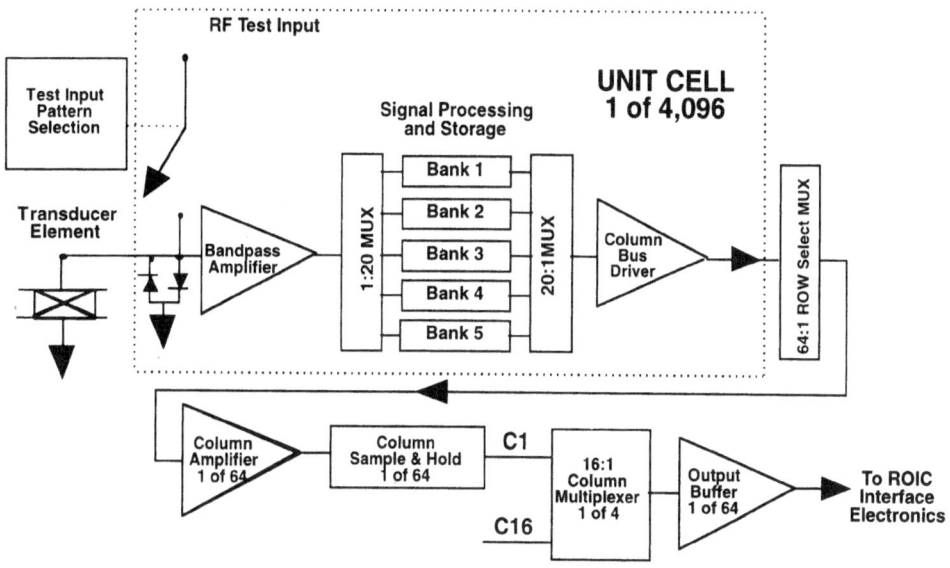

Figure 8. Functional Block Diagram of the ROIC.

INITIAL MEASURED PERFORMANCE

A 16x16 element 3.1 MHz test array was hybridized to a single ROIC for initial testing (Fig. 9). The average measured sensitivity was -188 (±6) dB re 1V/μPA with a broadband electronic noise level of 84 μV. Using relatively uniform direct insonification, typical responses were measured as shown in Figs. 10-13. In these figures, the voltage response of each element is plotted as a surface over a 2D grid representing the element location. The lower response in the corners of the array was found to be caused by a matching layer defect. In addition, for this initial test a less than perfect ROIC was used, resulting in some non-functional rows and columns which are omitted from the plots.

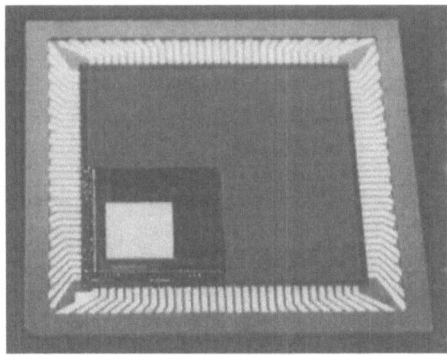

Figure 9. A 16x16 test array hybridized to a single ROIC for initial testing.

With a 60 μs long gated sine pulse, activity may be seen in each of the 5 planes (Fig. 10). Figure 11 demonstrates the range resolution of the THA by placing a short acoustical pulse in one range plane. Without any acoustical input (Fig. 12), the "fixed pattern" of the THA may be observed. Since these gain and offset variations are constant for each element of the array, they may be compensated for later in the system electronics. Finally, the system noise level is demonstrated in Fig. 13, where the results of subtracting data from the same plane in two successive frames is presented. This is equivalent to having corrected the fixed pattern variations, leaving only the random electronic noise.

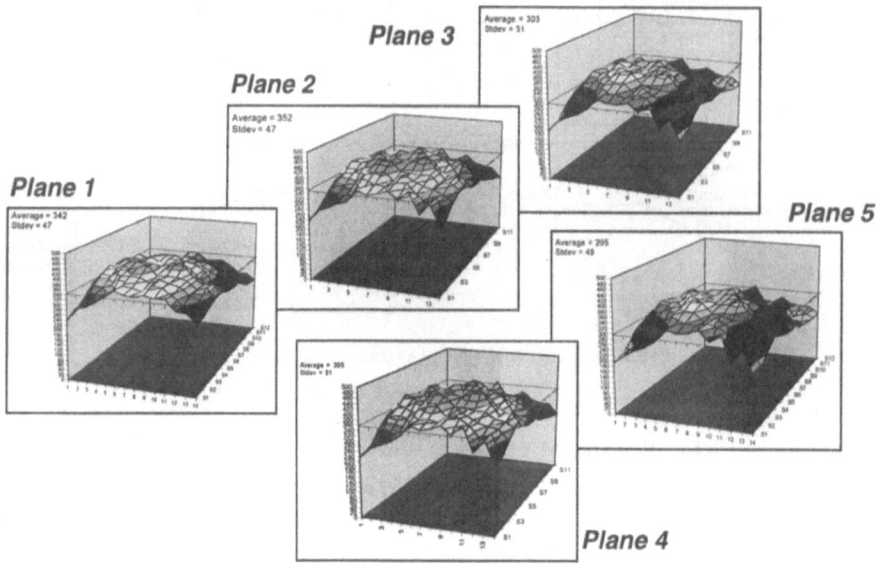

Figure 10. Response of the test THA to a 60 µsec gated sine pulse of low amplitude ultrasound. The range gates were set to capture the full acoustical pulse.

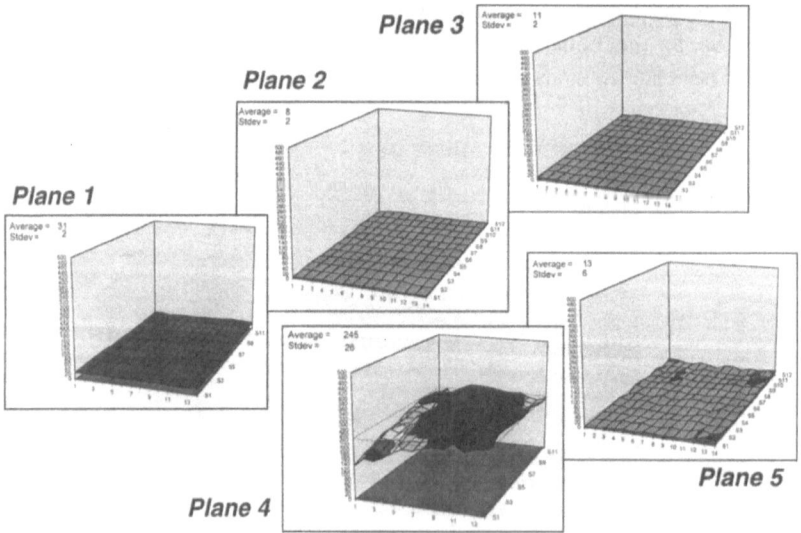

Figure 11. THA Range resolution. A short acoustical pulse (4.0 µs) appears only in range plane 4. The range gates are 3.25 µs long and are spaced by 0.75 µs, corresponding to pulse-echo ranges of 2.4 mm and 0.5 mm respectively.

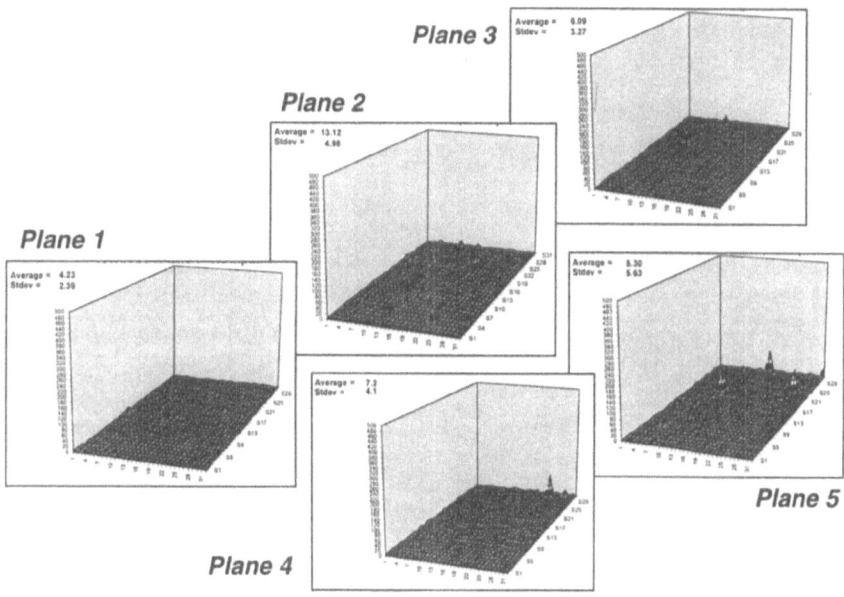

Figure 12. With no acoustical input, the "fixed pattern" noise of the 16x16 THA is revealed.

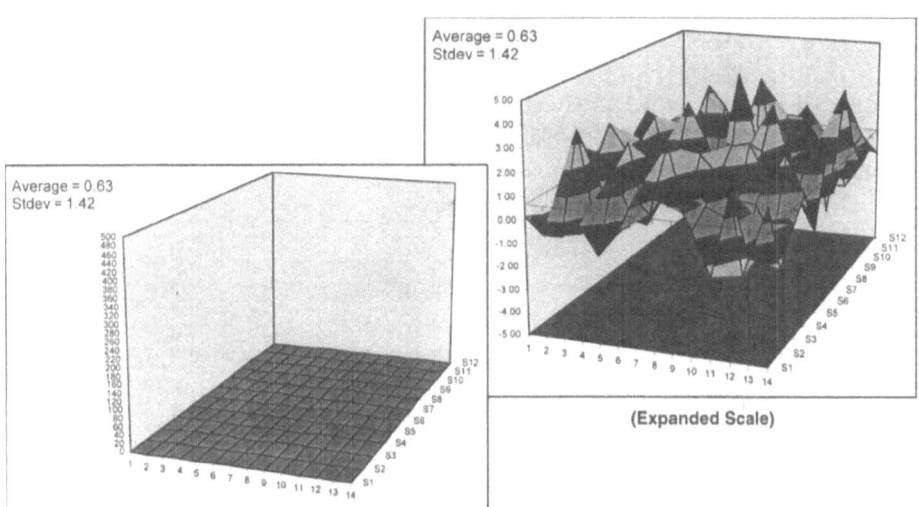

Figure 13. With no acoustical input, THA frame-to-frame repeatability is demonstrated by subtracting the same plane in one frame from the next frame. The difference is equal to the system electronic noise level. The right plot (b) is an expanded version of the left plot (a).

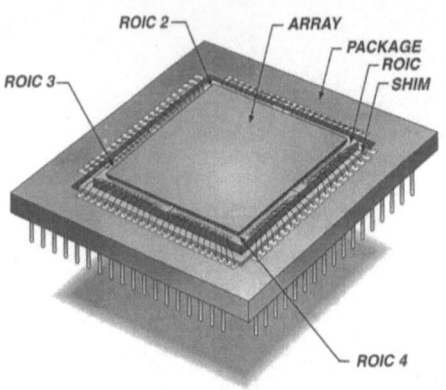

Figure 14. Schematic drawing of the 64x64 (or 128x128) THA.

Figure 15. A 64x64 element 3.1 MHz THA.

CONCLUSION

Fabrication and testing of 64x64 and 128x128 THA's is underway. Figure 14 is a schematic drawing of the THA, and Fig. 15 is a photograph of one such large array in the PGA package.

We have demonstrated the basic functionality of our THA design and shown that the initial versions meet the system performance goals.

ACKNOWLEDGMENTS

This work is supported by ARPA Grant DAMD17-94-J-4511 and US Navy Contracts N00014-96-C-0349 and N00174-96-0070 together with Lockheed Martin IR&D funding. The support and encouragement of Col. Richard Satava, M.D., Wallace Arden Smith, Ph.D. and Mr. Bruce Johnson is gratefully acknowledged.

REFERENCES

1. K. Erikson and R. Zuleeg, "Integrated Acoustic Array," *Acoustical Holography*, Vol. 7, ed. L. Kessler, Plenum Pub. Corp., pp. 423-444, 1977.
2. K. Erikson, S. Ishrak, W. Reckwerdt and R. Spratt, U.S. Patent No. 5301674, April 12, 1994.
3. O. Von Ramm and S. Smith, U.S. Patent No. 4694434, September 15, 1987.
4. American Medical Design, Paso Robles, CA, 93446
5. W. A. Smith and B. A. Auld: "Modeling 1-3 Composite Piezoelectrics: Thickness-Mode Oscillations," *IEEE Trans. Ultrasonics, Ferroelectrics and Frequency Control*, **UFFC-38**: 40-47 (1991).
6. D. Horvat, J. Bird and M. Goulding, "True Time-Delay Bandpass Beamforming," *IEEE J. Oceanic Eng.*, 17: 185-192 (1992).

DIVER-HELD IMAGING SYSTEM MODELING

D. Folds[1], B. Johnson[2], B. Kamgar-Parsi[3]

[1] Ultra-Acoustics, Inc., 4072 Hickory Fairway Dr. Woodstock, GA 30188

[2] Naval Explosive Ordnance Disposal Technology Division, Code 50A15, Indian Head, MD 20640

[3] Naval Research Laboratory, Code 5585, Washington, D.C. 20375

INTRODUCTION

Diver-held ultrasonic imaging is currently in the experimental development stage. Experiments are being carried out under controlled conditions to test various components and to evaluate critical parameters. The system is intended primarily for identifying explosive ordnance on the sea bottom where optical viewing is often impractical due to suspended materials stirred up by wave action or diver presence near the bottom. The objective is to provide a 3 MHZ pulse-echo system having range and cross-range resolution of 1.0 cm at a distance of 3 meters with a conical field of view of 20 degrees. This translates to beamwidths near 0.2 degrees and pulse lengths near 2 cm.

This research has concentrated on utilizing the beamforming capabilities of ultrasonic lenses to form an image on an array of sensors in the image plane of the lens. Modeling of lenses and lens-based imaging system performance has played an important role in guiding the research. This paper discusses results of recent modeling and simulation. Experimental results have been reported in reference 1.

ULTRASONIC LENS DESIGN

Lens designs have improved due to application of modern optical lens design tools such as ZEMAX™ (Focus Software, Inc., Tucson, AZ), and careful attention to material selection. Early designs[2] used crystal polystyrene, selected for its low attenuation. However, with polystyrene, multiple reflections between the retina and lens surfaces contribute to high amplitude 'ghost' images. Two factors contribute to ghost images: 1) impedance mismatches between lens surfaces and the surrounding medium; and 2) lens surface curvatures which cause reflections to refocus back onto the retina. The least favorable lens surface curvature is one which focuses reflected sound back to sensors at the image plane. Unfortunately, designs optimal in other respects frequently have the least favorable rear lens surface curvature. This fact forces the designer to seek materials with acoustic impedance near water to diminish reflections from lens surfaces. Combinations of solid materials such as TPX and low density polyethylene, and low velocity fluids such as Fluorolube FS-5, are currently favored. Figure 1 compares the reflectivity at interfaces of various combination of materials and provides the rationale for selecting the materials cited above. Low density polyethylene is included in this group in spite of its high attenuation because, as noted in reference 2, aperture apodization can be achieved using a lens (thin in the center and thick at the edges) with a high attenuation coefficient.

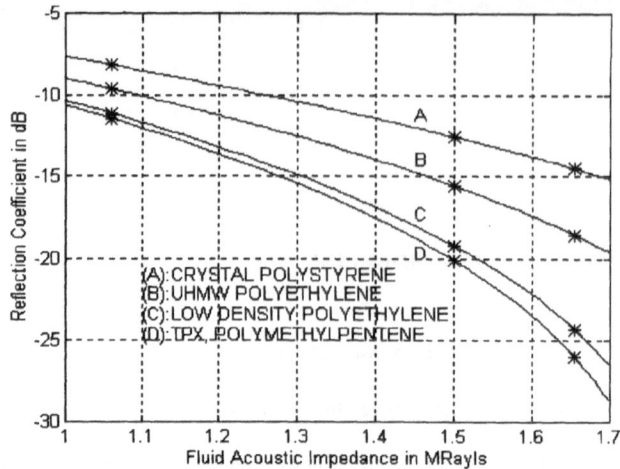

Figure 1. *Reflection coefficient of plastics immersed in fluids; Asterisks denote FC-75 (3M, ρc=1.06), water, and FS-5 (Occidental Chemical Corp., ρc=1.65).*

IMAGING ISSUES

The acoustic dynamic range of reflections from a typical scene of interest (e.g., a cylinder) may be greater than 50 dB. Regardless of the image processing and display method, image quality is strongly influenced by dynamic range. The ability to preserve a large dynamic range depends on system spatial and temporal impulse response. In the time domain, pulse shaping and large dynamic range low noise amplifier technology are normally used with success to control the temporal response of a system. In the spatial domain, the composite transmit/receive diffraction pattern of the system is the limiting factor in preserving image dynamic range. System diffraction patterns can be designed to meet stringent requirements by: 1) aperture apodization (shading); or 2) transmitting and receiving on a single beam (monostatic operation). Transmitting and receiving on a single beam reduces the -3 dB beamwidth and, more importantly, significantly reduces the sidelobe structure. In the unshaded single beam case of the monostatic system, the diffraction pattern is represented by $D1 = [2 \cdot J_1(x)/x]^2$ versus $D2 = 2 \cdot J_1(x)/x$ for the unshaded bistatic case.

Figures 2(a-d) illustrate the difference between bistatic and monostatic configurations by comparing simulated 2-D images of a rough surface cylindrical target on the bottom obtained with diffraction patterns D1 and D2 above. Figure 2(a) was generated under the condition where the target was broadly insonified and scanned by beam D2 (bistatic operation). In Figure 2(b), the same target was insonified by the scanning beam D1(monostatic operation). The source was positioned 70 cm above the bottom and the beamwidth in both cases was 0.28°. These images were obtained by ray tracing and summation procedures described in a following section. The spider-like arms most prominent in Figure 2(a) are due to multiple reflection modes between the cylinder and the bottom resulting from broad-beam, simultaneous insonification of the scene. Their intensity is a function of bottom reflectivity assumed here to be equal to the target reflectivity. The superior image in Figure 2(b) results from transmitting and receiving on a single beam to obtain reduced sidelobes and a reduction in observed multipath due to insonifying only a portion of the scene at any one time. However, single beam scanning results in low frame rates that are impractical for diver-held imaging systems. Thus, other approaches such as the beam multiplexing methods described in this paper are being examined. Beam multiplexing permits transmission-reception on mutiple beams simultaneously by separating the beams in angle or frequency or both angle and frequency. This method leads to higher frame rates and also the potential for a dramatic reduction in receiver channel electronics.

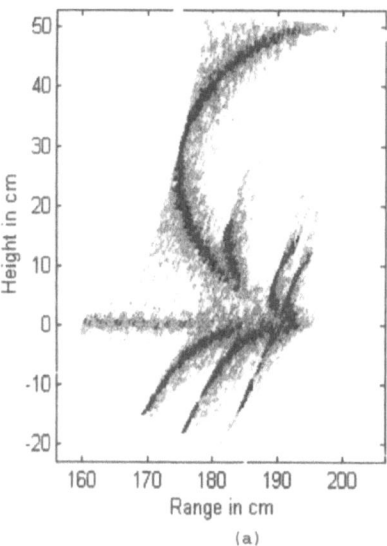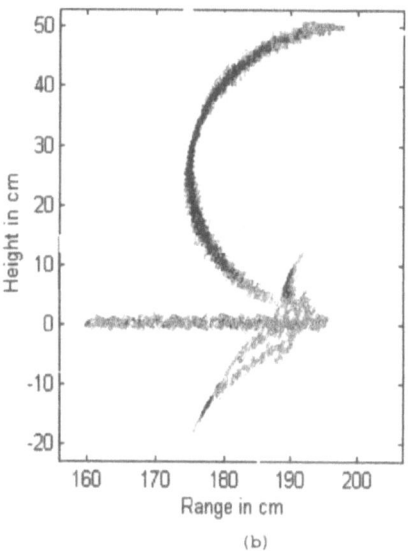

Figure 2. Simulated image profile of a rough surface 50 cm diameter cylinder resting on the bottom: (a) scanning beam of the form $[2 \cdot J_1(x)/x]$; (b) scanning beam of the form $[2 \cdot J_1(x)/x]^2$

In an attempt to approximate monostatic operation and to reduce the number of electronic channels required for the multi-element retina, we have examined various means of generating multiple independent transmit-receive beams simultaneously while retaining the characteristics of monostatic operation. Monostatic operation depends upon maintaining independence of each beam by some combination of angular, temporal, and frequency separation. Beams transmitted simultaneously at the same frequency must be spatially separated by several beamwidths to avoid excessive coherent interference and to achieve any significant advantage over bistatic operation. Separating beams by two or more beamwidths raises the issue of how to fill in the gaps between beams while maintaining an acceptable frame rate. This can be done by forming additional beams operating at different frequencies and/or transmitting at different times. For the short ranges required by the system under consideration, several transmission/reception cycles are permissible in generating an image frame. The range resolution requires a bandwidth of 75 kHz, but operating in three or four frequency bands does not place unreasonable bandwidth constraints on a 3MHz transducer design. Figure 3(a) shows one example of frequency-time allocation for a retina configuration which would provide a 9:1 multiplexing ratio using three different frequency bands and three different transmit/receive cycles to generate a single frame. In this configuration, beams are separated by three beamwidths. Figure 3(b,c,d) show images of a single point reflector (spatial impulse response) for a system where the transmit-receive beams are separated by 2, 3, and 6 beamwidths. Analysis shows that while two beamwidths separation reduces the -3 dB beamwidth, there is little gain in sidelobe reduction. Both beamwidth and sidelobes are reduced if beams are separated by three or more beamwidths as shown in Figures 3(c) and 3(d).

Figure 4(a) shows a single row of the matrix of transmit beams formed by a 46 by 46 element retina similar to that in Figure 3(a). This figure shows the relative pressure distribution of a single row of beams formed simultaneously by one of the three operating frequencies. Note that the levels of the nulls are near -15 dB; therefore, the entire scene is insonified but at a reduced level when compared to a bistatic system where a broad beam projector is used. For this reason, multiplexing in this manner is only an approximation to single beam monostatic operation. An example of an image based on the frequency-time multiplexing is shown in Figure 4(b). A target (a cylinder on the bottom) is insonified in three sequential angle shifted frames by the projector distribution in Figure 4(a). The three frames are interlaced and shown in Figure 4(b).

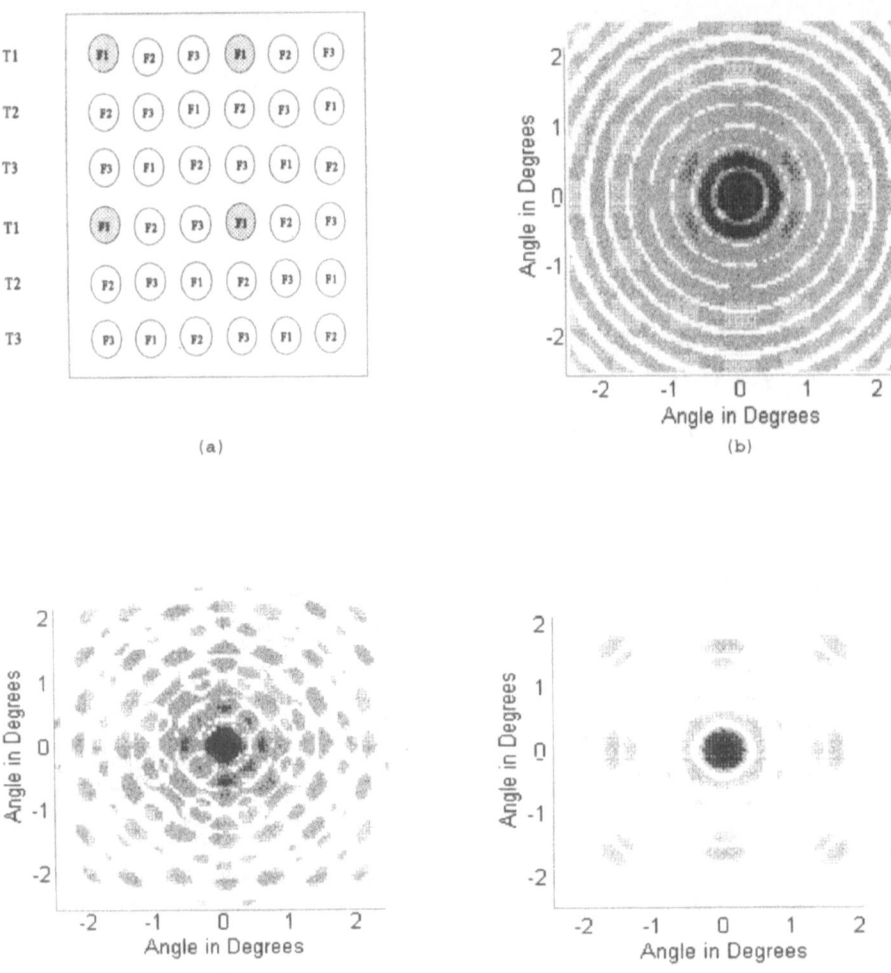

Figure 3. *(a) Example of a 3:1 multiplexed array with beams separated by 3 beamwidths; Image of point reflector with beamwidth separation of: (b) two beamwidths; (c) three beamwidths; (d) six beamwidths*

In this modified monostatic configuration the echos due to target/bottom multipaths are significantly reduced compared to Figure 2(a) (bistatic operation). Also, due to the reduced sidelobe structure, the flair in the cylinder image is absent. It should be noted that the influence of aperture apodization has been examined and found to be ineffective in reducing these multipath echos since the echos result from insonification by other beams and are entering in the main beam response. However, in spite of the remnant multipath echos, the multiplexing scheme described here appears to offer many advantages in terms of reduced electronics and the ability to achieve reduced -3 dB beamwidths. For example, in a 3:1 multiplex system as in Figure 3(a), a 100 by 100-element retina is served by a 10,000/3 processing channels. The effects of multipath appear to be the only price paid to achieve reduced sidelobe structure and significant reductions in system complexity, and this may be a small one depending on bottom reflectivity. Although measurements are planned for the near future, no data currently exists regarding bottom reflectivity in the 3 MHZ range and the analysis has assumed worst case, i.e., bottom reflectivity equal to target reflectivity.

498

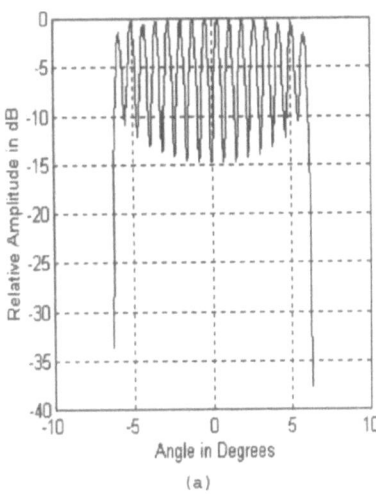 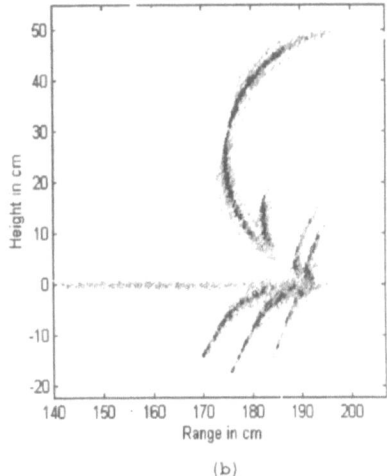

(a) (b)

Figure 4. *(a) Projector beam pattern for 3 beamwidth separations; (b) Simulated image of 50 cm cylinder resting on the bottom using frequency-time multiplexing and three beamwidth separation with composite transmit-receive beampattern shown in Figure 3(c).*

THREE DIMENSIONAL MODELING

Figure 5 shows an example of continuing modeling efforts. This surface rendered image is a section of a 50 cm diameter cylinder modeled by a grid of coordinate points with a spacing of 0.2 cm. A 10 cm diameter disk is recessed into the cylinder wall. Six 2 cm diameter bolt heads are distributed at equal angles around the disk circumference. Surface roughness of the target was modeled by a random normal perturbation of one wavelength (0.05 cm) in grid coordinates. The imaging system aperture was modeled as a bistatic system with an unshaded receiving aperture diameter of 15 cm, frequency of 3MHz, and a pulse length of 1 cm. Target distance was 200 cm.

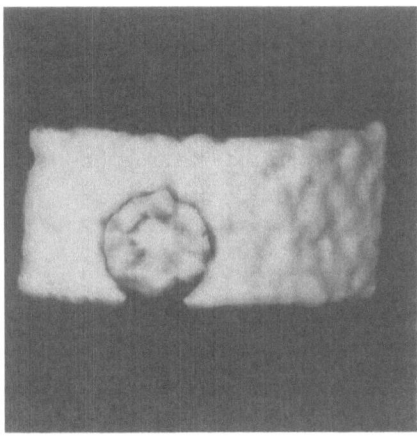 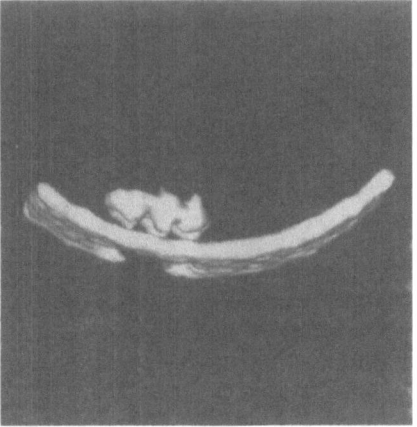

Figure 5. *Two views of a bistatic image of a 50 cm diameter cylinder section with recessed disk*

SIMULATION METHOD

In the image system model, a target is defined either by a grid of coordinate points or in the case of canonical surfaces, defined by a general equation of the second degree. Surfaces normal direction cosines are assigned to permit computation of ray reflection angles. A cone of rays is launched from the source location and traced to the target scene computing and storing intercept coordinates and reflected angles through a series of multiple reflections. Reflectivity coefficients and roughness are assigned to target, surface and bottom. Roughness is generated by randomizing the direction cosines of the surface normals and small perturbations in coordinate values. Contributions from each point are amplitude weighted by the directivity patterns and pointing angle of projector and receiver and cumulative reflection losses. Phase, or time of arrival information, is retained by storing the path lengths from the projector to each reflection point and return. For each beam pointing angle reflection contributions from each point in the scene are coherently summed within time intervals equal to the transmitted pulse length.

Data have been displayed in three forms: 1) in two-dimensions where each image pixel is assigned the maximum amplitude level in each beam; 2) in two dimensions, where time of arrival of the first echo exceeding a preset threshold is assigned a color or gray shade; 3) in pseudo three dimensions where time of arrival and amplitude data are used in gradient-based surface detection algorithms. The first method is generally unsatisfactory under all conditions. The second display method has been successfully employed for targets suspended in the water column, but may not be suitable for targets on a hard surface where extensive multiple reflections occur. Although the third method is computer intensive, it results in a superior image display. Selection of the image processing and display method awaits acquisition and analysis of image data under realistic operating environments.

DISCUSSION

The results of simulation and analysis demonstrate that frequency-time multiplexing provides a significant reduction in the number of electronics channels required to achieve a useful diver-held imaging capability. Depending on the multiplexing ratio, it is possible to achieve performance approaching that of single monostatic beam scanning. If bottom reflectivity is high, frequency-time multiplexing with widely separated beams will reduce the image degradation caused by multiple reflections. However, if bottom reflectivity is low and multiple reflections can be ignored, then monostatic and bistatic image quality will be essentially equivalent if sidelobe levels and beamwidths are comparable. For the bistatic case, achieving sidelobe levels and beamwidths equivalent to the monostatic case will require a larger aperture and aperture shading, but will not require separate transmit-receive circuitry for each element of the array.

REFERENCES

1. B. Kamgar-Parsi, B. Johnson, D. Folds and E. Belcher, *High-resolution Underwater Imaging with Lens- based System*, International Journal of Imaging Systems and Technology (in press).
2. D. L. Folds, Focusing Properties of Solid Four-Element Lens, *J. Acoust. Soc.*, **58**, (1975).

LITTORAL TARGET FORWARD SCATTERING

Brad Gillespie,[1] Ken Rolt,[1] Geoff Edelson,[1] Rob Shaffer,[1] and Paul Hursky[2]

[1] Sanders, A Lockheed Martin Company
Advanced Systems Directorate
P.O. Box 868
Nashua, NH 03061-0868

[2] Lockheed Martin Aeronautical Systems
Building 641, 52990 Gatchell Road
San Diego, CA 92152-7314

INTRODUCTION TO FORWARD SCATTERING AND BARRIER TECHNOLOGY

Bistatic sonar[i] can be employed for target regions outside an ellipse surrounding the source and receiver. However, it is particularly hard to obtain useful detection inside the ellipse (i.e. directly between the source and receiver) due to the very strong "direct blast" from the source. Using traditional processing techniques, this "direct blast" overwhelms any signature from a target located inside this ellipse, and the target is masked.

One means of overcoming this problem is to make use of the perturbations in the received signal due to the forward scattering from the target. Forward scattering is a special case of bistatic sonar, where there is large acoustic gain in the forward direction around the target. The goal of this work is to exploit this strong forward scattering response to detect the target inside the ellipse. Intuitively, if a stable underwater sound field is maintained in an underwater environment, one should be able to detect the perturbations in the sound field caused by an object entering this field (J. Underwater Acoustics, 1995; Robinson and Greenleaf, 1983). The only requirement is that the perturbation be "strong enough" to be detected at the receiver. It has been shown (Stenzel, 1938; Clay and Medwin, 1977) that the forward scatter strength for a fixed rigid sphere in the far field, for ka values greater than approximately 2, is significant compared to scattering in other directions.

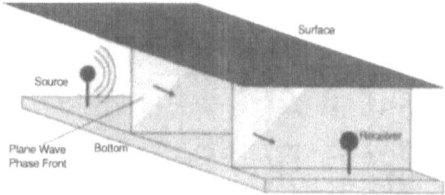

Figure 1. Source receiver configuration for a barrier system in a shallow-water waveguide. For this system to be most effective it should be deployed in an area (such as a harbor entrance) that forces the target to pass between the source and receiver. Thus this system acts as a "barrier".

---

[i] Bistatic sonar refers to a system with a spatially separated acoustic source and receiver.

In this paper, forward scattering examples from computer modeling in a time-varying littoral environment are generated and two fundamentally different techniques are applied in an attempt to detect the passage of the target. These techniques are outlined and results of applying them to modeled data are shown.

OVERVIEW

The acoustic barrier is shown in Figure 1. It is important to note that Figure 1 (and Figure 2) are illustrative, and are not meant to fully portray the scenario; no environmental variability nor multi-path structure are shown.

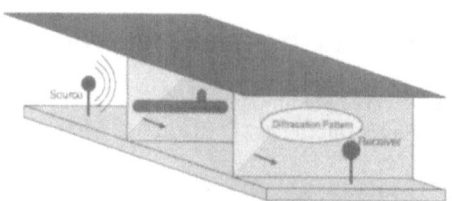

Figure 2. Passage of a target through the acoustic barrier system.

The continuous acoustic source transmits either pulsed FM or CW. Under ideal conditions (i.e. perfectly stable environment with no noise) the received waveform will be perfectly stable as well. Only when a perturbation in the environment occurs, such as the passage of a target between the "tripwire"[ii], can a variation in the direct field be observed. This variation is caused by the target diffraction pattern, which may be visualized as "punching a hole" in the plane wave as shown in Figure 2. As the plane wave progresses past the target the "hole" begins to be filled in, due to forward diffraction. However this leaves a measurable variation in the plane wave, which can be detected.

There are two problems that must be overcome for this system to work. First, the environment is not ideal. There are variations in the received waveform due to surface waves, variable water temperature, internal waves, marine life etc. Second, the variations in the direct field are weak compared to the aforementioned environmental variability.

We have considered two proposed detectors to exploit aspects of the received waveform. The first is based on analysis of "target signatures" extracted by careful filtering, and the second uses the variability in the modal arrivals.

TECHNIQUE #1

The first technique is to directly evaluate the modulation in the envelope of the direct field. A candidate algorithm to accomplish this is shown in Figure 3. The first filter is used to isolate the

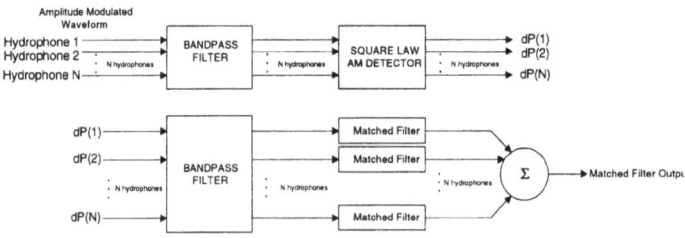

Figure 3. Candidate processing algorithm to detect the modulations in the received waveform due to the passage of a target through the acoustic tripwire.

[ii] This denotes the region between source and receiver, where the forward scattering effects can be observed.

modulation in the direct field due to the passage of the target from other environmental sources of perturbation. The envelope of the signal is extracted using a simple AM detector. The second bandpass filter was found to be necessary due to interference from internal waves in the frequency band of interest. It was found that internal waves modulated the envelope of the received waveform. Therefore a bandpass filter with appropriate cutoff frequencies was implemented to extract only the "clean" portion of the received envelope. The resulting waveform is individually matched filtered and the outputs summed.

To evaluate the proposed system, modeling and simulation was performed. The chosen scenario is displayed in Figure 4. The target is traveling perpendicular to the "tripwire" at approximately mid-water depth. A vertical receive array is located near the bottom of the ocean channel 15 km from the source (also located near the bottom of the channel). The source transmits CW.

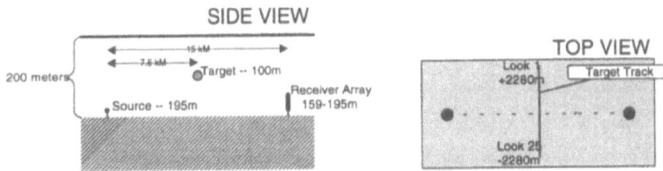

Figure 4. Scenario to evaluate technique #1.

To initially evaluate the system performance received data was generated for a case where the ocean surface and bottom are flat, the sound speed profile is iso-velocity (1500 m/s) and the target was traveling at 5 knots. The processing shown in Figure 3 was applied to this data set and the results are shown in Figure 5 and Figure 6. Figure 5 details the results of processing on one particular hydrophone (#11). The uppermost plot in this figure shows the envelope of the received waveform (mean removed). In Figure 5 the filtered signal represents the envelope after the second bandpass filter. In this case the matched filter replica (shown in the third plot from the top) is ideal, i.e. the replica is the same as the received signal. The final plot in Figure 5 shows the output of the matched filter on hydrophone #11. Similar results were achieved on all other hydrophones. Figure 6 is the result of summing all matched filter outputs together. This case is ideal, therefore this result is considered to be optimal.

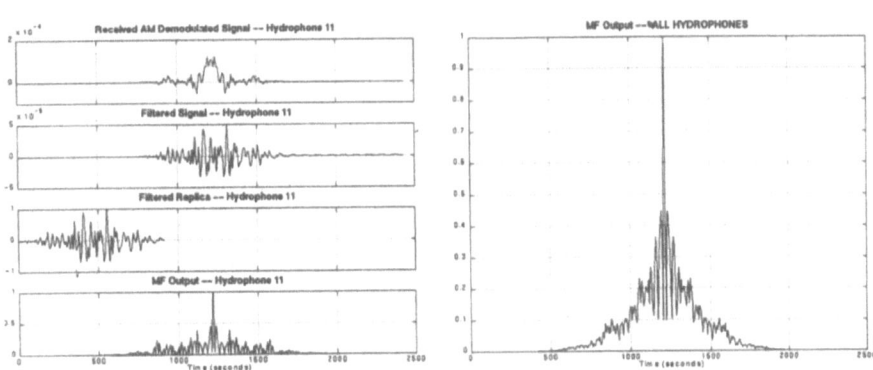

Figure 5. Processing results on Hydrophone #11. No noise and no internal waves.

Figure 6. Result of summing all matched filter outputs. No noise and no internal waves.

To investigate the robustness of this algorithm ocean variability was added to the received scenario in the form of internal waves and ambient noise. Internal waves are modeled as bandpass filtered white Gaussian noise. Ambient noise is modeled as additive white Gaussian noise (AWGN) scaled to yield a signal to noise ratio of 9.5 dB. The matched filter for this example are the same as those used in the previous example. Therefore, this no longer represents the "ideal" case.

The performance of this system is shown in Figure 7 and Figure 8. On an individual element basis detection performance is marginal, however when all elements are combined detection performance is significantly improved.

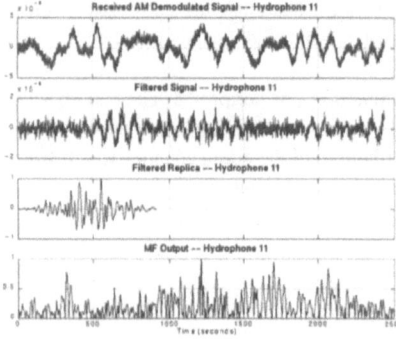 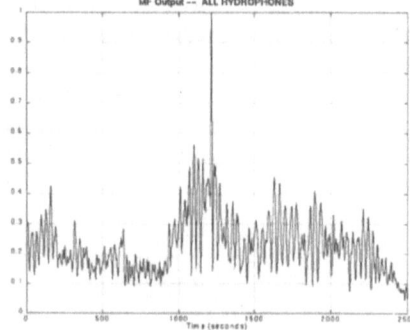

Figure 7. Processing results on Hydrophone #11. With noise and internal waves.

Figure 8. Result of summing all matched filter outputs. With noise and internal waves.

Clearly from our initial analysis this technique holds promise. However, direct field amplitude variability due to environmental conditions requires further investigation. Direct field variability must be identified as a function of time over a range of conditions and environments. Further clarification of amplitude variability due to target scattering is required. In addition, matched filter performance in the presence of strong direct field amplitude variability (that is not separate in frequency from target scattering) must be investigated.

TECHNIQUE #2

The second technique involves tracking the modal arrivals from the received waveform. Due to the brevity of this paper, the details of the processing will be omitted. The hypothesis of the processing is that in the absence of a target the modal arrivals should be relatively stable. The structure of these arrivals should change dramatically in the presence of a target. To evaluate the potential of this technique modeled data from the environment shown in Figure 9 was generated. As in Figure 4 the target is traveling perpendicular to the "tripwire" at approximately mid-water depth. A vertical receive array is located near the bottom of the ocean channel 15 km from the source, also located near the bottom of the channel. In addition the source is continuously transmitting a linear FM sweep one second in duration, at a higher frequency than in technique #1. Again, the target

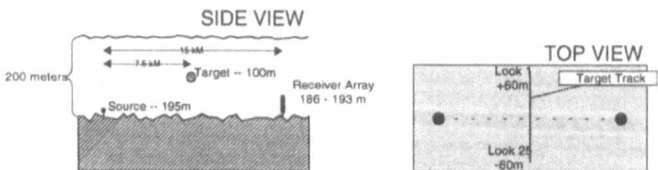

Figure 9. Scenario to evaluate technique #1.

velocity is 5 knots and the sound speed profile is iso-velocity (1500 m/s). To simulate environmental variability bottom and surface roughness was added. This case is worse than a downward refracting environment because of the multiple interactions with the varying surface. In this scenario, a shorter modeled target track than in the first technique. This is because the width of the main forward scattering lobe narrows considerably as the frequency is increased.

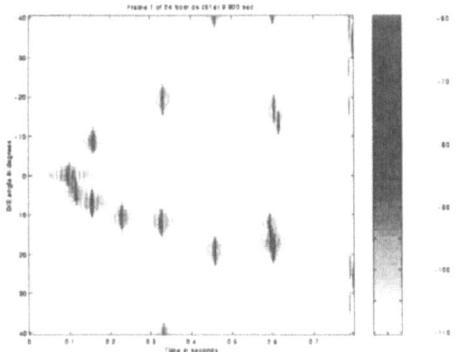

Figure 10. Typical arrival structure with no target present.

A typical arrival structure when no target is present is shown in Figure 10. Notice the clean well defined parabolic arrival structure and the lack of arrival at shallow angles in the later time periods. While the lack of noise in this case contributes to the clean nature of the arrivals, the overall structure will not differ dramatically in the presence of noise. Compare these results to those shown Figure 11 and Figure 12. As the target approaches the tripwire the arrival structure becomes

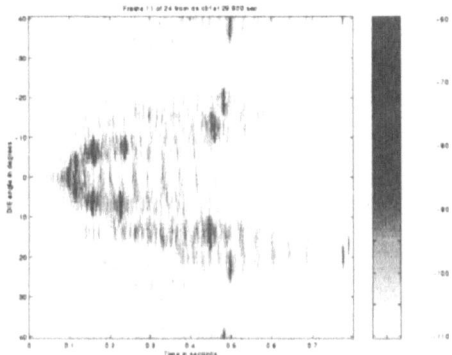

Figure 11. Arrival structure with a target 20 meters from the tripwire.

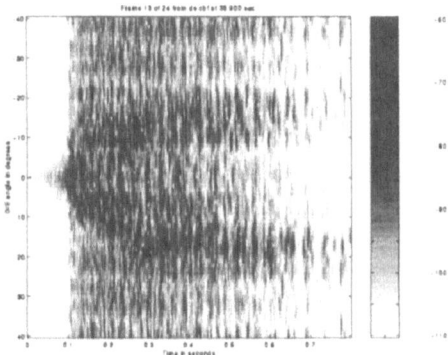

Figure 12. Arrival structure when the target crosses the tripwire.

increasingly cluttered. In addition when the target is near the tripwire a number of strong arrivals occur at very shallow angles much later in time than the direct arrival. This effect can be directly attributed to the presence of the target in the water column.

Another means of visualizing this data is to peak detect and threshold each snapshot/fm sweep, time align the first arrivals and plot them as a function of the target location. Therefore the delay time (from the first arrival) is plotted on the x-axis while the overall time is plotted on the y-axis. The result of this analysis is shown in Figure 13. There is a significant increase in the number of arrivals as the transmit time approaches 34 seconds (the time at which the target is directly on the tripwire). This increased density of arrivals can be used as a basis for target detection.

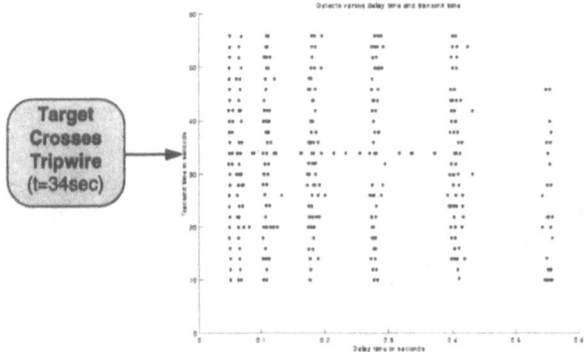

Figure 13. Arrival structure as a function of target crossing time.

CONCLUSIONS

Clearly these results show there is significant potential for target detection by exploiting the forward scattering phenomena. Two fundamentally different techniques have been demonstrated, both of which provide encouraging results. However, a great deal of work is still to be completed. The variability of the received waveform must be fully understood. Further refinements in the processing algorithm are necessary, to combat the environmental variability effects. In addition, localization analysis must be performed to understand the capability of these systems.

REFERENCES

Clay, C.S., and Medwin, H., *Acoustical Oceanography*, J. Wiley & Sons, New York, (1977), pp. 192-193.

U.S. Navy Journal of Underwater Acoustics, *Special Issue on Forward Scattering*, **45**(1), 1995.

Robinson, B.S. and Greenleaf, J.F., "Measurement and Simulation of the Scattering of Ultrasound by Penetrable Cylinders", *Acoustical Imaging volume 13*, Plenum Press, New York, (1983).

Stenzel, H., "On the disturbance of a sound field brought about by a rigid sphere," in German, *Elektr. Nachr. Tech.* **15**, (1938). Translated by G. R. Barnard and C. W. Horton, Sr., and republished as Technical Report No. 159, Defense Research Laboratory, University of Texas, Austin, 1959.

AUTOMATICALLY COMBINING A GEOMORPHOLOGIC MAP USING 2-D IMAGING SONAR DATA

GUAN Hao[1], TIAN Tan[1], ZHANG Dianlun[1], and SHI Zhenlong[2]

[1] Department of Underwater Acoustics Engineering,
Harbin Engineering University, Harbin,
150001, P.R.China
[2] JiangSu Raysound Electronic Equipment Plant, WuXi,
214061, P.R.China

INTRODUCTION

By using a multibeam image sonar, the return signals from the targets on seabed and background can be stored and displayed in the form of distance-azimuth, so that the location and classification for underwater targets are realized. When the sonar is to be evaluated or tested on a specified area, it is generally needed to know the feature of the seabed in the tested area. This is usually done by a side scan sonar[1]. However, when the ship used in the experiment is fixed, we can't employ the side scan sonar to get the geomorphologic map. In this case, this task can also be performed by the imaging sonar itself.

When the array is rotated with an angle step horizontally, a set of sonar images with a certain sector width and overlapping region between them will be obtained, which can cover the expected sector test area. For lack of the means to accurately measure the azimuth of the array, it can only be ensured that the angle step is less than the sector width in successive detect periods. If these images are simply combined, it is difficult to get a good geomorphologic map. A new method to combine the images is suggested in this paper, in which the correlativity between the adjacent images is used to calculate the overlapping width and automatically combine the set of images into a big one. The precision of the method only depends on the azimuth resolution of the sonar, so the accurate direction of the array does not need to be measured and the requirement for the measuring equipment in the test ground is relaxed. The new algorithm and relative results are presented in this paper.

ALGORITHM

Correlation Coefficient

Let the adjacent two images to be combined be A, B, their pixel values are stored as a rectangular format of distance-azimuth with dimensions $H \times W$. A, B can be written as

$$A = (a_{ij}) = (A_j)$$
$$B = (b_{ij}) = (B_j)$$

$$i=0,1,.....,(H\text{-}1),\ j=0,1,.....,(W\text{-}1) \qquad (1)$$

where a_{ij} and b_{ij} are the pixel amplitudes in row i and column j of A, B respectively. A_j, B_j are the column vectors of column j of A, B respectively with dimension of H. Suppose that during the data are collected, the array is only rotated horizontally, that is, there is no displacement in range between each two images. The linear correlation coefficient between the columns of A and B is defined as [2]

$$\rho_{mn} = \frac{\displaystyle\sum_{i=0}^{H-1}(a_{im} - \overline{A_m})\cdot(b_{in} - \overline{B_n})}{\sqrt{\displaystyle\sum_{i=0}^{H-1}(a_{im} - \overline{A_m})^2}\cdot\sqrt{\displaystyle\sum_{i=0}^{H-1}(b_{in} - \overline{B_n})^2}}$$

$$= \mathrm{corr}(A_m, B_n) \qquad m=0,1,....,(W\text{-}1),\ n=0,1,....,(W\text{-}1) \qquad (2)$$

where A_m and B_n are the mean value of data samples of column m in A and column n in B respectively. Equation (2) represents the elements of $W \times W$ matrix. If there is some overlapping region between A and B, there must be a common region corresponding to the echoes reflected from the same region of the seabed. In this case, one or more correlation coefficients ρ_{pq} over certain threshold may be found in the common region, where p and q are two column numbers and correlativity between column p and q is much greater than the others. It can be considered that these two columns most possibly represent the same region of the seabed. Therefore, the overlapping width between the two images is determined as $d=(W\text{-}q+p)$. In principle, by the same processing for each two adjacent images, an entire combined image can be obtained.

In the ideal condition, the overlapping regions between two images are completely correlative. If the width is d, the correlation coefficient $\rho_{W\text{-}d+l,l}$ between $A_{W\text{-}d+l}$ and B_l is equal to $1 (l=0,1,...,d\text{-}1)$. This implies that the column number $W\text{-}d\text{-}l$ of A and the column number l of B satisfies a linear relationship for the maximum correlation. Due to time-variance and complexity of underwater sound propagation environment, even two images obtained from the same detected area still emerge some difference. Therefore in practical situation, the linear relationship between two column numbers with stronger correlativity will not always be accurately held and some wild points will occur around the straight line. In this case, using a single maximum value of ρ_{pq} to determine d can not ensure the precision for the combined image.

Rejecting Wild Point

The existence of wild points shows that the column numbers, in which the maximum correlation coefficients appear between A and B, deviate from the linear relationship mentioned above. Because overlapping region really exists, it can be sure that there is an "apparent" linear relationship. The question is how to approach to the "apparent" relationship and how to obtain the best estimation value of d.

This problem can be solved by using the method called "match template". Let

$$Q_m = max(\rho_{mn}) \qquad n=0,1,....,(W\text{-}1) \qquad (3)$$

Q_m are column numbers in B corresponding to column number m in A with the maximum correlativity between them. Take the "match template" as a set of straight lines

$$f(k) = \begin{cases} m - k & m \geq k \\ 0 & another \end{cases} \qquad k=0,1,....,(W\text{-}1),\ m=0,1,....,(W\text{-}1) \qquad (4)$$

Let D be matrix of the correlation coefficient between Q_m and match template, the element of D is

$$d_k = corr(Q_m, f) \quad m \geq k \quad k = 0,1,....,(W\text{-}1) \tag{5}$$

So the subscript k corresponding to the maximum correlativity in D is the number of match template most similar to apparent line, and hence

$$d = M - 1 - k \tag{6}$$

is just the overlapping width of the two images.
  Another method to get overlapping width is the method of histogram. Let

$$T_m = Q_m + W - m \quad m = 0,\cdots,W-1 \tag{7}$$

then in overlapping region, the apparent line with the slope of 1 will change to a horizontal line. The height of the horizontal line is just the overlapping width of two images. By contrast, in un-overlapping region, the distribution of T_m is in a stochastic state. The values of T_m corresponding to the maximum number of distribution in histogram is the overlapping width.
  The match template method needs more calculation, but with this method, the maximum correlation coefficient can quickly be found by using adaptive method with variable steps; the histogram method only makes use of additions to get result, so this method is fast and simple but sometimes this method is liable to make mistakes.

Selecting Threshold

  Suppose acoustic images A and B have relationship in time and space, that is, A and B are sequences of the image data sampled at different time with a part of overlapping region. Those pixels within overlapping region would have stronger correlativity than another part.
  If the main lobes of some beams in overlapping region of A and B are aimed at each other in space, then maximum correlation coefficient must exist between corresponded columns. Otherwise the beams of the second image will be fallen into some places of each two beams of the first image, and the correlation coefficient will be decreased. The worst situation is that the second image's beams place at the middle of first ones. Let A_m and A_{m+1} are adjacent columns of image A; A_m and A_{m+1} are statistically independent. In overlapping region, column B_n of B has the direction between A_m and A_{m+1}, then

$$B_n = \frac{aA_m + bA_{m+1}}{a+b} \quad 0 \leq a \leq 1, 0 \leq b \leq 1 \tag{8}$$

where a, b is the scale factor. When A and B have the same distribution function, the covariance of A_m and B_n is

$$Cov(A_m, B_n) = \frac{a}{a+b} \cdot Cov(A_m, A_m) \tag{9}$$

Because the variance of B_n is

$$D(B_n) = \frac{a^2 + b^2}{(a+b)^2} \cdot D(A_m) \tag{10}$$

we can find through equation (2) that the correlation coefficient of A_m and B_n are

$$\rho_{mn} = \frac{\rho_{mm}}{\sqrt{1 + \frac{b^2}{a^2}}}$$

(11)

It is obvious that when $a=b$, ρ_{mn} reaches to minimum, the value is

$$\rho_{mn} = \frac{\rho_{mm}}{\sqrt{2}} = \frac{1}{\sqrt{2}}$$

(12)

then the minimum coefficient is 0.707 of maximum coefficient ρ_{mm}. If the coefficient is not smaller than 0.707, it can be considered that the two images have overlapping region, and the overlapping width can be calculated. In this case, the estimation error of overlapping width is less than half of the beamwidth.

In practical situation, the correlation coefficient between columns in a overlapping region can be less than 0.707 because of signal distortion and channel fluctuation. We can use the equivalent noise sequence N to represent above effects, that is

$$B_n = \frac{A_m + A_{m+1}}{2} + N$$

(13)

It can be derived out that

$$\rho_{mn} = \frac{\rho_{mm} + 2 \cdot \frac{\sqrt{D(N)}}{\sqrt{D(A_m)}} \cdot \rho_{A_m N}}{\sqrt{2 \cdot (1 + 2 \cdot \frac{D(N)}{D(A_m)} + 4 \cdot \frac{\sqrt{D(N)}}{\sqrt{D(A_m)}} \cdot \rho_{A_m N})}}$$

(14)

where $\rho_{A_m N}$ is the correlation coefficient between A_m and N. If A_m is independent to N, we have

$$\rho_{A_m N} = 0$$

(15)

then

$$\rho_{mn} = \frac{\rho_{mm}}{\sqrt{2 + \frac{D(N)}{D(A_m)}}} = \frac{\rho_{mm}}{\sqrt{2} \cdot \sqrt{1 + 2 \cdot \delta}}$$

(16)

where

$$\delta = \frac{D(N)}{D(A_m)}$$

(17)

represents the ratio of the variance of equivalent noise to the variance of signal. When the distortion and fluctuation are small, the correlation coefficient in overlapping region will be

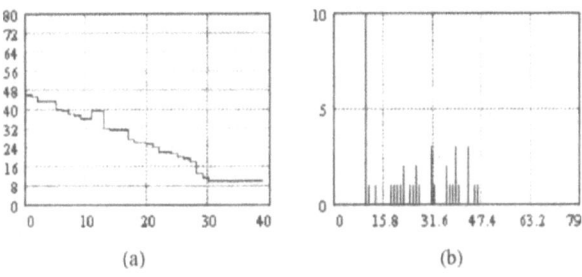

Figure 1. some intermediate result

approach to the value defined in equation (12). In practical calculation process, SNR can be increased by medium filtering in the direction of columns in image[3], and as a result the correlation coefficient can be improved about 10% or more.

The geomorphologic map can be combined after overlapping width is determined. Several original images can be combined to a large rectangular image. To avoid geometric distortion, the conversion from rectangular coordinate to polar coordinate is needed to make a non-distortion sector image.

EXAMPLE OF APPLICATION

We need two criteria for references: (1) the maximum correlation coefficient larger than 0.707, and (2) the existence of the apparent line. Otherwise, we think these two images are not overlapped. Some image data from an experiment on lake have been processed by using above algorithm.

The experimental system has 40 beams with the beamwidth of 0.17 degree in range from 100 meters to 160 meters with 6 degree sector in each detection period. The depth of water is 25 meters. Figure 1(a) is one of calculation results of equation (7), figure (b) is the histogram of figure (a). The result shows that the overlapping width of two adjacent images is 1.5 degree (about 10 beams). Figure 2(a) shows four adjacent original images, figure 2(b) shows a geomorphologic map combined and converted into polar coordinate by multiple images. The total width of combined sector is about 100 degrees. The raised rectangular section in figure 2(b) is a submerged farmland.

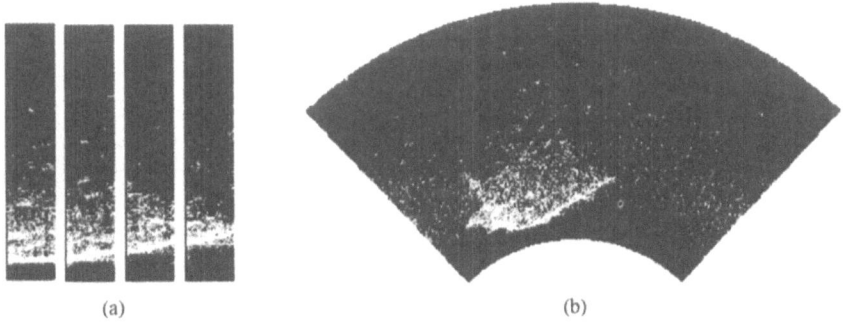

Figure 2. (a) original images; (b) an example of combined image with polar coordinate conversion

CONCLUSIONS

By using the algorithm mentioned above, a geomorphologic map at specified area can be automatically combined by 2-D imaging sonar data. The precision of the method only depends on the azimuth resolution of the sonar. The algorithm relaxes the requirement for experimental equipment. Further research on the algorithm can process more complex image combination. For example, when space movement exists not only in azimuth, but also in range, we can use two dimension correlation method to calculate displacements both in azimuth and range. The results can be used in geomorphologic map combination, data integration and display stabilization etc.

REFERENCES

1. John William Nicholson, Registration and variability of side scan sonar imagery, *AD-A201 895*(1988).
2. MAO Shisong, *Modern Engineering Mathematical Handbook*, Vol. IV, Part 59, Publisher of Hua Zhong Industrial Institute(1987).
3. ZHOU Xinlun, LIOU Jian, LIOU Huazhi, *Digital Image Processing*, Publisher of National Defense Industry(1988).

SHEAR SURFACE ACOUSTIC WAVES
LOCALIZED NEAR A PLANAR DEFECT

Arnold M.Kosevich, Eugenii S.Syrkin, and Andre V.Tutov

Institute for Low Temperature Physics
47 Lenin Avenue, 310164, Kharkov, Ukraine

INTRODUCTION

One-component acoustic waves localized near a planar defect in a crystal with FCC lattice are studied on the basis of the microscopic equations using the vector model. It is shown that due to the subsurface distortion in an area of the interface the eigen solutions of such a boundary problem are divided into symmetrical and antisymmetrical relatively to the plane of the defect. The properties of the high and low frequency vibrations of the both types of eigen solutions are studied.

The surface shear waves of horizontal polarization (SH-waves) have been studied for a long time but they remain of the special interest now [1]. These waves are absent in local and linear elasticity theory. The surface or the bulk of the elastic media should have an additional factors which make the existence of the surface waves possible. Such a factor could be the electrosound field coupling [2,3] or nonlinear properties of the elastic media [4-7] or the spatial dispersion of the media [8-10]. The waves of such a type can exist also near the interface between two elastic media. The data of vibrational characteristics' changes yield the information about such processes as the crystal growth, the stacking fault and Guniar-Preston zones creature and also about the great number of phenomena concerned with the interfaces where the number of atoms in the subsurface region is the same order of value as the number of the bulk atoms.

SYMMETRICAL AND ANTISYMMETRICAL SHEAR LOCALIZED WAVES

The equation of motion for an atom which has the small displacement from the equilibrium state in the framework of the vector model can be written in the following form in the harmonic approximation

$$m\frac{\partial^2 u_i(\vec{n})}{\partial t^2} = -\sum \Phi^{ik}(\vec{n} - \vec{n}_0)u_k(\vec{n}_0), \tag{1}$$

where m is the atom mass, $u_k(\vec{n})$ is the kth component of the displacement of the atom $(\vec{n})$ in the crystal. The matrix $\Phi^{ik}(\vec{n} - \vec{n}_0)$ account the symmetry of the crystal and possesses the number of properties which ensure the translative and rotatory invariance of the crystal energy. The components of the matrix $\Phi^{ik}(\vec{n})$ of the FCC crystal with account of the central interaction between the nearest neighbors have the form [11]

$$\Phi^{ik}(0) = 8\alpha\delta_{ik}, \tag{2}$$

$$\Phi^{ik}(\vec{n}_0) = -\alpha n_i^0 n_k^0, \tag{3}$$

where $\vec{n}_0$ is one of the vectors which connect the nearest neighbors in the crystal, α - is the force constant. The rest of the matrix components $\Phi^{ik}(\vec{n})$ can be obtained using the symmetry operation of the O_h group.

Let us suppose that the plane of the defect coincides with the plane (001) of the crystal and the shear wave of the horizontal polarization propagates along the direction [110] in the plane (001). The displacement of the atoms in this wave is in the direction [1 $\bar{1}$ 0] and has the following harmonic dependence on time t:

$$u_n = u(n_3)\exp(ik_\gamma n_\gamma)\exp(-i\omega t), \gamma = 1,2, \tag{4}$$

where ω is the frequency of vibrations. The factor $u(n_3)$ determines the dependence of the amplitude of the wave on the coordinate normal to the defect plane. We write down this factor $u(n)$ in the form for the localized vibration

$$u(n) = \{ \begin{matrix} u_0 q^n, n = 0, 1, 2, 3... \\ u_{-1}q^{-n-1}, n = -1, -2, -3... \end{matrix} \tag{5}$$

when $|q| < 1$ and n numerates the atom planes parallel to the defect plane. The equation (5) demonstrates the decrease of the amplitude of the wave into the bulk of the crystal starting from the defect plane.

Using the displacement in the form (5) and equations (1-4) one can obtain the dispersion relation for the bulk atom $(|n| > 1)$:

$$\lambda = 4 - 2\left(q + \frac{1}{q}\right)\cos k, \tag{6}$$

where $\lambda \equiv \frac{m\omega^2}{4\alpha}, k = k_x = k_y$. The lower and the upper edges of the bulk vibrational spectrum are described by the dependences

$$\lambda_{min} = 4(1 - \cos k), \lambda_{max} = 4(1 + \cos k). \tag{7}$$

In order to study the problem which was analyzed for the long wave lenghtes without account the effects of discreteness we consider the simple model of the planar defect in the FCC crystal. In this model of the planar defect we make the supposition that the force constant of the interaction between the atoms on the opposite sides of the defect differs from the force constant of the interaction between the bulk atoms. This model can be applied for the planar defect description when we pay the main attention to the properties of localized vibrations stipulated for the properties of the interface (not for the properties of the crystals creating such an interface). We show that even such a simple model of the plane defect makes it possible to describe a new type of waves localized near the plane defect and these waves are absent in the limit of the long wave lengths (elasticity theory).

Let us introduce a new force constant β. Suppose that the interface is created by the atom planes $n = 0, n = -1$. Let us write down the equations of motion (1) for these atom layers (such equations are the boundary conditions for the bulk equation of motion):

$$\lambda u(0) = 2(1 + \tfrac{\beta}{\alpha})u(0) - 2u(1)\cos k - 2u(-1)\tfrac{\beta}{\alpha}\cos k$$
$$\lambda u(-1) = 2(1 + \tfrac{\beta}{\alpha})u(-1) - 2u(-2)\cos k - 2u(0)\tfrac{\beta}{\alpha}\cos k. \tag{8}$$

Using the form of the localized wave (5) and introducing the definition $\epsilon = 1 - \tfrac{\beta}{\alpha}$ we obtain the set of the boundary conditions on the defect plane in the form

$$\lambda u_0 = (4 - 2\epsilon - 2q\cos k)u_0 - 2(1 - \epsilon)u_{-1}\cos k$$
$$\lambda u_{-1} = (4 - 2\epsilon - 2q\cos k)u_{-1} - 2(1 - \epsilon)u_0\cos k. \tag{9}$$

The parameter $\epsilon = 1 - \tfrac{\beta}{\alpha}$ characterizes the relative change of the force interaction in the interface. The reinforcement (slackening) of the interaction between the atoms on the opposite sides of the interface corresponds to the values $\epsilon < 0 (\epsilon > 0)$. The dispersive relation for the characteristics of the localized waves has the form:

$$\left| \epsilon - \frac{\cos k}{q} \right| = (1 - \epsilon)\cos k. \tag{10}$$

Two solutions of equation (10) are

$$q_s = \frac{\cos k}{\cos k + \epsilon(1 - \cos k)} \tag{11}$$

$$q_a = \frac{\cos k}{\epsilon(1 + \cos k) - \cos k} \tag{12}$$

The solution (11) corresponds to the symmetrical vibrations relatively to the defect plane, i.e. to the case when the displacements on the opposite sides of the interface coincide $u_0 = u_{-1}$. Respectively, the solution (12) corresponds to the antisymmetrical vibrations $u_0 = -u_{-1}$. The vibrations (11), (12) might have the frequencies as below as above the edges (7) of the continuum spectrum of the bulk vibrations.

The rapture of the bonds through the interface corresponds to the value $\epsilon = 1$. We have the free surface (001) of the FCC crystal in this case where the surface shear wave has the known characteristics [12]:

$$q = \cos k, \lambda_s = 2\sin^2 k. \tag{13}$$

The frequency of the localized near the free surface wave chipps from the edge of the bulk vibrational spectrum as

515

$$\Delta\lambda = 2(1-q)^2 \tag{14}$$

and such value is of the order $\Delta\lambda \sim k^4$ in the limit of the long wave lengths $k \to 0$. Such small chipping off does not allow us to describe SH waves in the framework of the linear non-dispersion elasticity theory and it is the reason of the large penetration depth of the interface SH waves for the long wave lengthes ("quasi-bulk" waves).

When $\epsilon > 0$ (slackening of the interaction constant between the interface atoms) the dumping parameter $q > 0$ and the amplitude of the localized near the interface waves decreases monotonically. The symmetrical vibrations (11) which correspond to the frequencies below the edge of the bulk vibrational spectrum occur with all the wave vectors k. The frequency of the localized vibrations (symmetrical vibrations) chipps from the edge λ_{min} as

$$\lambda_{min} - \lambda_s = \frac{2\epsilon^2(1 - \cos k)^2}{\cos k + \epsilon(1 - \cos k)}. \tag{15}$$

The antisymmetrical vibrations (12) which correspond to the frequencies below the edge λ_{min} occur for the values of the wave vector $k > k_*^H$ where $\cos k_*^H = \frac{\epsilon}{2-\epsilon}$. These vibrations have the frequency

$$\lambda_s = 4(1 - \cos k) - \frac{2(2 \cos k - \epsilon(1 + \cos k))^2}{\epsilon(1 + \cos k) - \cos k}. \tag{16}$$

The critical point $(k_0, \lambda(k_0))$ corresponds to the curve's getting into the bulk vibrational spectrum. The frequency of the localized wave is described by the formula near the critical point:

$$\lambda_s = 8\frac{1-\epsilon}{2-\epsilon} - T(k - k_0)^4, k \to k_0, \tag{17}$$

where $T = \frac{16}{\epsilon^2}(1 - \epsilon)(2 - \epsilon)$.

When $\epsilon < 0$ (slackening of the interaction constant between the interface atoms) the dumping parameter is negative ($q < 0$) and the decrease of the wave's amplitude into the bulk occur with oscillations. The antisymmetrical vibrations (12) with the frequencies above the edge λ_{max} of the bulk vibrational spectrum occur with all the wave vectors k and have the frequency

$$\lambda_s = 4(1 + \cos k) + \frac{2\epsilon^2(1 + \cos k)^2}{\cos k + |\epsilon|(1 + \cos k)}. \tag{18}$$

The symmetrical vibrations (11) which correspond to the frequencies above the edge λ_{max} of the vibrational spectrum occur for the values of the wave vector $k > k_*^H$ where $\cos k_*^H = \frac{|\epsilon|}{2+|\epsilon|}$. These vibrations have the frequency

$$\lambda_s = 4(1 + \cos k) + \frac{2(2 \cos k - |\epsilon|(1 + \cos k))^2}{|\epsilon|(1 - \cos k) - \cos k}. \tag{19}$$

The behavior of the dispersive relation near the critical point is described by the formula (17).

When the value of the parameter ϵ decreases from the limit value $\epsilon = 1$ (free surface) to the value $\epsilon = 0$ (homogeneous crystal) the dispersive curves of the localized vibrations tend to the lower edge λ_{min} of the bulk vibrational spectrum. Only the symmetrical vibrations (11) do occur in the limit case $k \to 0$ for the fixed value of the parameter $\epsilon > 0$. The curves of the both types of vibrations (11), (12) degenerate in the limit $k \to \frac{\pi}{2}$ and describe the vibrations which are localized in one atom layer. The dispersive curves (11), (12) have the following behavior near the point $k = \frac{\pi}{2}$:

$$\lambda_s = 2(2 - \epsilon) \pm 2(1 - \epsilon)\left(k - \frac{\pi}{2}\right). \tag{20}$$

When $\epsilon = 1$ the dispersive curves coincide with the lower edge $\lambda_{\min}$ of the bulk vibrational spectrum of the crystal.

When $\epsilon < 0$ the localized vibrations of both types (11), (12) have the high frequencies and are described by the curves which lie above the edge of the bulk vibrational spectrum. Only the antisymmetrical vibrations (12) do occur in the limit case $k \to 0$ for the fixed value of the parameter $\epsilon < 0$. The curves of the both types of the vibrations (11), (12) degenerate in the limit $k = \frac{\pi}{2}$ and have the following behavior near this point:

$$\lambda_s = 2(2 - \epsilon) \mp 2(1 + |\epsilon|)\left(k - \frac{\pi}{2}\right). \tag{21}$$

In conclusion we would like to mention that some of the results can be obtained in the framework of the elasticity theory accounting the capillary parameters [13].

REFERENCES

1. S.D.Bodar, M.K.Hinders, *Interface Effects in Elastic Wave Scattering*, Springer-Verlag, Heidelberg (1995).
2. Yu.V.Gulyaev, Surface electrosound waves in solids. *Pisma v JETP.* 9:63 (1969).
3. J.L.Bleustein, A new surface wave in piezoelectric materials. *Appl.Phys.Lett.* 13:412 (1968).
4. V.G.Mozhaev, A new type of surface acoustic waves in solids due to nonlinear elasticity. *Phys.Lett.* A139:333 (1989).
5. V.I.Gorenzveig, Yu.S.Kivshar, A.M.Kosevich, E.S.Syrkin, Nonlinear surface elastic modes in crystals. *Phys.Lett.* A144:479 (1990).
6. Yu.A.Kosevich, Nonlinear shear surface waves at interface and planar defects of crystals. *Phys.Lett.* A146:529 (1990).
7. A.S.Kovalev, E.S.Syrkin, Surface solitons in nonlinear elastic media. *Surf.Sci.* 346:337 (1995).
8. G.P.Alldredge, Shear horizontal surface waves on the (001) face of cubic crystals. *Phys.Lett.* A41:281 (1972).
9. B.Djafari-Rouhani, L Dobrzynski, V.R.Velasco, F.Garcia-Moliner, Acoustic waves in sandwich ABC. *Surf.Sci.* 110:129 (1981).
10. I.M.Gelfgat, Account of the spatial dispersion for study of non-Rayleigh surface waves in crystals. *Fiz.Tverd.Tela.* 22:2815 (1980).
11. G.Leibfried, *Gittertheorie der Mechanischen und Thermischen Eigenschaften der Kristalle*, Springer, Berlin-Gottindhem-Heidelberg (1955).
12. I.M.Gelfgat, E.S.Syrkin, Surface waves in layered crystals. *Low.Temp.Phys.* 4:141 (1978).
13. Yu.A.Kosevich, E.S.Syrkin, Capillary phenomena and elasic waves localized near a plane defect in crystal. *Phys.Lett.* A122:178 (1987).

BIOMIMETIC SONAR FOR OBJECT RECOGNITION

Roman Kuc

Intelligent Sensors Laboratory
Department of Electrical Engineering
Yale University, New Haven, CT 06520-8284

INTRODUCTION

Man-made sonar has had limited success for object recognition because the echo waveform varies significantly with the location of the object within the acoustic beam pattern. The variation of the echo from the same object located in different parts of the beam is typically significant [1]. This variation was minimized in previous experiments by careful placement of the objects in a stationary sonar field and by using objects that are large compared to the wavelength and that have shapes (spheres, cubes, pyramids and cones) that exhibit very different scattering properties [2, 3]. The system described in this paper exploits the important biological principle of mobility by translating and rotating the sonar in order to standardize the view of the object by positioning it at a constant location. The sonar design is motivated by bats, whose ears react by rotating to the direction of the echo source, and by dolphins, who appear to move as if to position the object at a standard location in the beam pattern [4, 5, 6].

An active sonar positioned at the end of a robot arm is described that adaptively changes its location and configuration in response to the echoes it observes. The sonar translates in a horizontal plane and rotates about vertical and horizontal axes to position an object at a standard location within the beam patterns. Recognition is accomplished by extracting 32 values from the binaural echo patterns and searching a data base. This paper demonstrates the following three points:

1. Echoes can be processed in the time domain for object recognition,

2. The localization problem of positioning a small object at a known location within the beam patterns can be solved without using memory, and

3. Object recognition can be accomplished directly from the echoes, without generating an intermediate map or image.

Acoustical Imaging, Vol. 23
Edited by Lees and Ferrari, Plenum Press, New York, 1997

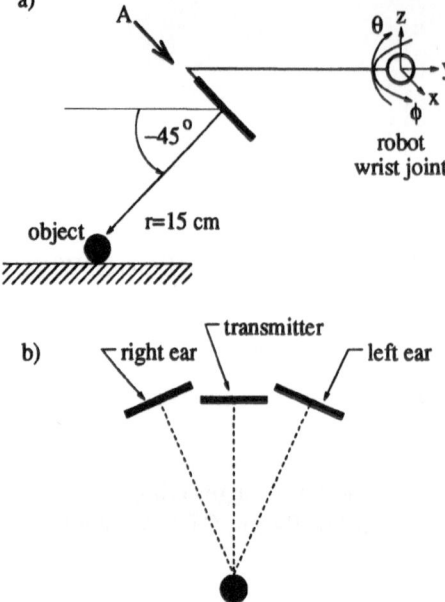

Figure 1: Biosonar configuration mounted on the end of a robot arm. a) Side view. b) View from direction A.

BIOMIMETIC SONAR SYSTEM

The sonar shown in Fig. 1 consists of three identical Polaroid electrostatic transducers that can act as either a transmitter or receiver [1]. Each transducer has a circular aperture with radius $a = 1.9$ cm and is resonant at 60 KHz with a nominal wavelength in air ($c = 343$ m/s) equal to $\lambda = 5.7$ mm. The beam has a full-width half-max amplitude spread of 12°. The center transmitter (mouth) is flanked by two receivers (ears), with the separation between the centers of the mouth and each ear being equal to $D = 4.5$ cm. The ears are rotated in opposite directions by cables connected to a stepper motor to accomplish a type of focusing operation. The entire assembly is mounted on the end of a Armdroid robot arm that provides three translational and two rotational degrees of freedom. The sonar can then be translated and rotated so that it can view an object from any desired aspect.

The sonar moves in a horizontal plane 11 cm above a smooth working surface and is directed downward at a 45° angle. The environment is scanned by transmitting interrogation pulses and processing the signals detected by the receivers while rotating the sonar about the vertical axis to sweep the sonar across the environment. The signal at each receiver is acquired using one of the two channels of an analog-to-digital converter (Gage CS1012) sampling at 10 MHz per channel with 12-bit resolution. A new interrogation pulse is transmitted as soon as the echoes have been detected and processed. Being controlled by a Pentium 120 processor, pulses are emitted as often as every 100 ms. Once echoes are detected, the sonar moves so as to maintain the object at a range of $r = 15$ cm. This range is in the far-field of each transducer ($r > a^2/\lambda$), thus avoiding the complicated near-field wavefronts, but is in the near-field of the three transducer system ($r < D^2/\lambda$) to obtain two novel views of the object.

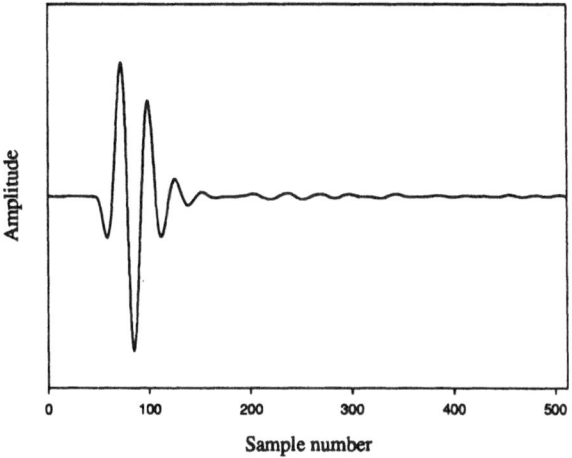

Figure 2: Echo waveform from plane reflector located at a range of 50 cm.

ECHO PROCESSING

An interrogation pulse is produced by impulsively exciting the transmitter. The echo waveform from a plane at 0.5 m range, indicative of the transmitted waveform, is shown in Fig. 2. The waveform has a short duration, making it useful for differentiating nearly identical objects. The usable bandwidth extends from approximately 30 kHz to 120 KHz [7].

The object is positioned in the center of the transmitter beam by driving the bearing to zero, to within the resolution of the robot arm wrist joint ±0.225°. While the object is being driven toward the transmitter axis, the receivers are rotated so that their axes point toward the object using the angle $\arctan(r/D)$. After these two positioning operations have been accomplished the sonar is viewing a particular object from a repeatable height, range and aspect. The sonar can then be instructed to perform one of two operations: learning or recognition.

In the **learning stage**, the sonar pitch is varied by rotating the wrist joint angle θ to observe the echoes from the object at a variety of elevations. Echoes are acquired and processed as the pitch is varied in 0.225° steps, from −43° to −47°. The features extracted from each echo pair, explained below, are stored in memory as a structured data base. This scan is necessary to compensate for the errors in elevation angle that could occur due to the discrete nature of the stepper motors and the backlash in the robot gears.

In the **recognition stage**, an object is viewed at a nominal pitch of −45°, but with an error within ±1.5° due to errors in the joints. There are no similar errors in the range and yaw because these are controlled by sensor feedback. A single interrogation pulse is emitted, the echoes are acquired and processed in the same manner as in the learning stage. The extracted features are compared with data base entries. The entry matching the observed features in the least squared error sense is used for identifying the object.

The steps in echo processing that occur in each ear are shown in Fig. 3. The echo

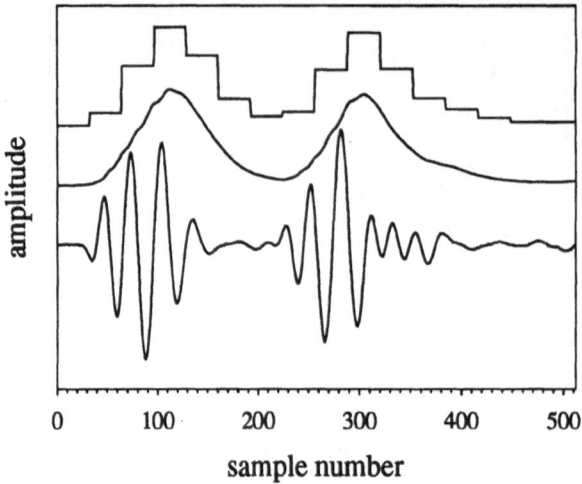

Figure 3: Processing of echoes. From bottom to top, the curves indicate the echo waveform from a 2.54 cm O-ring, the envelope and the 16 values that are used to represent the echo.

waveform consists of 512 samples, starting 50 samples prior to the time the threshold is exceeded. This allows low-level signals at the start of the echo packet to be included in the analysis. This data window duration was chosen to include all the echoes from the largest object in our collection. The waveform data are rectified and applied to a second-order low-pass filter to produce the echo envelope. Logarithmic compression has also been applied to reduce the dynamic range of the echoes and emphasizes the important small-amplitude structure that is present in the echo packet. Since the envelope exhibits a slow variation in time, its 512 samples can be reduced to a smaller number without losing significant information. This is done by segmenting the envelope into 32-sample segments and computing the mean for each segment. This segment size corresponds to approximately one period at the resonant frequency of the transducer. This data reduction allows the echo packet to be represented by an *envelope vector* containing 16 values. These values are normalized to have a constant energy to aid in computing matches. An object is then characterized by combining the vectors from each ear to produce a 32-value vector. These vectors were stored in a data base that is searched for identifying the object from the observed echo features.

EXAMINING A PAPER CLIP

To illustrate the sonar behavior on asymmetric objects we consider recognizing the bent-wire paper clip shown in Fig. 4. The clip was initially positioned with its long axis lying along the axis of the transmitter and its smaller opening closer to the sensor, defining the 0° pose. Typical echo waveforms are shown. The echo vectors from the 0° pose were entered into the data base in the learning stage.

Without moving the clip or the sonar, a series of 20 recognition trials was conducted immediately after completion of the recognition phase. At each trial, the observed vector was compared with those in memory. For this set of 0°-pose trials, the squared

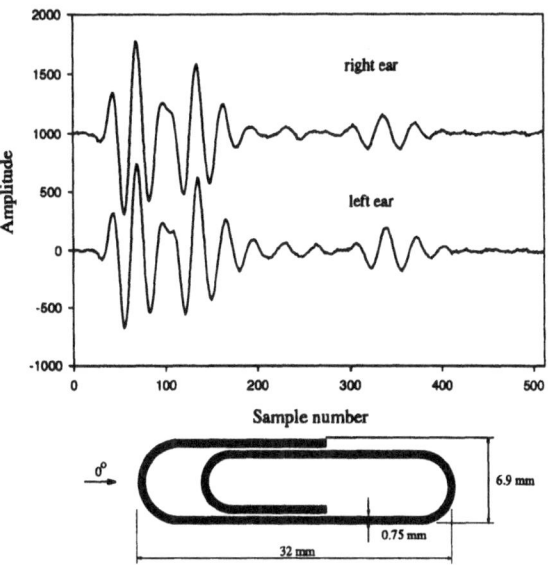

Figure 4: Paper clip used in recognition experiment and echoes observed from $0°$ pose.

error was due only to noise and jitter effects. The clip was then rotated in $1°$ increments upto $10°$ and the set of 20 recognition trials was repeated at each angle. The clip was finally rotated back to the $0°$ pose to provide and indication of repeatability. At each trial, the observed vector pair was compared with the clip-$0°$ vectors in memory. The errors observed in these trials are shown in Fig. 5. The mean of the squared errors at the initial $0°$ pose is 250, shown in dashed line, with a standard deviation of 50. With the envelope vector containing 32 values this corresponds to approximately an average error of 3 in each vector value. With the largest vector values being 300, this is a 1% error. As the pose deviates from $0°$, the error increases. When the clip was examined in the $180°$ pose, the squared error was observed to have a mean of $60,000$. When the clip is returned to the $0°$ pose the error mean was observed to be approximately 600, or approximately 2.5 times the error before the arm moved. This increase was due to errors in the robot arm in placing the sensor at the position where the learning phase was conducted. These result indicate that the pose of the clip can be determined within $4°$.

CONCLUSIONS

This paper describes an adaptive sonar system that is mounted on the end of a robot arm to detect and identify objects using echolocation. For reliable operation, the sonar must standardize the echo waveform and maximize the signal-to-noise ratio. The sonar does this by extracting low-level (time-of-flight) information from the echoes. Recognition is accomplished by using higher-level processing of envelope extraction.

ACKNOWLEDGMENT

The research described in this paper was supported by the National Science Foundation under grant IRI-9504079.

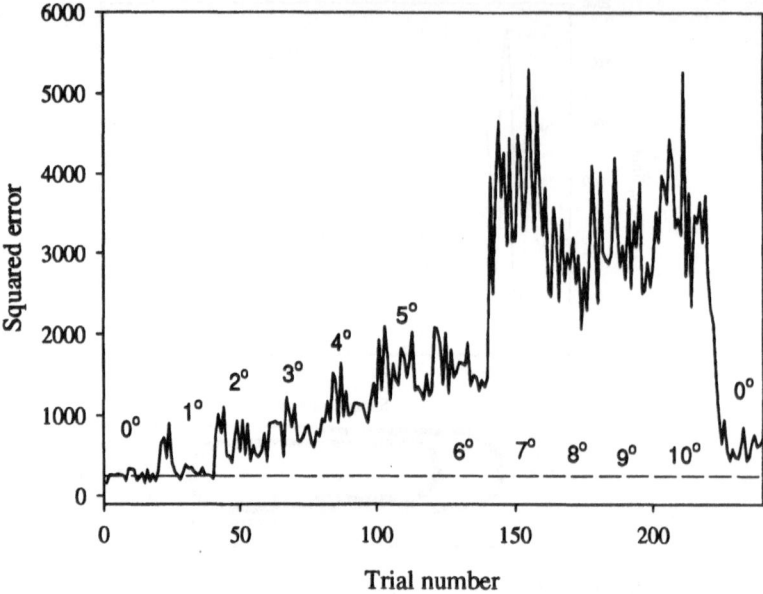

Figure 5: Squared error as a function of paper clip pose.

References

[1] R. Kuc. Biologically-motivated adaptive sonar system. *J. Acoust. Soc. Amer*, 100(3):1849–1854, 1996.

[2] J.M. Richardson, K.A. Marsh, D. Gjellum and M. Lasher. Acoustic recognition of objects in robotics: Determination of type, pose, position and orientation. In L. Kessler, editor, *Acoust. Imaging, vol. 16*, pages 613–620. Plenum Press, New York, 1988.

[3] H.L. Roitblat, P.W.B. Moore, P.E. Nachtigall, R.H. Penner, and W.W.L. Au. Natural echolocation with an artificial neural network. *Int. J. Neural Networks*, 1:239–248, 1989.

[4] P. E. Nachtigall and P. W. B. Moore. *Animal Sonar: Processes and Performance*. Plenum Press, New York, 1988.

[5] J. A. Simmons, M. B. Fenton and M.J. O'Farrell. Echolocation and pursuit of prey by bats. *Science*, 203:16–21, January 1979.

[6] W. W. L. Au. *The Sonar of Dolphins*. Springer-Verlag, New York, 1993.

[7] R. Kuc. Biomimetic sonar recognizes objects using binaural information. *J. Acoust. Soc. Amer*, 101(8), 1997.

DATA FUSION BETWEEN LOW-RESOLUTION BATHYMETRY AND HIGH-RESOLUTION IMAGES FOR THE 3D RESTITUTION OF THE SEA BOTTOM

Eric P. Maillard,[1] Didier Guériot[2]

[1]Gesma, BP 42
 29240 Brest-Naval, FRANCE
[2]Télécom Bretagne
 Brest
 FRANCE

INTRODUCTION

This paper deals with data fusion between high-resolution images from a sidescan sonar and low-resolution bathymetry from a multiple-beam echo sounder. Most of the open literature on this subject deals with the association between image and bathymetry from the same sensor [1] (usually for deep sea applications). While our approach is derived from an original idea based on data from a sector scanning sonar [2], we deal here with a multi-resolution and multi-sensor problem. The experimental results concern a tanker wreck: ahigh-resolution sidescan sonar imaged the wreck from both sides while navigating at a constant altitude, a multiple-beam echo sounder mounted on the hull of a ship acquired the relief of the wreck.

Depending on sensors characteristics acoustic data exhibit different properties. An algorithm which registrates directly all these data would be very complex and hardly evolutive. Two crucial points must be addressed. First, the kind of data are different, so the symbolic data extraction process must look for features that are present on both data sets. Second, the navigational conditions between the acquisitions are dramatically different, so the matching algorithm must be robust. In order to overcome these problems, we propose a two-step registrating system: its first stage is acquisition dependent and extracts information, after projection, through processes tailored for each sensor. Then, this symbolic knowledge feeds a generic system whose goal is to perform the matching task. The matching algorithm has been designed to handle a large variety of data ranging from coarsely detected objects to accurately delimited homogeneous sea-bed regions.

The final restitution consists in two steps. First the projected images are gathered after matching, then the resulting reflection map is merged with the bathymetry map. For each step, the results of the matching process constitute a set of transformations to be applied to the projected data. When shapings are achieved, a 3D representation of the sea bottom is proposed which associates the acoustic images with the sea bottom profile.

IMAGE AND BATHYMETRY GEOREFERENCING

Rough sonar images and bathymetry scans are given in the local reference system of the carrier, georeferencing (fig. 1) consists in transfering sonar images and bathymetry scans in a global geographic system, according to the navigational information provided by the attitude and positioning sensors.

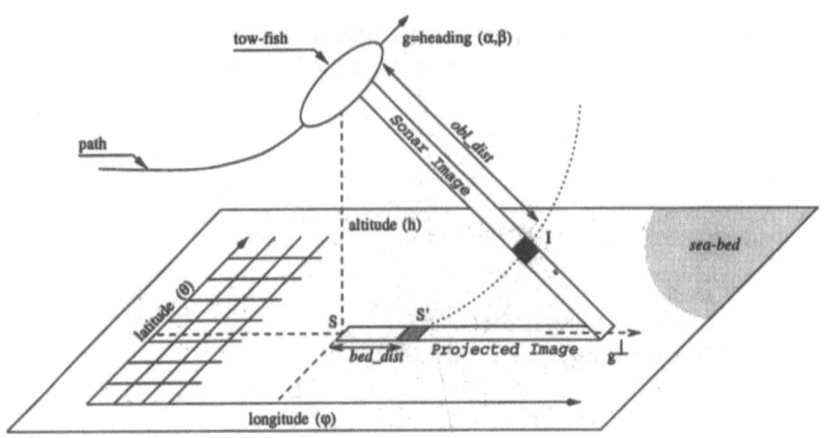

Figure 1: Sonar in-situ schematic representation

Definitions

We note $\Im$, the function which returns the real distance SS' on the sea-bed for a point I in the raw data. Assuming $\Re(bed\_dist)$ gives the altitude of the point S' located at the distance bed_dist from S in a direction $g^{\perp}$, $bed\_dist = \Im(obl\_dist) = \sqrt{obl\_dist^2 - (\Re(bed\_dist) - h)^2}$. Assuming the flatness of the sea-bottom due to the lack of the relief information at this stage of the process, $\Im(obl\_dist) = \sqrt{obl\_dist^2 - h^2}$. Considering the earth as a geoid and assuming that we actually work with neighbouring points, the local average radius of the geoid is $\rho_{A,B} = r_{Eq}\left(1 - \alpha \sin^2\left((\theta_A + \theta_B)/2\right)\right)$ where α is the geoid flattening ratio $\alpha = (r_{Eq} - r_{Po})/r_{Eq}$, r_{Po} is the earth radius at the poles, r_{Eq} is the earth radius at the equator, and A, B are two points defined by φ_i, θ_i. The distance $\delta_{A,B}$ on the geoid between A and B is:
$\delta_{A,B} = \rho_{A,B} \cdot \arccos(\cos(\theta_A)\cos(\theta_B)\cos(\Delta(\varphi_A, \varphi_B)) + \sin(\theta_A)\sin(\theta_B))$

We define the function Ψ which returns the coordinates of the point $S'(\varphi', \theta')$, given the point $S(\varphi, \theta)$, the distance $d = SS'$ and the perpendicular normalized vector $g = (\alpha, \beta)$ to (SS').

$$S' = \Psi(S, d, g) \ with \ \begin{cases} \varphi' = \varphi \mp |\beta d| \\ \theta' = \theta \pm |\alpha d| \end{cases}$$

Sidescan Sonar

For each pixel (i,j) of a recurrence, the vertices of the projected polygon C have the following coordinates (fig. 2):

$$C_{(i,j)} = \Psi(S_i, p_{i,j}, g^{\perp}) \qquad C_{(i+1,j)} = \Psi(S_{i+1}, p_{i+1,j}, g^{\perp})$$
$$C_{(i,j+1)} = \Psi(S_i, p_{i,j+1}, g^{\perp}) \qquad C_{(i+1,j+1)} = \Psi(S_{i+1}, p_{i+1,j+1}, g^{\perp})$$

with $S_i = \Psi(S, d_i, g)$, $d_i = \Im(pixel\_resolution * i)$, and $p_{i,j} = \frac{(p_d + i*p_e - p_d I)}{J} * \left(j - \frac{J}{2}\right)$.

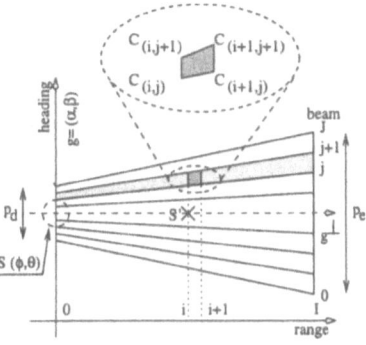

Figure 2: Sidescan sonar beam representation

Bathymetry

For each sample m of the depth during a scan n, the projected measurement S is calculated according to the position (φ, θ) and attitude (κ, μ, ν) of the ship:

$$S_{(m,n)} = \zeta(m, \varphi, \theta, \kappa, \mu, \nu)$$

The function $\zeta(.)$ corresponds to a series of rotations. Once the exact position of each sample of the bathymetry is estimated, a regular grid is defined by its longitude step $(d\varphi)$ and latitude step $(d\theta)$. For each vertices of the grid (fig. 3), the height is computed according to:

$$B_{(i,j)} = \frac{1}{\sum_{m_{ij}, n_{ij}} d(B, S)} \times \sum_{m_{ij}, n_{ij}} d(B, S) S_{(m_{ij}, n_{ij})}$$

with $d(.)$ euclidian distance and $S_{(m_{ij}, n_{ij})} = \{S, \ i < m < i + d\varphi \text{ and } j < n < j + d\theta\}$

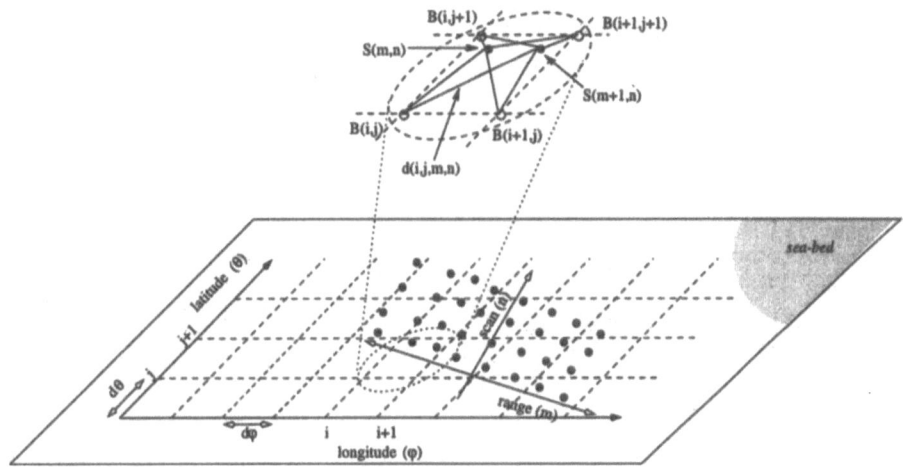

Figure 3: Bathymetry map generation through scan fuzzy discretization

IMAGE AND BATHYMETRY MATCHING

Genetic principle of matching

Genetic algorithms provide a powerful stochastic engine to find solutions to non-linear problems. Operating on binary encodings of the search variables or chromosomes, the task consists in minimizing a cost function which measures the accuracy of a given matching, defined by a composition of transformations (rotation, translation and scaling). Final wrapping between the two sets of symbolic data is obtained through convergence to the function optima, while current solutions offer interesting approximations. For this application, a chromosome represents directly such a composition of transformations (fig. 4) suited to match contours or isobaths [3].

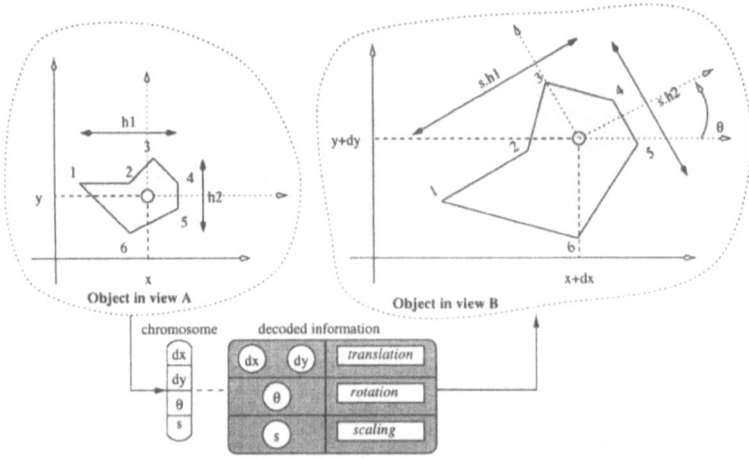

Figure 4: Genetic encoding of the transformation between two symbolic data

Sonar/bathymetry matching

The first step of the process consists in defining the common set of symbolic features. The high level of back-scattered energy in a sidescan sonar image corresponds to a high gradient of the sea relief. Detection of such phenomena provides sea marks that are encoded through a symbolic representation. Once the symbolic maps extracted, the matching process proposes a possible transformation that should be applied to one map to compute the final association. According to the encoding scheme of the matching problem, an homogeneous sampling is not required. This property is useful when the two contours are obtained through processes working on data coming from different sensors.

EXPERIMENTAL RESULTS

The proposed approach has been applied to the 3D reconstruction of a wreck. The Lagadmor system, consisting of two high-resolution sidescan sonars and a gap filler, imaged the wrech while following successively a North-South starboard track and a South-North port track. The high-resolution sidescan sonar images are presented in fig. 5. In order to extract symbolic information, the image is first low-pass filtered to eliminate the speckle, then an adapted thresholding isolates the echos. The results are presented in fig. 6. A RESON Seabat 9001 was used to provide bathymetry data of the area. Due to a bad sea state, the scans are not uniformly spaced. While the projection provides a better representation (fig. 7), the final map is far from perfect since the experimental setup did not allow an accurate recording of the platform attitude. The first matching process is applied to the two images. Once the transformation has been proposed by the genetic algorithm, one image is wrapped according to the estimated parameters. A max-value selection process provides the final reflection map. This map (fig. 8 left) is then matched with the isobath (fig. 8 right) of the bathymetry map. The 3D restitution is presented in fig. 9.

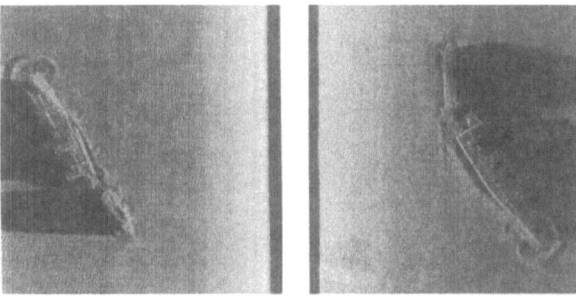

Figure 5: Raw images of the wreck: left) port, right) starboard

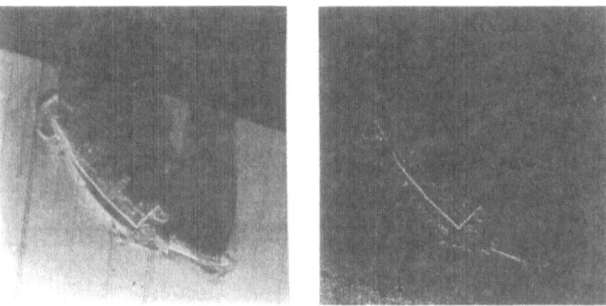

Figure 6: Processing of one image: left) low pass filtering , right) segmentation

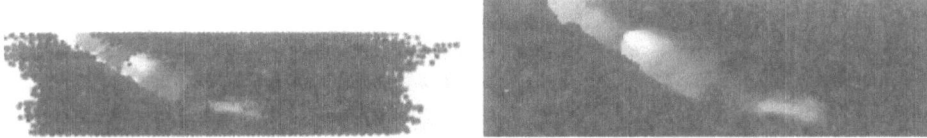

Figure 7: Bathymetry data: left) projected samples, right) discretized grid

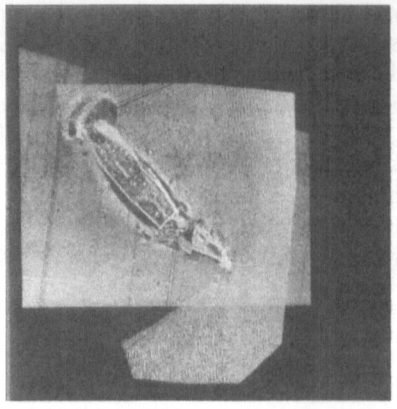 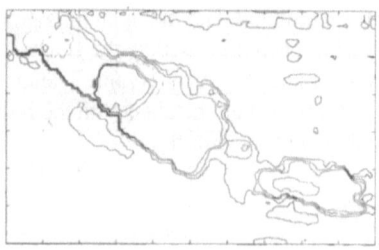

Figure 8: Final data before association: left) reflection map, right) isobath map

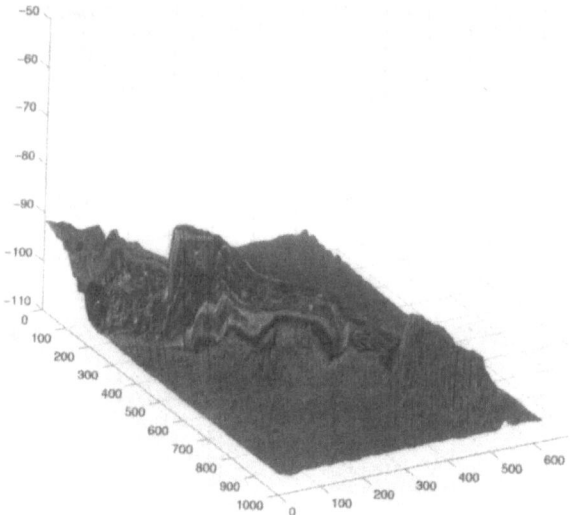

Figure 9: 3D restitution of the wreck

REFERENCES

[1] C. de Moustier et all. Simultaneous operation of the sea-beam multibeam echosounder and the seamarc ii bathymetric sidescan sonar system. *IEEE Journal of Oceanic Engineering*, 15(2):84–94, 1990.

[2] Benoît Zerr and Bjarne Stage. Three-dimensional reconstruction of underwater objects from a sequence of sonar images. In *IEEE International Conference on Image Processing*, 1996.

[3] Didier Guériot, Eric P. Maillard, and Jean-Pierre Kernin. Sonar image registration through symbolic matching: a fuzzy local approach using genetic algorithms. In *Oceans MTS / IEEE*, 1996.

INVERSION OF ACOUSTIC REFLECTION DATA
FOR SEA-BOTTOM RECONSTRUCTION

Panagiotis J. Papadakis

Foundation for Research and Technology-Hellas (FORTH)
Institute of Applied and Computational Mathematics (IACM)
P.O.Box 1527, 711 10 Heraklion, Crete, GREECE

INTRODUCTION

A problem of current interest with applications to geophysical science is the reconstruction of the top-most layers of the sea-bottom. Acoustic techniques are preferred to more classical methods - such as the use of seismic devices or coring samples- since they are more efficient, less time consuming and less expensive.

A set of codes have been developed in the past at FORTH-IACM for this problem in the case of signals emitted from parametric sources. The methods used in these codes were designed to be applied to particular models of the sea floor structure. Models of one layered bottom consisting of either fluid, or elastic material, as well as, models for two layered bottoms consisting of fluid-fluid, elastic-fluid, or elastic-elastic materials have been examined. The codes developed work well in these cases and they have been applied to experimental data either from laboratory experiments or from real experiments in the open sea. The input to these algorithms were the emitted and reflected from the sea-floor signals at several (sometimes 10-15) angles of incidence. These requirements were not always able to be satisfied in a real experiment restricting the application of the methods in such cases.

In this work an inversion technique is presented for the general case of a two-layered bottom consisting either of fluid materials, elastic materials or both kind of materials.The input to the algorithm is the emitted acoustic signal and the bottom echoes at two or three different angles of emission. These requirements can be easily satisfied in a sea experiment. The use of two only angles of incidence makes possible to recover all the properties of a one or two layered sea bottom except the elastic properties of the last layer of the sea-floor. The parameters estimated are the densities of the bottom layers, the compressional and shear wave velocities, the thickness of the top layer and the attenuation coefficients of both layers. There are a maximum of 11 parameters to be recovered and the complexity of the mathematical formulation of the problem makes impossible the use of direct inversion methods with reliable results in all cases.

The new inversion method presented in this study is based in the fact that after some manipulation the problem in hand can be split into several steps where only a few of the unknown parameters are estimated at a given step. The values of the parameters found in one step may be used in the following steps to facilitate the estimation of the rest of the unknowns, improving their own accuracy at the same time. Thus the method is accurate and fast. Another advantage is that it requires a minimum amount of data since only two or three angles of incidence are needed. The only additional requirement is that the bottom echoes at the different layers of the sea floor as they appear in the reflected signal are distinct and they can be isolated from the rest of the signal. If this is not the case then only an average of the combination of the layers can be obtained.

In the first part of the paper the models under examination and the mathematical formulation will be presented. The different approaches for the one or two-layered case will be explained and discussed. In the second part the inverse method will be presented and in the third part error induced synthetic data will be presented in order to demonstrate the accuracy and efficiency of the method.

THE MATHEMATICAL FORMULATION

The model under consideration is shown in Figure 1. A plane wave is emitted incident on the bottom with incident angle θ. From the reflected signal the reflection coefficient is calculated for two different angles of incidence.

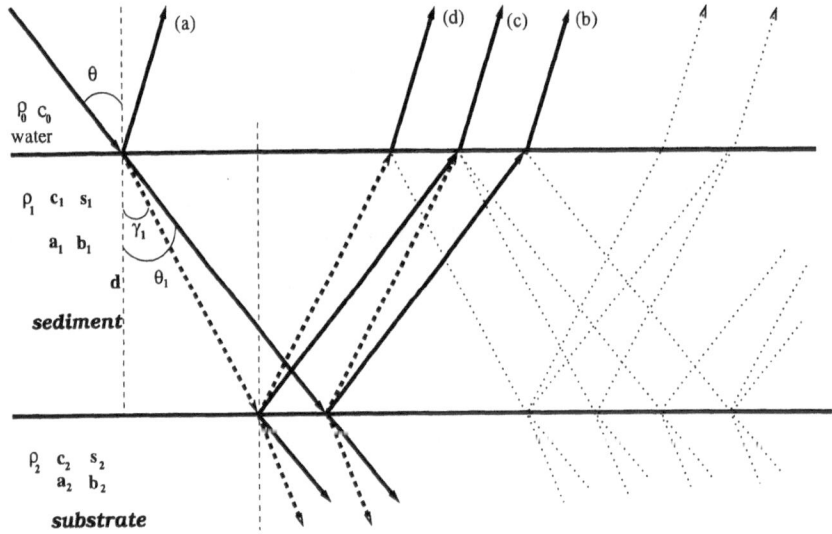

Figure 1. The 2-layer bottom model

A sample incident and reflected signal (for an angle of incidence 25 degrees) are shown in Figure 2. The properties to be recovered are the densities (ρ_1, ρ_2), the compressional sound velocities (c_1, c_2), the sear velocities (s_1, s_2) and the attenuation coefficients (a_1, a_2, b_1, b_2) of the compressional and shear waves respectively, of the bottom layers, as well as the thickness d of the sediment layer. We assume that the density ρ_0 and sound speed c_0 of the water are known and given. The rays denoted by (a), (b), (c) and (d) in figure 1 correspond to the recorded echoes (a), (b), (c) and (d) in figure 2. In the case of a one layered bottom the model is simpler (figure 3) and the unknown parameters are only those of the first layer. If a layer does not consist of an elastic material then all the rays corresponding to shear waves (denoted by dashed lines in figures 1 and 3) do not exist.

The mathematical formulation of the reflection coefficient of a plane wave for the general case of two elastic layers is a complex nonlinear equation of the form: $R = R(f, \theta, \vec{x})$ where $\vec{x}$ is the array of all the parameters described in the previous section. It will take a couple of pages in order to write down this equation (see for example Brekhovskikh (1982)). The reflection coefficient is a complex number which depends on both frequency and angle. An attempt to use this formulation in the inverse process will result to inaccurate results due to the strong non-linearity of the equation.

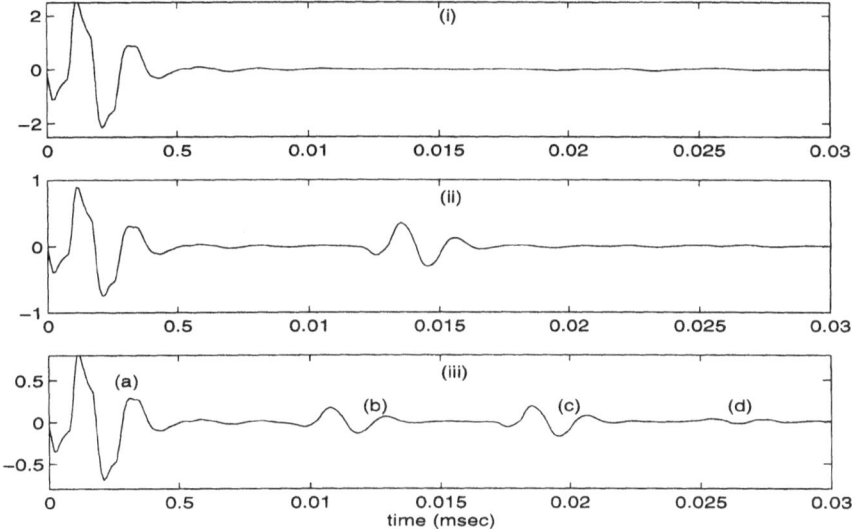

Figure 2. Reference and reflected signals

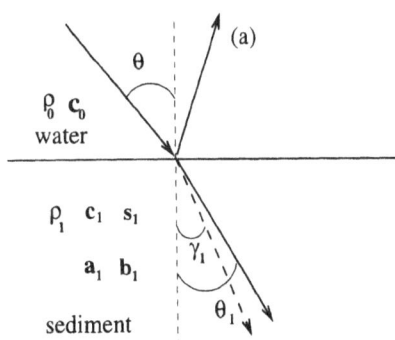

Figure 3. The 1-layer bottom model

In this work a different approach will be applied which will facilitate the use of the reflected data for inversion purposes. Let us emphasize that previous equation for R takes into consideration not only the first few (main) returns but also all the multiple returns from the bottom (denoted as dotted lines in figure 1). However in realistic situations almost none of these returns will be visible in the recorded signal. Looking at the equations which describe each one of the bottom echoes (a), (b) etc, it can be seen that the formulas are much simpler. So, for the four main (not multiple) returns in the general case the reflection coefficient becomes:

$$R(f,\theta) = R_1(\theta) + R_2(\theta)\exp(2\omega Mi) + R_3(\theta)\exp(\omega(M+N)i) + R_4(\theta)exp(2\omega Ni) \quad (1)$$

where, $\quad \omega = 2\pi f, \quad M = (d/c_1)\cos\theta_1, \quad N = (d/s_1)\cos\gamma_1$

The R_1, R_2, R_3, R_4, are complex numbers independent of the frequency and the terms of equation correspond to the bottom echoes (a), (b), (c), (d) respectively. In the case of a one layered bottom only the first term (R_1) is present and is given by the formula (assuming elastic material):

$$R(\theta) = R_1 = \frac{Z_l\cos^2 2\gamma_1 + Z_t\sin^2 2\gamma_1 - Z}{Z_l\cos^2 2\gamma_1 + Z_t\sin^2 2\gamma_1 + Z} \quad (2)$$

where, $\quad Z = \rho_0 c_0/\cos\theta \,, \quad Z_l = \rho_1 c_1/\cos\theta_1 \,, \quad Z_t = \rho_1 s_1/\cos\gamma_1$

In the two-layer case where the first layer is fluid, terms 3 and 4 do not exist. Equation (1) as well as simpler equations obtained from this equation (such as (2)) will be used in our inversion scheme.

REMARK: Actually in laboratory or sea experiments the emitted signal is a narrow or wide beam and not a plane wave. However if the source is far away from the bottom surface it can be considered as plane wave.

THE INVERSE METHOD

The one-layered bottom case

In this case the reflection coefficient is independent from the frequency as it can be seen from equation (2). In the case however of a fluid material the equation is even simpler and becomes:

$$R(\theta) = R_1 = \frac{Z_l - Z}{Z_l + Z} \quad (3)$$

If there are data available for only two angles of incidence θ and ϕ equation (3) is used since it is not possible to determine if the bottom is elastic. The main steps of the inverse procedure are:

1) Dividing the spectrum of the reflected signal by the spectrum of the reference signal we obtain a complex number which is the reflection coefficient R_1. This number should be constant for all frequency values. So we have two complex numbers $R(\theta)$ and $R(\phi)$.

2) From Equation (3) we calculate the ratio $\rho_1 c_1/\cos\theta_1$ and $\rho_1 c_1/\cos\phi_1$

3) Dividing the two previous expressions we calculate the compressional sound velocity c_1. Dividing the real and imaginary parts of c_1 we calculate the attenuation coefficient a_1.

4) From one of the expressions from step 2 we calculate ρ_1

If there are three angles of incidence then we can find out if the bottom layer is elastic and estimate its properties using the following anditional steps:

5) Using one of the previous angles and the new angle ω we repeat the previous steps. If the results are the same then the material is indeed fluid and no further processing is required. If not we continue with step 6

6) Using all three angles and equation 2, we form a system of three non-linear complex equations in 5 unknowns: $c_1, a_1, \rho_1, s_1, b_1$. We solve it with a library nonlinear solver.

534

The two-layered bottom case

Once more if there are data available for only two angles of incidence θ and ϕ we should assume that the substrate is fluid. The equation used in this case is (1) but the formulas for R_1, R_2, R_3 are even simpler. The steps in this case are:

1) From the reflected signal in the time domain we keep only the first pulse which corresponds to the first bottom echo (a). We obtain the spectrum of the new signal and divide it by the spectrum of the reference signal. The resulted number (which is independent from the frequency) corresponds to R_1 (equation 2).

2) Taking the first two returns in the signal (corresponding to echoes (a) and (b) and dividing again its spectrum by that of the reference signal we obtain the quantity

$$R_1 + R_2 exp(2\omega M i) \tag{4}$$

3) Since R_1 and R_2 are independent from the frequency if we divide equations (4) for two different frequencies we can calculate the ratio: $(d/c_1)\cos\theta_1$

4) Applying steps 1-3 for angle ϕ we calculate again the ratio $(d/c_1)\cos\phi_1$. Dividing these two ratios we obtain the c_1 and a_1. Then from one of them we calculate the thickness d.

5) We repeat the previous steps keeping in the signal the third return as well. Thus we obtain the s_2 and b_2. Then from R_1 we caclulate the ρ_2 and the first layer is completly recovered.

6) From the equation for R_2 we can then calculate analytically the c_3 and ρ_3.

Now if a third angle ω is available we can obtain the elastic properties of the substrate (if it is not fluid). The steps required are exactly the same as steps 5 and 6 from the one layer case.

REMARK: The attenuation coefficent b_2 of the shear wave does not contribute sufficiently in the calculation of the reflection coefficient and thus is hard to estimate accurately. So, if no a-priori estimates are known, we assume it is zero. Also when using only two angles of incidence the resulting values for the last layer are meaningless. this is an indication that the last layer is not fluid but elastic.

RESULTS

The code was tested with synthetic error induced data. The results from the three phases of the algorithm are presented in table 1. Three angles were used in the code:

Table 1: Results for error induced data

	exact	$15-19$	$15-22$	$15-19-22$
ρ_1	1290	1291	1291	1291
c_1	2110	2108	2107	2107
s_1	940	937	936	936
a_1	0.83	0.76	0.75	0.75
b_1	0.6	0.69	0.79	0.79
d	1.8	1.79	1.79	1.79
ρ_2	1600	—	—	1477
c_2	2900	—	—	3060
s_2	1600	—	—	1734
a_2	0.3	—	—	0.15
b_2	0.6	—	—	0.47

15°, 19°, 22°. The third and fourth column of the table contain the results with angles 15-19 and 15-22 respectively using steps 1-6 of the procedure described in the previous section. The values of the second layer were not physically acceptable values. The last column contains the results for the itteration method using all three angles. Note that the values of the parameters of the first layer were not recalculated but instead, the values obtained from column four were used. In figure 4 we present the exact reflection coefficient (dotted line) and the calculated reflection coefficient (solid line) using the values from the last column of table 1 and for the angle 22°.

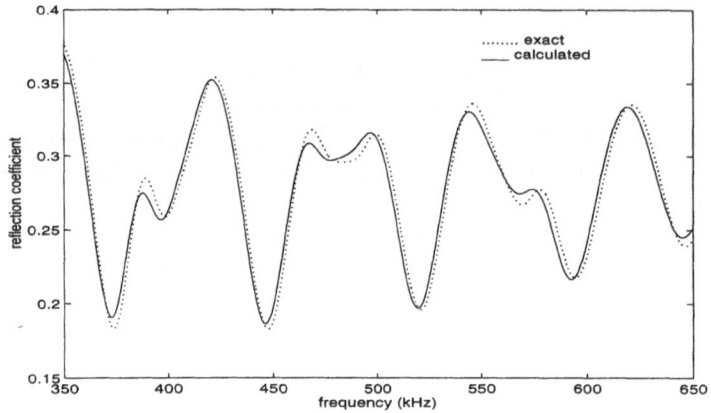

Figure 4. The exact and calculated reflection coefficient (22°)

CONCLUSIONS

The inverse method discussed in this study was developed in order to solve the problem of bottom reconstruction for one or two layered bottom. The sea–bottom may consist from either fluid or elastic materials. The input to the algorithm is the emitted acoustic signal, and the reflected signal at two or three oblique angles of incidence. This makes possible to use this method with experimental data since minimum amount of data is needed. Another requirement is that at least the first three pulses in the reflected signals are distinct and do not overlap.

Under the above assumptions the algorithm works very fast and is accurate as was shown in the tests even when the data are not error free. Subjects for further research is the extension of this method for cases where the pulses from the different layers are not distinct and the application of the method to laboratory and sea experimental data.

ACKNOWLEDGMENT

This work was partially supported by the European Union under contract CI1-CT94-0093 (**ACUSTICA** project)

REFERENCES

L.Brekhovskikh *Waves In Layered Media* , Springer Verlag (1982)

LITTORAL COHERENCE LIMITATIONS ON ACOUSTIC ARRAYS

Kenneth D. Rolt[1] and Philip A. Abbot[2]

[1] Sanders, A Lockheed Martin Company
P.O. Box 868, Nashua NH 03061-0868

[2] Ocean Acoustical Services and Instrumentation Systems, Inc.
5 Militia Drive, Lexington, MA 02173

INTRODUCTION

An initially planar acoustic wave traveling in the ocean has phase fronts which become wrinkled as the wave travels because the ocean is inhomogeneous from a spatial viewpoint and because the ocean is nonstationary from a temporal viewpoint. The coherence of the wave, and the ability to perform acoustical imaging with the wave, is thereby reduced. In our context, imaging could be: direction finding for a passive towed line array, range-depth-azimuth localization in a passive matched-field-processing (MFP) system, or any of the more usual imaging concepts for an active sonar. The mechanisms for temporal and spatial coherence loss are many, but include internal waves, ocean surface motion in shallow water acoustic waveguides, thermal microstructure and others. The coherence loss of a wavefront, as described above for the ocean case, also applies to the case for biomedical ultrasonic imaging, with similar results: blurred images.

The notion, that coherence of a wave is reduced by either space or by time, suggests an imaging system have an acoustic aperture length smaller than the spatial coherence limit, and a signal processing window smaller than the temporal coherence limit. However, long arrays built for deep ocean use may be too long, by exceeding the spatial coherence limit, when the same arrays are used in much shallower water. Likewise, long synthetic arrays may suffer the same performance degradation for the very same reason. These issues relating the length (or size) for real or synthetic arrays with the spatial and temporal limits for coherence in the ocean motivated our study.

COHERENCE, DIMENSIONLESS PHASE RATE AND EXPERIMENTAL DATA

Coherence for an array requires that the phase error across the array be less than $\pi/2$ (radians). Coherence, fluctuations, and the mechanisms and consequences are discussed in greater detail in companion papers from the *1974 Low-Frequency Noise and Propagation Workshop* held at Woods Hole Oceanographic Institution. Ocean environmental studies of temporal stability are usually conducted by measuring the phase stability of the environment, for a given geometry (source and receiver depths and separation), frequency and ocean conditions (surface effects, internal waves, $c(z)$ profile etc). The phase stability is often presented as a standard deviation of the phase, averaged over a time period. One quantitative measure of temporal stability is the complex temporal correlation function of a multipath received signal $r(\tau)$ defined per Dyer $et\ al.$ (1987) and Dahl $et\ al.$ (1988):

$$r(\tau) = \exp[-2\pi^2 v^2 \tau^2 - i2\pi f_D \tau] , \qquad (1)$$

where τ is the time delay for the correlation, f_D is the Doppler shift frequency of the signal radiated at frequency f, and v is the fluctuation parameter known as the rms single-path phase rate between the source and the receiver. The phase rate v is dependent on frequency, range and the changing sound speed profile, and is an effective measure of ocean dynamics on acoustic propagation. The phase rate v is, from a theoretical basis (Chernov, 1960; Urick, 1967 and 1982), is dependent on the type of scattering mechanism involved in sound propagation, and is modeled as $v \sim f\,R^{3/2}$ for R small compared to the scatterer size, and $v \sim f R^{1/2}$ for R large compared to the scatterer size.

Table I summarizes $ocean\ temporal\ stability\ experiments$ from the last 30+ years for deep ocean fluctuations, and for synthetic aperture sonar (SAS) stability measurements in deep and shallow water. The $dimensionless\ phase\ rate$, v/f_c , provides a comparison among these experiments which were done at different frequencies, ranges, water depths and conditions, and with different levels of control. The DEFINITIONS are at the Table bottom, including the relation of v, σ_{phase} and T. Figure 1 plots v/f_c vs. range. Highly stable environments tend to be plotted near the page bottom. Two distinct regions are noted: the lower-left region comprises SAS-type experiments which were well controlled (no source-receiver motional errors) and had single propagation paths in relatively shallow water. Also plotted on the lower-left are two dark lines for $v/f_c \sim R^{3/2}$, to depict the expected behavior from theory. The second region appears on the upper-right side of Fig. 1, and it consists of mostly deep-water experiments, some with few controls (there are errors in the measurements making the data appear worse than it actually should be). Also plotted on the upper-right are two dark lines having $R^{1/2}$ slope, and having intermediate shading; the lower of the two lines (Dyson $et\ al.$, 1976) is based on internal wave-induced acoustic scattering. The $R^{1/2}$ behavior is also consistent with theory for R >> scatterer size.

A similar study of $ocean\ spatial\ stability\ experiments$ was made, per Figure 2, only now the results are shown for $dimensionless\ aperture\ length\ L/\lambda$ versus range, where L is the spatial coherence length (versus the temporal coherence T used above). The data in this figure is much more sparse, but there is a trend: deep water measurements show better spatial coherence (larger L/λ) than shallow water, and for a given experiment (Wille & Thiele, 1971 under severe sea state conditions), lower frequencies provide larger L/λ (20 at 400 Hz versus 2 at 3.2 kHz). From the spatial stability data we conclude that L/λ from 30 to 100 are possible in shallow water, noting that L/λ from 100-400 have been observed for horizontal beamforming with long towed arrays in deep water.

In an attempt to reconcile the $temporal\ stability$ data from the $spatial\ stability$ data, especially from the context of an array, real or synthetic, moving horizontally through a shallow water area, we chose a realistic mid-frequency example: a 3 kHz acoustic wave and an receiver moving at U~5 m/s We converted the temporal scale data to a spatial scale through $L = U\,T$, and then plotted the results per Figure 3 on a L/λ versus range. Figure 3 shows that the assumed $30 < L/\lambda < 100$ zone from the spatial coherence data is below the L/λ zone from the temporal data, until 40 km where the zone begin to overlap.

We conclude that the $spatial\ coherence\ scale\ dominates\ in\ shallow\ water$, which in turn means: a) arrays longer than the spatial coherence length for shallow water will not provide the full array gain expected by theory, and b) arrays shorter than the spatial coherence length for shallow water should meet the array gain expected by theory, and c) both real and synthetic arrays could be fashioned to exceed the spatial coherence length provided they use signal processing to correct for the change in phase. This type of corrective processing has been used with success for passive towed synthetic arrays (Edelson and Sullivan, 1992) and is a form of autofocusing.

TABLE I - SUMMARY OF OCEAN STABILITY EXPERIMENTS

Authors	f_c, Hz	R, m	σ_{phase}	Time	ν (phase rate), Hz	ν/f_c	Comments
Christoff, Loggins & Pipkin, 1982	100 k / 100 k	48	.04 rad @ 3 m, bottom / .31 rad @ 9 m, surface	20 min / 2 min	5.31×10^{-6} / 4.11×10^{-4}	5.31×10^{-11} / 4.11×10^{-9}	St. Andrews Bay, 12m ± 0.6m, Tidal, Rail Based Platform
Gough & Hayes, 1989	15-30 k	130	$\leq 10°$	60 sec	4.63×10^{-4}	3.09×10^{-8} to 1.54×10^{-8}	From a pier on Loch Linnhe (Scotland), 6 kts Tidal Flow, Rail Based Platform
Stowe, et al., 1974	10 k	2500	3.5°	60 sec	1.62×10^{-4}	1.62×10^{-8}	AUTEC. Bottom Moored Hydrophone, Low Platform Motion
Williams, 1976	400	107-495 km	49°	148 to 450 sec	9.2×10^{-4} to 3.02×10^{-4}	2.3×10^{-6} to 7.56×10^{-7}	Deep water, Bermuda Plateau L/λ = 93 to 300 for SAS Applications
Williams & Battestin, 1976	400	270-1200 km	90°	120 to 480 sec	2.08×10^{-3} / 5.21×10^{-4}	5.21×10^{-6} / 1.3×10^{-6}	Bermuda-Eleuthera Island, RSR paths
Fitzgerald, et al., 1976	10.04	65-71 km	15°	1000 sec	4.17×10^{-5}	4.15×10^{-6}	Hatteras Abyssal Plain
Spindel, Porter, Jaffee 1974	406	210 km	15 cycles / 7.5 cycles	> 3 hrs @ d = 1500 m / > 3 hrs @ d = 305 m	1.39×10^{-3} / 6.94×10^{-4}	3.42×10^{-6} / 1.71×10^{-6}	Deep Ocean for SAS Applications
Dyer, et al., 1987	25-200	100 km			0.4 to 2.4×10^{-3}	$1.15 \pm 24\% \times 10^{-5}$	Marginal Ice Zone (MIZ) Tests
Clark, 1974	406	550-1250 km			2.8×10^{-3} / 4.0×10^{-3}	6.9×10^{-6} / 9.9×10^{-6}	Eleuthera-Bermuda (Project MIMI)
Georges, Boden, Palmer, 1994	57	9.1 Mm			1.3×10^{-4} (Flat Ray) / 3.9×10^{-4} (Steep Ray)	2.3×10^{-6} / 7.0×10^{-6}	HIFT: Ascension Island
Birdsall, Metzger, Dzieciuch, 1994	57	5.51 Mm			2×10^{-3} (max dev)	3.5×10^{-5} (max)	HIFT: Christmas Island
ARPA SAS Brief	20-150 k	91				6.6×10^{-9}	From Brief Notes
Russian SW Tests (Bunkin, et al.)	300	70 km				3×10^{-6}	Water Depth = 200 m
Raytheon High Freq. SAS Tests (Ciany et al.) 1994	100-300 k	300				2×10^{-9}	Near Bottom, Water Depth = 180 m
Parvulescu	400	36.5 km	1000 sec x 400 oscillations = $T f_c$*			6.25×10^{-7}	Tongue of the Ocean & AUTEC, MESS Processing **

DEFINITIONS: f_c = Center Frequency σ_{phase} = Standard Deviation of Phase   T = period

R = Range   Time = Measurement Interval of σ_{phase} * order of magnitude

λ = Wavelength $\nu = \sigma_{phase}/T$ ** Assume T = 360ν

NOTE: Doppler Shifts Removed From Data

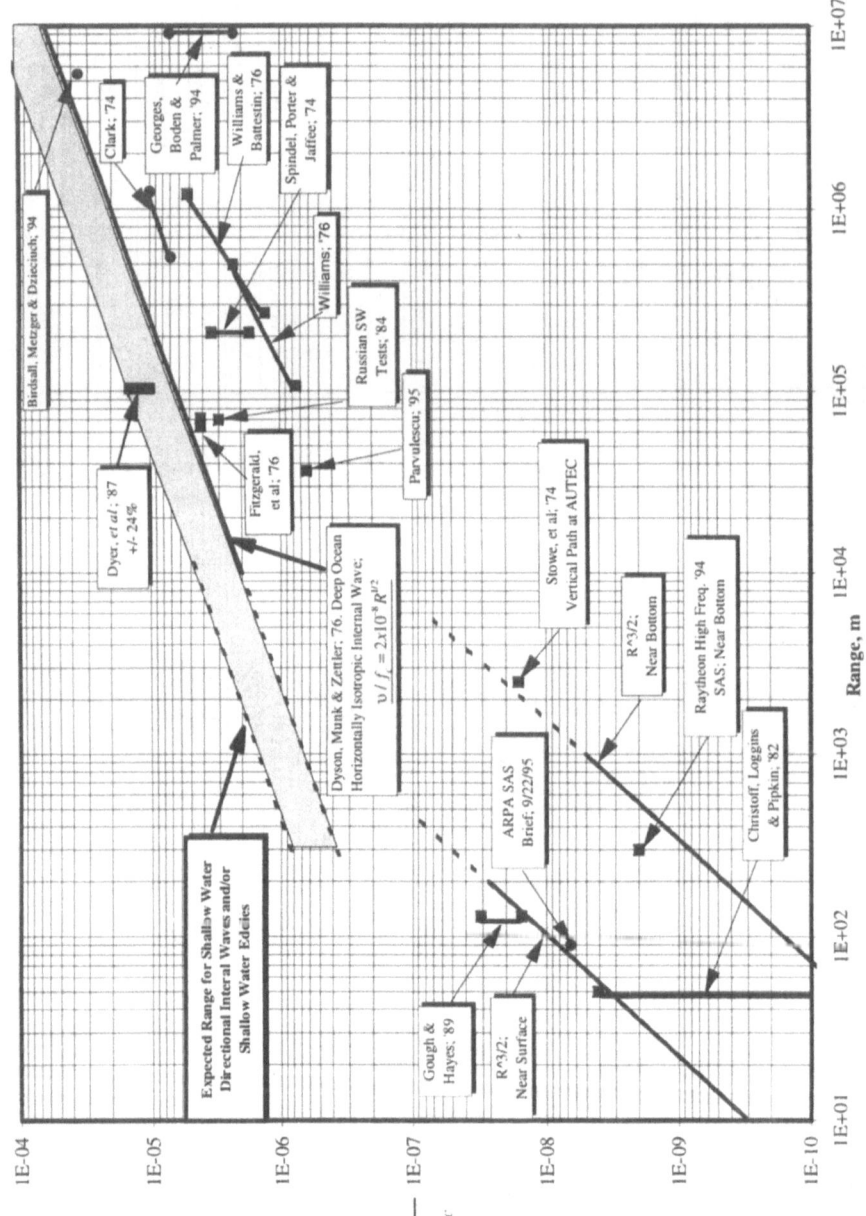

Figure 1 - Summary of Ocean Stability (Temporal) Experiments

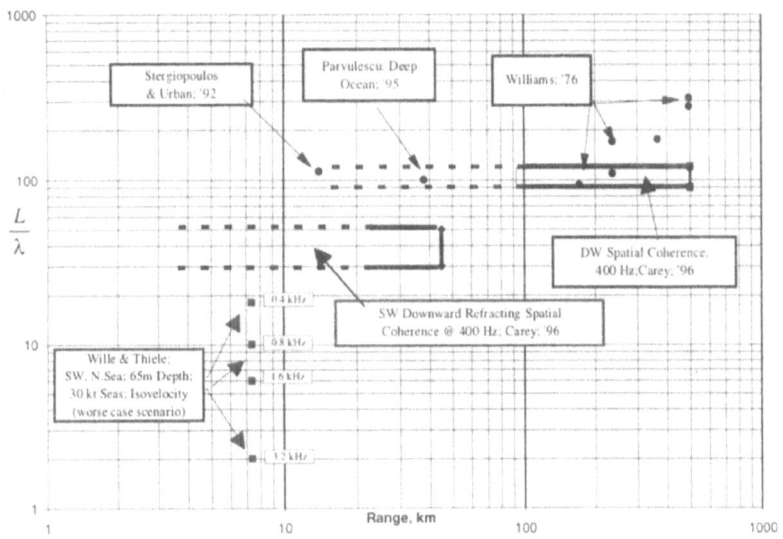

Figure 2 - Spatial Coherence for Deep ● and Shallow ■ Water

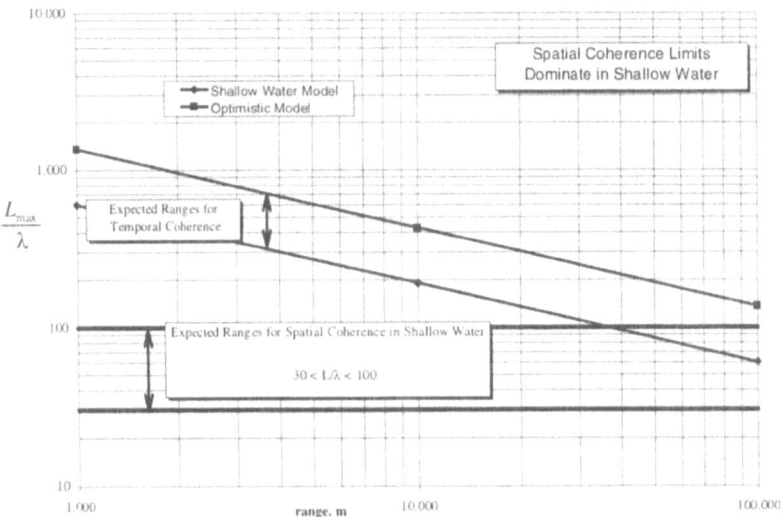

Figure 3. Lmax/λ vs range for temporal and spatial stability data,
U=5 m/s and f=3kHz

ACKNOWLEDGMENTS

This paper is an outgrowth of synthetic aperture sonar work described in Celuzza *et al.* elsewhere in this volume. Dr. Ira Dyer and Dr. William Carey, both of MIT, provided information and suggestions on coherence and medium stability. This work was supported by SBIR N95-207 via the Naval Underwater Warfare Center.

REFERENCES

Birdsall, T.G., 1974. Coherence, in *Int. Workshop on Low-Frequency Propagation and Noise*, 357-364, held at Woods Hole, Massachusetts, Oct. 1974. U.S.G.P.O. O-244-456 (1977).

Birdsall, T.G., *et al.*, 1994. *J. Acoustic Soc. Am.* 96(4): 2353-2356.

Bunkin, F.V., *et al.*, 1985. Preliminary results of a study of the space-time variability of a shallow sea along a stationary acoustic test range, *Sov. Phys. Acoust.* 30(5): 351-354.

Carey, W., 1996. The determination of signal coherence length base on signal coherence and gain measurements in deep and shallow water, *Institute of Acoustics International Conference on Arrays and Beamforming in Sonar*, 23-25 July 1996, University of Bristol, UK.

Chernov, L.A., 1960. *Wave Propagation in a Random Medium*, McGraw-Hill, NY.

Christoff, J.T., *et al.*, 1982. Measurement of the temporal phase stability of the medium, *J. Acoust. Soc. Am.* 71(6):1606-1607(L).

Ciany, C.M., *et al.*, 1994. Propagation medium impact on sonar coherent processing for high frequency synthetic aperture imaging, *Symposium on AUV, 1994 IEEE Technology*.

Clark, J.G. and Kronengold, M., 1974. Long period fluctuations of CW signals in deep and shallow water, *J. Acoustic Soc. Am.* 56(4):1071-1083.

Dahl, P., *et al.*, 1988. Measurement of temporal fluctuations of CW tones propagated in the Marginal Ice Zone, *J. Acoust. Soc. Am.* 83(6): 2175-2179.

Dyer, I., 1987. Ocean dynamics and acoustic fluctuations in the Fram Strait Marginal Ice Zone, *Science*, 236: 435-436.

Dyer, I., 1974. Fluctuations, in *Int. Workshop on Low-Frequency Propagation and Noise*, 365-386.

Dyson, F., Munk, W. and Zetler, B., 1976. Interpretation of multipath scintillations Eleuthera to Bermuda in terms of internal waves and tides, *J. Acoust. Soc. Am.* 59(5): 1121.

Edelson, G. 1994. Supplemental bibliography of acoustical synthetic aperture (U), *U.S.N. J. Underwater Acoust.* 44(4):1057-1064. *Contains many papers not cited, for brevity sake, in this reference list.*

Fitzgerald, R.M., Guthrie, A.N. and Shaffer, J.D., 1976. Low-Frequency coherence transverse to the direction of propagation, *J. Acoust. Soc. Am.* 60(3):752-753(L).

Flatte, S.M., 1979. *Sound Transmission Through a Fluctuating Ocean*, Cambridge Univ. Press, England.

Georges, T.M., *et al.*, 1994. Features of Heard Island signals received at Ascension Island, *J.Acoustic Soc. Am.* 96(4):2441-2447.

Gough, P.T. and Hayes, M.P., 1989. Measurement of acoustic phase stability in Loch Linnhe Scotland, *J. Acoust. Soc. Am.* 86(2):837-839(L).

Kooij, T., 1995. *ARPA SAS Program Brief.*

Lathrop, J.D., 1995. High area rate reconnaissance (HARR) and mine reconnaissance hunter (MR/H) exploring development programs, *SPIE* 2496:350-356.

Parvulescu, A., 1995. Matched-Signal ("MESS") processing by the ocean, *J. Acoust. Soc. Am.* 98(2):943.

Rolt, K. 1993. A bibliography of acoustical synthetic aperture imaging (U), *U.S.N. J. Underwater Acoust.* 43(3):889-904. *Also contains many papers not cited, for brevity sake, in this reference list.*

Spindel, R.C., 1974. Phase fluctuations, coherence and internal waves, in *Int. Workshop on Low-Frequency Propagation and Noise*, 423-464.

Spindel, R.C., *et al.*, 1974. Long-range sound fluctuations with drifting hydrophones, *J. Acoust. Soc. Am.* 56(2):440-446.

Urick, R.J., 1967. *Principles of Underwater Sound for Engineers*, 149-154, McGraw-Hill, NY.

Urick, R.J., 1982. *Sound Propagation in the Sea*, Peninsula Publishing, Los Altos CA. Ch. 12 and 13.

Wille, P. and Thiele, R., 1971. Transverse horizontal coherence of explosive signals in shallow water, *J. Acoust. Soc. Am.* 50:348-353.

Acoustic sounding the statistics of fluctuating fluid

Bernhard Schwarz-Röhr, Volker Mellert

Dept. of Physics, University of Oldenburg
D-26111 Oldenburg
FAX (+49) 441 798 3698

Abstract

The lateral correlation of a spherical wave propagating outdoors and in a model atmosphere was measured in order to determine statistical parameters of the propagation medium. Wind- and temperature data are monitored at the same time, and the structure constants and outer scale of tubulence are calculated. The acoustic correlation is evaluated using the parabolic equation method and assuming a von Kármán spectrum of turbulence. The acoustic measurements yield comparable results to the statistical parameters derived independently from the wind and temperature measurements. The deviations are due to the fact that in the present theory a von Kármán spectrum of turbulence is assumed which is not always met in the real experimental setup. On the other hand the local determination of the time history of wind and temperature is less representative than the spatial average which is performed by the tomographic sensing of the acoustic wave.

1. Introduction

Randomly fluctuating properties of a fluid are in general locally measured by appropriate sensors assuming that the spatial representation of the statistics is passing the sensor location by an average flow of the fluid (e.g. "frozen" turbulence). A typical example is the spatial field of temperature and wind in the atmosphere which is measured in samples of large separation with local time histories of temperature and wind. But it is desirable to registrate the average statistics of the fluid properties within a certain volume independently of the special sensor location which might reflect properties of a local artefact. Moreover, the "frozen" hypthesis is not always fulfilled and depends on the average flow, or the spatial representation of the fluctuating structure is such large that the average motion is not capable to map it into the time history of one sensor. Therefore it is useful to perform a tomography of the statistical properties of the fluid volume.

The imaging method is based on the fact that the wave fronts of a sound wave are changed due to the fluctuations of velocity and temperature in a fluid, in this case the turbulent atmosphere. The aim of the present study is to evaluate the fluctuations in amplitude and phase of a sound wave in order to derive first and second moments of velocity and temperature in a certain segment of the turbulent atmosphere. The most important parameters for characterizing the turbulent atmosphere are the structure constants C_T^2, C_v^2, and outer scale $L_0 = 1/\kappa_0$. Based on the "parabolic equation method" (PEM) it is possible to calculate the propagation of the first and second moment of a plane and spherical wave. Analytical results are known for certain spectral representations of the fluctuations of wind and temperature[1]. One objective of the present study is to test the theoretical predictions experimentally. The other objectiive is to provide a new remote sensing measurement tool to determine acoustically the structure constants and outer scale.

2. Theory

The statistical properties of wind and temperature are assumed to be homogeneous. Velocity and temperature fluctuate around an average value which is constant in time and space. Both assumptions do not hold for the real atmosphere but are approximately true for a time intervall of some minutes and in the stratified boundary layer over a distance of some hundred meters. Profiles are not investigated. The wave is assumed to propagate through a statistically homogeneous part. The fluctuations of wind and temperature are modeled by a von Kármán spectrum, modifying the infinite cascade of turbulence in the Kolmogorov spectrum by an outer and an inner scale of largest and smallest turbulent structures. In this investigation, only the large scale L_0, representing the largest structures of wind and temperature fluctuation within the time interval of measurement is of interest.

The sounding wave is reduced in amplitude due to geometrical spreading and to usual atmospheric absorption but not due to turbulence (provided the wave is not radiated as a narrow beam). The turbulent structures cause phase- and magnitude-fluctuations of the wave and some scattering. Hence, the energy is not reduced but spreaded over a larger volume. The amplitude is regarded as a complex quantitiy taking magnitude and phase informations into account. Spherical wave propagation is assumed in order to compare theoretical predictions with experimental results, but the theory could deal with other wave geometries as well.

The theoretical description is based on the PEM developed by *Klyatskin* and *Tatarskii* in the late sixties. Within the scope of this paper only the results of the theory will be given. Starting from a lossless wave equation isotropic, homogeneous and frozen turbulence is assumed. Two essential approximations enter into the theory. First, one assumes, that the wave propagation can be described in the parabolic approximation. The second approximation is to replace the correlation function of the medium by an effective one with vanishing correlation length in the direction of the propagation path ("Markov approximation"). Introducing Cartesian coordinates $(x, \mathbf{r}) = (x, y, z)$, where the x-axis is directed along the propagation path, the acoustical pressure $p(x, \mathbf{r}, t)$ is written as

$$p(x, \mathbf{r}, t) = A(x, \mathbf{r})e^{ikx}e^{i\omega t}. \tag{1}$$

Here $A(x, \mathbf{r})$ denotes the complex amplitude, k the acoustical wavenumber ω/c_0. For any set of points in the yz-plane $\mathbf{r}_1 \ldots \mathbf{r}_n, \mathbf{r}_1' \ldots \mathbf{r}_m'$ the parabolic equation method leads to closed differential equations for arbitrary statistical moments of the form

$$\left\langle A(x,\mathbf{r}_1)A(x,\mathbf{r}_2)\dots A^*(x,\mathbf{r}_1')A^*(x,\mathbf{r}_2')\dots\right\rangle.$$

The mutual coherence function

$$\Gamma(x,\mathbf{r},\mathbf{r}') := \left\langle A(x,\mathbf{r})A^*(x,\mathbf{r}')\right\rangle \tag{2}$$

depends on the wave geometry and analytical solutions are known only for special cases. Assuming spherical waves the mutual coherence function is given by[1]

$$\Gamma(x,\mathbf{r},\mathbf{r}') = \frac{1}{4\pi x^2}\exp\left\{ik\frac{[\mathbf{r}-\mathbf{r}'][\mathbf{r}+\mathbf{r}']}{2x} - \pi^2 k^2 x \int_0^1 dt \int_0^\infty d\kappa\ \kappa[1-J_0(\kappa\rho t)]\ \Phi(\kappa)\right\} \tag{3}$$

Here k denotes the acoustical wavenumber ω/c_0, $\Phi(\kappa)$ is a function of the spatial power spectrum of turbulence, $\rho = |\mathbf{r}-\mathbf{r}'|$ is an abbreviation for the lateral distance of the two points $\mathbf{r}$ and $\mathbf{r}'$. Note, that up to a phase factor the mutual coherence function decays like $\exp(-k^2 x\dots)$ independent of the form of the turbulent power spectra. For von Kármán spectra the the following expression for the coherence is obtained:

$$\Gamma(x,\mathbf{r},\mathbf{r}') = \frac{\exp\left(ik[\mathbf{r}+\mathbf{r}'][\mathbf{r}-\mathbf{r}']/2x\right)}{(4\pi x)^2}\Gamma_T(x,\rho)\Gamma_v(x,\rho)$$

where $\Gamma_T(x,\rho)$ and $\Gamma_v(x,\rho)$ are contributions to $\Gamma(x,\mathbf{r},\mathbf{r}')$ due to temperature and velocity fluctuations, given by

$$\Gamma_T(x,\rho) = \exp\left[-2\gamma_T\ x\ R_T\left(\kappa_0\rho\right)\right],$$

$$\Gamma_v(x,\rho) = \exp\left[-2\gamma_v\ x\ R_v\left(\kappa_0\rho\right)\right].$$

γ_T and γ_v are abbreviations for

$$\gamma_T = Bk^2\frac{C_T^2}{T_0^2}\kappa_0^{-\frac{5}{3}}, \qquad \gamma_v = 4Bk^2\frac{C_v^2}{c_0^2}\kappa_0^{-\frac{5}{3}}.$$

here $B = \pi/(12\ \Gamma(1/3)) \approx 0.098$, T_0 is the mean value of temperature corresponding to the mean sound speed c_0. $R_T\left(\kappa_0[\mathbf{r}-\mathbf{r}']\right)$ and $R_v\left(\kappa_0[\mathbf{r}-\mathbf{r}']\right)$ are abbreviations for

$$R_T(\mu) = 1 - \frac{2^{\frac{1}{6}}}{\Gamma(\frac{5}{6})}\frac{1}{\mu}\int_0^\mu d\beta\ \beta^{\frac{5}{6}}K_{\frac{5}{6}}(\beta) \tag{4}$$

$$R_v(\mu) = 1 - \frac{2^{\frac{1}{6}}}{\Gamma(\frac{5}{6})}\frac{1}{\mu}\int_0^\mu d\beta\ \beta^{\frac{5}{6}}\left[K_{\frac{5}{6}}(\beta) - \frac{\beta}{2}K_{\frac{1}{6}}(\beta)\right] \tag{5}$$

where $K_{5/6}(\beta)$ and $K_{1/6}(\beta)$ denote modified Bessel-functions.

Measurable quantities are obtained from the moment equation (3) as follows. The acoustical pressure at a microphone located at $(x,\mathbf{r})$ is given by Eq. (1). Its Fourier transform is obviously

$$\tilde{p}(x,\mathbf{r},\omega) = A(x,\mathbf{r})e^{ikx}. \tag{6}$$

Therefore, the averaged cross spectrum of the two pressures turns out to be the mutual coherence function

$$\left\langle\tilde{p}(x,\mathbf{r},\omega)\tilde{p}^*(x,\mathbf{r}',\omega)\right\rangle = \left\langle A(x,\mathbf{r})A^*(x,\mathbf{r}')\right\rangle = \Gamma(x,\mathbf{r},\mathbf{r}'). \tag{7}$$

The parabolic equation method does not account for atmospheric absorption. Furthermore, it is desirable to eliminate source and receiver sensitivities from the measurements. Within the scope of the PEM the intensity is not affected by medium fluctuations, as can be seen by setting $\rho=0$ in Eq. 3. Therefore the transfer characteristics are normalized with the averaged intensity leading to

$$\frac{\Gamma(x,\mathbf{r},\mathbf{r}')}{\sqrt{\Gamma(x,\mathbf{r},\mathbf{r})\Gamma(x,\mathbf{r}',\mathbf{r}')}} = \exp\left(ik\frac{[\mathbf{r}+\mathbf{r}'][\mathbf{r}-\mathbf{r}']}{2x}\right)\exp\left(-2x\left[\gamma_v R_v(\kappa_0\rho) + \gamma_T R_T(\kappa_0\rho)\right]\right) \quad (8)$$

The complex phase factor $\exp(ik[\mathbf{r}+\mathbf{r}'][\mathbf{r}-\mathbf{r}']/2)$ does not carry any information about the turbulence. Therefore, multiplying Eq. (8) with its complex conjugate leads to

$$\frac{\left|\Gamma(x,\mathbf{r},\mathbf{r}')\right|^2}{\Gamma(x,\mathbf{r},\mathbf{r})\Gamma(x,\mathbf{r}',\mathbf{r}')} = \exp\left(-4x\left[\gamma_v R_v(\kappa_0\rho) + \gamma_T R_T(\kappa_0\rho)\right]\right)$$

Using relationship (7) to express $\Gamma(x,\mathbf{r},\mathbf{r}')$ by the averaged spectra of the acoustic pressure leads to

$$\frac{\left|\left\langle\tilde{p}(x,\mathbf{r},\omega)\tilde{p}^*(x,\mathbf{r}',\omega)\right\rangle\right|^2}{\left\langle\tilde{p}(x,\mathbf{r})\tilde{p}^*(x,\mathbf{r})\right\rangle\left\langle\tilde{p}(x,\mathbf{r}')\tilde{p}^*(x,\mathbf{r}')\right\rangle} = \exp\left(-4x\left[\gamma_v R_v(\kappa_0\rho) + \gamma_T R_T(\kappa_0\rho)\right]\right) \quad (9)$$

Within the context of signal processing the left hand side of the last equation is defined as the so called "coherence" η^2. This coherence is directly evaluated by any Fast-Fourier-Transform signal analyzer. The coherence (9) is used in the data exploition, as it carries all information of the medium fluctuations that is hidden in the second moment of the sound amplitude.

3. Outdoor measurement setup for fluctuations of wind

Measurements were made over a flat lawn site surrounded by some medium height trees. Weather conditions were neutral with (low) average windspeed up to about

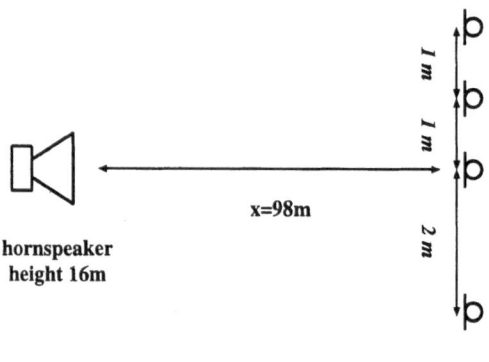

Fig. 1: Sketch of measurement setup outdoors.

1 m/s, clouded sky and practically no temperature fluctuations. Fig. 1 gives a sketch of the measurement setup. A wide angle horn loudspeaker mounted on a mast at 16 m height is driven with sinusoidal pulse-trains of appropriate length allowing to eliminate undesired reflections. At a distance of 98 m a lateral, horizontal arrangement of four microphones allows to pick up the spatial correlation of the wave in 10.5 m height. All microphone signals are stored on digital tape recorders together with synchronisation signals of the pulse generator. The pulse-train was set up by five sinusoidal pulses with frequencies 650 Hz, 1.08 kHz, 1.8 kHz, 3 kHz, 5 kHz, and 100 ms duration. They were Hanning-switched with a frequency dependent rise- and fall-time from 40 ms to 20 ms. The repetition rate of the pulse-train was about 0.5 Hz. An ensemble of 150 successive indiviual measurements at each frequency representing a time window of about 5 min was registrated.

The meteorological data were monitored in parallel with sensors mounted on the same mast. Wind was registered with a three-component ultrasound anemometer. Temperature was monitored with a hotwire probe. A typical time series of the wind is shown in Fig. 2 over a period of 35 min.

3.1 Determination of C_v^2 and κ_0 from anemometer data

Considerable care has to be taken in the exploition of the anemometer measurements due to the vector character of the velocity, because even if one assumes isotropic and homogeneous turbulence, for any separation vector $\mathbf{r}$ several correlation functions of the form

$$B_{ij}(\mathbf{r}) = \big\langle \mathbf{v}(\mathbf{r}_0) \cdot \mathbf{e}_i \quad \mathbf{v}(\mathbf{r}_0 + \mathbf{r}) \cdot \mathbf{e}_j \big\rangle$$

can be defined. Here $\mathbf{v}$ denotes wind velocity fluctuations. $\mathbf{e}_i$, $\mathbf{e}_j$ are unit vectors in a any arbitrarily choosen directions. Taking both $\mathbf{e}_i$ and $\mathbf{e}_j$ in the direction of the

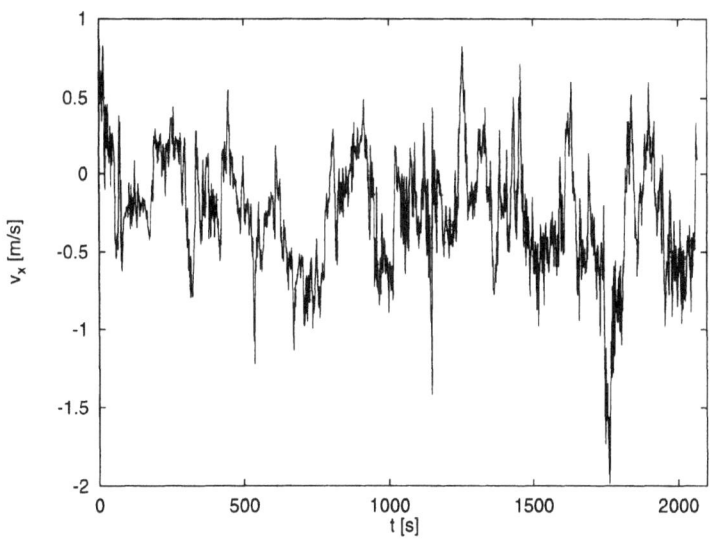

Fig. 2: Time series of one windvector component during a measurement intervall of about 35 min, registrated with an ultrasound anemometer.

separation vector $\mathbf{r}$ leads to the longitudinal correlation function

$$B_{rr}(r) = \left\langle \mathbf{v}(\mathbf{r}_0) \cdot \mathbf{e}_r \quad \mathbf{v}(\mathbf{r}_0 + \mathbf{r}) \cdot \mathbf{e}_r \right\rangle$$

where $\mathbf{e}_r = \mathbf{r}/r$. For isotropic turbulence this function depends only on the magnitude r of the separation vector. The Fourier transform of B_{rr} is called one-dimensional, longitudinal power spectrum $F_1(k)$

$$B_{rr}(r) = \int_{-\infty}^{\infty} dk \ e^{-ikr} F_1(k).$$

In order to calculate $F_1(k)$ from a locally measured time series of wind velocities $\mathbf{v}_m(t)$, the following steps must be taken. First the average wind vector $\bar{\mathbf{v}}$ is calculated. Next the time function of the velocity component in direction of $\bar{\mathbf{v}}$ is given by

$$\tilde{v}_r(t) = (\mathbf{v}_m(t) - \bar{\mathbf{v}}) \cdot \mathbf{e}_{\bar{v}},$$

here $\mathbf{e}_{\bar{v}}$ denotes $\bar{\mathbf{v}}/|\bar{\mathbf{v}}|$. With frozen turbulence, $\tilde{v}_r(t)$ is the spatial function of the velocity fluctuation component in direction of the mean wind $\mathbf{v}(\mathbf{r}) \cdot \mathbf{e}_v$ at the position $\mathbf{r} = \bar{\mathbf{v}} \, t$:

$$\mathbf{v}(r\mathbf{e}_r) \cdot \mathbf{e}_v = \tilde{v}_r(r/\bar{v}).$$

Denoting the power spectrum of $\tilde{v}_r(t)$ by $\tilde{F}(f)$, the spatial power spectrum $F_1(k)$ is therefore given by

$$F_1(k) = \frac{\bar{v}}{2\pi} \tilde{F}\left(k \frac{\bar{v}}{2\pi}\right).$$

It must be emphasized that due to the need of calculating the wind velocity component in direction of the mean wind vector, hot-wire or cup-anemometer data, that give

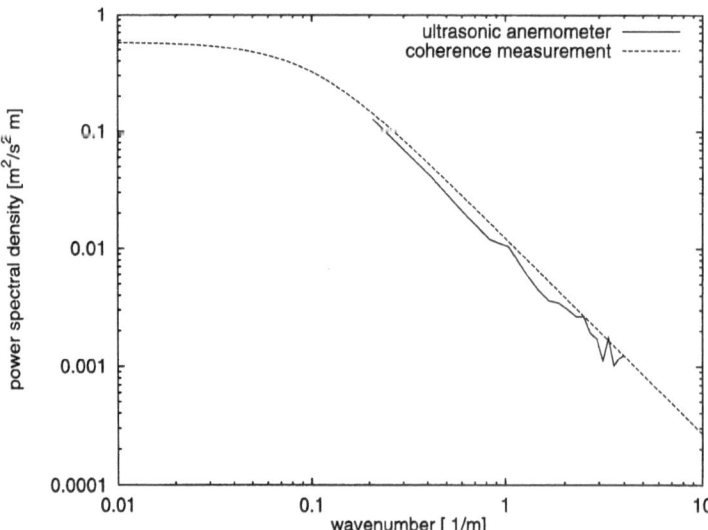

Fig. 3: One-dimensional power spectral density derived from the ultrasonic anemometer data compared with the estimated spectrum from the coherence measurement. ($C_v^2 = 0.1$ m^2/s^2 m$^{-2/3}$, $\kappa_0 = 0.1$ m^{-1})

only the magnitude of the wind vector, are generally not sufficient to calculate the longitudinal spectrum correctly. The one-dimensional power spectrum derived from the ultrasonic anemometer data is plotted in Fig. 3 (solid line). The lower bound of the wavenumber k is determined by the largest structure that crosses the anemometer within the registration interval of each time history used for ensemble averaging. Since time intervals of 40 s were used, no information can be achieved below k= 0.2 m^{-1}.

The dashed line in Fig. 3 corresponds to the one dimensional von Kármán spectrum

$$F_1(k) = \frac{C_v^2}{3\ \Gamma\left(\frac{1}{3}\right)} \frac{1}{(k^2 + k_0^2)^{5/6}} \tag{10}$$

using the parameters $C_v^2 = 0.1$ m^2/s^2 m$^{-2/3}$, $\kappa_0 = 0.1$ m^{-1} that are deduced from the *acoustic* coherence measurements elucidated below. Both curves coincide with reasonable accuracy.

3.2 Acoustic determination of C$_v^2$ and κ_0

The results of the coherence measurements are plotted in Fig. 4 together with the theoretical prediction (Eq. 10) for $C_v^2 = 0.1$ m^2/s^2 m$^{-2/3}$, $\kappa_0 = 0.1$ m^{-1}, showing reasonable agreement. Thus the parabolic theory seems to be applicable in this experiment.

During the experiments the neutral meteorological condition was characterized by neglegible temperature fluctuations, therefore $C_T^2 = 0$ is assumed. Thus the structure constant of the velocity fluctuations C_v^2 and the outer scale L_0 –or equivalently $\kappa_0 = 1/L_0$– remain to be determined. These two parameters are combined in the coherence (Eq. 9). In the special case of two microphones at propagation distance x from the

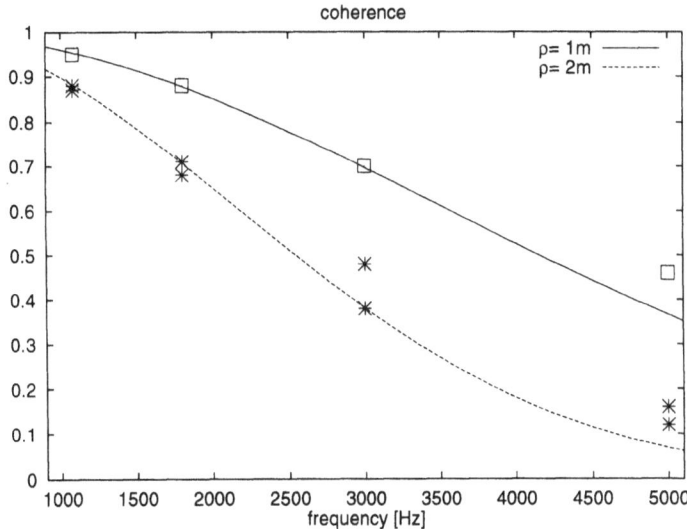

Fig. 4: Measured coherence at lateral distances of 1 m and 2 m compared with the predicted coherence(Eq. 12, $C_v^2=0.1$m^2/s^2 m$^{-2/3}$, $\kappa_0=0.1$ m^{-1}).

source with lateral displacement ρ, this equation might be written as

$$\eta^2(x,\rho) = \exp\left(-4x \; \mathcal{B} \; k^2 \; \frac{C_v^2}{c_0^2} \; \kappa_0^{-\frac{5}{3}} \; R_v(\kappa_0\rho)\right). \tag{11}$$

Of course the determination of two parameters requires two measured values or at least three microphones. Denoting the two coherences measured at lateral distances ρ_1 and ρ_2 by η_1^2 and η_2^2 respectively, the following trick allows to separate the parameters. First determine κ_0 by computing the ratio

$$\frac{\ln \eta_1^2}{\ln \eta_2^2} = \frac{R_v(\kappa_0\rho_1)}{R_v(\kappa_0\rho_2)}. \tag{12}$$

For fixed values of ρ_1 and ρ_2, the right hand side of the last equation turns out to be a monotonic function of κ_0. Thus for each frequency the value of $\ln \eta_1^2 / \ln \eta_2^2$ provides an unique estimate for κ_0. The present measurements give values of κ_0 in the range of 0.09 m^{-1} to 0.105 m^{-1}. Knowing κ_0, the structure constant C_v^2 can be calculated for each measured coherence by Eq. (11), leading to values in the range of 0.083 to 0.12 m^2/s^2 m$^{-2/3}$.

The predictions of the PEM for the coherence using the average values κ_0=0.1 m^{-1} and C_v^2= 0.10 m^2/s^2 m$^{-2/3}$ are plotted in Fig. 4. Obviously these parameters agree well with the measurements, proving the proposed evaluation procedure on the basis of the PEM.

The von Kármán spectrum of velocity fluctuations (Eq. 12) is calculated using the same parameter values and is plotted in Fig. 3. The assumed von Kármán spectrum is in excellent agreement with the anemometer data.

4. Laboratory measurement setup for fluctuations of temperature

Since meteorological parameters are difficult to control outdoors a model atmosphere was established in a laboratory scale. Electrical heating elements warm up air in an open box of size $0.8 \times 0.8 \times 0.5$ m^3. The convecting warm air mixes with cold air and is additionally whirled by a grid covering the opening of the box. Temperature is measured with a hot-wire anemometer used as a temperature probe. Correlation with a second temperature probe gives the average convection velocity, ranging from 0.7 m/s

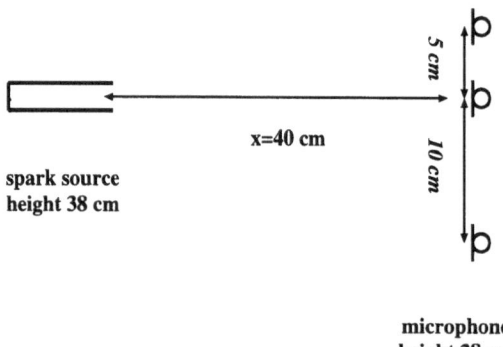

Fig. 5: Sketch of laboratory measurement setup.

to 1.4 m/s depending on heating. The fluctuations are more or less homogeneous within an area of about 0.3×0.4 m^2 but depend strongly on height. The temperature spectra are registrered for a time intervall of 150 min.

The wave propagation was measured by radiating a short acoustic impuls through the convecting air about 0.5 m above the grid. The wave front is picked up at about 0.4 m distance by three 1/4" microphones, 0.05 and 0.15 m apart (Fig. 5). The acoustic impulse is generated by a spark sound source which is ignited in a tube of 1/2" diameter. The impuls travels within the tube forming a shockwave with a reproducable wavefront. The velocity step function is radiated from the opening of the tube as a short δ-impulse (bandwidth > 1 MHz). The short acoustic impulse allows for elimination of reflections in the laboratory setup by appropriate windowing. The other advantage is the broad band measurement, of course. A frequency range from 5 kHz to 25 kHz was evaluated. The repetition rate of the pulses is about 0.5 Hz. The statistical information is derived from a number of 300 successive registrations corresponding to a measuring period of 10 min.

The spectrum of the turbulence is controlled to some extent by the amount of electrical heating power and the outlet of the convecting air. Three cases (I, II, III) of measured turbulence spectra are shown in Fig. 6. In case I low heating power and a reduced outlet yield a diminished convection velocity. This results in smaller spectral density at low wavenumbers and a slower decay at high wavenumbers in comparison to the other two cases. In case II the convection velocity was increased by enlarging the outlet. Therefore the time for the redistribution of turbulent power to smaller structures was reduced, resulting in lower spectral densities at higher wavenumbers. In case III the heating was increased and the same outlet size as in case II was used. The

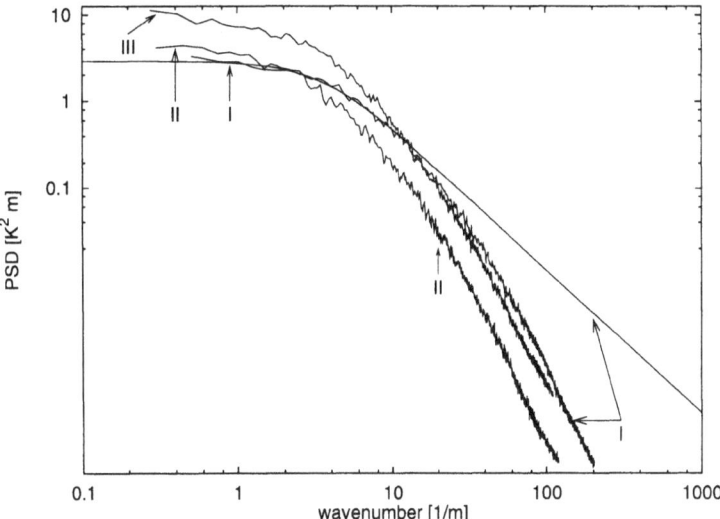

Fig. 6: Measured spatial power spectra of temperature fluctuations and best fit of a von Kármán spectrum for case I.

turbulence is enhanced at all scales. A fit of a one dimensional von Kármán spectrum

$$F_1^T(\kappa) = \frac{C_T^2}{T_0^2} \frac{1}{3\,\Gamma\left(\frac{1}{3}\right)} \frac{1}{(\kappa^2 + \kappa_0^2)^{-5/3}}$$ (13)

to the measured spectrum of case I gives the values $C_T^2 = 196$ K^2 m$^{-2/3}$, $\kappa_0 = 3.6$ m^{-1} (cf. Fig. 6 fitted curve). Obviously the power spectra of the convecting air do not obey a von Kármán law, i.e. the turbulence is not fully developed. At high wavenumbers the structures decay like $\kappa^{-\beta}$, $\beta = 3.2 \ldots 3.5$ in contrast to the theoretical 5/3-law. The parameters that are obtained by fitting a von Kármán spectrum to the measured spectra are of minor physical relevance.

4.1 Acoustic determination of C_T^2 and κ_0

The theory of Sec. 2 is based on the assumption of a von Kármán spectrum. As this condition is not met in the laboratory experiments, the parameters C_T^2 and κ_0 of Eq. 13 are not defined in a strict sense. Nevertheless a technique analogous to the description in Sec. 3.2 yields unique estimates for these parameters by replacing R_v by R_T (Eq. 4) in Eq. 12. The following values are obtained:

		I	II	III
C_T^2	[K^2 m$^{-2/3}$]	376	398	507
κ_0	[m^{-1}]	19	22	9

The two spectra in case I and II yield nearly the same parameters, whereas the increased turbulent power in case III leads to higher estimates for C_T^2.

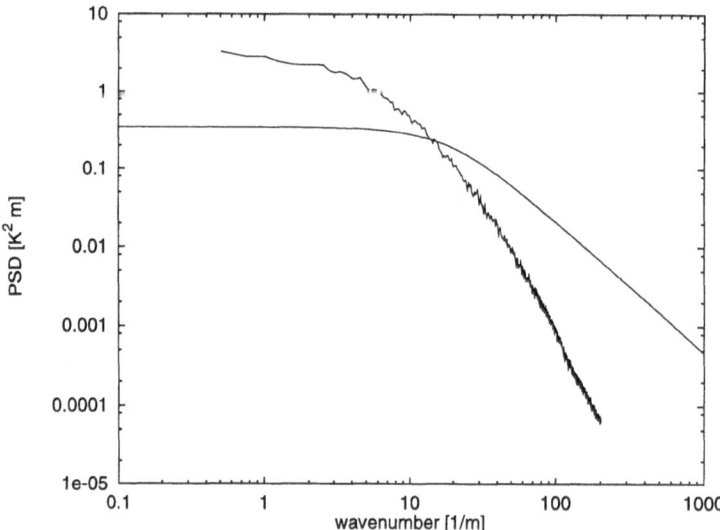

Fig. 7 Comparison of the measured spectrum of temperature fluctuations with the estimated spectrum of the acoustical coherence measurement for case I.

Fig. 7 gives the von Kármán spectrum using the acoustically determined parameters in comparison with the measured spectrum of turbulence for case I. The acoustically estimated von Kármán spectrum behaves as if there were less contributions for small wavenumbers and more contributions for higher wavenumbers to the decorrelation of the acoustic amplitudes. This behaviour seems reasonable, because the wave statistics is determined qualitatively by integrating all structure sizes of turbulence (Eq. 3). Since the asymptotic region at high values of κ is determined by the 5/3-law in the von Kármán spectrum, the low κ region is underestimated. Fig. 8-10 show the measured coherences for the three cases. The measured coherences are reasonably well fitted by $\exp(-\alpha\ f^2)$, indicating that the PEM is applicable in the proposed acoustic sounding.

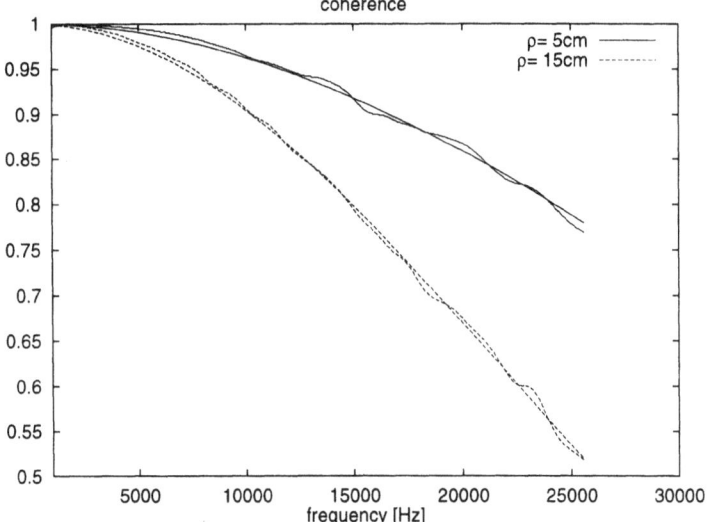

Fig. 8: Coherence of experiment I and fits of exp(-a f^2)

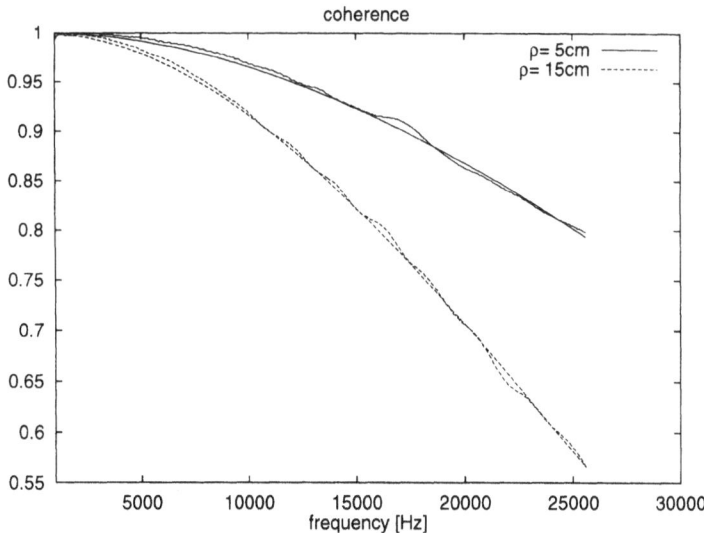

Fig. 9: Coherence of experiment II and fits of exp(-a f^2)

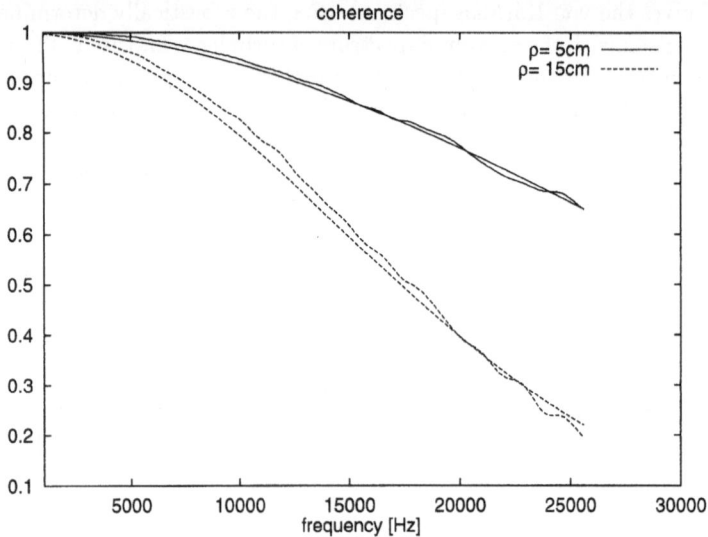

Fig. 10: Coherence of experiment III and fits of exp(-a f^2)

5. Conclusions

One objective of the present study was to verify the predictions of the PEM experimentally. In both setups the general exp(α f^2)-law described the measured coherences. It was shown how to determine the two parameters of the von Kármán turbulence by a three microphone measurement of coherence. In the measurement outdoors the acoustically determined von Kármán spectrum agreed well with the anemometer data. In the laboratory setup the turbulence was not fully developed, the asymptotic roll off was stronger than $\kappa^{-5/3}$. Therefore the acoustically estimated von Kármán spectrum deviated from the temperature measurements.

Generalizing the procedure of determining the two parameters of the von Kármán spectrum will allow for the evaluation of additional parameters (which might be necessary for describing more complex spectral shapes) in refining the statistical image by adding further microphones in the lateral plane of measurement, n+1 microphones for n parameters.

Acknowledgement

The experiments were supported by members of the acoustics group, University of Oldenburg: D. Englich, R. Matuschek, A. Schomburg.

References

1. V.E. Ostashev, B. Brahler, V. Mellert, and G.H. Goedecke, Coherence function and mean sound field in moving random media with the von Kármán spectrum of medium inhomogeneities., accepted by Journ. Acoust. Soc. Am., 1997.

TRAJECTORY ESTIMATION OF MOVING TARGET IN THE MEDIUM WITH A STRONG SCATTERING

V.D.Svet[1], T.V.Kondratieva[1], N.V.Zuikova[1]

[1] Federal Science Center "N.N.Andreyev Acoustical Institute"
Shvernik St.-4, Moscow, 117036, Russia

INTRODUCTION

The coordinate estimation of moving target in sea medium (in active or passive mode) is based on the measurement of angle spectrum by 2D beamforming . The term "beamforming" automatically implies that hydroacoustical system executes the solution of reverse wave problem. In other words at first the system is forming the 2D- angle spectrum of incident waves on array what is the same that the reconsruction of source function. Then one can estimate the angle coordinates from maximum of angle response using different procedures. In free space beamforming is based on Fourie-Frenel transformations and on a priory estimation of Green's function for inhomogeneous medium. The last procedure is known as matchfield signal processing.

Unfortunately the real effectivity of Fourie processing and matched filtering in the medium with a strong scattering and multiray propagation is very small. The strong phase scattering leads to catastrophic expansion of angle spectrum and ambiguity of bearing. At the same time matched filtering procedures make no sense too since it is impossible to estimate Green's function or channel impulse response with sufficient accuracy in practice.

The new method of attack is originating from answer on the following question: is it always necessary to solve the reverse problem for coordinate estimation if it is granted that it is a point source ?

As one can see furthermore such formulation of the question leads to a new method of angle estimation for source placed in the medium with a strong scattering and multiray sound propagation [1,2,3].

MEDIUM WITH A STRONG SCATTERING

The developed method is based on the following main assumption: the rate of target moving is greater than rate of space-time variations of scattering structure.

Under this condition scattering fields from moving source will be unchanged or correlated for some time. When such is the case we can compare two (or more) correlated angle spectrums of scattering fields with different positions of moving source in successive moments of time. Since the angle scattering spectrums are correlated their dissimilarity from each other is responsible for angle variation of incident wave. That is the reason why it is possible to extract the information about angle displacements without solution of reverse problem in the scattering medium.

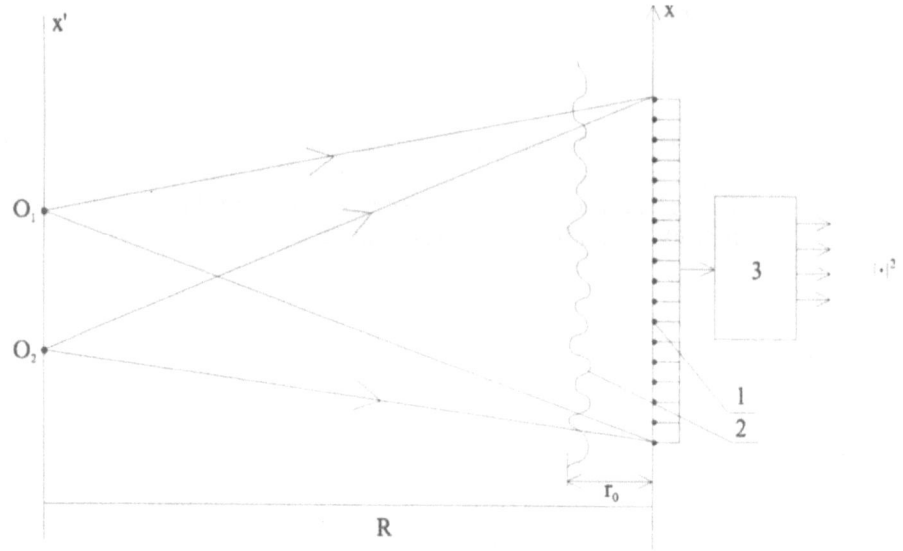

Figure 1. The point source in the scattering medium: 1 - line array, 2 - inhomogeneous layer, 3 - beamforming.

Let place in a free water layer (Figure 1) point sound source 0 of frequency f and line array with N receivers and length L. The array spacing is $\Delta = \lambda/2$. Let place the source in far field zone on the distance $R \geq 2 L^2/\lambda$. The point source is moving from point O_1 to point O_2 in parallel to array. Beamforming is performed by acoustical lens or Fourie signal processing. Inhomogeneous medium represents the layer of scatters or layer of transparent phase structure $\psi(x)$ with random variations from 0 to 2π and more. If many intervals of space correlation function of $\psi(x)$ is placed on the length L the incident plane wave on the rear of the layer will be absolutely distorted and angle information will be disappeared from Fourie beamforming (Figure 2a, 2b). Note that these images look like speckle-structures in coherent optics.

Let's represent point source field as $P(x) = A(x) \, exp \, \{jk\varphi(x)\}$. After the inhomogeneous layer and beamforming the angle space spectrum $\Phi(y)$ is

$$\Phi(y) = \Im[P(x)exp\,\{jk\psi(x)\}], \qquad (1)$$

where $\Im$ - Fourie-operator, x - array coordinate. Now let form a new function

$$I(y) = |\Im(y)|^2, \qquad (2)$$

and fix $I_1(y)$ and $I_2(y)$ in sequential moments of time t_1 and t_2. In a time $\Delta t = t_2 - t_1$

the point source moved from point *01* to point *02*.

As this takes place, the phase structure did not change during this time interval, i.e. $\psi(x, t1) = \psi(x, t2)$.

Now let us sum $I1(y)$ and $I2(y)$ and take the Fourie- transformation from this sum again. Leaving aside interface mathematical treatments it is easy to show that $/\Phi_1(y)/ = /\Phi_2(y + \Delta y)/$ and since

$$\Im_u [\Phi_2(y + \Delta y)] = \Im_u [\Phi_1(y)] \exp(j\Delta yu),$$

$$/G(u)/ = /\Im_u[I1(y) + I2(y)/ = /\Im_u [I1(y)]/ \sqrt{\{2[1 + \cos(\Delta y\, u]\}} \qquad (3)$$

and

$$/G(u)/ = 2 / \Im_u[I1 (y)]/ \overset{2}{[} 1 + \cos(\Delta yu)]. \qquad (4)$$

From the last relation it is evident that function $G(u)$ is modulated with cosine factor with frequency Δy proportional to the source angle shift α which we want to estimate.

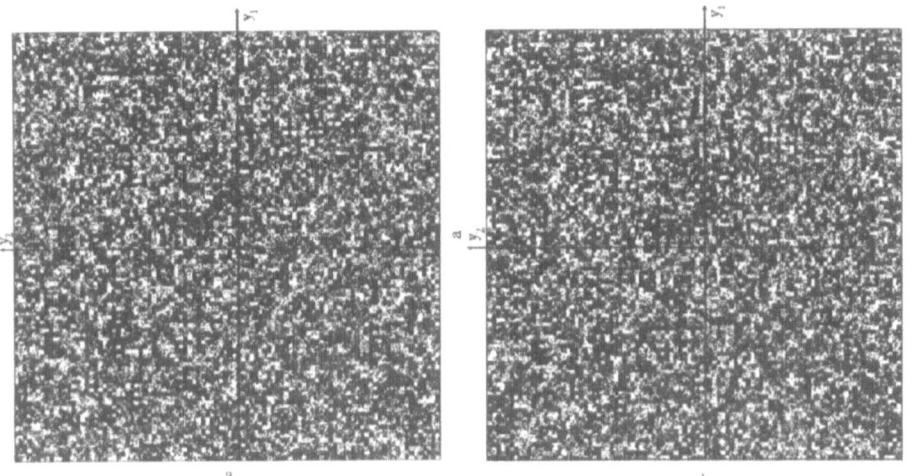

Figure 2. Angle spectra after 20 beamforming: a - point O_1 , b - point O_2.

The similar method of estimation of star angle dimensions in the turbulence medium for the first time has been described in optics by Laiberry[4] and derived its name as "speckle-interferometry".

Relation (4) permits to estimate only projection of source displacement on x-axis of line array. If we have *2D* array such procedure (4) leads to the occurrence of interference stripes, Figure 3. Their angle orientation permits to estimate two angle projections of source shift.

Naturally the correlation procedure is more preferable algorithm for such processing instead of procedure (4), i.e.

$$covar\{I1\ I2\} = I1\ (y)\ \otimes I2\ (y), \qquad (5)$$

where symbol $\otimes$ means integral of correlation.

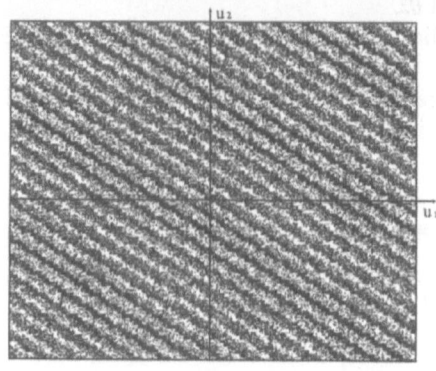

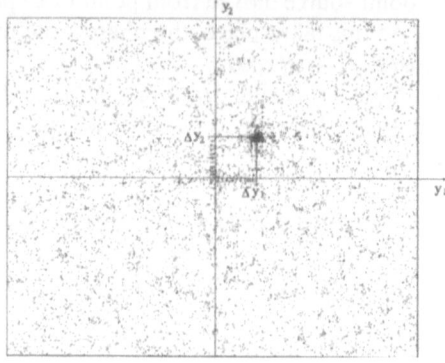

Figure 3. Interference strips.   **Figure 4.** Image of "Covar function".

The results of correlation processing for two-dimensional beamforming are shown on the Figure 4. It is seen that angle position of source is estimated with high resolution although phases of signal are changing arbitrarily in each array sensor from 0 to 2π.

If the space spectrum of phase function $\psi(x)$ is uniform and many intervals of space correlation of this function are placed on array correlation procedure (5) gives good results. But if the space spectrum of phase inhomogeneties contains some periodic components the correlation processing can lead to indefinite estimation.

In this connection one can suggest another algorithm of shift estimation. Let us take Fourie -transformation from (2) before their summation, i.e. transfer the shift in phase inclination again:

$$\Lambda_1 = \Im_\eta \left[|\Phi_1(y)| \right]^2,$$

$$\Lambda_2 = \Im_\eta \left[|\Phi_2(y)| \right]^2 \tag{6}$$

and form the ratio $B\ (\ \eta\ )$. It is evident that

$$B(\eta) = \Lambda_1(\eta)/\Lambda_2(\eta) = exp\ (j\ \Delta\xi\eta\ ). \tag{7}$$

The opposite Fourie transformation gives

$$Div[\ I_1(y),\ I_2(y)\ ] = \Im[\ B\ (\ \eta\ )\ ] = \delta\ (\ y - \Delta y\ ), \tag{8}$$

i.e. the single peak independent of the type of the function $\psi\ (\ x\ ),\ \psi\ (x,\ t_1\ ) = \psi\ (\ x,\ t_2\ )$.

Some simulating and experimental results of signal processing (8) are described in[1, 2].

SEA WAVEGUIDE WITH MULTIRAY SOUND PROPAGATION

The optimal signal processing for sound bearing and ranging in a multiray medium is a matched filtering processing based on the knowledge (calculation or measurement) of Green's function for sound propagation channel. Unfortunately for the most practical situations it is impossible to estimate Green's function with a required accuracy. So, these methods keep serviceability only for very low sound frequencies.

One can consider the multiray sound propagation in a sea waveguide as a peculiar kind of random speckle structure. Really, 2D angle spectum after beamforming for unknown hydrology will look like random structure of peaks. If this multiray structure remains constant with movement of the source it is possible to apply the same ideology considered above for scattering medium for angle estimation of source in a waveguide[3].

The signal in a ocean waveguide can be represented as a group of rays or normal modes. Beamforming of such signals for long horizontal array leads to discrete angle spectrum. We'll consider the situation when multiray effect shows itself at the most: sound source radiates in the face of line horizontal array, Figure 5.

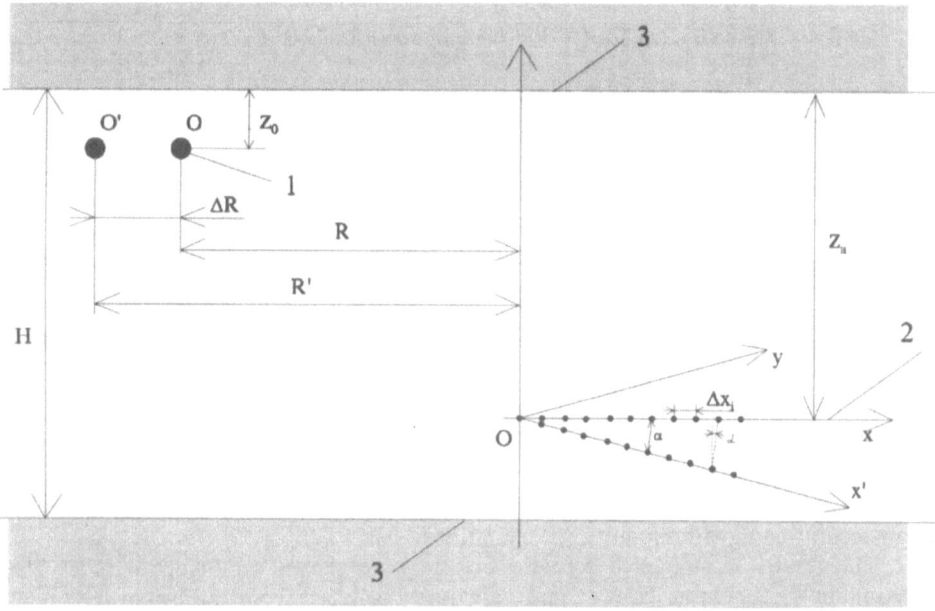

Figure 5. The point source in the waveguide: 1 - point source, 2 - line horizontal array, 3 - bounds of waveguide.

The received field represents the sum of normal modes:

$$P(\,xi\,) \approx \sum Am\ exp\ [\,j\ \xi m\ (\,R + xi\,)\,]\,, \tag{9}$$

where Am - amplitude of m-th mode, $\xi m = 2\pi/\lambda \bullet am$ - the horizontal component of wave vector, xi - coordinate of i-th sensor in array, the sum is taken from $m = 1$ to $m = M$, M is the number of propagating modes. Let us suggest that $am = 1 - bm$, where bm - small additives, but not vertical wavenumbers, $bm << 1$. In this case we can rewrite (9) as:

$$p(xi) \cong exp[jk\ (\,R + x\,i\,)\,]\sum A\,m\ exp\ [\,jkbm(r + xi)\,]. \tag{10}$$

The first phase multiplier before sum describes the propagating incident wave on array. If the second multiplier (sum) in (10) is not changing significantly with source's shift from distance $R1$ to distance $R2$ and its angle α relating to array, one can believe this complex function as constant and consider it as a speckle-structure. Therefore it is possible to apply the developed algorithms *"covar"* and *"div"* for one-valued angle estimation of the point source in a waveguide without matched filtering processing. Quantitative analysis of possibility of application of such methods for waveguide were performed for ideal sea waveguide[3] since horizontal wavenumbers are easily obtainable. Let us assume that point source moved from point $O1$ to point $O2$, Figure 5, and the distance from source to first sensor in array changed as $\Delta R = R - R^{'}$. The variation of the source angle is equivalent to rotation of array on the angle α in the plane (x,y). For small angles α the changing of distance relative to *i-th* sensor is determining by ΔR and $\Delta x i = x i \, \alpha_e^2$. The phase multiplier in (10) before sum, $exp[R^{'} + x i \, (1 - \alpha^2)]$, can be extracted by algorithms *"covar"* and *"div"*, (5,8) and this multiplier contains information about plane incident wave from point source. But it is true only in the case when all terms in the sum (10) do not change sign, i.e.

$$2\pi/\lambda \bullet bm(R + xi) - 2\pi/\lambda \bullet bm(R^{'} + x^{'}i) = 2\pi/\lambda \bullet bm\Delta R + 2\pi/\lambda \bullet bm \, \alpha \, xi \leq \pi. \qquad (11)$$

The analysis of this relation shows that it is true for many real waveguides. For straticified waveguides $bm \sim 0,001....0,0001$, so variations of distance can be $\Delta R \sim 1000\lambda$. Such possible changes of distance are notably sizable. The second term in (11) is too small and has an order $2\pi bm$, if α is equal to diffraction array resolution α_a and $xi = x_{max} = L$, since $\alpha^2 = \alpha_a = \lambda/L$, where L is the length of array.

The results of simulation[3] show that suggested methods are working successfully for many real waveguides when relation (11) is nonhighly accurate.

Some results are shown on the Figure 6abc. Figure 6a depicts the result of array beamforming (horizontal angle spectrum) for ideal waveguide with $M = 50$, $N = 256$, spacing $\Delta X = \lambda/2$ and source angle shift $\alpha = \sqrt{6} \, \alpha_0$.

It is evident that these angle spectrums have many components due multiray sound propagation and the estimation of source angle is ambiguous. Note that the images of these spectrums are similar in appearance, although qualitative relation (11) was received at $bm \ll 1$ and under simulating the quantities bm are calculated with high accuracy and without any approximation.

The results of suggested signal processing are shown on Figure 6 b, c. The maximums of functions *"covar"* and *"div"* are displaced from the origin by preset value and bearing is uniquely determined. The angle resolution is equal to diffraction array limit (λ/L).

DISCUSSION

The specialists always have considered the sound scattering in combination with mulitray propagation as interference and noise and tried to suppress its influence. Such suppressing is a very hard problem because for the most part this noise has multiplicative character.

Twenty years ago in holography the opticians have called attention to a very high sensitivity of laser diffraction patterns generated by scattering screens to very small random shifts. Such shifts have caused interference and degraded the images. Subsequently it occurred to somebody that it is possible to use such interference for shift measurement of the image with a very high accuracy without reconstruction of image in itself. In such a manner the speckle-interferometry made its appearance.

The developed methods of acoustical bearing in inhomogeneous sea medium are similar to speckle-interferometry and permit to estimate the coordinates of source without reconstruction of source function.

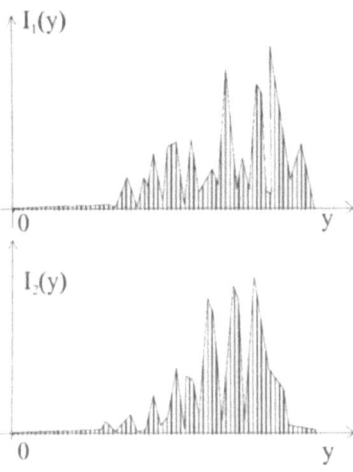

Figure 6a. Angle spectra after beamforming: 1 - point O_1, 2 - point O_2.

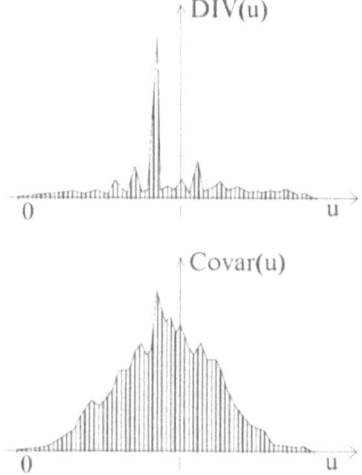

Figure 6 b, c. "Covar" (b) and "Div" (c) functions.

It is very main for different applications because the majority of real sea channels are uncomputable.

Moreover such signal processing is a quadratic in signal after beamforming. This factor leads to a sizable contraction of software and hardware.

One would expect that developed methods will receive a wide acceptance not only in hydroacoustics, but also in seismology , ultrasound medicine diagnostics and so on, i.e. in any mediums, where the position of random scattering structures is stable during long time.

ACKNOWLEDGMENTS

This work was supported by Russian Research Fund, *GRANT 96-02-16372.*
The presentation of this work at the 23rd International Symposium on Acoustical Imaging is supported by ONREUR

REFERENCES

1. V.D.Svet, T.V.Kondrat`eva, N.V.Zuikova, Application of Speckle Interferometry to Problems of Ocean Acoustics, Acoustical Physics. v.42, # 2 (1996).

2. V.D.Svet, T.V.Kondrat`eva, N.V. Zuikova, Estimation of the Distance to the Source Placed Under Scattering Layer, Acoustical Physics. v.43, # 2 (1997).

3.V.D.Svet, T.V.Kondrat`eva, N.V.Zuikova, Estimation of Angle Source Movement in a Multimode Waveguide, Acoustical Physics. (1997) (to be published)

4. Franson M. Speckle-Optics, Mir, Moscow (1980).

A STUDY OF DETECTING AND IMAGING SMALL TARGETS ON SEABED

TIAN Tan, GUAN Hao, LIU Guozhi, SUEN Dajun, ZHANG Dianlun

Department of Underwater Acoustics Engineering
Harbin Engineering University
Harbin 150001, Heilongjiang, P.R.China

Key words: Underwater target, Detection, Acoustical imaging

INTRODUCTION

The most effective approach to detect underwater targets is to use the sound wave. The reason for this is that the propagating absorption of sound wave in water is much smaller than the other kinds of waves, and a greater working distance can be reached. The problems faced with detecting the small targets on seabed, are fully different from that in detecting underwater target for the general purpose. First, because of the small size of targets, the target strength is small and received echo amplitudes are unlikely great. Second, to decide the presence of small targets and classify them, both the azimuth and range resolution are required to be high for a system. Third, it perhaps is the most important that the main background is the bottom reverberation. All of these imply that there are some particularities in detecting and imaging small targets on seabed. The purpose of this paper is to present such a kind of system. The signal to reverberation ratio and the reverberation to noise ratio related to detecting small targets on seabed, beamforming and focusing in the near field, signal processing and display technique are discussed. Some experiments have been done on lake with the developed system and the obtained results are satisfactory.

THE SIGNAL TO REVERBERATION RATIO AND REVERBERATION TO NOISE RATIO

The high resolution sonar used to detect and classify the small targets on seabed mainly works in the bottom reverberation background. The echo amplitudes from targets and the reverberation level depend on the transmitting sound level, the sensitivities of array elements, propergation losses. The echoes from targets are also effected by the scattering strength of targets, and on the other hand, the reverberation level is a function of the backscattering strength of the seabed and azimuth resolution (or beamwidth) of the sonar. Both the echo amplitudes from targets and the reverberation are degreased as the range increases. The signal to reverberation ratio at the receiving point is[1]

$$EL - RL = TS - Sa - 10\lg(\Theta \cdot R \cdot \frac{c\tau}{2})$$ (1)

where TS is the target strength, for a rigid sphere with the size of less than $1m$ or a cylinder with the length of less than $2m$ and the diameter of about $0.5m$, the value is between $-12dB$---$20dB$. Θ is the horizontal beamwidth of receiving beam (at -3dB), R is the slant range from the transducer array to the target area, and c, τ are the sound velocity in water and the duration of the transmitting signal respectively. Sa is the backscattering coefficient of the sea bottom, which is related to the grazing angle, frequency and the type of the seabed. Increasing the transmitting sound level has no help for improving the signal reverberation ratio. In fact, this value is the signal to reverberation ratio after beamforming. Based on the required level of correctness, the detecting index is determined in the receiver operating curve (ROC), then the minimum input signal to reverberation ratio $(E/R)_{min}$ is calculated according to a practical processing system. If the value in eq. (1) is greater than $(E/R)_{min}$, the system can detect the target.

To form a shadow of a target and to classify the target, the horizontal size of the target should at least be greater than the horizontal spacing spanned by two beamwith, so that whatever azimuth the target is illuminated from, at least one beam is included in the target dimension. In this case, after the target echo is received, it is ensured not to receive the bottom reverberation, and a shadow will be formed. In other words, the necessary condition to form a target shadow is that at least one beam does not receive the bottom reverberation after receiving the echo from target. As we know, the background amplitude without reverberation depends on the noise level. Therefore, to form a visible shadow, the reverberation to noise ratio at a given range must exceed some threshold. The noise is composed of environment noise, and themer noise of the receiver. For a typical sea state, ship velocity and a certain receiver, the noise can be considered as constant. Therefore, increasing the transmitting sound level and the incident angle of the sound ray to the seabed (or depression angle of the array) is helpful for increasing the reverberation to noise ratio. However, to get a shadow long enough, the depression angle of the array has to be less than some value. According to the experience, an angle between 15° and 45° is available in general.

BEAMFORMING AND FOCUSING IN NEAR FIELD

A wide beam is generally employed to insonify a certain sector and the multiple pre-formed beams are used for receiving. The receiving beams must be very narrow (such as, less than 0.2°) to ensure the azimuth resolution high enough. The resistance phase shifted network is still employed in such a kind of high resolution sonar. Its advantage is that beamforming can be realized at the small signal stage following by mixer circuit, so that the complexity with hardware is avoided and the system is easily stabilized.

To meet the requirement for shadow classification, the sidelobes in receiving beampattern must be low enough, so that a good shadow classification contrast is ensured. The contrast can be defined as

$$C = 10\lg\frac{I_1}{I_2} \qquad (2)$$

where I_1 is the total received energy by some receiving beam, which is returned from the bottom in a transmitting sector, I_2 the difference of the total returned energy above and the energy in the area of the receiving beamwidth forming the target shadow (see Fig.1). It can be seen that the lower the sidelobes and the higher the contrast. The sidelobes of receiving

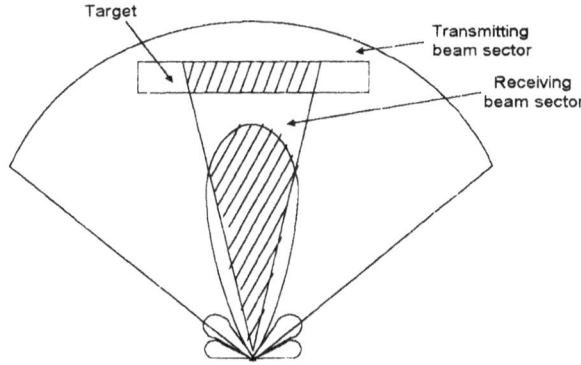

Fig. 1   The shadow classification contrast

beams are reduced to below -20dB by chabishev shading in our system. The working distance in classification mode is not too far (in general, less than 200m). It can be known from the far field condition

$$R > L^2/\lambda \qquad (3)$$

where L is the size of the array, λ is the wave length, that the system wholly works in the near field. Hence the focusing is needed, otherwise the wider mainlobe and the higher sidelobes will occur. When the hardware beamformer is employed, it is difficult to focus to each point in space[2]. The focusing is generally performed for a few points at some distances in a certain direction[3][4]. In this paper we adopt the focusing design for each beam at several distances. The focusing geometry is shown in Fig .2. It can be seen from this figure that, for the beam of

the azimuth θ, when the distance from the focusing point to the center of the array is R, the phase shift to be compensated for the i th element of the array is

$$\varphi_i = 2\pi f \cdot \frac{R_i - R}{c} \qquad (4)$$

where

$$R_i = \left[R^2 + (\frac{N-1}{2} - i)^2 \cdot d^2 + (\frac{N-1}{2} - i) \cdot d \cdot 2R \sin \theta \right]^{1/2}$$ (5)

and d is the spacing between two elements. The amplitude weighting needed for suppress sidelobes together with the phase compensation required for focusing are realized in the phase shifted network. Several sets of weighting resistors are set for each beam in this system. When the system is working, the dynamic focusing for each beam in different distances are realized.

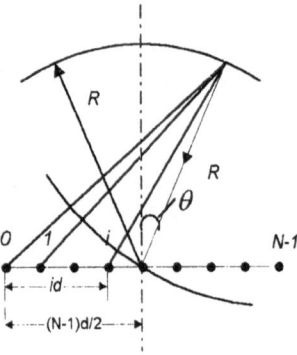

Fig.2   The focusing geometry for near field

SIGNAL PROCESSING AND IMAGE DISPLAYING

The integral processing is generally employed after detecting the received signal. For a single point target, the integral time should not exceed the duration of the transmitting pulse. When there is target spreading, the integral time can be longer to obtain a better processing gain.

A limited number of beams realized by hardware in the system will make the pixel number on screen rather fewer in azimuth dimension. To improve the visible sensation effect, the pixel number in azimuth dimension must be increased. The simple way is to display the output data of each beam at the same distance for several times to extend the width of displayed picture. However, it will make the target in the same azimuth to emerge a short horizontal line. The method adopted in this paper is to interpolate digitally among the output data of beams. The three-point parabola interpolation is a better approach to extend the displayed picture, so that the obtained image will become smother.

The contrast of the displayed image not only relates to the data record directly but also relates to the mapping relationship of the data amplitudes and the brightness in displaying. In general, the level of received signals are of the shadow (or noise), the bottom reverberation and target's echo from small to large in order. Based on the obtained data, the contrast can be enhanced further using some nonlinear amplitude-brightness mapping. As shown in Fig. 3, when the different amplitude-brightness maps are used in the echo classification mode and the shadow classification mode, the ratio of echo to reverberation and the ratio of reverberation to shadow can be increased on screen. This mapping process must be performed for each pixel, and hence the table-looked method is adopted to ensure the processing required for image display in real time.

566

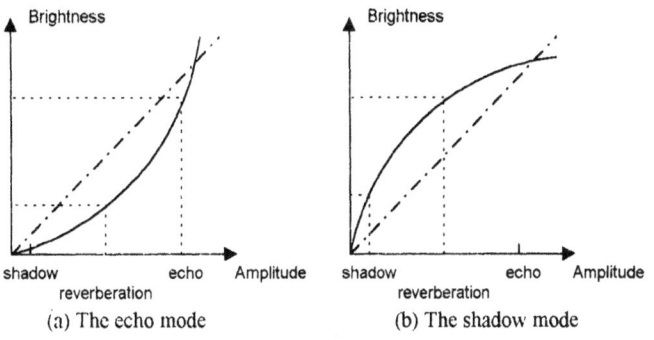

(a) The echo mode   (b) The shadow mode

Fig.3   The different amplitude-brightness mapping relationship used in differnt modes

THE EXPERIMENTAL SYSTEM AND SOME RESULTS

A complete experimental system has been built, which composed of the transmitting array, equally spaced line receiving array, transmitter, receiver, beamformers, and an embedded microcomputer for data displaying and system controlling. The transmitting and receiving array can be rotated and depressed by a controlling unit. The tasks of data acquisition and beam interpolation and so on are performed by a DSP of TMS320c30[5]. The principle block-diagram is shown as in Fig. 4.

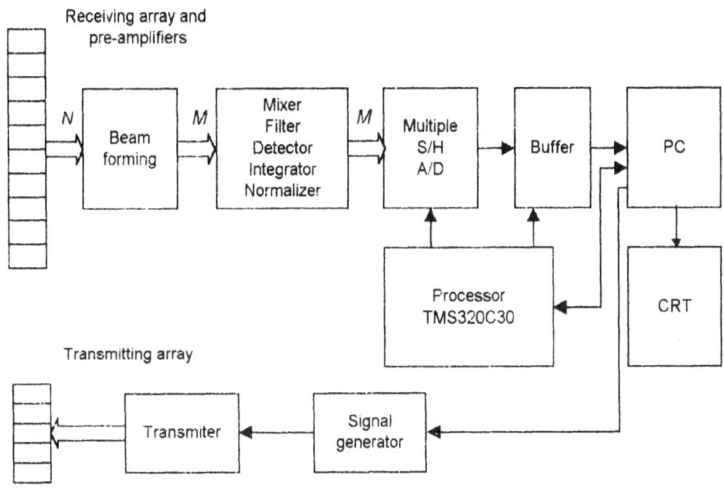

Fig.4   The principle block-diagram of the built system

Some experiments with the system have been done in Xin-an lake, China, in recent times. The used target is a rigid air-core cylinder with the length of about 2*m* and the diameter of about 0.5*m*. The arrays are located at about 2*m* underwater. The target is suspended on to the bottom of the lake (15*m*--25*m* water depth) from a little boat using ropes and a buoy is left on water surface so as to get the target back easily. Fig.5 shows some experimental results. (a) is the image with the target lying on the bottom of the lake, (b) and (c) are the obtained images when the target is being raised from the bottom ,(d) is a digitized

A-mode display of (c) at the direction indicated by the arrow. It can been seen from this pictures that the results are satisfactory.

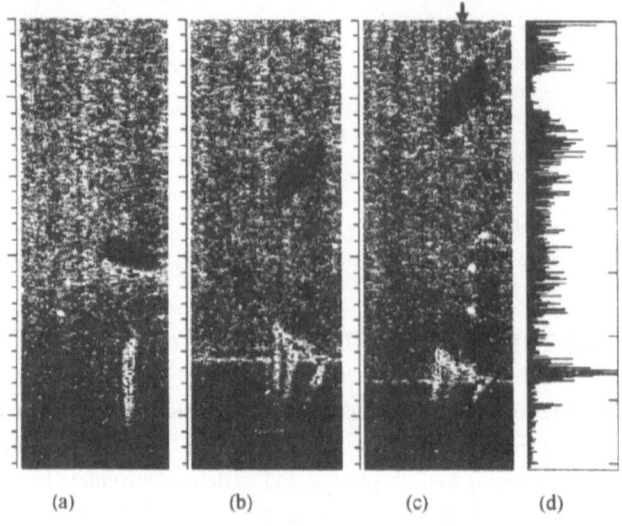

Fig.5   Some experimental results

REFERENCES

1.   R.J. Urick, "Principles of underwater sound", New york, Mac Graw Hill, 1983.

2.   J.L. Sutton, "Underwater acoustic imaging", Proc. of IEEE, 1979, 67(7), pp554-566.

3.   T. Tian, S. Fan and H. Guan, "A real time imaging sonar with single beam using electronic scanning", Proceedings of the 20th International Symposium on Acoustic Imaging, 1993 Plenum Press, New York, pp 715--721.

4.   T. Tian, H. Guan and J. Zhang, "A new within-pulse sliding DFT Beamscanner", proc. of the 14th International Acoustical congress, 1992.

5.   "Digital Signal Processing Application with the TMS320 Family", Texax Instrument Incorporated, 1987

3D SONAR IMAGING IN REAL-TIME USING INVERSE BEAMFORMING (IBF)

James H. Wilson, Josette Paquin Fabre, David Fabre, Christopher Dubea, and Ken Meyers

Neptune Sciences, Inc.
150 Cleveland Ave.
Slidell, LA 70458

ABSTRACT

SubSea International, Inc.'s (SubSea) three dimensional SONARGRAPHICS from robotically operated platforms have been an unqualified success in diver field operations all over the world and such services are currently in great demand based on the successful results. An example of SubSea's 3D SONARGRAPHICS sonar performance in dislodging a stuck dredge head in a reservoir in the Andes in Ecuador was presented at the 23rd International Symposium on Acoustic Imaging to illustrate the power of 3D sonar graphics. These color figures are not presented in this short version of the paper due to space limitations. A full report including color figures is available upon request.

Two improvements to 3D SONARGRAPHICS sonar are discussed in this paper that would make the demand for these services increase significantly. First, the 3D SONARGRAPHICS sonar can be made to run in real-time, or near real-time by the integration of state-of-the-art high speed array processing technology developed and tested recently by Neptune Sciences, Inc. (NSI). Second, NSI's Inverse Beamforming (IBF) technology will be incorporated into the 3D SONARGRAPHICS sonar's beamformer and IBF's well tested performance gains of 10 dB or more will enhance the 3D images significantly.

INTRODUCTION

Sonar Imaging has become a major tool in oceanographic survey, construction and diving projects. The hardware that is presently available is quite fast but it is archaic compared to what can be done with faster layered systems. Standard sonars presently in use are mechanical devices that physically scan and return data at a snails pace compared to what can be done with electronic phase delay beamforming. Dr. James H. Wilson of

Neptune Sciences, Inc. in San Clemente, CA, is developing new devices, hardware and software for deep ocean and shallow water sonar investigations.

NSI has developed the engineering capability to custom design remotely operated vehicles (ROVs). An example of using ROV to unstick a dredge head in a reservoir in Equador in the ANDES was presented at the conference, but the figures had to be omitted in this paper due to length limitations and cost of publishing color figures. The NSI ROV may be fitted with various side scan sonars and subbottom profilers. Details on acoustic imaging results from past ROV based exercises may be obtained from Dr. J.H. Wilson of NSI.

REAL-TIME PROCESSING

Neptune Sciences, Inc. (NSI) has a long and highly successful track record in designing and building real-time sonar systems for the Navy. Recent submarine real-time exercises include RANGEX 1-92 (Dec 91), RANGEX 1-95 (Dec 94), SHIMKAME (Mar 95), and SHIMKAME (Jan 95). NSI developed an Inverse Beamforming (IBF) real-time sonar system and delivered such a significant increase in sonar performance (> 10 dB in detection of signals of interest) that the Navy is considering buying an IBF real-time sonar system for every submarine in the Fleet. NSI is now starting real-time sonar system designs for Navy surface ship passive and active sonars, minehunting high frequency sonars, and Navy air active sonars. In addition, NSI has begun the design of real-time radar systems using IBF and its high speed computer technology. The remainder of this presentation will focus on NSI's effort to upgrade the SONARGRAPHICS 3D Sonar to real-time operation and integrate IBF into the existing sonar.

Phase I is to take the existing IBF real-time system (see figure 1) and implement this technology into sonars of interest. A VME super computer is being built that has an interface board to the SubSea sonar and several Quatro 860™ 320 mflops compute engines developed by NSI through its hardware vendor Universal Computing. The Quatro 860™ is commercially available through NSI and has two distinct advantages over competing hardware. First, its significantly less expensive because of NSI's very low overhead operational philosophy. Second, the design is superior for applications with sonar or radar systems. NSI is implementing both the post-processing 3D graphics algorithm and a new real-time beamformer to enhance the real-time performance of both the SONARGRAPHICS 3D Sonar and other sonars that are capable to be installed on an ROV.

Real-time software is the key to the success of any real-time sonar or radar system and in this area NSI has nationally recognized expertise and experience. All high speed computers have a "theoretical" or design processing speed. For a single Quatro 860™ board it is 320 mflops and for other computers the design computing speed is specified. The **actual** speed for specific software, benchmarks can vary dramatically below the design speed of the computer. NSI has developed several innovative real-time software development techniques to optimize computer speed.

Thus, the conclusion of this section is that the IBF/SONARGRAPHICS 3D Sonar real-time system will be similar in hardware design to the pilot VME bus/i860 IBF real-time system. It will consist of commercial off-the-shelf (COTS) hardware, have an inexpensive, modular design, and have a small footprint. This system will be capable of being back fit into the current SONARGRAPHICS 3D sonar system or forward fit into any current or next generation sonar system. This real-time sonar system will be capable of running current conventional SONARGRAPHICS 3D sonar system algorithms or the IBF algorithms described in the next section in real-time. This will be demonstrated in a separate hardware system. The real-time operation of the system will be demonstrated satisfactorily in the laboratory by real-time operations using data tapes, and then demonstrated at-sea in the field.

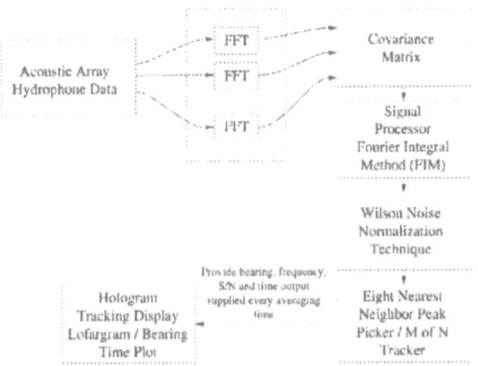

Figure 1. NSI Inverse Beamforming (IBF) Processing Schematic

INVERSE BEAMFORMING (IBF) OVERVIEW

It is not possible to describe IBF's theory and performance gains in detail in this paper and the reader is referred to numerous references to IBF (Wilson, 1980; Wilson, 1983; Wilson, 1995; Wilson and Donald, 1995a and 1995b; Nuttall and Wilson, 1991; Fabre and Wilson, 1995a and 1995b; Nuttall 1972; Nuttall, 1983). IBF has provided large performance gains to produce sharper images of underwater objects for the following two reasons: (1) The IBF beamformer, the Fourier Integral Method (FIM) has a 3dB array gain (AG) advantage over conventional beamformers currently in use. Thus, FIM produces sharper images of underwater objects; and (2) The IBF post processor tracks weak, but persistent, objects using FIM beams. Thus, buried oil pipelines, scoured oil platforms, etc. will be detected and imaged far better than current conventional beamformers and post processors. In the next two subsections some of IBF's underlying theory producing the above gains is summarized.

Beamforming

During several recent real-time, at-sea exercise, IBF algorithms resulted in 10 to 15 dB of detection gain over the performance of conventional sonar systems. More importantly, IBF provided high bearing resolution and tracking solutions that resulted in vastly improved images even when, the array aperture was as small as approximately one acoustic wavelength. IBF consists of three separate algorithms: (1) beamforming, performed by the Fourier Integral Method (FIM); (2) data thresholding, performed by the eight nearest neighbor peak picker (ENNPP); and (3) post-processing, performed by the four-dimensional "M of N' tracker. Implementing FIM into the SONARGRAPHICS 3D sonar system will result in 3 dB less reverberation as is shown in this section.

As the name implies, Inverse Beamforming has been historically defined as a beamforming method only, and is the solution to the IBF integral equation:

$$C(j,k,f_i) = \int_0^\pi FIM(\theta,f_i) \bullet e^{i(\bar{r}_j - \bar{r}_k)} d\theta \qquad (1)$$

where $C(j,k,f_i)$ is the covariance matrix element between hydrophones j and k in frequency bin i, $FIM(\theta, f_i)$ is the acoustic field plane-wave density as a function of bearing θ and frequency bin i, and $\bar{r}_j$ is the position vector for hydrophone j, $\bar{r}_k$ is the position vector for hydrophone k, and $\bar{k}$ is the incoming acoustic wave vector. Here, IBF will refer not only to the solution to equation (1) but also the specific multi-dimensional peak picking and post-processing tracker algorithms.

In the 1970's, extensive efforts to solve equation (1) failed and the solution was thought to be mathematically ill conditioned. Wilson (1983) obtained a robust solution called the Fourier Series Method (FSM) that has resulted in performance gains when applied to measured data. Recently, the solution to equation (1) was obtained for linear, planar and volumetric arrays and was called the Fourier Integral Method (FIM). FIM was shown to be mathematically equivalent to FSM and the solution to equation (1) will be known as FIM for the remainder of this report. As is implied by "Fourier," FIM is the most fundamental plan-wave beamforming algorithm in that FIM, not conventional beamforming, minimizes the error (in the least squares sense) in estimating the acoustic plane-wave density.

The solution to equation (1) for an equally spaced, horizontal line array of M elements is given by:

$$FIM(\theta,f) = \frac{1}{2M-1} p = \sum_{-(M-1)}^{+(M-1)} \bar{C}(p,f) e^{i2\pi fpd\cos\theta/c} \qquad (2)$$

where $\bar{C}(p,f)$ is the Toeplitz averaged covariance matrix. For comparison with CBF (θ, f), the corresponding expression for conventional beamforming is given by:

$$CBF(\theta,f) = \frac{1}{2M-1} p = \sum_{-(M-1)}^{+(M-1)} \bar{C}(p,f) e^{i2\pi fpd\cos\theta/c} \qquad (3)$$

where f is frequency in Hz, c is the speed of sound in m/s, d is the hydrophone spacing in m, p is the hydrophone pair separation $|j-k|$ $j,k = 1,2,3 ... M$, and θ is the azimuth measured counterclockwise from the positive x-axis. For an equally spaced line array, IBF (θ, f) gives approximately 3 dB more array gain (AG) or array noise gain (ANG) than CBF (θ, f) as shown below.

Under ideal signal (a single perfectly coherent plan-wave) and noise fields (isotropic, spatially incoherent noise that is statistically independent between any pair of hydrophones), AG for CBF and for FIM are given by:

$$AG_{CBF} = 10\log\frac{1}{1/M} = 10\log M \qquad (4) \text{ and}$$

$$AG_{FIM} = 10\log\frac{1}{1/(2M-1)} = 10\log(2M-1) \qquad (5)$$

$$\frac{AG_{FIM}}{AG_{CBF}} = 10\log\left(\frac{2M-1}{M}\right) \approx 3dB \qquad (6)$$

for large M, QED. For planar and volumetric arrays, the AG advantage using FIM in the place of CBF in ideal conditions is approximately 6 and 9 dB, respectively. For non-ideal ocean conditions, FIM's AG advantage over CBF can be much greater than for the ideal conditions. From an acoustic viewpoint, the ideal assumptions of classical detection theory concerning either the spatial coherence of the signal or the noise field are not even approximately valid in actual ocean environments. Equations (2) through (6) show that there will be 3 dB less reverberation using FIM than using CBF, the current beamforming algorithm used in nearly all commercial sonars.

Equations (2) and (3) are normalized to unity response in the signal direction and the AG enhancement really comes from the reduced reverberation term, because there is approximately 3 dB less area (i.e., less reverberation or ambient noise) under the FIM beam pattern curve than there is under the CBF beam pattern curve. Figure 2 shows the FIM and CBF beam pattern curves plotted near broadside ($\theta = 90°$) for a plane-wave arrival form broadside. The FIM beamwidth is approximately 2/3 narrower than the CBF beamwidth and the overall FIM sidelobe level is less than the CBF sidelobe level. The FIM beam pattern is negative in some azimuthal intervals outside the main lobe; this is negative intensity and is due to a basic property of the Fourier Series (Wilson, 1983). Given an arbitrary, continuous (actually piecewise continuous) function, the Fourier Series is the orthogonal basis that minimizes the error, in the least squares sense, between the arbitrary continuous function and the finite series of orthogonal basis sine and cosine functions. The FIM solution can be thought of as the special case where the arbitrary continuous function is the plane-wave acoustic density as a function of azimuth. FIM is the spatial Fourier Series approximation of the plane-wave acoustic density determined with a finite number of terms in the Fourier Series once the number of array elements, the spacing, and acoustic frequency are specified. In fact, Wilson (1983) showed that the Fourier Series solution automatically truncates at a spatial frequency where there is approximately one full cycle within the FIM beam width. It was also shown that the spatial Fourier coefficients of the acoustic field automatically go to zero beyond this point (Wilson, 1983).

Normally, the trade-off for FIM's AG advantage and narrower beam width for any array is higher peak sidelobe levels. To overcome this potential problem, NSI's IBF team has developed four-dimensional (beam level versus azimuth, range and time) post-processing algorithms known as the eight nearest neighbor peak picker (ENNPP) or twenty-six nearest neighbor peak picker (TSNNPP) when timewidth (TW) is added as an independent variable and the "M of N" tracker to threshold data and automatically recognize or detect a signal of interest (e.g. buried pipeline, etc.) in a reverberation background. No constant threshold is necessary in the ENNPP data thresholding technique as there is in classical detection theory. Thus, the ENNPP/TSNNPP allows the detection of extremely low SNR peak echoes in beam level relative to the beam levels in all neighboring cells without thresholding the data. The 4-D "M of N" tracker greatly reduces false targets by being "trained" to look for 4-D signal and reverberation characteristics of interest that are addressed by 16 spatial/temporal/spectral tracker parameters.

The IBF principles that have been so successful in achieving vastly improved detection and extremely high spatial resolution in real-time during at-sea exercises are now being applied to commercial active sonar systems to optimize bearing accuracy and image quality in a shallow water, time dependent, multipath environment. Environmental acoustic (EVA) properties of echo and reverberation are used to design the IBF "M of N" tracker.

CONCLUSIONS

There are two potentially very significant conclusions from this paper: (1) NSI and SubSea International are designing and building real-time sonars such as the SONARGRAPHICS 3D sonar, and will be able to provide in the very near future REAL-TIME underwater 3D sonar imaging services; and (2) NSI will add IBF beamforming and post-processing improvements to SubSea's underwater sonar products and the performance of these sonars will be improved by a "quantum leap."

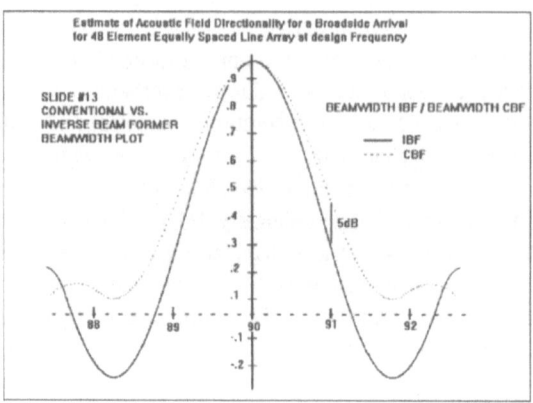

Figure 2. IBF and Conventional Beampattern Comparison

REFERENCES

Fabre, J. Paquin and J.H. Wilson, (1995a). "Minimum Detectable Level (MDL) Evaluation of Inverse Beamforming (IBF) Using Outpost SUNRISE Date," J. Acoust. Soc. of Am., Vol. 98, No. 6, Dec 1995, pp. 3262-3275.

Nuttall, A.H., (1972). "Estimation of Noise Directionality Spectrum," NUSC Technical Memorandum TC-211-71, Naval Underwater Systems Center, New London, CT, 29 October 1971; also NUSC TR-4335, 1 September 1972.

Nuttall, A.H., (1983). "Estimation of Noise Directionality Spectrum Extensions and Generalizations," NUSC Technical Memorandum TC-6-73, Naval Underwater Systems Center, New London, CT, 7 May 1983.

Nuttall, A.H. and J. H. Wilson, (1991). "Estimation of the Acoustic Field Directionality by Use of Planar and Volumetric Arrays via the Fourier Series Method and Fourier Integral Method," J. Acoustic. Soc. of Am., vol 90, no 4, October 1991.

Wilson, J.H., (1983). "Signal Detection and Localization Using the Fourier Series Method (FSM) and Cross-Sensor Data," J. Acoust. Soc. of Am., vol 73, no 5, May 1983, pp. 1648-1656.

Wilson, J.H., (1980). "Solution of Integral Equation Important in Signal Processing of Radar and Sonar Hydrophone Array Outputs," SIAM, J. Appl. Mathe., 93 (1), August 1980.

Wilson, J.H. (1995). "Applications of Inverse Beamforming," J. Acoust. Soc. of Am., Vol. 98, No. 6, Dec. 1995, pp. 3250-3261.

Synthetic Aperture Computed Tomography

Iwaki Akiyama and Kiyoshi Yano

Department of Electrical Engineering
Shonan Institute of Technology
Fujisawa 251 Japan

INTRODUCTION

This paper describes a novel tomographic reconstruction technique for a phased array based on a pulse-echo imaging[1]. One dimensional phased array forms an focusing beam to improve lateral resolution. For the normal direction to the image plane an acoustic lens is also used for the improvement of resolution. Since the acoustic lens is generally designed to have the fixed focal length, it is less resolution for the defocal region. On the other hand, if one uses a divergent beam instead of focusing beam, one could obtain the high resolution at any range by using a back projection algorithm of computed tomography. Because all the wavelets which are returned from scatterers at a circular arc is simultaneously received by the array. It means that the echo signals, which are expressed as summation of wavelets from the scatterers at the circular arc is considered as a projection of a back projection algorithm for the computed tomography. In this paper we investigate the feasibility of this technique by some experiments and compare the spatial resolution for the resultant images. Also we show the capability of three dimensional imaging.

THEORY

Now we consider a simple model of the scattering field with an object which consists of a number of point scatterers in a linear, lossless, and homogeneous medium[2]. We assume each element of the array as a point transmitter and receiver. Fig.1 shows the geometry of the model. The array is assumed to be located at an origin along the x-axis. When ultrasonic pulses are insonified at an i-th element in the array, the echoes which are received by j-th element in the array is expressed as a convolution of transmitted waveform of $f(t)$ to an impulse response of the system including scatterers distribution of $e_{ij}(t)$;

$$e_{ij}(t) = w(t) * s_{ij}(t), \qquad (1)$$

where a symbol of '$*$' represents convolution.

The impulse response of $s_{ij}(t)$ is expressed as the signal which is received by the j-th element when an impulsed wave is insonified at the i-th element.

$$s_{ij}(t) = \int \int \int_V \frac{\delta(t - \frac{R_i}{c} - \frac{R_j}{c})}{R_i R_j} f(r, \theta, \psi) r^2 \cos\theta \, r d\theta d\psi, \qquad (2)$$

where $f(r, \theta, \psi)$ is a coefficient of back scattering, $R_{i,j}$ is a distance between a scatterer and i-th or

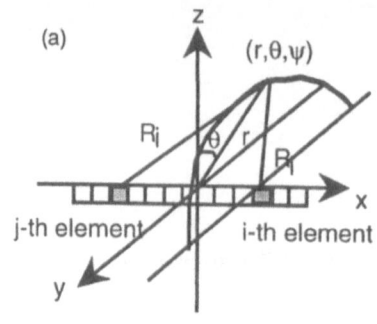

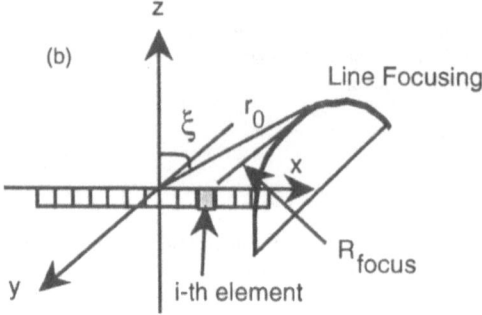

Figure 1: Coordinates system and location of the array(a) and ultrasonic waves are focused on the circular arc (b)

j-th element in the array. $R_{i,j}$ is expressed as,

$$R_{i,j} = \sqrt{x_{i,j}^2 + r^2 - 2x_{i,j}r \sin\theta}. \tag{3}$$

Note that $R_{i,j}$ is not a function of ψ. Thus the impulse response of $s_{ij}(t)$ is given by

$$s_{ij}(t) = \int_{-\pi/2}^{\pi/2} \int_0^\infty \frac{\delta(t - \frac{R_i + R_j}{C})}{R_i R_j} r^2 \cos\theta \, I(r,\theta) dr d\theta, \tag{4}$$

where

$$I(r,\theta) = \int_{-\pi/2}^{\pi/2} f(r,\theta,\psi) d\psi. \tag{5}$$

Now we focus on the circular arc of $r = r_0, \theta = \xi$ shown in Fig.1(b) by phasing of the received signals from each element. The distance between the arc of $(r_0, \xi, 0)$ and the i-th element is given by

$$R_{focus}(x_i; r_0, \xi) = \sqrt{x_i^2 + r_0^2 - 2x_i r_0 \sin\xi} \tag{6}$$

$$= r_0 + \Delta r_i. \tag{7}$$

Note that Δr_i is a function of x_i, ξ, r_0. The time delay of τ_i is thus given by,

$$\tau_i = \frac{\Delta r_i}{C} = \frac{1}{C}(R_{focus} - r_0). \tag{8}$$

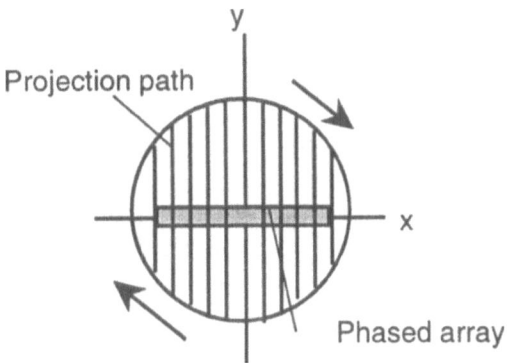

Figure 2: The schematic illustration of phased array and a projection path.

The delayed impulse response is

$$s_{ij}(t + \tau_i) = \int \int I(r,\theta) \frac{r^2}{R_i R_j} delta(t - \frac{R_i + R_j - R_{focus} + r_0}{C}) \cos\theta dr d\theta. \tag{9}$$

When a focusing beam is formed by m elements in the array, the signal which is received by the j-th element is expressed as,

$$s_j(t) = \sum_{i=0}^{m} s_{ij}(t + \tau_i)$$

$$= \int \int I(r,\theta) \cos\theta \sum_{i=0}^{m} \left[\frac{r^2}{R_i R_j} \delta(t - \frac{R_i + R_j - R_{focus} + r_0}{C}) \right] dr d\theta. \tag{10}$$

Let the circular arc on which the ultrasonic waves are focused be 'σ' and let the other region be 'E'. The signal received by the j-th element is expressed as a summation of the waveforms scatterered at the arc of σ because they are simultaneously received by the j-th element. The echoes from 'E' is not simultaneously received by the j-th element. Thus the echoes from σ grow higher than the echoes from E. The impulse response of the echoes received by the j-th element is given by

$$s_j(t) = \int \int_\sigma I(r,\theta) \cos\theta \delta(t - \frac{R_j + r_0}{C}) \sum_{i=0}^{m} \left[\frac{r^2}{R_i R_j} \right] dr d\theta$$

$$+ \int \int_E I(r,\theta) \cos\theta \sum_{i=0}^{m} \left[\frac{r^2}{R_i R_j} \delta(t - \frac{(R_i - R_{focus}) + R_j + r_0}{C}) \right] dr d\theta. \tag{11}$$

$$s_j(t) = \int \int_\sigma I(r,\theta) \cos\theta \delta(t - \frac{R_j + r_0}{C}) \sum_{i=0}^{m} \left[\frac{r^2}{R_i R_j} \right] dr d\theta \tag{12}$$

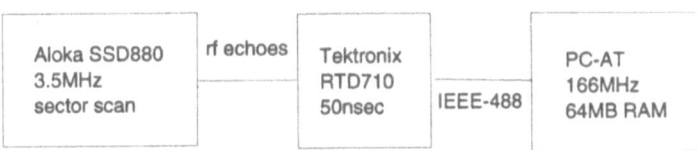

Figure 3: Block diagram of experimental system.

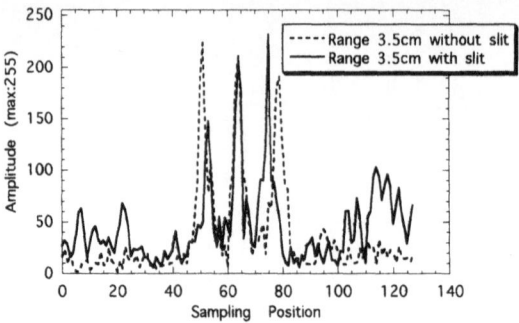

Figure 4: Projection at range of 3.5cm which is obtained by probe with slit and without slit.

The summation of the echoes received by each of m elements is done to form a focusing beam at the circular arc. Let the summed signal be $e(t)$. It is expressed as a convolution of the transmitting waveform of $w(t)$ to the impulse response of $b(r_0, \xi; t)$. $b(r_0, \xi; t)$ is given by

$$b(r_0, \xi; t) = \sum_{j=0}^{m} s_j(t + \tau_j)$$

$$= \delta(t - \frac{2r_0}{C}) \int \int_{\sigma} \sum_{j=0}^{m} \sum_{i=0}^{m} \left[\frac{r^2}{R_i R_j} \right] \cos\theta I(r, \theta) dr d\theta \tag{13}$$

$$\tag{14}$$

where we assume the following approximation.

$$\sum_{j=0}^{m} \sum_{i=0}^{m} \left[\frac{r^2}{R_i R_j} \right] = m^2 \tag{15}$$

The integration of σ is larger than those of E. It becomes

$$b(r_0, \xi; t) = m^2 \delta(t - \frac{2r_0}{C}) r_0 \cos\xi I(r_0, \xi), \tag{16}$$

where $I(r_0, \xi)$ is the integration on the σ for $f(r, \theta, \psi)$. Since $I(r_0, \xi)$ is a projection for back projection algorithm, a hemi-spherical cross sectional image is reconstructed. The projection of $p_\phi(\xi)$ which is

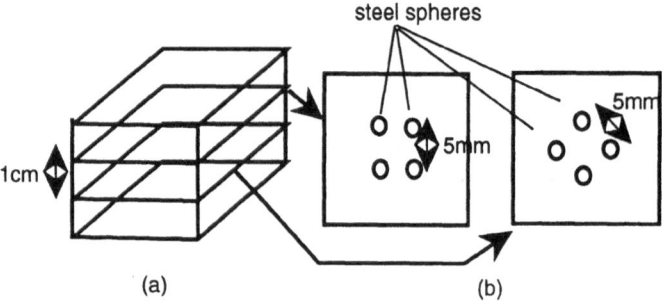

Figure 5: Layer phantom made of agar-gel and small spheres of 2mm in diameter. Thickness of each layer is about 1cm (a). There are four spheres at each layer, where they are arranged in square (b).

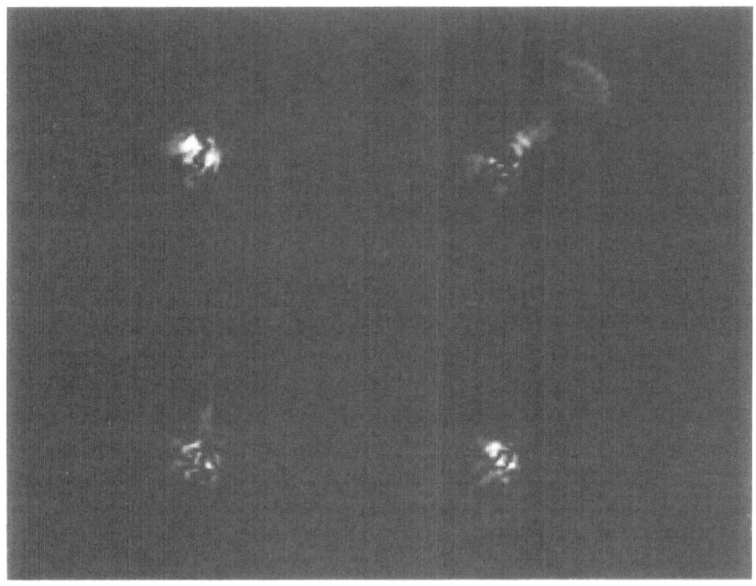

Figure 6: Reconstructed image of layer phantom at partially spherical surface at range of every 0.3mm. (a) is top surface and (b),(c) and (d) are lower surface, respectively.

Cylindrical hole

(b)

Figure 7: Picture of silicon rubber block.

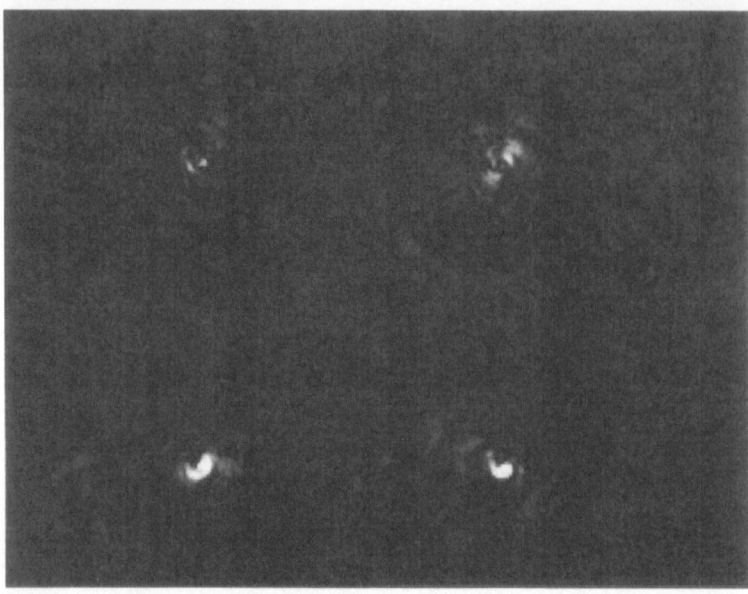

Figure 8: Reconstructed image of silicon rubber block at partially spherical surface. (a), (b), (c) and (d) are of every 0.3mm in range.

measured by the phased array mentioned above is finally given by,

$$p_\phi(\xi) = \frac{b(r_0, \xi; \frac{2r_0}{C})}{m^2 r_0 \cos \xi}. \qquad (17)$$

Figure 2 shows the schematic illustration of the relation of phased array and a projection path for filtered back projection algorithm.

EXPERIMENTS

A block diagram of experimental system is shown in Fig.3. An ultrasonic diagnosis equipment is Aloka SSD880 3.5MHz sector scanner system. A slit is arranged in front of the face of the array to form the divergent beam. The Tektronix RTD710 waveform digitizer is interfaced to a computer of PC-AT via IEEE-488 bus. Sampling interval and total length of acquired data is 50 nano seconds and 637,440 points, respectively. The acquired rf waveforms is correspondent to a sector format of 90 degrees of one frame. A rotation stage is manually rotated at every 3 degrees. Sixty projections are totally acquired. The algorithm which is used for computed tomography is a filtered back projection method. The results are shown in 128×128 images which is reconstructed from 60 projections with 128 samples per projection.

First, we show the experiments to test the resolving capability of the system. A test target is made of steel spheres of 2mm in diameter. Three of them are put on a agar-gel to keep an image of the targets separated from the boundary. Fig.4 shows the projection at range of 3.5cm by the probe with and without slit. The width of the peak on left and right sides of center peak which is obtained by the probe with the slit are narrower than those without the slit. It means that the image reconstructed by the proposed mehtod has higher resolution than the conventional B-scan image because the relation of projection to reconstructed image is linear.

Second, we show the experiments to test the capability of three dimensional imaging. A layer phantom is made of steal spheres of 2mm in diameter and an agar-gel as shown in Fig.5. Thickness of each layer is about 1cm. There are four spheres at each layer and they are arranged in square as shown in Fig.??. Fig.6 shows the reconstructed images of the layer phantom which is correspondent to the scatterer distribution at partial spherical surface. Fig.6 (a) is an image at the partially spherical surface of top layer of the phantom. Fig.6 (b), (c), and (d) are images of every 0.3mm in range. Two spheres are found in (a) and (d), however the other two spheres are not found in those figures. Because the targets are not perfect sphere, the reconstructed image of each target has a different level of magnitude.

Finally, we show examples of applying the method to silicone rubber block. There are a cylindrical hall and flaws in the block as shown in Fig.7. Fig.8 shows the reconstructed images at the partially spherical surface. Fig.8 (a) is the image at the range of about 5mm lower from the top the block. (b), (c), and (d) are the lower images from the range of (a) at every 0.2mm. A couple of flaws are found in those images.

CONCLUSION

The novel method for ultrasonic computed tomography using synthetic aperture technique is described in this paper. This method is based on the fact that echoes are express as a circular arc line-integration, when a divergent beam is formed by the phased array with an acoustic lens or a slit. The features of the technique is the capability of higher resolution than of conventional B-mode image and the capability of three dimensional imaging. The experimental results show that the higher resolution was achieved by the slit phased array and the tomographic imaging at arbitrary range was performed by shifts of time gate for the echoes.

References

[1] I.Akiyama and K.Yano:"Tomographic reconstruction technique for pulse-echo imaging", J.Acoust.Soc.Am.,100,2794 (1996)

[2] K.A.Dines and S.A.Goss:"Computed ultrasonic reflection gomography", IEEE Trans. on UFFC, Vol.UFFC-34, No.3 (1987)

SIMULTANEOUS SPATIAL AND VELOCITY VECTOR MAPPING WITH DIFFRACTION TOMOGRAPHY

Michael P. André,[1,2] Helmar S. Janée,[1,2] Todd K. Barrett,[3] Brett A. Spivey,[3] and Peter J. Martin[3]

[1]Department of Radiology, Veterans Affairs Medical Center, La Jolla, CA
[2]School of Medicine, University of California, San Diego, CA 92093-9114
[3]ThermoTrex Corporation, San Diego, CA 92121-4339

INTRODUCTION

Current Doppler ultrasound technology privides only two-dimensional projections of vessels or velocity patterns, is highly operator-dependent and is unable to accomplish accurate maps of the vessel lumen and flow velocity at the same time. Doppler techniques are limited in their ability to detect or measure small or deep vessels especially in the presence of slow flow, although new contrast media may enhance sensitivity. Even in superficial vessels such as the corotid artery, a very small number of pixels defines the flow channel. The aim of this study was to explore a novel diffraction tomography technique for producing quantitative, high-resolution cross-sectional Doppler images. Using this approach, a flow channel or vessel may be mapped in sequential tomograms providing a unique representation of the three-dimensional patterns of complex, curvilinear, multi-phasic or turbulent flows. The method is based on instrumentation initially developed for breast imaging[1] where the value of Doppler flow data is proving valuable for breast cancer diagnosis.

Our method of Doppler CT employs a circular array of transducers which surrounds and defines the field of view (Figure 1). A separate transducer emits continuous-wave sound into the area defined by the ring. Tissue and moving blood scatter sound waves into the ring of transducer elements which record the complex amplitude of the scattered sound waves as a function of time. The Doppler-shifted information is extracted from this time series and reconstructed to form a map of flow velocities in the form of a tomographic slice.

THEORY

A continuous acoustic wave (single frequency) transmitted from a transducer into human tissue as shown in Figure 2, may be described by $p_0(t) = \cos(\vec{k} \cdot \vec{r} - \omega_0 t + \phi)$ where

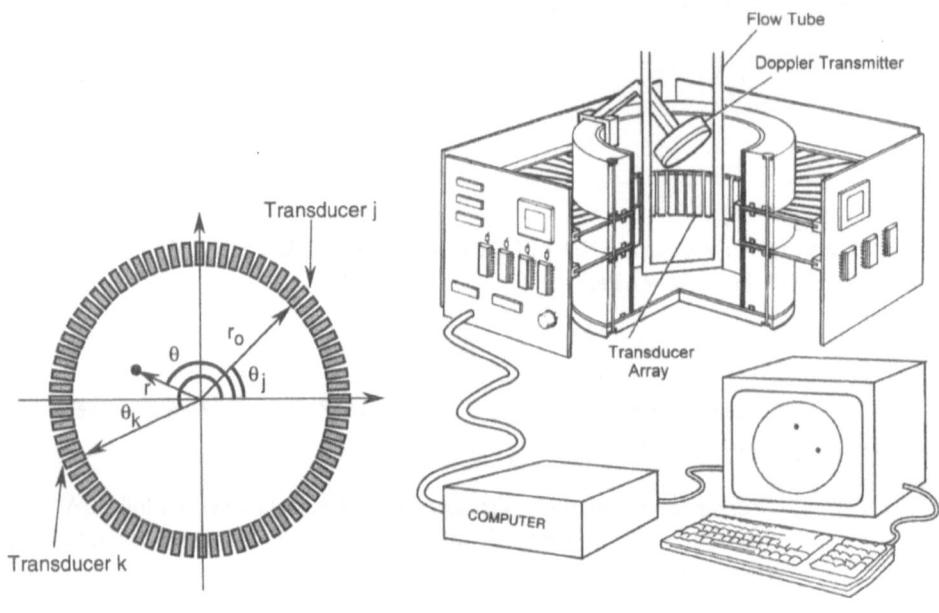

Figure 1   Figure 2

$k=2\pi/\lambda$, $\omega_0=2\pi f_0$, f_0 is the operating frequency and ϕ is a constant phase delay. The tissue may contain a number of vessels that carry flowing blood cells that scatter sound waves in all directions and shift the frequency by the Doppler effect in an amount which is dependent on motion relative to the transmitter. The amount of Doppler shift caused by a blood cell with velocity $\vec{v}$ relative to the transmitter is $\omega = \vec{k} \cdot \vec{v}$, corresponding to a shift frequency $f=\omega/2\pi$. For typical blood flow, $v \leq 100$ cm/sec and $f \leq 1000$ Hz for $f_0=2$ MHz (wavelength $\lambda=0.75$ mm). Thus, the Doppler shift frequency $f<<f_0$.

The acoustic wave incident on receiver j can be written $p_j(t) = m_j(t) e^{i\omega_0 t}$, a form which separates the signal into its carrier component and the complex amplitude $m_j(t)$ containing the Doppler shift. The complex amplitude $m_j(t)$ contains a constant part due to a combination of acoustic waves scattered from the static tissue to receiver j and a time-dependent part due to Doppler-shifted acoustic waves scattered from the flowing blood.

The measured complex amplitude $m_j(t)$ at receiver j can be written as:

$$m_j(t) = \int f(\vec{r},t)\, D_j(\vec{r})\, d^2\vec{r} \qquad (1)$$

where $D_j(\vec{r})$ is the antenna pattern for receiver j. The source function $f(\vec{r},t)$ is the complex amplitude of the transmitted acoustic wave scattered from a volume element located at position $\vec{r}$. The volume element can contain static tissue, represented by static $f(\vec{r})$ or flowing fluid, time-dependent $f(\vec{r},t)$. The discrete Fourier transform in time of Equation 1 yields

$$X_j(\omega) = \int F(\vec{r},\omega)\, D_j(\vec{r})\, d^2\vec{r} \qquad (2)$$

where $F(\vec{r},\omega)$ is the Fourier transform of the source function. Next, we correlate all pairs of receivers by forming the correlation matrix

$$\langle X_j(\omega) X_k^*(\omega) \rangle = \int D_j(\vec{r}) D_k^*(\vec{r}') \langle F(\vec{r},\omega) F^*(\vec{r}',\omega) \rangle d^2\vec{r}\, d^2r' \qquad (3)$$

where $\langle...\rangle$ denotes an ensemble average in the vertical direction over different realizations of $X_j(\omega)$. In the case of flowing fluid, a realization of $X_j(\omega)$ consists of the Doppler spectrum from a specific sample of discrete scatterers which is defined by the pixel size and slice width. The algorithm relies on the fact that the scattering particles in the blood are randomly distributed in the blood vessel. Furthermore, since the particles are moving, the distribution of particles in the scattering volume varies randomly as a function of time. Averaging the moving particles over time is equivalent to the ensemble average in Equation 3. The algorithm relies on the fact that the spatial part of the complex source function $f(\vec{r},t)$ consists of a sum of localized point-like scattering objects (with diameter $<<\lambda$), each multiplied by a random complex amplitude $e^{i\xi}$, where ξ is a random phase of Gaussian distribution. Thus, we can replace the ensemble average on the right-hand side of Equation 3 with the form appropriate to random incoherent radiators: $\langle F(\vec{r},\omega)F^*(\vec{r},\omega)\rangle = \delta(\vec{r}-r')S(\vec{r},\omega)$. In this equation, $S(\vec{r},\omega)$ is the acoustic power radiated at Doppler frequency ω. Combining this with Equation 3 and defining $\langle X_j(\omega)X_k^*(\omega)\rangle \equiv R_{jk}(\omega)$ we arrive at[2]

$$R_{jk}(\omega) = \int D_j(\vec{r})S(\vec{r},\omega)D_k^*(\vec{r})d^2\vec{r} \tag{4}$$

Obtaining an image of the flow medium consists of inverting Equation 4 for the source power distribution, $S(\vec{r},\omega)$.

The antenna pattern $D_j(\vec{r})$ represents the response of the receiver j to incident acoustic radiation. In the laboratory system used for this study, the transducers were designed to closely resemble dipoles in the x-y plane, although any measured pattern may be implimented. In this simplified form, $D_j(\vec{r})$ may be given by:

$$D_j(r,\theta) = \sum_{m=-\infty}^{m} e^{im(\theta-\theta_j)} H_m'(k_0 r_0) J_m(k_0 r) \tag{5}$$

In the above equation, J_m and H_m' are the Bessel function and first derivative of the Hankel function, respectively.

In theory, we could reconstruct the source power distribution $S(\vec{r},\omega)$ from Equation 4 as a continuous function of the Doppler shift frequency ω at each spatial point r. In practice, we will reconstruct $S(\vec{r}_\alpha,\Delta\omega_\beta)$ for an average of frequency bands $\Delta\omega_\beta$ and finite spatial pixels $\vec{r}_\alpha$ (Figure 3). Our initial method of inverting Equation 4 exploits the homomorphic relationship to the reconstruction of coherent ultrasound tomography. Diffraction tomography solves for a

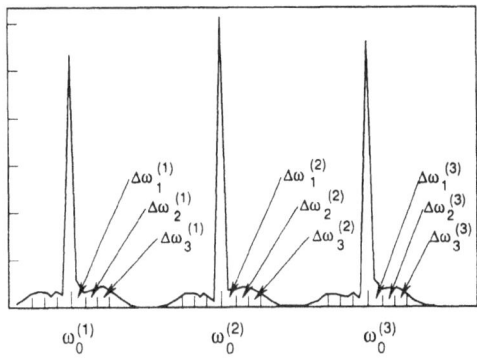

Figure 3

source function given measurements $M_{jk} = \int D_j(\vec{r})S(\vec{r})D_k(\vec{r})d\vec{r}$, where M_{jk} is the complex amplitude of sound received by transducer k from emissions of transducer j, $D_j(\vec{r})$ is the antenna pattern of transmitter j and $D_k(\vec{r})$ is the pattern of receiver k. This equation is identical to the Doppler Equation 4, with $S(\vec{r},\omega)$ substituted for $S(\vec{r})$. An inverse solution for either form of S in terms of M_{jk} (or R_{jk}) was developed in our previous work on diffraction tomography.[3]

While the CT approach can yield high quality images, it is difficult to implement in real time and is computationally intensive. As a way to potentially simplify these problems, Equation 4 may be recast into matrix form and a beamforming approach used. Writing Equation 4 as $\mathbf{R} = \mathbf{DSD}^{T} + \mathbf{N}$ allows S to be obtained in a number of different forms depending upon the approximations and or statistical model used in the inversion ($\mathbf{N}$ is the noise matrix). In particular, directly inverting this equation yields S as $\mathbf{S} = \mathbf{D}^{T}\mathbf{RD}$ and by using the maximum likelihood (ML) approach to improve the resolution at the expense of increased computation S can be written as $\mathbf{S} = (\mathbf{D}^{T}\mathbf{R}^{-1}\mathbf{D})^{-1}$. Other approaches include maximum entropy (ME) and iterative estimators (IE) which have the potential for higher resolution but at the expense of significantly increased computation and sensitivity to noise.[4,5] Unlike the CT method and any of these statistical estimators (ML, ME and IE), the direct inversion approach offers the possibility of real-time operation since there is no matrix inversion or series summations required. It is expected, however, that there would be a reduction of about a factor of one-third in resolution compared to CT and ML methods. This more efficient approach could be used for real-time interactive applications during the ultrasound exam, after which high-quality images for diagnosis could be reconstructed off line in a few minutes using the other, higher resolution approaches. In addition, since statistical estimators such as ME allow the inclusion of *a priori* knowledge in the reconstruction process, it is possible that such approaches may be useful in aberration correction.

A map of the source term S represents the power radiated at Doppler frequency ω from a volume element located at position r of dimension of the order of $\lambda/2$ x $\lambda/2$ x slice thickness, which corresponds to 0.5 x 0.5 x 8 mm for our experimental arrangement. We can form a map of the average velocity of fluid passing through point r by calculating:

$$V_z(\vec{r}) = C\int S(\vec{r},\omega)\frac{\lambda\omega}{2\pi\cos\theta}d\omega \tag{6}$$

where V_z is the average value of the velocity perpendicular to the plane containing the receiver transducers, C is a constant set by the transducer configuration of the device, and θ is the angle between the fluid velocity vector $\vec{v}$ and the wave number vector $\vec{k}$, controlled precisely by the directivity of the transmitter transducer.

EXPERIMENTAL RESULTS

A small laboratory system (Figure 2), originally developed for a breast imaging research program, was used to test the feasibility of the Doppler CT method. The laboratory device consists of 256 lead zirconate titanate (PZT) transducers, backed with air-filled foam, arranged in a four-inch cylinder. The receiving elements are 0.9 mm wide, 25.4 mm long with 1.2 mm spacing, have a center frequency of 530 kHz with approximately 7.5% bandwidth. The transducer length is about eight times the acoustic wavelength in water to assure a collimated beam pattern in the vertical direction. A separate 25.4 mm annular array transducer (530 kHz center frequency) was used as the transmitter placed at a known angle to the imaging plane.

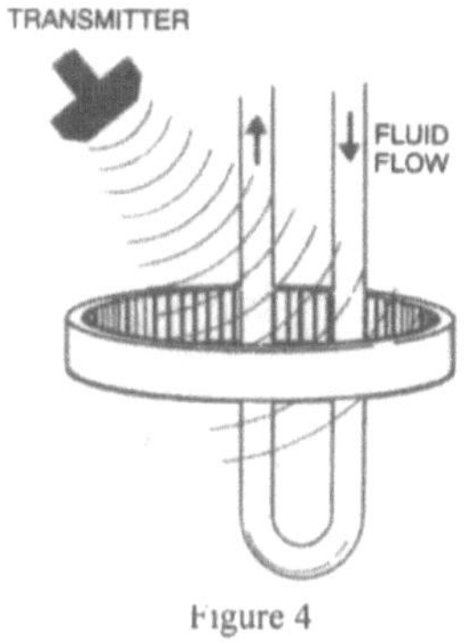

TRANSMITTER

FLUID FLOW

Figure 4 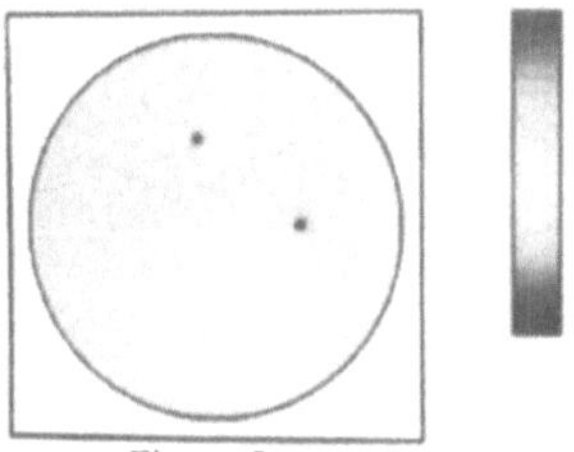   Figure 5 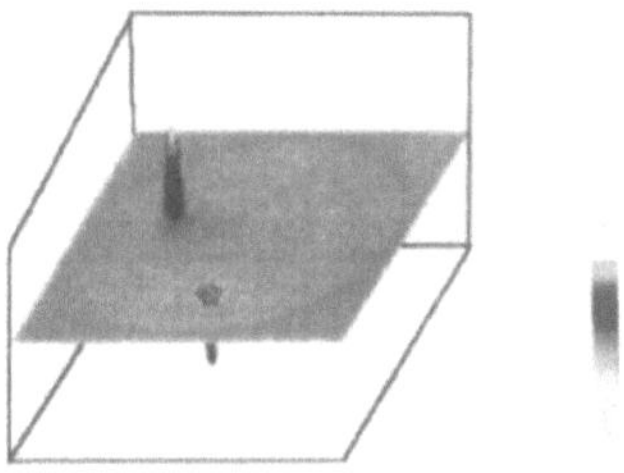   Figure 6

The area inside the cylinder was filled with water to couple sound to scattering fluid (corn starch in water) which flowed through latex rubber tubes.

The transmitter was activated with a Wavetek Model 23 function generator at ten discrete frequencies from 405-630 kHz. Quadrature data was collected on each of the 256 receivers at 3.7 msec intervals for 1.9 sec using a high-speed, parallel, 16-channel custom multiplexing network and a fast 12-bit analog-to-digital converter through a VME backplane. The data was then reconstructed by the CT method to produce an image of the medium in the form of a tomographic slice. Reconstruction of the 256 x 256 pixel image from single-frequency data required approximately 1 minute on a PC-compatible with an i860 accelerator.

In the first experiment shown in Figure 4, a pair of latex tubes (inner diameter 5 mm) were gravity fed under constant pressure to produce constant flow with a mean velocity of approximately 11 cm/sec. The Reynolds number was very low (~110) and laminar conditions were approximated. The axes of the tubes were perpendicular to the image plane and the flow within the tubes was equal in magnitude but opposite in direction; one flowing downward through the plane, the other upwards. The power spectrum $|X_i(\omega)|^2$ has two sharp peaks, one at $+\omega$ for the tube flowing towards the transmitter and one at $-\omega$ for the tube flowing away. Each signal also contains $X_i(\omega)$ static complex Fourier amplitude which indicates position r of the moving Doppler radiator. First, the covariance matrix corresponding to a positive Doppler shift, $R_{jk}(\omega)=\langle X_i(\omega)X_i^*(\omega)\rangle$, was used to reconstruct the image of the flow within the tube carrying the liquid upwards out of the reservoir. Next, the data corresponding to a negative Doppler shift $R_{ij}(-\omega)$ was used to reconstruct the flow within the other tube. The two reconstructions were combined to form the final image. The Doppler data is shown in image form (Figure 5) and in a surface plot in Figure 6. The parabolic velocity profile, characteristic of laminar flow, is apparent in this presentation.

Figure 7 shows a second experiment of a larger single latex tube with inside diameter of 1.5 cm and mean velocity of 5 cm/sec. The flow within this tube is downwards into the reservoir away from the transmitter. The parabolic flow profile is evident in this data and readily apparent in Figure 9. The highest velocities are in the center of the tube, as is expected for laminar flow. The flow velocities are normally color-coded for display with a linear scheme similar to that of conventional color duplex Doppler. The grey-scale representations in these figures are more difficult to interpret.

CONCLUSIONS

These preliminary results demonstrate feasibility of the method for mapping velocity using diffraction tomography techniques. Practicality of the idea, however, depends on several

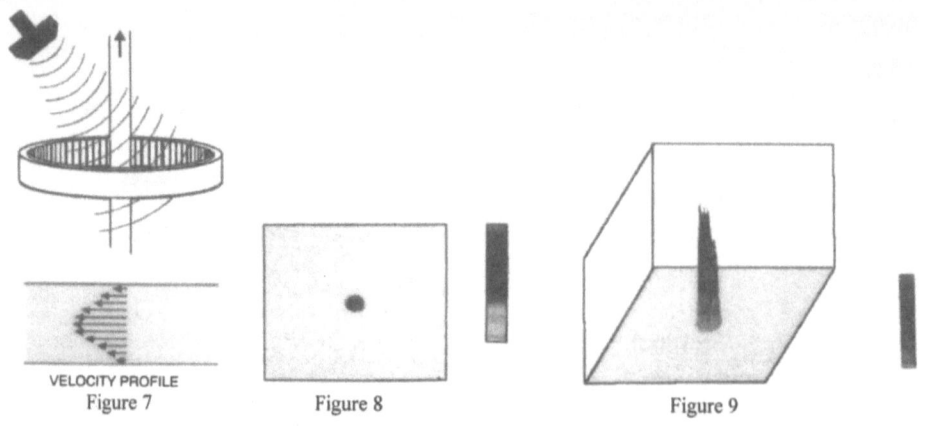

VELOCITY PROFILE
Figure 7   Figure 8   Figure 9

issues that remain to be resolved. The effects of clutter and stationary scatterers may be large enough to significantly reduce the signal-to-noise ratio and subsequent image quality. These preliminary experiments did not include significant background scatter. The system can operate either in cw or pulse mode so reflections from bone and the effects of the large backscatter from tissue compared to blood may be dramatically reduced by appropriately range gating (time delaying) the receivers in burst mode. The system sensitivity to slow flow and the accuracy of the Doppler shift measurements need to be established. Certainly, higher frequencies than those used in this pilot study will be needed to improve Doppler accuracy, although uncertainty of the Doppler angle ω may be as much of a problem as it is for conventional Doppler ultrasound. If the Doppler tranmsitter is placed within the transducer ring or arc, accurate angles between the transmitter and receivers can be established from the tomographic reconstruction using the entire array. In this approach, a high resolution tomogram may be acquired from which the cross-sectional Doppler reconstruction may be guided (à la duplex Doppler). Recent experiments and simulations have shown that this method may be accomplished with very sparse arrays of as few as 16 receiving elements thereby reducing system complexity. The optimum number of receiver pairs and their spacing are important design considerations to be determined.

ACKNOWLEDGEMENTS

This work was supported in part by the California and U.S. Army Breast Cancer Research Programs.

REFERENCES

1. M.P. André, H.S. Janée, G.P. Otto, P.J. Martin, B.A., et al, High-speed data acquisition in a diffraction tomography system using large-scale toroidal arrays, *Intl. J. Imaging Systems Technol.* 8(1):137 (1997).
2. S. Johnson, M.L. Tracy, Inverse scattering solutions by a sinc basis, multiple source, moment method -- Part I: Theory, *Ultrasonic Imaging* 5:361 (1983).
3. M.P. André, H.S. Janée, G.P. Otto, P.J. Martin, Reduction of phase aberration in a diffraction tomography system for breast imaging. *Acoustical Imaging* 22:151 (1996).
4. R.O. Schmidt, Multiple emitter location and signal parameter estimation. *IEEE Trans. Ant. & Prog.* 14:276 (1986).
5. S.F. Gull, G.J. Daniell, Image reconstruction from incomplete and noisy data. *Nature* 272:686 (1978).

MULTI-DIMENSIONAL ACOUSTICAL TOMOGRAPHY BY INCOMPLETE DATA

V.A.Burov, A.L.Konjushkin and O.D.Rumiantseva

Moscow State University, Faculty of Physics, Department of Acoustics,
Moscow, 119899, Russia

The reconstruction problem of three-dimensional biological structures, as determined by the purposes of medical tomography, faces a number of basic and technical difficulties. The main difficulties are the following. Firstly, biological medium in real conditions appears to be inhomogeneous with regard to three components simultaneously (phase velocity $c(r)$, density $\rho(r)$ and amplitude coefficient of absorption $\alpha(r, \omega)$). Each of these components should be reconstructed. Secondly, for the majority of problems the inhomogeneities of medium belong to a class of strong scatterers, i.e. they bring in an essential distortion to an internal field in comparison with an incident field. Thirdly, some spatial-geometrical positions of sources and receivers might be unavailable in experiment. Absence of these positions leads to incompleteness of the scattering data.

In this connection an attempt of the complex consideration of the mentioned set of difficulties in the solution of multi-dimensional acoustical inverse scattering problem is undertaken in the report. Iterative methods [1] (if necessary with introduction of a procedure of "gradual account" of inhomogeneity into algorithm [2]) seem to be the most expedient for solution of the problem because of incompleteness of the scattering data. These methods have been developed to date concerning highly contrast inhomogeneities for which the Born approximation fails. However solution of the inverse scattering problem in the Born approximation could be taken as a basis for each fixed iterative step. By this reason the main attention in the report is concentrated on the Born problem as a key moment for organization of iterative cycle.

Reconstruction of Two-Dimensional Scatterer in Monochromatic Mode for Complete Data

The scatterer characteristics $c(r)$, $\rho(r)$, $\alpha(r, \omega)$ can be evaluated by reconstruction of function $v(r, \omega)$:

$$v(r,\omega) = \omega^2 \left(c_0^{-2} - c^{-2}(r) \right) + \sqrt{\rho(r)} \nabla^2 \left(1 / \sqrt{\rho(r)} \right) - 2i\omega\,\alpha(r,\omega)/c(r) , \qquad (1)$$

where the time factor is $\exp(-i\omega t)$, c_0 is the sound velocity in background homogeneous lossless medium. An estimation $\hat{v}(r,\omega)$ of this function is calculated by classical algorithm [3] admitting a generalization on pulse mode [4]. For two-dimensional problem $r = \{x, y\}$,

$$\hat{v}(r,\omega) = \frac{k_0^2}{2} \int\limits_0^{2\pi} d\varphi \int\limits_0^{2\pi} d\varphi'\; f(\varphi, \varphi', \omega)\, \left| \sin(\varphi - \varphi') \right|\, \exp\left\{ ik_0 \left[x(\cos\varphi' - \cos\varphi) + y(\sin\varphi' - \sin\varphi) \right] \right\}$$

This relationship supposes that the scatterer is irradiated by plane waves $\exp(ikr)$, $|k| \equiv k_0 = \omega/c_0$, and the *complete* set of the scattering data - the scattering amplitude $f(k, l, \omega) \equiv f(\varphi, \varphi', \omega)$ - is available. Angels φ and φ' define direction of the probing incident wave and direction of the scattered wave, correspondingly. In practice the similar schemes are usually realized by synthesis of the scattering data received from the points-like sources surrounding the scattering area rather than by the use of receivers and sources of large extent.

Results of computer modeling for the solution of the reconstruction problem for two-component scatterer have been obtained. In this case the real part of the function $v(r,\omega)$ is formed by inhomogeneities of the phase velocity $\Delta c \equiv c(r) - c_0$ and the imaginary part - by the absorption coefficient $\alpha(r, \omega)$ at the invariable density $\rho(r) \equiv const$. Various scatterers of the Gauss form are used as the test objects to estimate the influence of spatial characteristics of the scheme and accuracy of measurement of the scattering data: $\Delta c(r) = c_1 \exp\{-(r-r_1)^2/a_1^2\}$; $\alpha(r, \omega) = \alpha_2 \exp\{-(r-r_2)^2/a_2^2\}$. Vectors r_1 and r_2 define coordinates of center for the refractive and absorbing components of the scatterer correspondingly. Numerical values of parameters of the model problem correspond to their values for problems of tomographic imaging of soft biological tissues: frequency of probing waves 500 kHz, the background velocity $c_0=1500$ m/s (wave length $\lambda=3\times10^{-3}$ m/s), maximal coefficient of absorption $\alpha_2=1$ dB/cm $\approx 11,5$ m^{-1}, maximal difference of velocity within the scatterer and in the background medium $c_1 = 10$ m/s. The choice of such the parameters is based on the fact that magnitude of the scattering data observed from two inhomogeneities with the same linear size - single absorbing inhomogeneity with $\alpha_2=1$ dB/cm and single refractive one with $c_1 = 10$ m/s - are approximately equivalent. Such a simple model allows to estimate a stability of the used procedure of reconstruction for addition of noise in the scattering data. Really, stability of reconstruction mainly depends on size of a scatterer and width of its spatial spectrum. Details of the scatterer's form are not so essential.

Quality of reconstruction remains acceptable at the following relative errors in the scattering data. In the case of a single smooth scatterer with characteristic sizes about half of wave length $(2a_1 = 2a_2 \approx 0,4\lambda$ for numerical modeling) errors about -6dB from the maximal amplitude of the scattered field still allow to reconstruct $\Delta c(r)$ and $\alpha(r, \omega)$ accurately enough. Such a high level of errors is admitted thanks to coherent accumulation of all the scattering data during reconstruction of a single *fine-scale* scatterer. In the case of a single smooth *large-scale* scatterer with characteristic sizes of several wave lengths $(2a_1 = 2a_2 \approx 6,6\lambda$ for numerical modeling) the acceptable noise level is restricted only by -30 dB. The same restricting level of errors takes place for reconstruction of the fine-scale scatterer on the background of the large-scale one (fig.1a,b) and for a combination of fine-scale scatterers with arbitrary location. For example, the section y=0 of scatterer reconstructed without noise (thick line) and for the noise level $\approx$-40dB (thin line) is pictured on fig.1a,b. The scatterer is the sum of a large-scale scatterer $(2a_1 \approx 6,6\lambda$; $c_1 = 10$ m/s; $x_1 \approx 17\lambda$, $y_1 = 0)$ and a fine-scale one $(2a_1 \approx 0,4\lambda$; $c_1 = 10$ m/s; $x_1 \approx 20\lambda$, $y_1 = 0)$ for the velocity inhomogeneity and a large-scale scatterer $(2a_2 \approx 6,6\lambda$; $\alpha_2=1$ dB/cm; $x_2 \approx -17\lambda$, $y_2=0)$ for the absorption inhomogeneity.

In such manner, the two-dimensional monochromatic algorithm of the solution of the inverse scattering problem allows to reconstruct qualitatively the Born scatterer if relative errors in the data are less than, in the general case, -30dB as compared to maximum.

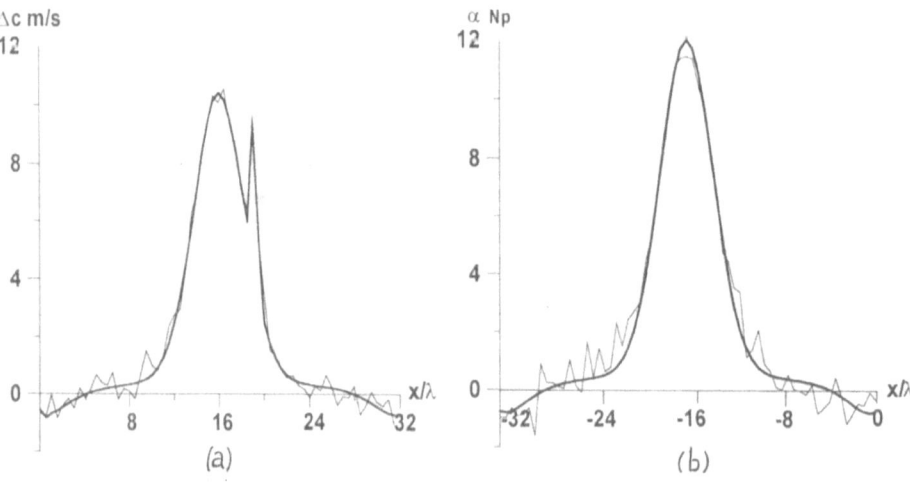

Figure1. Two-dimensional velocity (a) and absorption coefficient (b) reconstructed by complete data without noise (thick line) and for noise level -40dB (thin line).

Reconstruction of Three-Dimensional Scatterer in Multiple-Frequency Mode for Incomplete Data

In tomographic experiment the situation frequently occurs when the complete (i.e. corresponding to all directions of radiation and reception) scattering data are impossible to be obtained. This situation takes place when receiver-sources elements forming array are located not over all sides of the tested object, by some reasons. In this connection point spread (PS) functions have been researched for an algorithm of reconstruction of Born three-dimensional scatterers by incomplete data. The function $v(\mathbf{r}, \omega)$ defined by (1) is estimated by the relationship [3]:

$$\hat{v}(\mathbf{r},\omega) = \frac{k_0^3}{\sqrt{2\pi}} \int_0^{2\pi} d\varphi \int_0^{2\pi} d\varphi' \int_0^{\pi} d\theta \int_0^{\pi} d\theta' \ f(\mathbf{k},\mathbf{l},\omega) \ \sin\theta \ \sin\theta' \times$$

$$\times \sqrt{1 - \sin\theta' \sin\theta \cos(\varphi - \varphi') - \cos\theta' \cos\theta} \quad \exp[ik_0[x(\sin\theta' \cos\varphi' - \tag{2}$$

$$- \sin\theta \cos\varphi) + y(\sin\theta' \sin\varphi' - \sin\theta \sin\varphi) + z(\cos\theta' - \cos\theta)]]: \quad \mathbf{r} = \{x,y,z\}.$$

Here φ and θ are azimuth and polar angles of the wave vector $\mathbf{k}$ of the incident wave in the spherical coordinate system; φ' and θ' are the angles of the wave vector $\mathbf{l}$ of the scattered field; $f(\mathbf{k},\mathbf{l},\omega)$ - the scattering amplitude.

PS functions correspond to different variants of ultrasonic tomographs in view of specificity of their medical application. The variants differ from each other concerning the range of values of the polar angles θ and θ' used in the experiment. It is supposed that for fixed θ and θ' the azimuth angles take all possible values $\varphi,\varphi' \in [0,2\pi)$. Then a two-dimensional section of PS function by a plane z=const is similar to PS function of two-dimensional problem for the complete data. At the same time, z-structure of the three-dimensional PS function is different depending on the range of the values θ, θ'. This fact is illustrated by fig.2a,b, when the scattering data are generated by very small model scatterer at the origin: $v(\mathbf{r},\omega) \sim \delta(\mathbf{r})$. Real (fig.2a) and imaginary (fig.2b) parts of each PS function are normalized to the maximal value of its real part. In the case of *complete* scattering data ($\theta,\theta' \in [0,\pi]$) the PS function is real (dashed line). In the case of *incomplete* data two following models of the three-dimensional tomograph have been considered. The first model corresponds to irradiation in the horizontal plane($\theta = \pi/2$) and to reception of the scattered waves in a part of the upper semispace ($\theta' \in [\pi/4, \pi/2]$). In both model the data incompleteness results in fragmentary information on spatial frequencies of scatterer function. This fact leads to appearance of the imaginary part of the PS function, and absolute values of the imaginary and the real parts are comparable to each other (fig.2: the first model -thick line, the second one - thin line). The information incompleteness is illustrated by fig.3, where z-component ξ_z of the space spectrum of the PS function for the first model (thick line) and for the second one (thin line) is presented. The z-component is normalized to its maximal value. Instead of desirable full information about the scatterer corresponding to all values of space frequencies $\bar{\xi}$ within $2k_0$-sphere, all the values in the plane $\xi_z=0$ and only a part of the values in the upper semispace $\xi_z > 0$ are accessible.

The reconstruction of the scatterer function therefore means not only to estimate values of the scattering components (velocity, density and absorption coefficient) as independent characteristics, but also to exclude an influence of the imaginary part of the PS function on these estimations. Solution of this double problem is possible with the help of pulse or multiple-frequency probing mode [5,2] because the type of dependence on frequency is individual for each the scattering component of the function $v(\mathbf{r},\omega)$ (see (1)). This regime should supply three independent sets of the scattering data corresponding to three different frequencies ω_j in the case of multiple-frequency mode: $f(\mathbf{k},\mathbf{l},\omega_j)$,

j=1,2,3. Really, supposing that the absorption coefficient is quadratic in frequency $\alpha(\mathbf{r},\omega) \cong \omega^2 \beta(\mathbf{r})$, the relation (1) is reduced to the form

$$W(\mathbf{r},\omega) \equiv (c_0^2 / \omega^2) v(\mathbf{r},\omega) = g_c(\mathbf{r}) + \omega^{-2} g_\rho(\mathbf{r}) + i \ \omega g_\alpha(\mathbf{r}). \tag{3}$$

Functions g_c, g_ρ, g_α are independent of the frequency and are connected with physical medium characteristics looked for:

$$g_c(\mathbf{r}) = \left(1 - c_0^2 / c^2(\mathbf{r})\right); \quad g_\rho(\mathbf{r}) = c_0^2 \sqrt{\rho(\mathbf{r})} \ \nabla^2\left(\rho^{-1/2}(\mathbf{r})\right); \quad g_\alpha(\mathbf{r}) = -2\beta(\mathbf{r}) c_0^2 / c(\mathbf{r}).$$

An estimation $\hat{v}(r, \omega_j)$ of the function $v(r, \omega_j)$ is calculated by (2) for each fixed $\omega = \omega_j$. Then an estimation $\hat{W}(r, \omega_j) \equiv (c_0^2 / \omega_j^2) \hat{v}(r, \omega_j)$ of the function $W(r, \omega_j)$ is known for the three frequencies. In the case $g_c(r) = \delta(r)$, $g_\rho = g_\alpha = 0$ the estimation $\hat{W}(r, \omega_j) \equiv \hat{g}_\delta(r, \omega_j) = Re(\hat{g}_\delta) + i \, Im(\hat{g}_\delta)$ is the PS function. The lower is the frequency, the wider is the main lobe of its real part $Re(\hat{g}_\delta(r, \omega_j))$. For example, for the first model of tomograph the semiwidth of the PS function for the level 0,7 is $\approx \lambda_j / 8$ in the plane (X,Y) and it's only $\approx 0,35\lambda_j$ along the vertical Z-axis. It follows from the fact that the space frequencies are accessible, for which $|\xi_x| \le 2k_j$, $|\xi_y| \le 2k_j$ and only $|\xi_z| \le k_j / \sqrt{2}$ because of the data incompleteness (fig.3).

For arbitrary scatterer the estimation $\hat{W}(r, \omega_j)$ is convolution ($\otimes$) of the functions $W(r, \omega_j)$ and $\hat{g}_\delta(r, \omega_j)$, then the relation (3) results in

$$\hat{W}(r, \omega_j) = W(r, \omega_j) \otimes \hat{g}_\delta(r, \omega_j) =$$
$$= g_c(r) \otimes \hat{g}_\delta(r, \omega_j) + \omega_j^{-2} \, g_\rho(r) \otimes \hat{g}_\delta(r, \omega_j) + i \, \omega_j \, g_\alpha(r) \otimes \hat{g}_\delta(r, \omega_j). \qquad (4)$$

The estimations looked for the functions g_c, g_ρ, g_α should be the same for any frequencies ω_j, so reference PS function $g_\delta^0(r, \omega_{low})$ is chosen. Its space spectrum $\tilde{g}_\delta^0(\xi, \omega_{low})$ should provide a low level of side lobes of the function $g_\delta^0(r, \omega_{low})$. At the same time the width of this space spectrum should correspond to space spectrum $\tilde{g}_\delta(\xi, \omega_j)$ of the PS function $\hat{g}_\delta(r, \omega_j)$ for the lowest frequency $\omega_j = \omega_{low}$. Then space spectra $\tilde{W}(\xi, \omega_j)$ of the estimations $\hat{W}(r, \omega_j)$ are corrected according to $\tilde{g}_\delta^0(\xi, \omega_{low})$:

$$\tilde{W}^{cor}(\xi, \omega_j) = \tilde{W}(\xi, \omega_j) \left[\tilde{g}_\delta^0(\xi, \omega_{low}) / \tilde{g}_\delta(\xi, \omega_j) \right] \qquad (5)$$

The corrected estimations $\hat{W}^{cor}(r, \omega_j)$ may be presented, according to (4), as

$$\hat{W}^{cor}(r, \omega_j) = g_c(r) \otimes g_\delta^0(r, \omega_{low}) + \omega_j^{-2} g_\rho(r) \otimes g_\delta^0(r, \omega_{low}) +$$
$$+ i \, \omega_j g_\alpha(r) \otimes g_\delta^0(r, \omega_{low}) \quad , \quad j=1,2,3. \qquad (6)$$

Separating the function $\hat{W}^{cor}(r, \omega_j)$ into its real and imaginary parts leads (6) to a system of 6 algebraic equations for each fixed point r. In the system terms $S_c = g_c \otimes Re[g_\delta^0]$, $S_\rho = g_\rho \otimes Re[g_\delta^0]$, $S_\alpha = g_\alpha \otimes Re[g_\delta^0]$ are informative and are taken as estimations of the functions g_c, g_ρ, g_α, correspondingly. Terms $N_c = g_c \otimes Im[g_\delta^0]$, $N_\rho = g_\rho \otimes Im[g_\delta^0]$, $N_\alpha = g_\alpha \otimes Im[g_\delta^0]$ are uninformative. Because of the system obtained is linear, it is possible to exclude N_c, N_ρ, N_α and to find S_c, S_ρ, S_α looked for by three remaining equations. As a result,

$$S_c = \left[A(\omega_1 / \omega_3)^2 \left(1 - (\omega_3 / \omega_1)^3\right) - B(\omega_1 / \omega_2)^2 \left(1 - (\omega_2 / \omega_1)^3\right) \right] / D \ ;$$

$$S_\rho = -\omega_1^2 \left[A(1 - \omega_3 / \omega_1) - B(1 - \omega_2 / \omega_1) \right] / D \ ;$$

$$S_\alpha = \left[\mathrm{Im}\, W_1 \left(\frac{(\omega_2/\omega_3)^2 - 1}{1 - (\omega_2/\omega_1)^2} \right) + \mathrm{Im}\, W_2 \left(\frac{(\omega_1/\omega_3)^2 - 1}{1 - (\omega_1/\omega_2)^2} \right) + \mathrm{Im}\, W_3 \right] \Bigg/$$

$$\Bigg/ \left[\omega_1 \left(\frac{\omega_3}{\omega_1} - \frac{\omega_2}{\omega_1} - 1 + \frac{1 + \omega_1\omega_2/\omega_3^2}{1 + \omega_1/\omega_2} \right) \right] , \qquad \text{where}$$

$$A = \mathrm{Re}\, W_2 - \mathrm{Re}\, W_1 \; (\omega_2/\omega_1); \quad B = \mathrm{Re}\, W_3 - \mathrm{Re}\, W_1 \; (\omega_3/\omega_1); \quad W_j \equiv \hat{W}^{\mathrm{cor}}\!\left(\mathbf{r}, \omega_j \right);$$

$$D = \left(1 - \omega_2/\omega_1\right) \left(1 - \omega_3/\omega_1\right) \left(\omega_1/\omega_3 - \omega_1/\omega_2\right) \left(1 + \omega_1/\omega_2 + \omega_1/\omega_3\right).$$

It should be noted that in case of complete volume of the scattering data two frequencies ω_j are sufficient to find S_c, S_ρ, S_α.

Numerical modeling has shown that it is expedient to reconstruct nondimensional values S_c, S_ρ/ω_m^2, $\omega_m S_\alpha$ (ω_m is the mean frequency from the three ones) because these are the values that are physically comparable with each other in medical tomography. Model scattering data for the frequencies 400kHz, 500kHz, 600kHz and $c_0 = 1500$ m/s have been formed by points-like inhomogeneities:

$$g_c(\mathbf{r}) = \delta(\mathbf{r} - \mathbf{r}_c) = \delta(x)\delta(y)\delta(z - z_c); \qquad g_\rho(\mathbf{r})/\omega_m^2 = \delta(\mathbf{r} - \mathbf{r}_\rho) = \delta(x)\delta(y)\delta(z - z_\rho);$$

$$\omega_m S_\alpha(\mathbf{r}) = \delta(\mathbf{r} - \mathbf{r}_\alpha) = \delta(x)\delta(y)\delta(z - z_\alpha), \qquad \text{where } z_c = 0, \; z_\rho = 0.5\lambda_{\min}, \; z_\alpha = -0.5\lambda_{\min},$$

the wave length $\lambda_{\min}$ corresponds to the highest frequency. Location of c-, ρ-, α-inhomogeneities in the plane (X,Y) is characterized (for simplification of illustration) by the same coordinates x=y=0. The Hemming spectrum is taken as the reference one $\tilde{g}_\delta^0\!\left(\vec{\xi}, \omega_{\mathrm{low}}\right)$. Result of the reconstruction of

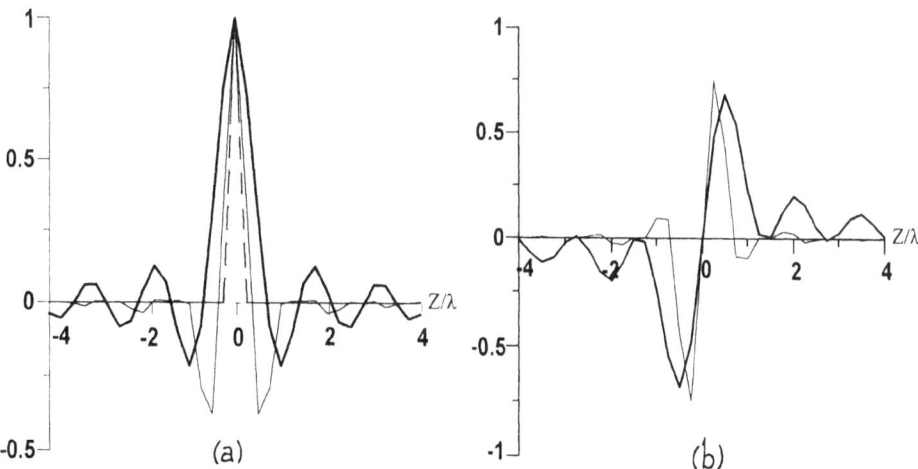

(a)   (b)

Figure2. Real (a) and imaginary (b) parts of three-dimensional point spread function normalized to maximal value of its real part for complete scattering data (dashed line) and incomplete data (the first tomography model -thick line, the second one - thin line).

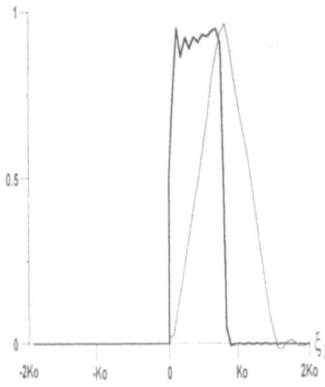

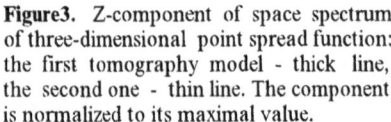

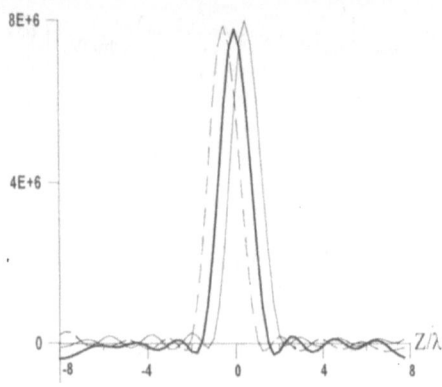

Figure3. Z-component of space spectrum of three-dimensional point spread function: the first tomography model - thick line, the second one - thin line. The component is normalized to its maximal value.

Figure4. Functions characterizing velocity (thick line), density (thin line) and absorption coefficient (dashed line) of three-dimensional points-like scatterer reconstructed by incomplete data in three-frequencies mode.

estimations S_c, S_ρ / ω_m^2, $\omega_m S_\alpha$ as functions of z-coordinate at the coordinates fixed x=y=0 is shown on fig.4.

It is shown, by these means, that the application of multiple-frequency mode for three frequencies allows to reconstruct complete spatial structure of inhomogeneities of velocity, density and absorption. The use of multiple-frequency (or pulse) mode is a unique way allowing to separate these three types of inhomogeneities in reconstructed scatterer function. However, it should be emphasized that conditionality of the method highly depends on the given ratio between the operating frequencies ω_j. While the relative interval between the frequencies decreases, conditionality becomes significantly worse. Additional problem is to set a priori the type of frequency dependence of absorption coefficient, since an error in this influences the result of estimations.

REFERENCES

1. S.A.Johnson, Y.Zhou, M.K.Tracy, M.J.Berggen, F.Stenger, Inverse scattering solutions by a sink basis, multiple source, moment method - Part 3: Fast algorithms, *Ultrasonic Imaging* 6:103 (1984).
2. V.A.Burov, M.N.Rychagov, A.V.Saskovets, Account of multiple scattering in acoustic inverse problems of tomographic type, in: *Acoustical Imaging-19*, H.Ermert and H.-P.Harjes, ed., Plenum Press, New York (1992).
3. A.J.Devaney, M.L.Oristaglio, Inversion procedure for inverse scattering with the distorted-wave Born approximation, *Phys. Rev. Lett.* 51:237 (1983).
4. V.A.Burov, O.D.Rumiantseva, Linearized inverse problem of scattering in monochromatic and pulse modes, *Acoust. Phys.* 40:34 (1994).
5. A.J.Devaney, Variable density acoustic tomography, *J. Acoust. Soc. Am.* 78:120 (1985).

BACK SCATTERED ULTRASONIC TOMOGRAPHY: EXPERIMENTS AND MODELIZATIONS

S. Delamare, J.P. Lefebvre, P. Lasaygues

Laboratoire de Mécanique et d'Acoustique
CNRS/LMA/PI
31 ch. Joseph Aiguier, 13402 MARSEILLE Cedex 20, FRANCE
E-mail : delamare@lma.cnrs-mrs.fr

INTRODUCTION

Ultrasonic tomography with back-scattered field is based on three assumptions: low contrast medium (biological medium), large frequency range signal (echographic signal), dense and complete sets of projection. If these are ideally verified, we can construct images of the impedance of the medium. This technique, which we called Acoustical Impedance Tomography [1], involves an algorithm based on First-Order Born Approximation.

To date, we have tested only the qualitative aspect of this imaging technique on our tomographic testing ground. It is composed of a support for the target, of a precise rotating and translating mechanical structure that can hold several transducers, and of a numerical chain to record the projections and to compute them.

Now we would like to obtain quantitative images. We begin this paper by a study of the First-Order Born Approximation in the case of a homogeneous circular cylinder. After, we fully simulate the reconstruction experiments on our tomographic testing ground, including the influence of the contrast, of the transducer frequency bandwidth, and of the shape of the target. Lastly, we discuss on the possibility to obtain quantitative imaging using the Acoustical Impedance Tomography technique.

STUDY OF THE FIRST-ORDER BORN APPROXIMATION FOR A CIRCULAR CYLINDER

Let a homogeneous cylinder of infinite length, radius R, density ρ and sound speed c be placed at the origin, and an insonifying plane wave coming from infinity in the propagating medium of density ρ_o and sound speed c_o. This simple case allows us to analytically calculate the Exact far field pattern u_s^∞ and the Born Approximation far field pattern $\phi_s^{B,\infty}$ for back-scattered measurements. After, we make a comparison between these two solutions.

Exact Far Field Pattern

The general solution for the back-scattered field is obtained by the method of separation of variables in the Helmholtz equation when polar coordinates are used. After, the expansion of the Hankel function for large arguments gives the field for large distance r to wavelength ratio $k_o r$ (far field) [2]:

$$P_s^\infty(r) = \sqrt{\frac{2}{\pi k_o r}} e^{j(k_o r - \frac{\pi}{4})} u_s^\infty \quad ; \quad u_s^\infty = \sum_{n \in \mathbb{Z}} \frac{N_n}{D_n} e^{jn\pi} \tag{1}$$

$$(\forall n \in \mathbb{Z}) \quad \frac{N_n}{D_n} = \frac{J_n(k_o R) \dot{J}_n(\frac{k_o R}{\gamma_c + 1}) - (\gamma_\rho + 1)(\gamma_c + 1) J_n(\frac{k_o R}{\gamma_c + 1}) \dot{J}_n(k_o R)}{(\gamma_\rho + 1)(\gamma_c + 1) J_n(\frac{k_o R}{\gamma_c + 1}) \dot{H}_n(k_o R) - H_n(k_o R) \dot{J}_n(\frac{k_o R}{\gamma_c + 1})} \tag{2}$$

where $k_o R$ is the frequency variable ($k_o = \frac{2\pi f}{c_o}$: wave number, f: frequency), $\gamma_\rho = \frac{\rho - \rho_o}{\rho_o}$ is the density contrast, and $\gamma_c = \frac{c - c_o}{c_o}$ is the speed contrast. J_n, H_n are the Bessel and first kind Hankel functions of order n.

We can't give the exact result for the infinite series in (1), so we have calculated an optimal truncature given by: $|n| \le k_o R + 10 \ln k_o R$.

Born Approximation Far Field Pattern

Using a Lippmann-Schwinger integral equation, the Born Approximation and the expansion of the Hankel function seen before, we obtain the Born Approximation far field pattern for back-scattered measurements [3]:

$$P_s^{B,\infty} = \sqrt{\frac{2}{\pi k_o r}} e^{j(k_o r - \frac{\pi}{4})} \phi_s^{B,\infty} \tag{3}$$

$$\phi_s^{B,\infty} = -\frac{j\pi k_o R}{2} \ln \left[(\gamma_\rho + 1)(\gamma_c + 1) \right] J_1 \left[2k_o R \right] \tag{4}$$

Comparison Between u_s^∞ and $\phi_s^{B,\infty}$

The conditions of the study are imposed by the measurement system of our tomographic testing ground. In figure 1(a), we consider a circular cylinder of diameter 10 mm. In the time-domain, the signal sampling rate is 20 MHz, so the frequency signal is known from 0 to 10 MHz according to the Shannon Theorem when a Fourier Transform is applied [4].

We also consider the case of low contrasts ($\gamma_\rho = \gamma_c = 0.01$) that corresponds to a biological medium, and high contrasts ($\gamma_\rho = 6.6, \gamma_c = 2.9$) in the case of a target made of steel. We compute the relative error $\left| \frac{u_s^\infty - \phi_s^{B,\infty}}{u_s^\infty} \right|$ in order to compare the Exact and the Born Approximation far field patterns.

In the low frequency domain ($k_o R \le 1$), the results are the same as in the litterature ([5] for example): the First-Order Born Approximation gives quite perfect results for low contrasts and diverges for high contrasts. In our case, the large frequency range ($k_o R \in [0, 210]$) includes high frequencies. The results are shown in the figure 1(b):

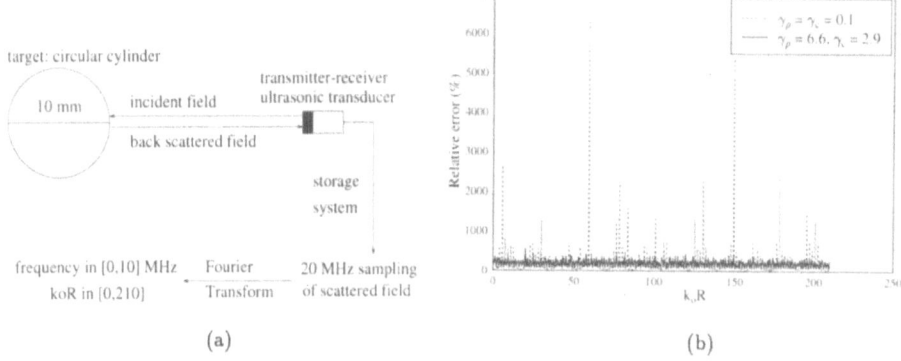

Figure 1. (a) The tomographic testing ground: principle, (b) Relative error: $\left| \frac{u_s^\infty - \phi_s^{B,\infty}}{u_s^\infty} \right|$

the First-Order Born Approximation gives far better results for high contrasts than for low contrasts. According to the theory, this result is surprising but it explains the quality of the reconstructions we performed in the following part.

SIMULATION OF THE BACK-SCATTERED TOMOGRAPHY EXPERIMENTS

Consider the system of the figure 1(a), and replace the circular cylinder by any cylindrical target, that is to say, the shape is unknown. In this case, similar calculations as those of the Born Approximation far field pattern give a close relationship between the back-scattered field and the Acoustical Impedance inside and outside the target:

$$\phi_s^{B,\infty}(\overline{n}_i) = -\frac{jk_o^2}{2}\hat{\xi}(-2k_o\overline{n}_i) \qquad (5)$$

where $\hat{\xi}$ is the two-dimensional Fourier Transform of the function $\xi = \ln\frac{Z(x,y)}{Z_o}$ that gives the natural logarithm of the normalized impedance for each point (x,y) of the image. $\overline{n}_i$ is the unit vector in the direction of the incident field.

Finally, we obtain the reconstruction formula using a two-dimensional Fourier Transform:

$$\xi(2\pi x, 2\pi y) = \frac{2j}{\pi^2}\int_0^\pi \left\{ \int_{-\infty}^{+\infty} \frac{A(|\chi|)}{|\chi|} u_s^\infty(|\chi|, \theta_i) e^{2j\pi\chi(x\cos\theta_i + y\sin\theta_i)} dx \right\} d\theta_i \qquad (6)$$

$\chi = 2k_o$ is the frequency variable. A is a filter due to the transducer frequency bandwidth. u_s^∞ is the measured back-scattered field, in place of $\phi_s^{B,\infty}$. This formula is composed of two integrals: the integral with respect to χ is an Inverse Fourier Transform, and the integral with respect to θ_i (angle of the incidence) is an Inverse Radon Transform [6].

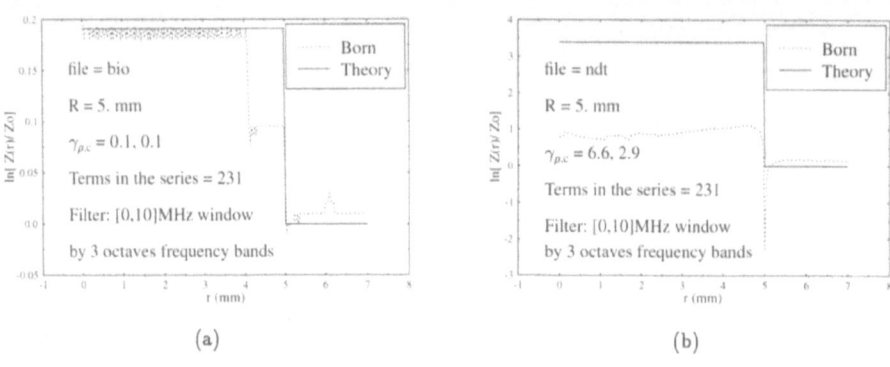

Figure 2. Reconstruction with: (a) low contrasts, (b) high contrasts

Influence of the Contrast

As a first result, we show the reconstruction of the relative acoustical impedance of a circular cylinder of radius 5 mm. We assume that $k_o R \in [0, 210]$. We consider the cases of low contrasts (figure 2(a)) and high contrasts (figure 2(b)). The circular symmetry of the target allows us to represent the reconstruction along any radius, from the center to the external medium. The theoretical reconstruction is represented by the solid line. The dotted line shows the simulation of our experiments.

In the case of low contrasts, the reconstruction is quantitative: see the closeness of the two curves. It is obvious everywhere but in a region localized between the radius 4 mm and 5 mm, near the surface of the cylinder. We haven't studied this phenomenon yet, but we observe that the drop is equal to the half of the higher value of the relative impedance. So, at first glance, the drop is due to the fact that we don't take into account a multiplying factor 2 on the surface of the cylinder, as we do in the case of impenetrable targets [7]. We must find the solution in the Distributions Theory. Turning now to the high contrasts case, we lose quantitative information but the shape is well defined. Note that when the impedance is high, that is to say, the sharp drop on the cylinder boundary is more visible, the numerical artifact disappears.

Influence of the Transducer Bandwidth

Until now, we only used the [0,10] MHz window. To give a more precise simulation of the experiments, let's consider the influence of a gaussian frequency filter. The frequency response is a gaussian curve centered at 500 kHz and the bandwidth is 550 kHz. In the case of high contrasts, the reconstruction is shown in the figure 3. Quantitative information is lost because the lower cutoff frequency is not zero [8], and the higher cutoff frequency is less than 10 MHz. But the external shape is well defined.

Influence of the Shape of the Target

Finally, we are interested in a multi-part target composed of two cylinders of radius 4mm separated by a distance of 10mm. The direct problem of finding the scattered field has been solved with a classical boundary element method [9].

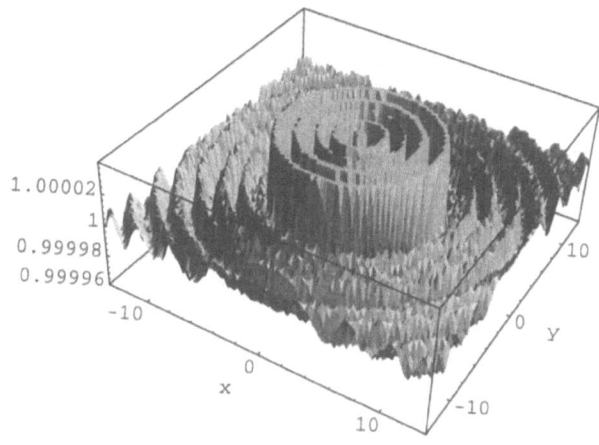

Figure 3. Reconstruction with the 500 kHz gaussian frequency filter: high contrasts

The reconstruction of the figure 4 shows the two cylinders: quantitative information is lost but the shape is again well defined and the reconstructed radius is exact. The only problem is the space between the two cylinders: the reconstruction shows a distance of 4mm instead of 10mm. It can be a default due to the Inverse Born Approximation when the target is not centered at the origin.

COMMENTS AND CONCLUSIONS

In a first part, we have studied the first-Order Born Approximation in a large frequency range and we found that in the case of a high frequency range, the Born Approximation gave far better results for high contrasts than for low contrasts. Judging from these observations, it is possible to extend the applications of Ultrasonic Tomography to a certain degree.

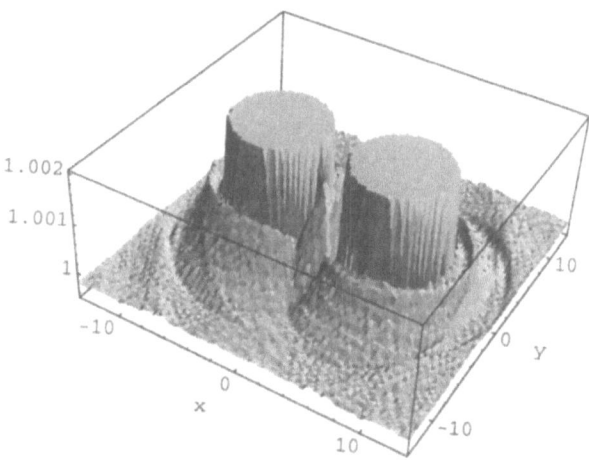

Figure 4. Reconstruction of a multi-part target: high contrasts

Next, we have simulated the reconstruction of the Acoustical Impedance. In the case of low contrasts, we found quantitative information in the low frequency domain, and a lack of information in high frequency domain. After, for high contrasts, quantitative information was lost but the shapes were far better defined in a large frequency range including high frequencies. Thus, the extend of Acoustical Impedance Tomography for high contrasts is possible but it must be considered as qualitative imaging: we observe the external shape of the target with a great accuracy.

Thereby, we know what can be expected from the experiments on our tomographic testing ground and we can say that the Inverse Born Approximation is a precise initial step for more complex iterative algorithms that can work in a large frequency range. It is a way to perform quantitative imaging in high contrasts and high frequency cases.

REFERENCES

[1] J.P. LEFEBVRE. La Tomographie d'Impédance Acoustique. *Revue du traitement du signal*, II(2):103 – 110, 1985.

[2] W.C. CHEW. *Inverse Waves and Fields in Inhomogeneous Media*. IEEE PRESS Series on Electromagnetic Waves. Donald G. Dudley, Series Editor, New York, 1995. Pages 14-15.

[3] P. M. MORSE K. U. INGARD. *Theorical Acoustics*. International Series in Pure and Applied Physics. Mc Graw-Hill Book Company, 1968. Pages 408-414.

[4] R.W. RAMIREZ. *The FFT: Fundamentals and Concepts*. PTR Prentice Hall, Englewood cliffs, New Jersey 07632, 1985.

[5] D. COLTON R. KRESS. *Inverse Acoustic and Electromagnetic Scattering Theory*, volume 93 of *Applied Mathematica Sciences*. Berlin : Springer-Verlag, 1992. Page 262.

[6] S.R. DEANS. *The Radon Transform and Some of its Applications*. Wiley-Interscience Publication, 1983.

[7] W. TOBOCMAN. Inverse Acoustic Wave Scattering in two Dimensions from Impenetrable Targets. *Inverse Problems*, 5.1131–1144, 1989.

[8] W. TOBOCMAN. In Vivo Biomicroscopy with Ultrasound. *Current Topics in Acoust. Res.*, 1:247–265, 1994.

[9] A. WIRGIN. Scalar Difffraction by N Cylinders - Boundary Element Method. Laboratoire de Modélisation en Mécanique, UPMC-CNRS, Paris, 1992.

NONLINEARITY PARAMETER TOMOGRAPHY
—— A NOVEL IMAGING METHOD IN TISSUE CHARACTERIZATION

Xiu-fen Gong, Dong Zhang

(Institute of Acoustics and State Key Laboratory of
Modern Acoustics, Nanjing University, Nanjing 210093, China)

The acoustic nonlinearity parameter B/A is a basic parameter in nonlinear acoustics and is defined as the ratio of the coefficients of quadratic term to linear term in the Taylor expansion of the state equation: $B/A=2\rho_0 c_0(\partial c/\partial p)_{o,s}$, where ρ_0 and c_0 are the density and velocity, p is static pressure and s is entropy. This parameter can describe the nonlinear ultrasonic phenomena in the range of biomedical frequencies and intensities, for example, wave distortion, harmonic generation, sound saturation and extra-attenuation etc. In recent years, many research works indicated that nonlinearity parameter B/A can provide more important information, such as structural features and pathological state of tissues than that linear parameters and may be a new parameter in biological tissue characterization. Several efforts have been made in nonlinear parameter imaging by T.Sato[1] and Y.Nakagawa[2] using phase shift method and finite amplitude method, respectively. In our previous works successful computer simulation of acoustic nonlinearity parameter tomography from calculated second harmonic data using conventional CT technique and some preliminary experimental results were reported[3,4]. In this paper we will present our recent experimental investigation of B/A tomography for normal and several pathological liver tissues.

PRINCIPLE AND METHOD

(1) When a finite amplitude ultrasonic wave with frequency ω is transmitted in a medium, it will be distorted along the propagating path and second harmonic wave will be generated. Assume the primary wave at the surface of the transmitter has the form $p_1=p_0\sin(\omega\tau)$, then considering the sound attenuation of the medium, the pressure amplitude of second harmonic wave at distance L is given by:

$$P_2(L) = \frac{\omega}{2} p_1^2(0) \int_0^L \beta_i(x) \exp[\int_0^x -2\alpha_1(x)dx - \int_x^L \alpha_2(x)dx]dx \qquad (1)$$

where $p_1(0)$ is the pressure amplitude of primary wave at x=0, α_1 and α_2 are the attenuation coefficients of primary wave and second harmonics, respectively. $\beta_i(x)=\beta(x)/(\rho_0 c_0^3)$, $\beta(x)=1+B/(2A)$. For distilled degassed water with known B/A=5.2, as a reference medium, $p_{20}(L)$ can be expressed by

$$P_{20}(L) = \frac{\omega_2 L}{4} p_1^2(0)\beta_{i0} \qquad (2)$$

where β_{i0} is the β_i of distilled water. From Eq.(1) and (2), the ratio of pressure amplitude of second harmonics in sample $p_{2x}(L)$ to that in water $p_{20}(L)$ can be obtained using comparative insert-substitution method as follows[4,5] :

$$\frac{P_{2x}(L)}{P_{20}(L)} = \frac{1}{\beta_{i0}(L)} \int_0^L \beta_i(x) \exp[\int_0^x -2\alpha_1(x)dx - \int_x^L \alpha_2(x)dx]dx \qquad (3)$$

(2) Using conventional CT method, the ratio of $p_{2x}(L)/p_{20}(L)$ is collected as projection data $p(u,\theta)$

$$p(u,\theta) = \frac{1}{\beta_{i0}L} \int_{-V}^{V} \beta_i'(x,y)dv \qquad (4)$$

where $\beta_i'(x,y) = \beta(x,y)C(x.,y)$, V=L/2, and C(x,y) is the attenuation matrix as :

$$C(x,y) = \frac{1}{2\pi} \int_0^{2\pi} \exp(\int_{-V_1}^{V} -2\alpha_1(x,y)dv - \int^{V_1} \alpha_2(x,y)dv)d\theta$$

$p(u,\theta)$ is the line integral of $\beta_i'(x,y)$ along the propagating path of ultrasound wave. Therefore, β' image can be reconstructed by using the filtered convolution method in CT technique. Then the nonlinearity parameter B/A tomography can be obtained by multiplying the attenuation matrix C(x,y) to compensate the attenuation of the sample.

EXPERIMENTAL ARRANGEMENT

The arrangement of experimental system is shown in Fig.1

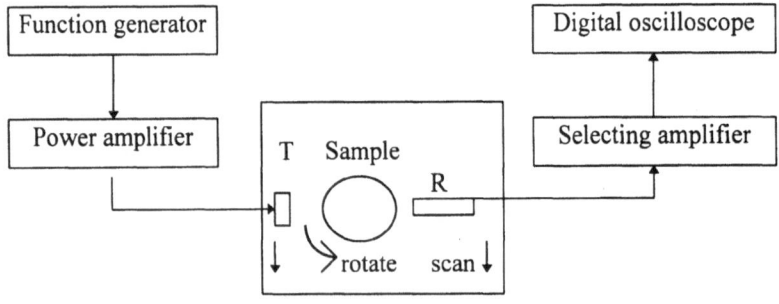

Figure 1. The block diagram of the experimental system for nonlinearity parameter tomography

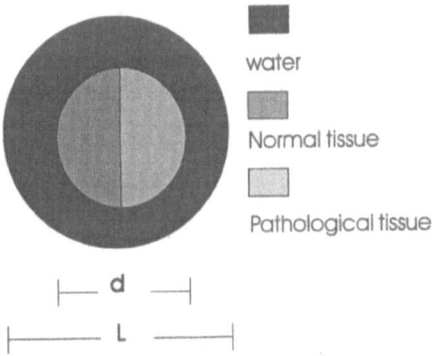

Figure 2. The sample model of nonlinearity parameter tomography for normal and pathological tissues

In experiment a 2MHz planar PZT transducer is used as the transmitter (T) and a broadband hydrophone is used as the receiver (R) to measure the signal of second harmonics. The T and R scan the sample synchronously step by step and sample rotates from 0° to 180° at interval 15°. The received signals are recorded by HP digital oscilloscope. After further processing the imaging of B/A is visualized on the monitor with 64 gray scales.

EXPERIMENTAL IMAGING FOR NORMAL AND PATHOLOGICAL TISSUES

Fresh porcine liver tissues in vitro were obtained from a slaughterhouse, stored in 0.9% saline solution and studied within 6h. A cylindrical sample holder was designed that surrounded the sample with a polythene membrane. The cross section of sample model is shown in Fig.2. The internal part consists of two half-cylindrical tissue samples with normal tissue in the left and pathological tissue in the right, the external part is water. In this paper, three kinds of pathological porcine livers including hepatocirrhosis liver, hepatitis and fatty liver are studied in experiments. The nonlinearity parameter tomography for hepatocirrhosis-normal livers, hepatitis-normal livers and fatty-normal livers are shown in Fig.3 (a), (b) and Fig.4 (a), respectively.

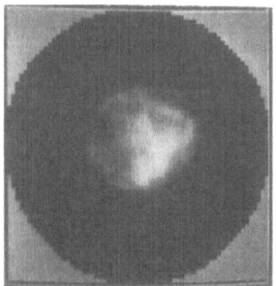
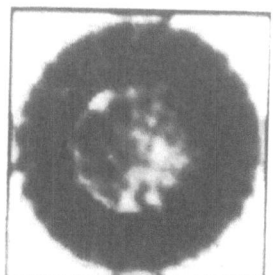

Figure 3. B/A tomography for (a) hepatocirrhosis-normal livers (b)hepatitis-normal livers

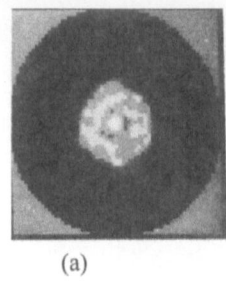
(a)

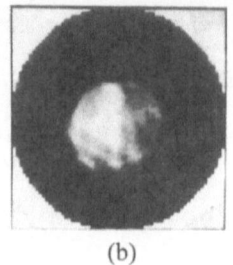
(b)

Figure 4. (a) B/A tomography for fatty-normal livers (b) Linear tomography for hepatitis-normal livers

For comparison, the ultrasound linear and nonlinear parameters of tissues are listed in a Table 1. and the linear parameter attenuation coefficient tomography for hepatitis-normal livers is given in Fig.4 (b).

Table 1. The ultrasound parameter of liver tissues

phantom	density g/cm^3	velocity m/s	attenuation (dB/cm) 2MHz	4MHz	B/A
normal liver	1.05	1605	1.08	2.20	6.8
hepatocirrhosis liver	1.05	1499	1.12	2.16	8.1
hepatitis liver	1.05	1600	1.11	2.18	7.4
fatty liver	1.05	1598	1.20	2.44	8.3

CONCLUSION

Results obtained in experiments indicate that values of nonlinearity parameter of pathological livers are higher than that of normal liver obviously. In nonlinearity parameter tomography, the difference between normal and pathological tissue is more sensitive than linear attenuation coefficient tomography. Therefore, the nonlinearity parameter B/A tomography may become a novel imaging method for biological tissue characterization.

This work is supported by National Natural Science Foundation of China and Natural Science Foundation of Jian-Su Province.

REFERENCES:

1. T.Sato, Y.Yamakoshi and T.Nakamura, "Nonlinear tissue imaging", *1986 Ultrasonics Symposium*, 889

2. Y.Nakagawa, W.Hou, A.Cai, etc., "Nonlinear parameter imaging with finite-amplitude sound waves", *1986 Ultrasonics Symposium*, 901
3. D.Zhang and X.F.Gong, "Computer simulation of acoustic nonlinear parameter tomography", *Chin.J.Acoust.*, 13(2), 169-175 (1994)
4. D.Zhang, X.F.Gong and S.G.Ye, "Acoustic nonlinearity parameter tomography for biological specimens via measurement of the second harmonics", *J.Acoust.Soc.Am.*, 99(4), 2397-2402 (1996)
5. X.F.Gong, Z.M.Zhu, T.Shi and J.H.Huang, "Determination of acoustic nonlinearity parameter using FAIS and ITD methods", *J.Acoust.Soc.Am.*, 86(1), 1-5 (1989)

EXPERIMENTAL INVESTIGATIONS OF THE CORRELATION
TOMOGRAPHY SPACE RESOLUTION

Chmill A.I., Gerasimov V.V., Guluaev Yu.V.,
Mirgorodsky V.I. and Peshin S.V.

Institute of Radio Engineering and
Electronics, sq. Vvedensky 1, Fryazino,
Moscow region, Russia, 141120.

The correlation location principle was for the first time proposed in the recent work[1]. It was shown that by the means of the correlation location obtaining of the spatial distributions of the incoherent source's intensity of radiation is possible. It was illustrated by the experiment with an artificial source of radiation.

In the given work theoretical consideration for an estimation of the spatial resolution correlation location for signals with a limited spectrum of frequencies and limited duration will be stated.

As was shown[1], the distribution of intensity $I(r)$ of incoherent radiation may be obtained from the processing of signals received by the 4 receivers arranged at different locations. The main procedure is based on the 4-th order correlation analysis, according to the following expression:

$$\langle I^2(r) \rangle = k_0 k_1 k_2 k_3 |r - r_0||r - r_1||r - r_2||r - r_3| C_{1234}(\tau_{01}, \tau_{02}, \tau_{03}) \exp(\alpha(|r - r_0| + |r - r_1| + |r - r_2| + |r - r_3|))$$

Where k_0, k_1, k_2 and k_3 - sensitivity factors of receivers of radiation, r_0, r_1, r_2 and r_3 - coordinates of receivers of radiation, $C_{1234}(\tau_{01}, \tau_{02}, \tau_{03})$ - 4-th order correlation function, α-factor of attenuation of radiation in environment.

Point of the analysis is determined by the values of delays τ_{01}, τ_{02} and τ_{03} 1-th, 2-th and 3-rd channels with respect to 0 (or to any of 4 receivers). Obvious, that the assigning of the delays determines hyperbolic surfaces of equal delays from each of two points, in which receivers are located. The point of the analysis is determined by intersection of these surfaces.

It is true for δ-correlated radiation. In practice, when signal spectrum is limited, the peaks of correlation functions have widths Δ, values of which are inverse proportional to the width of a spectrum. In this case, the surfaces of equal delays will be transformed to layers, which consist of a pair of surfaces that are appropriate to delays $\tau_{01} - \Delta/2$ and $\tau_{01} + \Delta/2$.

Intersection of such layers will form a spatial resolution cell, shape of which depends on the position of receivers of radiation and the position of a point of sounding.

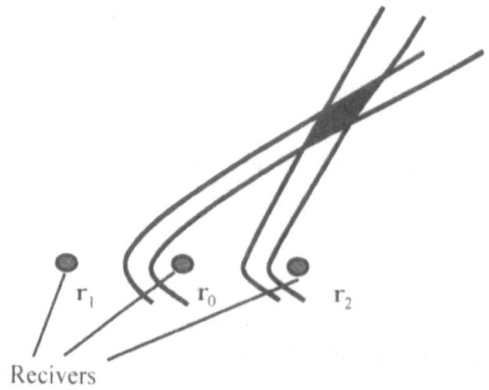

Figure 1. Forming of the resolution cell.

It is illustrated on Fig. 1, for two-dimensional case; r_0, r_1 and r_2 - points of an arrangement of receivers of radiation. It can be seen, that the cell of the spatial resolution has rhombus-like form. Its longitudinal and transverse sizes may be different. Such approach allows to explain, why in the first experiments[1], when the receivers of radiation were placed in one plane, the longitudinal spatial resolution was appreciably (in 5 times) worse than transverse. Estimation, made for this particular geometry results in the same ratio between the longitudinal and transversal spatial resolutions.

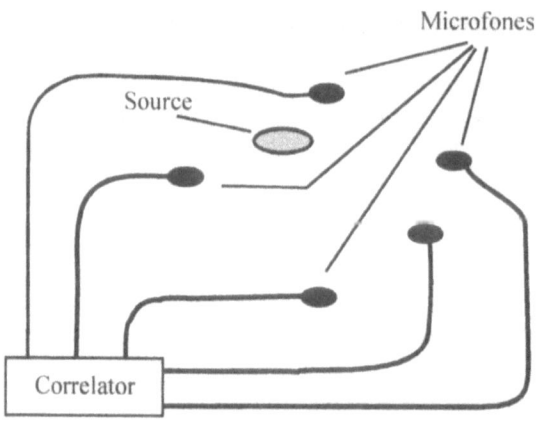

Figure 2. Experimental setup.

Similar consideration allows to propose the arrangement of receivers for getting an identical on different directions spatial resolution. The analysis shows that for this purpose the receivers must be arranged near the tetrahedron tops, and sources must be arranged near to the tetrahedron center. For check of correctness of the made assumption, the appropriate experiments were carried out.

The experimental set up is represented on Fig. 2. As the source of acoustic incoherent radiation was used the nozzle, from which compressed air flowed out. In the base configuration five microphones was used, forming two tetrahedrons: upper and lower. The size of the tetrahedrons' sides was equal about 6 m.

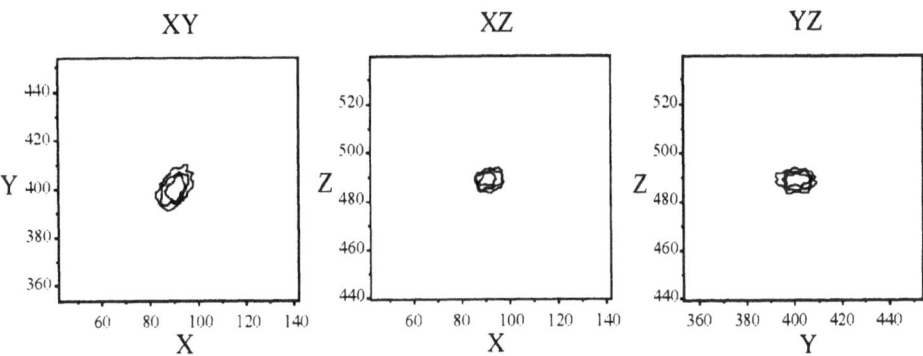

Figure 3. Sections of the recovered image by XY, XZ and YZ planes.

Thus we had an opportunity to recover few (5) images from the different points of view for improvement of the resulting final image. The experimental results obtained on this installation have confirmed the assumption of the practically equality in such configuration of the spatial resolution on different directions. It is illustrated on Fig. 3, where are represented three mutual perpendicular sections of the image of a noise source of a signal.

The outer contours at the Fig. 3 correspond to a level of a signal 0.5. Apparently reached spatial resolution is about 20 cm in all (XY, XZ and YZ) sections. It is close to the coherence length of received radiation with the bandwidth - 1.2 kHz.

Other circumstance, which is necessary for taking into account at extending of the theory[1] conclusions, developed for δ-correlated signals, on real situations is finiteness of measurement time. It is convenient to illustrate the basic laws determining opportunities of a correlation location method on the examples of few point sources. The analysis becomes especially evident with Gauss type signals. In this case, as it is known, one may calculate the 4-th order correlation function on the basis of the 2-th order correlation functions:

$$C_{0123}(\tau_{01}, \tau_{02}, \tau_{03}) = C_{01}(\tau_{01})*C_{23}(\tau_{03}-\tau_{02}) + C_{02}(\tau_{02})*C_{13}(\tau_{03}-\tau_{01}) + C_{03}(\tau_{03})*C_{12}(\tau_{02}-\tau_{01})$$

As the first step, we shall consider trivial case with one point-like source of radiation. Let we have registered a signal in N>>1 time points. 2-nd order correlation functions have one peak, which is distant from a beginning of coordinates at the interval, which is proportional to a difference of distances from a source up to the appropriate receivers, and noise-like part without correlation. The ratio of value of a root-mean-square deviation of correlation function where correlation is absent to correlation function maximum value will be equal to $1/\sqrt{N}$.

The picture as a whole will be legible if the correlation peaks will be appreciable on a background of noise without the correlation. It is obvious that it may be at N>10 at least. If the requirements of accuracy of determination of intensity of a source and (or) its coordinates are assigned, the number of points of registration may be accordingly calculated. For example, estimations show, that for achievement of relative accuracy of the intensity measurement 10^{-2} number of points N should be more than 10^4. The more number "M" of sources the more number of maximums at paired correlation functions.

In the simple case of equal source's intensity the maximum values of the correlation function, appropriate to each source will be equal. Relation of value of maximums to the root-mean-square deviation of correlation function in the part of absence of correlation, as it is not difficult to calculate, will be $\sqrt{N}$ /M.

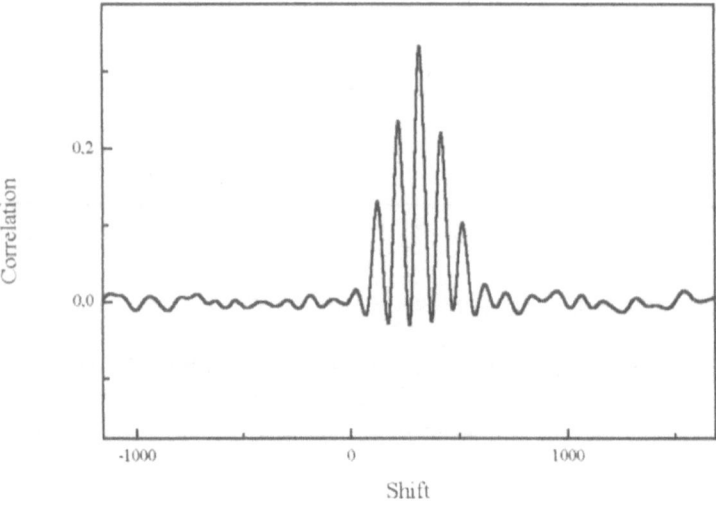

Figure 4. Example of the correlation function for the case of computer simulation radiation of the 27 point sources.

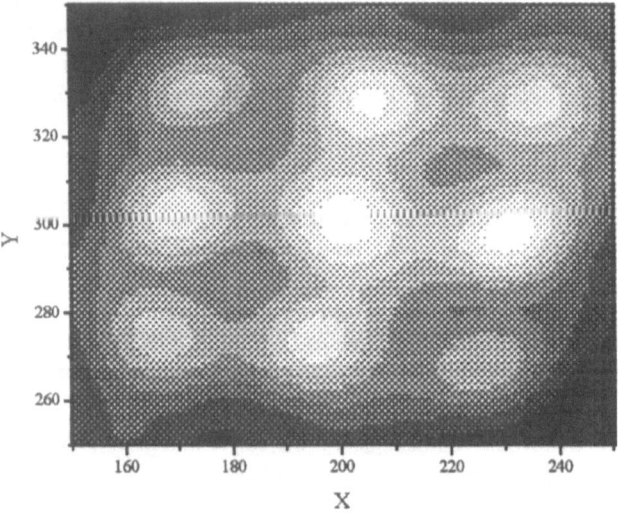

Figure 5. Section of the recovered image by XY plane for the case of computer simulation radiation of the 27 point sources.

The example of such function (one of 10), obtained at the case of 27 independent sources disposed inside of cube with the distance between them 30 cm, is represented on Fig. 4. Number of points in time series used at such modeling was equal to 100000. On Fig. 5

section by XY plane of the image of the 27 sources is submitted. On Fig. 5, one can see the image of the 9 sources, located at the one layer. Other 18 sources are located in other layers: one of them is on 30 cm higher, and other is on 30 cm lower than submitted one.

This work had financial support of RFFI, grant 96-02-17734a.

Conclusions

1. On the basis of geometrical considerations the basic relationship between the longitudinal and transversal resolutions was interpreted.
2. It is shown, that at an arrangement of receivers of radiation near to the tetrahedron tops, the spatial resolutions of sources arranged near to center of the tetrahedron are equal on all directions. The size of the spatial resolution is close to the coherence length of received radiation.
3. Basic relationship between signal - noise ratio, number of sources and number of the time points in measured signals is represented.

REFERENCES

1. V.I. Mirgorodsky, V.V. Gerasimov, S.V. Peshin, Three-dimensional ultrasonic imagine of temperature distributions, in: *Acoustical Imaging-22*, P. Tortoli and L. Masotti ed., Plenum Press, New York (1996).

MULTIPLE FREQUENCY DISTORTED BORN ITERATIVE METHOD FOR TOMOGRAPHIC IMAGING

Osama S. Haddadin and Emad S. Ebbini

Department of Electrical Engineering and Computer Science
University of Michigan
Ann Arbor, MI 48109

INTRODUCTION

The goal of ultrasound tomography is to produce a two-dimensional (2D) map of the acoustical properties of a region of interest (ROI) from measurements of the acoustic field taken outside the ROI. This map or image can prove very useful in medical diagnosis, non-destructive evaluation, and earth sciences [2, 5, 8, 12]. One solution of the tomographic imaging problem is by solving the inverse scattering problem [2, 4, 6, 10]. The inverse scattering problem is dependent on the total field inside the ROI which is unknown and is dependent on medium properties. This makes the inverse scattering problem nonlinear, and difficult to solve. A linearized, Born-approximate wave equation is usually inverted instead [4, 12].

A powerful procedure for solving the inverse scattering problem is the distorted-wave Born iterative (DBI) approach [1, 2, 4, 9, 10]. In this technique the inverse scattering problem is solved by incrementally correcting some initial estimate of the spatial structure of the ROI. This requires that we compute estimates of the total field and the inhomogeneous Green's function at every iteration. The procedure is repeated until convergence. The Born inverse solution is normally used as the initial solution [1, 3, 4, 9, 10]. This approach has been shown to significantly improve the reconstruction quality. One major limitation is that the algorithm diverges for strongly scattering media. This is observed to be due to an erroneous Born-based initial estimate of the spatial structure of the region. The validity of the Born assumption can be improved by increasing the wavelength (decreasing the operating frequency). This, in effect, makes the inhomogeneous medium small relative to the wavelength, thus, reducing scattering. Numerically, we show that the DBI algorithm converges by simply reducing the operating frequency. However, a solution at very low frequency suffers from poor spatial resolution.

We present a multiple frequency approach in which we solve for the strongly scattering medium at sufficiently low frequency to achieve convergence. Then, we use this low frequency solution as the initial estimate for the DBI method at a higher frequency. Convergence is achieved at the higher frequency since the initial solution is not erroneous and has a contrast that is relatively close to the true contrast. This process can be repeated for multiple frequencies until the required spatial resolution is attained. This algorithm, which is unlike other multiple frequency approaches, extends the applicability of the DBI method to the levels of dealing with biological tissue.

THE SCATTERING PROBLEM

Consider an infinite space containing a homogeneous loss-free acoustic medium with a known constant background wavenumber, k_0. Embedded in this space is a constant density scatterer with space-varying wavenumber, $k(r)$. An integral wave equation describing this system is the Lipmann-Schwinger equation given as [4, 6, 10]

$$\Phi(r) = \Phi_i(r) + \int G_0(r, r')f(r')\Phi(r')\, dr', \tag{1}$$

where Φ is the total field, Φ_i is the incident field, $f(r) = k(r)^2 - k_0^2$ is the scattering function, and G_0 is the 2D scalar homogeneous Green's function. The second term on the right hand side of Equation 1 is the scattered field which is usually measured at some observation points outside the ROI. The 2D homogeneous background Green's function in cylindrical coordinates is known and can be defined in terms of the zero-th order cylindrical Hankel function of the first kind. In general the inhomogeneous Green's function in unknown, however, it considered as the impulse response of a line source in the presence of medium with space varying wavenumber, $k(r)$.

We define the inverse scattering problem as that of solving Equation 1 for the scattering function f. This requires knowing the total field inside the ROI which is impossible. In addition, the total field itself is dependent on the medium properties, which makes the inverse scattering problem nonlinear in f. We resort to using the Born approximation, which assumes that the inhomogeneity is simply a weak perturbation and that the scattered field it produces is weak compared to the incident field. Therefore, the total field in the ROI, Φ under the integral in Equation 1, can be replaced by the incident field, Φ_i, [7, 4, 10].

Upon discretizing the resulting integral equation, we obtain the linearized matrix representation of the inverse scattering problem,

$$Qf = \phi_s, \tag{2}$$

where $Q_{m,n} = \int G_0(r_m, r')g_n(r')\Phi_i(r')\, dr'$ for $m = 1, \cdots, M$ and $n = 1, \cdots, N$ are the kernel matrix elements, f is a $N \times 1$ vector containing the unknown scattering function coefficients of the expansion $f(r) = \sum_{n=0}^{N} f_n g_n(r)$, and ϕ_s is a $M \times 1$ vector containing measurements of the scattered field at M observation points. The function g_n is a pulse basis function used in the expansion of the unknown scattering function.

To produce a reasonable image of the ROI, the inverse scattering problem is solved for multiple transmit positions. A matrix system of the form shown in Equation 2 is constructed and solved for each source position. The final representation of the scattering function is computed as the vector average of the individual solutions. The inverse scattering problem produces an ill-conditioned, underdetermined matrix system, which we solve by computing a regularized pseudoinverse operator [11]. Unfortunately, an image obtained from a Born approximated wave equation exhibits two types of errors. First, for moderately scattering media, the solution is biased due to the assumption that the medium is weakly scattering. Second, for strongly scattering media, the algorithm produces an incorrect solution. The second drawback is more fundamental and it occurs when the Born assumption is strongly violated. Numerical simulations indicate that the Born inverse solution can be reasonable if the phase shift introduced by the inhomogeneous media is smaller than π. Therefore, the validity of the assumption depends on the size (in wavelengths) of the inhomogeneity and its contrast [12].

The DBI method, which improves the Born solution dramatically, requires solving the inverse and forward scattering problems iteratively. We define the forward scattering problem as that of solving Equation 1 for the total field. In matrix form,

$$P\phi = \phi_i, \tag{3}$$

where $P_{m,n} = \{\delta_{mn} - \int G_0(\boldsymbol{r}_m, \boldsymbol{r}')f(\boldsymbol{r}')g_n(\boldsymbol{r}')\,d\boldsymbol{r}'$ for $n, m = 1, \cdots, N$ are the forward problem kernel elements, f is the true (or estimated) scattering function, ϕ is a $N \times 1$ vector containing the unknown total field coefficients of the expansion $\Phi(\boldsymbol{r}) = \sum_{n=0}^{N} \phi_n g_n(\boldsymbol{r})$, and ϕ_i is an $N \times 1$ vector of elements from the incident field evaluated at $\{\boldsymbol{r}_m\}_{m=1}^{N}$. Once the matrix P is computed, it may be inverted to solve for the coefficients of the total field. Note that we have used the same pulse function in the expansion of f and Φ.

The transmit/receive array configuration and ROI are shown in Figure 1. A low frequency ultrasound wave is transmitted from one source at a time. This wave interacts with the inhomogeneities in the medium to produce a scattered wave. Both the incident and scattered waves are measured by the M receive elements. The measurements and system model make up the matrix system of Equation 2. This process is repeated for multiple transmit position. For simplicity, the final image is produced as the vector average of the individual scattering function solutions.

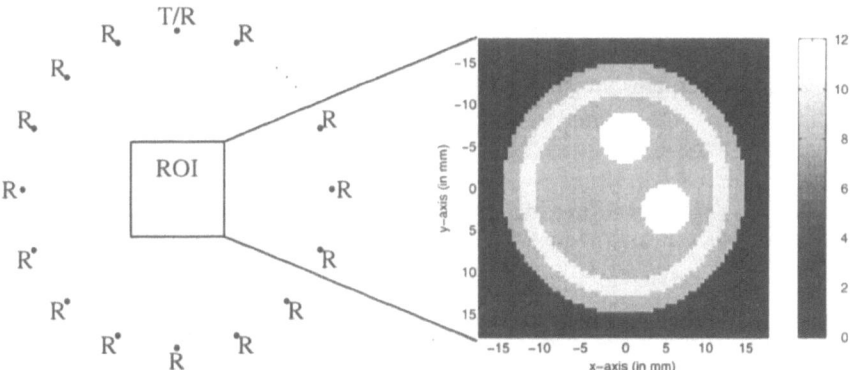

Figure 1: The circular transmit/receive array and region of interest shown as a percentage increase in speed of sound over the background speed of sound.

THE DBI METHOD

A solution obtained by simply inverting the Born approximated wave equation is not a reasonable method for imaging biological tissue. However, it provides an initial solution for Born-based iterative techniques, namely, the Born iterative and the DBI methods [3, 4, 10].

In the DBI method, we solve

$$\Phi_s(\boldsymbol{r}) - \hat{\Phi}_s^b(\boldsymbol{r}) = \int G^b(\boldsymbol{r}, \boldsymbol{r}')\Delta f(\boldsymbol{r}')\Phi^b(\boldsymbol{r}')\,d\boldsymbol{r}' \tag{4}$$

for the incremental correction $\Delta f(\boldsymbol{r}) = f(\boldsymbol{r}) - f_b(\boldsymbol{r})$, where $f_b(\boldsymbol{r})$ is the initial scattering function estimate. This algorithm requires that we compute the inhomogeneous Green's function and the total and scattered fields associated with f_b. In effect, we solve an inverse scattering function in which some initial solution is used as the background. This process is repeated until convergence is achieved. The Green's function can be computed using the forward scattering problem, since it satisfies the Helmholtz equation. In matrix form, we solve $PG = G_0$, where G and G_0 are $N \times M$ matrices of the sampled inhomogeneous and homogeneous Green's function, respectively.

In brief, the DBI algorithm alternates between (1) incrementally updating an estimate of the spatial structure of the inhomogeneity (the scattering function) given a current best estimate of the inhomogeneous Green's function and the total acoustic field in the ROI (an inverse scattering problem) and (2) updating the total field and Green's function given the previously generated estimate of the inhomogeneity (a forward scattering problem).

MULTIPLE FREQUENCY DBI METHOD

Consider the ROI shown in Figure 1. The image displays a gray scale representation of the percentage increase in the speed of sound over the background speed of sound. The ROI is of size $12\lambda_0 \times 12\lambda_0$, which is discretized using a pixel dimension of $\frac{\lambda_0}{6} \times \frac{\lambda_0}{6}$ (producing N = 5184 pixels). The contrast and size of the inhomogeneities in this region are chosen to emphasize the limitations of the DBI method in the presence of highly scattering media. The transmit/receive array consists of 128 elements that are equally distributed along a circle of diameter $20\lambda_0$. All elements are used for reception (M = 128). Only 32 equally spaced elements are used individually for the sequential transmission. The size of this region is slightly smaller than realistic regions, however the contrast is increased to emulate scattering that is typical to breast tissue. A DBI algorithm operating at 300 kHz is used to reconstruct this region. The initial estimate and the 4-th iteration images are shown in Figure 2(a) and 2(b). Obviously, the algorithm diverges quickly, which is due to an erroneous initial solution.

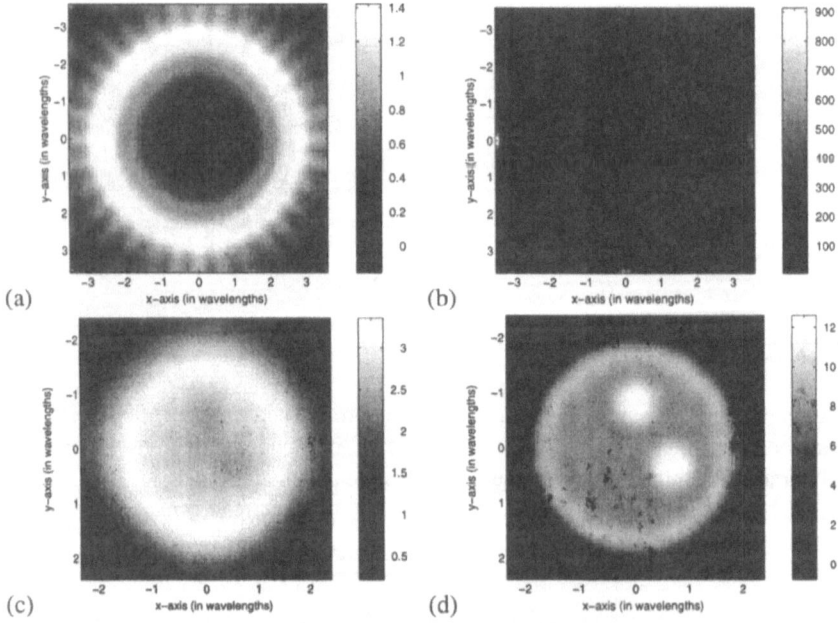

Figure 2: (a) Born inverse solution and (b) 4-th iteration of the DBI method operating at 300 kHz. (c) Born inverse solution and (d) 20-th iteration of the DBI method operating at 200 kHz.

616

Since the DBI method is initialized by the Born solution, it suffers due to the second drawback of the Born assumption. The validity of the Born assumption is dependent on the contrast level and size (in wavelengths) of inhomogeneities in the ROI. That is, for a constant wavelength, the assumption is improved by decreasing the contrast or decreasing the size of the inhomogeneous regions. These are parameters the are beyond our control. However, we may reduce the operating frequency, thus increasing the wavelength which, effectively, reduces the size of the inhomogeneities. This results in a convergent solution. The initial estimate and the 8-th iteration DBI estimate at 200 kHz are shown in Figure 2(c) and 2(d). This is a convergent solution, however, the spatial resolution is very poor.

The validity of the Born approximation and the Born inverse solution and their dependence on the size and contrast of the inhomogeneous media has been evaluated for simple structures [4, 12, 13]. Consider a region consisting of a single cylindrical scatterer of constant contrast and of diameter, d. Then, the phase change across this object is proportional to $\omega \left(1/c - 1/c_0 \right) d$, where ω is the operating frequency, c is the speed of sound in the cylinder and c_0 is the speed of sound in the background medium. The results on the Born-based inversion indicate that, in general, the reconstruction of cylinders that introduce a phase change greater than π show severe artifacts near the center of the cylinder [4, 12, 13]. That is, the diameter and contrast must satisfy $\omega \left| \frac{1}{c} - \frac{1}{c_0} \right| d < \pi$, which can be written as

$$\left| \frac{c_0 - c}{c} \right| d < \frac{\lambda_0}{2} \tag{5}$$

where $\lambda_0 = \frac{2\pi c_0}{\omega}$ is the wavelength in the background medium. It should be obvious from Equation 5 that increasing λ_0 extends the applicability of the method. This is however, at the expense of a loss in spatial resolution which is proportional to the wavelength of the operating frequency of the system.

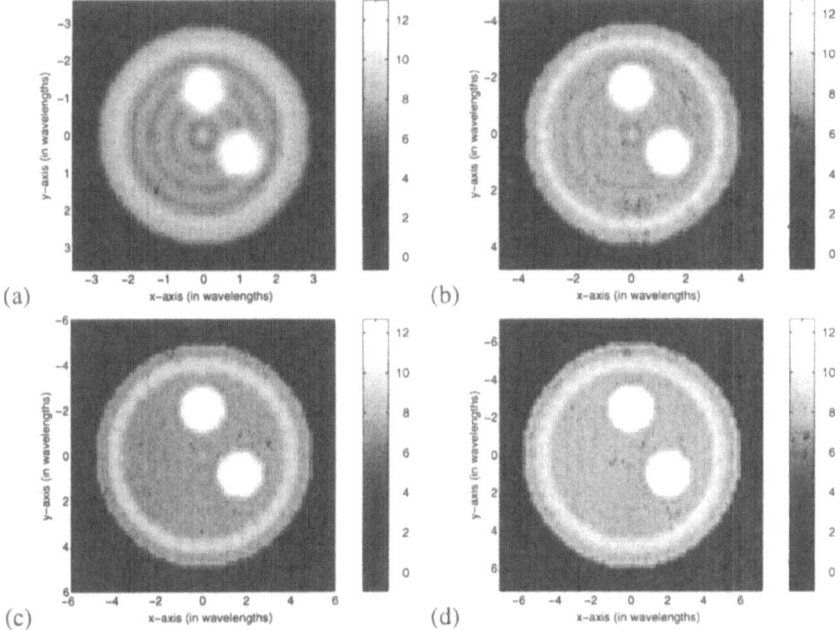

Figure 3: Reconstructions using the multiple frequency DBI method: (a) 8-th iteration at 300 kHz, (b) 8-th iteration at 400 kHz, (c) 8-th iteration at 500 kHz, and (d) 8-th iteration for the final 600 kHz stage.

In the multiple frequency approach, we iteratively solve for the ROI at sufficiently low frequency. This process does not need to provide an accurate representation of the region. Then, this solution is used as the initial estimate of the region in a higher frequency DBI method. Since we initialize the high frequency DBI method with a solution that has contrast close to the true contrast, and the new operating frequency is chosen properly, the algorithm will continue to converge. This process can be repeated by incrementally increasing the frequency until a required spatial resolution is achieved. The algorithm does not necessarily have higher computational complexity. If the pixel dimension (in wavelengths) is chosen constant, the number of pixels required for the ROI in the early stages of the algorithm is small. Therefore, the added computational complexity is low. At latter stages, the DBI method does not require many iterations to converge since it starts with a contrast level close to the true contrast.

A safe strategy to choose the starting frequency is to approximately satisfy the criterion of Equation 5 for an inhomogeneity that is as large as the ROI and contrast as the maximum expected contrast level of the tissue under consideration. For example, in the above example, the largest possible inhomogeneity is 36 mm in diameter, and a typical contrast is 20%. Thus, Equation 5 can be satisfied by choosing a frequency of 200 kHz ($\lambda_0 = 7.42$ mm). The maximum frequency is determined according to two factors. First, knowledge of the underlying tissue structure can indicate a reasonable spatial resolution. Second, the size of the problem, since by keeping the pixel dimension (in wavelengths) reasonably small, the number of pixels in the ROI must be increased.

We apply our approach to the problem by using the solution shown in Figure 2(d) as the initial scattering function for the 300 kHz reconstruction. This is repeated up to a final frequency of 600 kHz, and a frequency step of 100 kHz. Figure 3 shows the final solution for each of the frequencies, 300 kHz, 400 kHz, 500 kHz and 600 kHz. The final result, presented in Figure 3(d), shows the advantage of using this multiple frequency DBI method.

CONCLUSION

Although, the distorted Born iterative method is very powerful, it diverges for strongly scattering inhomogeneous media. It is apparent that the cause of divergence is the initial Born inverse solution, which is erroneous for strongly scattering media. In our multiple frequency approach, we solve for the region at sufficiently low frequency. This solution becomes the initial step for a higher frequency DBI method. The DBI method operating at a low frequency can converge since the wavelength in large. The converges is conserved at high frequencies, since the initial step is closer in contrast to the true ROI. Therefore, this approach does not diverge and it achieves the appropriate spatial resolution. Simulation results show the advantages of this technique.

References

[1] G. Beylkin and M. Oristaglio, "Distorted-wave Born distorted-wave Rytov approximations," *Optics Comm.*, vol. 53, pp. 213–215, 1985.
[2] H. Blok and M. Oristaglio, *Wavefield Imaging and Inversion in Electromagnetics and Acoustics*, Delft University of Technology, report number Et/EM 1995-21, June 1995.
[3] W. C. Chew and Y. M. Wang, "Reconstruction of two-dimensional permittivity distribution the using the distorted Born iterative method," *IEEE Trans. Medical Imaging*, vol. 9, no. 2, pp. 218–225, June 1990.
[4] W. C. Chew, *Waves and Fields in Inhomogeneous Media*, Van Nostrand Reinhold, 1992.
[5] Z. H. Cho, J. P. Jones, and M. Singh, *Foundations of Medical Imaging*, John Wiley and Sons, 1993.
[6] D. Colton and R. Kress, *Inverse Acoustic and Electromagnetic Scattering Theory*, Springer-Verlag, 1992.
[7] A. J. Devaney, "Inversion formula for inverse scattering within the Born approximation," *Optical Lett.*, vol. 7, pp. 111–112, 1982.

[8] A. J. Devaney, "Geophysical diffraction tomography," *IEEE Trans. Geological science, Special Issue on Remote Sensing*, vol. 22, pp. 3–13, January 1984.

[9] A. J. Devaney and M. Oristaglio, "Inversion procedure for inverse scattering within the distorted-wave Born approximation," *Phys. Rev. Lett.*, vol. 51, pp. 237–240, 1984.

[10] T. M. Habashy and R. Mittra, "On some inverse methods in electromagnetics," *J. Electromagnetic Waves and Appl.*, vol. 1, no. 1, pp. 25–58, 1987.

[11] O. S. Haddadin and E. S. Ebbini, "Adaptive Regularization of a Distorted Born Iterative Algorithm for Diffraction Tomography," *Int. Conf. Image Processing*, pp. 725–728, September 1996.

[12] A. C. Kak and M. Slaney, *Principles of Computed Tomographic Imaging*, IEEE Press, 1988.

[13] M. Slaney, A. C. Kak, and L. E. Larson, "Limitations of imaging with first order diffraction tomography," *IEEE Trans. Microwave Theory and Tech.*, vol. 32, no. 8, pp. 860–873, 1984.

METHOD OF FORMAL PARAMETER EXPANSION FOR ACOUSTICAL INVERSE SCATTERING PROBLEMS

Zhen-Qiu Lu

Department of Physics, Nankai University
Tianjin, 300071, People's Republic of China

INTRODUCTION

In acoustical diffraction tomography, the inverse scattering perturbation theory, especially the first-order Born perturbation approximation has its advantages: comparatively simple calculations. That is why it has been used in diffraction tomography in many fields, such as in medical and seismic imaging [1-2]. But this method has its disadvantages : severe limitations on scatterers, i.e., objects to be imaged[3-5]. These limitations are impracticable in the most cases, such as in medical imaging and petroleum exploration. The use of high-order, for example, second-order Born perturbation algorithms can reduce these limitations to a certain extent. But they also failed to reconstruct the object with good accuracy in many cases [5]. In such cases, the third- or even higher-order Born approximation must be taken into account. This will result in more and more tedious calculations. Can and how do we find a method which needs comparatively simple calculations and has not severe limitations on objects ?

A FUNDAMENTAL IDEA

Why does the inverse scattering perturbation theory have such severe limitations on objects? This is because the inverse scattering perturbation expansion has infinitive terms. If a parameter ϵ is introduced into the generalized scattered wave or its angular spectrum, set an expansion of the object or its Fourier spectrum in series of ϵ and only finite terms are left with the other terms being zero, a non-perturbation analytical solution is obtained. In such a case, the parameter is no perturbation one, and I call it a formal parameter. Therefore this method can be anticipated to be applicable to many practicable cases, including intermediate and strong scattering.

The key problem is how to choose an appropriate field transform that can lead to a formal expansion. In the case of the Helmholtz equation which is commonly used in the acoustical medical imaging and petroleum exploration, Rytov transform is the only one required.

THEORETICAL RECONSTRUCTION ALGORITHMS USING THE METHOD OF FORMAL PARAMETER EXPANSION

Acoustical diffraction tomography has been formulated as an approximate inversion of the 2-dimensional Helmholtz equation

$$\nabla^2 \psi(\mathbf{x}) + k_0^2 \psi(\mathbf{x}) = -k_0^2 f(\mathbf{x}) \psi(\mathbf{x}) \tag{1}$$

where

$$f(\mathbf{x}) = (\frac{k(\mathbf{x})}{k_0})^2 - 1, \quad k(\mathbf{x}) = \begin{cases} k_0 & \mathbf{x} \in D^* \\ k_i(\mathbf{x}) & \mathbf{x} \in D \end{cases} \tag{2}$$

in which $k(\mathbf{x})$ is the circular wave number, D and D^* are domains occupied by an object to be imaged and its surrounding medium, respectively.

After the Rytov transform

$$\chi = \psi_0 ln(\frac{\psi}{\psi_0}) \tag{3}$$

Eq.(1) becomes[6-7]

$$\nabla^2 \chi + k_0^2 \chi = -k_0^2 f \psi_0 + \psi_0 (\nabla \frac{\chi}{\psi_0})^2 \tag{4}$$

where $\psi_0 = A e^{i k_0 \cdot \mathbf{x}}$ (with the time-dependent factor $e^{-i\omega t}$) is a single-frequency incident wave and $\chi(\mathbf{x})$ is called a generalized scattered wave.

Introducing a rotation parameter θ of the object, Eq.(4) with the object oriented at the angle θ becomes[7]

$$\nabla^2 \chi(\theta; \mathbf{x}) + k_0^2 \chi(\theta; \mathbf{x}) = -k_0^2 \psi_0(\mathbf{x}) f(\Theta \mathbf{x}) + \psi_0(\mathbf{x}) [\nabla \frac{\chi(\theta, \mathbf{x})}{\psi_0(\mathbf{x})}]^2 \tag{5}$$

where

$$\Theta(\theta) = \begin{pmatrix} \cos\theta, & \sin\theta & 0 \\ -\sin\theta, & \cos\theta & 0 \\ 0 & 0 & 1 \end{pmatrix}$$

Taking the Sommerfeld radiation conditions into account, Eq.(5) becomes the integral equation[7]

$$
\begin{aligned}
\chi(\theta, \mathbf{x}) &= \int G(|\mathbf{x} - \mathbf{x}'|) \{ -k_0^2 \psi_0(\mathbf{x}') f(\Theta \mathbf{x}') + \psi_0(\mathbf{x}') [\nabla \frac{\chi(\theta, \mathbf{x}')}{\psi_0(\mathbf{x}')}]^2 \} dx' dy' \\
&= G(|\mathbf{x}|) \otimes_{\mathbf{x}} \{ -k_0^2 \psi_0(\mathbf{x}) f(\Theta \mathbf{x}) + \psi_0(\mathbf{x}') [\nabla \frac{\chi(\theta, \mathbf{x})}{\psi_0(\mathbf{x})}]^2 \}
\end{aligned} \tag{6}
$$

where $\mathbf{x} = (x, y)^T$, $\mathbf{x}' = (x', y')^T$, T denotes transpose, $\otimes_{\mathbf{x}}$ convolution in $\mathbf{x}$ and $G(|\mathbf{x}|) = H_0^{(1)}(k_0|\mathbf{x}|)/4i$. After Fourier transforming of Eq.(6) with respect to y, i.e., after performing its plane wave angular spectrum, we obtain[7]

$$\chi_a(\theta, x, \nu_y) = A \int dx' G_a e^{ik_0 x'} \{ -k_0^2 f_a(\Theta, x, \nu_y) - [(\nabla \frac{\chi}{\psi_0})^2]_a \} \tag{7}$$

or formally

$$\chi_a = \Gamma F \Psi_{0a} + Z \chi_a \tag{8}$$

where the subscript a denotes angular spectrum,

$$\chi_a(\theta, x, \nu_y) = \int \chi(\theta, x, y) e^{-i2\pi \nu_y y} dy \tag{9}$$

$$\Psi_{0a} = \int \psi_0(x,y)e^{-i2\pi\nu_y y}dy = Ae^{ik_0 z}\delta(\nu_y) \tag{10}$$

$$f_a(\theta, x, \nu_y) = \int f(\Theta x)e^{i2\pi\nu_y y}dy \tag{11}$$

and

$$G_a(x,\nu_y) = \int G(|\mathbf{x}|)e^{-i2\pi\nu_y y}dy = \frac{e^{i2\pi\xi(\nu_y)|z|}}{4\pi i\xi(\nu_y)} \tag{12}$$

is the plane wave angular spectrum of the 2-D free-space Green function, in which

$$\xi(\nu_y) = \sqrt{\overline{k}_0^2 - \nu_y^2 + i0}$$

$$\frac{1}{\sqrt{\overline{k}_0^2 - \nu_z^2 - \nu_y^2 + io}} = \lim_{q \to 0} \frac{1}{\sqrt{\overline{k}_0^2 - \nu_z^2 - \nu_y^2 + iq}}$$

is a singular generalized function, q is a positive definite quadratic form and $\overline{k}_0 = k_0/2\pi = 1/\lambda_0$, F is the Fourier transform of f, $L = \Gamma F$ is a linear integral operator

$$\Gamma F = -k_0^2 \int\int\int dx'd\nu_z'd\nu_y' G_a(x - x', \nu_y)e^{i2\pi\nu_z' x'}F(\Theta(\nu_z', \nu_y - \nu_y')^T) \tag{13}$$

and Z is a nonlinear integral operator

$$(Zp)(x,\nu_y) = \int dx'G_a(x - x', \nu_y)e^{-ik_0 x'}\{[ik_0 p(x', \nu_y) - \frac{\partial}{\partial x'}p(x', \nu_y)]$$

$$\otimes_{\nu_y}[ik_0 p(x', \nu_y) - \frac{\partial}{\partial x}p(x', \nu_y)] + 4\pi^2[\nu_y p(x', \nu_y)] \otimes_{\nu_y} [\nu_y p(x', \nu_y)]\} \tag{14}$$

Now introducing a parameter ϵ into the angular spectrum of the generalized scattered wave in Eq.(8) and setting an expansion of the Fourier spectrum of the object in series of ϵ

$$F = \sum_{n=1}^{\infty} F_n \epsilon^n \tag{15}$$

we have

$$\chi_a = (\Gamma F_1)\Psi_{0a} \tag{16}$$

$$(\Gamma F_2)\Psi_{0a} + Z\chi_a = 0 \tag{17}$$

$$(\Gamma F_n)\Psi_{0a} = 0 \quad n \geq 3 \tag{18}$$

i.e., we obtain the formal parameter expansion

$$F = F_1 + F_2 \tag{19}$$

and then

$$f = f_1 + f_2 \tag{20}$$

According to Eq. (16), we have the reconstruction formula of the first term in the formal parameter expansion[7]

$$\chi_a(\theta, x, \nu_y) = (\Gamma F_1)\Psi_{0a} = -Ak_0^2 G_a(x, \nu_y)F_1(\Theta\nu_t) \tag{21}$$

where

$$\nu_t = (\xi(\nu_y) \quad \overline{k}_0, \quad \nu_y)^T$$

From Eq. (21) we are able to obtain F_1 and then f_1 [7] Canceling χ_a in Eq. (17) using Eq. (16) leads to

$$[(\Gamma F_2) + Z(\Gamma F_1)]\Psi_{0a} = 0 \tag{22}$$

Using Eq. (21) we have

$$(\Gamma F_n)\Psi_{0a} = -Ak_0^2 G_a(x, \nu_y) F_n(\Theta\nu_t) \qquad n = 1, 2 \tag{23}$$

Substituting them into (22) results in

$$-Ak_0^2 G_a(x, \nu_y) F_2(\Theta\nu_t) + Z(-Ak_0^2 G_a(x, \nu_y) F_1(\Theta\nu_t)) = 0 \tag{24}$$

or according to Eq.(14),

$$G_a(x, \nu_y) F_2(\Theta\nu_t) = Ak_0^2 Z(G_a(x, \nu_y) F_1(\Theta\nu_t)) \tag{25}$$

Equation (25) gives F_2 and then f_2. Thus we obtain the reconstruction formula of the second term in the formal parameter expansion, which is very closely analogous to the reconstruction formula of the first term. The only calculation subject to change is that the angular spectrum χ_a in the reconstruction formula of the first term should be replaced by $-Z\chi_a$ or by $-A^2 k_0^4 Z(G_a F_1)$.

DETAILED RECONSTRUCTION ALGORITHMS OF THE SECOND TERM IN THE FORMAL PARAMETER EXPANSION

Now we calculate $AZ(G_a(x, \nu_y) F_1(\Theta\nu_t)$ and then F_2, the second term in the formal parameter expansion of the object's Fourier spectrum. According to Eq. (13), we have

$$A(2\pi)^2[Z(G_a F_1)](x, \nu_y) = \int d\nu_y' F_1(\Theta\nu_t(\nu_y - \nu_y')) F_1(\Theta\nu_t(\nu_y'))[G_1(x, \nu_y, \nu_y') + G_2(x, \nu_y, \nu_y')] \tag{26}$$

where

$$G_1(x, \nu_y, \nu_y') = (2\pi)^2 \int dx' G_a(x - x', \nu_y) e^{-ik_0 x'} \cdot$$

$$[ik_0 G_a(x', \nu_y - \nu_y') - \frac{\partial}{\partial x'} G_a(x', \nu_y - \nu_y')] \cdot$$

$$[ik_0 G_a(x', \nu_y') - \frac{\partial}{\partial x'} G_a(x', \nu_y')] \tag{27}$$

and

$$G_2(x, \nu_y, \nu_y') = (2\pi)^4 (\nu_y - \nu_y')\nu_y' \int dx' G_a(x - x', \nu_y) G_a(x', \nu_y - \nu_y') G_a(x', \nu_y') e^{-ik_0 x'} \tag{28}$$

According to the theory of generalized functions[8]

$$\frac{\partial}{\partial x} G_a(x, \nu_y) = \frac{1}{2} e^{i2\pi\xi(\nu_y)|x|} \tag{29}$$

and then

$$G_1(x, \nu_y, \nu_y') = \frac{1}{8\xi(\nu_y)}\left(\frac{k_0}{\xi(\nu_y - \nu_y')} - 1\right)\left(\frac{k_0}{\xi(\nu_y')} - 1\right) G_3(x, \nu_y, \nu_y') \tag{30}$$

$$G_2(x, \nu_y, \nu_y') = -\frac{\pi(\nu_y - \nu_y')\nu_y'}{4\xi(\nu_y)\xi(\nu_y - \nu_y')\xi(\nu_y')} G_3(x, \nu_y, \nu_y') \tag{31}$$

624

where

$$G_3(x, \nu_y, \nu_y') = -2\pi i \int_{-\infty}^{\infty} dx' e^{i2\pi\phi} \tag{32}$$

in which

$$\phi(x, x', \nu_y, \nu_y') = [\xi(\nu_y - \nu_y') + \xi(\nu_y')]|x'| + \xi(\nu_y)|x - x'| - \overline{k}_0 x' \tag{33}$$

In the following we assume that $x > 0$. The x is the position where a receiver array is located. Let G_3 be decomposed into three parts.

$$G_{31} = -2\pi i \int_{-\infty}^{0} e^{i2\pi\phi} dx' = -2\pi i e^{i2\pi\xi(\nu_y)x} \int_{-\infty}^{0} e^{i2\pi\phi_1 x'} dx' \tag{34}$$

where

$$\phi_1(x', \nu_y, \nu_y') = \xi(\nu_y - \nu_y') + \xi(\nu_y') + \xi(\nu_y) + \overline{k}_0 \tag{35}$$

According to[8]

$$\int_0^{\infty} e^{-i2\pi\nu x} dx = \frac{1}{i2\pi\nu} + \frac{1}{2}\delta(\nu) \tag{36}$$

$$G_{31} = i2\pi e^{i2\pi\xi(\nu_y)x} \left[\frac{1}{-i2\pi\phi_1} + \frac{1}{2}\delta(\phi_1)\right] = -\frac{1}{\phi_1} e^{i2\pi\xi(\nu_y)x} \tag{37}$$

This is because $\phi_1 > 0$, and hence $\delta(\phi_1) = 0$

$$G_{32} = -2\pi i \int_0^x e^{i2\pi\phi} dx' = \frac{1}{\phi_2} e^{i2\pi\xi(\nu_y)x} (1 - e^{i2\pi\phi_2}) \tag{38}$$

where

$$\phi_2(x', \nu_y, \nu_y') = \xi(\nu_y - \nu_y') + \xi(\nu_y') - \xi(\nu_y) - \overline{k}_0 \tag{39}$$

and

$$
\begin{aligned}
G_{33} &= -2\pi i \int_x^{\infty} e^{i2\pi\phi} dx' = -2\pi i e^{i2\pi(\phi_3 - \xi(\nu_y))x} \int_0^{\infty} e^{i2\pi\phi_3 x'} dx' \\
&= -2\pi i e^{i2\pi(\phi_3 - \xi(\nu_y))x} \left[\frac{1}{-i2\pi\phi_3} + \frac{1}{2}\delta(\phi_3)\right] = -\frac{1}{\phi_3} e^{i2\pi(\phi_3 - \xi(\nu_y))x} \tag{40}
\end{aligned}
$$

where

$$\phi_3(x', \nu_y, \nu_y') = \xi(\nu_y - \nu_y') + \xi(\nu_y') + \xi(\nu_y) - \overline{k}_0 \tag{41}$$

This is because equation $\phi_3 = 0$ has not any real root, and hence $\delta(\phi_3) = 0$. Thus we have

$$G_3 = G_{31} + G_{32} + G_{33} = \left[\frac{1}{\phi_1} + \frac{1}{\phi_2}(1 - e^{i2\pi\phi_2}) - \frac{1}{\phi_3} e^{i\phi_3}\right] e^{i2\pi\xi(\nu_y)x} \tag{42}$$

Finally we obtain

$$F_2(\Theta\nu_t) = \frac{\overline{k}_0^2}{G_a(x, \nu_y)} \int d\nu_y' F_1(\Theta\nu_t(\nu_y - \nu_y')) F_1(\Theta\nu_t(\nu_y'))[G_1(x, \nu_y, \nu_y') + G_2(x, \nu_y, \nu_y')] \tag{43}$$

Analogous to the reconstruction formulas of the first term in the formal parameter expansion, we are able to obtain f_2 from F_2.

DISCUSSION AND CONCLUSION

We have presented the reconstruction algorithms using the method of formal parameter expansion which is a non-perturbation analytical method. These algorithms need comparatively simple calculations and have not severe limitations on scatterers. Therefore they can be anticipated to be applicable to many practical cases, including intermediate and strong scattering.

Why can Eq.(15) be reduced to two terms ? This is the result produced by Rytov transform which separates the scatterer from the scattered field in acoustical imaging.

The f_1 is the solution of an inverse source, the scattered field to which is the same as the scattered field to the inverse scattering problem concerned. The f_1 is called an equivalent source and f_2 the modified term.

The work to give some numerical examples is currently under way.

ACKNOWLEDGMENT

This work was supported in part by the National Natural Science foundation of China.

References

[1] M. Kaveh, R.K. Mueller and R.D. Inversion, Ultrasonic tomography based on perturbation solution of the wave equation, Computer graphics and image processing, 9:105(1979).

[2] N. Bleistein, J.K. Cohen, and F.G. Hagin, "Two and one-half dimensional Born inversion with an arbitrary reference," Geophysics, 52(1):26(1987).

[3] M. Kaveh, M. Soumekh, and R.K. Mueller, A comparison of Born and Rytov approximations in acoustic tomography, in:" Acoustical Imaging, vol.11," J. Powers, ed. Plenum, New York (1981).

[4] M. Slaney, A.C. Kak, and L.E. Larsen, Limitations of imaging with first-order approximations in acoustic tomography, IEEE Trans. Microwave Theory Tech., 32(8):860(1984).

[5] Zhen-Qiu Lu and Yan-Yun Zhang, Acoustical tomograpy based on the second-order Dorn transform perturbation approximation, IEEE Trans. Ultrason. Ferroelectrics, Freq. Contr., 43(2):296(1996).

[6] M. Kaveh, M. Soumekh, Zhen-Qiu Lu, R.K. Mueller, and J.F. Greenleaf, Further results on diffraction tomography using Rytov approximation, in: "Acoustical Imaging, vol.12," E.A. Ash and K. Hill, eds. Plenum, New York (1982).

[7] Zhen-Qiu Lu, JKM perturbation theory, relaxation perturbation theory and their applications to inverse scattering: theory and reconstruction algorithms, IEEE Trans. Ultrason. Ferroelectrics, Freq. Contr., 33(6):722(1986).

[8] I.M. Gel'fand and G.E. Shilov, "Generalized Functions (Vol.1)," Academic, New York (1964).

MODEL-BASED OBJECT RECOGNITION
BY 3-DIMENSIONAL SPARSE ULTRASONIC ARRAYS

Bernhard Menz

University of Karlsruhe, Institut für Meß- und Regelungstechnik
P.O. Box 6980, 76128 Karlsruhe, Germany

ABSTRACT

A new model based imaging technique is presented producing accurate images with small expense of hardware and time. High efficiency of the procedure is achieved by utilizing a priori information about the reflector models in the reconstruction algorithm. The main requirement for applying this technique is a limitation of the number of possible object shapes, each of which can be described by a sufficiently simple geometric model. By applying 1-, 2- or 3-dimensional arrays with few transmitters and receivers it is possible to obtain the required information by one or few measurements. The model-based reconstruction algorithm adapted to the specific measurement problem enables the classification of the objects and the estimation of the corresponding model parameters. Since the detailed description of the reconstruction algorithm was presented in a former paper, this article demonstrates the application and practical advantages as well as the limits of the method. The efficiency and flexibility of the method are demonstrated by applications within robot navigation, defect recognition (non-destructive testing), inspection of pipelines and level measurement of bulk material.

1 INTRODUCTION

Simple methods of ultrasonic imaging (e.g. 1- or 2-transducer systems without further reconstruction) usually provide only limited information about the object under consideration, e.g. distance or approximate size, or their application is restricted to the examination of objects with one particular shape[1-3]. On the other hand, high-resolution techniques (e.g. tomography, holography) generally require a great expense of sensorics and measurement time[4-6]. Given an expected image quality, for many applications the expense of the measuring system can be reduced by modeling the possible objects and by utilizing this a priori information in the reconstruction algorithm. A geometrical modeling essentially based on the shape of the objects proved to be suitable due to its simplicity and flexibility. Such a model in a simple form has already been successfully applied in robotics applications of ultrasound[7,8]. In this article it is extended to a measurement system, which is suitable to a wider field of applications. In contrast to conditions so far, no general restrictions exist on the parameter range of the objects as well as on the maximum number of transducers and their arrangement in the array and also models of more complex shape can be introduced[9].

In order to receive the required information from the object, the transducers which are operating in impulse echo mode are arranged to a 1-, 2- or 3-dimensional array. The array may be implemented with synthetic array techniques or, as only few transducers are needed, by means of a real array of fixed transmitters and receivers. Due to the generally applicable principle of the reconstruction algorithm described below, the arrangement of the transmitters and receivers can be easily adapted to the shape of the measuring volume to be examined and to the types of expected objects which leads to great flexibility of the measuring system.

2 OBJECT MODELING

In many practical applications a priori information about the possible object shape is available and the number of object types is strongly limited. Provided that the objects may be approximated by simple geometrical classes, they can be described by a model of virtual transmitters. The virtual transmitter $T_{V_{ij}}{}^k$ describes the effect of the ultrasonic field originating from the i-th real transmitter and influenced by the k-th reflector R upon the j-th receiver. Thus the object can be alternatively described by the real reflectors or approximately by the set of all virtual transmitters. Fig. 1 summarizes simple models with mainly geometrical parameters (a-f). Extensions of these models enable the description of geometrically more complicated objects (g) and the consideration of other object parameters like the object roughness (h) or the classification of interfaces with a negative gradient of the acoustic impedance[10]. Each object model ω is defined by a system of generally nonlinear equations

$$\mathbf{g}_\omega (\mathbf{M}_m, \mathbf{R}_m) = \mathbf{0}, \tag{1}$$

where the vector $\mathbf{M}_m$ includes the parameters of the model and $\mathbf{R}_m$ includes the theoretical (relevant) parameters of the received impulses which correspond to the object. With simple geometrical models, Eq. (1) is derived from the relation between object parameters, location of the virtual transmitters and the arrangement of the receivers[9]. For simplicity, total reflection is assumed and diffraction effects due to the boundaries and local curvatures of the objects are neglected. Thus, the relative positions of the virtual transmitters contain the information on the object type; the absolute location and the parameters of the virtual transmitters describe the parameters (location, orientation, characteristics) of the objects.

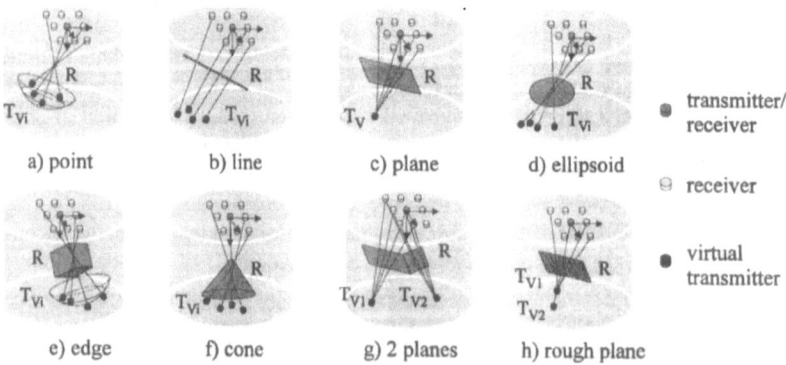

Figure 1. Modeling objects by virtual transmitters T_V.

3 RECONSTRUCTION ALGORITHM

Due to the models introduced, the measuring process can be reduced to the determination of type and possible location of virtual transmitters, which, by the reconstruction algorithm, are assigned to the individual reflectors. Fig. 2 schematically shows the principle of the measuring system.

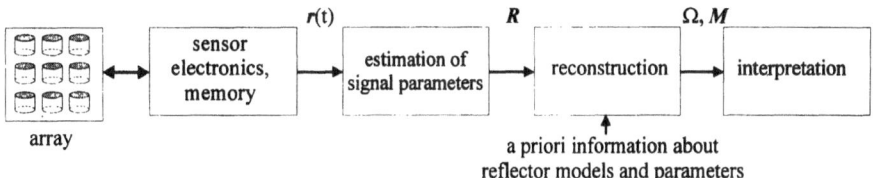

Figure 2. Measuring system.

From the received signals $r(t)$ which are components of the vector $\mathbf{r}(t)$ the relevant parameters $\mathbf{R}$ are extracted. Assuming that the relevant information of each received signal is included in few parameters, the dataset and therefore the expense of the following reconstruction algorithm can be reduced strongly without reducing necessarily the quality of the reconstruction result. In the following we use the times of flight τ_{ij} and the amplitudes a_{ij} of the received impulses as components of the matrices $\mathbf{R}_\tau = (\tau_{ij})$, $\mathbf{R}_a = (a_{ij})$, where index i relates to the i-th receiver and index j to the j-th echo. The signals received from an object with 3 simple objects (1 transmitter, 3 receivers) and the corresponding extracted parameters are shown in Fig. 3.

The reconstruction process consists of 3 tasks:

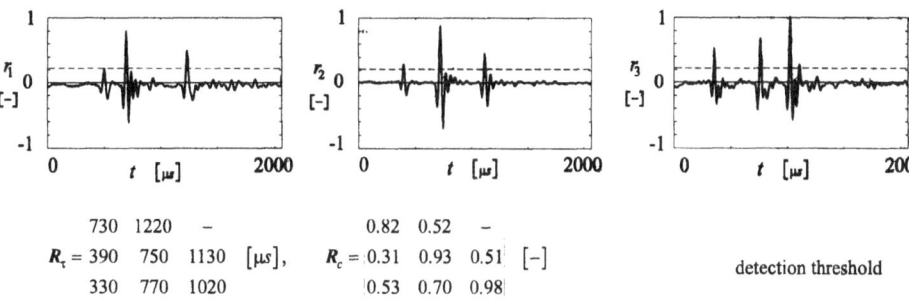

Figure 3. Received signals and extracted parameters.

I.   Assignment of the impulses to the n_0 individual objects,
II. Determination of the object models $\Omega = (\omega_1, \omega_2, \ldots, \omega_{n_0})$ (e.g. point, plane),
III. Estimation of the model parameters $M = (m_1, m_2, \ldots, m_{n_0})$ (e.g. location, orientation).

Generally the tasks I-III are coupled and therefore they have to be computed jointly. Due to the incorrect estimation of $\mathbf{R}$ and deviations from the ideal object models, a stochastical approach has to be applied to solve the problem successfully. The maximum of the a posteriori probability $p(\Omega, M | \mathbf{R})$ of the assumed object models Ω and the model parameters M, given the signal parameters $\mathbf{R}$, yields the objects[11]:

$$p_I = \max\{p(\Omega, M|R)\} \Rightarrow \hat{\Omega}, \hat{M} \qquad (2)$$

With a high number n_I of received impulses, the solution of the inverse problem, which corresponds to the multidimensional optimization of Eq. (2) is numerically exhaustive. In particular the solution is complicated because the number of reflectors, i.e. the dimensions of Ω and M, are a priori unknown. Therefore a simplified reconstruction algorithm was developed which is adapted to this problem: the maximum a posteriori estimation is reduced to combinations C_i of n_J impulses,

$$n_{min} \leq n_J \leq n_I , \qquad (3)$$

with the relevant parameters r and to the consideration of one object of unknown type ω with the corresponding model parameters m:

$$p_J = \max\{p(\omega, m|r)\} \Rightarrow \hat{\omega}, \hat{m} . \qquad (4)$$

The minimum number of impulses n_{min} depends on the defect models and prevents trivial ambiguities within the solution of Eq. (4) by ensuring overdetermination of Eq. (1). If p_J is smaller than the given threshold p_{min}, the combination of the n_J impulses is rejected. If p_J equals or exceeds p_{min} a reflector is assumed and the class ω_{max} that maximizes p_J is selected:

$$\begin{aligned} &p_J < p_{min} : C_i \text{ rejected,} \\ &p_J \geq p_{min} : C_i \text{ accepted, } \omega = \omega_{max} \end{aligned} \qquad (5)$$

The estimation and the detection of Eq. (4) and Eq. (5) are applied to all combinations C_i (excluding the impulses which have been already used in the determination of accepted objects). The combination of all accepted objects is finally filtered by joining objects with identical class and approximately the same parameters and by eliminating geometrical contradictions (e.g. a point which was detected behind a plane). Further details of the reconstruction algorithm as well as the specification of p_J and p_{min} can be found in [12].

4   APPLICATIONS

In many applications in which a priori information about the objects under consideration is available, the additional expense of the ultrasonic array and the model based reconstruction is worthwhile, because a more accurate image can be obtained. In the following the system is adapted to different industrially relevant applications where conventional ultrasonic systems are already used. The practical advantages and limits of the system are discussed by the results of simulations and measurements.

4.1 Obstacle Detection and Robot Navigation

Due to low costs and high axial resolution, acoustical sensors are often applied to obstacle detection and navigation of robots. The performance of these systems which are mostly built by a set of serially or parallelly excited 1- or 2-transducer systems is limited to the detection of objects with particular orientation. Artifacts, which result from multiple reflections are very hard to eliminate[13,14]. The low lateral resolution of these systems results from the wide directivity pattern of the transducers but a narrow directivity pattern would increase the number of not detected objects. The presented model based recognition can be easily applied by modeling the environment with 2-dimensionally (horizontal) structured reflectors (planes, lines/corners, cylinders, edges) and by applying a 2-dimensionally spherical array.

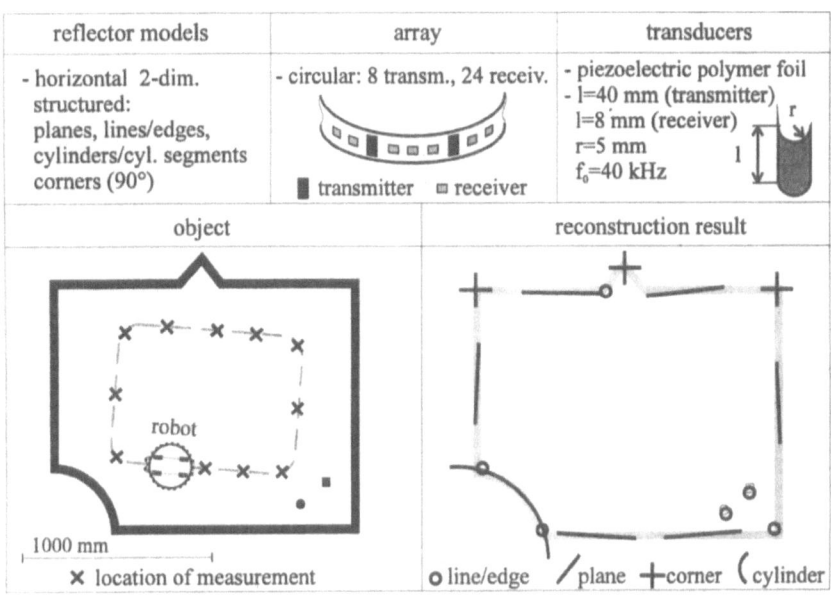

reflector models	array	transducers
- horizontal 2-dim. structured: planes, lines/edges, cylinders/cyl. segments corners (90°)	- circular: 8 transm., 24 receiv. ▮ transmitter ▯ receiver	- piezoelectric polymer foil - l=40 mm (transmitter) l=8 mm (receiver) r=5 mm f_o=40 kHz

object	reconstruction result
robot	
1000 mm ✗ location of measurement	o line/edge / plane + corner (cylinder

Figure 4. Application to obstacle detection and robot navigation.

In order to reduce the expense of the test equipment the experimental set up was built with reduced scale (scale ≈1:3 in comparison to technically relevant environments). In Fig. 4 the parameters of the set-up and the result of the reconstruction are shown. The robot moves on a rectangular path in a room with different obstacles. With the model based method all reflectors are detected and only the corner on the bottom right is classified incorrectly, which results from the partial shading by the two line objects. The two 'corners' on the bottom left are not shaped by two rectangular planes and therefore they are classified as lines. The reflector parameters are estimated with an accuracy of 0,8 mm (location) and 3° (orientation) relative to the position of the vehicle.

4.2 Defect Recognition within Non-Destructive Testing

The application of the method to defect recognition is demonstrated by measurements with a specimen of polymeric methylmetacrylate (Perspex) containing artificial point and plane shaped aluminium defects (point: small circular discs, d=5 mm, planes: a=20 mm, b=30 mm; Fig. 5). A synthetic array realized by 1 transmitter and 1 receiver with wide angle radiation patterns ($\varphi_0 = \pm30°$) is contacted with the specimen. The applied a priori information consists of the point and plane shaped defect models with arbitrary orientation and maximum dimensions of $a,b \leq 30$ mm.

With the minimum number of impulses ($n_J = n_{min} = 4$) all defects are recognized, but one point is classified as a plane and one artifact plane is detected. The false classification and the artifact are eliminated by combinations of $n_J = 5$ impulses. The defect parameters are determined within an accuracy of 2 mm (location) and 5° (orientation). Using 6 impulses, two existing point defects are not detected because only 5 impulses originating from each of the two missed point defects were received. This example demonstrates the importance of n_J to the quality of the measurement (in particular to the probability of detection and the suppression of artifacts). The problem generally can be solved by applying a sufficient number of transducers and by using a small set of impulses ($n_J=n_{min}+1$)[12].

The standard deviation of the TOF estimation $\sigma_\tau \approx 1,2$ mm is much higher than the standard deviation of the TOF estimation with known reflector ($\sigma_{TOF} \approx 0,2$ mm) because it is mainly determined by the approximations within the modeling which neglects the diffraction due to the

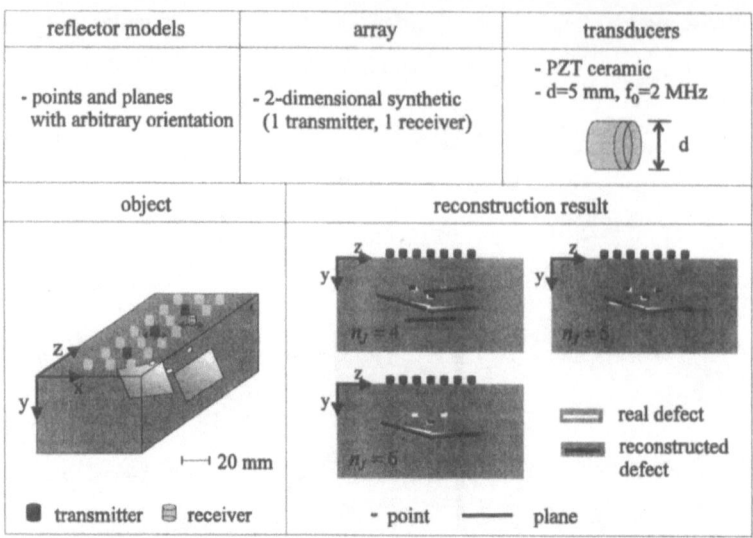

Figure 5. Application to defect recognition (non destructive testing).

boundary and the local curvature of the defects. If in addition to these approximations deviations from the assumed models appear, σ_τ increases further. So it may be advantageous to apply an extended model which includes or at least approximates the deviations.

4.3 Inspection of Pipelines

In addition to the frequently applied optical inspection of sewage pipes using man controlled TV-cameras the application of ultrasonic sensors may be advantageous. As they are able to produce

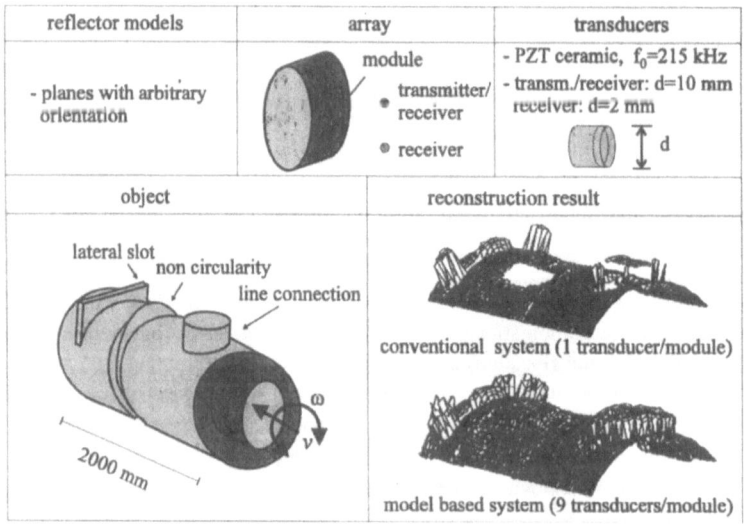

Figure 6. Application to the inspection of pipelines.

quantitative images of the inner geometry of the pipe, defects can be detected and assessed automatically[15-17]. In order to apply the model based system the array is adapted to the expected inner geometry of the pipe and designed modularly (Fig. 6). A module consists of 1 transmitter/receiver and 8 receivers and is used one at a time. By modeling the inner pipe wall by segments of local tangent planes, sufficiently large anomalies can easily be detected and their size can be determined with a high accuracy (0,5 mm location and 2° orientation, smooth PVC-pipes, ∅300 mm). With rough surfaces e.g. of concrete pipes the accuracy decreases (1 mm location and 7° orientation). Most of the measurement failures and artifacts which frequently occur with conventional 1-transducer systems are eliminated. In order to reduce the measurement time of the system, the transmitters can be excited simultaneously by code multiplex techniques, which enables an increases of the velocity of the inspection vehicle[18].

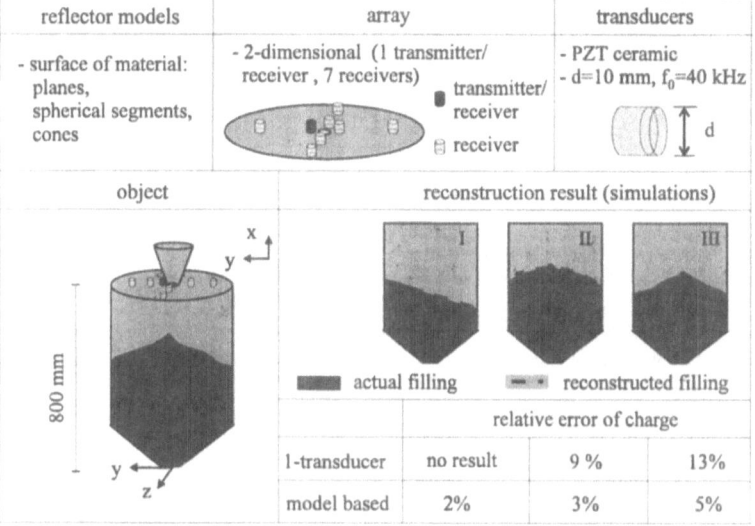

Figure 7. Application to level measurement of bulk material.

4.4 Level Measurement of Bulk Material

The accuracy of ultrasonic sensors which are applied to measure the charge of bulk material in silos or on band conveyors is limited due to the unknown shape of the bulk material and the disturbances, which result e.g. from multiple reflections or from depositions on the silo wall[19, 20]. These disadvantages can be eliminated by applying the model based system, which is able to estimate the shape of the material charge. Fig. 7 demonstrates the set-up of a silo which is built at reduced scale. The shape of the material charge is modeled by planes, spherical segments and by cones with unknown location and inclination angle. Due to the modeling the measurement no longer is limited to the estimation of one local filling height. In contradiction to the applications presented so far, no measurements have been taken yet but simulations demonstrate that the model based system enables an increase of accuracy in the determination of the charge. Furthermore the effect of disturbances can strongly be reduced.

REFERENCES

1. Gericke, O.R.: Determination of the Geometry of Hidden Defects by Ultrasonic Pulse Analysis Testing. *The J. of the Acoust. Soc. of America* 35 3/1963, pp. 364-368.

2. Sasaki, K.; Takano, M.: Classification of Objects' Surface by Acoustic Transfer Function. *Proceedings of the 1995 IEEE/RSJ Intern. Conf. on Intelligent Robots and Systems*, Vol. 1, pp. 821-828.

3. Yamani, A.; Al Akhdhar, S.Z.: A Novel Technique for Defects Classification from their Ultrasonic Pulse Echoes. *Advances in Signal Processing for Nondestructive Evaluation of Materials*, Kluwer Acad. Pub. (1994), pp. 371-384.

4. Berkhout, A.J.: Resolution Limits of Acoustical Echo Systems. *Acoustical Imaging Vol. 14*, Plenum Press; New York 1986, pp. 19-31.

5. Langenberg, K.J.; Brück, D.; Fischer, M.: Invers Scattering Algorithms. *New Procedures in Nondestructive Testing*, Springer Verlag 1983, pp. 381-391.

6. Capineri, L.: Time-of-flight Diffraction Tomography for NDT Applications. *Ultrasonics* Vol. 30 No. 5 1992, pp. 275-288.

7. Kleeman, L.; Kuc, R.: Mobile Robot Sensor for Target Localization and Classification. *The Intern. Journal of Robotics Research*, Vol. 14 No. 4 (1995), pp. 295-318.

8. Peremans, H.; Audenaert, K.; Van Campenhout, J.M.: A High-Resolution Sensor Based on Tri-aural Perception. *IEEE Trans. on Robotics and Automation*, Vol. 9 No. 1 (1993), pp. 36-48.

9. Menz, B.: Ultraschallarrays zur modellbasierten Objekterkennung. Sensoren und Meßsysteme, *VDI-Berichte 1255* (1996), pp. 357-362.

10. Ogilvy, J.A.: Model for the Ultrasonic Inspection of Rough Defects. *Ultrasonics*, Vol. 27 (1989), pp. 69-79.

11. Van Trees, H.L.: *Detection, Estimation, and Modulation Theory*, Part I, JohnWiley & Sons Inc. New York (1968)

12. Menz, B.: Model-Based Defect Recognition by 3-Dimensional Sparse Arrays. *Proc. of the 1996 IEEE Ultrasonic Symposium* (to be published).

13. Lawitzky, G.; Feiten, W.; Möller, M.: Sonar Sensing for Low-Cost Indoor Mobility; *Robotics and Autonomous Systems* 14 1995, 149-157.

14. Jörg, K.-W.: *Echtzeitfähige Multisensorintegration für autonome Roboter*; BI-Wissenschaftsverlag Mannheim 1994.

15. Haffner, H. u.a.: Fortschrittliches Abwasserkanal-Inspektionssystem; *KfK Nachrichten* Kernforschungszentrum Karlsruhe Jahrgang 26 2/94, 90-97.

16. Richardson, J.M.; Marsh, K.A.; Martin, J.F.: Techniques of Multisensor Signal Processing and Their Application of Vision and Acoustical Data. *Proc. of the 4th International Conference on Robot Vision ans Sensory Controls*, Bedford UK 1984, 395-408.

17. Clark, J.J.; Yuille, A.L: *Data Fusion for Sensory Information Processing Systems*. Kluwer Academic Publishers, Massachusetts 1990.

18. Menz, B.; Rappold, J.: Optimization of Scalar and Vectorial Codes for Codemultiplex Tecniques in Ultrasonic Multi-Tranmitter Systems. *Proc. of the World Congress of Ultrasonics 1997*, Yokohama (to be published).

19. Stock, J.R.: Schall, Ultraschall und elektromagnetische Wellen: Anwendungen bei Echosystemen und Schranken für die Füllstandsmessung. *Füllstands-Meßtechnik, VDI-Berichte 231* (1975). 103-111.

20. Kötzle, G.: Exakt definierte Grenzen - Berührungslose Füllstandsmessung bei Schüttgütern. *mpa-Messen, Prüfen, Automatisieren*, Nr.9 (1996), 38-44.

ULTRASONIC REFLECTION TOMOGRAPHY OF POST-DISCOTOMIC SCARRING

A. Pesavento[1], H. Ermert[1], J.Grifka[2], E. Broll-Zeitvogel[2]

[1]Dept. of Electrical Engineering, Ruhr University
[2]Orthopaedische Universitaetsklinik, St. Josefs Hospital
44780 Bochum, Germany

INTRODUCTION

Up to 40% of all back surgery patients suffer from significant post-operative relapsing complaints frequently caused by scarring. Despite modern MR imaging systems there is still al lack of reliable screening diagnostics. Furthermore the image quality of conventional B-scans suffers from low contrast and shadows.

In the past the ultrasound image quality of organs which allow multidirectional scanning like breast, thyroid, and testicles was improved significantly by ultrasonic reflection tomography. With this concept, B-mode images of the same cross sectional object area but from different aspect directions (subimages) are averaged incoherently. While moving the B-scan probe a short distance from the patient's back has to be kept. Therefore, only laterally displaced subimages can be acquired. In order to obtain highly overlapping subimages a sector B-scan system is used.

This way speckle effects and other artifacts are reduced and the image contrast is improved significantly. However, the improvement strongly depends on the decorrelation of the used subimages, whereas the calculation, data acquisition and storage effort rises with every used subimage. Since theoretical derivations of the correlation function of the subimages suffer from approximations that are only valid in the focal region the correlation of the subimages was measured over a wide depth range.

An adaptive compounding system has been developed using the measured correlation functions. The image which is to be compounded is divided into small segment. The segments are compounded using the optimum displacement depending on the considered depth and available subimages. The compounding system was optimized to find a good compromise between the calculation, storage and data acquisition effort and the SNR improvement of the system. The image quality of the system is demonstrated for both, phantom and in vivo images.

SYSTEM SETUP

Figure 1 shows the main components of the compounding system. A conventional B-scan system (Siemens Sonoline SL-2) with a 5 MHz sector scanner is used. The applicator is laterally moved along a line over the patients back by an stepping motor. The RF echo data is acquired by a 30 MHz A/D converter and stored in a PC. TGC and demodulation is done by the compounding software in the PC.

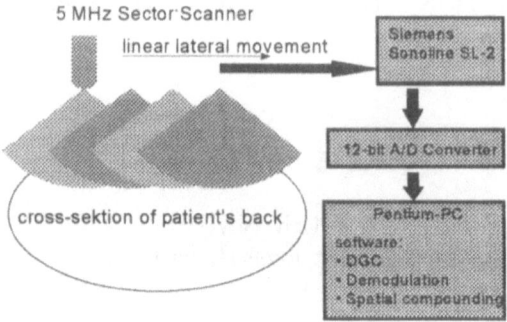

Figure 1. System setup of the applied spatial compounding system

CORRELATION OF SUBIMAGES

Theoretic considerations

The correlation of speckle imaged from different transducer positions has often been subject of research. Burkhardt[1] and later Gehlbach[2] derived the correlation between images formed from different transducer orientations. Subsequent work by Wagner et al.[4,7] following Gehlbach with minor modifications derived a correlation function as a function of lateral transducer shift b using the lateral point spread function $p(x)$,

$$\rho(b) = c \left[\left| FT\left\{ |p(x)|^2 \right\} \right|^2 \right]_{f=2b/\lambda_0 z_0} \qquad (1)$$

with c a constant to assure $\rho(0) = 1$. The result was obtained using monofrequency analysis at the systems center frequency. It assumes that those scatterer which cause echoes are randomly distributed on a line parallel to the transducer instead of being randomly distributed over the resolution cell of the system. Furthermore it assumes the point spread function to be separable into lateral and axial components[4]. Using an approximation of

$$p(x) = sinc\left(\frac{Dx}{\lambda_0 z_0} \right) = \frac{\sin(\pi Dx / \lambda_0 z_0)}{\pi Dx / \lambda_0 z_0} \qquad (2)$$

with the aperture size D, $\rho(b)$ is depth independent and only depends from the fractional aperture shift b/D. O'Donnell[3] showed that assuming

$$p(x) = A\left(\frac{Dx}{\lambda_0 z_0}\right) \mathrm{sinc}\left(\frac{Dx}{\lambda_0 z_0}\right) \qquad (3)$$

with any beam divergence function $A(z)$ the correlation function $\rho(b)$ in Equation (1) can be analytically solved to a composition of polynomial parts of third order, which is independent from $A(z)$. Despite of the used approximations measurements done by Trahey[5] show good correspondence with this prediction in the focal and far field region. However, the approximations are not valid for a constant focus sector scanner used in our system. Particularly the separation of the point spread function in axial and lateral multiplicative components is only valid for the focal region (because of the rotation of the single transducer). For the same reason the approximation of a sinc function is only valid for sector geometry.

Measurement of the correlation function

Consequently we measured the correlation function over a wide depth range. The measurement was made with a commercially available homogenous speckle phantom.

The correlation of the (sector) image viewed from two different transducer position (1) and (2) was calculated for different image depths d, lateral displacements Δx and aspect angles β (see Figure (2)). The correlation function was found to depend only on the effective lateral displacement $\Delta x' = \Delta x \cos\beta$. This is obvious since the echoes received from transducer position (2) are the same as from an imaginary transducer position (2)' (for small

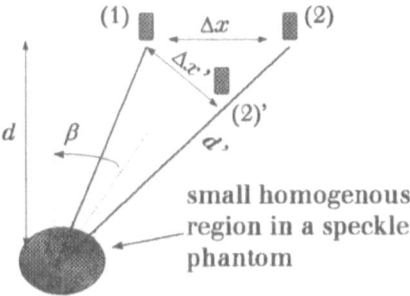

Figure 2. Measurement of speckle correlation

Δx, except of an phase shift, which is irrelevant for the demodulated signals). Considering this, for symmetry reason the correlation function only depends on the distance between the transducer and the imaged region $d' = d / \cos\beta$, not on d and β independently. This agrees with the measurement results. The measured functions were found to be very closely related to exponential functions of the form

$$\rho(\Delta x', d') = \exp(-\Delta x'/\alpha(d')) \qquad (4)$$

with an depth depended correlation length $\alpha(d')$. However, the correlation length in the focal region of the transducer meets the theoretical predictions of Wagner[4] and O'Donnell[6]. Since the correlation length is only slightly dependent on the depth we used the polynomial fit described in Equation (5) for $\alpha(d')$

$$\alpha(d') = a_0 + a_1 d' + a_2 d'^2 + a_3 d'^3 \qquad (5)$$

The polynomial coefficients a_0, a_1, a_2 and a_3 were estimated using a least square estimator Therefore the correlation length for the measured depth d' had to be estimated using an minimization of the summed absolute differences. The result is presented in Figure 3 showing the single measured correlation lengths and the polynomial fit.

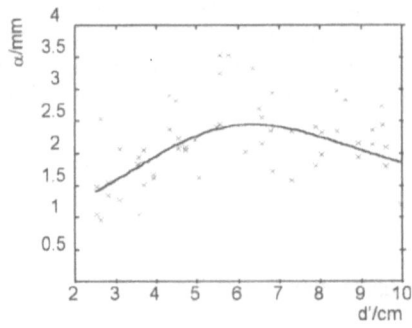

Figure 3. Correlation length α as a function of depth d'

The assumption of a sinc-function for the lateral component of the PSF underestimates the lateral resolution in front of the constant focus of the transducer. Applying Equation (1) this leads to a smaller correlation length, which we could confirm with our measurements. Behind the focus the correlation length decreases slightly. This results meets the observations of Trahey[5] and O'Donnell[6].

AN EFFICIENT COMPOUNDING CONCEPT

The SNR of a single ultrasound image with fully developed speckle is known[3] to be approximately 1.91. With spatial compounding this value can be improved[6] to

$$SNR \approx 1.91 N / \sqrt{\sum_{i=1}^{N}\sum_{j=1}^{N}\rho(x_i,x_j)} \qquad (6)$$

where N is the number of used subimages and $\rho(x_i,x_j)$ is the correlation of the subimages viewed from positions x_i, and x_j whereas the SNR improvement for uncorrelated images is proportional to $1/\sqrt{N}$. Due to the speckle correlation, assuming an constant and relatively large acoustical window on the patients back (ca 10 cm) a very dense compounding scheme leads to high acquisition and calculation time but does not lead to a significantly higher SNR than a slightly wider compounding scheme. For very dense compounding schemes the SNR can even fall under an reachable maximum[6,7]. However, since the calculation, data acquisition and storage effort increases with the number of used subimages, a compromise between the effort and the merit of a compounding scheme must be found, meaning an increase of used subimages should lead to an significant increase of the SNR. This compromise can be found by searching the compounding positions x_i that minimizes the cost function

$$f(x_i) = N^k / SNR. \qquad (7)$$

Because of the dependence of the SNR on the square root of the number of uncorrelated images, k must be significantly smaller than 0.5. If an value of k=0.5 is taken the cost function is constant for uncorrelated images. In this work a value of $k = 0.35$ was chosen.

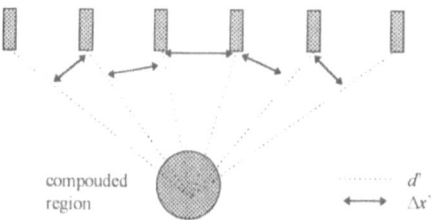

compouded region d

$\Delta x'$

Figure 4. Lateral displacements $\Delta x'$ and transducer distance d' for an compounding scheme with constant grid

The problem can not satisfactorily solved by a compounding scheme with constant grid, because the lateral displacements $\Delta x'$ and transducer distances d' are not constant for neighboring transducer positions. Especially the images viewed from transducer positions with a large angle β are more correlated because of a smaller effective lateral displacement $\Delta x'$. For this reason we choose a parametric model for the transducer positions of $x_i = b_0 + b_1 i + b_2 i^2 + b_3 i^3$. The polynom coefficients where found using MATLAB's simplex search method[8], minimizing the cost function Equation (7). Therefore the compounded image was divided into small rectangular segments. For each segment the optimal images to be used where found by the described method. The result was not very much affected by the unavoidable discretisation of the transducer positions which was chosen near the minimum of all the lateral displacements found by the optimization.

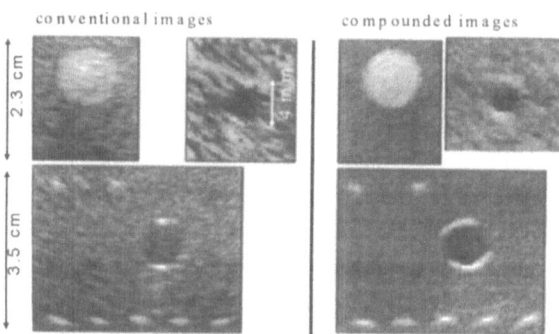

conventional images   compounded images

2.3 cm

3.5 cm

Figure 5. images of a speckle phantom

RESULTS

The SNR improvement of the optimized compounding concept, optimizing cost function Equation (7), was found to be 12 dB in the center segment of the image, with 45 acquired images, assuming an acoustical window of 10 cm. Optimizing the SNR with constant displacement between the subimages (standard compounding concept) a SNR improvement of 12.6 dB can be obtained, with 125 images to acquire. Since some of the acquired images are not used for some segments in the optimized concept, the calculation time increases by a factor of 6.5 for the standard compounding concept. Figure 5 compares images of a speckle phantom produced with the optimized compounding concept with conventional B-mode images of the same areas. Despite of the obvious speckle reduction borders are significantly improved due to the varity of aspect angles.

Figure 6 compares a compounded in vivo image of a human with a conventional one. The images show the back of a post dicotomy patient with scarring on the right side. In contrast to the conventional single images the perturbation of the facia due to the operation is well defined in the compounded image (low echogeneity in region 1) and the asymmetry in the echogeneity of the left and the right side of the processus spinosus with higher echogeneity on the right side indicates the scarring (region 2).

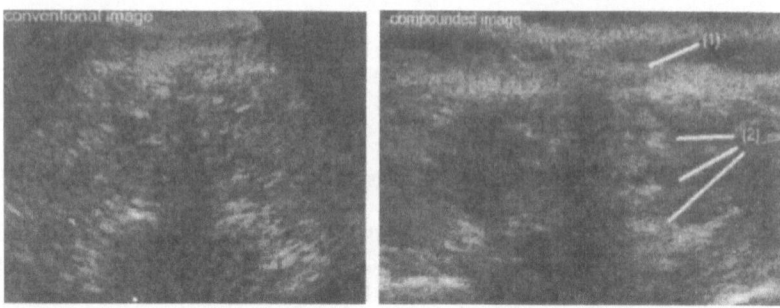

Figure 6. in Vivo images of post discotomic scarring

CONCLUSIONS

Our measurements have shown that in ultrasonic reflection mode tomography the speckle correlation function is depth dependent for a constant focus transducer. For efficient compounding, with a large acoustical window, a segmentwise optimized compounding concept with unequal displacement of the single images minimizes data acquisition and calculation effort by a factor of 6.5 without significant loss of SNR improvement. With the optimized compounding scheme a SNR improvement of approximately 12 dB could be obtained in the center of the image.

ACKNOWLEDGMENTS

This work was supported by the Deutsche Forschungsgemeinschaft, grant Gr 1506/1-1.

REFRENCES

1. C. B. Burckhardt, Speckle in ultrasound B-mode scans, *Trans Sonics Ultrason.* 25:1-6 (1978)
2. S. M. Gehlbach, *Pulse reflection imaging and acoustic speckle*, Ph. D. dissertation, Stanford University, Stanford, CA (1983)
3. R. F. Wagner, S. W. Smith, J. M. Sandrik and H. Lopez, Statistics of Speckle in Ultrasound B-Scans, *Trans Sonics Ultrason.* 30:156-163 (1983)
4. R. F. Wagner, M. F. Insana and S. W. Smith, Fundamental Correlation Lengths of Coherent Speckle in Medical Ultrasonic Images, *Trans. Ultrason. Ferroelec. Freq. Contr.* 35:34-44 (1988)
5. G. E. Trahey, S. W. Smith and O. T v. Ramm, Speckle Pattern Correlation with Lateral Aperture Translation: Experimental Results an Implications for Spatial Compounding, *Trans. Ultrason. Ferroelec. Freq Contr.* 33:257-264 (1986)
6. M. O'Donnell and S. D. Silverstein, Optimum Displacement for Compound Image Generation in Medical Ultrasound, *Trans. Ultrason. Ferroelec. Freq Contr.* 35:470-476 (1988)
7. A. Lorenz, L. Weng and H. Ermert, A Gaussian Model Approach for the Prediction of Speckle Reduction with Spatial and Frequency Compounding, *Ultrasonic Symposium* (1996)
8. W.H. Press, S.A. Teukolsky, W. T. Vetterling, B.P. Flannery, Numerical Recipes in C, Cambridge Univerity Press, Cambride (1992)
9. G. Röhrlein, H. Ermert, Limited angle reflection-mode computerized tomography, *Acoustical Imaging* 14:413-424 (1986)
10. D. Hiller, H. Ermert, System analysis of ultrasound reflection mode computerized tomography, *IEEE Trans. Sonics. Ultrason.*, 31:240-250 (1984)

RENORMALIZED BORN INVERSION

Rebecca B. Shuman and Paul E. Barbone

Department of Mechanical Engineering
Boston University
Boston, MA 02215

INTRODUCTION

Typical ultrasound B-scan processing relies on one dimensional Born inversion. That is, the images displayed represent grey-scale maps of Born inversion profiles: impedance changes are proportional to backscattered amplitude, and depth $x = c_0 t/2$. These approximations yield useful images. By systematically exploiting the implicit assumptions that underlie the success of traditional imaging, however, significant improvements can be made to better image quality with little or no additional processing expense.

In the context of perturbation theory, Born inversion represents a regular perturbation expansion in powers of ϵ, the scale of medium inhomogeneity. Such an expansion, however, is nonuniform in distance and time. It is this nonuniformity that can be easily corrected to yield a method that is both simple and accurate.

There are many methods by which one might attempt to improve a first order Born approximation. An obvious one is to continue the expansion to $O(\epsilon^2)$; i.e. to use a second order Born approximation. This requires significant additional processing time, however, and still fails to overcome the non-uniformity in the expansion. Another approach is the distorted wave Born approximation [3]. Here, the inversion is linearized about each $c(x)$ in the iteration. In comparison to first or second order Born inversion, this method is very computationally expensive, and can be applied only to higher order corrections of the first order approximation.

The approach we shall explore is renormalization. We redefine the order zero operator to remove non-uniformity from the expansion. The first term is as simple to obtain as in regular Born scattering, but the approximation is uniformly valid in space and time. Of course, iterating the process might improve the approximation further, but we are concerned here with methods that can be applied in real time. The renormalization that we choose is consistent with WKB ansatz. Similar or identical renormalization procedures have been suggested for use in seismic imaging by Hagin [4], Gray [1], and Raz [7], for the special case of constant density media. After renormalization, our order zero operator depends implicitly on the unknown sound speed $c(x)$. In order to evaluate this dependence, we must assume that changes in $c(x)$ are directly correlated to changes in the impedance. Soft tissue data shows such a correlation. We note that regular Born inversion makes such an approximation implicitly with the special case of constant sound speed.

MATHEMATICAL MODEL

Problem Formulation

In creating our mathematical model, we assume the tissue to be nearly homogeneous. Since the absorption length of ultrasound in tissue is typically smaller than the dimensions of the body, we shall

Acoustical Imaging, Vol. 23
Edited by Lees and Ferrari, Plenum Press, New York, 1997

treat the medium as infinite. We write the stiffness $E(x)$ and density $\rho(x)$ as weak perturbations from the background values:

$$E(x) \;=\; E_0 + \epsilon\,E_1(x) \qquad \epsilon \ll 1 \tag{1}$$

$$\rho(x) \;=\; \rho_0 + \epsilon\,\rho_1(x) \qquad \epsilon \ll 1. \tag{2}$$

We shall further assume that the medium is homogeneous in $x < 0$. Thus $E_1(x) = 0$ for $x \le 0$, and $\rho_1(x) = 0$ for $x \le 0$. In all that follows, we shall consider one dimensional wave propagation and scattering.

We will solve the one dimensional wave equation for an inhomogeneous medium:

$$\frac{\partial}{\partial x}\left(E(x)\frac{\partial u}{\partial x}\right) - \rho(x)\frac{\partial^2 u}{\partial t^2} \;=\; 0. \tag{3}$$

$$u(x,t) \;=\; w(t - x/c_0) \qquad \text{for } t < 0 \tag{4}$$

$$w(\eta) \;=\; 0 \qquad \text{for } \eta < 0. \tag{5}$$

Here, $u(x,t)$ represents the displacement, acoustic pressure, or velocity potential. $c_0 = \sqrt{E_0/\rho_0}$ is the speed of sound in the background medium, and $w(t)$, the incident wave shape, is assumed to be known. Our goal is now to solve equations (3-5) to obtain a relation between $u(0,t)$ and $\rho_1(x), E_1(x)$.

Born Inversion

Here, we briefly describe regular Born inversion so that we may later compare it to our renormalized inversion. Consistent with the Born scattering approximation, we write the field as a regular perturbation expansion:

$$u(x,t;\epsilon) \;\sim\; u_0(x,t) + \epsilon\,u_1(x,t) + O(\epsilon^2) \qquad (\epsilon \to 0; x,t \text{ fixed}): \tag{6}$$

$$u_0(x,t) \;=\; w(t - x/c_0). \tag{7}$$

We substitute (6) and (7) into equation (3) and collect terms to $O(\epsilon)$ to find that u_1 satisfies:

$$c_0^2\frac{\partial^2 u_1}{\partial x^2} - \frac{\partial^2 u_1}{\partial t^2} = \left(\frac{\rho_1}{\rho_0} - \frac{E_1}{E_0}\right)w''(t - x/c_0) + \frac{E_1'(x)}{\rho_0 c_0}w'(t - x/c_0). \tag{8}$$

We now solve equation (8) using the Green's function, $g(x,t;x_0,t_0)$, which satisfies

$$c_0^2\frac{\partial^2 g}{\partial x^2} - \frac{\partial^2 g}{\partial t^2} \;=\; \delta(x - x_0)\,\delta(t - t_0), \tag{9}$$

$$g(x,t;x_0,t_0) \;=\; 0 \qquad \text{for } t < t_0 \tag{10}$$

In terms of $g(x,t;x_0,t_0)$, u_1 may be written as

$$u_1(x,t) \;=\; \int_{-\infty}^{\infty} \int_{-\infty}^{\infty} \left[\left(\frac{\rho_1}{\rho_0} - \frac{E_1}{E_0}\right)w''(t_0 - x_0/c_0)\right.$$

$$\left. + \frac{E_1'(x_0)}{\rho_0 c_0}w'(t_0 - x_0/c_0)\right] g(x,t;x_0,t_0)\,dx_0\,dt_0. \tag{11}$$

We note that:

$$g(x,t;x_0,t_0) \;=\; \frac{1}{2c_0}H\big((t - t_0) - |x - x_0|/c_0\big). \tag{12}$$

Here, $H(\xi)$ is the Heaviside step function. Using (1),(2),(5) and (12) in (11), allows us to perform the time integration, yielding:

$$u_1(x,t) \;=\; -\frac{1}{2c_0}\int_0^{\infty}\left(\frac{\rho_1(x_0)}{\rho_0} - \frac{E_1(x_0)}{E_0}\right)w'(t - |x - x_0|/c_0 - x_0/c_0)\,dx_0$$

$$- \frac{1}{2c_0}\int_0^{\infty}\frac{E_1'(x_0)}{\rho_0 c_0}w(t - |x - x_0|/c_0 - x_0/c_0)\,dx_0. \tag{13}$$

Step function Incident

Equation (13) is a linear integral equation relating $u_1(x, t)$ to ρ_1 and E_1. This relation assumes its simplest form for a step function incident wave. Thus, we let:

$$
\begin{align}
w(t) &= A H(t) \tag{14}\\
w'(t) &= A \delta(t) \tag{15}
\end{align}
$$

Substituting (14) and (15) into (13), and assuming that $x \leq 0$, we find:

$$
\begin{align}
u_1(x, t) &= -\frac{A}{4}\left[\rho_1((x + c_0 t)/2)/\rho_0 + E_1((x + c_0 t)/2)/E_0\right] \tag{16}\\
&= -\frac{A}{2}z_1((x + c_0 t)/2)/z_0 \qquad \text{for } x \leq 0. \tag{17}
\end{align}
$$

In (17), $z = z_0 + \epsilon z_1 + O(\epsilon^2)$ is the impedance of the medium, and is defined by

$$
z = \rho c = \sqrt{\rho E}. \tag{18}
$$

Equations (17) and (18) show that, given z_0, measurements of $u_1(0, t)$ directly yield $z_1(c_0 t/2)$:

$$
z_1(x) = \frac{2z_0}{A} u_1(0, t) \tag{19}
$$

with $x = c_0 t/2$. That is, the reflected amplitude is directly proportional to impedance changes and distance is proportional to elapsed travel time.

RENORMALIZED BORN INVERSION

Here we renormalize the order zero operator which is the basis of our perturbation expansion. The unperturbed operator depends implicitly on the unknown properties of the medium. Inspired by a WKB ansatz, we introduce a travel-time variable instead of a physical distance. That is, we measure distance from the origin by how long it takes a wave disturbance to reach a point. We first define the speed, $c(x)$, as

$$
c(x) = \sqrt{\frac{E(x)}{\rho(x)}}. \tag{20}
$$

The travel time ξ is then given by:

$$
\xi(x; \epsilon) = \int_{-\infty}^{x} \frac{1}{c(x')} dx'. \tag{21}
$$

We now introduce the impedance function, $z(\xi)$, such that

$$
z(\xi) = \rho(x)c(x), \tag{22}
$$

where ξ and x are related by (21).

Reformulation

With the definitions (21) and (22), equation (3) can be rewritten as

$$
\frac{\partial^2 u}{\partial \xi^2} + \epsilon \frac{z_1'(\xi)}{z(\xi)} \frac{\partial u}{\partial \xi} - \frac{\partial^2 u}{\partial t^2} = 0. \tag{23}
$$

Equations (1), (2), (21) and (22) give us

$$
\begin{align}
z(\xi) &= z_0 + \epsilon z_1(\xi) + O(\epsilon^2) \tag{24}\\
z(\xi) &= z_0 \qquad \text{for } \xi \leq 0 \tag{25}
\end{align}
$$

Equation (23) represents the renormalized wave equation. The order zero operator $(\partial_{\xi\xi} - \partial_{tt})$ depends implicitly on ϵ through the definition of ξ in (21). This dependence keeps corrections uniformly bounded in space and time.

Solution

We again assume the field can be expanded:

$$u(x, t; \epsilon) \sim u_0(\xi, t) + \epsilon u_1(\xi, t) + O(\epsilon^2), \qquad \epsilon \to 0; \ \xi, t \text{ fixed} \tag{26}$$
$$u_0(\xi, t) = w(t - \xi). \tag{27}$$

Combining equations (23 - 26) yields the following equation for u_1:

$$\frac{\partial^2 u_1}{\partial \xi^2} - \frac{\partial^2 u_1}{\partial t^2} = \frac{z'(\xi)}{z(\xi)} w'(t - \xi). \tag{28}$$

We note that $z(\xi)$ in the denominator on the right hand side of (28) can formally be replaced by z_0 to the same order of accuracy, but it costs very little effort to include the (slightly) greater accuracy expressed in (28). Following precisely the same procedure as in the previous section, we obtain the solution

$$\epsilon u_1(\xi, t) = -\frac{1}{2} \int_0^\infty \frac{z'(\xi)}{z(\xi)} w(t - |\xi - \xi_0| - \xi_0) \, d\xi_0. \tag{29}$$

As before, we assume the incident wave is a step function. Thus,

$$w(\xi) = A H(\xi) \tag{30}$$

We substitute (30) into (29) and integrate with $\xi \leq 0$ to obtain

$$\epsilon u_1(\xi, t) = -\frac{A}{2} \Big\{ \log \big[z((t + \xi)/2) \big] - \log \big[z(0) \big] \Big\}. \tag{31}$$

Specializing this result to the case $\xi = 0$ allows us to solve for $z(\xi)$, with $\xi = t/2$:

$$z(\xi) = z(0) \exp \left\{ \frac{-2\epsilon u_1(0, t)}{A} \right\}. \tag{32}$$

Thus, measurements of the scattered field at the origin again determine $z(\xi)$. In this case, z is determined as a function of the travel-time, ξ.

Determining $\xi(x)$:

From (21), we have

$$\frac{d\xi}{dx} = \frac{1}{c} \tag{33}$$

We note that with knowledge of $z(\xi)$ only, we cannot solve equation (33) for $\xi(x)$. Thus we must make an assumption regarding the mutual dependence of the material properties. Therefore, we shall assume (for the purposes of reconstruction) that fluctuations in $E(x)$ and $\rho(x)$ are correlated. That is:

$$E = E(\rho). \tag{34}$$

Expanding to $O(\epsilon)$:

$$E(x) = E_0 \left(1 + \epsilon \gamma \frac{\rho_1(x)}{\rho_0} + O(\epsilon^2) \right). \tag{35}$$

For soft tissue imaging, $\gamma \approx 2$. Equations (33) and (35) (with (20) and (22)) now give:

$$x(\xi) = \int_0^\xi \left\{ c_0 + \frac{\gamma - 1}{\gamma + 1} (z - z_0) \right\} d\xi' \qquad (\xi > 0). \tag{36}$$

The function $\xi(x) \equiv x^{-1}(\xi)$. Therefore, the spacial impedance function is $\hat{z}(x) = z(\xi) \circ \xi(x)$. Thus, once we have determined $z(\xi)$, we can determine $x(\xi)$, and so in principle we can likewise determine $\xi(x)$ and $c(x)$.

EXAMPLES

We consider a piecewise constant medium for which the exact solution is easily obtained in the frequency domain. To obtain our simulated data, we compute the reflection coefficient for each of several

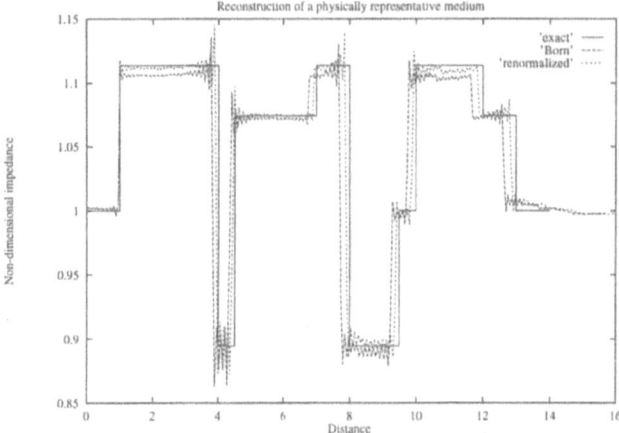

Figure 1: Comparison of inversion methods: material properties represent soft tissue.

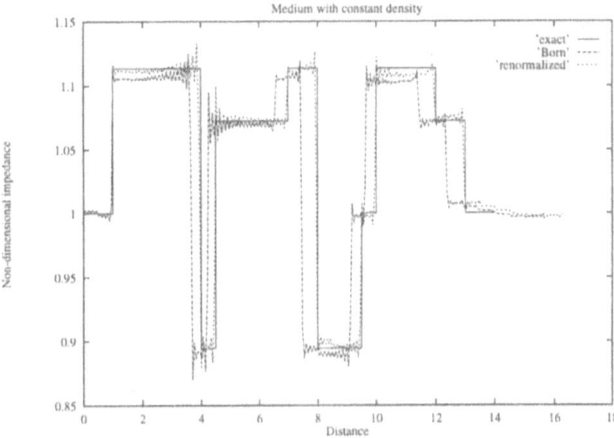

Figure 2: Comparison of inversion methods: constant density material.

discrete frequencies. We multiply the reflection coefficients by the Fourier amplitudes of the incident wave. Taking the inverse transform of this product then yields the reflected wave at $x = 0$. Thus we obtain the reflected field at discrete times:

$$\epsilon\, u_1(0, t_i) = u_i \qquad t_i = i * \delta t. \tag{37}$$

Given the reflected field at discrete times, our inversion scheme obtains ordered pairs (x_i, z_i).

Born Inversion Method

We use equation (19) to obtain our Born inversion results. For $c_0 = 1$:

$$z_i = z_0 + \epsilon\, z_1(x_i) = z_0 - 2 * u_i \tag{38}$$

$$x_i = \frac{t_i}{2} \tag{39}$$

Renormalized Born

We implement our renormalized Born method inversion using a trapezoidal rule to evaluate the integral in equation (36). Thus, equations (32) and (36) yield (for $c_0 = 1$):

$$z_i = z(\xi_i) = \exp\left(-2u_i\right) \tag{40}$$

$$x_i = x_{i-1} + \left\{1 + \frac{\gamma - 1}{\gamma + 1}\left(\frac{z_i + z_{i-1}}{2} - z_0\right)\right\}\frac{\delta t}{2} \tag{41}$$

Results and Conclusions

In figure 1 and figure 2, we show the results of the two methods compared to the exact impedance profiles. In figure 1, the data correspond to a medium with material properties similar to that of soft tissue [6]. For those tissue properties, $\gamma \approx 2$. In the reconstruction, therefore, we used $\gamma = 2$. The data in figure 2 correspond to $\rho(x) = constant$. The renormalized reconstruction was obtained with $\gamma = 100$.

The reconstructions based on the renormalized Born perturbation expansions are more accurate than the regular Born inversions. The interface positions are more accurate due to the use of the renormalization. The impedance of each layer is more accurately predicted because of the use of $z(\xi)$ instead of z_0 in equation (28).

The renormalized equations can recover the Born approximations as special cases. This can be seen by expanding (32) for $\epsilon \ll 1$ and noting that $\xi = x/c_0 + O(\epsilon)$.

In order to apply our renormalized expansions, we needed to assume that sound speed and impedance fluctuations are correlated. This is often the case in soft tissue imaging, and the correlation constant is known in advance. To evaluate each new x_i, only three new floating point multiplications are required, which is certainly within real-time processing capabilities. Thus the renormalized reconstruction is very simple and efficient to apply, and improves the accuracy of the images created.

Acknowledgments

The authors are grateful to Ms. Echo Miller who wrote the program used to generate our simulated reflection data. Ms. Miller's effort was financially supported by the NSF "Research Experiences for Undergraduates" program. The authors also gratefully acknowledge the financial support of NSF through grant BES9410218.

References

[1] S.H. Gray, "A second-order procedure for one-dimensional velocity inversion," *SIAM J. Appl. Math.*, vol. 39, pp. 456-462, 1980.

[2] S.H. Gray and N. Bleistein, "Imaging and Inversion of Zero-Offset Seismic Data," *Proc. IEEE*, vol. 74, no. 3, pp. 440-456, 1986.

[3] O.S. Haddadin and E.S. Ebbini, "Multiple Frequency Distorted Born Iterative Method for Tomographic Imaging," *Acoustical Imaging*, Vol. 23, S. Lees and L.A. Ferrari, eds. Plenum Press, New York, 1997.

[4] F.G. Hagin, "A stable approach to one-dimensional inverse problems," *SIAM J. Appl. Math.*, vol. 40, no.3, pp. 439-453, 1981.

[5] E.J. Hinch, *Perturbation Methods*. Cambridge University Press, New York 1991.

[6] R.A. Lerski, ed., *Practical Ultrasound*. IRL Press, Washington, D.C. 1988.

[7] S. Raz, "A direct profile inversion: Beyond the Born Model," *Radio Sci.*, vol. 16, pp. 347-353, 1981.

AUTHOR INDEX

SUBJECT INDEX

The items in the index were taken from a list of key terms provided by the authors. The referenced page number is the first page of the article.